2nd International Innovative Mining Symposium (Devoted to Russian Federation Year of Environment)

E3S Web of Conferences Volume 21 (2017)

Kemerovo, Russia
20 - 22 November 2017

Editors:

Maxim Tyulenev
Sergey Zhironkin
Alexey Khoreshok
Stefan Voth
Michal Cehlar
Demirel Nuray

ISBN: 978-1-5108-5193-1

Printed from e-media with permission by:

Curran Associates, Inc.
57 Morehouse Lane
Red Hook, NY 12571

CURRAN ASSOCIATES INC.
proceedings
.com

Some format issues inherent in the e-media version may also appear in this print version.

Printed by Curran Associates, Inc. (2017)

For additional information, please contact EDP Sciences – Web of Conferences
at the address below.

EDP Sciences – Web of Conferences
17, Avenue du Hoggar
Parc d'Activité de Courtabœuf
BP 112
F-91944 Les Ulis Cedex A
France

Phone: +33 (0) 1 69 18 75 75
Fax: +33 (0) 1 69 28 84 91

contact-edps@webofconferences.org

Additional copies of this publication are available from:

Curran Associates, Inc.
57 Morehouse Lane
Red Hook, NY 12571 USA
Phone: 845-758-0400
Fax: 845-758-2633
Email: curran@proceedings.com
Web: www.proceedings.com

TABLE OF CONTENTS

edp sciences

E3S Web of Conferences

Statement of Peer review

In submitting conference proceedings to E3S Web of Conferences, I certify to the Publisher that I adhere to the **Policy on Publishing Integrity** of the journal in order to safeguard good scientific practice in publishing.

1. All articles have been subjected to peer review administered by the proceedings editors.
2. Reviews have been conducted by expert referees, who have been requested to provide unbiased and constructive comments aimed, whenever possible, at improving the work.
3. Proceedings editors have taken all reasonable steps to ensure the quality of the materials they publish and their decision to accept or reject a paper for publication has been based only on the merits of the work and the relevance to the journal.

Title, date and place of the conference

IInd International Innovative Mining Symposium (Devoted to Russian Federation Year of Environment), 20-22 November 2017, Kemerovo, Russian Federation

Proceedings editor(s):

Maxim Tyulenev, PhD, Kuzbass State Technical University

Sergey Zhironkin, DSc, National Research Tomsk Polytechnic University

Alexey Khoreshok, DSc, Kuzbass State Technical University

Stefan Voth, Dr.-Ing, TFH Georg Agricola

Michal Cehlar, PhD, Technical University of Kosice

Demirel Nuray, PhD, Middle East Technical University, Turkey

Date and editor's signature

Date

17.09.2017

Innovative Competencies of Mining engineers in Transition to the Sustainable Development

Andrey Krechetov[1,*], *Alexey* Khoreshok[2], Valery Blumenstein[3]

[1]T.F. Gorbachev Kuzbass State Technical University, Acting Rector, 650000 Kemerovo, 28 Vesennya st., Russian Federation

[2]T.F. Gorbachev Kuzbass State Technical University, Director of Mining Institute, 650000 Kemerovo, 28 Vesennya st., Russian Federation

[3]T.F. Gorbachev Kuzbass State Technical University, Machinery Technology Department, 650000 Kemerovo, 28 Vesennya st., Russian Federation

Abstract. The transition to the sustainable development posed new challenges to the system of mining higher education. They are determined by the acceleration of scientific and technological progress and widespread introduction of innovations, convergence of technologies from various industries. On the one hand, globalization and rapid technology development are constantly increasing quality requirements for the labor resources of the mineral and raw materials complex and constant improvement of their skills. On the other hand, the transition to the sustainable development provides the necessity for rational use of raw materials and environmental protection. This requires the improvement of staff support system for mining operations and the interaction of enterprises with universities training mining engineers, aimed at the innovative competencies development of future miners.

1 Introduction

Today, many branches of the mineral resource complex are experiencing the need for highly skilled managers and specialists, which is not only a deterrent for the development of national economies, but also one of the main reasons for the loss of life and large-scale pollution of environment caused by industrial accidents.

Along with this, the development of mining complex, the basic one for the Russian economy, has always depended on the latest development results in mining, engineering, and power industry. Mining and processing enterprises have always been high-technology facilities, where the rapid introduction of the latest developments in the production process took place. Kuzbass is not an exception; it is a region with an ultrahigh concentration of mining operations, which exert a significant load on environment. The sustainable development of such mining region is impossible without staff support, which can be guaranteed only by the existing educational system being ready to train innovative competencies for future mining engineers.

* Corresponding author: kuzstu@kuzstu.ru

The leader of engineering education in Kuzbass – T.F. Gorbachev Kuzbass State Technical University - has old traditions and unique "know-how" in the field of resource base and power industry [1]. It is responsible for creating the prerequisites for the transition of the extracting region to the sustainable development changing the quality of human capital at mining and processing enterprises.

2 Materials and Methods

The challenges that technical universities involved in training of mining engineers will inevitably face in the first half of the 21st century are the following:

1. The global inter-industry and inter-disciplinary integration, which enhances the development of individual industrial complexes, scientific and technological progress and structural changes in the economy, which together can dramatically change the structure of demand for engineering staff in the near future.
2. The trend towards aging of the population in industrialized countries, which is going to lead to employable population reduction and, as a result, the decrease in number of highly skilled staff which provide the operation of mineral resource and raw materials complex facilities and conduct scientific research.
3. The development of geo-information technologies and unmanned mining of mineral resources that will transfer the profession of a mining engineer to a qualitatively new level.
4. The development of aviation, space remote sensing and navigation systems, which will result in the formation of integrated control systems for the search, exploration, extraction, transportation and processing of minerals. This will greatly enhance the role of IT in training future mining engineers.
5. The development of NBIC-convergence (nano-, bio-, information-, cogno) technologies will have a revolutionary impact on the demand structure of the extractive industry in future. There will be the replacement of fossil fuels with synthetic and biomass, wide distribution of biological underground leaching of metals. Particular attention will be paid to post-mining and restoration of ecosystems by using bioremediation of soils and sub-soils contaminated with hydrocarbons. As a result, the professional competences of a mining engineer are substantially altered, as are the tasks set for him/her by the innovative development of the mineral and raw materials complex [2].

3 Results and Discussion

In order to meet the requirements set by the transition to the sustainable development for the higher mining education, a modern technical university should provide innovative activity in the following key areas of mining training:

- influencing on the policy in the field of mining and technical education at the regional and national levels and drawing up the recommendations and proposals for the innovative development of mining enterprises;
- implementing the concept of ideas and innovations' laboratory in the field of natural-and-scientific and engineering education, as well as promoting global dialogue between universities, finding new ways for innovative development of mining education, strengthening their importance in the business community of the mineral resource complex;
- providing the scientific-and-technical cooperation with enterprise-stakeholders to create an effective system for interaction between the research and production sector ensuring the demand for research and development and affecting the independence and competitiveness of the mineral and raw materials complex;

- creating favorable conditions for the global mobility of students, post-graduate students, teachers and scientists at inter-regional, national and international levels, expansion of inter-university cooperation in the field of academic and research programs;
- developing the continuous professional education system, including all the necessary knowledge, skills and experience of innovative activity and facilitating full integration of mining specialists into the international professional environment.

To achieve the goals of innovative development based on the strategic vision of sustainable development, the technical university must complete a number of long-term tasks:

1. The development of a unified approach to assessing the quality of training and minimum threshold knowledge of mining specialists through engaging stakeholders into assessing the level of professional skills of future mining engineers.
2. The development of joint educational programs and training plans together with employers that meet modern challenges and trends in the development of mining, and use, as the multiplier, the synergy of the best international achievements of the mineral resource universities' community.
3. Conducting joint research with innovative companies in strategic scientific areas of development of the mineral and raw materials sector and making promising proposals and recommendations in the field of mining, transport and processing of mineral raw materials.
4. Equitable access for students studying at the expense of the state, own means and programs of targeted training to the educational and research infrastructure, including available scientific centers, laboratories, installations and instruments for fundamental and applied mining education.
5. Holding joint multi-university educational and scientific sessions and practices, creating joint effective systems for advanced training of teaching staff, as well as specialists and managers of the mineral and raw materials complex.

To complete a number of interrelated tasks set before the university by the imperative of sustainable development, the target area in which innovative competences will be trained should include the following:

1. Manufacturing new equipment which provides a condition for the industrial development of non-traditional sources of hydrocarbon resources.
2. The development of systems and methods for increasing the extraction of mineral raw materials, including depleted deposits of energy carriers.
3. The introduction of systems for complex and deep processing of mineral raw materials for increasing the extraction of both main and associated components at the sites of minerals.
4. The discovery of new genetic types of deposits, as well as the expansion of the geography of search and exploration of mineral deposits.
5. The development of resource-saving technologies for complex coal enrichment for involving off-balance reserves in processing and profitable use of waste from concentrating plants.
6. Introduction of new developments that contribute to reducing the level of environmental pollution, including minimization of areas for storage and disposal of waste in the areas of mining baffles of coal mines and open pits.
7. The introduction of equipment for pre-enrichment, working on different principles, at a quarry or in a mine.

The areas of application of innovative competences of mining engineers in the transition to the sustainable development include Intelligent Mine, Green Mining, integrated development of mineral resources and energy-saving technologies for their processing, including the associated extraction of components.

Taking into account all mentioned above, the initial range of innovative competences of the mining engineer, from which the upgrading of his university training should begin, is the following:
- understanding the basic positions of IT embedding in mining.
- the ability to rationalize the production of high-quality end products due to a new look at the international ISO quality standards;
- the ability to treat man-made waste as a resource for restoring the economy and solving environmental problems in the regions;
- the skill in applying and improving technologies for increasing the efficiency of processing resources, which will allow creating a mechanism for regulating the distribution of mineral resources among regions;
- a systematic understanding of the rational use of raw materials in the entire technological cycle of their extraction and complex processing obtaining products with high added value;
- understanding the basic positions of IT and the Internet of Things-technologies.

Training of these innovative competences requires a review of the functions of a technical university, ensuring its transformation into a regional center for innovative development. In the transition to the sustainable development, the functions of Alma Mater should include the following:

1. Increasing the relevance of higher mining education as an element of high-quality and inclusive education for all throughout life and participation in the formation of mechanisms that promote the sustainable development in the mineral and raw materials complex.

To achieve that, modern technical university will need to develop concepts and mechanisms for involving stakeholders (consumers of innovations) in planning, managing, drawing up curricula and training programs for miners' training, developing and assessing their professional skills, and also for cooperation between universities and mineral resource enterprises in the region for training in the workplace.

Along with this, the university should develop and present relevant mechanisms and tools for identifying current and future mining engineers' competencies being in demand in the mineral complex in transition to the sustainable development. The success of this lies in ensuring the relevance and quality of training programs for mining engineers in the context of changes in the labor markets and innovative development of the economy.

2. Activation of scientific research at the mineral resource technical university for setting up an information basis for the development of breakthrough innovations that ensure the reduction of environmental damage from the increasing extraction of mineral raw materials.

In turn, this requires the establishment of an effective system of interaction between scientific research and educational activity of the technical university, ensuring the correspondence of higher education training programs to the real needs and prospects for the development of enterprise-stakeholders of the mineral and raw materials sector.

Technical university, accordingly, should play the role of ideas and innovations' laboratory in the field of engineering mining education, as well as promote dialogue between company-stakeholders, and search for new ways for the conceptual development of the mining and processing industry.

3. Promotion of processes and results of mining engineers' training focused on the acquisition of innovative competencies, the exchange of knowledge and the promotion of ideas and values of engineering education in mineral and raw materials sector of the economy.

To realize this function, the technical university - the center for innovative development of the region - should strive to create favorable conditions for the global mobility of students, post-graduate students, teachers and scholars and the expansion of inter-university

 DOI: 10.1051/e3sconf/20172100001

cooperation in the field of educational and scientific-research programs for mineral resource and raw materials complex.

The establishment of an inter-regional center for the engineering practical training sessions within the framework of international cooperation for scientific and staff support of the extracting region in transition to the sustainable development means inclusive access to the educational and scientific infrastructure, including functioning scientific centers, laboratories, unique facilities and instruments for fundamental and applied education.

4 Conclusion

All mentioned above will fully determine the vector of development of scientific and educational activities at T.F. Gorbachev Kuzbass State Technical University as it is being transformed into a center for innovative development of the region and organization of engineering practical training sessions. The goal of the next decade for Alma Mater will be to increase the prestige and attractiveness of getting mining engineer profession focused on training innovative competencies in the mineral and raw materials sector.

References

1. A. Krechetov, E3S Web Conf., **15**, 00001 (2017)
2. A.A. Khoreshok, S.A. Zhironkin, M.A. Tyulenev, G.A. Barysheva, V.Y. Blumenstein, M.C. Hellmer, S.V. Potyagailov, IOP Conf. Ser.: Mater. Sci. Eng., **142: 1**, 012122 (2016)

DOI: 10.1051/e3sconf/20172100002

From Mining Innovations to Sustainable Development: Keynote Speakers of the First to the Second International Innovative Mining Symposium

Michal Cehlár[1], *Juraj* Janočko[1], *Nuray* Demirel[2], *Seroni* Anyona[3], *Stefan* Vöth[4], *Maxim* Tyulenev[5,*], and *Sergey* Zhironkin[6]

[1]Technical University of Kosice, Faculty of Mining, Ecology, Process Control and Geotechnologies, Letná 9, 04200 Kosice, Slovak republic
[2]Mining Engineering Department, Middle East Technical University, Üniversiteler Mah., Dumlupınar Bulvarı, No: 1, 06800 Çankaya, Ankara, Turkey
[3]Jomo Kenyatta University of Agriculture & Technology, Mining, Materials & Petroleum Engineering Department, P.O. Box 62000 – 00200, Nairobi, Kenya
[4]Technische Hochschule Georg Agricola, Centre for Drive and Lifting Technology ZAFT, Bochum, Germany
[5]T.F. Gorbachev Kuzbass State Technical University, Department of Open Pit Mining, 28, Vesennyaya, 650000, Kemerovo, Russia
[6]T.F. Gorbachev Kuzbass State Technical University, Belovo Branch, 652644 Ilyicha st. 32, Inskoy, Kemerovo oblast, Russia

Abstract. 24-26 April, 2017 the Scientific Practical Conference "International Innovative Mining Symposium (in memory of Prof. Vladimir Pronoza)" was successfully organized at T.F. Gorbachev Kuzbass State Technical University - KuzSTU (Kemerovo Russia). More than 20 participants submitted their papers in presence, and more than 40 authors sent their papers for virtual participation. The main participants of the conference – Keynote Speakers – paid special attention to the development of international cooperation of technical universities as a form of accelerating the transfer of innovations in mineral resources mining and processing.

1 Introduction

The problems of training global competences of mining engineers [1], resource universities' graduates international cooperation and working abroad [2], old industrial territories development [3], achievement of new ecologically-oriented goals [4] are the issues of discussions of Kuzbass scholars. So honorable Keynote Speakers initiated fruitful discussion on the Plenary Session, which was continued on Pannel Sections, devoted to perspective mining technologies, environment saving development of mining, innovative mining

* Corresponding author: tma.geolog@kuzstu.ru

equipment, economic and social development of mining regions in general and Kuzbass (Western Siberia) in particular [5].

2 Results and Discussions

2.1 The Keynote Speaker Prof. Michal Cehlár, Dean of Faculty of Mining, Ecology, Process Control and Geotechnologies of Technical University of Košice (Slovak Republic): Assessment process of concept for mining and its impact on the region

In his Symposium Report Prof. Dean Michal Cehlár noted that the raw material policy of the Slovak republic reflects the all-society interest and is essential for the sustainable functioning of modern societies. Criticality of different metallic and mineral raw materials has gained attention in recent years. Consequently, the systematic evaluation of raw material criticality has been the subject of numerous studies in recent years.

Presenting their report Dean M. Cehlár initiated the discussion of the principles of evaluation of reconnaissance mineral resources, such as:

- reconnaissance mineral resources are estimated to the depths accessible to the current or future mining in accordance with current economic, mining, technical, technological and other conditions, while considering possible changes of these conditions in the future. The estimation is carried out in units of mass or tonnage in a limited reconnaissance area;
- the quantity of reconnaissance mineral resources is defined on the basis of expert estimations within the limits of large regions, basins or parts of deposit units;
- reconnaissance mineral resources are divided into Reconnaissance mineral resources P1 and P2, based on the positive evaluation of deposit indications and anomalies, and based on favourable stratigraphic, litologic, tectonic and paleographic assumptions.

Special interest of all participants was attracted to the position of mining studies and projects in the process of deposits evaluation which provides suitable information before verifying actual deposit (Fig. 1).

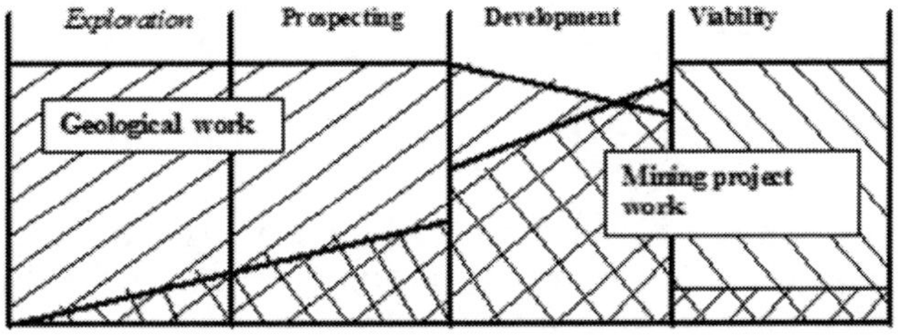

Fig. 1. Schematic display of consequences between geological and mining work in an industrial mineral deposit.

2.2 The Keynote Speaker Prof. Dr. Juraj Janočko, Vice Dean of Faculty of Mining, Ecology, Process Control and Geotechnologies of Technical University of Košice (Slovak Republic): Faculty BERG and T.F. Gorbashev Kuzbass State Technical Uniretsity: Perspective ways of international cooperation

Prof. Dr. Juraj Janočko described the place of Faculty of Mining, Ecology, Process Control and Geotechnologies of TUKE in the row of the word mineral resource universities. He presented F BERG TUKE as one of the most experienced mining Alma Mater in Europe,

 DOI: 10.1051/e3sconf/20172100002

which historically refers to the Mining Academy in Banská Štiavnica, promoted to the academic level by Maria Theresa, the Empress of the Austria-Hungarian Monarchy, in 1762.

Now Faculty of Mining, Ecology, Process Control and Geotechnologies of Technical University of Košice provides sufficient base for innovative development of underground and open pit mining, surveying, raw material processing.

It was achieved by the development of strong research and educational system of F BERG that comprises both traditional mining (geology; mining; raw materials processing, environmental technologies, mining mechanization and automation, blasting, transport etc.) and innovative disciplines (deep drilling, GIS, geothermal exploration, 3D seismic interpretation, photogrammetry and remote sensing etc.).

Privileged place of F BERG TUKE in Slovakian system of education is proved by the high confidence of the Government. The Faculty of Mining, Ecology, Process Control and Geotechnologies acts as a strategic partner of the Ministry of Economy and Ministry of Environment of the Slovak Republic in the field of defining the new Slovak Raw Material Policy. The Faculty established and maintains a long-term cooperation with numerous mining and raw material processing companies, regional and state authorities dealing with mineral resources extraction issues

But for the development of F BERG TUKE and KuzSTU inter-universities cooperation the most important feature of Faculty of Mining, Ecology, Process Control and Geotechnologies is its membership in Slovak Raw Material Platform (Fig. 2).

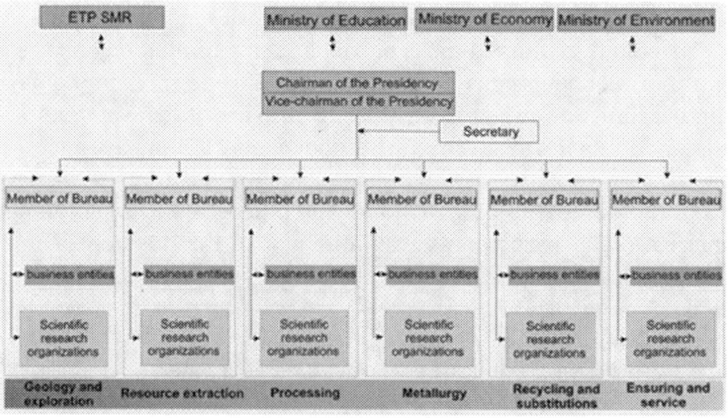

Fig. 2. F BERG – a member of Slovak Raw Material Platform

2.3 The Keynote Speaker Assoc. Prof. Nuray Demirel, Mining Engineering Department of Middle East Technical University (Turkey): Mining Projects in Turkey and METU Mining Engineering Department

In her Symposium report Lady Demirel presented one of the biggest and well-known technical universities of Asia - Middle East Technical University (Ankara, Turkey). Despite METU as a state university was founded not so long ago (November 15th 1956), it currently has 26,500 students, executes 43 undergraduate programs within 5 faculties, 5 Graduate Schools with 107 masters and 69 doctorate programs. 15 undergraduate programs and 3 graduate programs are offered in connection with METU Northern Cyprus Campus. The language of instruction at METU is English, and University has "School of Foreign Languages" which includes the English Preparatory Department and hosts over 1,700 international students from nearly 94 different countries studying toward myriad of academic degrees.

METU initiated a long-term program to have all its engineering programs accredited by the United States Accreditation Board for Engineering and Technology (ABET). Each year, among the students taking the National University Entrance Examination, over 1/3 of the 1000 applicants with the highest scores enters METU. Due to high demand towards METU, most of the departments accept only the top 1% of approximately 1.5 million applicants taking the National University Entrance Examination. Over 40% of METU's students go on to graduate school. The campus area is 4500 hectares and the forest area is 3043 hectares, including Lake Eymir, which is about 20 kilometers from the centrum of Ankara.

UK-based Times Higher Education (THE) ranked METU the 12th in the "Asia University Rankings" for 2015 on June 10, 2015, the 85th among the world's best universities in October 2016. In 2012, 2013, 2014 The World Reputation Rankings, ODTÜ was also named among the top 100 most reputable universities in the world.

During the discussion all the participants of the Symposium were agree that the important key for METU and KuzSTU cooperation development is the steady growth of mineral resource extracting in Turkey for the last decade (Fig. 3).

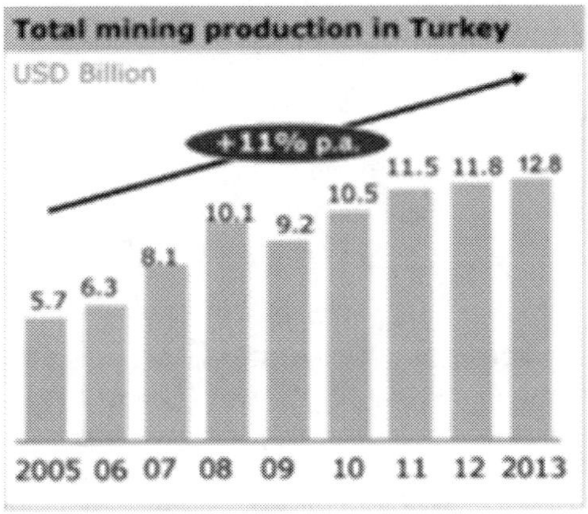

Fig. 3. Growth of mineral resource mining in Turkey.

2.4 The Keynote Speaker Assoc. Prof. Seroni Anyona, Mining, Materials & Petroleum Engineering Department of Jomo Kenyatta University of Agriculture & Technology (Kenya): The ways of establishing cooperation between JKUAT and KuzSTU

Mr. Anyona presented very interesting information about Jomo Kenyatta University, located more than 7,000 km from Kuzbass. Participants were pleased to know that JKUAT staff counts more than 2,250 people among which 761 are academic and 1489 non-academic. Academic structure of Jomo Kenyatta University of Agriculture & Technology includes:

1. College of Engineering and Technology (CoETEC).
2. College of Pure and Applied Sciences (CoPAS).
3. Faculty of Agriculture.
4. School of Architecture and Building Sciences (SABS).
5. College of Human Resource Development (COHRED).
6. College of Health Sciences (COHES).
7. Seven research centers: Institute of Biotechnology Research (IBR), Institute of Energy and Environmental Technology (IEET), Sustainable Materials Research and Technology Center (SMARTEC), Water Resource and Research Center (WARREC), Center for Urban Design and Planning.

8. Six Production and Innovation Centers. The students' body of JKUAT accounts for 35,000 students being taught 234 programmes.

Mr. Anyona demonstrated map of University linkages, including KuzSTU (Fig. 4).

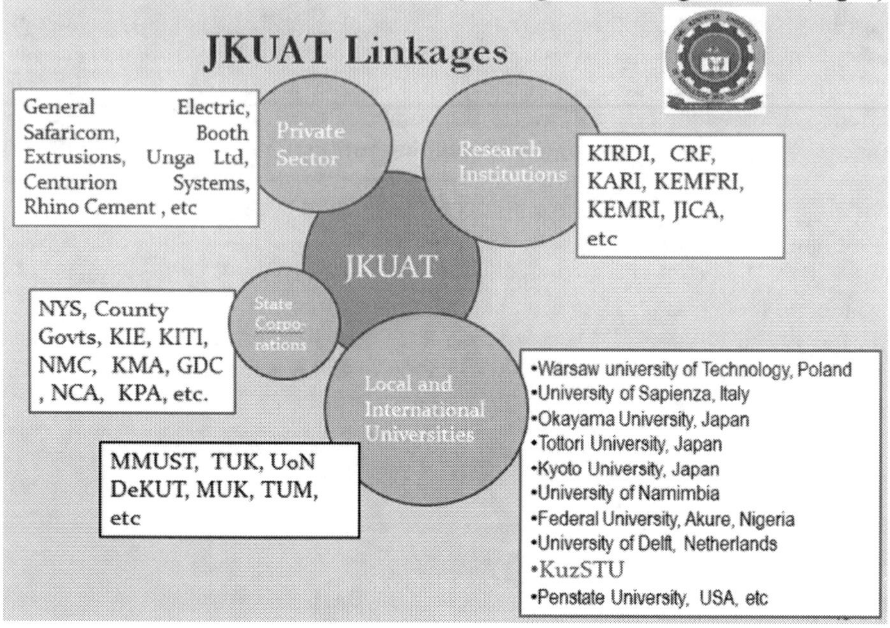

Fig 4. Map of inter-universities connections of JKUAT

2.5 The Keynote Speaker Dr.-Ing. Prof. Stefan Vöth, Technische Fachhochschule Georg Agricola, Bochum (Germany): Safety is a priority for innovations

Prof. Vöth introduced TFH Georg Agricola to the participants and visitors of Symposium as a private, state-recognized university in Bochum, founded in 1816 as a mountain school. After years it developed into a modern European university, with nearly 2300 students enrolled at the winter semester 2016/2017. 88% of them are male, 9% are students with a different nationality and 49% are graduates. Since 1995 this Technische Fachhochschule has received the name of the universal scholar Georgius Agricola, who was honored as a pioneer of mining and mineralogy.

Pof. Stefan Vöth told to the listeners that mining has an old tradition in the Ruhr area. In order to train the management staff for mining, a special school – Bergbauschule – was established 1816, and its goal was to provide people with a broad basic technical knowledge. The graduates of the Bochum miming school deserved at that time a good reputation that came ahead. In 1864, the Westfälische Berggewerkschaftskasse (WBK), which remained until 1990, became the bearer of the school through the merger of the Märkische and the Essen-Werden mining cooperative. In addition to the training of the experts, the Bergschule had to cope with further tasks, such as carrying out scientific examinations and making seam maps. Now TFH Georg Agricola specializes in mining machine technology and electrical engineering, surveying, mineral resource process engineering.

Substantial part of Prof. Vöth's report was devoted to Multifunctional Testing Rig for Machinery Safety Equipment. The attention of KuzsTU members – mining machinery engineers – attracted the drivetrain of testing rig (fig. 5).

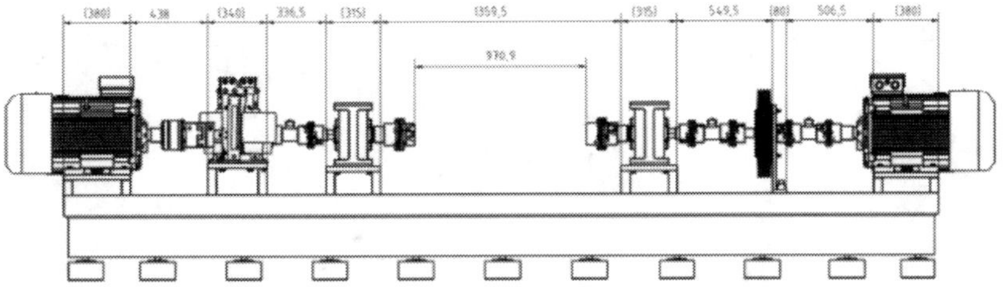

Fig. 5. Drivetrain of testing rig for Machinery Safety Equipment.

2.6 The Keynote Speakers Prof. Dr. Sergey Zhironkin and Prof. Maxim Tyulenev, Department of open pit mining, T.F. Gorbachev Kuzbass State Technical University (Russian Federation)

Prof. Zhironkin and Prof. Tyulenev reported about the man in whose honor and memory the First International Innovative Mining Symposium was named after - Dr. Prof. Vladimir Pronoza (1939-2015). The most famous books of Prof. Pronoza are the following: "Scientific bases of open pit mining (direct dumping method)", Kemerovo, 1984; Fundamental work "Typical technological schemes at coal open pits", Moscow, 1982; "Development of coal-bearing zones of quarry fields by excavator-and-transport complex", Kemerovo, 2010; "Open pit mining technology for complex-structured coal deposits development", Kemerovo, 2010.

Prof. Sergey Zhironkin described the leading role of Western Siberian mining cluster in Russian coal industry (Fig. 6).

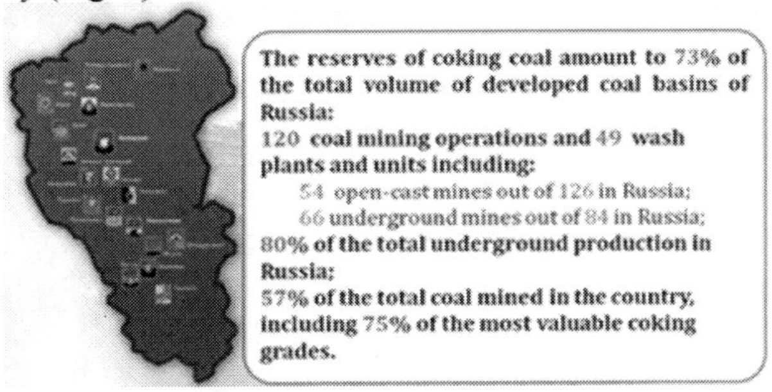

Fig. 6. Mining engineers – the future of Coal Regions Kuzbass.

Prof. Maxim Tyulenev emphasized that the competitive advantage of KuzSTU on higher education market is a unique platform for establishing business and scientific-technical cooperation with leading enterprises of Kuzbass. A relatively small area of the Kemerovo region has more than 5,500 industrial enterprises, in which the majority of specialists and managers are KuzSTU graduates who maintain a close relationship with their Alma Mater. Moreover, currently KuzSTU has 76 strategic agreements with enterprises of the Kemerovo region.

DOI: 10.1051/e3sconf/20172100002

Fig 7. The Keynote Speakers of the First International Innovative Mining Symposium (in memory of Prof. Vladimir Pronoza) at T.F. Gorbachev Kuzbass State Technical University. From the left to the right: Prof. Maxim Tyulenev, Dr.-Ing. Prof. Stefan Vöth, Prof. Dr. Juraj Janočko, Assoc. Prof. Nuray Demirel, Prof. Dr. Sergey Zhironkin, Prof. Dr. Oleg Misnikov (guest), Assoc. Prof. Seroni Anyona, Assoc. Prof. Alexandr Nikolaev (guest), Prof. Dean Michal Cehlár (photo courtesy of S. Zhironkin).

3 Conclusion

At the end of the First International Innovative Mining Symposium and following its results there was signed a Communique of International Cooperation of Universities.

Acknowledgement

We express their great gratitude to EDP Sciences and Isaline Augusto personally. Publishing the Symposium Papers with E3S Web of Conferences appeared to be beyond our expectation. We also would like to extend thanks to the members of EDP Sciences Isabelle Houlbert and Catherine Nissen for their utmost attitude to our issues and questions.

References

1. A.A. Khoreshok, S.A. Zhironkin, M.A. Tyulenev, G.A. Barysheva, V.Y. Blumenstein, M.C. Hellmer, S.V. Potyagailov, IOP Conf. Ser.: Mater. Sci. Eng., **142: 1**, 012122 (2016)
2. I. Pevneva, O. Gavrishina, A. Rolgayzer, M. Agienko, A. Myaskov, E3S Web of Conf., **15**, 04016 (2017)
3. V. Trifonov, O. Loyko, D. Nesteruk, S. Zhironkin, E. Strekovtsova, AIP Conf. Proceed., **1800**, 050009 (2017)
4. A. Krechetov, E3S Web Conf. **15**, 00001 (2017)
5. M.A. Gasanov, K.A. Kolotov, K.A. Demidenko, E.A. Podgornaya, O.V. Kadnikova, IOP Conf. Ser.: Earth Environ. Sci., **50:1**, 012025 (2017)

Gas Hydrates of Coal Layers as a Methane Source in the Atmosphere and Mine Working

Valery Dyrdin[1], *Sofya* Shepeleva[1*], and *Tatiana* Kim[1]

[1] T.F. Gorbachev Kuzbass State Technical University, 650000, 28 Vesennyaya St., Kemerovo, Russia

Abstract. Living conditions of gas hydrates of a methane in a coal matrix as one of possible forms of finding of molecules of a methane in coal layers are considered. However, gas hydrates are formed not in all mineral coals even under the thermobaric conditions corresponding to their equilibrium state as the minimum humidity and the corresponding pore width are necessary for each brand of coal for formation of gas hydrate. It is shown that it depends on electric electrical dipole moment of a macromolecule of coal. Coals of brands K, D, Zh were considered. The electric field created by the surface of coal does not allow molecules of water to carry out three-dimensional driving, and they keep on an internal surface of a time. By means of theoretical model operation a dipole - dipole interaction of molecules of water with the steam surface of coal values of energy of fiber interaction for various functional groups located in coal "fringe" which size for the first and second layers does not allow molecules of water to participate in formation of gas hydrates are received. For coals of brands K, Zh, D, considering distribution of a time on radiuses, the percent of moisture, which cannot share in education solid coal of gas solutions, is calculated.

1 Introduction

By underground mining of gas-bearing coalfields in some cases there are sudden emissions of coal and gas – the catastrophic natural phenomena accompanied with emission of hundreds of tons of the crushed coal and tens of thousands of cubic meters of natural gas. Not only destruction of excavations, cars and mechanisms, but also a deadly traumatizing of miners follow them. In spite of the fact that coal layer from a particular depth can be carried to threatened or dangerous on sudden emissions of coal and gas, emissions happen irregularly and not on all territory of the mine field, and in particular points. Apparently, it is bound to padding factors, which still were not considered when developing the theory of sudden emissions of coal and gas and methods of their prevention.

For practice of mining and environment protection against harmful emissions it is important to know not only total of the methane which is contained in coal, but also its phase state, that is a ratio between amounts of methane in the form of the free gas and a methane in other phase states. The role of various forms of existence of a methane in coal substance in formation of gas of dynamic processes is up to the end not clear, but from practice, it is known that the volume of the methane, which is allocated at the same time, exceeds natural gas

* Corresponding author: shepelevasa@kuzstu.ru

content. The thermodynamically coordinates assuming existence in coal layers of solid solutions of natural gases as gas hydrates are characteristic of the existing development depths in mines of Kuzbass. The factors promoting existence of gas in the form of gas hydrates in natural system are existence in reservoir gases of mixes of hydrocarbons and impurity of other gases, natural moisture, particular thermo baric parameters and also high mechanical tensions which at a depth of 500 m are about 12,5 MPa (and in zones of tectonic violations of tension can exceed this size [1]). It is possible to consider the following forms of finding of a methane in coal layers [2]:

- the free state in cracks and a time of coal. In this case the coal-gas system submits to laws of a condition of real gas;
- condition of the surface (physical) adsorption; this case the system submits to the law of Lengmyur, i.e. the amount of stripped gas is proportional to its pressure;
- the macromolecule of coal forms a chemical bond with molecules of gases, that is between an adsorbent and an adsorbate electron transfer takes place. Such type of communication is called chemical adsorption, or a chemosorption;
- crystallohydrate or just gas hydrate of a methane; fragile crystal connections which are usually unstable also exist in a coal matrix under particular thermo dynamic conditions[3].

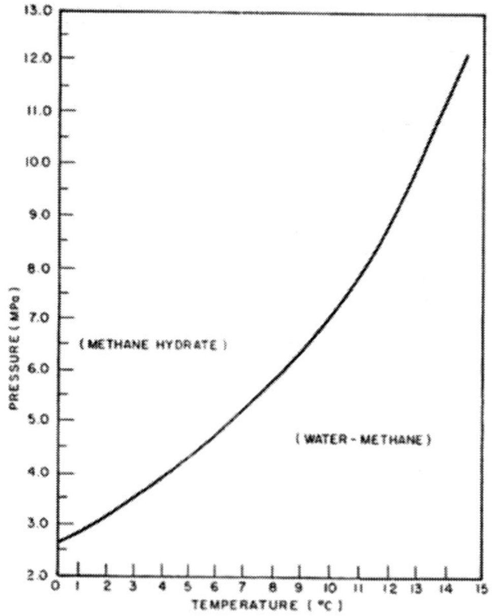

Fig. 1. Thermobaric zones of possible decomposition of hydrate of a methane on phases gas/ice and gas / the overcooled water at a depressurizing to various values [8].

Therefore, the gas balance of sudden emissions has to be presented by the sum of the free gas concluded in a coal pore space; the gas occluded on a surface of cracks and macropores, and gas, which is emitted at dissociation of gas hydrates.

2 Materials and methods

Earlier A.G. Zenin, 1973; I.L. Oettinger, 1974; Yu.F. Makogon, 1973 suggested that a methane in mineral coal under particular thermo dynamic conditions can be in a type of the gas hydrates representing the solid connections formed by a crystal lattice of molecules of water which cavities are busy with methane molecules. On average, six molecules of water are the share of one molecule of a methane in gas hydrate. The water necessary for formation

 DOI: 10.1051/e3sconf/20172101001

of hydrate of a methane is in a coal time and is connected by electrostatic forces with the surface of coal. Gas which is in the free state and water under particular thermobaric conditions practically is always formed by gas hydrates. But in a coal matrix of a molecule of gas and water are polarized under the influence of electrical dipole moment [4-6]. In the course of formation of gas hydrate water molecules which energy exceeds bond energy with the surface of coal can participate. Therefore, for the solution of a question of formation of gas hydrates in a coal matrix it is necessary to compare interaction energy of molecules of water with a surface of a coal matrix and energy of thermal motion.

Depending on thermodynamical parameters system moisture – methane can be in a state gas – hydrate (fig. 1) or gas – moisture, and states are divided by an equilibrium curve [7]. At a temperature about 10 °C transition of a methane from a hydrated state in gaseous is possible at sharp pressure drop from values above of 8 MPa (over a curve of phase change) to values of smaller 6 MPa (below a curve).

It is experimentally established that gas hydrates are formed not in all mineral coals, even under the corresponding thermobaric conditions. For coals of different extent of metamorphism there is a minimum humidity at which gas hydrates are formed [9].

The crystal lattice of coal represents graphite similar grids (fig. 2) consisting of hexangular rings. Electrical dipole moment of coals depends on functionality, contained in "fringe" of a macromolecule of coal, their concentration and the location on a surface. A task of this article is establishment of a possibility of formation of gas hydrates of a methane in coals of a different stage of metamorphism depending on their electrical dipole moment and pore size.

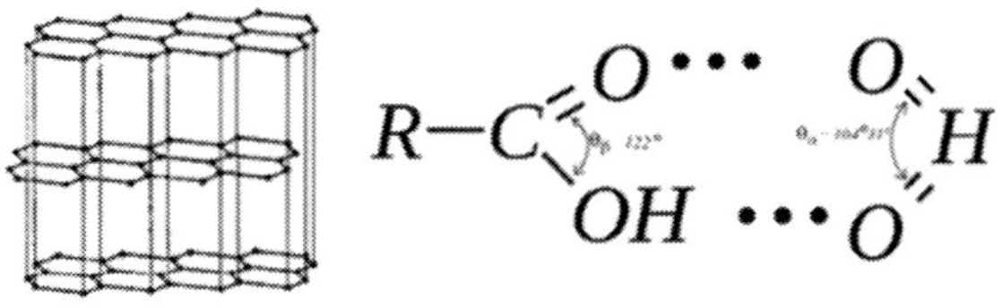

a *b*

Fig. 2. Model of interaction of the functional groups on the surface of coal with a water molecule: *a* – a coal crystal structure, a general view; *b* – an illustration of designations at record of potential of Shtokmayer on the example of the functional group of R-COOH and a molecule of water.

Let's consider model of interaction of molecules of water from one of the functional groups which is contained on a carbon surface. The model is limited to consideration of pair interaction of the molecule H2O and one of the functional groups, for example, by COOH (fig. 2). Molecules of water are caught by the surface of coal and lose an opportunity to carry out a three-dimensional Brownian motion. The mobility of the adsorbed molecules is limited to their twodimensional movement on a surface; molecules keep on an internal surface of a time, but do not get into a field of force of atoms and molecules of coal [10,11]. In most cases at the quantitative description use so-called potentials of an intermolecular interaction. One of widespread potentials of pair interaction of the molecules having constant electrical dipole moment is potential of Shtokmayer [12]:

$$U(r,\theta_\alpha,\theta_\beta,\varphi) = 4\varepsilon\left(\left(\frac{\sigma}{r}\right)^{12} - \left(\frac{\sigma}{r}\right)^{6}\right) - \frac{\mu_\alpha \mu_\beta}{r^3} g \ ,$$

$$g = 2\cos\theta_\alpha \cos\theta_\beta - \sin\theta_\alpha \sin\theta_\beta \cos\varphi \ ,$$

 DOI: 10.1051/e3sconf/20172101001

where ε – characteristic energy, J; σ – characteristic length, nanometer; r – center distance of the interacting molecules, nanometer; μ_α, μ_β – electrical dipole moments of molecules, Kl·m; θ_α, θ_β – polar angles of molecules, degree; φ – difference of lateral angles of molecules, degree.

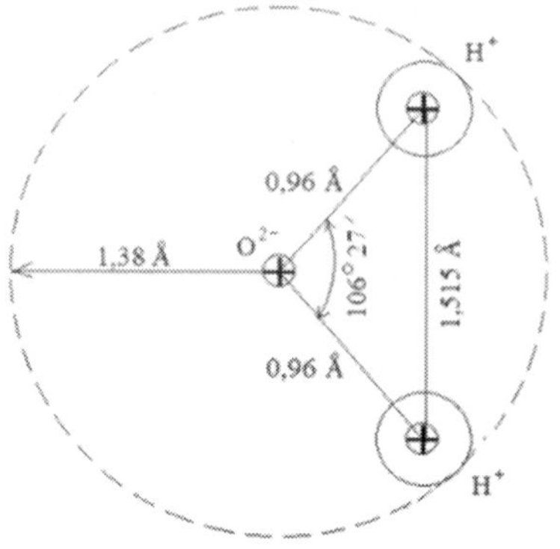

Fig. 3. Water molecule mode.

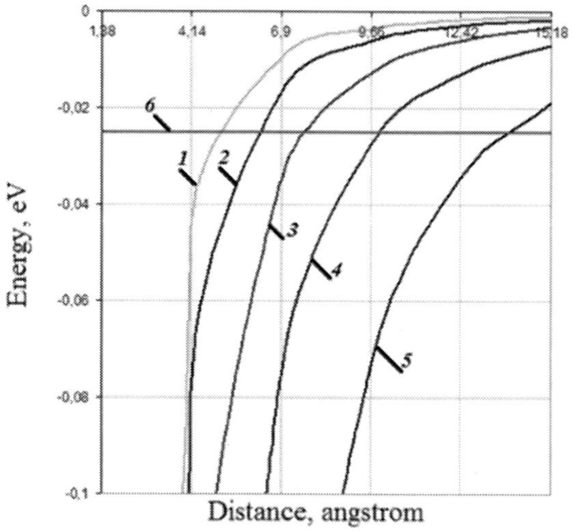

Fig. 4. Dependence of interaction energy of the functional groups on the surface of coal with water molecules from distance: 1 – with the C-O group; 2 – with the CH_3 group; 3 – with the C-H group; 4 – with the COON group; 5 – with the C=O group; 6 – thermal energy at T = 286 K

For the qualitative description of interaction of a molecule of water with the surface of coal we use model of Bernal and Fowler (fig. 3). According to the considered model, the center of the negative charge is located in the center of the sphere which radius is equated to the efficient radius of the molecule H_2O and makes size about 0,138 nanometers. As a first approximation, the water molecules which are immediately near the surface of coal are polarized along the field created by this surface. And at some distance from a surface polarization of molecules of water will be defined by a density fluctuation of distribution of volume negative charge or influence of the next polarized molecules.

3 Results and discussion

 DOI: 10.1051/e3sconf/20172101001

With use of potential of Shtokmayer was calculated energy of fiber interaction for each functional group with water molecules. In fig. 4 it is visible that for the first coat of molecules of water located apart 4,14 angstrom, energy of thermal agitation modulo is less than energy of dipole interaction of molecules of water with the functional groups on the surface of coal. Therefore, it is impossible to tear off the water molecules located in the first coat of rather coal surface.For different functional groups dipole-dipole energy will be various. The method ^{13}C of a nuclear magnetic resonance spectroscopy obtained the data reflecting changes in structure of coals of various extent of metamorphism [13, 14]. Several functional groups are allocated, we will consider only those which have electrical dipole moment: CH_3 (electrical dipole moment 0,4 D); OCH_3 (1,3 D); C-H (0,4 D); C-O (2,5 D); COOH (1,63 D); C=O (2,7 D). Using data [13] was calculated that for brand D coals electrical dipole moment can change within 0,04 – 0,17 of D; for coals of the Zh 0,03 - 0,28 D brand; for coals of the K 0,02 - 0,28 D brand.

Comparing the radius of molecules of water and cross sectional dimension of a time, it was received that in micropores (cross sectional dimension up to 0,6 nanometers) there can be only two molecules of water that corresponds to one layer; in the transitional time (cross sectional dimension of 1 - 10 nanometer) of 1 - 18 layers of molecules of water; in macropores (cross sectional dimension of 10 - 100 nanometers) of 1 - 180 layers of molecules of water. Considering distribution of a time, for coals of different brands the percent of moisture which cannot share in formation of gas hydrates is calculated.

4 Conclusion

1. It is established that for coals of the considered brands the moisture which is in micropores cannot take part in these processes. For brand D coals from transitional and macropores of a time it is not possible to use 33% of moisture. For coals of brand Zh and K up to 83%.

2. It is shown that at the uniform distribution of water on a time at natural humidity of 2% in coals with porosity of 10% education up to 132 g of hydrate of a methane on 1 t of coal is possible. In coal layers the congestion of gas hydrates represents not a monolithic piece, but distribution in structure of coal layer in the form of the particles comparable to pore sizes. Decomposition of hydrate in a porous medium or in selfcontained volume leads to the substantial increase of gas pressure. At the same time it can be allocated up to 160 – 180 volumes of pure gas that asks decrease in a share of a methane as greenhouse gas in gas balance of sudden emissions.

References

1. D. R. Hanson, Mining Engeneering, **37:1**, 54 (1985)
2. A.D. Alexeev, T.A. Vasylenko, SOLID STATE COMMUN, **130:10**, 669 (2004)
3. S. A. Shepeleva, V. V. Dyrdin, Journal of Mining Science, **47:5**, 660 (2011)
4. Dimo Kashchieva, Abbas Firoozabadib, Journal of Crystal Growth, **243**, 476 (2002)
5. P. Claudio, Jr. Ribeiro, L.C. Paulo, Chemical Engineering Science, **63**, 2007 (2008)
6. Matthew Clarke, P. Ray Bishnoi, The Canadian Journal of Chemical Engineering, **79**, 143 (2001)
7. V. G. Smirnov, V. V. Dyrdin, Z. R. Ismagilov, Fuel, **166**, 188 (2016)
8. Christophe Dicharry, Pascal Gayet, Gerard Marion, Alain Graciaa and Anatoliy Nesterov, J. Phys. Chem., **109**, 17205 (2005)
9. T. Kim, V. Dyrdin, V. Smirnov, E3S Web of Conferences, **15**, 01022 (2017)

10. I. L. Ettinger, Russian Chemical Reviews, **34:7**, 496 (1965)
11. E. R. Khabibulina, Z. R. Ismagilov, N. V. Zhuravleva, Workshop on Black Carbon, 20 (2016)
12. N. I. Fedorova, T. S. Manina, Z. R. Ismagilov, Solid Fuel Chemistry, **48:4**, 245 (2014)

The Mine Working's Roof Stress-strain State Research in the Perspective of Development of New Coal Deposits of Kuzbass

Svetlana Kostyuk[1], *Nikolay* Bedarev[2] , *Oleg* Lyubimov[1,*], and *Arthur* Shaikhislamov[3]

[1]T.F. Gorbachev Kuzbass State Technical University, Vesennyaya str., 28, Kemerovo, 650000

[2] Prokopyevsk branch of the T.F. Gorbachev Kuzbass State Technical University, Nogradskaya str., 19a, Prokopyevsk, 653033

[3] West Siberian Industrial Engineering Ltd., Kutuzova str., 39, Novokuznetsk, 654041

Abstract. The present now normative and information base is regulating of the Kuzbass coal seams treatment but is not considering of the mining-geological and mining-engineering conditions for new coal deposits. The analysis of works for the research of the rock pressure manifestation shows that in many cases numerous results require of the practical confirmation in mine conditions directly, and also confirmation by the physical models. This work reflects one of the stages of research on changing the stress-strain state of the massif with the formation of unloading zones, increased rock pressure, and recovery. As an example, the results of the information analysis obtained by means of contour and depth benchmarks on the ventilation drift in the course of the 34 seam treatment at the "Tagaryshskaya" mine are presented. The differences of the analyzed results from the results obtained in the conditions of other mines are established. The values of the drift's roof stratification on the contour and at the distance from the contour of 1.0 to 4.0 m are given. The revealed maximums of the rock pressure and pressure changes in the hydraulic supports of the complex used for movement are presented. Recommendations on the choice of the anchor's length taking into account the roof stratification size are given. The further research stages on models from equivalent materials at various geometric scales are proposed.

1 Introduction

In the "Guidelines for the rational treatment of sequence of closed steep and steeply inclined seams of Kuzbass", developed by VNIMI, which have actuality this time, the mining-geological and mining-engineering conditions for new coal deposits (Erunakovskoe, Kazankovskoe etc.) are not taken into consideration. That is why, to compose of the same normative document for these deposits, it's necessary to have mine and laboratory researches in accordance with terms of similarity under simulated testing of the seams on the physical models, including of the stress-strain state researches for the preparatory mine workings. It's

* Corresponding author: oleg_lyubimov@mail.ru

possible to use the results of the mine preparatory working's roof stress-strain state researches for the choice of the anchor's length, which exceeds the roof stratification size, and also for the practical registration of the rock pressure and pressure changes in the hydraulic supports applied to specific conditions.

There are no clear recommendations on the choice of the anchor's length taking into account the formation of zones of the increased rock pressure in domestic and foreign practice now.

In this regard the rock pressure manifestation researches in domestic conditions and abroad are carried out by complex method, including analytical part, the mine development modeling and the full-scale mine researches.

For example, the article [1] describes the mathematical modeling of a system linking the movement drift over of the longwall.

In the time, when the process of formation of zones of the rock pressure reference ahead of the coalface is studied, the modeling of the extraction area [2-5] is described.

In publications [6 – 11] with respect to local field studies, it was noted that the length of the zones of the rock pressure reference does not exceed 15…20% of the depth of development, and the values of the reference pressure can exceed hydrostatic 2…4 times.

In articles [12-13] the attempt to determine the influence of the coalface moving speed on the nature of the rock pressure manifestation is made. But, for justification of the obtained dependencies, additional researches are needed.

In works [14-15] the opportunity of the connection of the rock pressure manifestation with the relative strain changes in rock massif during treatment of a single seam and seam suits.

So, to obtain new information correcting and justifying of the modern instructions similar named above, the aim is tasked to study the rock pressure manifestations (for example, in the "Tagaryshskaya" mine) with the task of obtaining new data and comparison them with previously known for mine conditions and also with data of the physical modeling researches.

2 Method of experimental researches

Researches were conducted on the ventilation and conveyor drifts during mining seam 34-3 (longwall 3) in the presence the undermined seam 35 (the thickness between seams – 80 m) and the overmined seam 33 (the thickness between seams – 90 m).

When developing seam 34 (m=2.8 m, a=5-8 deg, N=320-340 m) in the longwall (L_v=250 m, Lpr=2500 m), the complex MCD was used. Excavation of coal was carried out by harvester KSW 460. The immediate roof strata is the siltstone fine-grained (m=20,0 m, f=3-4). The average speed of the movement of the longwall was 2.5-3.5 m/day.

Conveyor and ventilation drifts in the roof were anchored with a length of 2.2 m for 6 anchors at the roof bar of channel No. 20 (l=4.4 m), including 2 key anchors (with malleable washers) at the side of the ventilation drift and the 1 key anchor at the side of conveyor drift. Sides of drifts were anchored by SHK anchors with 1.8 m length, one anchor on the sides with the distance along strike and 1.0 m in alignment with the roof bars.

The displacement measurements were carried out on the contour and on the distance up to 4 m from the contour of drifts, on the distance 400 m from the mounting chamber. The depth benchmarks GR1, GR2, GR3 were placed in alignment with the contour benchmark KR at the distance: GR1 – 1.0 m; GR2 – 2.5 m; GR3 – 4.0 m.

Measuring of the depth and contour benchmark's displacements values was performed according to the VNIMI method. The measurement circuit is represented on fig. 1.

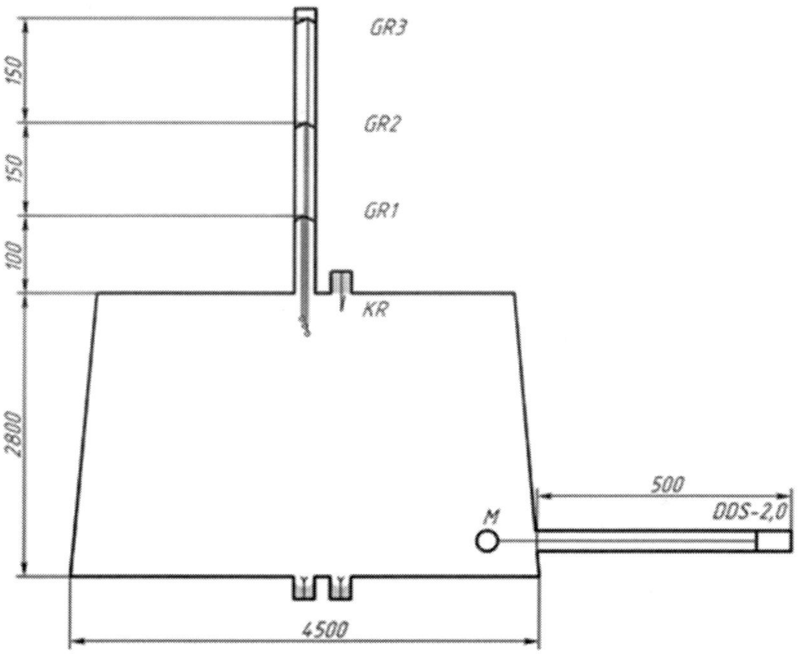

Fig. 1. The measurement circuit in the course of researches.

The contour benchmark was created by drilling the 0.2 m long hole, a hook was inserted into the wooden wedge from the side of the working's roof for the measuring tool clamping, and a pin in the working's bedrock was inserted into the wood with a recess for fixing the conical part of the measuring tool.

From the depth benchmarks fixed in the bore by spacers, the numbered flexible links with loops for engaging the measuring tool are withdrawn, and there is also a pin in the bedrock with a recess for fixing the conical part of the measuring tool.

Simultaneously the formation of the reference pressure at the bedrock by means of the hole pressure detector DDS-2,0, and also the character of the pressure manifestation in the hydraulic supports at the middle of the longwall were studied. Changes in the reference pressure were recorded with a manometer M.

3 Results and discussion

The displacements of contour KR and depth GR1 – GR3 benchmarks on the ventilation drift are represented in table; on the fig. 2 the displacement diagrams are shown. Results are lower by 5-7% on the conveyor drift. It is quite explainable by lack of availability of the developed area behind of the conveyor drift.

Analysis of the table and diagrams shows, that the contour benchmark's displacements are fixed at the approach of the coalface at 50 m to observation station and whey are reaching a maximum in alignment with the line of the coalface movement.

Table. 1. The values of the contour and depth benchmark's displacements on the ventilation drift in dependence of the coalface movement.

Names of benchmarks	Distance from the coalface, m											
	50	45	40	35	30	25	20	15	10	5	0	-5
	Displacements of benchmarks											
KR	1.9	5	10	20	29	31	49	77	94	115	119	120

DOI: 10.1051/e3sconf/20172101002

GR1		2	3	7	12	14	15	35	47	52	55	56
GR2			2	3.3	5.4	6.6	7.7	17	24	27	28	29
GR3				1.6	2	3.3	3.6	8	11	14	14	14

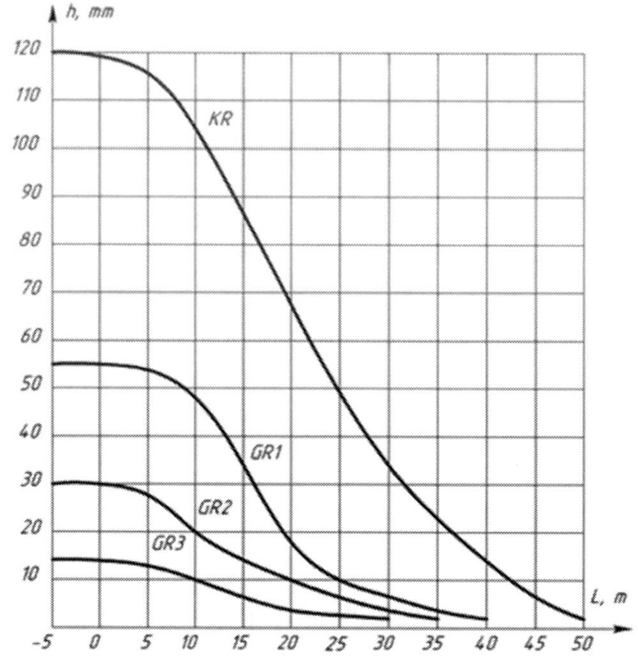

Fig. 2. Diagrams of displacements of contour (KR) and depth (GR) benchmarks.

Displacements of the GR1 depth benchmark on the contour distance 1.0 m had begun manifest in approach of coalface on 37 m to measuring point and had reached 55 mm with the line of the coalface.

Displacements of the GR2 depth benchmark had begun manifest in approach of coalface to measuring point on distance 33…35 m and had reached 28…29 mm with the line of the coalface.

Displacements of the GR3 depth benchmark had begun manifest in approach of the coalface on distance 30 m and had reached maximum in 14 mm.

The character of change of pressure in hydraulic supports oscillated in limits 10.5…17.5…10.5 MPa. These oscillations can be explained by the fact that the periodical growth of maximum occurred as secondary caving steps of the immediate roof every 6…8 m (fixed in another publication about developing of this longwall).

Maximum of pressure with concentration factor K=2.1 had manifested at the coalface approach to DDS-2,0 detector on 1.8…2.0 m. The small concentration factor value compared with the another author's data (K=2.5…4.0) can be explained by the fact that in this case developed space was constantly filled by collapsed rocks of the immediate roof represented by the siltstone (fulfilled the role of the backfill array).

The existing experience of simulation the treatment of seams on physical models shows that it is practically impossible to simulate the work of the supports with different configurations on small scale models. But it is possible to describe the change in the stress-strain state of the rock massif and the change in the character of the redistribution of the rock pressure in moving of the coalface. Therefore, the idea which has been put for forward researches is: after the modeling on a small scale (1:100, 1:200) to use the obtained results of the formation of unloading zones, zones of increased rock pressure and damping zones on large-scale models (1:20, 1:30) and simulate step-by-step functioning of supports for further comparative analysis.

4 Conclusion

1. As a result of researches, carried out at the "Tagaryshskaya" mine, new data about the stress-strain state of the rock massif around of the preparatory mine workings on the single seam with the presence of overlying and underlying seams are received.

2. Researches of the rock pressure manifestations shows, that the established data significantly (2…3 times) lower than results, which were obtained at other mines with the hard-to-fall roofs.

3. The contour benchmark's displacements can be the basis of recommendations that anchor's length in the same conditions would exceed maximum of the roof stratification and would be about 2.5…2.6 m.

4. During researches of character of change in pressure in the hydraulic supports should include the pressure increase as steps of roof collapse (primary and subsequent), applied to specific conditions.

5. In order to clarify the nature of the redistribution of the rock pressure around preparatory mine workings, when they are overmined or undermined, there are not enough the obtained data. To obtain the missing data, an imitation of the treatment of the suit of seams on models from equivalent materials can be provided, with obtaining correction coefficients for the transition from models to real conditions.

6. In developing of normative documents for new Kuzbass deposits it is necessary to spend more additional researches (both full-scale and laboratory on the physical models), so as the separate researches in mine conditions (complex, time-consuming and expensive) are insufficient to justify of the rock pressure manifestation regularity with a wide variety of the mining-geological and mining-engineering conditions.

References

1. Q. Chen, K. Zhou, T. Long, F. Gao, Journ. of Inst. of Sc. Tech. Inf. of China, **5**, 200 (2008)
2. Y. Zhang, Zh. Wan, B. Gu, Ch. Zhou, Therm. Sc., **20**, 2149 (2016)
3. L. Barron, *Proc. Of 9th Intern. Conf. On Ground Control in Mining*, 142 (1990)
4. V. Sotskov, I. Saleev, Min of miner. depos., **4**, 165 (2013)
5. V. Sotskov, G. Gusev, Min of miner. depos., **6**, 197 (2014)
6. K. Stefan, Gluckauf **1**, 50 (2004)
7. K. Erer, A. Heidareh – Zadeh, Min. Sc. and Tech. **3**, 191 (1985)
8. P. Lane, Rept. Invest Bur Mines us Dep. Intez **89**, 29 (1985)
9. *Prediction of shears zones associated with outbursts in coal mines* (Lama R.D.S., 1987)
10. L. Fu-Chen, Zh. Wei, S. Ai-hua, El. J. of Geotech. Eng. **18**, 1221 (2013)
11. P. Peng-Zhi, J. Rutquist, F. Xia-Ting, Y. Fei, J. Quan, Rock Mech. And Rock Eng. **47**, 6 (2014)
12. Z. Bieniawski, *Rock mechanics design in mining and tunelling* (Rotterdam, 1984)
13. D. Oyler, C. Mark, W. Gale, J. Chen, *Performance of Roof Support Under High Stress in U.S. Coal Mine* (Littelton, 2004)
14. S. Kostyuk, N. Bedarev, O. Lyubimov, A. Shaikhislamov, E3S Web of Conf. **15**, 01009 (2017).
15. R. Scigala, Gosp. Sur. Min., **29**, 191 (2013)

Carbon-Containing Waste of Coal Enterprises in Magnetic Sorbents Technology

Ekaterina Kvashevaya[1], *Elena* Ushakova[1], and Andrey Ushakov[1*]

[1] T.F. Gorbachev Kuzbass State Technical University, 650000, 28 Vesennyaya St., Kemerovo, Russia

Abstract. The article shows the issues state of coal-mining enterprises carbonaceous wastes utilization, including by obtaining oil-sorbent. The characteristics of the feedstock are presented; experiment methods of obtaining a binder based on the livestock enterprises waste, of forming binder with filler (sawdust, coal waste); of pyrogenetic processing to obtain a sorbent are described. Possible options for the introduction of magnetite (a magnetic component) in the composition of the oil sorbent are considered: on the surface, in the volume of the granule and the magnetite core. In the course of the work it was found that the optimum content of coal dust in the sorbent granules is 75% by weight, and the most effective way of obtaining the magnetic sorbent is to apply the carbon material directly to the "core" of magnetite. However, in this case, the problem of finding an effective binder for magnetite arises. The option of applying magnetite on the surface of a carbon sorbent is not effective. Thus, at present, we use a mixture of coal waste, which binds to the uniform distribution of magnetite in the volume. The developed magnetic sorbents can be used in various weather conditions, including strong winds and icing of water bodies, as well as for small and medium currents.

1 Introduction

To date, one of the most large-scale sources of environmental pollution is mining, including coal – 92% of the total production of waste production according to 2015 [1]. Methods of utilization of coal waste are diverse; the choice of one or another method is based on the composition of the mass, the particle size distribution, and the volume of education.

There are technologies for the use of coal waste in construction, when obtaining rare-earth elements and granulated fuel, as raw materials for gasification, etc. [2, 3]. One of the promising directions may be the production of sorbents for cleaning water from petroleum products together with organic binders [4].

According to GREENPEACE in recent years, oil spills are the order of several million tons per year. In this case, there are no clear signs of a reduction in this volume [5]. The International Tanker Owners Pollution Federation (ITOPF) reports a decrease of the amount of oil that has entered water bodies, but it should be noted that spills with a volume of more than 700 tons are included in the statistics [6, 7].

* Corresponding author: brels@list.ru

 DOI: 10.1051/e3sconf/20172101003

Oil is a complex of substances consisting of more than 3000 elements, most of which are easily oxidized, so oil has a rather large toxic effect on plants and living organisms. Another important circumstance is the ability of oil, and especially its light fractions, to spread at high velocity along the water surface, thus forming a thin film of a large area [8, 9]. Because of this oil spills on the water are considered more dangerous than on the soil, where it is kept for some time by soil particles.

The severity of the consequences of accidents in the water areas is associated with the high rate of oil spread along the water surface, as well as the complexity of the technical means necessary to eliminate the consequences. [10] To reduce the consequences of accidents of spilled oil products, it is necessary to conduct an oil spill operation promptly and with maximum efficiency.

Mechanical methods of removing oil and oil products from the water with sorbents is the most perspective. However, the sorbents currently used do not always meet the requirements for effective liquidation of oil spills. In particular, powder sorbents are difficult to apply to spills, especially in windy weather.

The aim of the work is to develop a composition for obtaining magnetic oil sorbents by medium-temperature pyrolysis based on carbonaceous wastes from industrial enterprises, in particular coal mining waste and coal processing.

2 Materials and methods

Experimental studies consisted of the following stages:

1. Analysis of raw components.
2. Binder obtaining based on agricultural waste.
3. Preparation of molded granules from binder, filler and magnetite component.
4. Pyrolysis of the molded granules to obtain a sorbent.
5. Analysis and testing of the finished oil-sorbent.

At the first stage, the properties of the raw materials were studied in accordance with GOST. Coal dust, waste livestock farm, magnetite are used as a raw material

The binder was obtained by anaerobic digestion of the biomass of livestock enterprises in laboratory plants - methane tanks by the methods described in [11].

The experiments on mixtures pelletizing were carried out on a drum-type laboratory granulator, which consisted of a polyethylene drum 0.27 m in diameter, 0.31 m in length, and an electromechanical part. The key parameters for mixture pelletizing in a drum were determined experimentally, but at the same time, the roll-over regime was to be maintained, and the yield of the target product fraction was more than 50% by weight.

The speed of drum rotation ranged from 40 to 140 turnover/min. The required size of the molded granules was achieved by varying the speed of drum rotation: to obtain granules with an average diameter of 5 mm – 80 turnover/min, 10 mm – 60 turnover/min, 15 mm – 40 turnover/min.

The process of granulation took place within 30 minutes. After completion of the granulation, the raw molded granules were dried to constant moisture in a granulator using fans and sent to pyrogenetic processing.

In addition to the "classic" filler – coal fines, the mixture was added magnetite in various ways. Perhaps three models of introducing magnetic particles into the sorbent (Figure 1): "on the surface", "in the bulk" and "magnetite core".

DOI: 10.1051/e3sconf/20172101003

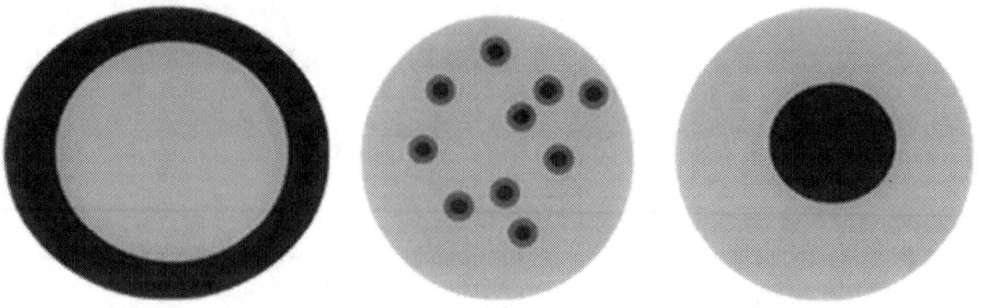

a *b* *c*

Fig. 1. Models of introducing magnetic particles into the sorbent: ● – mixture of filler and binder; ● – magnetic particles.

The first model provides the location of magnetite on the surface of the molded granules (Figure 1a). With this method, magnetite is not added to the pelletizing mixture, but is introduced in the last stage of granulation by dusting.

The second model (Figure 1b) is the distribution of magnetite particles in the volume of the sorbent. This model is realized at the mixing stage of the binder and filler, where an additional amount of powdered magnetite is additionally introduced.

In the third model (Figure 1c), a mixture of binder and filler is applied to the already formed core of magnetite by the rolling method, that is, the magnetite particles are the centers of granule formation.

The pyrogenetic processing of molded granules consisted in successive pyrolysis and cooling in air or inert gas. The experiments were carried out on an experimental pyrolysis plant (Figure 2).

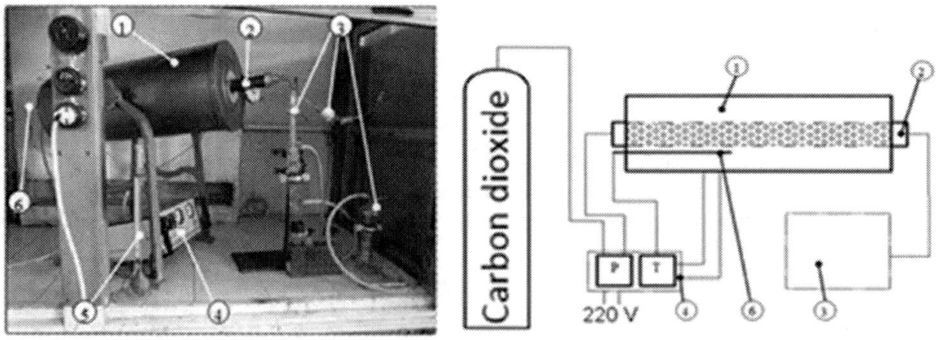

Fig. 2. Experimental pyrolysis plant for oil-sorbent preparing: 1 – tube furnace; 2 – pyrolyser reactor; 3 – cooling and gas cleaning system; 4 – control unit; 5 – foam flowmeter; 6 – temperature sensor.

The heating rate of the furnace is 15-25 °C, the maximum temperature is 1000 °C. The reactor of the pyrolysis furnace is a cylindrical steel vessel (internal diameter 27 mm, length 820 mm) with a useful volume of 4.7×10^{-4} m^3, equipped with fittings at both ends: for introducing inert gas and for removing the vapor-gas mixture.

Loading of the reactor-pyrolyzer took place in 2/3 of the volume, and then it was placed in a tube furnace for heating for 30 minutes. After the end of pyrolysis, the plant was turned off; an inert carbon dioxide-containing agent was supplied to the reactor for cooling.

The magnetic sorbent was tested as follows: oil was applied to the water surface in such a way that oil occupies the entire surface. A granule of a known mass was placed on the imitated spill and the sorption process was observed. After 24 hours, the oil -sorbent pellets were weighed, dried for 3 hours at a temperature of up to 35 °C (an increase in temperature is not permissible, since the gasoline fraction escapes from oil). The moisture capacity and oil capacity of the magnetic oil-sorbent were determined in experiment on the "oil-water" phase section. In a similar way, a separate determination of the moisture capacity and oil capacity in conditions of one phase, or "water", or "oil," is determined.

3 Results and discussion

The main properties of used raw materials are shown in Table 1.

Table 1. Properties of used raw materials

Component	Humidity, %	Ash content, %	Density, кg/m^3
Biomass	80.6	21.1	1,048
Coal dust (cyclone)	0.6	24.5	760
Magnetite	3.0	100.0	2,800

The moisture content of the biomass in anaerobic digestion was selected in such a way that combustible biogas with a methane content of more than 65% was released.

Thus, for molded granules producing we use a binder with the following characteristics: humidity – 89.2% wt., ash content – 24.2% wt., density – 1021 kg / m^3.

According to laboratory tests, to obtain the optimum sorbent composition, it is necessary to use 21% wt. binder in the mixture for pelletizing. With a given amount of binder, the maximum yield of the desired product after pyrolysis corresponds to the processing of molded granules of diameter 2-10 mm. This is the minimum amount of binder necessary for the granulation process. With a lack of a binder, a cellular structure is formed. It should be noted that with an overabundance of the binder, the mixture adheres to the internal surface of the apparatus, which reduces the efficiency of the rolling process.

There are advantages and disadvantages in various ways of magnetite introduced. The method of powdering the finished shaped pellets with magnetite is the simplest in the implementation and allows the components to be evenly distributed over the granule. However, in this case, the "magnetite shell" peeled off by mechanical action also creates an obstacle to the penetration of oil into the granule sorbing carbonaceous material. Therefore, this method of introducing magnetite is not recommended.

The option of introducing magnetite into the granulation mixture causes the need for more thorough mixing to distribute the components evenly. Advantages of the method used: magnetite particles do not interfere with the sorption of oil and oil products, remain in the oil sorbent granule throughout the entire period of use.

Theoretically, "magnetite core" is the most effective, because the most expensive component is initially represented in the core of granules, and, consequently, its losses are minimized. Secondly, such a nucleus is easy to reuse (the spent sorbent is burned using heat, and the magnetite cores are separated from the ash by means of a magnet). However, in order to realize this method, it is necessary to search for a binder that does not do the magnetite heavier, is stable at temperatures up to 600 °C without air access (during pyrolysis), and does not burn at temperatures up to 1,000 °C. Since such a binder is not found, it is considered optimal to use magnetite distributed over the volume of granules.

In result of pyrolysis, the granules do not lose their original form, although the strength of the granules decreases by a factor of 2 (compressive strength of the initial granules is 1,500 g, the finished sorbent – 680 g.). However, it is worth noting that the introduction of magnetite has virtually no effect on the strength of the sorbent (when its content in the mixture is up to 15% wt.).

Within the framework of the laboratory study, an experiment of the dependence attractive force of the obtained oil-sorbents with different magnetite content from the distance to the magnet was performed (Figure 3).

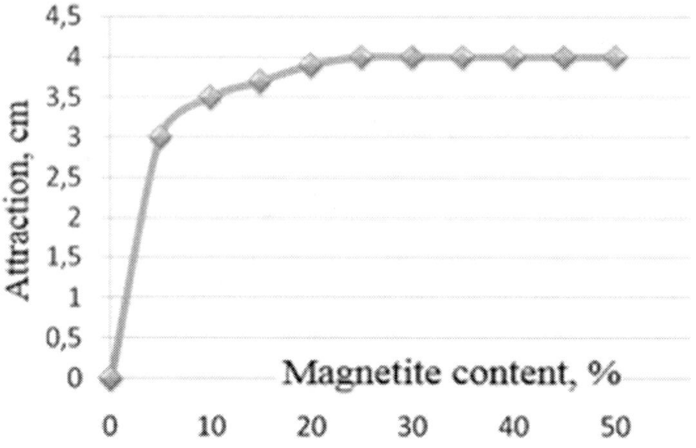

Fig. 3. The graph of the force of attraction (the distance from the magnet to the sorbent) dependence on the content of magnetite in the oil-sorbent.

As seen with 5-10% wt. containing of magnetite, attraction to the magnet is observed at a distance of 37 mm, after additional introduction of magnetite this value practically does not change. In practice, to increase the distance from the magnet to the oil-sorbent, it is necessary to increase the magnetic field strength or choosing another magnet material.

Below is a comparison of the characteristics of the oil-sorbent with the introduction of magnetite and without it (Table 2). From practical data it can be seen that the magnetic sorbent is not inferior in performance to the usual one. A significant difference can be seen in the parameter ash content, since magnetite enters the inorganic part of the sorbent and during combustion passes into ash.

Table 2. Comparison of magnetic and conventional sorbent characteristics

Parameter	Conventional sorbent	Magnetic sorbent
Humidity, %	2.0	2.8
Ash, %	22.4	30.6
Oil capacity, g/g	3.5	3.0
Moisture capacity, g/g	2,1	1.6
The yield of volatile substances , %	35.2	33.3
Density, кг/м^3	195.0	345.0
Compressive strength, g/granule	520.0	550.0
Buoyancy, days	20	20
Bulk density, кg/m^3	151.0	165.0
Adsorption capacity for iodine, g/g	4.3	4.2
Additional characteristics	-	Magnetically controlled

The possibility of extracting magnetite from the spent sorbent after combustion is very important: the resulting ash is placed in water and thoroughly mixed to "wash" magnetite from the ash; the pure magnetite is subsequently collected by an electromagnet and sent for reuse.

In case of using of magnetic sorbent in large reservoirs, the use of existing single-hull magnetic vessels is recommended, the basic scheme of which is shown in Figure 4 [12].

DOI: 10.1051/e3sconf/20172101003

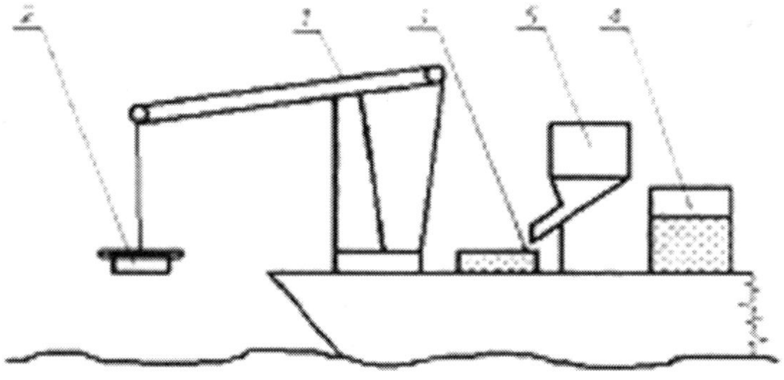

Fig. 4. Scheme single-hull vessel: 1– lifting device; 2 – an electromagnetic washer; 3 – initial adsorbent; 4 – assembled tank; 5 – bunker with initial sorbents .

The device for collecting the saturated oil-sorbent works as follows: the electromagnetic washer attracts a predetermined amount of the initial sorbent from the container 3 and places it on a layer of spilled hydrocarbons for 5-7 minutes. After the sorbent is saturated with a magnetic washer, it is removed from the oil product layer and transferred to a collecting container 4 above which the electric power supply of the washer is disconnected and the saturated sorbent falls into the collecting container under its own weight. Then the cycle repeats.

4 Conclusion

The use of coal mining waste, in particular coal dust, in the production of sorbents is possible, while its content can vary in a wide range. However, the optimal composition is: 75% wt. coal mining and coal processing waste, 20% wt. binder, and 5% magnetite.

References

1. *State report "On the state and on the protection of the environment of the Russian Federation in 2015"* 639 (M.: Ministry of Natural Resources of Russia; NNA-Nature. – 2016)
2. Andrey Ushakov, Elena Ushakova, Gennadiy Ushakov, Taishan Academic Forum – Project on Mine Disaster Prevention and Control, 445 (2014)
3. Andrey Ushakov, Alternative energy and ecology, **7**, 106 (2011)
4. Elena Briukhanova, Gennadiy Ushakov, Andrey Ushakov, Alternative energy and ecology, **5**, 71 (2010)
5. Andrew Nikiforuk, *Coastal Science and Societies, A Brief Overview of the Oil Spill Problem in Russia* (2016)
6. *Oil tanker spill statistics, Annual report of International Tanker Owners Pollution Federation Limited (ITOPF)* (2016)
7. Mervin Fingas, *Oil Spill Science and Technology*, 1078 (Gulf Professional Publishing, 2016)
8. Stephanie E. Chang, Jeremy Stone, Kyle Demes, Marina Piscitelli, Ecology and Society, **19:2**, (2014)
9. E.R. Gundlach, Journal of Coastal Research, **39**, 39

DOI: 10.1051/e3sconf/20172101003

10. Elena Briukhanova, *Processes for oil-sorbent obtaining by pyrolysis of pellets based on wood waste and organic binder in layered apparatus*, 152 (Tomsk, 2013)
11. Flores Arias, *Development of a sorbent with magnetic properties based on iron oxides and metallurgical waste for liquidation of oil spills*, 137 (Belgorod, 2012)

Predicting the Possibility for Deep Hydroprocessing of Some Kuzbass Coals

Ivan Petrov [1*], and *Boris* Tryasunov [2]

[1] Institute of Coal Chemistry & Chemical Materials Science, Federal Research Center of Coal & Coal Chemistry, Siberian Branch of the Russian Academy of Sciences, 650000, Sovetsky 18 Ave, Kemerovo, Russia

[2] T.F. Gorbachev Kuzbass State Technical University, 650000, 28 Vesennyaya St., Kemerovo, Russia

Abstract. Using thermal analysis and photoacoustic IR spectroscopy techniques, the abilities to be hydrogenated for some low-metamorphosed Kuzbass coals have been estimated in the aim of studying the possibilities of their deep industrial processing to obtain liquid fuels and valuable chemical products. Two independent parameters for quick assessment of solid fuel suitability for hydrogenation processes under high pressure of hydrogen have been proposed: 1) the relative maximal rate of coal pyrolysis in the 300-500°C temperature range, $[(1/m_o) \cdot dm/dT]$ (calculated from thermal analysis data), and 2) the intensity of aliphatic IR absorption bands (the sum of their optical densities), normalized to the optical density of an IR absorption band associated with aromatic ring C=C-bonds [i.e., $(D_{2920}+D_{2860})/D_{1600}$] (obtained from photoacoustic FTIR spectra of coal materials). It has been shown that for a number of low-metamorphosed Kuzbass coals (Barzas sapromixite, Itatsky lignite and Karakansky long-flame coal) the above mentioned parameters were in good consistency with generally accepted hydrogenation ability parameters of solid fossil fuels – atomic H/C ratios and volatile matter values.

1 Introduction

One of the most important directions for non-fuel use of coals is their hydrogenation processing to produce artificial liquid fuels and valuable chemical components. Processes of direct coal hydrogenation are usually conducted at the temperatures of 400-600°C under high hydrogen pressure (15-20÷70-80 MPa) [1-7]. For intensifying hydrogenation reactions and providing less severe conditions of coal conversion, these processes are often carried out in the presence of hydrogen-donor solvents (tetralin, alkyltetralins, phenanthrene and anthracene hydro-derivatives, high-temperature fractions of oil and coal-derived products) with using relatively inexpensive catalysts (compositions on the base of iron, molybdenum, tungsten, zinc, nickel and cobalt compounds or their mixtures) [2-7]. Analysis of literature data shows [2-5] that for effective occurring of hydrogenation processes, it is most expedient to use coals containing 65-85% of carbon, ≥5% of hydrogen (mass C/H ratio is not higher

* Corresponding author: ipetrov@kemcity.ru

than 16-17), ≥30% of volatile substances (on dry, ash-free basis), ≤5.5% of ash content for hard coals and ≤10% for brown coals (if ash content exceeds these values, then coal should be enriched), with fusenised micro components content up to 15% and humidity up to 12%.

Among the coals that present within the Kuznetsk coal basin, the following low-metamorphosed solid fuels fully correspond to the above criteria: 1) Barzas sapromixites; 2) brown coals of the Itatsky deposit (lignites); and 3) long-flame and gas coals from the deposits of Belovo and Leninsk-Kuznetsk areas of Kuzbass, in particular, long-flame coals of the Karakansky-Western coal mine. In this paper, using the methods of thermal analysis and photoacoustic IR spectroscopy, an attempt has been made to estimate the possibility of using the Barzas sapromixite, Itatsky lignite and Karakansky long-flame coal as raw materials for direct hydrogenation processes under high hydrogen pressure.

2 Materials and methods

Proximate and ultimate analysis of coals was carried out using conventional methods (GOST standards). Derivatograms of coals in an inert medium were recorded with a Q1500D instrument (Hungary), which construction was modified by adding an ADC module I-7018P, produced by ICP DAS company (Taiwan), for analog-to-digital conversion of T, TG, DTA and DTG signals, followed by their registration on a personal computer using a SIMP Light program software (SIMP Light SCADA Company, Russia); the final mathematical processing of the data obtained was performed with an Origin Pro program complex (OriginLab Corp., USA). Sample charges in thermal analysis experiments were of ~200-220 mg; heating rate of ~10°C/min; α-alumina precalcined at 1200°C was used as the reference standard; the pyrolytic experiments were conducted in a flow of helium (~450 He cm^3/min). Photoacoustic infrared Fourier transformed spectra (FTIR spectra) of the coals under investigation were registered with a Perkin Elmer 2000 FT-IR spectrometer (USA) equipped with a photoacoustic MTEC detector (Model 200); high purity electrode graphite was used as the reference standard.

3 Results and Discussion

Basic characteristics of three types of Kuzbass coals chosen (proximate and ultimate analysis data) are listed in Table 1. It is considered [2, 8] that the most reliable characteristics of coals hydrogenation abilities are their atomic H/C ratio and volatile matter (V^{daf}) values; reactivity in the process of hydrogenation increases with enhancing these parameters. However, as it was noted in [9], in a number of cases, an important characteristic of coal hydrogenation ability is the relative maximal rate of coal pyrolysis in the 300-500°C temperature range, i.e., the maximal rate of mass loss per 1 g of coal in an inert medium, $[(1/m_o) \cdot dm/dT]$, which value also correlates to the yield of liquid products in the process of coal hydrogenation. In order to evaluate this parameter in the present paper, we recorded the derivatograms of three types of low-metamorphosed Kuzbass coals in a flow of helium.

A typical view of such derivatograms (specifically, for Barzas sapromixite) is presented in Fig. 1. Three main temperature ranges of coal thermal decomposition are observed in these derivatograms: 1) 20-25÷150-170°C (removal of adsorbed water), 2) 300-500°C (a primary or quick pyrolysis stage) and 3) >500-600°C (a secondary or slow pyrolysis stage). On the whole, the above described picture is characteristic for the processes of non-isothermal pyrolysis of solid fuels [10, 11]. In the primary pyrolysis stage, CO_2, CO, H_2O, methane and its homologues, as well as light unsaturated hydrocarbons, are the prevailing components among the gaseous products released; these products are formed during the decomposition of aliphatic structures and elimination of lateral substitutes from carbon-based framework of coal organic matter [10, 11].

Table 1 Characteristics of Kuzbass coals studied.

Coal sample	Proximate analysis (wt.%)			Ultimate analysis (wt.%, *daf* basis)			H/C atomic ratio
	W^a	A^a	V^{daf}	C	H	(O+N+S)	
Barzas sapromixite	3.4	40.5	66.2	84.1	9.5	6.4	1.36
Itatsky lignite	12.1	15.8	52.4	71.4	4.9	23.7	0.82
Karakansky long-flame coal	7.7	13.6	41.3	81.9	4.6	13.6	0.67

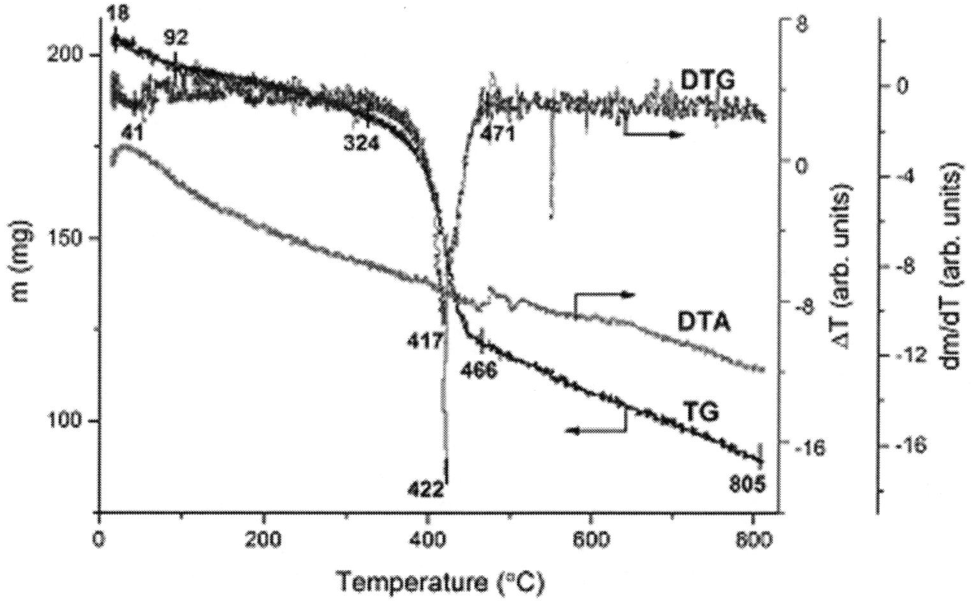

Fig. 1. A typical derivatogram of Barzas sapromixite in a flow of helium.

In the secondary (high temperature) pyrolysis stage, bond breaking and structural rearrangements of coal carbon-based matrix occur; in this process, the main gaseous products released are hydrogen and CO, as well as (in small quantities) CS_2 и HCN [11]. Calculated from derivatograms the relative maximal rates of pyrolysis in the 300-500°C temperature range for the Kuzbass coals studied, $[(1/m_o) \cdot dm/dT]$, are listed in Table 2. As it can be seen from these data, the values of these rates increase in the order of: Karakansky long-flame coal < Itatsky lignite < Barzas sapromixite.

To study the functional composition of coals under investigation, we used a method of photoacoustic IR spectroscopy. When recording IR spectra of solid and pasty materials, the main problem is that most of these samples are opaque in the IR spectral range and, therefore, are not available for measurement in transmitted light. As an alternative, various methods of analysis in reflected light are offered, but they also have their drawbacks. Thus, in order to register spectra of diffuse reflection, the sample often has to be strongly crushed (leading to destroy its crystalline structure), and, besides that, should be diluted using KBr; on the other hand, when using the method of disturbed total internal reflection (IRR), it is necessary to

 DOI: 10.1051/e3sconf/20172101004

take care of the reproducible optical contact with the surface of the sensor crystal [12, 13]. Photoacoustic spectroscopy (PAS) offers a simple and fast method of analysis for the majority of solid and powder samples. Since the photoacoustic effect produces a signal to be caused only by the energy absorbed by the sample, there are no here any undesirable obstacles, or they are very small. Therefore, the PAS method becomes more and more widespread when studying the structure of coals and carbonaceous materials [12, 14].

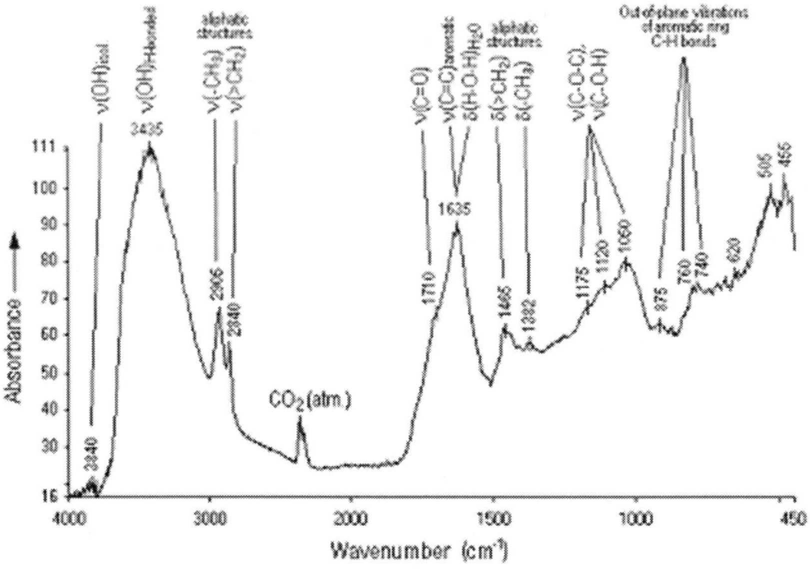

Fig. 2. A typical photoacoustic FTIR spectrum of Barzas sapromixite.

A typical photoacoustic FTIR spectrum of Barzas sapromixite is depicted in Fig. 2. Spectra of coals studied point out to the presence of broad IR absorption bands attributed to hydrogen-bonded OH groups (3435-3485 cm^{-1}), as well as weak signals related to isolated OH groups (~3840 cm^{-1}); the spectrum of Barzas sapromixite is characterized by the increased (compared to spectra of Itatsky and Karakansky coals) intensity of IR absorption bands associated with aliphatic CH_3- and CH_2<-groups (stretching vibrations at 2800-2900 cm^{-1} and deformation vibrations in the 1380-1460 cm^{-1} region) [2]. This is not surprising if we take into account the increased content of hydrogen in the elemental composition of the Barzas coal. The spectrum of this coal also clearly shows a shoulder with a maximum at 1710 cm^{-1} (stretching vibrations of carbonyl C=O groups) against the background of the intense absorption band of aromatic ring C=C bonds (~1600-1635 cm^{-1}). It should be noted that this intense IR absorption band related to aromatic structures partly overlaps with a band of deformation vibrations of water at ~1630-1650 cm^{-1}. The presence of O-bridged C-O-C and C-O-H groups in the spectra under investigation is reflected by a group of IR absorption bands in the 1000-1200 cm^{-1} region. In the 700-900 cm^{-1} range out-of-plane vibrations of C-H bonds of aromatic rings are manifested [1, 2].

Table 2 Various hydrogenation ability parameters for the coals studied.

Coal sample	**H/C atomic ratio**	**Volatile matter, V^{daf} (wt.%)**	**$[(1/m_o) \cdot dm/dT]^*$ $[(1/g_{samp.}) \cdot mg/°C]$**	**$[(D_{2920}+D_{2860})/D_{1600}]^{**}$**

Barzas sapromixite	1.36	66.2	2.236	1.3447
Itatsky lignite	0.82	52.4	0.848	0.8125
Karakansky long-flame coal	0.67	41.3	0.742	0.7481

*Relative maximal rate of coal pyrolysis in the 300-500°C temperature range.
**A sum of optical densities of IR absorption bands related to aliphatic CH_2< and CH_3 groups, normalized to the optical density of an IR absorption band associated with aromatic ring C=C-bonds.

It is well known [1-3, 10, 11] that the solid fuel primary pyrolysis stage in the ~300-500°C temperature range corresponds to mainly thermal decomposition of aliphatic structures in coals. Since aliphatic structures can easily be identified in the IR spectra of coals (by IR absorption bands of CH_3- and CH_2< group stretching vibrations at 2840-2860 cm^{-1} and 2900-2920 cm^{-1}, respectively [1, 3]), then the relative content of aliphatic structures in the coals studied (estimated from the data of their IR spectra) can serve as an additional independent parameter characterizing coal hydrogenation ability, along with the relative maximal rates of coal pyrolysis in the range 300-500°C and volatile matter values (V^{daf}) for them. In particular, a sum of optical densities of IR absorption bands attributed to C-H stretching vibrations in aliphatic structures ($D_{2920}+D_{2860}$), normalized to the optical density of an IR absorption band at ~1600-1630 cm^{-1} associated with stretching vibrations of aromatic ring C=C-bonds (D_{1600}), can be used as the above mentioned independent parameter for evaluation of aliphatic structure contents in solid fuels. The $(D_{2920}+D_{2860})/D_{1600}$ values calculated from the photoacoustic FTIR spectra of coals studied are listed in Table 2.

In Figures 3 and 4, the proposed coal hydrogenation ability parameters, $[(1/m_o) \cdot dm/dT]$ and $(D_{2920}+D_{2860})/D_{1600}$, for the coals studied in the present work are compared to H/C atomic ratios and volatile matter values (V^{daf}) for the given materials. Analysis of these results shows that very good correlations (practically ideal linear dependences) are observed between the relative maximal rates of pyrolysis of low-metamorphosed Kuzbass coals in the 300-500°C temperature ranges, optical density ratios of aliphatic and aromatic IR absorption bands in these coals and H/C atomic ratios in the given materials (Fig. 3). Somewhat less satisfactory (but quite acceptable) linear dependences are observed between the proposed hydrogenation ability parameters and volatile matter values in the coals studied (Fig. 4). This means that the above described $[(1/m_o) \cdot dm/dT]$ and $[(D_{2920}+D_{2860})/D_{1600}]$ parameters proposed for characterization of solid fuel reactivities, along with generally accepted parameters of coal hydrogenation ability (H/C atomic ratios and volatile matter values), can successfully be applied for an express assessment of the possibility of coal using as raw materials for hydrogenation processes under high hydrogen pressure.

DOI: 10.1051/e3sconf/20172101004

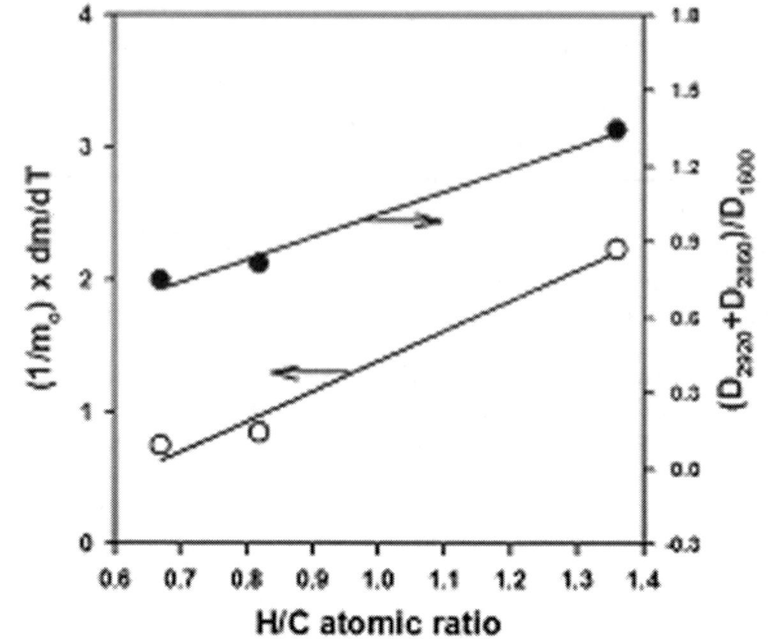

Fig. 3. Relationships between the H/C atomic ratios in Kuzbass coals, relative maximal rates of their pyrolysis in the 300-500°C temperature range (O) and $(D_{2920}+D_{2860})/D_{1600}$ optical density ratios of aliphatic and aromatic IR absorption bands (●) in these materials.

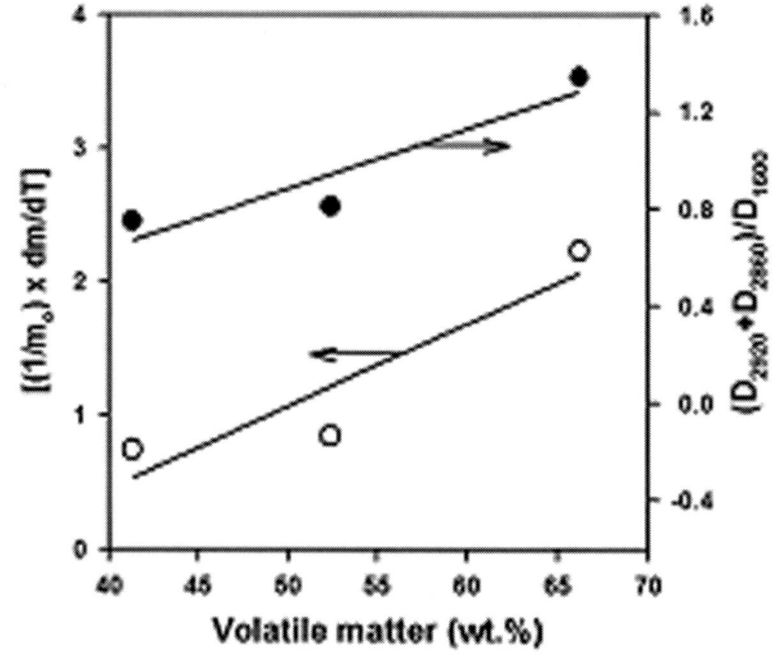

Fig. 4. Relationships between the volatile matter values in Kuzbass coals, relative maximal rates of their pyrolysis in the 300-500°C temperature range (O) and $(D_{2920}+D_{2860})/D_{1600}$ optical density ratios of aliphatic and aromatic IR absorption bands (●) in these materials.

DOI: 10.1051/e3sconf/20172101004

4 Conclusions

The relative maximal rates of coal pyrolysis in the 300-500°C temperature range, [(1/m_o) • dm/dT] (calculated from thermal analysis data) and (D_{2920}+D_{2860})/D_{1600} optical density ratios of aliphatic and aromatic IR absorption bands (obtained from photoacoustic FTIR spectra of coal materials) have been shown to be quite applicable parameters for preliminary express assessment of coals hydrogenation abilities (i.e., for checking possibility of using solid fuels as feedstocks in the processes of their direct hydrogenation under high hydrogen pressure).

References

1. J.G. Speight, *The Chemistry and Technology of Coal. 3rd Edition. – Boca Raton*, 845 (New York: CRC Press, Taylor & Francis Group, 2013)
2. V.G. Lipovich, *Coal Chemistry and Processing* 336 (Moscow: Khimiya, 1988)
3. *Coal Liquefaction Fundamentals*, (Washington, DC: ACS, 1980)
4. A.S. Maloletnev, M.Ya. Shpirt, Russ. J. Gen. Chem, **79:11**, 2499 (2009)
5. S. Vasireddy, B. Morreale, A. Cugini, C. Song, J.J. Spivey, Ener. Env. Sci, **4**, 311 (2011)
6. I. Mochida, O. Okuma, S.H. Yoon, Chem. Rev, **114:3**, 1637 (2014)
7. J. Xu, Y. Yang, Y.W. Li, Fuel, **152**, 122 (2015)
8. D. Gray, G. Barrass, J. Jezko, J.R. Kershaw, Fuel, **59:3**, 146 (1980)
9. M.M. Ibrahim, M.S. Seehra, R.A. Kegh, Fuel. Process. Technol., **25**, 215 (1990)
10. S.K. Janikowski, V.I. Stenberg, Fuel, **68**, 95 (1989)
11. S. Ma, J.O. Hill, S. A. Heng, J. Therm. Anal., **37**, 1161 (1991)
12. K. Michaelian, *Photoacoustic IR Spectroscopy. Instrumentation, Applications and Data Analysis. 2nd Edition.* 385 (Weinheim, Germany: Wiley-VCH, 2010)
13. J.F. McClelland, S. Luo, R.W. Jones, L.M. Beaverson, Photoacoustic and Photothermal Phenomena (Springer-Verlag: Berlin-Heidelberg, 1992)
14. J.A. Orrego-Ruiz, R. Cabanzo, E. Mejía-Ospino, Int. J. Coal Geol., **85**, 307 (2011)

Development and Substantiation of Parameters of Environmentally Friendly Technology for Filling the Vertical Mine Workings with Autoclaved Slag-Concrete

Andrey Uglyanitca[1*], and *Kirill* Solonin[1]

[1] T.F. Gorbachev Kuzbass State Technical University, 650000, 28 Vesennyaya St., Kemerovo, Russia

Abstract. The environmentally friendly technology for filling the vertical mine workings with autoclaved slag-concrete, prefabricated on the surface of the mine is presented in the article; the optimal parameters of filling technology are proposed. The developed technology for filling the abandoned vertical mine workings allows ensuring the environmental safety of the territories adjacent to the abandoned mine, utilizing slag dumps and providing the possibility of shaft recovery, if necessary, with minimal labor and material costs.

1 Introduction

According to the requirement of normative documents of the Russian Federation, vertical shafts and pits with unsatisfactory support, put down in unstable rocks, must be completely filled with waterproof non-shrink material up to the level of the earth's surface. The main requirements for the waterproof non-shrink filling mass: the filtration coefficient of the mass is less than 0.001 m/day; absence of the filling mass compression in the most loaded lower part of the opening.

The methods for filling the vertical mine workings with hardening mixtures in layers are developed in the Russian Federation., The expensive cement concretes or hardening mortar mixtures based on floured burnt rocks containing a large amount of cement-lime binder and special additives are used to create the filling waterproof and non-shrink mass in these methods. The proposed methods have not found practical application, due to the high cost of filling operations and the exhaustion of burned rock stocks in coal mines. In addition the filling of mine workings by hardening concrete mixtures completely eliminates the mine and excludes the possibility of its recovery and use in certain period of time after the termination of its operation.

2 Materials and methods

* Corresponding author: uav@kuzstu.ru

The autoclave slag concrete, produced on the basis of ash and slag waste from fuel and energy enterprises [1-9], will be an effective filling material for the abandonment of vertical mine workings.

It is known that the autoclave curing of slag-lime mixture mix depending on the parameters of the mixture and its autoclave treatment allows producing the low water permeability and compression autoclave slag-concrete. Therefore, by autoclaving the slag-lime mixture it is possible to produce the water-proof and non-shrink autoclave slag-concrete to fill the abandoned vertical mine workings based at cheaper mortar mixes containing less binding and special additives,. The use of ash and slag wastes to fill the vertical mine workings will allow to utilize of these wastes, which will be beneficial for the environmental state of the territories.

The experimental laboratory tests of compression and filtration properties of autoclave slag concrete, carried put in KuzSTU, showed that it is possible to use the autoclaved floured slag-lime mixtures for filling the vertical mine workings, but the production of waterproof and non-shrink filling mass is possible only with certain weight ratios of the components of the mixture, its chemical composition and autoclave curing modes [10, 11,12].

Table 1 shows the compositions of slag-lime mixtures and the parameters of their autoclave curing for producing the waterproof and non-shrink slag-concrete, laid down by the present research.

Table 1. Parameters slag-lime mixes and their autoclave curing, at which samples didn't contract and proved to be waterproof.

Coefficient of mixture basicity $K_{осн}$	**Slag and lime fractions**	**Water-binding ratio BBO**	**Pre-autoclave exposure duration, hour**	**Duration of pressure rise, hour.**	**Duration of autoclave curing under pressure 0,9 МПа, hour**	**Duration of pressure descent, hour**
0,3	0,08	0,5	4	0,75	6	5
0,5	0,08	0,5	4	0,75	6	5
0,7	0,08	0,5	4	0,75	6	5
0,7	0,16	0,5	4	0,75	6	5

The method for filling the mine opening with cylindrical autoclave slag-concrete blocks, previously manufactured on the surface of the mine (RF patent No. 2449129, 2012) is developed in T.F. Gorbachev Kuzbass State Technical University (KuzGTU). Filling the vertical mine workings according to the developed method is carried out as follows (see Figure 1).

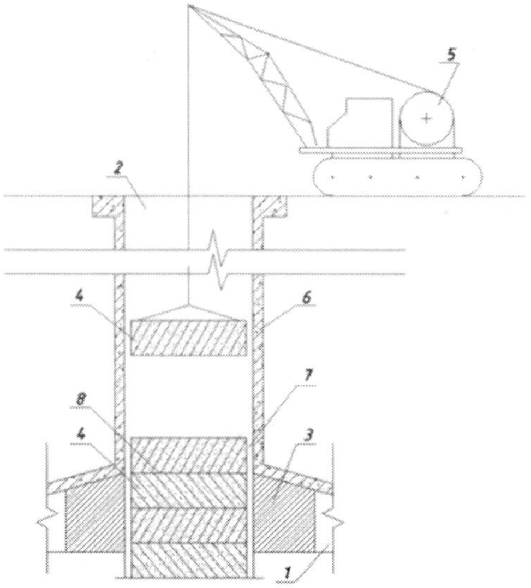

Fig. 1. Scheme of filling the vertical mine workings by the large-size cylindrical autoclave slag blocks.

The level openings 1 are filled in the areas of their junction with the vertical opening 2 by building the wall 3.

The cylindrical autoclave slag-concrete blocks 4 are manufactured using an autoclave mold on the surface and stored near the collar. The diameter of the cylindrical autoclave slag-concrete block 4 is taken less than the internal diameter of the concrete support of the shaft, taking into account the unevenness of its walls, from the condition that the cylindrical block does not jam when it is lowered into the shaft.

The shaft filling with cylindrical block is performed by pockets. The projected number of slag-concrete blocks is lowered down into the shaft within the limits of the pocket, and then the grouting container is lowered down into the shaft, from which the calculated amount of the grout is poured into the space 7 between the concrete support of the shaft 6 and the cylindrical autoclave slag-concrete blocks 4 and plugged. In this case, the solution also plugs the horizontal slits between the stacked cylindrical blocks 4. They pass to the laying by cylindrical cinder blocks of the overlying ridge. The blocks and и grouting container are lowered into the shaft using either jib (self-propelled or tower) or gantry cranes 5.

The method for the production of cylindrical large-sized autoclave slag-concrete blocks according to which the slag-concrete blocks are manufactured in a cylindrical autoclave form using steam wells (RF patent No. 2562307, 2015) is developed in KuzGTU. The autoclave curing of the concrete mixture in the autoclave form is carried out both through the open surface of the concrete mixture in the form and through the steam wells, which are located in a concrete mixture at a given distance from each other.

The distance between the steam wells in the mold is taken depending on the magnitude of the of the autoclave curing propagation radius from the steam well to the slag-lime mixture.

3 Experimental research methods

The experimental laboratory radial autoclave was developed to perform studies on establishing the relationship between the propagation radius of the autoclave curing of the slag-lime autoclave mixture from the steam well, the parameters of the slag-lime autoclave mixture and its autoclave curing.

Fig. 2 shows the design of the experimental laboratory radial autoclave.

 DOI: 10.1051/e3sconf/20172101005

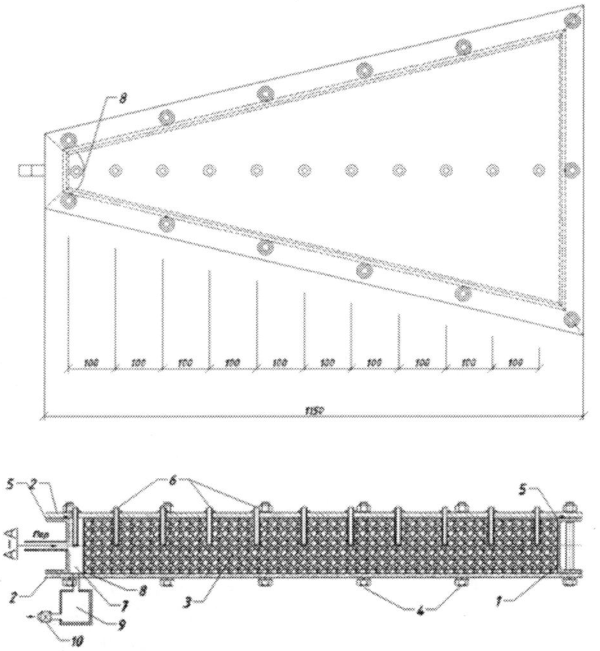

Fig. 2. Experimental laboratory radial autoclave.

The experimental laboratory radial autoclave is a prefabricated metal structure made in the form of a sector with an angle at the apex of 30° (see Figure 2) consisting of a frame 1, two 8 mm thick sheet metal covers 2. In the whole, the frame 1 and its two covers 2 form a radial autoclave chamber, which is filled with the slag-lime autoclave mixture 3. The lower cover is fixed and connected to the frame by the welded joint along the inner contour. The top cover is removable for the direct filling of the radial autoclave chamber with filling material. To ensure the integrity of the autoclave chamber in working condition, the top cover is connected to the frame by bolted connections 4. The paronit cushion is laid on the high-temperature sealant 5 between the frame and the cover. There are eleven wells 6 for placing thermometers in the top lid. The distance between the thermometers is 100 mm. The alcohol thermometers were filled with the glycerin before the installation.

The 76 mm diameter steam well 7 is made in the form of a sector with an angle at the apex of 30^0. The radial wall of the steam well is made of a steel mesh 8. During the autoclaving process, the steam condensate (water) is removed by gravity from the steam well 7 to the of steam condensate collection tank 9 equipped with a drain cock 10.

The essence of the experimental research method was to establish the effective propagation radius of the autoclave curing from the steam well, within which the non-shrink and water-proof autoclave slag-concrete was being produced in radial autoclave. The autoclave curing radius was determined from the temperature change of the slag-lime mixture in an autoclave form as it moved away from the steam well. The maximum water vapor pressure in the steam well was taken as $P_{скв}$ = 1,2 МПа, which corresponded to the vapor temperature in the well ≈187.0°C. The radius at which the temperature in the autoclave form was ≈174.5 ° C, which corresponds to a pressure of 0.9 MPa, was taken as the limit of the radius of propagation of effective autoclave curing from the steam well. The thermometer readings were recorded at the end of the autoclave curing - "duration of autoclaving at maximum pressure". The parameters of the slag-lime mixture and its autoclave curing were assigned from Table 1. In addition, in order to study the effect of the composition of slag-lime mixtures on the radius of propagation of autoclave curing from the steam well, the experiments with slag and lime fraction «-0,16» and $К_{осн}$ = 0,3 и 0,5were additionally performed.

4 Results and discussion

The results of the experimental studies are presented in Table 2 (the experiments, in which the waterproof and non-shrink autoclave slag-concrete for filling the laying of vertical mine workings was formed with the specified compositions of the slag-lime mixture and its autoclave curing, are presented in bold).

As a result of the performed studies, the ratios of the propagation radius of the effective autoclave curing of the filling mixture from the steam well are established depending on the parameters of the slag-lime mixture and its autoclave curing. As follows from Table 2, the maximum radius of autoclave curing of the slag-lime mixture equal to R = 0.53 m, at which water-proof and non-shrink autoclave slag-concrete for filling the vertical mine workings is produced, is attained at $К_{осн}$ = 0,3, slag and lime fractions "-0.08" the water-binding ratio of the mixture BBO = 0.5 and the "medium" autoclaving mode.

Table 2. Radiuses of autoclave curing propagation from the steam hole depending on mix compositions and autoclave curing parameters.

Radius of autoclave curing, м	$К_{осн}$	Slag and lime fractions	Water-binding ratio BBO	"Medium" autoclaving mode, hour
0,61	0,3	(-0,16)	0,5	4+0,75+6+5
0,53	0,3	(-0,08)	0,5	4+0,75+6+5
0,49	0,5	(-0,16)	0,5	4+0,75+6+5
0,34	0,5	(-0,08)	0,5	4+0,75+6+5
0,42	0,7	(-0,16)	0,5	4+0,75+6+5
0,21	0,7	(-0,08)	0,5	4+0,75+6+5

5 Conclusion

The developed technology for filling the abandoned vertical mine workings with waterproof and non-shrink autoclave slag-concrete allows ensuring the environmental safety of the territories adjacent to the abandoned mine, utilizing slag dumps and providing the possibility, if necessary, of shaft recovery with minimal labor and material costs.

References

1. L. I. Dvorkin, O. L. Dvorkin, *Building materials made of industrial wastes* (Rostov-on-Don, Feniks Publ., 2007)
2. A. A. Mechay, Ye. I. Baranovskaya, S. V. Lasankin, BGTU Proceed., **3**, 40 (2011)
3. A. A. Mechay, Ye. I. Baranovskaya, BGTU Proceed., **2**, 369 (2011)
4. D. M. Kuzmenkov, A. A. Sacovich, *Materials of the International Scientific and Technical Conference "Resource and energy saving technologies and equipment, ecologically safe technologies"*, 42 (2010)
5. M. A. Micheenkov, Cement and its application, **5**, 81 (2009)

 DOI: 10.1051/e3sconf/20172101005

6. P. Dolzhikov, K. Kiriyak, E. Ivlieva, *Progressive Technologies of Coal, Coalbed Methane and Ores Mining,* (Moscow, MMU. 2014)
7. Liu Hongjun, Yuan Feng, Yang Donghai., *The strenghth varieties of the seibsurface made of lime and fine coal ash of the Hingwaj from Changba to Baichengt. Dongbei linye daxue xuehao* (J. Nort-East Forest. Univ. 2000)
8. Chen Xiaotong, Shao Jiexicn, Zhang Jun, Chen Rongsheng, Don Younian, Zhang Fan. *Dongnan daxue xuebao. Ziran kexue ban*. (J. Southeast Univ. Natur. Sci. Ed. 2001)
9. P. Lahtinen, Kunnallisteknukka, **52:5** (1997)
10. A.V. Uglyanica, T.I. Khmelenko, K.D. Solonin, IJAER, **9:22**, 16837 (2014)
11. A.V. Uglyanica, K.D. Solonin, E.A. Strukova, Res. J. Appl. Sci. **10:2**, 84 (2015)
12. A.V. Uglyanica, K.D. Solonin The 8th Russian-Chinese Symposium. Coal in 21st Century: Mining. Processing and Safety, **16**, 66 (2016)

Rock Deformation Behavior Near Excavations Under the Influence of High Tectonic Stress in Coal Seam V-12, "Severnaya" Mine, JSC "Urgalugol"

Pavel Grechishkin[1]*, *Evgeny* Razumov[1], *Olga* Petrova[1], *Alexey* Kozlov[2], and *Evgeny* Aushev[3]

[1]Kuzbass State Technical University named after T.F. Gorbachev, branch in the city of Prokopyevsk, 19A, Nogradskaya Str., Prokopievsk, 653033, Russia.

[2]JSC "Urgalugol", 2 Magistralnaya Str., Khabarovsk region, Verchnebureinsky District, Chegdomyn, 682030, Russia.

[3]"RANK 2" Ltd., 119 Kuznetsky Ave., Kemerovo, 650055, Russia.

Abstract. The article is devoted to determination of roof bolt parameters in coal mine excavations when there is rather high tectonic stress. The work was complicated by the presence of considerable number of fractures in the adjacent rock of seam V-12 in "Severnaya" mine (JSC "Urgalugol") as well as variable stress field, low roof rock strength and heterogeneity, considerable rock shift during excavations development and maintenance. The article presents the results of horizontal and vertical stress measurements in the rock mass, physical and mechanical properties, roof rock structure and fracturing. According to the results of boreholes study by means of borehole endoscope the authors fully confirmed the forecasted parameters of areas with roof rock discontinuity, the boundaries of unstable rock. The article presents the measurement results of roof rock shift during development face advance. Based on the research, recommendations on excavations development and proper support using were given under conditions of seam V-12 in "Severnaya" mine (JSC "Urgalugol").

1 Introduction

Within last decade mines have begun to use intensive coal mining technology it turned out that current regulatory documents can be poorly adaptive to the new organizational and technical conditions of the working mines, and that due to frequent accidents and downtime in longwalls and development faces does not make it possible to enhance labor efficiency sufficiently and increase coal output using underground mining method [1]. As a result, some companies carry out scientific work aimed at studying mining, geological and technical conditions of intensive mineral resources extraction to gain and generalize practical experience and study results [2-4]. As a rule, different scientific and research companies do

* Corresponding author: pv_grechishkin@mail.ru

such research. The main goal of these companies is to implement the results of their studies in practice in coal mines and classify the guidelines for correction of the current regulatory documents.

The authors of the article have done the same work under conditions of "Severnaya" mine (JSC "Urgalugol") during seam V-12 mining on level +200 (about 230 m deep, maximum mining depth was 340 m). According to the results of preliminary survey and evaluation of mine excavations operation conditions the authors determined the following differences from the current regulatory documents:

- shifting of ribs, roof and floor of the excavations, especially at the intersection of the excavations and longwall, can reach 20% of the design size;
- falling in of roof rock in the development faces when they head to each other leads work stoppage and downtime up to accident response and recovery;
- sufficient variety in shifting of excavation ribs at their different positioning in reference to the meridian, which shows that there is heterogeneous tectonic stress field;
- vertical deformations of roof lead to the failure of the main support (rock bolts);
- in high rock pressure areas there are failures of the secondary support cable bolts.

The factors having negative effect on stability of excavations include geological faults, laminated and fractured roof rock, local zones with low rock strength, variation in coal seam thickness and properties of coal and rock, heterogeneous tectonic stress field.

The authors determined the goal and objectives of the study considering the importance of such work at present moment and influence of the factors mentioned above. They developed the sequence of procedures and chose proper methods to carry out an experiment under conditions of "Severnaya" mine (JSC "Urgalugol").

2 Materials and methods

The goal of the studies is to develop methodological guidelines which secure operational conditions of the excavations under conditions of "Severnaya" mine (JSC "Urgalugol").

According to the goal of the study the authors have set and met the following scientific and practical objectives:

- to determine the trends in physical and mechanical properties distribution in different parts of the mine field using rock samples taken out at boreholes drilling in the roof of the excavations;
- to determine the intensity of fracture distribution in the roof rock according to the results of borehole studies by means of borehole endoscope;
- to determine the direction and value of tectonic stress vector by means of partial coal mass relaxation in the face and ribs);
- to test the input parameters of the software using finite elements method according to the results of the experiment;
- to perform numerical modeling of geomechanical processes in a wide range of mining, geological and technical conditions and establish the parameters of supports to be used in excavations according to estimated stress and potentially unstable rock mass around an underground excavation;
- to reveal the parameters of tectonic field stress vector based on the results of numerical modeling and determine its influence on excavations which are drifted in different locations and parameters of the supports;
- to confirm the compliance of the forcasted parameters with real deformation parameters of the rock enclosing the excavations according to the results of the geomechanical processes monitoring.

These objectives were included in the research program. The results of the research are given below.

3 Results and discussion

The authors determined the trends in physical and mechanical properties distribution in different parts of the mining field according to rock samples testing. They found out that roof rock thickness and structure of seam V-12 are very heterogeneous. Immediate roof rock thickness varies from 2,0 up to 7,0 m. To make quantitative assessment of immediate roof parameters the authors did additional mine research which included borehole drilling in the roof, taking samples (drill samples), analysis of roof rock composition using borehole video endoscope.

11 drill samples were taken out of seam V-12 roof within three extraction areas. Results of measurements and tests of these samples were used as basic data at software testing and project documents correction.

According to the results of studies using borehole endoscope the authors estimated the intensity of fracture spread in the rock roof. They determined considerable discontinuity and heterogeneity of rock and coal mass. They include argillite and siltstone, carbon-bearing interbeddings and thin layers of sandstone and tuffite.

To determine directions and values of tectonic stress vector the authors carried out nine experiments to perform partial relaxation of coal mass in the face and ribs of excavations with changes in the deformation values.

The authors used the following scheme of the research: they installed dial gauge on the rib of the excavation in three directions; drilled a relaxation hole and determined deformations, installed a hydraulic jack into the borehole and gradually loaded the seam edge till coal broke down. The authors wrote down manometer and dial gauge readings [5]. After the results of the experiment were processed the authors determined the elasticity modulus, coal compression strength, values and direction of main stress vector (Table 1).

Table 1. Results of stress measurement in coal seam V-12.

Test No	Stress, MPa		Angle φ*, degree.	Elasticity modulus, MPa
	σ_1	σ_2		
1	8,31	20,58	23	2797
2	10,35	6,85	44	2174
3	11,9	4,13	24	3678
4	7,44	-8,74	21	3787
5	7,59	-3,84	-7	3636
6	7,56	1,64	1	726
7	7,28	9,75	-23	1282
8	6,57	19,38	38	2161
9	9,52	4,34	-19	2770

φ* – angle between X-axis and main maximum stress vector.

According to Table 1, stress field in the coal and rock mass under seam V-12 conditions of JSC "Urgalugol" is very heterogeneous. That is due to the high level of rock mass fracturing and complex pattern of tectonic stress field. Some tests revealed that in some cases horizontal stress can be 2,5-3,0 times higher than vertical stress. According to tests 4 and 5, there is tension stress in the right and left ribs of the conveyer roadway. That can be explained by the fact that the measurements were taken in area where there is influence of various rupture-type small-amplitude fractures.

The results of parameter measurements of strain-stress behavior and experiments were used to adjust input parameters of the software using finite elements method.

The authors determined the trends in display of rock pressure intensity on the direction of development faces advance and their position in relation to the line perpendicular to geological faults (Figure 1).

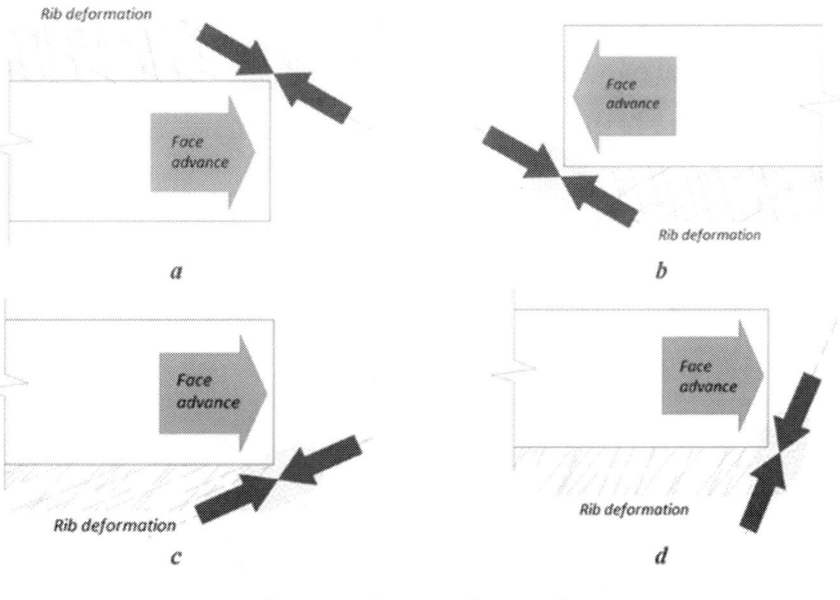

Fig. 1. Type of coal destruction in the ribs of the excavations and roof rock falls depending on the direction and maximum horizontal stress vector (red arrows): a – airways and conveyer roadways (mining direction – from the main excavations to the lateral excavations); b – airways and conveyer roadways (mining direction – from the lateral excavations to the main excavations); c – airways and connection excavations (mining direction – up-dip); d – installation chambers (mining direction – down-dip).

The most intensive rock and coal crashing takes place in the ribs of the excavation when main horizontal stress vector is directed almost at a right angle to the axis of the excavation (Figure 1b). When the angle between the direction of the face advance and main horizontal stress vector is acute coal distruction in the ribs of the excavations is less intensive. But in this case roof rock shift and volume of coal fall increase in the area adjacent to the face, especially in areas where there is maximum stress concentration (Figure 1 a-c) [6-9]. The authors found considerable damage of rock bolts in these areas of the excavations. They found deformations on the support elements, regular breaks of the rock bolts and cable bolt shafts, nuts tearing off the first level cable bolts. These phenomena took place in roof and ribs of the excavation depending on its advance direction.

According to the results of numerical modelling made using finite elements method, the authors determined the parameters of rock destruction adjacent to mine excavations. The models were developed considering information given in sources 10-15. As for areas outside high stress zone and longwall influence on the excavation, the authors forecasted two factured zones: arched zone (with depth up to 2,4 m from the roof) and a local one (at the depth of about 4-4,3 m).

To determine the causes of such an abnormal display of rock behavior in tectonic stress field the authors took measurements of roof rock shifts at drift advance using multiwire extensometers. The installation diagram of three-wire extensometer (RG3 type) in connection #8 near drifting face is given in Figure 2. The sets of three-wire extensometers RG3-1 and

DOI: 10.1051/e3sconf/20172101006

RG3-2 were placed at a depth of 8,0 m, 4,0 (4,1) m and 1,85 m in two boreholes in the crosscut of the excavation. The research results are given below.

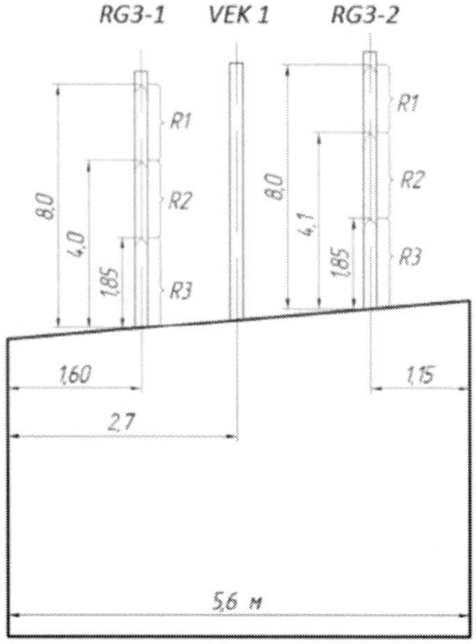

Fig. 2. Scheme of multiwire extensometers placement.

Measurements of multiwire extensometer were taken 1, 6, 10 and 14 m away from the face. The diagram in Figure 3 gives the multiwire extensometer data.

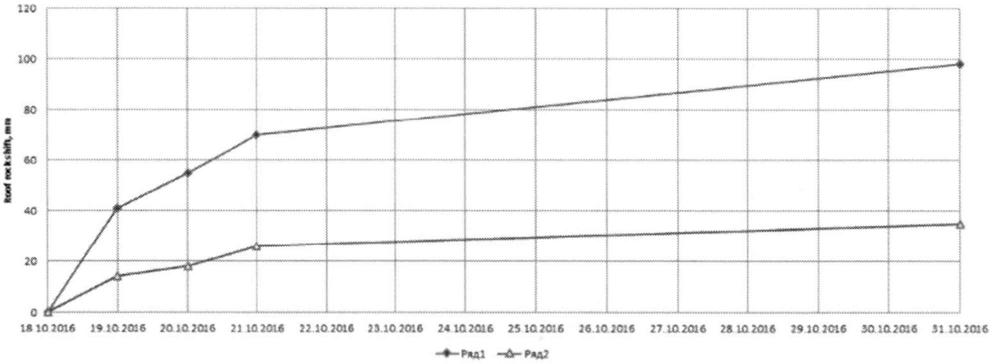

Fig. 3. General roof rock shift during observation (stations RG3-1 and RG3-2).

According to the results of borehole VEK 1 study using borehole endoscope in the roof of the drift when the distance between the borehole and face was 1 m the authors obtained the data characterizing rock discontinuity. The data fully comply with the readings of the three-wire extensometer when borehole depth was equal to 0-0,11 m (vertical fracture); 0,11-0,20 m (stratification, fractured rock); 0,32-0,41 m (vertical fracture); 2,05-2,11 m (slant fracture); 2,35-2,36 m (slant fracture); 3,95 m (slant fracture); 5,48-5,67 m (fractured rock). This confirmed high forecast accuracy of the developed numerical models (up to 95%).

4 Conclusions

Driving procedures and technology of using supports in development excavations imply coal mining by means of road header. The authors recommend to make excavations using road header with cutting boom and installation of roof bolts not closer than 0,8 m from the face line to avoid considerable roof rock stratification and rock falls.

It is possible to enhance the efficiency of supports in the excavations using arched or trapezium-shaped cross-cut when roof coincides with rock bolts and cable bolts in conjunction with channels and beams.

Enclosing rock intense relaxation zone is 20-30 m from face line. At the same time roof rock shifts can reach 100 mm.

To avoid overloading and rupturing of the rock bolts during roof rock relaxation it is necessary to use primary and secondary support with yielding not less than 100 mm.

Installation of secondary support should be made 15-20 m away from the face. It is necessary to take roof rock bolts with yielding which corresponds to relaxation of roof rock.

The developed numerical models let make an accurate forecast of roof rock deformation parameters in mine excavations located under conditions of V-12 (JSC "Urgalugol") when there is high tectonic stress.

References

1. I. G. Tarazanov, Ugol', **3**, 36 (2017)
2. Z.T. Bieniawski, *Bieniawski strata control in mineral engineering*, 212 (NewYork: Wiley, 1987)
3. D.H. Loubscher, Trans. Instn Min. Metall., **93** (1984)
4. D. H. Laubscher, Trans. S. Afr. Inst. Min. Metal., **9:10** (1990)
5. A. D. Sashurin, N. P. Vlokh, A. V. Zubkov, V. M. Tsaregorodtsev, Soviet Mining Science, **1:42** (1968)
6. G.M. Molinda, K.A. Heasley, D.C. Oyler, J.R. Jones, *Effects of horizontal stress related to stream valleys on the stability of coal mine openings: U.S. Bureau of Mines Report of Investigations*, 26 (1992)
7. C. Mark, T.P. Mucho, U.S. Bureau of Mines Technology Transfer Seminar, **94:01**, 53. (1994)
8. T.P. Mucho, C. Mark, Proceedings of 13th International Conference on Ground Control in Mines, Morgantown, **2:4**, 277 (1994)
9. J.F.T. Agapito, L. Gilbride, W. Koontz, Proceedings of the 24th International Conference on Ground Control in Mining. Morgantown, 196 (2005)
10. J. A. Hudson, J. P. Harrison, *Engineering Rock Mechanics: An Introduction to the Principles*, (Elsevier Ltd. 1997)
11. T.G. Sitharam, V. Vijay Kumar Geotechnical & Geological Engineering, **16:2**, 129 (1998)
12. S.S. Peng, *Longwall Mining. 2nd ed. Morgantown*, (WV: West Virginia University, 2006)
13. G. G.Shtumpf Russian Journal of Mining Sciences, **4:30**, 43 (1994)
14. R.O. Davis, A.P.S. Selvadurai, *Plasticity and Geomehanics*, (Cambridge: Cambridge University Press, 2002)
15. H. Kose, C.O. Aksoy, *Numerical Modelling for the Recovery of Safety Pillars*, (Lambert Academic Publishing, 2010)

New Technical Solution for Vertical Shaft Equipping Using Steel Headframe of Multifunction Purpose

Elena Kassikhina[1*], *Vladimir* Pershin[1], and *Yurij* Glazkov[1]

[1] T.F. Gorbachev Kuzbass State Technical University, 650000, 28 Vesennyaya St., Kemerovo, Russia

Abstract. The article reviews a novel approach to the design of steel angle headframe for vertical shafts of coal and ore mines on the basis of rational design solutions. Practice of construction of coal and ore mines provides application of various designs for steel angle headframes which are divided into separate large assembly blocks and constructive elements during assembling operations. Design of these blocks and elements, their weight and dimensions effect the chose of the method of assembling on which economic and technological indicators, as well as duration of down-time, depend on during performance of construction operations in shaft. The technical solution on equipment provision for mine vertical shaft using headframe of multifunctional purpose will allow changing the management construction of vertical shaft. The constructive design of the headgear allows application of the effective method of assembly and thus to provide improvement of the technical and economic indexes, and high calendar time rate of the shaft construction due to reduction of duration of works on equipment provision for the shaft and to refurbishment of the shaft in order to carry out horizontal mining.

1 Introduction

The construction project of the vertical mine shaft consists of the following main stages: provision of equipment for mine surface and sinking of the technological part of the shaft (including shaft collar); provision of equipment for surface for a shaft sinking, including assembling of the sinking headgear; shaft work; shaft walling; refurbishment of the shaft to carrying out horizontal and inclined developments, including dissembling of headgear unit during sinking operations and, construction of the permanent headframe [1 –10].

2 Materials and methods

The practice of construction of the coal and ore mines provides various designs for the steel angle headframes, both sinking headgear and permanent headframe: 4-post headframe, 4-

* Corresponding author: kalena-07@mail.ru

post headframe with two stay legs, tent headframe and two-post headframe which are divided into separate large assembly blocks and constructive elements during assembly operations [1–15]. Namely the assembly time of the headframe depends on the design peculiarity of the setup blocks, and, as a consequence, the total time of work stoppage in the shaft.

Method of erection is selected depending on the structural design of the setup blocks and parts of the headframe, their weight and dimensions, that affects the economic and technological data, as well as the duration of idle time during operations in the shaft.

Construction of Underground Structures and Mines' department at KuzSTU developed technical solution regarding furnishing of the vertical shaft with equipment utilizing steel headframe of multifunctional purpose [16, 17, 18] what will allow to change mining management on some of the hereinabove mentioned stages.

3 Results and discussion

As the proposed headframe (fig. 1) due to assembly/disassembly of the replaceable functional blocks combines the functions of both sinking and operation, then such stages as disassembly of gear complex during sinking operations and construction of the permanent headframe will be eliminated from the schedule.

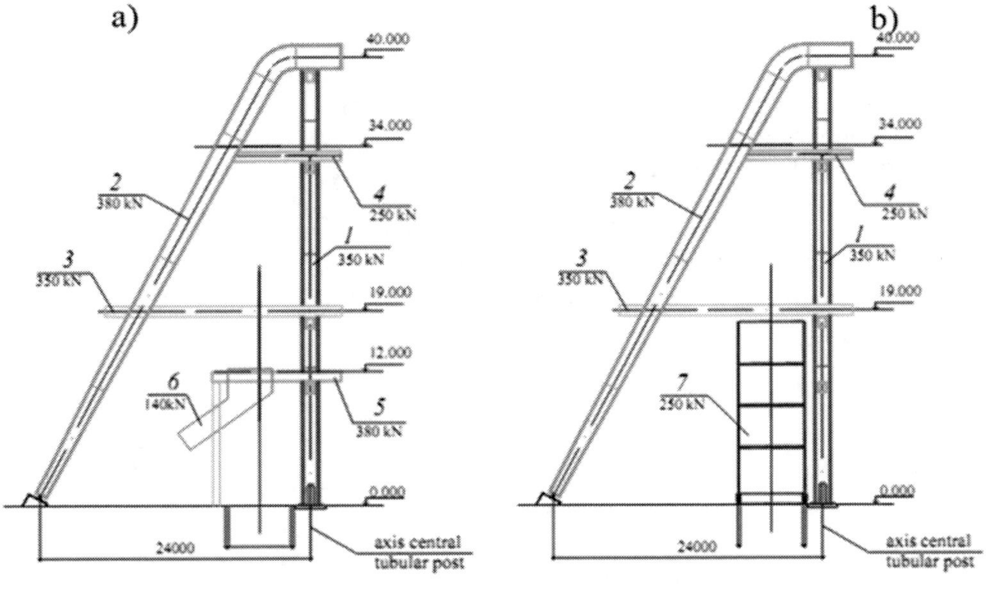

Fig. 1. Breakdown of the headgear on blocks: a) –scheme for sinking period; b) – scheme for operation; 1 – central tubular post; 2 –stay leg; 3 – sheave wheels platform; 4 – permanent sheave wheels platform; 5 – unloading machine; 6 – reception bunker; 7 – rig.

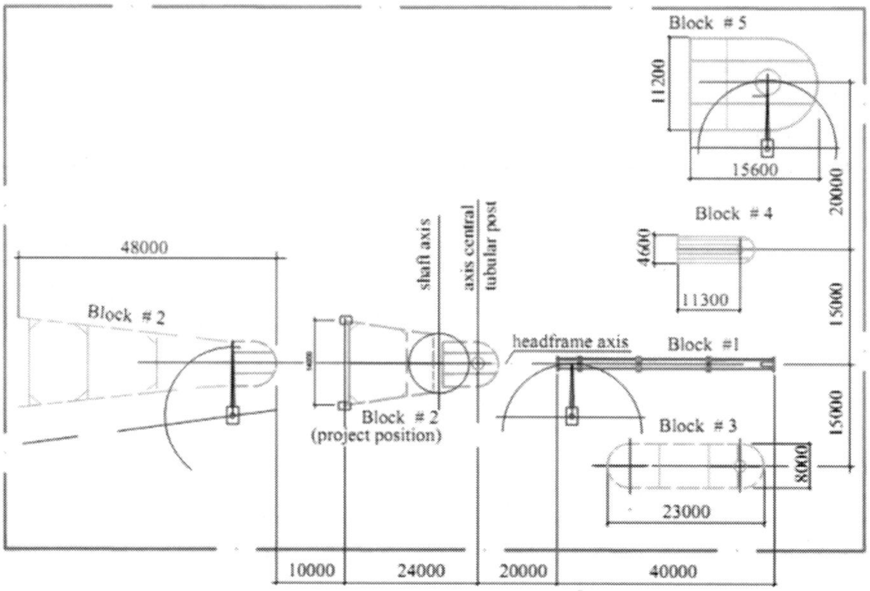

Fig. 2. Assembly schedule for blocks of the headframe.

The assembly of the headframe is carried out in the following consequence. On the assembling stand (fig. 2) located in the close proximity from the constructing shaft, the assembly is performed in horizontal position of the central tubular post (block #1). The assembled post is pulled to the shaft by sliding using two valves working as masts, erected to the vertical position and installed on the foundation as per the project position (fig. 3, a).

On the assembly stand (see fig. 2) they perform pre-assembly of stay leg (block #2) in horizontal position with the subsequent movement by sliding to the shaft by two cranes, there it is erected to the project position and fixed on the central tubular post *1* (fig. 3, b).

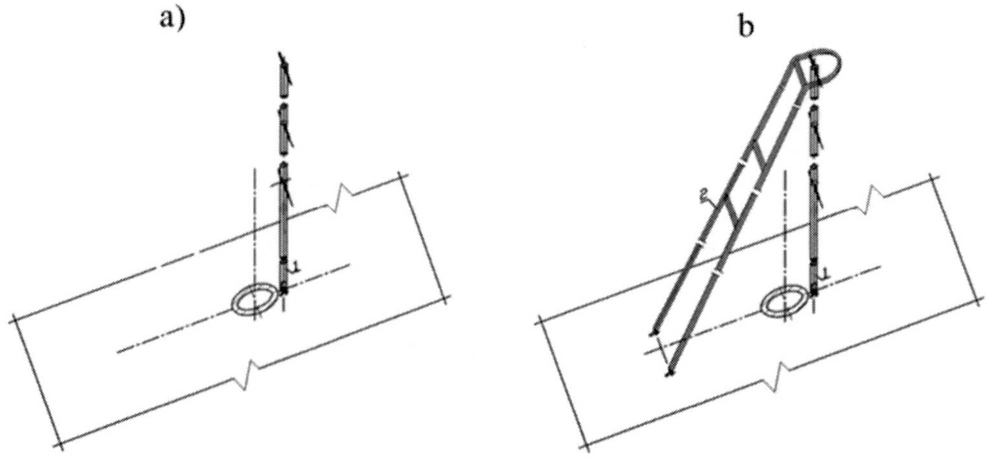

Fig. 3. Scheme of assembly of pulley device: a) – erection of central tubular post; б) – assembly of stay leg; *1* – central tubular post; *2* – stay leg.

The driving sheave wheels platform (block #3) is put together on the assembly stand in the form of two blocks, each of them is separately moved to the position 'on the ground' on the level ±0.00 m where they are connected to each other (fig. 4, a). By two cranes using the central tubular post *1* as a guide, the driving sheave wheels platform *3* is erected to the project position (fig. 4, b).

Fig. 4. Scheme of assembly of the driving sheave wheels platform: a) – assembly on the ground; b) – lifting to the design position; 1 – central tubular post; 2 – stay leg, 3 – the driving sheave wheels platform.

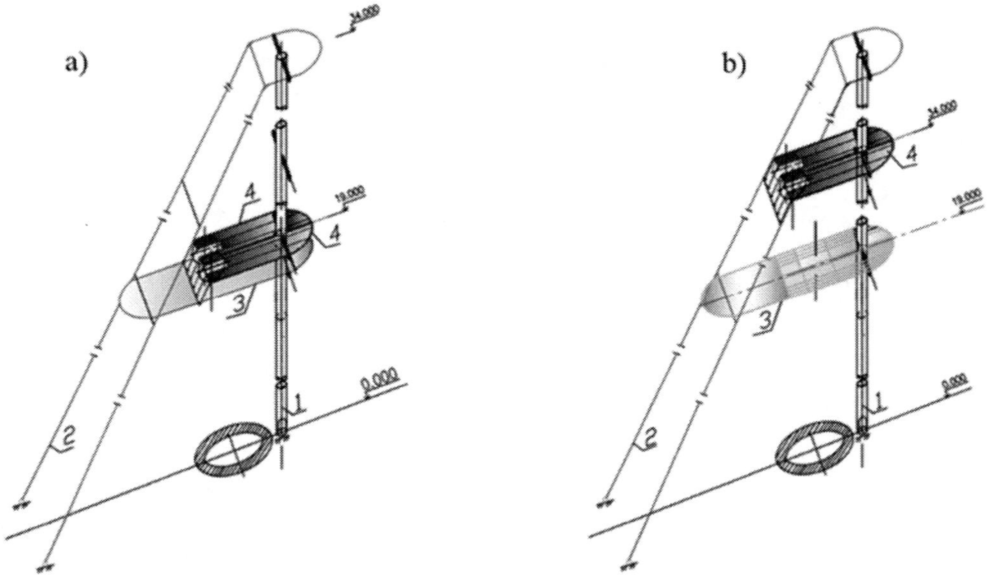

Fig. 5. Scheme of assembly of permanent sheave wheels platform: a) – assembly on driving sheave wheels platform; б) – erecting to project position; *1* – central tubular post; *2* – stay leg, *3* – driving sheave wheels platform, *4* – permanent sheave wheels platform.

The permanent sheave wheels platform (block #4) is put together on the assembly stand also as of two blocks, each block is moved and placed on the driving sheave wheels platform ***3*** by two cranes (fig. 5, a). On the driving sheave wheels platform ***3*** they perform the final assembly of the permanent sheave wheels platform ***4*** from the two enlarged blocks and erect

 DOI: 10.1051/e3sconf/20172101007

it up to the project position (fig. 5, b). The lifting is carried out by two cranes using the central tubular post ***1*** as a guide.

The assembly of the enlarged block #5 (the driving unloading platform) is performed as per analogy with the assembly of the block #3 (the driving sheave wheels platform) with the subsequent erection of the bunker (block #6). At this stage of assembling of the winch on the level 40.000 m is already possible for the elevator moving inside of the tubular post *1* which can be in turn used for needs of assemblers.

After a long break when mining and walling of the shaft are complete, it is necessary to dissemble the unloading machine and the receiving bunker, part of the driving sheave wheels platform and the mining equipment to assemble the unit (block #7) (see fig. 1).

The unit 7 [17] consists of ring elements *8*; each element is separately moved and stack on the driving sheave wheels platform *3* for the subsequent assembly by the 'from up to down' method over the shaft collar *9* (fig. 6). Thus, the driving sheave wheels platform is used again as assembly and installation platform already during transition to the operation mode.

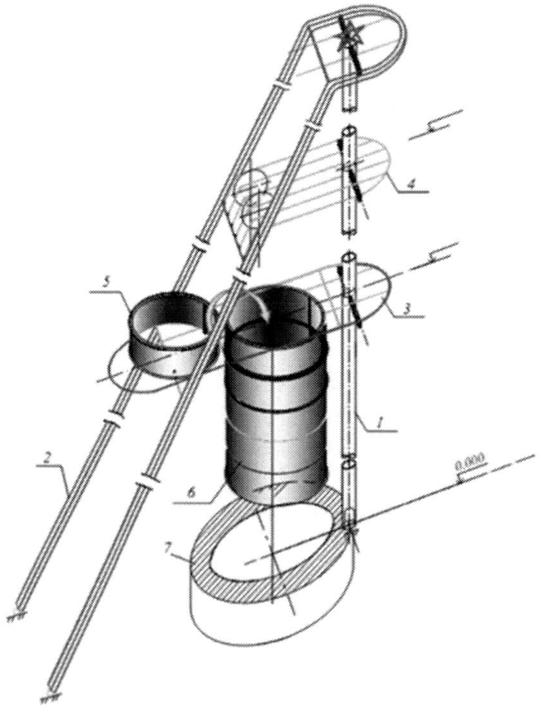

Fig. 6. Scheme of assembly of driving sheave wheels platform: *1* – central tubular post; *2* – stay leg, *3* – driving sheave wheels platform, *4* – permanent sheave wheels platform, 7 – rig, *8* – ring element, 9 – shaft collar.

The efficiency of the suggested method of assembly of the headframe of multifunctional purpose is provided by:

– Reduction of the total time for assembling of the headframe and, in this regard, considerable reduction of shutdown during operations in shaft, considerable reduction of the overall labor costs and the cost of headframe assembly;

– Reduction of labor input and cost of installation due to breakdown of the headframe on the assembly blocks of small weight (250 kH - 350 kH);

– Application of the sheave wheels platforms for assembly and erection;

– Use of the central tubular post as a guide what considerably reduces the number of operations on aligning and adjustment of structures during assembling;

– Reductions of labor costs on rigging up of the unit which is put together separately from the sheave wheels platform and other constructive elements of the headframe;

– Lack of labor costs on arrangement of the rolling track to deliver ring elements of the rig to the place of installation due to their small weight;

– Minimizing of works at height.

The suggested option of the vertical shaft construction utilizing headframe of multifunctional purpose versus the traditional technical solution [1-15] allows to reduce labor input of assembly operations in 1.5 times, duration of works in 1.22 times, steel consumption in 1.5 times (table 1).

Table 1. Option of the vertical shaft construction utilizing.

Index	**Sinking headgear 'Sever-1' (excluding dissemble)**	**Perma-nent angle head-frame**	**Headframe of multifunc-tional pur-pose**
Labor input of as-sembly MK, man/hr.	95.9	153	174
Assembly duration as per time-table, days	20	25	15
Steel consumption, kN	1200	1890,5	2060

4 Conclusion

Equipping of the vertical shaft using headframe of multifunction purpose allows to optimize the arrangement of mining operations in terms of construction of the mine, as the proposed headframe combines the functions of shaft-sinking and operation, which eliminates the cost of assembly / disassembly of the derrick.

The constructive design of the headgear allows application of the effective method of assembly and thus to provide improvement of the technical and economic indexes, and high calendar time rate of the shaft construction due to reduction of duration of works on equipment provision for the shaft and to refurbishment of the shaft in order to carry out horizontal mining.

References

1. D. W. Butler, A. C. Schneyderberg, Min. Cong. Jour., Ref, **7**, 15, (1982)
2. F. A. Edwards, *Hoisting systems, SME Mining engineering handbook, Society for Mining, Metallurgy, and Exploration*, (New York, Willey, 1992)
3. J. I. Rojas-Sola, J. M. Montalvo-Gil, M. Castro-Garcia Dyna, **80:181**, 118 (2013)

4. A. Ye. Nechytailo, Ye. V. Horokhov, V. N. Kushchenko, 18th International Conference on the Application of Computer Science and Mathematics in Architecture and Civil Engineering. Weimar, **1** (2009)
5. I. Ross, *Goldcorp going electric with Chapleau gold mine. Northern Ontario Business* (Detroit, UGHHJ Pres., 2016)
6. J. I. Rojas-Sola, I. Palomares-Muñoz, Dyna, **90:6**, 602 (2015)
7. F. Wei, I. F. R. Wang, Key Engin. Mater. **455**, 340 (2010)
8. I. Ross, Northern Ontario Business, **33:12**, 14 (2013)
9. L. Kelly, Northern Ontario business, **34:3**, 31 (2014)
10. E. G. Kassikhina, Qiao Weiguo, V. V. Pershin, N. O. Butrim, Taishan Academic Forum – Project on Mine Disaster Prevention and Control, **1** (2014)
11. E. G. Kassikhina, Gornyi Zhurnal, **8**, 56 (2017)
12. E. G. Kassikhina, GIAB, **4**, 28 (2014)

Research of the Quality of Quarry Dumpers Engine Crankshafts Sliding Bearings of Various Manufacturers

Alexander Korotkov [1*], *Lidiya* Korotkova [1], and *Denis* Vidin [1]

[1] T.F. Gorbachev Kuzbass State Technical University, 650000, 28 Vesennyaya St., Kemerovo, Russia

Abstract. Sliding bearings are an important part of many large and critical components. They are widely used in power equipment, high-capacity pumps, compressors, electric motors and internal combustion engines (ICE). As a rule, sliding bearings include an antifriction bushing, part of the shaft surface (bearing journal), and a layer of oil between them. These are complex and critical parts in which there may occur dangerous defects, and which directly affect the durability, accuracy and reliability of the entire unit. To ensure high reliability of the equipment with sliding bearings applied in complex equipment, it is necessary to provide the quality control and sufficient level of monitoring of the technical condition, as well as diagnosis of emerging defects. This paper presents a comparative analysis of the internal combustion engines sliding bearings quality of various manufacturing companies. It gives operational properties of bearings depending on the compositional composition. The results of chemical analysis of the base, the cover and intermediate layers of the ICE liners are presented here. We have also made recommendations to increase the operational performance of sliding bearings.

1 Introduction

Sliding bearings are critical parts of an internal combustion engine, failure of which leads to its complete stop and often to the following expensive repairs. Bushings are high-tech products made of composite material, having a complex construction and precise dimensions.

Internal combustion engines are characterized by cyclic loading of bearings due to variable pressure in the cylinder as well as inertial forces caused by the moving parts.

The vast majority of the crankshafts of the engines of modern cars rotate in the sliding bearings, so called bushings.

Long and reliable work of the sliding bearing is achieved by providing a number of indicators, including fatigue strength, wear resistance, cavitation resistance, adaptation, resistance to setting, absorption capacity.

Rotating components of internal combustion engines are equipped with sliding bearings, performing different functions [1]:

* Corresponding author: korotkov.a.n@mail.ru

- *The main bushings* support the crankshaft and rotate it. They are installed in the cylinder block, and each bushing consists of the upper and lower semi-rings. As a rule there is a groove for lubrication and an opening for the supply of oil on the inner surface of the upper semi-ring.
- *Connecting rod bushings* ensure the rotation of the crank pin that rotates the crankshaft. Connecting rod bushings are installed in the lower end of the connecting rod.
- *Thrust rings* prevent axial movement of the shaft. In many cases, the thrust rings are part of one of the main bushings. Such combined bearings are called collar liners.
- *The bushings* of the connecting rod upper head ensure the rotation of the piston pin connecting the piston with the connecting rod.
- *The camshaft* bushings support and rotate it. They are installed on the top of the cylinder block head.

The sliding bearings are lubricated with engine oil constantly supplied in sufficient quantity to their surface and providing hydrodynamic friction mode. Fluid (hydrodynamic) friction is characterized by a pressurized oil film separating the bearing and shaft surfaces.

There is no direct contact between the surfaces in the hydrodynamic mode. This is ensured by the force of the hydrodynamic pressure emerging in the oil, which is forced through the converging gap (oil wedge) formed by the bearing and shaft surfaces.

The working conditions of the sliding bearings are classified as heavy stress. They operate in a mixed friction mode, when direct contact occurs during operation alternating with hydrodynamic friction [2].

The oil film contributes to the distribution of the force applied to the bearing, thus preventing local load concentration [3].

However, under certain conditions, hydrodynamic friction mode is changed to a mixed one. These conditions are as follows [1]:

- Insufficient oil flow;
- High loads;
- Low viscosity of oil;
- Overheating of the oil, further reducing its viscosity;
- High roughness of bearing and shaft surfaces;
- Oil contamination;
- Deformation and geometric defects of the bearing, its nest or shaft.

A mixed friction mode can lead to scuffing, increased wear of the bearing and even gripping the shaft [4].

Cyclic bearing loads can lead to its destruction due to fatigue of the material.

The following requirements apply to materials for sliding bearings [1]:

- Fatigue strength (maximum load) is the maximum cyclic load that the bearing can withstand for an unlimited number of cycles. Exceeding this load leads to the formation of fatigue cracks in the material.
- Resistance to setting (compatibility) - the ability of the bearing material to resist welding to the shaft material during direct physical contact between them.
- Wear resistance - the ability of the bearing material to maintain its dimensions despite the presence of abrasive particles in the oil as well as under conditions of metal contact with the shaft.
- Adaptation - the ability of the bearing material to compensate for small geometric defects in the shaft and the seat due to minor local wear or plastic deformation.
- Absorption capacity - the ability of the bearing material to capture small foreign particles that circulate with the oil.
- Corrosion resistance - the ability of the bearing material to resist the chemical influence of the oxidized or contaminated oils.

- Cavitation stability - the ability of the bearing material to withstand shock loads produced by collapsing cavitation bubbles (bubbles are formed because of a sudden drop in the pressure of the oil flow).

2 Materials and methods

Depending on the operating conditions, the design of the sliding bearings is chosen. It is possible to distinguish two basic designs of bearings used in heavy stress conditions - these are bimetallic and trimetallic bearings [5, 6, 7].

Properties of bearing materials

Comparative properties of the most popular bearing materials, depending on the structure composition, are shown in Table 1 [8].

As we may see from Table 1, properties that characterize strength and softness are combined in different proportions for different materials.

Sufficient soft antifriction properties of trimetal are limited by the coating thickness, which should be no more than 12 μm. If a geometric defect or foreign particles exceed the thickness of the coating, its antifriction properties drop sharply [9, 10].

The soft properties of bimetal are somewhat lower than those of trimetal, but they are not limited to coating thickness, so bimetallic bearings are capable of adapting to various geometric defects [6].

On the other hand, the fatigue strength (maximum load) of bimetallic bearings is lower (40-50 MPa) than for trimetallic bearings (60-70 MPa).

To control the quality of sliding bearings (bushings), a methodology has been developed to meet the standard samples. This includes the following steps:

• layer-by-layer control of the chemical composition of the researched alloys;

• carrying out metallographic tests to determine the number and depth of layers of applied coatings;

• layer-by-layer control of the applied coatings and the base of the bushing microhardness distribution.

Table 1 Operational properties of bearings depending on the composition of the composition

Property	**Bimetal Without Si**	**Bimetal With Si**	**Trimetal with lead coating**
Fatigue strength, MPa	40	55	60-70
Compatibility with the steel shaft	very good	good	excellent (until the coating has been worn off)
Compatibility with the cast iron shaft	satisfactory	very good	excellent (until the coating has been worn off)
Wear resistance	good	very good	satisfactory
Absorption capacity for particles larger than 12 mm	very good	good	bad
Absorption capacity for particles smaller than 12 mm	very good	good	excellent
Adaptation to geometric defects larger than 12 μm	very good	good	bad

Property	**Bimetal Without Si**	**Bimetal With Si**	**Trimetal with lead coating**
Adaptation to geometric defects smaller than 12 μm	very good	good	excellent

Conducting chemical analysis of alloys

To conduct a layer-by-layer chemical analysis of metal coatings and the base of the bushings, we used an *Q4 TASMAN* optic emission spectrometer.

Atomic emission spectrometers apply the principles of atomic emission spectroscopy (AESA). These devices are the most common analytical instruments in the world and in our country. They are designed for analysis (most often elemental analysis) of the composition of various substances in different aggregate states.

Metallographic studies

These included preparation of samples and direct study the microstructure.

Preparation of samples for metallographic studies

It is one of the most important stages in carrying out metallographic studies of the metals' structure and mechanical properties, and it is very important in metallography. Metallographic preparation of samples is a process of obtaining quality samples for research. The quality of sample preparation directly affects the quality of research results [8, 9].

We have to provide a number of conditions to conduct metallographic studies in order to determine the number and depth of layers of coatings applied on bearings:

• to cut samples with minimal residual deformations;
• not to damage surface layers during the preparation of thin sections;
• to form a high quality of the researched surface of the samples.

Determining the number and thickness of the layers

To assess the number and thickness of the applied layers of metal coatings on the bushing, we used a special inverted optical metallographic microscope, EPIQUANT, which allows accurate measurements of both the depth of the applied coating and the quantitative phase size by the quantitative metallography method at X1500 magnification.

Measuring the thicknesses of the sliding bearing layers was carried out with the help of an eyepiece micrometer, with glass with a printed scale inserted in it.

Measurement of microhardness

Microhardness was measured by pressing a diamond pyramid according to Vickers method. The PMT-3 microhardness meter allowed placing the imprint in a separate grain, phase or layer.

Fig. 1 and Fig. 2 show the determination of the hardness of individual bearing layers. The microhardness meter allows to make prints and determine microhardness at any point of the layer. Microhardness is determined here at a load of 100 g. To determine the microhardness of a thin upper layer, we used the load of 50 g.

3 Results and discussion

This paper covers the results of studying internal combustion engines trimetal bushings of different sizes, types and manufacturers. A comparative evaluation of the quality of the bushings was carried out basing on the results of chemical analysis of the layers, their thickness and microhardness.

The research was conducted on the samples of sliding bearings, the list of which is given in Table 2.

The results of the studies are summarized in Tables 3 - 5. Based on the presented results, it is possible to carry out a comparative analysis of the bearings' quality, based on their structure, chemical composition of layers and microhardness.

The conducted metallographic studies made it possible to notice that all the samples under research (Figures 1 and 2) include three basic layers: a steel base, a copper alloy bushing (intermediate layer), and a whitemetal cover layer. In the design of the examined bearings, we found protective tin layers of a shallow depth of several micrometers (coating), as well as a nickel gasket (barrier) between the cover layer and the bearing shell. Thus, the design of all bearings is the same.

Table 2. List of researched samples

No. of researched sample	Size of the bearing	Manufacturer's code
1	205160	Cummins POL
2	205160	Cummins CCEC
3	205140	Cummins CCEC
4	205840	WETZ
5	205840	WETZ

Table 3. Results of studying bushings bases

№	Steel chemical analysis										
	C	Si	Mn	P	S	Cr	Ni	Cu	Al	Sn	HV, Kgf / mm2
1	0,053	<0,0015	0,275	0,0032	0,0046	0,025	0,0053	0,036	0,028	0,003	164-210
2	0,124	0,0020	0,580	0,0010	0,0010	0,037	0,038	0,053	0,016	0,017	183-194
3	0,126	0,0030	0,557	0,0010	0,0015	0,037	0,026	0,072	0,016	0,017	170-183
4	0,158	0,014	0,363	0,0052	0,0054	0,0093	0,0015	0,013	0,032	0,002	186-214
5	0,140	0,020	0,371	0,0049	0,0053	0,0098	0,0045	0,031	0,035	0,034	159-190

Table 4. Results of studying bushings coating layers

№	Whitemetal chemical analysis						
	Pb	Sn	Fe	Ni	Cu	HV, Kgf / mm2	h, μm
1	>26,40	>26,40	0,310	20,24	26,28	16-25	25
2	>26,40	>26,40	0,284	27,62	18,87	22,4-20,5	16
3	26,40	5,707	0,047	2,588	65,09	21-20,5	16
4	26,40	2,460	0,035	5,209	65,75	18,5-19,3	16
5	26,40	1,709	0,033	4,014	67,69	22,4-21,9	16

 DOI: 10.1051/e3sconf/20172101008

Table 5. Results of studying bushings intermediate layer

№	Bronze chemical analysis						
	Pb	**Sn**	**Fe**	**Ni**	**Cu**	**HV, Kgf / mm2**	**h, µm**
1	21,61	2,828	0,198	0,018	75,32	86-107	520
2	22,57	<2,000	0,045	0,078	75,28	106-128	360
3	25,95	2,921	0,079	0,017	71,00	77,2-100	360
4	26,27	1,680	0,021	0,597	71,30	110-135	320
5	26,40	1,397	0,024	0,465	71,58	122	304

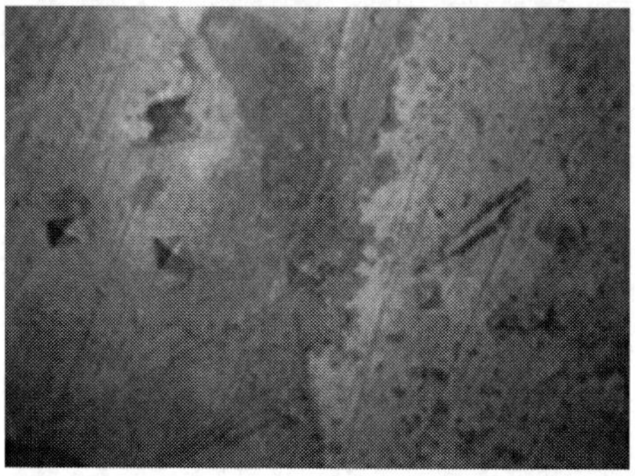

Fig. 1. Imprint of an indenter when measuring the microhardness of a layer and a substrate.

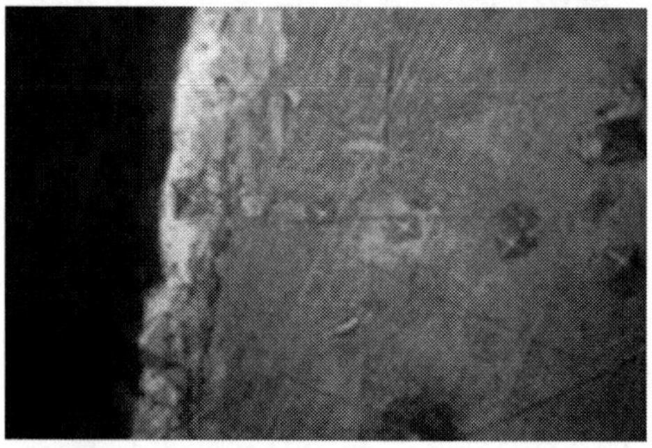

Fig. 2. Imprint of the indenter of the microhardness meter when measuring a thin layer of whitemetal.

The base for all studied bushings is made of low-carbon, high-quality steel. The method of manufacturing is cold-rolled steel with fine-grained ferritic microstructure. Its microhardness is 160 - 210 HV (see Table 3).

The conducted research made it possible to identify some differences in the chemical composition, depth, and microhardness of the working layers (see Tables 4 and 5).

Thus, in the production of the coating layer, they use lead-tin plies (samples No. 1, 2) or leaded whitemetals (samples No. 3, 4, 5). In all the samples studied, the depth of the cover layer is in the range of admissible values of 12-25 µm. The chemical composition of whitemetals does not make any significant influence on microhardness (see Table 4).

The intermediate layer in all the samples studied is a tin lead-containing cast bronze. By the chemical composition, bronze can be identified according to State Standard (GOST) 613-79 as brands approximating BrO3C22 and BrO5S25 [15]. This shows that the chemical composition of the intermediate layer of the ICE bushings of various manufacturing companies is very close. The uniformity of the alloys has been proved by metallographic studies

 DOI: 10.1051/e3sconf/20172101008

and similar microhardness values in the range of 100-160 HV (two-phase microstructure) (Table 5).

Thus, basing on the comparative analysis of the quality of the ICE bushings of different manufacturers, it can be concluded that they are close in structure and chemical composition, namely:

• Cummins POL: lead-tin-based whitemetal - nickel interlayer - tin lead-containing bronze.

• Cummins CCEC: lead-tin-based whitemetal - nickel interlayer - tin lead- containing bronze.

• WETZ (domestic manufacturer): lead-based whitemetal - nickel interlayer - tin lead-containing bronze.

As a result, we can conclude that all bearings are suitable for operation, and it should be noted that the ICE sliding bearings of domestic manufacturer (WETZ) are not inferior in quality to foreign analogues (Cummins POL, Cummins CCEC).

It should be borne in mind that the service life of ICE bushings is determined not only by their quality, but, to a large extent, by the conditions of their operation. In this regard, it is reasonable to formulate conditions for the non-failure operation of sliding bearings.

4 Recommendations to improve the performance of bearings

To ensure high performance of the internal combustion engine bushings in the course of operation, a number of conditions related to the quality of the bearing parts in the delivery state and during operation must be fulfilled. This will make internal combustion engines work for a long time (up to 1.5 million km). However, to ensure such a service life, the following requirements must be met:

1. Correct installation of bearings.
2. High quality of a journal contours crankshaft processing.
3. Ensuring the required routine maintenance.
4. Meeting the requirements on the recommended oil change intervals and the use of oils with the suitable grades of SAE viscosity and the API quality standard.
5. Prevent the coolant and fuel from entering the engine oil.
6. Monitoring of engine operating conditions to avoid overloads, high speed and engine overheating.

How long is the period of engine bearings trouble-free operation depends on the implementation of the quality standard, which should regulate the fault detection tests.

5 Conclusion

1. There has been carried out a patent-literature analysis on the standard designs of sliding bearings bushings, in terms of chemical composition, technology of production of crankshaft bushings for combustion engines.

2. We have studied typical designs of ICE crankshaft bushings of various manufacturers: materials for sliding bearings, requirements for these materials and properties of bearing materials.

3. We have developed the methodology for quality control of the ICE bushings. It includes chemical analysis of alloys, metallographic studies, measurement of microhardness, measurement of the layers depths.

4. As a result of approbation of the developed technique for quality control of internal combustion engines crankshafts bearings, it has been revealed that all the researched crankshaft bushings are trimetallic in structure.

5. The comparative analysis of the ICE bushings of various manufacturing companies (Cummins POL, Cummins CCEC, WETZ) has shown that they differ from each other in chemical composition, depth and microhardness of the working layers:

5.1.Cummins POL: lead-tin-based whitemetal - nickel interlayer - tin lead-containing bronze.

5.2.Cummins CCEC: lead-tin-based whitemetal - nickel interlayer - tin lead-containing bronze.

5.3.WETZ (domestic manufacturer): lead-based whitemetal - nickel layer - tin lead-containing bronze.

6. On the basis of a comparative analysis of the quality of the internal combustion engines bushings of different manufacturers, it has been concluded that all of them are suitable for operation.

7. Domestic manufacturer\s (WETZ) sliding bearings for internal combustion engines are not inferior in quality to foreign analogues (Cummins POL, Cummins CCEC).

References

1. D. Kopeliovich, *Inserts for the engine - details are critical.* (ABS-Auto. The first autoservice magazine.)
2. S.S. Kim, H.N. Yu, I.U. Hwang, D.G. Lee, S.N. Kim, K. Suzuki, H. Sada, Tribology Letters, **35:3**, 211 (2009)
3. O.G. Zavyalov, Y.V. Alyabyeva, O.S. Savchenkova, Journal of Friction and Wear, **30:4**, 235 (2009)
4. A. Zeren, Materials and Design, **28:8**, 2344 (2007)
5. *GOST 28813-90 Bearings of sliding. Metal multilayer materials for thin-walled plain bearings.* Publishing Standards, **11** (1990)
6. GOST *ISO 4383-2006 Bearings of sliding. Multilayer materials for thin-walled plain bearings,* **8** (2009)
7. L. I. Simma Chemical and Petroleum Engineering, **45:9:10**, 649 (2009)
8. F.H. Burumkulov, S.A. Velichko, P.V. Chumakov, Proceedings of GOSNITI, **111:2**, 142 (2013)
9. A. N. Korotkov, L. P. Korotkova, D. V. Vidin, S.V. Lashchinen, O.V. Frolova, KuzGTU, Bul. **1:113**, 119 (2016)
10. B. Leszczyńska-Madej, M. Madej, Archives of Metallurgy and Materials, **56:3**, 805 (2011)

New Opportunities to Expand Information on Intense-Strained State of the Earth's Crust in the Areas of Development Mineral Resources During Monitoring Creation

Vladimir Pershin[1*], and *Aleksander* Solovitskiy[1]

[1] T.F. Gorbachev Kuzbass State Technical University, 650000, 28 Vesennyaya St., Kemerovo, Russia

Abstract. It is established that new safe and pollution-free technologies of development of Kuzbass coal deposits should be based on new knowledge of development geodynamic and technogenic processes in exploration of mineral resources. Such information is impossible without formation of new models of deformations of the earth crust blocks. Now in traditional technologies of geomechanical ensuring development of mineral resources the main characteristic is the information about the kinematics of these processes. A comprehensive approach which beginning is development of the theory for justification of scale of the explored territory and establishment of uniform integral parameters of a strained state of blocks of crust. Justification of scale of the explored territory defines effectiveness of expenses. Establishment of uniform integral parameters of a strained state of crustal blocks characterizes the new level of information exchange between sciences about Earth and geomechanics. Practical use of the specified theory consists of assessment of geodynamic danger at development of coal fields.

1 Introduction

Traditional technologies of mineral resources development are focused on studying their geomechanical conditions by improving the information on the intense-strained state of the rock mass [1-3]. Therefore, in practice the dominating development is gained by geophysical and surveying methods. The new stage of development of studying intense-strained state of a block of rock mass is not observed after the fundamental research 80-ties and 90-ties. There are some reasons.
1. The absence of a unified technological policy.
2. Training Costs.
2. The pursuit of profit.
All three specified reasons are interrelated, therefore, obtaining new information and new knowledge of an intense strained state of crust is restrained. However, it should be noted that a tendency of development of the modern technologies of Earth Sciences is to improve the accuracy, efficiency and automation of processes such as the measurement of

*Corresponding author: pvv.spssh@kuzstu.ru

DOI: 10.1051/e3sconf/20172101009

production and processing of the received information [4-6]. Therefore, they are in demand for studying of natural and technogenic geodynamic processes in the development of the coalfields, information about which is the most important in the aspect of forecast catastrophic geodynamic phenomena (life threatening situations). However, despite the material and human losses, systemic monitoring of natural and technogenic geodynamic processes at development of coalfields is not developed. Occasionally re-performed monitoring of technogenic geodynamic polygons (GFC) is ambiguous, time-consuming, not sufficiently detailed and operational. Therefore, the development of theory and methods of geodetic monitoring of the stress-strain state of the crust (GMDSS) in the areas of development of coal deposits, ensuring the safety of their development, is a scientific issue with important socio-economic importance. Its development is constrained by lack of new mathematical models of geodetic constructions on the geodynamic polygons allowing registering; evaluating and predicting the stress-strain state of the earth's crust [7-12]. The main condition for such models is the relationship between changes in the coordinates of points obtained from repeated observations, with parameters of the stress-strain state (SSS) of the earth's crust, as they are not available by direct measurement of geodetic methods. The objective of this monitoring is the collection, storage, processing, and analysis and forecast information on the stress-strain state of the crust during the development of coal deposits to reduce the risk of manifestations of geodynamic phenomena (life threatening situations). The statement allows to consider the actual development of theory and methods GMDSS in the areas of development of coal deposits.

2 Results and discussion

Acquiring new knowledge about the state of a block of the earth's crust based on the results GMDSS is necessary for its evaluation, analysis and use [13]. This will require the following:

• establishing one number of the earth crust block;

• the establishment of common integral parameters of intense-strained state of the earth crust blocks;

• assessment of the scale of the investigated area.

A modern approach to the use of the results GMDSS is based on information technology, therefore, the following classification is proposed which includes the region code, year, and method of execution, as well as the hierarchy structure of the crust. The proposed single number of block of the earth's crust has the following structure:

• the region code;

• the year of the;

• the method of implementation;

• a rank unit;

• hierarchical room

The region code corresponds to the political-administrative division of Russian Federation and determines its subject. The procedure is determined by table 1.

Table 1. The code of the method of geodynamic zoning.

Method	Code
Geological	01
Geophysical	02
Geomorphologic	03
Cartographic	04
Remote sensing	05
Integrated	06

The rank of a block is determined by the adopted hierarchy structure of the crust, so at geodynamic zoning of Kuzbass six rank system is applied, so the hierarchical number will be a 12-digit number, for example, 42:1987:04: 03: 000001000000.

The establishment of common integral parameters of intense-strained state of the earth crust blocks eliminates the ambiguity of the kinematic characteristics of the crustal blocks of different ranks used in traditional technologies. Quantitative characteristics of vertical and horizontal movements of the earth's crust are the amplitudes $\Delta A_j\,[t\!-\!t_0]$ [6] determined by the difference of coordinates $\Delta A_1[t\!-\!t_0] = (X[t] - X[t_0])$, $\Delta A_2[t\!-\!t_0] = (Y[t] - Y[t_0])$ и $\Delta A_3[t\!-\!t_0] = H[t] - H[t_0]$ of the same items at different times t and t_0 according to the formulas:

$$\Delta A[t\!-\!t_0] = A[t] - A[t_0], \tag{1}$$

where $j=x, y, z$.

The most common presentation of quantitative characteristics of movements of crustal blocks on the results of the analysis is shown in the table 2.

Table 2. Representation of the quantitative characteristics of the movements of the earth's crust

Type of presentation of quantitative characteristics	
Tabular	Tables
The amplitudes	Charts
Speed	Vector schemes
Gradients	Maps

To substantiate these parameters, the authors carried out the following experiment. For blocks of the crust of the studied grades in the areas of development of deposits using the programme "BM" the time variation of deformations of crustal blocks of Kuzbass is modeled based on their kinematics, which are shown in the figure 1.

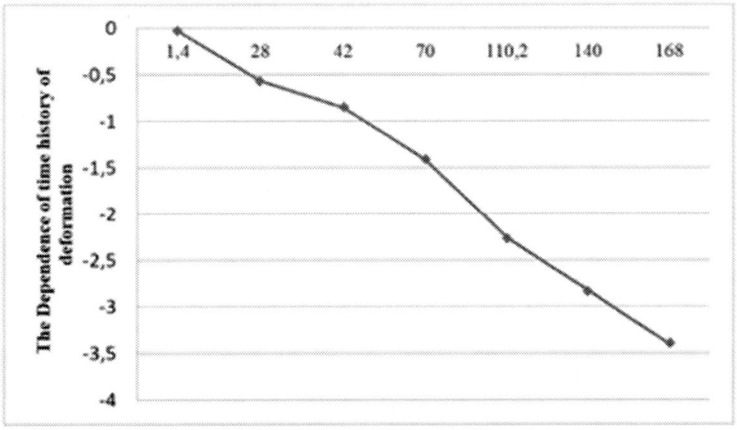

Fig. 1. The Dependence of time history of deformation of the block of crust of III rank on the speed of its velocity: the vertical axis-strain variation in time blocks of the crust ($1\cdot10^{-6}$); the horizontal axis is the velocity of motion of the earth blocks (mm / year).

The received dependences by results of researches confirm that the same values of time variation of the deformation of crustal blocks of different grades correspond to different velocities of their movements. Because of this difference in the velocities of the movements of the blocks of the earth's crust of different ranks corresponding to the same values of the time variation of their deformations, the kinematic characteristics of the blocks of the earth's crust can not be accepted as dynamic parameters of the crustal block. A more acceptable option is the time variation of deformations of blocks of the earth's crust, which are integral characteristics. Thus, the dynamic parameters are the changes over time of deformation (stresses) and the potential energy determining the kinematics of earth crust block. To determine the dynamic parameters of the earth's crust, the author optimized and compiled the program "BM" [13].

The integral nature of these parameters is to provide a new level of information interaction between the Geosciences and geomechanics. In addition, changes in time of the deformation of the crustal blocks are expressed analytically depending on the corresponding changes of stress and potential energy, that is, of the entire range of considered parameters, which quantitatively characterize the manifestation of geodynamic phenomena (life threatening situations). The obtained dependences allow assessing the scale of the studied area of coal companies. Defined by the authors of the program "BM" based on changes over time of the potential energy and density from the ranks of the crustal blocks of Kusbass are shown in the figures 2.

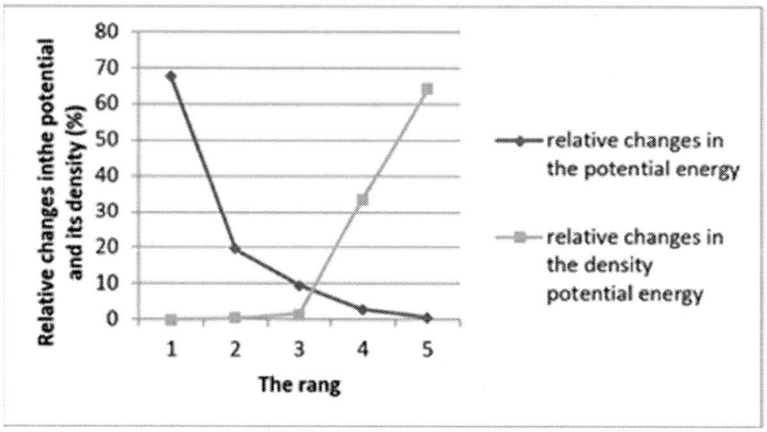

Fig. 2. The dependence of the change in the time of the potential energy and its density from the ranks of the crustal blocks

The results which are shown in the figure 2 demonstrate that the variation in time of the potential energy of deformation of crustal blocks of Kuzbass are in inverse proportion to their rank (of 67% in unit II rank and 1% in unit VI rank), and the corresponding change of its density (64% in unit VI rank and 0.5% in units II and III rank) in a straight line. Such dependencies can be obtained for a block of the earth's crust of any field. The received dependences regulate the scale of studying of dynamic parameters of blocks of crust around development of the field. According to the results shown in the figure 2, the most popular territory for the study will be the territory of the crustal blocks of the V-th and VI-th rank.

3 Conclusions

Based on the results of the research the author has developed a theory that defines the basics of creating GMDSS, and made the following conclusions:

1. Justification and establishment of dynamic block parameters of the earth's crust provide: disambiguation of kinematic characteristics; a new level of information interaction

between the Earth Sciences and geomechanics; to assess the scale of the studying area of coal companies.

2. It is established that the variation in time of the potential energy of deformation of crustal blocks of Kuzbass is in inverse proportion to their rank and the corresponding change of its density in a straight line.

3. Quantitative changes over time of the potential energy of deformation of the crustal blocks and the corresponding change of its density indicate that the most popular for studies in the areas of development of deposits are areas of crustal blocks of the V-th and VI-th rank.

4. Proposed a unified numbering blocks of the earth's crust in the area of development creates GMDSS and eliminates the uncertainty of the crustal structure.

References

1. A. A. Kozyrev, I. E. Semenova, V.V. Rybin, I. M. Avetisyan, Journal of Mining Science, **51**, 4 (2015).
2. B. G. Saksin, I. Y. Rasskazov, B. F. Shevchenko, Journal of Mining Science, **51,** 2 (2015).
3. G. G. Kocharyan, A. A. Ostapchuk, Journal of Mining Science, **51**, 1(2015).
4. A. A. Mylnikova, Yu. V. Yasyukevich, E. V. Kunitsyu, Results in Physics. **5, (**2015).
5. A. P. Karpik, I. G. Ganagina, S. N. Kosarev, D. N. Goldobin, Gyroscopy and Navigation, **7**, 1(2016).
6. A. N. Solowizkij, Journal of Shandong Mining Institute. **16,** (1997).
7. C. DeMets, R. G. Gordon, D. F. Argus, Geophys Journal Int., **181**, (2010)
8. D. F. Argus, R. G. Gordon, M. B. Heflin, C. Ma, R. J. Eanes, P. Willis, W. R. Peltier, S. Owen, Geoophys Journal Int., **180, (**2010).
9. R. McCaffrey, Journal Geophys. Res.,**110**, B07401(2005)..
10. J. C. Savage, W. Gan, J. L. Svarc, Journal Geophys. Res., **106**, (2001)
11. Z. Shen, D. D. Jackson, B. X. Ge, Journal Geophys. Res., **101**, (1996).
12. A. Solovitskiy, 8th Russian-Chinese Symposium. Coal in the 21st Century: Mining, Processing and Safety, **2**, 255-260 (2016)

DOI: 10.1051/e3sconf/20172101010

The Extent of Destruction Zones Within Protective Pillars in Jsc "Suek-Kuzbass" Underground Mines

Natalya Pirieva[1*], and *Inna* Ermakova[2]

[1]JSC "SUEK-Kuzbass", 652507, Vasilyeva St., 1, Leninsk-Kuznetsky, Russia

[2] T.F. Gorbachev Kuzbass State Technical University, 650000, 28 Vesennyaya St., Kemerovo, Russia

Abstract. To justify an effective way of extracting protective pillars, which have lost their primary function, we need to establish the extent of destruction zones within them. This problem has been tackled using the finite element method allowing for their relative position, the depth of cover, the structure, and the deformation and strength properties of the strata.

1 Introduction

Pillar size calculation is one of the classic problems of geomechanics, whose solution has been the subject of multiple publications [1]. In terms of definition and solution methods, these publications can be divided into two categories: (1) engineering solution methods based on re-distribution of rock stress, and (2) methods based on the mechanics of a deformed solid body. Being the most physically stringent, the methods based on the mechanics of a deformed solid body include the finite element method, a boundary element method and analytical methods [2-11]. This article is based on a finite element method [12] which takes into account deformation and strength properties of the complete coal compression diagram [13].

2 Materials and methods

The analysis of the strength of protective pillars is shown using an example of pillars for slopes in Rubana Mine, JSC "SUEK-Kuzbass". This mine is extracting three gently dipping seams:

1) Polysaevsky-II Seam, depth of cover 80 – 124 m, angle of dip 6 – 8^0, thickness from 4.40 to 5.4 m;

2) Nadbaikaimsky Seam, depth of cover 210 – 290 m, angle of dip 3 – 7^0, thickness from 2.20 to 2.70 m;

3) Baikaimsky Seam, depth of cover 245 – 285 m, angle of dip 5 – 12^0, thickness from 2.20 to 2.85 m.

The schematic of the relative position of these seams allowing for planned mining operations in all three seams is shown in Fig. 1.

* Corresponding author: PirievaNN@suek.ru

The solution of the problem takes into account the structure of the strata and properties of each layer specified for wells 21577, 21567, and 11724.

The computational region of the problem was defined taking into account the depth and the thickness of the existing layers. The properties of the rocks were set in the following way:

For topsoil with alluvium the deformation module is $E = 4\cdot10^9$ N/m^2; Poisson's ratio is $\nu = 0.3$; bulk density is $\gamma = 18000$ N/m^3.

For siltstone: $E = 1\cdot10^{10}$ N/m^2; $\nu = 0.25$; $\gamma = 25000$ N/m^3.

For sandstone: $E = 2\cdot10^{10}$ N/m^2; $\nu = 0.2$; $\gamma = 27000$ N/m^3.

For coal seams: $E = 1\cdot10^9$ N/m^2; $\nu = 0.25$; $\gamma = 13000$ N/m^3.

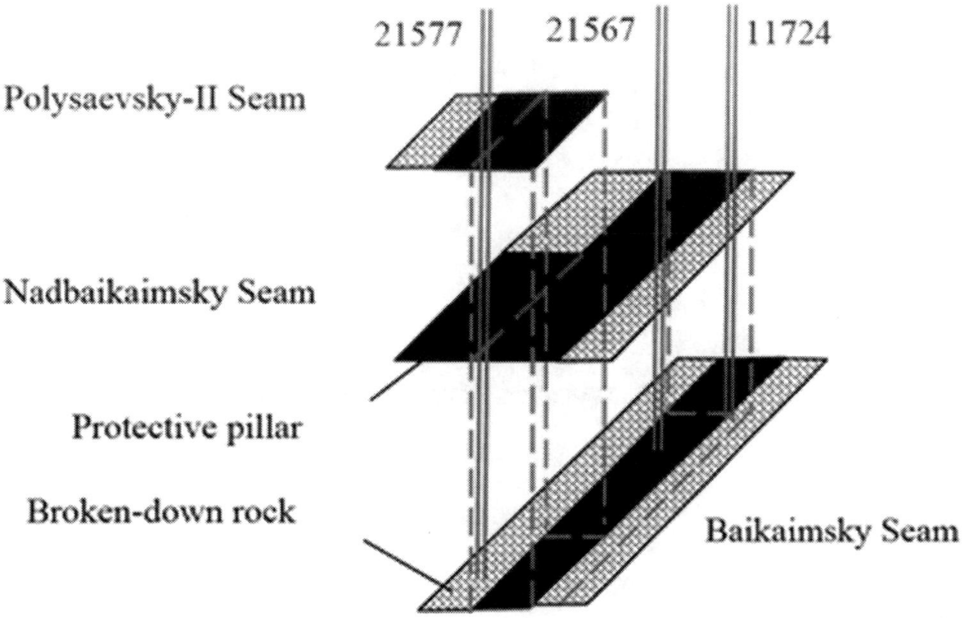

Fig. 1. Scheme of location of protective pillars for coal seams with the indication of wells – border of protective pillars

Fig. 2 shows vertical profiles and calculation schemes for the stressed protective pillars for slopes through wells 11724 (a), 21567 (b), and 21577 (c) respectively.

DOI: 10.1051/e3sconf/20172101010

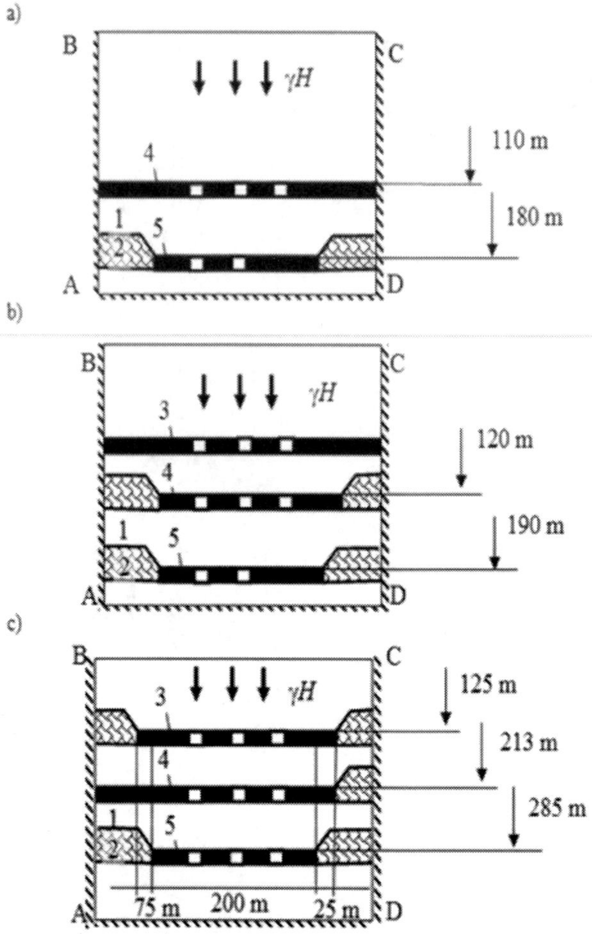

Fig. 2. Vertical profiles and schemes for calculating the strength of protective pillars through wells 11724 (a), 21567 (b), 21577 (c) respectively: 1 – host rock; 2 – goaf; 3 – Polysaevsky-II Seam; 4 – Nadbaikaimsky Seam; 5 – Baikaimsky Seam .

The problem at hand was solved using finite element method with ELCUT licensed software package. The number and size of grid elements were selected in such a way that boundary conditions are fulfilled with adequate accuracy: up to 0.001 m for movements and up to 1 N/m^2 for stresses.

The condition of the pillars was modelled based on a developed methodology for numerical calculation of the condition of pillars taking into account superlimiting deformation [14].

A criteria introduced in V.A. Gogolin's publication [13] was used for assessing pillar strength. The ultimate strength of a gently dipping seam which is in a planar deformation state is calculated using the following formula:

$$\sigma_S^P = \gamma H + \frac{\sigma_S}{1+\nu}, \quad (1)$$

where ν – lateral deformation ratio; H – depth of cover (m); γ – bulk density of rocks (N/m^3); σ_s – ultimate monoaxial compressive strength of rock (MPa).

So the section of the seam which meets this requirement: $\left|\sigma_{yy}\right| > \sigma_S^p$, loses its strength, partly breaks down, thus goes into a superlimiting state.

The following values were used in the calculations: ν = 0.25; σ_s = 10 MPa.

Fig. 3 shows distribution of vertical stresses within the section of well 11724 (upper section). Within this profile a protective pillar in Baikaimsky seam is being analysed (see Fig. 2a). The maximum compressive stresses at which the coal seam retains its strength were calculated using this formula (1): $\sigma_S^p = -14{,}5$ MPa. Sections of the seam where $\sigma_S^p < -14{,}5$ MPa and which lose their strength are shown in purple. The length of the destruction zone both sides of the pillar is 3.7 m.

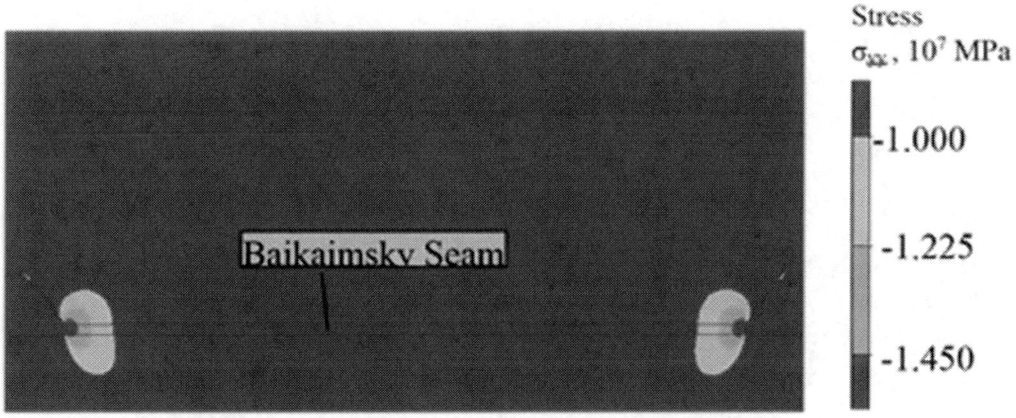

Fig. 3. – Distribution of vertical stresses σ_{yy} in section of well 11724.

Fig. 4 shows distribution of vertical stresses within the section of well 21567 (middle section). Two protective pillars are located within this section.

For Baikaimsky Seam the ultimate strength is $\sigma_S^p = -15$ MPa. Destruction zones in this seam are shown in purple and amount to 2.2 m on the left-hand side and 2.7 m on the right-hand side. For Nadbaikaimsky Seam $\sigma_S^p = -14$ MPa. The destruction zone is shown in purple and blue and amounts to 0.8 m on the left-hand side and 0.6 m on the right-hand side.

Fig. 5 shows distribution of vertical stresses within the section of well 21577 (see Fig. 2b). Three protective pillars are located within this section.

For Baikaimsky Seam the ultimate strength is $\sigma_S^p = -17$ MPa. Destruction zones in this seam are shown in purple and amount to 8.2 m on the left-hand side and 8.8 m on the right-hand side. For Nadbaikaimsky Seam $\sigma_S^p = -15$ MPa. The destruction zone is shown in purple and blue and amounts to 2 m.

 DOI: 10.1051/e3sconf/20172101010

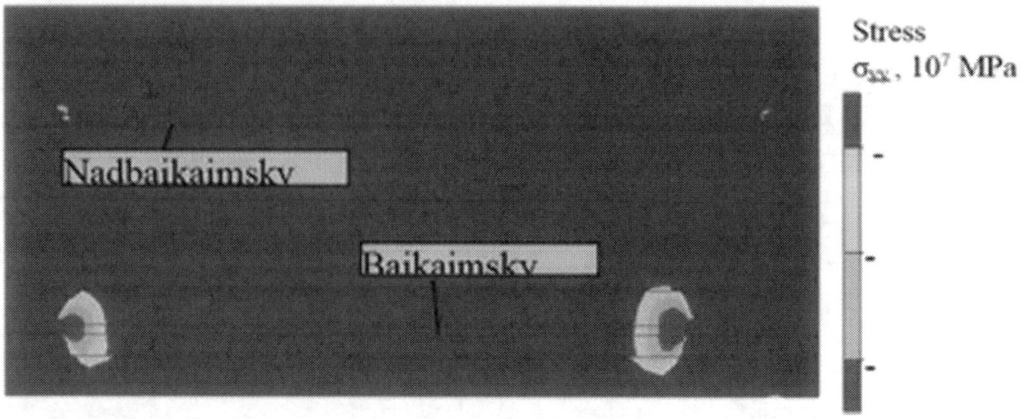

Fig. 4. –Distribution of vertical stresses σ_{yy} in section of well 21567

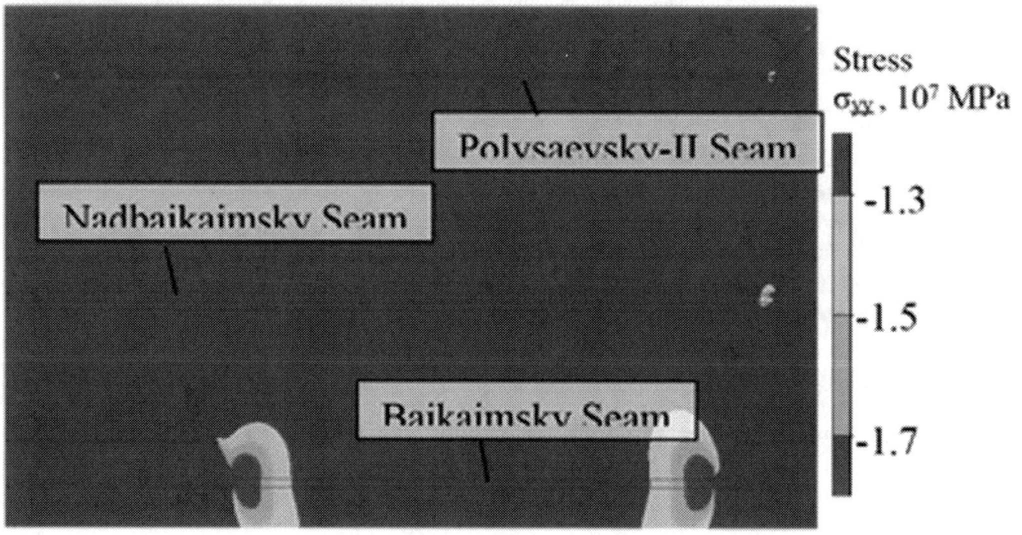

Fig. 5. – Distribution of vertical stresses σ_{yy} in section of well 21577.

For Polysaevsky-II Seam the ultimate strength is $\sigma_S^p = -13$ MPa. The destruction zone in Polysaevsky-II Seam is shown in purple, blue and yellow. Its sizes are 1.1 m on the left-hand side and 2.6 m on the right-hand side.

4 Conclusions

The deduced sizes of destruction zones within the boundary parts of protective pillars for slopes should be considered during design of a technology for extracting them. In particular, the amount of coal within these zones will be accounted for as operational losses. For pillars in Nadbaikaimsky and Polysaevsky-II seams these losses will amount to 1-1.5% while in Baikaimsky seam the losses will reach 3.7%, 2.5% and 8.5% in the upper, middle and lower part of the pillar respectively.

References

1. A.M. Lin'kov, J. Min. Sci. **37:1**, 10 (2001)

2. Isachenko A.A., Rib S.V., Voloshin V.A., Frjanov V. N., GIAB, **1**, 297. (2016)
3. Kurlenja M.V., Serjakov V.M., Eremenko A.A., *Technogenic geochemical stress field* , 264 (Novosibirsk, Nauka, 2005)
4. Dyagileva A.V., Gogolin V.A., Ermakova I.A., GIAB, **6**, 13 (2011)
5. Kornev E.S., Pavlova L.D., Fryanov V.N., KuzSTU Bul. **2**, 20 (2015)
6. Stavrogin A.N., Protosenya A.G., *Prochnost' gornyh porod i ustojchivost' vyrabotok na bol'shih glubinah*, 271 (M., Nedra, 1985)
7. Cherdancev N. V., KuzSTU Bul. **1**, 15 (2012)
8. Komissarov S.N. *Upravlenie massivom gornyh porod vokrug ochistnyh vyrabotok*, 237 (M., Nedra, 1983)
9. Kornev E.S., Pavlova L.D., Fryanov V.N., KuzSTU Bul. **2**, 65 (2013)
10. Kurlenya M.V., Mirenkov V.E., Shutov V.A., J. Min. Sci. **50:6**, 1001 (2014)
11. Zienkiewicz O.C., Taylor R.L., Zhu J.Z., *The Finite Element Method: Its Basis and Fundamentals (7th Edition)*, 756 (Butterworth-Heinemann, 2013),
12. Gogolin V.A., KuzSTU Bul. **3**, 3 (2016)
13. Ermakova I.A., Pirieva N. N., KuzSTU Bul. **4**, 3 (2016)

Prerequisites for the Establishment of the Automated Monitoring System and Accounting of the Displacement of the Roof of Underground Mines for the Improvement of Safety of Mining Work

Alexandr Abramovich [1*], *Evgeniy* Pudov [1], and *Evgeny* Kuzin [1]

[1]T.F. Gorbachev Kuzbass State Technical University, 653039, Nogradskaya str., 19a, Prokopievsk, Russia

Abstract. In the article the necessity of continuous control over the condition of the roof of mine workings is considered, to increase the safety in the conduct of mining operations. Provided the rationale for monitoring in complex mining and geological conditions, as well as in areas prone to rock blows and sudden coal emissions. The existing methods for controlling the displacement of the roof rocks are described, and their shortcomings are given. An idea is given of an automated system for monitoring the displacement of the workings. The stages of the system as a whole are considered, including the choice of a linear displacement sensor, a platform for software development, and a programming language. In order to ensure integration into other systems and subsequent analysis of the results, it is envisaged to output data to spreadsheets. Are shown the interfaces of the program and the output of the readings from the sensors to the monitors of the mining manager.

1 Introduction

Increasing the safety of coal production remains an urgent task given the growing concentration of equipment and mining. One of the factors of possible accidents is the development of areas dangerous for mountain impacts and sudden emissions. In this case, the task is to determine the zone of destruction of the rock massif in the vicinity of the mine and the ends. The shape and size of the fracture zone depend on the actual stresses and strength of the rocks, on the depth of the workings, on the presence of geological disturbances.

Sudden geodynamic manifestations in the roof of the excavations are caused by a number of factors: geological disturbances of rocks, overmoistening of the massif, exceeding the load on the anchor support, simultaneous development of several layers and inadequate support for given geological conditions, these are just some of the many of these factors. A big step towards solving this problem is to identify the causes of collapse. Sometimes the causes of

* Corresponding author: abramovich_sanja@rambler.ru

roof collapse can not be determined, but the study of the fall of exfoliated and decomposed rocks, the state of the surrounding and geotechnical environment can reveal the causes of collapse [1].

An increasing amount of mined coal, an increase in the depth of excavations require improved methods for monitoring the condition of the roof, with high reliability of results, speed and continuity of control.

The Regulation on the Safe Management of Mining [2] regulates the conduct of research at each deposit, prone or dangerous for mining impacts, to improve the forecasting and prevention of mountain impacts. In this case, the forecast of the impact hazard of sections of the rock massif should be carried out, both according to the systems of continuous automated control of the stressed state of the array, and by changing the values (speeds) of deformations and displacements.

The variety of mining-geological and mining-technological factors continues to create complexity in predicting the manifestations of rock pressure.

In order to identify the patterns of rock pressure, timely detection of complex situations for managing rock pressure and crepe work, taking measures to eliminate the accident rate of mining operations, it is necessary to accumulate information and create a database with data on the magnitude and speed of displacement of the roof sections. Effective work is possible using a system of continuous monitoring of roof movements, analyzing the influence of various factors, and issuing information on dangerous sections of excavations, where it is necessary to take measures to strengthen the fastening.

2 Materials and methods

At present, methods of numerical simulation of geomechanical processes are being conducted, cores are selected, the structure of the roof rocks is analyzed using a video endoscope and GPR method [3, 4].

In addition, today, enterprises that mine coal underground, to control the displacements of rocks in the excavations used reference depth stations, which allow for visual contact to detect the displacement of rock mass. The method of visual control with the help of deep reference stations is to observe the rock shift by recording the readings when moving the frames fixed in the hole at the appropriate distance [5, 6].

Nowadays, this method is obsolete and has a number of shortcomings [7-9]:

1. For specific, clear information, it is necessary to monitor the frames continuously, that is, one directly needs to continuously record the shifts that occurred.
2. The uncertainty of the exact time of occurrence of the displacement of the rock mass, as well as the lack of information on the rate of its displacement.

Proceeding from the above, it becomes clear that the engineer and technical worker of the site, checking the state of the reference deep stations during the shift, with a rapid and large displacement of the roof rocks may not be in time to warn other employees of the mine. As a rule, this is done by the mining master once a shift, and it is not possible to say for what time the movement occurred.

The foreign researchers results [10-12], show the urgency of creating an automated monitoring system, with various ways of implementation.

Thus, our main task is to create a full-fledged automatic monitoring and accounting system for the displacement of the roof of underground workings, to improve the safety of mining, which will directly include:

1. Sensors of displacement;
2. Transmitting devices;
3. Communication channels;
4. Software;

5. Output of data to the monitors of the mountain dispatcher.

The work on the creation itself will consist of several stages, mentioned earlier:

1. Select a displacement bias sensor.
2. Select the transmitter.
3. Determination of the method of data transmission (selection of data transmission channels).
4. Definition of the software development platform.
5. Software development.

The displacement sensor is a device designed to determine the magnitude of the linear or angular mechanical displacement of an object. It should be noted that all displacement sensors can be divided into two main categories - linear displacement sensors and angular displacement sensors (encoders). Of all the existing types of sensors, we are most suited to linear displacement sensors.

Having analyzed the merits and demerits of various converters, relative to the necessary parameters, it can be concluded that the most acceptable as a displacement sensor for the roof are capacitive linear displacement sensors with a variable plate area. The main advantages of capacitive converters are the simplicity of the device, high sensitivity, low power consumption, no moving contacts, comparative ease of manufacture, small dimensions and weight, long service life.

At the stage of approach to software development, we have constructed a general scheme for representing the system as a whole, a diagram (see Figure 1).

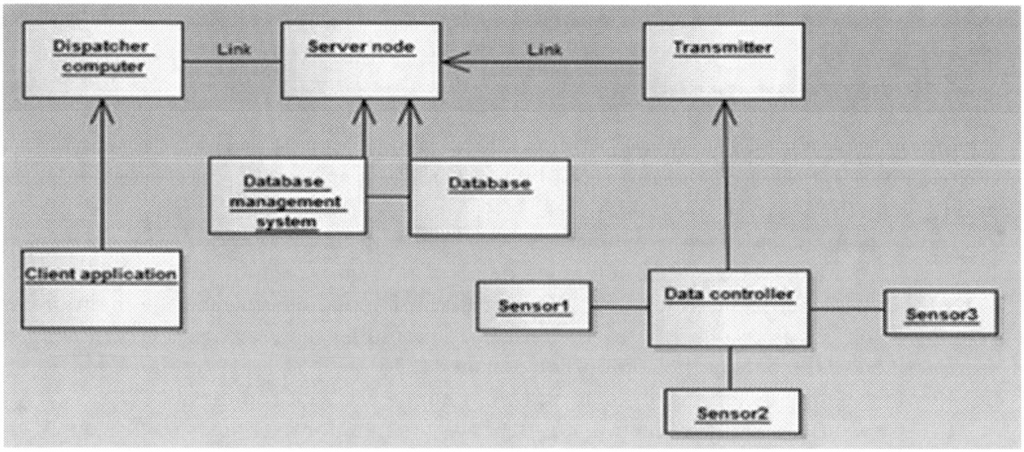

Fig. 1. System representation diagram.

3 Results and discussion

At the software development stage, the subject area was first analyzed. The most important point to which we drew attention is very large volumes of information, and it was decided to store all non-recurring recorded rock mass readings taken from each displacement sensor.

The architecture of the information system being developed is the client-server architecture. This architecture is implemented by the database management system (DBMS) of MSSQL Server.

Using this DBMS will provide the following functions:

- Ensures the integrity of the database (DB);
- provides fast recovery after various failures (hardware and software);
- provides backup;
- provides high operational reliability;

 DOI: 10.1051/e3sconf/20172101011

- provides high performance.

During the development of the database it was decided to create a table for each sensor and thus the following information objects were identified:

• Sensor 1
• Sensor 2
• Sensor n

Integrity requirement: to table contents Sensor 1, Sensor 2, Sensor n must necessarily refer to level 1, level 2, date and time of reading, for two-level sensors. As a result, we got a rational database for storing information, a database diagram (see Fig. 2).

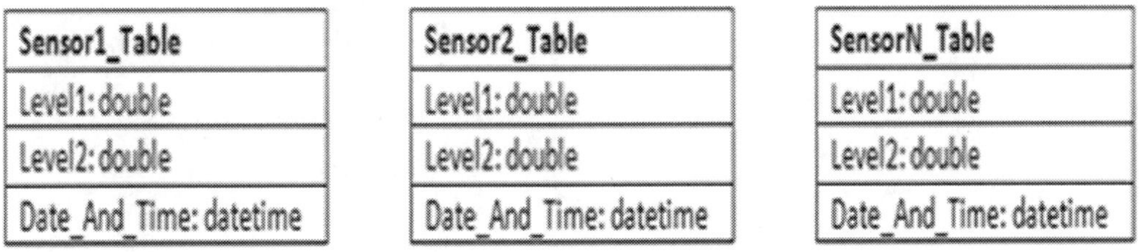

Fig. 2. Resulting ER diagram (entity diagram-connection) of the database.

After implementing the data warehouse, we went on to the process of creating the program itself.

As technology of interaction of user-defined components with data, ADO.NET technology is chosen - the main model of data access for applications based on Microsoft .NET. The choice of technology is based on the choice of operating system. As a programming environment, choose Visual Studio, and as the language of C #. This language and environment are universal programming tools, so they are suitable for solving the set task of creating a system. The development of an implementation model (program code) can be represented as a description of the components that implement the boundary object classes and entity classes.

The implementation model includes the components integrated into the subsystems that implement the objects identified in the design process. The component model is shown in Fig. 3.

DOI: 10.1051/e3sconf/20172101011

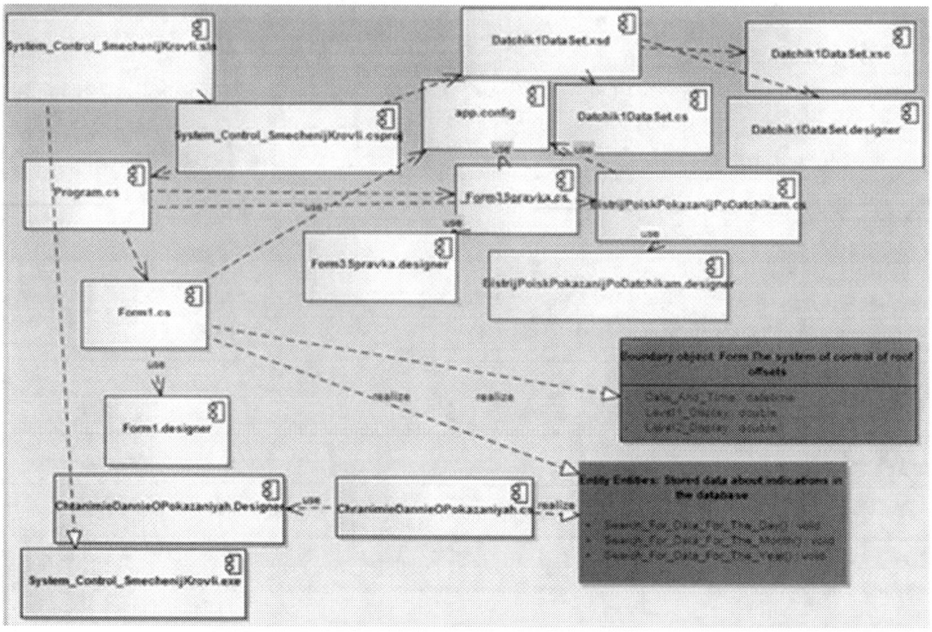

Fig. 3. Component model.

For each form and bookmark on the form, the composition of interactions with the user is determined, thanks to which the process of receiving and accounting of the readings of the sensors is carried out, as well as notification of the controller about the displacement of each sensor in a particular area.

To fulfill each functional requirement, we select according to a separate scenario. Scenario for fulfilling the functional requirement of finding the readings of the selected sensor for the entire time since the introduction of the system and installing the sensor:

The dispatcher has the ability to plot the selected sensor for the entire time, from the moment of the sensor introduction, as well as for the day, week, month, year.

To implement this scenario, two stored procedures were created in the database, which will be called in a program created on the Windows platform. As part of the system being developed, the client component and the server component are distinguished. The server component implements scenarios related to access to the data and its main processing. The client component implements the presentation of data in a user-friendly form and selects the actions to be performed.

The developed interface of the program is presented in Fig. 4

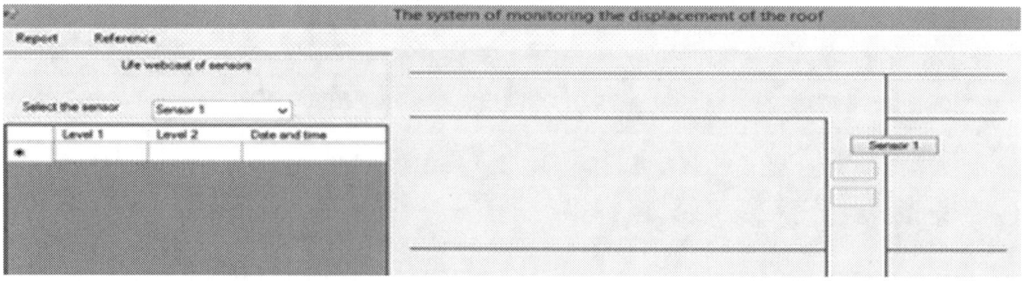

Fig. 4. Main page program interface.

The interface of the program for each separate sensor was also developed. The interface is shown in Fig. 5 below.

DOI: 10.1051/e3sconf/20172101011

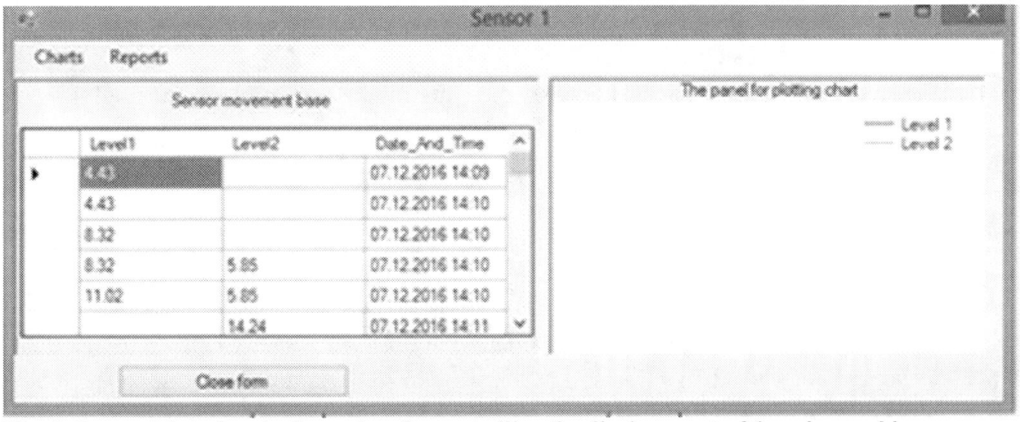

Fig. 5. Sensor 1 interface in the system for controlling the displacement of the mine workings.

In connection with the possible emergence of the need to work with the received data, including processing and analysis, it was decided to comprehensively export the received data to a Microsoft Office Excel document.

To implement the output of data in an Excel document, an Excel Report button was created on the form.

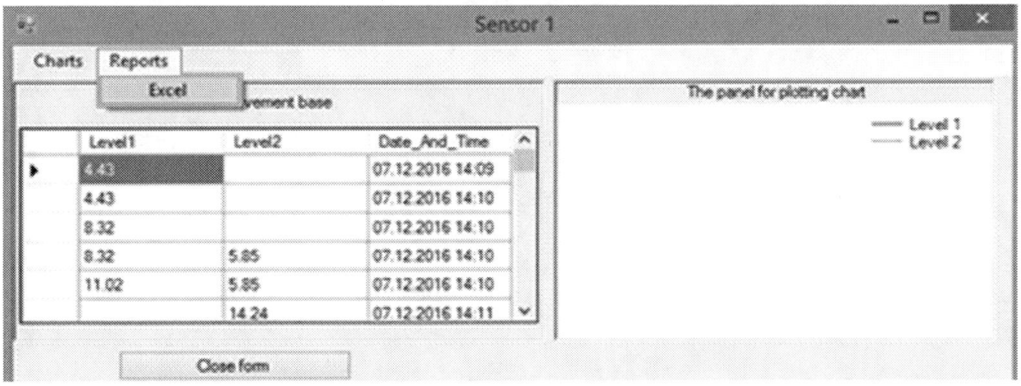

Fig. 6. Report interface in a spreadsheet.

4 Conclusions

As a result of the work, the first version of the system for monitoring and recording the displacement readings in the mine workings roof was created, which will allow to record changes in the state of the roof of underground mine workings in real time. Based on the accumulated information, it is possible to predict the behavior of the roof in such mining and geological conditions, and also to analyze the causes preceding the change in the roof state (displacement, stratification, decompression).

References

1. G. Molinda, C. Mark, Electron Journal Geotech Engineering, **15**, 547 (2010)

2. *The order of Rostechnadzor from 02.12.2013 N 576 «About approval of Federal norms and rules in the field of industrial safety «Regulations for the safe conduct of mining works on deposits prone and dangerous mountain blows»* (Registered in Ministry of justice of Russia, 2014).
3. G.L. Feofanov, Coal, **3**, 18 (2017)
4. Haijun Zhao, Fengshan Ma, Jiamo Xu, Jie Guo, International Journal of Rock Mechanics & Mining Sciences, **53**, 20 (2012)
5. M.I. Alvarez-Fernandez, C. Gonzalez-Nicieza, A. Menendez-Dıaz, Alvarez-Vigil., Eng. Geol., **80**, 1 (2005)
6. *Federal occupational safety standards and regulations «Guidelines for computation and application of coal mines rock support»*, (Moscow, JTC NTS Industrial Safety Publ., 2015).
7. Chunbin Zhao, Yubao Zhang, Genu Zhang, Janji Li, Shuki Ma., **17:5**, 1044. (2017)
8. P. James Doherty, Allsize Hasan, H. Gonzalo Suazo, A. Andy Fury, Canadian Geotechnical Journal, **2:12**, 1901 (2015)
9. E Lawson, T. Douglas, M. K. Larson, 35th International Conference on Ground Control in Mining (2016)
10. Yiming Zhao, Nong Zhang, Guangyao Si., Sensors (Basel), **16:10** (2016)
11. Z.Z. Zhang, J.B. Bai, Y. Chen, S. Yan, Rock Mech. Min. Sci., **80:1**, (2015)
12. A. Catalano, F.A. Bruno, M. Pisco, A. Cutolo, A. Cusano, Sensors. **14**, 18268 (2014)

Increasing Stability of Mine Surface Facilities on the Fill-Up Ground

Michail Sokolov [1*], and *Sergey* Prostov[1]

[1] T.F. Gorbachev Kuzbass State Technical University, 650000, 28 Vesennyaya St., Kemerovo, Russia

Abstract. The emergency condition of the mine surface facilities including a coal wash plant, a crusher, a bunker and three galleries, is considered. As a result of a complex of engineering-geological and geophysical surveys, it is established that the significant deformations of the foundations of the structure are caused by the zones of weakened (unconsolidated) soil in the bases of the foundations formed due to landslide developments in the bulk ground mass. The article describes the techniques for justifying the technological parameters of the consolidation grouting of soil bases including computer modeling, forecasting of deformations of the natural and consolidated ground mass, analysis of the stress-strain state of soils in the most dangerous area. The effectiveness of the adopted scheme of consolidation grouting of soil bases according to the approved methodology for elimination of the emergency condition of the mine surface facility was confirmed. The limit values of the calculated horizontal deformations in the soil bases of one of the supports of the galleries are determined. To prevent the indicated deformations, it was recommended to provide an additional row of injection wells with a depth below the border of the loosened soil zone.

1 Introduction

Stability of surface mine structures is an important condition of a safe technological process for the development and extraction of minerals. In some cases, in the foundations of such structures there are technogenic soils with different degrees of compaction and changeable deformation properties, which contribute to the development of uneven vertical displacement and are the causes of accidents. To improve the stability of the foundations of surface mine structures, methods of injection consolidation are used the most effective of which are methods of pressure injection [1, 2] and electrochemical consolidation [3, 4]. The choice of approach to improving the stability of structures need to be supported by rock mechanics forecast, study of the stress-strain state of rock mass, including those with consolidating elements [5-7]. The use of software programs for geotechnical calculations using the finite element method significantly increases the efficiency, accuracy and speed of forecasting [8, 9].

* Corresponding author: ea-sokolov42@mail.ru

 DOI: 10.1051/e3sconf/20172101012

2 Materials and methods

A seasonal coal wash plant with separators and galleries at Krasnobrodsky coal open-pit mine, the branch of OAO UK "Kuzbassrazrezugol" is an example of the above mine surface facilities. This facility includes a number of technological buildings and 3 galleries (Fig. 1). All structural elements have standard solutions, including post reinforced concrete foundations.

The facilities are located on the territory with a complex relief, formed as a result of mining operations and having a completely technogenic (man-made) character. On the site under consideration, the rock mass at a depth of 8 to 20 m is represented by man-made fill-up soils in the form of a mixture of land waste, crushed stone, lumps, loam and sandy material compactly rolled. The fragments and fine aggregate are represented by sedimentary rocks of weathered sandstone type of low strength. Loam contains local lenses and nests of medium fine gravel with loamy hard, dense aggregate up to 35-45 %. In the hydrogeological respect, the site is characterized by the presence of an aquifer of groundwater in weathered and destroyed bedrock (formation-fractured waters).

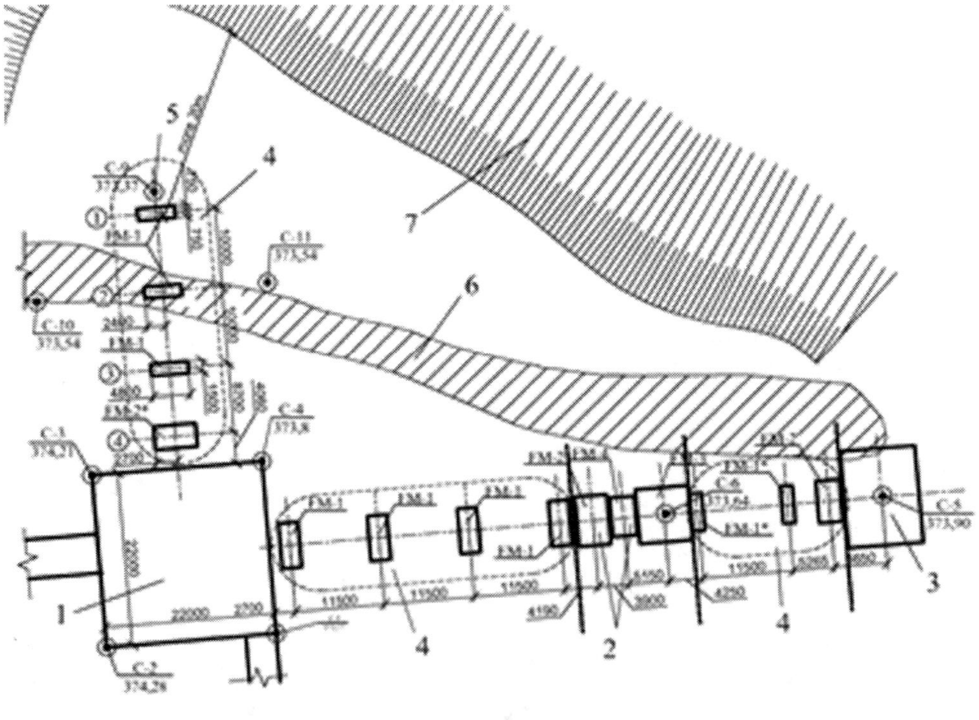

Fig. 1. Scheme of the territory and the plan of the investigated object: 1 – coal washing building; 2 - crusher and crushing section; 3 - receiving hopper; 4 - galleries No. 11, No. 5, No. 9; 5 - geological wells; 6 - zone of soils decompaction; 7 - slope of fill-up foundation.

According to the data of visual surveys and engineering and geological surveys conducted by OOO NOOCENTR, the structure is in an emergency condition, it has significant uneven subsidence of the supports of the conveyor galleries. During the work on the surface of the site in the northeastern part, closed cracks of a landslide character (cleavage cracks) were found, and zones of decompaction were observed, which develop as a result of a partial displacement of the ground masses towards the unconsolidated slope of fill-up soil mass. Since a part of the receiving hopper and gallery No. 11 are located above the zones of decompaction, additional work has been carried out aimed at identifying and outlining these zones by geoelectric and radar control methods [10]. Analysis of the data of survey showed the need to perform measures to increase the load-bearing capacity of the soil base in the northeastern part by pressure injection methods, in particular under the foundations of the receiving hopper and supports of gallery No. 11, located directly above the zone of soil decompaction.

3 Results and discussion

The first stage of the analysis of the geomechanical state of the base of gallery No. 11 was aimed at predicting the vertical displacement of the surface along the longitudinal axis of the structures. The forecasting was carried out by mathematical modeling on the basis of the finite elements method and the modern geotechnical software package "Alterra" of OOO "InzhProektStroy", the Russian program developer. The obtained distribution graphs of the vertical displacement Δh of the supports of gallery from the relative coordinate along the axis of the object L_i /L_{all} (Fig. 2), where the total length L_{all} = 33.05 m, show that the displacement has maximum values mainly at intervals located above the decompaction zones: in the ranges of L_i /L_{all} = 0.3...0.4 and L_i /L_{all} = 0.6...0.7.

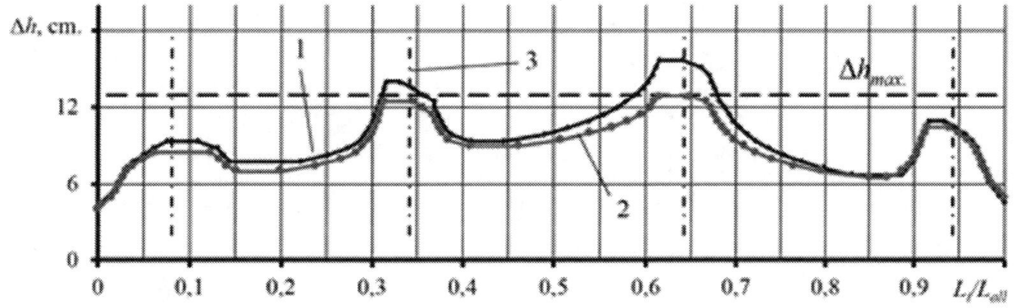

Fig. 2. Distribution of vertical displacement Δh along the axis: 1 - with natural base; 2 - when consolidated according to the approved scheme; 3 - axes of foundations symmetry.

Taking into attention the maximum permissible size of subsidence of Δh_{max} = 14 cm, determined according to SP 23.13330.2011, it can be concluded that the proposed consolidation scheme allows us to achieve the required stability of the structure, however, the subsidence in the above ranges is close to the maximum permissible values, so it is advisable to conduct a more detailed study of individual supports of the gallery.

The second stage of modeling was aimed at studying and analyzing the stress-strain state of the base of potentially the most dangerous support of the gallery in the longitudinal and transverse directions. The schemes of basic models of the soil base of the gallery support are shown in Fig. 3.

The base was simulated from the top mark of the foundation with the specification of equivalent vertical pressure forces P, the horizontal load T and the bending moment from deformation of the support in the transverse direction M. Foundations of FM-1 type are a slab with a width of b_f = 1.5 m, a length of l_f = 4.8 m and a height of h_f = 0.6 m, made of heavy monolithic reinforced concrete (density ρ_f = 2500 kg/m^3, elastic modulus E_f = 2000 MPa, Poisson's ratio υ = 0,15). The depth of foundation from the lowest mark of the surface was d_f = 3.090 m. In the base of the foundation there is concrete preparation with a height h = 0.1 m. The foundations are in open pits backfilled with large clastic technogenic deposits with a thickness of 0.8 m. To solve the problem of the various kinds, the decompaction zones were averaged and divided into two identified characteristic regions.

The physical and mechanical properties of the elements of the model are presented in the Table.

Based on the data given and the algorithms, described in the work [11], the main parameters of the model for the case under consideration were assigned: the width in the longitudinal direction $b_{m1} = 41.0$ m, in the transverse direction - $b_{m2} = 31.0$ m (taking into account the size of the pit and the foundation); height of the model is $h_m = 21.0$ m. The boundary conditions in the form of restrictions of displacements are superimposed similarly to the models adopted in [11, 12]. For the model in the longitudinal section, the load P was set by two forces of magnitude $P/2$ with a displacement of $l = 0.6$ m on either side of the side faces of the foundation, simulating the position of the support elements. In the cross-section, the loads P, T and M are given along the axis of symmetry of the foundation. In both cases, the upper edge of the foundation (cutoff) is considered the point of application of the load.

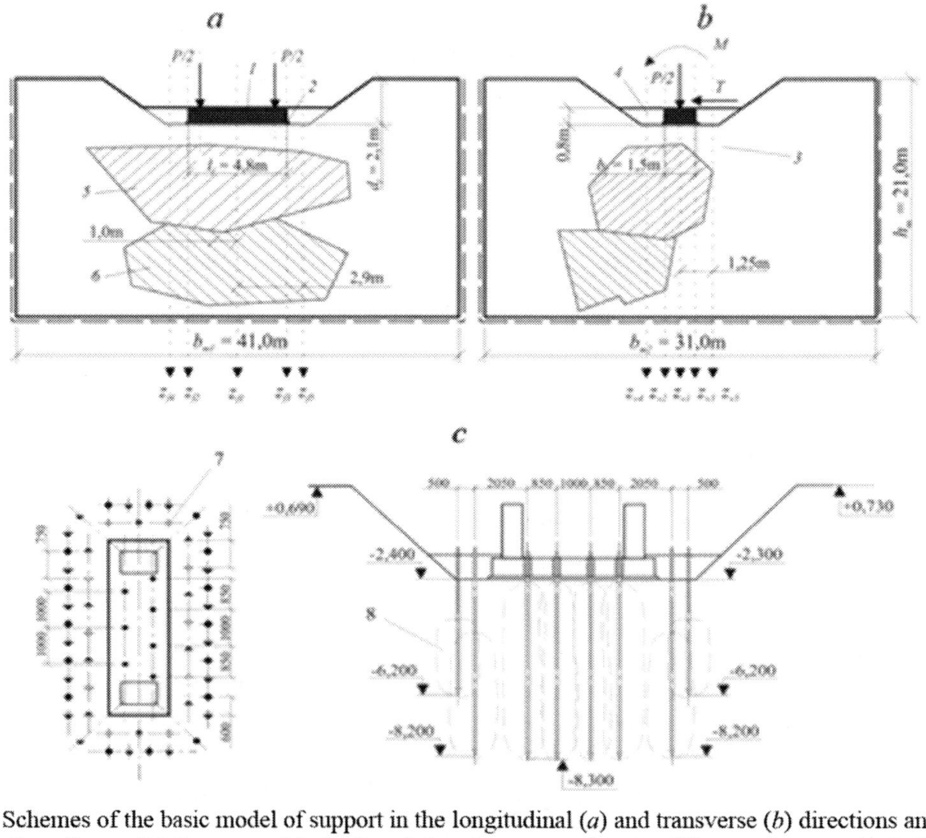

Fig. 3. Schemes of the basic model of support in the longitudinal (*a*) and transverse (*b*) directions and the location of consolidation zones (*c*): 1 - foundation; 2 - preparation; 3 – fill-up rock mass; 4 – technological backfilling; 5 - decompaction zone No. 1; 6 - decompaction zone No. 2; 7 - injector; 8 - consolidation zone.

Table 1. Physico-mechanical properties of the elements of the model

#	**Name of the model element**	**Density of the material (soil) ρ, kg/m^3**	**Modulus of deformation *E*, MPa**	**Poisson's ratio υ**	**Angle of internal friction φ, deg.**	**Bond *C*, kPa**
1	Reinforced concrete mat foundation	2500	2000	0.15	–	–

 DOI: 10.1051/e3sconf/20172101012

2	Concrete preparation for foundation	2400	1800	0.15	–	–
3	Backfilling of the excavation	1950	20.0	0.25	20.0	–
4	Soil bases	2050	38.0	0.29	23.7	73.6
5	Decompaction zone No. 1	1900	17.0	0.36	25.0	25.0
6	Decompaction zone No. 2	1800	12.5	0.37	25.0	15.0

The results of the second stage of modeling are reflected in the form of fields of isolines of the stress-strain state of the soil mass of a separate support in the longitudinal and transverse directions, shown in Fig. 4.

In the longitudinal direction the stress distribution due to the influence of the decompaction zones occurs asymmetrically with the presence of significant distortions at the boundaries of these zones. The zone of stress concentration remains in the area under the foundation, without significant displacements described in the work [12], and the introduction of the consolidation zones into the calculation considerably reduces the level of stresses and the area of their concentration. The influence of the consolidation zones can be traced in the form of redistribution of deformations from decompaction zone No. 1 to the decompaction zone No. 2 and the soil mass, the value of vertical displacement are greatly reducing. In the cross section, the stresses are distributed predominantly in the left support part of the rock mass due to the asymmetric load and the location of the decompaction zones in the rock mass relative to the foundation. Consolidation increases the asymmetry, loading mainly the zone of decompaction No. 2 (at a depth of $z_2 = 8$ m), but significantly reduces the horizontal stresses under the foundation, thereby preventing horizontal displacements.

 DOI: 10.1051/e3sconf/20172101012

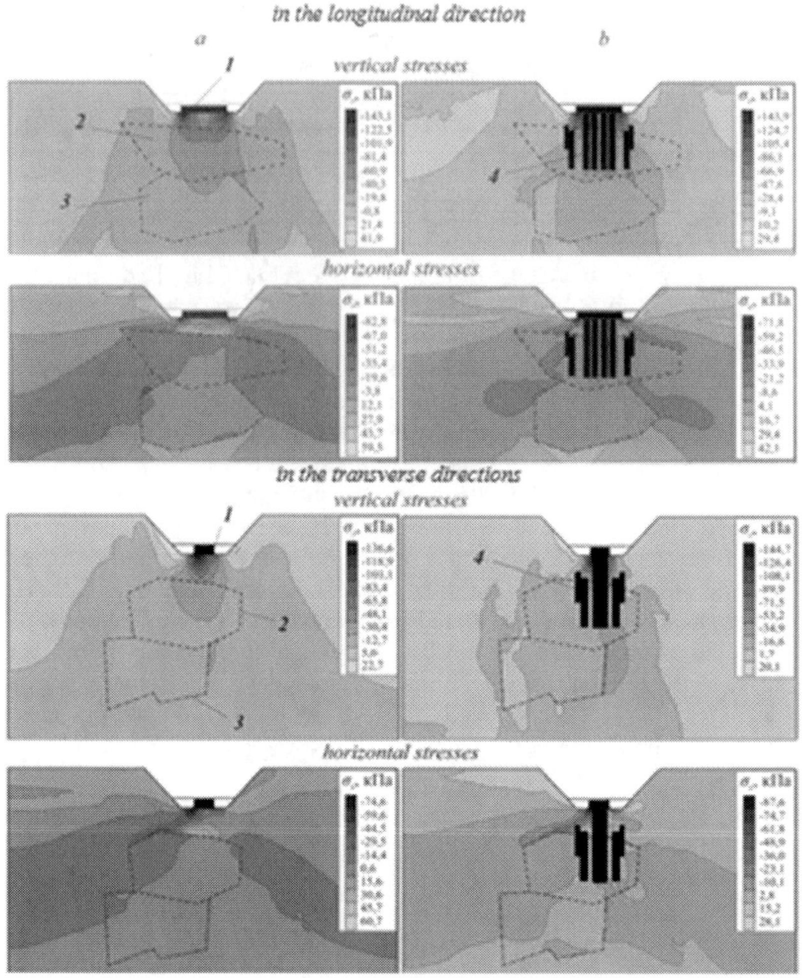

Fig. 4. Distribution of stresses in the natural and artificial soil mass in the longitudinal (*a*) and transverse (*b*) cross-sections: 1 - foundation; 2 - decompaction zone No. 1; 3 – zone of decompaction No. 2; 4 - consolidation zones.

The characteristic regularities in the distribution of vertical and horizontal deformations are shown in Fig. 5. The distribution of vertical deformations ε_z has a similar character both in natural and in artificial soil mass. In the range of the near-surface part of the soil, the reduction of deformations is noted, and within the consolidation zones the deformation ε_{zf} is much lower than in the natural soil mass; within decompaction zone No. 2 an increase in deformation ε_{zf} is noted. Similar regularities can be traced in the cross section, however, additional concentrations of maximum deformations ε_{zv} and redistribution of deformations into the soil mass are noted at a greater depth in the range of z_{vi} = 12.0...17.0 m.

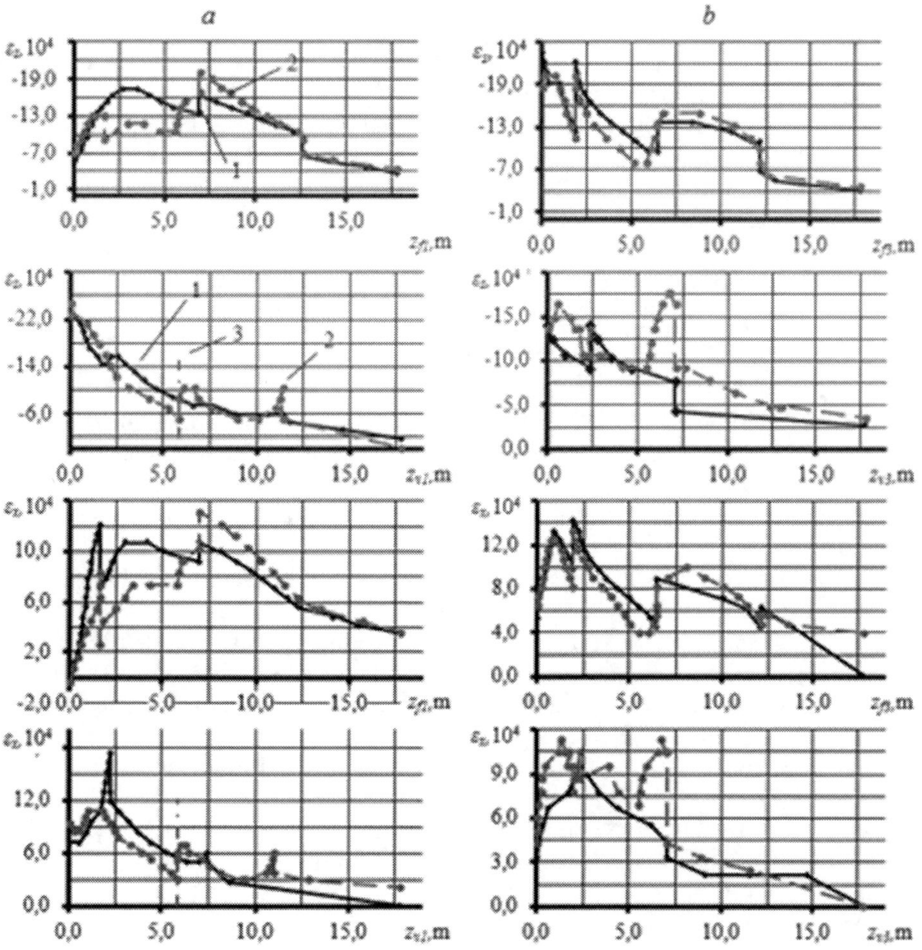

Fig. 5. Dependences of vertical ε_z and horizontal ε_x deformations at the depthьof model z_i in longitudinal (a) and transverse (b) directions: 1 - natural soil; 2 – consolidated soil; 3 - boundaries of the consolidation zone.

The distribution of horizontal deformations ε_x in both sections occurs mainly monotonically with the formation of separate local "bursts," mainly at the borders with decompaction zones. In the longitudinal section, there is a slight decrease in deformations ε_{xf} in the artificial soil mass, as well as their noticeable increase in the regions below the consolidation zones. When calculating the artificial soil base it should be taken into account that in the longitudinal direction the decompression zone is subject to considerable horizontal deformation. In the transverse profile there are similar patterns, at the same time, the following anomalies are recorded: the formation of an additional deformation difference ε_{xv} at $z_{v1} > 11.3$ m; change in the nature of the distribution and increase in the values of deformations ε_{xv} in the interval $z_{v3} = 0.0...7.2$ m; local increase in deformation within the consolidation zone. Thus, the consolidation zones located in the direction of the action of the horizontal load can have significant transverse deformations.

4 Conclusions

Geomechanical analysis has proved the effectiveness of the method of injection grouting in managing the properties of weak soils and the stability of structures. Methods of mathematical and computer modeling allowed us to predict with sufficient detail the effect obtained from consolidation and to identify the most problematic areas of the soil. In particular, in the presence of additional horizontal load and decompaction zones it is necessary to increase the number and depth of injection wells to prevent increased vertical and horizontal deformations. Based on the modeling data, the effectiveness of the consolidation scheme developed by OOO NOOCENTR has been proved, however, in order to achieve the greatest effect, it is recommended to place additional consolidation zones in the most loaded part of the soil mass at an asymmetric load, immersed in the ground to the depth of decompaction zone No. 2. Such an approach will allow us to redistribute vertical and horizontal stresses to a stronger soil mass and reduce the foundation subsidence, its tilt displacement and transverse deformations of the consolidation zones themselves.

References

1. M. N. Ibragimov, Rock Mechanics and Foundation Engineering, **50:5**, 200 (2013)
2. M. N. Ibragimov, Rock Mechanics and Foundation Engineering, **52:2**, 100 (2015)
3. Shao-Chi Chien, Fu-Chen Teng, Chang-Yu Ou, Acta Geotechnica, **10:6**, 813 (2015)
4. A. Flora, S. Gargano, S. Lirer, L. Mele, Geotechnical and Geological Engineering. (2017)
5. A. V. Azarov, M. V. Kurlenya, A. V. Patutin, S. V. Serdyukov, Journal of Mining Science, **51:6**, 1063 (2015)
6. T. V. Lobanova, S. V. Moiseev, Journal of Mining Science, **45:3**, 227 (2009)
7. S. G. Bezvolev, Soil Mechanics and Foundation Engineering, **45:3**, 77 (2008)
8. Z. G. Ter-Martirosyan, A. S. Abdulmalek, Soil Mechanics and Foundation Engineering, **44:6**, 195 (2007)
9. D. M. Shapiro, N. N. Mel'nichuk, Soil Mechanics and Foundation Engineering, **44:2**, 62 (2007)
10. S. M. Prostov, N. Yu. Nikulin, Journal of Mining Science, **51:5**, 908 (2015)
11. S. Prostov, V. Sokolov, Taishan Academic Forum - Project on Mine Disaster Prevention and Control, 350 (2014)
12. M. Prostov, V. Sokolov, V. Pokatilov, International Journal of Applied Engineering Research, **10**, 45297 (2015)

E3S Web of Conferences **21**, 01013 (2017) DOI: 10.1051/e3sconf/20172101013
The Second International Innovative Mining Symposium

Logistic Principles Application for Managing the Extraction and Transportation of Solid Minerals

Alexey Tyurin[1*]

[1] T.F. Gorbachev Kuzbass State Technical University, 650000, 28 Vesennyaya St., Kemerovo, Russia

Abstract. Reducing the cost of resources in solid mineral extraction is an urgent task. For its solution the article proposes logistic approach use to management of mining company all resources, including extraction processes, transport, mineral handling and storage. The account of the uneven operation of mining, transport units and complexes for processing and loading coal into railroad cars allows you to identify the shortcomings in the work of the entire enterprise and reduce resources use at the planned production level. In the article the mining planning model taking into account the dynamics of the production, transport stations and export coal to consumers rail transport on example of Krasnoyarsk region Nazarovo JSC «Razrez Sereul'skiy». Rolling planning methods use and data aggregation allows you to split the planning horizon (month) on equal periods and to use of dynamic programming method for building mining optimal production programme for the month. Coal mining production program definition technique will help align the work of all enterprise units, to optimize resources of all areas, to establish a flexible relationship between manufacturer and consumer, to take into account the irregularity of rail transport.

1 Introduction

Logistic approach application to management of material and accompanying information, financial and service flow has been successfully implemented in various of economy sectors [1-2], including in the area of solid minerals mining [3-6].

Using the basic logistic principles (rationality, integrity, consistency, hierarchy, and integration) in the field of mining allows you to link all production, transport and other service processes in a single system [7-9].

The use of resource-saving technologies not only during mining, but also in its dissemination to the end customer allows you to get a superior competitive advantage - the reduction in the price of the finished product and potential market expansion, profitability of the enterprise.

To achieve this, must reduce all logistics costs in the chain «producer-consumer» due to the review of the production, transport and storage processes consistency in the field of mining solid minerals [10-12].

* Corresponding author: alexturin07@rambler.ru

2 Materials and methods

The object of the study was JSC «Razrez Sereul'skiy», Nazarovo of Krasnoyarsk region. The coal mine is carried out extraction of brown coal. The main consumer of the coal mine is Novosibirsk CHPP-3 of JSC «SIBEKO». JSC «Razrez Sereul'skiy» includes the following production areas:
- production area, which produces overburden and mining of the existing excavator fleet;
- technological transport area, which transports overburden on the inside, the outside blade (the fleet of BelAZ dump trucks), transportation of coal mined at the warehouse "Central" (the fleet of BelAZ dump trucks);
- motor transport area, which carries out transportation of coal from the warehouse «Central» to warehouse «Transshipment» Volvo dump trucks;
- technological complex of coal loading, which provides the receipt and storage of coal at the warehouse «Transshipment», and performs the loading of coal into railroad cars to the consumer.

Coal mining is carried out by excavators HITACHI-470, Liebherr 984 loading of coal into dump trucks BelAZ 7555D, as well as attracting overburden dump trucks BelAZ 7540, 75473, hauling to the warehouse «Central». Warehouse organization is allowed:

-eliminate downtime for transportation of coal due to climatic conditions;

-exclude access to transport third-party organizations and consumers working in the mining sector (loading coal in the mine), ensuring safety of mining operations;

-remove the hard interaction of production processes, transportation and commercial products shipment, which increases the productivity of mining shovels.

Coal loading at the warehouse «Central» is carried out by a front loader DRESSTA - 534C. The transportation of coal from the warehouse «Central» to Nazarovo station railway deadlock on warehouse «Transshipment» is carried out by dump trucks VOLVO BCM-51.1 with a transportation distance of 39.5 km. the General scheme of extraction and transportation of coal to consumers is presented in figure 1.

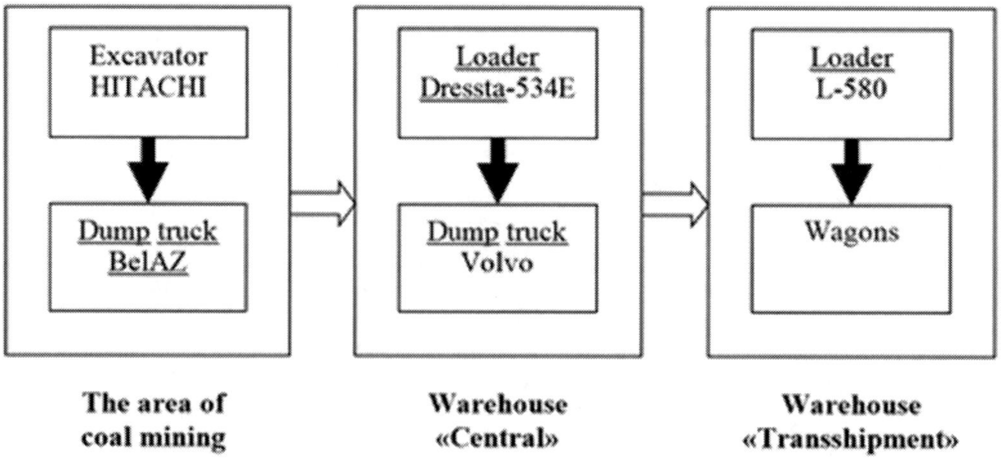

Fig. 1. General diagram of the extraction and transportation of coal.

The uneven supply of wagons for loading to the warehouse «Transshipment» (figure 2) affects the irregularity of the excavators work (figure 3) and trucks at the coal mining area and warehouse «Central». All this leads to equipment downtime, reduced productivity of production area and, as a consequence, to higher costs in the chain «producer-consumer» and the prices of finished products. Thus, the resources of the enterprise are used inefficiently.

The elimination of these disadvantages is possible through the logistic approach use to management of all coal mine processes and accounting of all costs for production, handling and transportation of coal to the consumer. For this, rolling planning method use is proposed. Planning period – a month, which is divided into 10 periods of 3-4 days in the period. The aggregation of time intervals will take into account the uneven supply of wagons (environmental conditions) and to adapt the coal mine operation (conditions of the internal environment) for the new parameters of the environment.

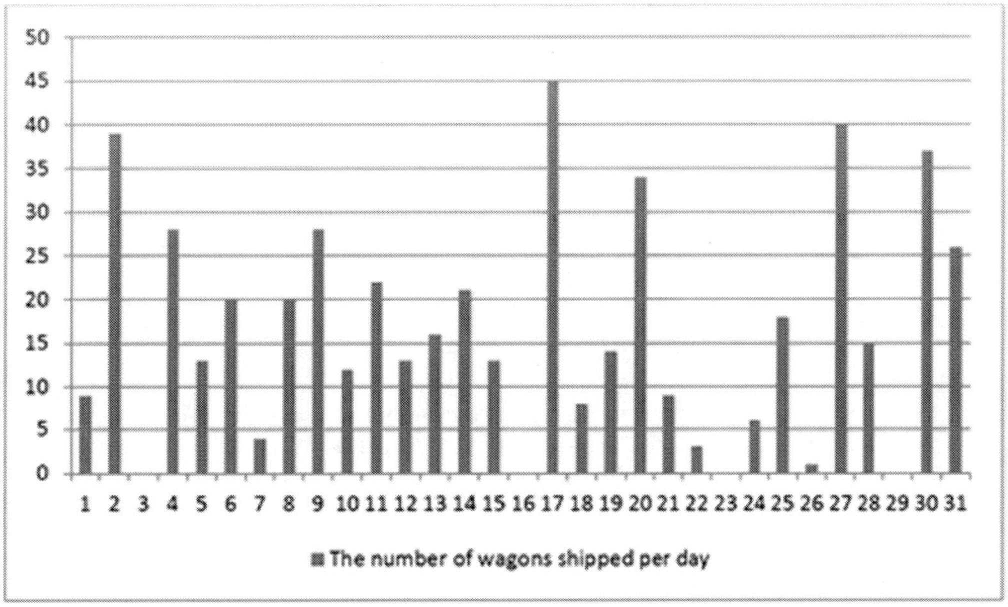

Fig. 2. Uneven loading of wagons in March 2017.

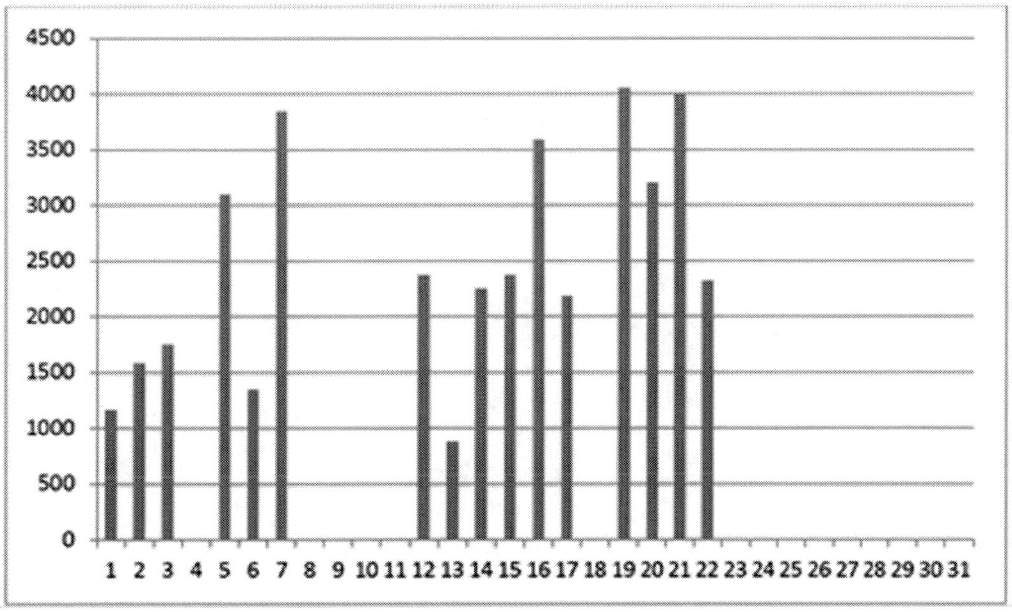

Fig. 3. Uneven production of coal in tons for March 2017.

The optimal production plan problem for March 2017 based on the production scheduling based on the analysis of product demand (wagons loading).

Consider the mathematical formulation of the problem. Denote by rk the demand, and using zk – the required capacity of the coal mine in k-th period, k = 1,m, where m is the number of orders received during the month. In this case z0 = c is some fixed initial level of production. For timely execution of orders requires that demand is always met, i.e., zk ≥ rk, k = 1,m .

The problem is to minimize the sum of two functions damages (increased costs towards the optimal costs) and penalties [13]:

а) $g_k(z_k - r_k)$ - damages and penalties in the k-th period, due to the fact that production exceeds demand and excess stocks appear;

б) $h_k(z_k - z_{k-1})$ - damages and penalties in the k-th period, caused by the unevenness of the production program by months.

Thus, the first function (g_k) determines the losses from overproduction of products, the second (h_k) - losses associated with inventories or services.

Then the objective function can be written in the form:

$$L(z_1, z_2, ..., z_m) = \sum_{k=1}^{m} \left[g_k(z_k - r_k) + h_k(z_k - z_{k-1}) \right] \to \min \quad (1)$$

under the constraints

$$z_k \geq r_k, \; k = 1,m \quad (2)$$

This problem can be solved by dynamic programming. Denote by f_k (c) total cost at the optimal production program for a month, if before the end of the planning period is k periods. Then, the estimated solution can be obtained using the following recurrence relations:

$$f_k(c) = \min_{z_k \geq r_k} \{ g_k(z_k - r_k) + h_k(z_k - c) + f_{k+1}(z_k) \}, \quad k = \overline{1, m}, \quad (3)$$

where

$$r_{k-1} \leq c \leq \max_k r_k, \quad k = \overline{1, m} \quad (4)$$

$$r_k \leq z_k \leq \max_k r_k, \quad k = \overline{1, m} \quad (5)$$

Initial conditions: $f_{m+1}(c) = 0$ и r0 = 0.

3 Results and discussion

Using a mathematical model (1-5) will calculate the planned performance indicators of the open pit given the intensity of export of coal to consumers.

The planning horizon (month) is divided into m=10 periods.

Count the total number of loaded wagons for periods (it will demand) (table 1).

Table 1. Dynamics of demand for March 2017

k	1	2	3	4	5	6	7	8	9	10
r_k	48	61	52	47	50	53	57	9	59	78

 DOI: 10.1051/e3sconf/20172101013

Given the cost of loading and transportation of coal parameter g_k in (1) will be 8925,74 RUB/wagon. The h_k option will affect the cost of unproductive downtime of excavators and dump trucks in the case of the uneven production during the month. Hence h_k in (1) will be 1789,46 RUB/wagon.

Using (3) with (4-5) we get the optimal production programme of coal production (table 2), which leads to extraction and transportation costs minimization of coal to consumers.

Table 2. Optimal production program of March 2017

k	1	2	3	4	5	6	7	8	9	10
r_k	48	61	52	47	50	53	57	9	59	78
z_k	61	61	60	60	60	60	60	60	60	78

In table 2 the production program presented in the wagons. Moving on from the wagons to the ton and breaking periods per day, and will receive the calculated optimal production programme of coal production for March 2017 (figure 4).

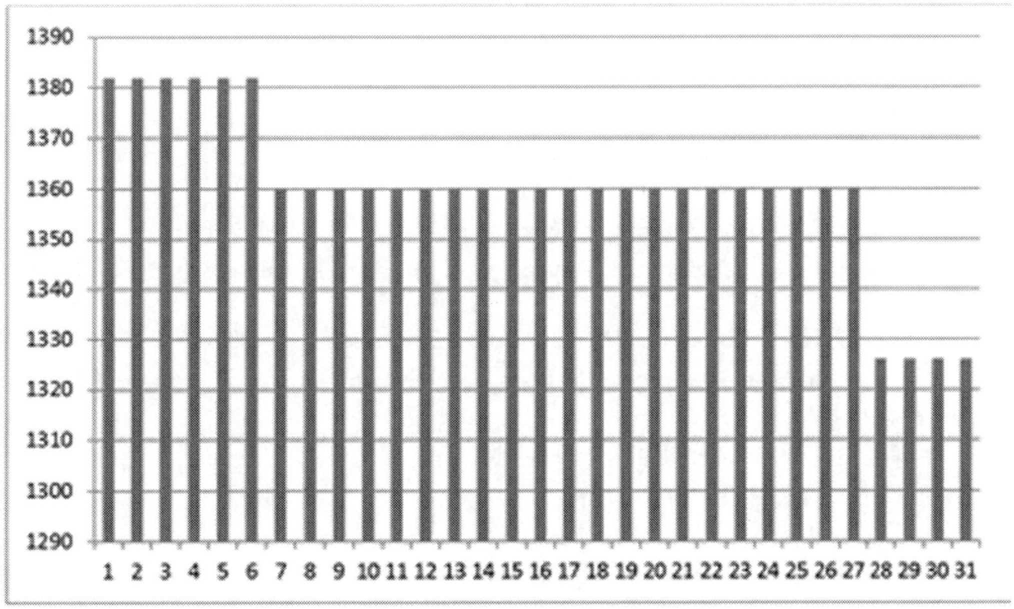

Fig. 4. Estimated optimal programme of coal production in tons for March 2017.

The results show that the level of coal production is expected to decline from the beginning of the month to the end. The use of such production policies will allow to align the work of production sites, transport and warehouses with a decrease in financial resources, as reflected in table 3.

Table 3. Analysis of actual and optimal costs on the production – warehouse «Transshipment»

No.	Area	Expenses for March 2017, RUB		Deviation, RUB
		actual	optimal	
1.	Area of coal production - warehouse "Central" including:	2 206229.64	1 923970.33	282 259.31
1.1.	Loading of coal by excavator HITACHI	1 037 332.18	857 430.67	179 891.51
1.2.	Transportation of coal by haul trucks BelAZ 7555D	1168 907.46	1 066 539.66	102 367.80
2.	Warehouse «Central» -	6 666 652.41	5 645 039.44	1 021 612.97

	warehouse «Trans-shipment», including:			
2.1.	Coal loading loader DRESSTA -534E	659 223.36	527 512.59	131 710.77
2.2	Coal transportation dump trucks VOLVO BCM-51.1	6 007 429.05	5 117 526.86	889 902.20
3.	Area of coal production - warehouse «Trans-shipment»	8 872 882.05	7 569 009.78	1 303 872.28

4 Conclusions

Coal mining production program definition technique will help align the work of all enterprise units, to optimize resources of all areas, to establish a flexible relationship between manufacturer and consumer, to take into account the irregularity of rail transport. Application of rolling planning methods (grouping by periods) to allow you to release unnecessary enterprise resources and redirect them to the needed objects or to realize in the market.

References

1. A. Yu. Tyurin, KuzSTU Bul. **3**, 106 (2005)
2. E. V. Khrapova, J. Rus. Entr., **11:2**, 61 (2010)
3. A. A. Tverdov, A. V. Zhura, S. B. Nikishichev, Coal, **8:1037**, 86 (2012)
4. R. Chicoisne, D. Espinoza, M. Goycoolea, E. Moreno, E. Rubio, Oper. Res. **60:3**, 517 (2012)
5. C. Cullenbine, K.Wood, A. Newman, Optimization Letters, **88:3**,365 (2011)
6. D. Espinoza, M. Goycoolea, E. Moreno, A. Newman, Annals of Operations Research, **206**, 93 (2013)
7. A. Yu. Tyurin, KuzSTU Bul. **1**, 40 (2000)
8. Ju. D. Pristupa, GIAB, **4**, 369 (2012)
9. Ju. D. Pristupa, KuzSTU Bul. **1**, 91 (2012)
10. H. Askari-Nasab, K. Awuah-Offei, H. Eivazy, Int. J. Min. Miner. Eng. **2**, 185 (2010)
11. E. Klotz, A. M. Newman, Surveys in Operations Research and Management Science, **18:12**, 18 (2013)
12. A. Newman, E. Rubio, R. Caro, A. Weintraub, K. A. Eurek, Interfaces, **40:3**,222 (2010)
13. A. Y. Tyurin, KuzSTU Bul. **1**, 94 (2004)

Promising Technologies of Mining and Processing of Solid Minerals

Sergey Shabaev[1*], *Seregey* Ivanov[1], and *Evgeniy* Vakhianov[1]

[1] T.F. Gorbachev Kuzbass State Technical University, 650000, 28 Vesennyaya St., Kemerovo, Russia

Abstract. The continuing growth in mineral extraction entails an increase in industrial waste, which in turn has a negative impact on the environment. Rubber-tired vehicles, in which the tires wear colossally, is mainly used as a transport for loading, unloading, transportation and other types of work in the extraction of solid minerals. The used tires are not disposed in any way, but are stored in special areas where harmful toxic substances are emitted under the influence of ultraviolet rays. Therefore, a decision was made to find a method for utilization and rational use of industrial waste in the road construction sector.The operating temperature of composite rubber-bituminous binders based on rubber crumb from the used automobile tires is estimated in this paper, which is necessary for assigning technological parameters of production and laying of asphalt-concrete mixtures produced on their basis. It is established that composite rubber-bituminous binders based on rubber chips from the used automobile tires, produced according to the two-stage technology, have the same viscosity as the original petroleum bitumen, at a temperature increased by 20°C.

1 Introduction

The increase in coal mining in Kuzbass leads to the ever-increasing number of used equipment. Rubber-tired vehicles are mainly used at open pits and other industries. Because of the heavy load on the tires of heavy equipment, there is a rapid wear of these units. In this regard, there is the acute problem of utilization of tires, which are currently stored in specially designated areas where toxic substances are released under the influence of ultraviolet rays.

At present, both in the Kuzbass and in Russia, plants for processing the used tires to produce the various fractions of rubber crumbs are being put into operation. However, due to the fact that the amount of used tires is many times greater than the real consumption of rubber crumbs, this type of waste continues to be accumulated in specially designated areas.

The most promising method of recycling rubber crumb, produced from the used automobile, is its use in road construction, as a modifier of bituminous binders.

* Corresponding author: Shabaev81@rambler.ru

There are various technologies for bitumen modification in the world practice: Asphalt Rubber [1-9], Terminal Blended [10] and their combinations [11, 12]. The rotational method for determining their viscosity is the most common. In this case, the highest kinematic viscosity is normalized, which should be 8.65 mm^2/s at 100 °C. At the same time, devices designed to determine the kinematic viscosity of rubber bituminous binders are designed to determine the viscosity of sufficiently thick materials, such as rubber bituminous binders, while they are practically unable to assess the viscosity of bitumen at operating temperature. However, rubber bituminous binders produced according to the two-stage technology have a significantly lower viscosity than rubber bituminous binders produced according to other technologies, therefore, the task of their viscosity comparing with the original bitumen is actual.

2 Materials and methods

When carrying out the studies, the relative viscosities of БНД 60/90petroleum bitumen from Omsk refinery and the composite rubber-bitumen binder produced on its basis according to the following two-stage technology were estimated:
a rubber crumb in an amount of 37.5% was added at room temperature into the dispersion medium of bitumen in an amount of 50% and petroleum oil in an amount of 12.5% heated with constant stirring to a temperature of 230±5°C;
the mixture was kept under constant stirring at a temperature of 230 ±5°C for 2.5 hours to produce a concentrated slurry of the rubber-bitumen composite;
the heating elements were switched off and bitumen, having a temperature of 105±5°C, in an amount of 47.5%, was added to the concentrated slurry of the rubber-bituminous composite in an amount of 52.5%, and the mixture was held for 0.25-0.50 hours to produce the composite rubber-bituminous binder.
To estimate the operating temperature, the flow time of 50 cm^3 of the binder through 5 mm diameter hole at the bitumen viscometer was measured in accordance with the test procedure. Planning matrix of the experiment is presented in Table. 1.

Table 1. Planning matrix of the experiment to determine the operating temperature of the composite rubber-bitumen binder.

Binder	Flow time of 50 cм^3 of binder, c, at test temperature, °C				
	100	**120**	**140**	**160**	**180**
60/90 feed bitumen	Opt_1	Opt_2	Opt_3	Opt_4	Opt_5
Composite rubber-bituminous binder	Opt_6	Opt_7	Opt_8	Opt_9	Opt_{10}

Note: Opt_i – optimization parameter.

3 Results and discussion

The results of the experiment for determining the operating temperature of the composite rubber-bituminous binder, produced according to the two-step technology, are presented in Table. 2. The graphical form of the obtained dependences is shown in Fig.

 DOI: 10.1051/e3sconf/20172101014

Table 2. The results of the experiment to determine the operating temperature of the composite rubber-bituminous binder.

Binder	**Flow time of 50 cм³ of binder, c, at test temperature, °C**				
	100	**120**	**140**	**160**	**180**
60/90 feed bitumen	100	50	15	15	15
Composite rubber-bituminous binder	-	180	60	17	15

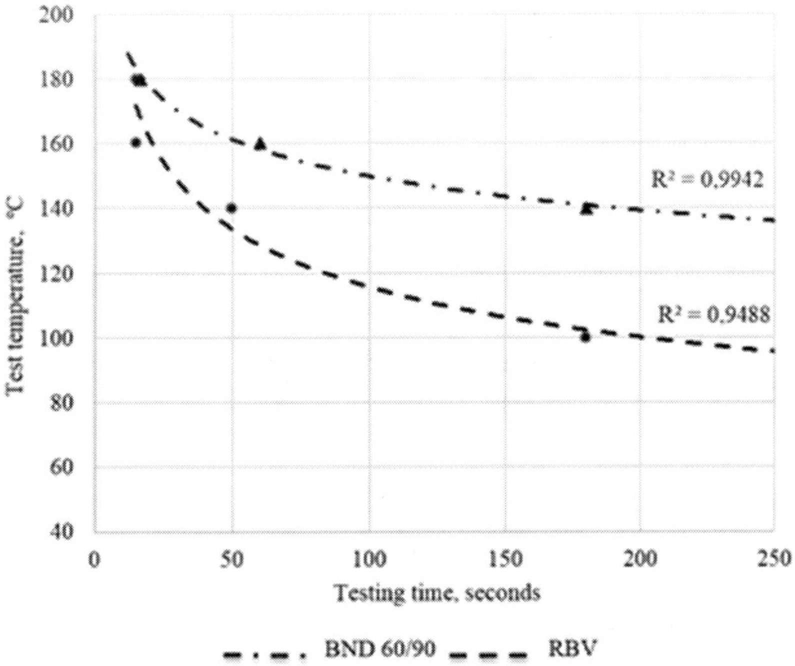

Fig. 1. Dependences of the relative viscosity of the feed bitumen and the composite rubber-bituminous binder, obtained on the basis of the feed bitumen at the two-stage technology

The analysis of the results of the experiment shows that the relative viscosity of the feed 60/90 petroleum bitumen at the operating temperature of the asphalt mixes of 145-155°C is 15 s. The composite rubber-bituminous binder has the same relative viscosity at a temperature of 165-175°C, hence, to ensure the comparable workability of asphalt-concrete mixtures based on the composite rubber-bituminous binders and the bitumen-based asphalt mixtures, their operating temperature should be increased by 20ºC.

4 Conclusions

The operating temperature of the composite rubber-bituminous binders, produced according to the two-stage technology, is 20°C higher than the operating temperature of the feed petroleum bitumen. In this regard, there will be a decrease in anthropogenic impact on the ecosystems of both mining areas and in other places of the composite rubber-bituminous binder applications.

DOI: 10.1051/e3sconf/20172101014

References

1. B. Abdulwarith, H. Norhidayah, M. Hanif, M. Jahangir, Materials and Structures, (2016)
2. M. A. Mull, K. Stuart, A. Yehia Journal of Materials Science, **37**, 557 (2002)
3. H. Kim, S. Lee, S. Amirkhanian, KSCE Journal of Civil Engineering, **14**, 839 (2010)
4. G. R. Morrison, S. A. Hesp, Journal of Materials Science, **30**, 2584 (1995)
5. X. Chuan, L. Tianqing, Q. Yanjun, Journal of Modern Transportation, **21**, 273 (2013)
6. S. Shakir, P. Jorge, M. Manuel, 7th RILEM International Conference on Cracking in Pavements, **4**, 1157 (2012)
7. M. Tao, Z. Yongli, H. Xiaoming, Z. Yao, KSCE Journal of Civil Engineering, **20**, 1347 (2016)
8. R.G. Hicks, D. Cheng, T. Duffy, *Evaluation of Terminal Blend Rubberized Asphalt in Paving Applications* (California Pavement Preservation Center, 2010)
9. H. Zhu, C. Liu, K. Tom, T. Norasit, Materials and Structures, **41**, 383 (2008)
10. G. Juan, A. Ana, G. Felice, Mechanics of Time-Dependent Materials, **20**, 389 (2016)
11. A.A. Khristoforova, M.D. Sokolova, S.E. Filippov, B.N. Zarovnyaev, International Polymer Science and Technology, **42**, 27 (2015)
12. A.A. Khristoforova, M.D. Sokolova, B.N. Zarovnyaev, A.N. Akishev, Mining Journal, **3**, 47 (2016)

Three-Dimensional Computer Simulation as an Important Competence Based Aspect of a Modern Mining Professional

Olesya Aksenova[1*], and *Anna* Pachkina[1]

[1] T.F. Gorbachev Kuzbass State Technical University, 650000, Vesennyaya 28 st., Kemerovo, Russian

Abstract. The article deals with the problem of necessity of educational process transformation to meet the requirements of modern miming industry; cooperative developing of new educational programs and implementation of educational process taking into account modern manufacturability. The paper proves the idea of introduction into mining professionals learning process studying of three-dimensional models of surface technological complex, ore reserves and underground digging complex as well as creating these models in different graphic editors and working with the information analysis model obtained on the basis of these three-dimensional models. The technological process of manless coal mining at the premises of the mine *Polysaevskaya* controlled by the information analysis models built on the basis of three-dimensional models of individual objects and technological process as a whole, and at the same time requiring the staff able to use the programs of three-dimensional positioning in the miners and equipment global frame of reference is covered.

1 Introduction

Nowadays progressive development of mining industry demands highly qualified professionals. It involves the necessity of adaptation of modern educational institutions to meet the requirements of educational services customer as well as of rapid and timely educational process transformation to achieve the qualitative system of analysis and control of the education programs efficiency.

With regard to timely prediction and taking into account the country's changing economic conditions requirements, the mining industry is aimed at the effective and profitable extraction and processing of solid mineral deposits which is to guarantee the underground operation safety and to save people's lives.

Field experience in many branches of industry, successful practices in computerization of work processes and potentialities of three-dimensional computer simulation facilities and of information technologies in general demonstrate the timeliness and urgency of cooperative developing of new educational programs and implementation of educational process taking into account modern manufacturability [1-3].

* Corresponding author: olesya_aksenova@mail.ru

2 Materials and methods

To create a modern mining enterprise it is necessary to have a completely new approach to the training of professionals. Working out educational programs one should take into consideration the study of three-dimensional models of surface technological complex, ore reserves and underground digging complex as well as creating these models in different graphic editors and working with the information analysis model or system obtained on the basis of these three-dimensional models [4-6].

Information analysis system enables, being based on the online data, to assist in making tactical, strategic and operative management decisions at the enterprise following on from immediate graphic representation of all the necessary data set to the users responsible for the work analysis. Information model as a constituent of information analysis system is an object model given in the form of information describing relevant for this model characteristics and variables of an object along with their connections, object inputs and outputs and allowing with the help of transmitting to the model information about input values changes to simulate possible states of an object; it is total information characterizing essential features and states of an object, process, phenomenon as well as interrelation with the outside. Information analysis system enables to find insights and implicit tendencies, to carry out end-to-end analysis of process steps, actual oncoming of critical developments and abnormal situations, to keep, cumulate and make use of implicit tendencies [7-12].

Information analysis system in operation and development of information analysis models on the basis of three-dimensional models of production facilities are used in the work of mine *Polysaevskaya* (SIBENCO) in Kemerovo region. Here, for the first time in Russia, the face enabling to carry out manless coal mining was put into operation. The system of manless coal mining is based on cooperative development of three companies specialists: OAO SIBENCO-Kuzbass, MARCO (Germany) and EICKHOFF (Germany).

The purpose of development and implementation of technological coal mining process which is realized and controlled by the use of information analysis models is to increase the enterprise profits, to enhance underground operations safety and to save people's lives.

The methods used to achieve the purpose are to control and timely make a decision in case of emergency, to increase employees work efficiency thanks to highly-professionally trained university graduates.

3 Results and Discussion

Manless coal mining involves coal winning technology without people's presence in a stoping face which solves not only technological but also the most important social and ecological problems – that is to take people out of the face and to release them from heavy manual operations. There are different ways of manless coal mining, each having its own advantages and disadvantages (fig.1). At present the most widely used is the timberless coal winning technology with the help of the following equipment: augers, scraper boxes, sawing machines, equipment system KMD-72, hole blasting, etc. [1, 2, 7].

The three-dimensional scale diagram – mine network at a producing mine forms a complex topological structure (fig. 2). The technological process and equipment control is performed by one operator. This came true thanks to the use of numerous information analysis detectors put on the equipment and integrated system of video monitoring.

For visualization of automatically controlled coal mining technological process three-dimensional models of mine openings, extraction and loading equipment, conveyers, transport etc. were made in different graphic editors and afterwards served as a basis for information models.

The given information models and computer performance analysis are able in a

 DOI: 10.1051/e3sconf/20172101015

completely automated mode to monitor operation facilities at any stage of coal mining technological process, to determine and realize the most effective combines movement, damper valves of powered support units, the work of winning-and-transport equipment (fig. 3).

Information system has a dynamic object data structure with an unlimited opportunity to increase functionality. This database in the system serves as an integrating basis of all the system operation modules corporate work. The most significant criteria of liveness and effectiveness of a corporate system are its in-built opportunities for information model development by the users themselves. Indeed, operating personnel of a mining enterprise works, as a rule, with real objects. The engineering technicians duties of chief surveyor's, geologist manager's, production manager's and other services are difficult to formalize and are changing all the time so there is always an opportunity to "misunderstand the task" and so-called "indispensable" professional can change the place of work.

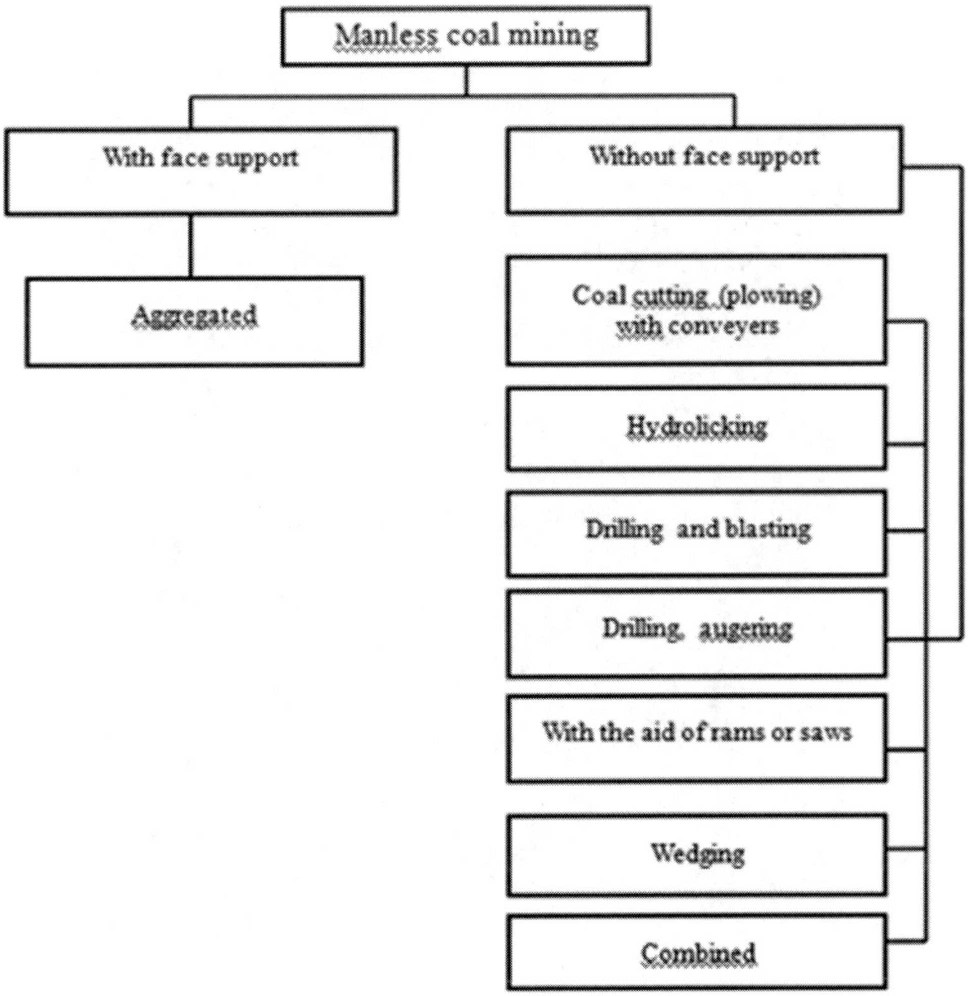

Fig. 1. Classification by the way of manless coal mining.

DOI: 10.1051/e3sconf/20172101015

Fig. 2. Three-dimensional system of mine openings.

Fig. 3. Information models of individual stages of coal mining technological process.

4 Conclusions

Technical solution based on modern information systems should provide:

1. Interactive visualization of production processes on the basis of 3D model of technological surface complex and underground mine facilities including dynamic lava changes.
2. Calculation and estimation of key enterprise management aspects: profit, output and mineral resources.
3. End-to-end operation control.

That is the reason why there is a necessity for a very flexible, powerful but at the same time user-available tool for self-study and development of data storage structure needed, getting answers to the unknown in advance requests in the informational space. On the other hand, there appears a demand in highly qualified personnel having not only deep theoretical knowledge in the coal mining field but also having a command of three-dimensional computer simulation language both of individual objects and of technological processes on the whole.

So, the development and the implementation of integrated control systems at the coal mining enterprises requires the alteration of the education and reeducation programs for engineering technicians, assistant mine rescue crews and stope miners, introduction into the course of learning the programs of three-dimensional positioning in the miners and equipment global frame of reference.

Implementation of information analysis models complex management systems studying into the educational programs of mining professionals will help to enhance underground operations safety and to save people's lives.

References

1. A.A. Khoreshok, S.A. Zhironkin, M.A. Tyulenev, M.C. Hellmer, S.V. Potyagailov, IOP Conference Series: Materials Science and Engineering, **142:1**, 012122. (2016)
2. D.V. Alekseev, G.A. Kazunina, A.V. Cherednichenko, Journal of Mining Science, **51:5**, 917 (2015)
3. S.P. Bakhaeva, V.A. Gogolin, I.A. Ermakova, Journal of Mining Science, **52:3**, 454 (2016)
4. M.A. Tyulenev, S.A. Zhironkin, E.A. Garina, International Journal of Mining and Mineral Engineering, **7:4**, 363 (2016)
5. T. Gvozdkova, M. Tyulenev, S. Zhironkin, V.A. Trifonov, Yu.M. Osipov, IOP Conference Series: Earth and Environmental Science, **50:1**, 012010 (2017)
6. M.A. Tyulenev, T.N. Gvozdkova, S.A. Zhironkin, E.A. Garina, Geotechnical and Geological Engineering, **35:1**, 203 (2017)
7. J. Kretschmann, A.B. Efremenkov, A.A. Khoreshok, IOP Conference Series: Earth and Environmental Science, **50:1**, 012024 (2017)
8. M. Behandish, T. Horea Ilieş, Computer-Aided Design, **70**, 100 (2016)
9. Ba Wenlan, Ren Ning, Cao Lixin, Graphical Models, **82**, 1 (2015)
10. Yang Hao, Zhang Hui, Computers & Graphics, **65**, 45 (2017)
11. M. Behandish, T. Horea Ilieş, Computer-Aided Design, **70**, 100 (2016)

Unmanned Mine of the 21st Centuries

Irina Semykina [1*], *Aleksandr* Grigoryev [1], *Andrey* Gargayev [1], and *Valeriy* Zavyalov [2]

[1] T.F. Gorbachev Kuzbass State Technical University, 650000, 28 Vesennyaya St., Kemerovo, Russia
[2] National Research Tomsk Polytechnic University, 634050, 30 Lenin ave., Tomsk, Russia

Abstract. The article is analytical. It considers the construction principles of the automation system structure which realize the concept of «unmanned mine». All of these principles intend to deal with problems caused by a continuous complication of mining-and-geological conditions at coalmine such as the labor safety and health protection, the weak integration of different mining automation subsystems and the deficiency of optimal balance between a quantity of resource and energy consumed by mining machines and their throughput. The authors describe the main problems and neck stage of mining machines autonomation and automation subsystem. The article makes a general survey of the applied «unmanned technology» in the field of mining such as the remotely operated autonomous complexes, the underground positioning systems of mining machines using infrared radiation in mine workings etc. The concept of «unmanned mine» is considered with an example of the robotic road heading machine. In the final, the authors analyze the techniques and methods that could solve the task of underground mining without human labor.

1 Introduction

The problem of improving the human labor safety in mining enterprises does not lose its relevance. One of the ways to solve this problem is the development and application of automation systems. The remote operation does not allow us to achieve maximum efficiency, as there is a problem concerned with the unmanned mining machines operators training. The productivity of the mining enterprise is reduced because the operator feels and coordinates the mining machine worse. Remote operation allows us to deal with the mining equipment at a considerable distance, that significantly increases the mining operations safety but still, it does not resolve the operator - "weak spot" and a source of errors and non-optimal actions [1]. The most reliable but also complex way to solve this problem (to eliminate this so called "weak spot") is the complete withdrawal of an operator from the mine opening and the creation of the fully automatic robotic systems that allow us to implement the "unmanned mine" concept.

2 Materials and methods

* Corresponding author: siyu.eav@kuzstu.ru

The development of algorithms, methods, and technologies for "unmanned mine" is a complex scientific and technical problem that has not been solved yet. The same situation is with the problems concerning mining machines energy and resources optimization, integrating various mine subsystems into a single system, planning and managing the production and assets of a mining enterprise.

The current state of the above set problem is characterized by the use of remote-operated autonomous complexes with an arrangement in a single operator complex [3-8]. Notable examples are the mines in the area of the Pilbara (Western Australia), Quarry Bingham Canyon (Salt Lake City, USA), Mine Kadi-Ridgeway (Australia). These facilities provide the use of the automatically operated excavators, bulldozers, dump trucks and other equipment in places with the unstable ground. There are also examples of the creation of fully automatic coal mining systems for example, in the mines "Tang Shan Gou" and "Nan Liang" (PRC). Monitoring the mining machines harvesting is carried out with the help of the real time cameras through the optical fiber transmission signal. In Russia, the main research focus on this issue is the direction of developing new automated machines and complexes equipped with sensors and video cameras that can be remotely operated (e.g. 2015 the Polysaevskaya mine).

It should be noted that while concerning the implementation of the above-mentioned automation systems there was not raised an issue of the completely automatic operation of road heading machines based on specified optimality criteria and the issue of providing methods and algorithms for the excavation operations. Nevertheless, modern automation systems can significantly improve the mine's productivity, power, and resource efficiency. In order to do this, it's necessary completely redefine the concept of coal mining underground methods and to develop new technological machines, means of rocks transportation to the surface, means of electricity and drainage.

3 Results and discussion

It is necessary to consider the problem of constructing the mine automation system implementing the concept of «unmanned mine» by the example of the robotic road heading machine. (Fig. 1). Such an automation system includes both global and local subtasks, work-related actuators and movement mechanisms.

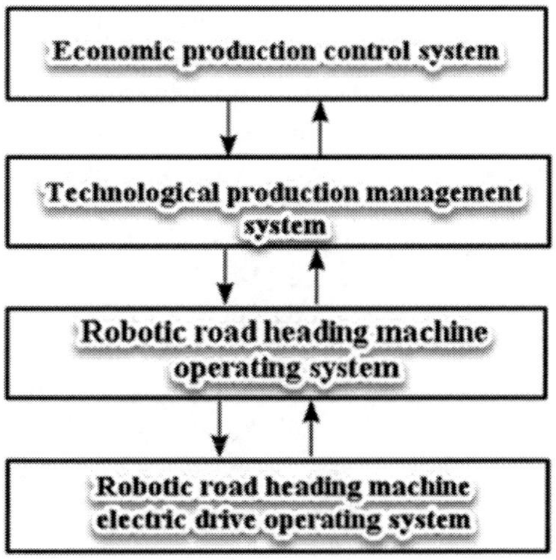

Fig. 1. Block diagram of the control system levels.

In order to solve local subtasks, it is necessary to form the set values of displacements, forces, and speeds of actuators and movement mechanisms from the upper-level control devices, so that the deviation from the set values is within the near-extreme error.

Systems that solve global subtasks in the operation of the robotic road heading machine can be built along a hierarchical chain from the top to bottom as follows: the economic production control subsystem (enterprise management system); the technological subsystem (the formation of common indicators and goals (for example, setting the productivity of the slaughter), robotic road heading machine control system (setting specific values of coordinates, velocities and trajectories of motion, as well as control time intervals), robotic road heading machine electric drive control system (working out the set values of technological parameters).

Integration of the presented subsystems will make the production process predictable, cost-effective and allow us to define the "neck stages" in production, to forecast and manage the production and assets of the enterprise.

One of the main tasks of development robotic road heading machine control system is the task of bottom hole space positioning. The task of controlling the movement of an object within a closed three-dimensional space bounded by a surface with apriori unknown shape and physical properties is the most difficult task of the modern robotics transferred to the robotic road heading machine.

When solving this problem, it is necessary to take into account a great number of restrictions imposed on the movement of the specified object in space: the restriction on the safety conditions of the mine or mine infrastructure, the limitations on the conditions of the minimum consumption of electric energy and the life of the elements of the robotic road heading machine, the limitations on the productivity of the transport infrastructure of the mine, unforeseen properties of the excavating rock mass (increased strength, increased explosive gases concentration going out from the layer seam, etc.). All these limitations lead to the fact that it is impossible to specify in advance the exact trajectory of motion and the robotic road heading machine speed. The organization of positioning in the underground space by known methods is difficult because of the poor admittance of the rock mass via electromagnetic waves. This task is further complicated by the lack of methods for creating landmarks (geodetic marks) in an unmanned mine, which are necessary for correcting positioning.

When creating or upgrading unmanned mining machines, the following requirements must be met: minimum infrastructure requirements; simplicity of installation and maintenance; navigation should be carried out without precise geometric mine maps; the positioning speed should be no less than with the operator; navigation should be carried out in a changing space geometry; there should be provided high work reliability with technical condition diagnostics; a high degree of integration with top-level automation systems [2].

Nowadays there are several mining machines positioning technologies in the mine space: by infrared radiation [9] using RFID (RFID) [10], using radar (Through-the-earth communication) [11] using the antennas in the form of transmission lines with radiating elements (leaky communication) [12], using a local wireless network (WLAN) [13], using a bar code applied to the surface of the mine [14]. However, the listed methods do not provide the necessary positioning accuracy and require external cues or external signals.

One of the effective solutions for the positioning of the robotic road heading machine is to use an inertial measurement unit (IMU) with information on the electric drive operating mode and the actuator's motion and movement parameters [15]. The inertial measuring unit includes a 3-axis accelerometer, a 3-axis gyroscope, and a 3-axis magnetometer – all of them are located on one board. The main IMU drawback is the measurement error presence, that is insignificant but in the working process, it is able to accumulate and become much worse.

This drawback leads to the appearance and subsequent increase in the discrepancy between the positioning system readings and the actual object coordinates. It is possible to restrict this error via the use of high-precision clock generator integrated into the IMU (Timing & IMU), and the use of the hybrid navigation system with further aggregation in the data from the various navigation systems. There may also occur the emergence and accumulation of errors caused by the track mover slippage. In order to compensate track mover slippage, it is necessary to determine this mode occurrence and the slippage amount by means of the variable speed electric drive. At the same time, there is required a computer model for moving the robotic road heading machine, that takes into account the track mover slippage, as well as the limited rigidity of the robotic road heading machine construction, that in its turn affects the readings of the onboard accelerometers and gyroscopes. In the noisiness and incompleteness conditions of the measurement data, there is needed a method to restore the information on the robotic road heading machine movement using data from 3 axial accelerometers and gyroscopes [16]. Approved devices for solving this problem are the advanced Kalman filter and the artificial neural network apparatus.

When mining a bottom hole space by a robotic road heading machine there is always raised a question of the excavation operations parameters rationality (the effector movement speed, its rotation frequency and the thickness of the removed chip load). On the one hand in case of these parameters irrational selection there can be enhanced the effector intensive wear and the operations increased power consumption. On the other hand, the reduction of these parameters settings increases the operation control time, reduces the job and the movement execution speed, i.e. affects the "unmanned mine" performance. In order to solve the robotic road heading machine power and resource efficiency problem, it is necessary to divide it and consistently solve the following three subtasks.

Firstly, it is necessary to develop methods for estimating the residual operation life of the robotic road heading machine effector and its expenditure speed. In order to solve this problem, we can analyze the existing models of machine parts destruction, modifying these models in terms of accounting forms change over time force generated in the robotic road heading machine element. Also synthesizing the resource consumption model of the robotic road heading machine \mining machine, combining the destruction model the robotic road heading machine parts, the actuator model, the electric drives model, the effector model and the model of the mechanism for moving the robotic road heading machine.

Secondly, it is necessary to implement the methods of controlling the robotic road heading machine, which allow us to reduce the resource consumption of the effector and the executive mechanism as a whole. In order to do it, it is necessary, via investigating the interconnection of the mining machine elements resources and the electric drives various operating modes efforts to develop the resource-saving management principles of the robotic road heading machine. As far as all technological operations are carried out to fulfill the main task - mining, a method must be developed for managing the robotic road heading machine, that allows us to achieve the task for a given time with the minimum possible resource consumption. Thus, the task of operating a robotic road heading machine is an optimization problem with two opposite criteria. Complementing the technological task formation subsystem model based on the current value of resource consumption to the model of a robotic road heading machine, it is proposed to solve this problem via methods of the optimal and synergetic control theories.

Thirdly, it is necessary to develop a technological parameters task forming method for a robotic road heading machine on the current data basis of the residual operation life, the rate of resource consumption and on the marketing and economic performance data of the mining enterprise. This problem solution is possible by synthesizing a mathematical model of economic activity aimed at extracting minerals by a robotic road heading machine, based on the residual operation life data analysis, the electrical energy consumption, the resource consumption rate, the useful yield and the extraction volume. Further, it is necessary to develop a method for determining the specification of the production parameters for the robotic road heading machine (the rotation speed of the effector, the movement speed of the effector in the bottom hole space, the chips thickness, the final movement coordinates, the task control time) on the basis of the above given model.

4 Conclusions

Despite the evolving technologies, robotic machines and autonomously operating equipment are not still used in the underground mining. The most important challenge for the «unmanned mine» are the issues of interconnected at different functioning levels of the complicated and hazardous process complex. The solution of these problems can be realized on the basis of the optimal and adaptive control theory methods and on the comprehensive technical and economical approach application.

References

1. T. B. Sheridan, R. Parasuraman, Hum Fact, **42**, 403 (2000)
2. J. Larsson, *A survey on autonomous robot navigation techniques. Technical report, Department of Technology*, (Orebro University, Orebro, Sweden, 2007)
3. J. Larsson, J. Appelgren, J. Marshall, In Proc. of the SME Annual Meeting and Exhibition (2010).
4. P. Ridley, P. Corke, In Proc. Australian Conference on Robotics and Automation, 26 (2001)
5. A. Saffiotti, *Autonomous Robot Navigation: a fuzzy logic approach. Ph.D. thesis* (Universite Libre de Bruxelles, 1998)
6. A. Saffiotti, K. Konolige, and E. H. Ruspini, Artificial Intelligence, 76:1:2, 481, (1995)
7. F. von Hundelshausen, M. Himmelsbach, F. Heckcr, A. Mueller, and H.-J. Wuensche. J. Field Robot, **25:9**, 640 (2008)
8. R.A. Luke, World mining Equipment, **6**, 24-28 (1992)
9. Q. Huang, Y.Lu, Z. Sun, Mining Safety & Environmental Protection, **31:5**, 13-16 (2004)
10. B.Tang, T. Xu., M. Luo, X. He, Modern Electronics Technique, **29:7**, 10 (2006)
11. J. Li, Express Information of Mining Industry, **10**, 94 (2004)
12. S. Outalha, R. Le, P. M.Tardif, CIM Bulletin, **93**:1044, 100 (2000)
13. J. Li, *Research on the UHF Radio Wave Propagation and Location Algorithms in Mine Tunnels. Ph.D. Thesis* (Tianjin University, 2006)
14. Yu Meng, Hongmei Wu, Li Liu, and Wenhui Li, AICI. CCIS, **237**, 414 (2011).
15. Wu Heping, Wu Ling, Zhang Yi, Yang Libing, J. Nonferrous Metals, **59:06**, 12 (2007)
16. Belkacem Abdous, Alexandre Berred. Journal of Statistical Planning and Inference, **132**, 3 (2005)

Determination of the Geometric Form of a Plane of a Tectonic Gap as the Inverse Ill-posed Problem of Mathematical Physics

Dmitry Sirota[1], *Vadim* Ivanov[1]

[1]T.F. Gorbachev Kuzbass State Technical University, 650000, 28 Vesennyaya St., Kemerovo, Russia

Abstract. Any mining operations influence stability of natural and technogenic massifs are the reason of emergence of the sources of differences of mechanical tension. These sources generate a quasistationary electric field with a Newtonian potential. The paper reviews the method of determining the shape and size of a flat source field with this kind of potential. This common problem meets in many fields of mining: geological exploration mineral resources, ore deposits, control of mining by underground method, determining coal self-heating source, localization of the rock crack's sources and other applied problems of practical physics. This problems are ill-posed and inverse and solved by converting to Fredholm-Uryson integral equation of the first kind. This equation will be solved by A.N. Tikhonov regularization method.

1 Introduction

The problem of stability of natural and technogenic massifs is bound to definition of geometry of the center of differences of mechanical tension which is a source of a seismic event (SE). According to D. L. Anderson's theory, the sources of SE is the some volume area of rocks surrounding the place of the arising gap and corresponding to a zone of nonreversible deformations and maximal differences of mechanical tension. In this definition gaps are violations of a structure of rocks. D. L. Anderson [1, 2] considers that in the massif of rocks there is a plane on which there are a gap and shift of blocks relatively each other. Let's consider that the plane has the form ellipse as at any explosive violation there has to be its attenuation towards the periphery of a gap. Using harbingers of seismic events we can define various characteristics of SE. In this article we shall use the electric harbingers of SE.

The search for electric harbingers of SE (rock cracks, earthquake) has led to the beginning of the nineties of the last century, to the discovery of the mechanism of electrification of pockets the forthcoming SE at the stage of the avalanche – unstable fracturing, i.e. at the final stage of preparation. It was shown both theoretically and experimentally that each crack in the process of the emergence and rapid spread bears at its apex a charge of the same sign, when multiple their accumulation in the focus leads to a strong electrification of rocks and causes severe abnormalities in the electrical field in the atmosphere and on the surface of the earth. The definition of the shape and size of the area of electrification of rocks on measurements on the earth's surface can serve as the basis of the forecast power of the impending seismic events and predict the time of occurrence of the event.

The problem of definition of the shape of the area is the inverse ill-posed problems of mathematical physics. We can formulate this problem as Fredholm-Uryson integral equation of the first kind, which will be solved by A. N. Tikhonov regularization method [3-14].

2 Results and discussions

2.1 Formulation and solution of the direct problem

Let us assume that the natural electric field is generated by flat source arbitrary shape S_p which is inclined at an angle $\gamma \in [0^0;70^0]$ and located at a certain depth $z_M = h$ (Fig 1) or $z_M = h + z_1$ (Fig 2). We also assume that the sources containing space is homogeneous and isotropic with one (Fig 1) or two plane-parallel borders (Fig 2).

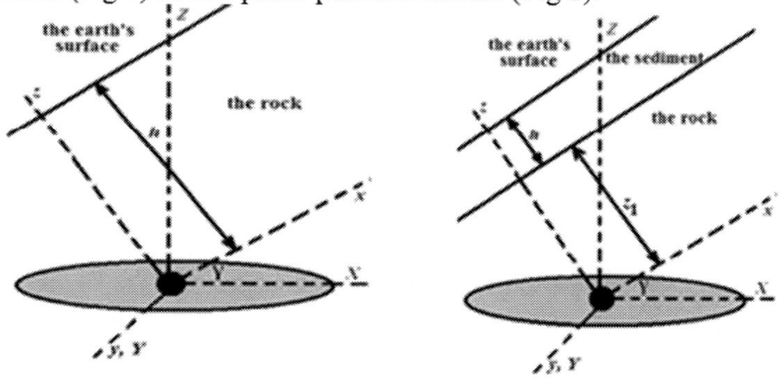

Fig. 1, 2. The schemes of the containing space and flat field source

The value of potential of such a source at an arbitrary measurement point M on the surface will be determined by the general formula [13]

$$U^M = \int_{S_P} u^M dS_P, \tag{1}$$

where S_P – flat source of electric field; u^M – the potential of point sources that determined by one of the two formulas (Fig 1 or Fig 2)

$$u_1^M = \frac{C}{\left[D + z_M^2\right]^{1/2}} \quad \text{or} \tag{2.1}$$

$$u_2^M = C\int_0^\infty \frac{J_0(mD)\exp(-mz_M)}{1 + W\exp(-2mh)}dm = C\sum_{k=1}^{11}\frac{q(k)}{\left[D + \left[z_M + 2(k-1)\right]^2\right]^{1/2}}, \tag{2.2}$$

where $J_0(mD)$ – Bessel function; $W = (\rho_2 - \rho_1)\cdot(\rho_2 + \rho_1)^{-1}$ – the reflection coefficient of the second layer; $\rho_{1,2}$ – electrical resistivity layers, (Om·m) $D = (x_P - x_M)^2 + (y_P - y_M)^2$; M – point of measurement of potential, P – the point located in the flat sources; $q(k)$ – the coefficients of the approximation formula; C – the coefficient, which characterizes the intensity of the field.

Since the source is located at an angle γ to the earth's surface, we will give the formula to calculate the coordinates of measuring points in the new system *OXY* $X_M = x_M \cos\gamma + z_M \sin\gamma$, $Z_M = -x_M \sin\gamma + z_M \cos\gamma$.

For integration in the region S_p let us use the polar coordinate system (ρ, φ) where $\varphi \in [0;2\pi]$, $r \in [0; \rho(\varphi)]$ and $X_P = r\cos\varphi$, $Y_P = r\sin\varphi$.

Additionally, let us turn to dimensionless quantities with formula $w \cdot h^{-1} = \overline{w}$, where w are all linear parameter with dimension [m]. Further, for simplicity, we will not write a line above the letters.

Thus, formula (1) will converts to $U^M = \int_0^{2\pi} d\varphi \int_0^{\rho(\varphi)} u^M(r,\varphi)dr$ and with (2.1) and (2.2) then converts to

$$U_1^M = C \cdot h \cdot \int_0^{2\pi} d\varphi \int_0^{\rho(\varphi)} \frac{rdr}{\sqrt{R}} \qquad \text{or} \qquad U_2^M = C \cdot h \cdot \int_0^{2\pi} \left[\sum_{k=1}^{11} q(k) \int_0^{\rho(\varphi)} \frac{rdr}{\sqrt{R}} \right] d\varphi, \tag{3}$$

where $R = r^2 + B \cdot r + A$, $B = -2(X_M \cos\varphi + Y_M \sin\varphi)$, $A = X_M^2 + Y_M^2 + Z_M^2$ for (Fig 1) and $A = X_M^2 + Y_M^2 + [Z_M + 2(k-1)]^2$ for (Fig 2). The inner integral in (3) we can calculate analytically: $\int_0^{\rho(\varphi)} \frac{rdr}{\sqrt{R}} = \left(\sqrt{R} - \frac{B}{2} \ln\left| r + \frac{B}{2} + \sqrt{R} \right| \right) \Big|_0^{\rho(\varphi)}$

Thus we have the Fredholm-Urysohn integral equation (IE) [3, 4] of the 1-st kind with unknown parameter $\rho(\varphi)$ to determine the shape of the source field with the Newtonian potential

$$\int_0^{2\pi} K(M, \rho(\varphi))d\varphi = U^*(M), \tag{4}$$

where K – nonlinear kernel of equation; U^* – values of potential measured on the earth's surface in the point M.

2.2 Solution of the inverse problem

The inverse problem is to determinate the function $\rho(\varphi)$ if we know potential value on the earth's surface. For the definition this function we consider the problem of finding the minimum of regularized functional of A.N. Tikhonov

$$\Phi(\rho) = \mathrm{T}(x_M, y_M, \rho(\varphi)) + \alpha\Omega(\rho(\varphi)), \tag{5}$$

where T – a measure of the error of measurement of the value of the potential; $\Omega = \int_0^{2\pi} \left(\rho(\varphi)^2 + \left(\frac{d\rho(\varphi)}{d\varphi} \right)^2 \right) d\varphi$ – the stabilizing functional; α – the regularization parameter.

In article [15] we assumed, that the measure of potential value is on a one-dimensional region ($y_M = 0$ for example) therefore, as the measure has been used the integral

$T(\rho)=\int_a^b\left[A(x_M,\rho(\varphi))-U^*(x_M)\right]^2dx_M$ where $A(x_M,\rho(\varphi)$ – non-linear integral direct modeling operator (3); U^* – the right part of the IE (4), which in practice is set experimentally, and for test problems is determined by solving a direct problem using formulas (3) with the addition of a random correction simulating inaccuracies in full-scale measurements.

In this article we will assume, that the measure of potential value will be on a two-dimensional region therefore, as the measure has been used the sum $T(\rho)=\sum_{k=1}^{N}\left[A(M_k,\varphi,\rho(\varphi))-U^*(M_k)\right]^2$ where N is the number of measurement points M.

We shall seek the minimum of the functional (5) by the method of conjugate gradients (CG) [16], whose general iterative scheme has the form

$$\rho_{q+1}=\rho_q+k_q\cdot I_q, \quad (6)$$

where $I_q=-\frac{d\Phi(\rho_q)}{d\rho}+\gamma_q\cdot I_{q-1}$, q – iteration index; k_q – minimization step; γ_q – CG-coefficient; $I_0=-\frac{d\Phi(\rho_0)}{d\rho}$ where $\frac{d\Phi(\rho_q)}{d\rho}=\Phi_\rho^{/}$ – the Freshet derivative of the functional (3); .

As well as know from the functional analysis the Freshet derivative of functionals T and Ω will be calculated by the following formulas

$$\frac{dT(\rho)}{d\rho}=2\sum_{k=1}^{N}\left[\left[A(M_k,\varphi,\rho(\varphi))-U^*(M_k)\right]\times\left(\frac{dK(M_k,\varphi,\rho(\varphi))}{d\rho}\right)\Bigg|_{\varphi=t}\right],$$

$$\frac{d\Omega}{d\rho}=2\left(\rho(\varphi)-\frac{d^2\rho(\varphi)}{d\varphi^2}\right)\Bigg|_{\varphi=t}, \quad (7)$$

where $\frac{dK(M_k,\varphi,\rho(\varphi))}{d\rho}=\frac{\rho(\varphi)}{\sqrt{R_\varphi}}$ for Fig 1 and $\frac{dK(M_k,\varphi,\rho(\varphi))}{d\rho}=\sum_{k=1}^{11}\frac{q(k)\cdot\rho(\varphi)}{\sqrt{R_\varphi}}$ for Fig 2, $R_\varphi=\rho^2(\varphi)+B\cdot\rho(\varphi)+A$ t – a fixed value of the argument φ, which defines the variable of differentiation $\rho(t)$.

3 Numerical realization

Numerical realization the above formulas consists of the following aspects.

The first is the numerical integration and differentiation functions in (5). For this we shall define the grid to the variable φ with step $u=\frac{2\pi}{N}$, N – the number of nodes on the grid, at each point where we shall compute $\rho(\varphi)$. To calculate the integral in (4), we shall use the Simpson formula. To calculate the second order derivative in (7), we shall use central deference relation $\frac{d^2\rho_k}{d\varphi^2}=\frac{\rho_{k-1}-2\rho_k+\rho_{k+1}}{u^2}$, $\frac{d^2\rho_1}{d\varphi^2}=\frac{\rho_{N-1}-2\rho_1+\rho_2}{u^2}$, $\frac{d^2\rho_1}{d\varphi^2}=\frac{d^2\rho_N}{d\varphi^2}$.

The second is to calculate three coefficients: k_q, γ_q, α. To find the step of minimizing k_q, we will use the "golden section" method for the objective function $\Phi(\rho+k\cdot I)$. Numerical experiments have shown that this function is unimodal on the interval $k\geq 0$. Since it is impossible in advance to determine the interval at which the minimum point is contained, we shall adhere to the algorithm described in [15]. To calculate the CG-coefficient γ_q we can use one of the multiple existing formulas [16]. In this article we shall use

$$\gamma_q=\max\left\{0;\min\left\{\frac{\Phi'_{q+1}\bullet\Phi'_{q+1}}{\Phi'_q\bullet\Phi'_q};\frac{\Phi'_{q+1}\bullet\left(\Phi'_{q+1}-\Phi'_q\right)}{\Phi'_q\bullet\Phi'_q}\right\}\right\}, \quad (8)$$

where $a\bullet b$ is the dot product [14]. To calculate coefficient α we shall use the iterative method [17] which is expressed in the formula $\alpha_q=\alpha_0(1+q)^{-0,25}$ where start value $\alpha_0=10^{-4}$ [17].

The third is the test program. Let us consider a test region in the form ellipse the dimensionless equation of which in the polar coordinate system has the form

$$\rho(\varphi)=\frac{ab}{h\sqrt{a^2\sin^2\varphi+b^2\cos^2\varphi}}, \quad (9)$$

where $a=50$, $b=15$, $h=150$ for Fig. 1 and $h=25$ meters for Fig. 2. The start region for (6) is the circle with the radius 40 m. To simulate errors in full-scale measurements, we will add a random correction to the potential value $U^*=U^{ex}+\beta\cdot(rand-0,5)$, where *rand* are uniformly distributed numbers on the interval [0; 1], U^{ex} – the exact calculated value of the potential according to the formula (3), $\beta=\{0,05;0,5\}$.

We will show in Fig. 3, 4 the graphic result of restoration the shape of flat sources for Fig. 1 and Fig. 2.

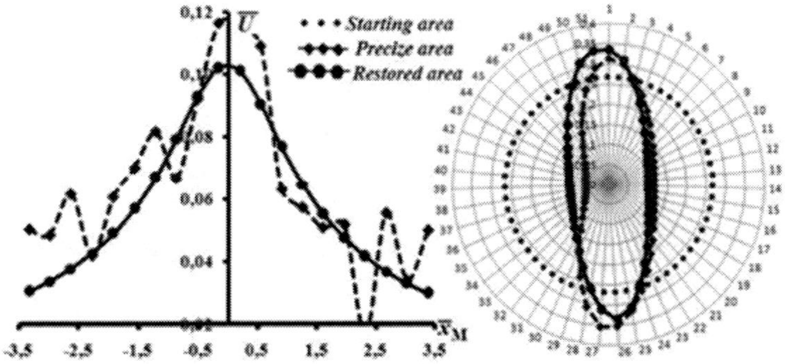

Fig. 3. Restoration of the area and potential value after 100 iterations CG-methods and $\beta=0,05$.

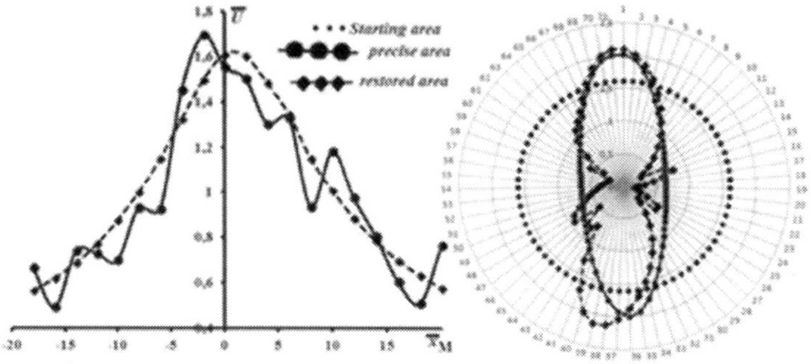

Fig. 4. Restoration of the area and potential value after 150 iterations CG-methods.

4 Conclusion

Let's give practical recommendations on the using of the developed methodology some aspects of which was introduced in article [15]. A coordinate grid is constructed on the surface with a step of about 25-50 meters. As we said above these values are used as the right part of IE (4). Then we use the developed program, which removed the random additive.

References

1. D.L. Anderson, *New Theory of the Earth*, (London, CUP, 2007)
2. D.L. Anderson, *Theory of the Earth*, (Hampshire, BSP, 1989)
3. A.N. Tikhonov, V.B. Glasko, J. Comp. Math. Math. Phys.,. **4**, 564, (1964)
4. A.N. Tikhonov, V.B. Glasko, J. Comp. Math. Math. Phys., **5**, 463, (1965)
5. J.P. Butler, J.A. Reeds, S.E. Dawson, J. Numer. Anal., **18**, 546 (1981)
6. V.A. Morozov, *Regularization methods fo ill-posed problems* (New York, CRC Press, 1993)
7. F.D.M. Neto, A.J.S. Neto, *An introduction to inverse problems with application* (New York, Springer, 2013)
8. H.W. Engl, M. Hanke, A. Neubauer, *Regularization of inverse problems* (London, CUP, 1996)
9. C.R. Vogel, *Computational methods for inverse problems* (Hamburg, SIAM, 2002)
10. Y. Wang, A.G. Yagola, C. Yang, *Optimization and regularization for computational inverse problems and applications* (London, Springer, 2011)
11. R.C. Aster, B. Borchers, C.H. Thurber, *Parameter estimation and inverse problems* (Cambridge, Elsevier Academic Press, 2005)
12. G. Chavent, *Nonlinear least squares for inverse problems* (New York, Springer, 2009)
13. M. S. Zhdanov, Geophysics, **76**, F77 (2011)
14. R. Kress, Inverse problems, **21**, 1207 (2005)
15. D. Sirota, E3S Web Conf, **15** 1012 (2017)
16. W.W. Hager, Pacific J. Optim. **2,** 35, (2006)
17. A.B. Bakushinsky, M.Yu. Kokurin, A. Smirnova, *Iterative methods for ill-posed problems* (Berlin/New York, Walter de Gruyter GmbH & Co. KG2011)

Parameters of Solidifying Mixtures Transporting at Underground Ore Mining

Vladimir Golik[1,2,*], and *Yury* Dmitrak[1]

[1]North-Caucasian State University of Technology, 44 Nikolaev street, Vladikavkaz, 362021, Russia[1]
[2]Vladikavkaz Scientific Center of the Russian Academy of Sciences and the Government Republic of North Ossetia-Alania, 22, Markusa street, Vladikavkaz, 362027, Russia

Abstract. The article is devoted to the problem of providing mining enterprises with solidifying filling mixtures at underground mining. The results of analytical studies using the data of foreign and domestic practice of solidifying mixtures delivery to stopes are given. On the basis of experimental practice the parameters of transportation of solidifying filling mixtures are given with an increase in their quality due to the effect of vibration in the pipeline. The mechanism of the delivery process and the procedure for determining the parameters of the forced oscillations of the pipeline, the characteristics of the transporting processes, the rigidity of the elastic elements of pipeline section supports and the magnitude of vibrator' driving force are detailed. It is determined that the quality of solidifying filling mixtures can be increased due to the rational use of technical resources during the transportation of mixtures, and as a result the mixtures are characterized by a more even distribution of the aggregate. The algorithm for calculating the parameters of the pipe vibro-transport of solidifying filling mixtures can be in demand in the design of mineral deposits underground mining technology.

1 Introduction

Concrete mixtures occupy a significant place in modern underground mining. An important characteristic of them is the moisture that determines its congealing, liquefaction and other parameters. Therefore, technologies using water as a vehicle are getting restricted up to the evicting, and technologies that can be applied without the addition of water as the transporting media are obtained [1-2].

The problem becomes particularly relevant for the underground mining of mineral deposits, especially in cases where adapted mining works are used for transportation, for example, with reverse bias.

To the technology of using concrete mixtures, the following requirements are set: the possibility of moving them to a considerable distance, reliability, minimal costs, loyalty to the environment,

* Corresponding author: v.i.golik@mail.ru

The peculiarity of the transportation of mixtures is the force acting on the pipe-line [3-4]. Vibration reduces the resistance to transportation and allows the flow under pressure to move concrete mixes.

The parameters of transportation are significantly influenced by the parameters of force impact. With the increase in the amplitude of the action at a constant frequency, the thickness of the water-cement film of the mixture increases due to the inertness of the mixture.

The peculiarity of the movement of the granular mixture through the pipes is the compliance of the pipeline diameter to the fineness of the mixture components, which must be at least 5/1. The grain size of the material should not exceed 15 mm, and the ratio "large – fine" filler must be provided within 2:3.

Next to the pipeline for concrete mix for its maintenance it is necessary to locate the main rifles of compressed air and water. Compressed air is supplied to the pipelines through ejectors mounted at an angle of 15-300 with an interval of 50-70 m.

2 Materials and methods

The main place in the study is given to industrial experiment.

The first use of this technology refers to the practice of developing deposits of the consern "Wismut" (Germany).

In the USSR, the technology was used in the development of the Shokpak-Kamyshovoye deposit in Kazakhstan to deliver hardening mixtures over a distance many times higher than the vertical height. When feeding the packing mixture for a distance of up to 2.5 km, the energy consumption was 0.15-0.22 kW/m3. The particular interest to this technological precedent is that during transportation the strength of the solidifying mixture has increased by 20-25%.

To increase transportability through pipelines, water is usually added to the concrete mixes before the cone 11 precipitating. With the new technology, the mixture is transported to the precipitation of cone 9 without the premix reducing the water-reducing strength.

The installation for solidifying filling mixtures transportation worked under the following conditions: diameter of air tapping is 40 mm, diameter of pipeline is 170 mm, the pressure of compressed air in air pipeline is 6 MPa, the capacity is 60 m^3/h.

Parameters of the vibro-transport installation are the following: compulsive force is 2-5 kN, amplitude of the pipeline oscillation is 1.2-2.0 mm, oscillation frequency is 6.0-13 Hz, influence of one exciter is 200-220 m. The range of the mixture supply was 2 times higher than possible for traditional technologies.

For delivery of filling mixes in underground conditions gravity, gravity-pneumatic and vibration schemes of pipeline transport are used. The first scheme provides stable delivery for a distance of 850 m, the second – for 1200-1500 m, and the possibilities of the third scheme have not been studied enough to estimate the quantitative parameters of delivery. It can only be asserted that the vibrators give the pipeline the functions of the mixer-activator [5-6].

The installation of vibro-pneumatic transport included a pipeline and vibrators. The system of vibro-transport of the concrete mix consisted of a vertical section of 175 m height and a horizontal section of 2500 m long (Fig. 1).

 DOI: 10.1051/e3sconf/20172101018

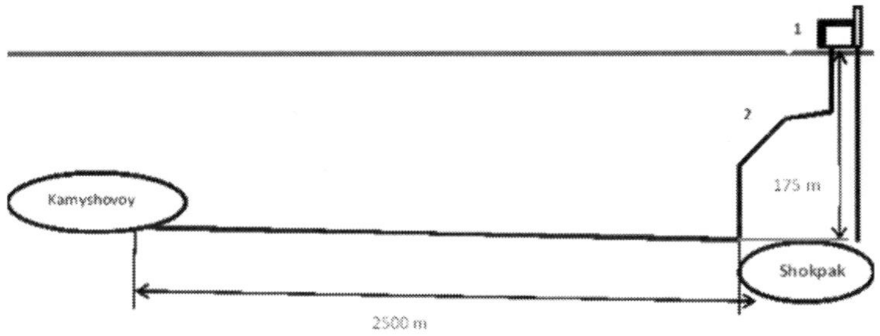

Fig. 1. Scheme of use of the filling complex of the Shokpak deposit for the delivery of hardening mixtures to the development of the Kamyshovoye deposit: 1-packing complex; 2-vertical part of the pipeline; 3 - horizontal part of the pipeline

If the length of delivery of the filling mixture according to the traditional scheme rarely exceeds 1.5 km at a ratio of the vertical and horizontal parts of the pipeline 1:5, then with the new technology this ratio was 1:20.

The composition of the transported mixture (kg/m^3) is the following: cement up to 100, blast furnace slags – 160-250, water – 380. Feeding range of the mixture: horizontal with a lift up to 14 m – 2500 m. Speed of the mixture through the pipeline is 1.2 -1.5 m/s. The maximum capacity of the system is 100 m^3/h.

The pressure drop over the length of the section during the feeding of the mixture was: in the self-cooling mode – 0.6-1.0 MPa, grad P = 3.0-5.0 kPa/M, with the application of vibration – 0.12-0.20 MPa, grad P = 0.8-1.0 kPa/M.

The transported hydro-mixture is characterized by plasticity within 1-7% (Tab. 1).

Table 1. Plasticity of solid part of hydro-mixture

Plasticity,%	Samples					Average
	1	2	3	4	5	
The limit of fluidity	39	37	39	38	37	38
The limit of unrolling	33	33	34	33	33	33
The indicator of plasticity	6	5	4	5	5	5

With traditional delivery technology, transportation of the mixture is carried out at an increased water flow, which reduces its strength. The use of vibration technology in combination with pneumatic gravity flow eliminates this drawback.

The accumulated experience allows the authors to develop an algorithm for the project of concrete mixtures supplying containing 0.1-0.35 dispersed particles, with a solids concentration in water of 0.10-0.85 and with settlement of a standard cone of 10-13 m.

3 Results and discussion

The main parameters of the vibrodelivery are the following: transport length (L), vertical line height (H), section length (l_1) and vibration exciter location within section (l_2) (Fig. 2).

DOI: 10.1051/e3sconf/20172101018

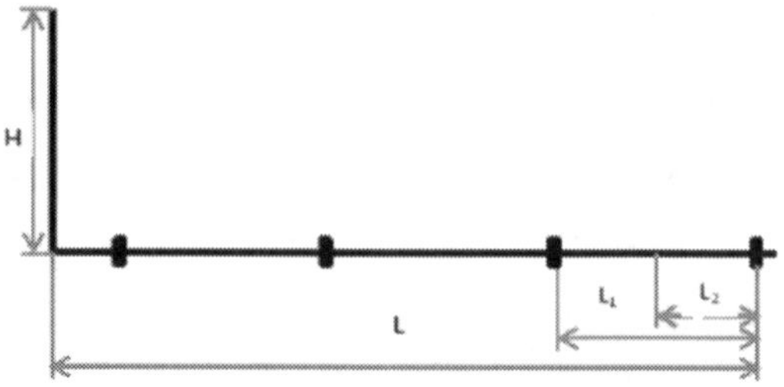

Fig. 2. Scheme for the calculation of vibrotransport: l- length of the pipeline, m; l_1 - distance between the vibrators, m; l_2-distance of installation of the vibrator on the section

The effective influence of vibration on the mixture is ensured by the oscillation of the pipeline with minimum amplitude A_m

$$A_m = \frac{\rho_{\tau} - \rho_o}{\omega^2 \rho_{\tau}} g \quad (1)$$

where: p_{τ} – the density of filling particles, kg/m^3;

p_o – the density of dispersed media, kg/m^3;

When the vibration exciter is located in the center of inertia of the section, its length is:

$$l_1 = b\frac{A - A_m}{a_1} \quad (2)$$

where: b is the coefficient considering the location of the vibrator exciter (2);

a_1- coefficient of attenuation of oscillations (0.007-0.008 mm/m).

Stratification of the mixture is eliminated at a speed of its movement of 0.5-0.7 m/s - for mixtures with aggregate size up to 5.0 mm and 0.7-1.0 m/s - for mixtures with a filler of 5.0- 40.0 mm. Herein the internal diameter of the pipeline is:

$$D = 24.45 V_{av} \times d_{av} \times \sqrt{\frac{\rho}{\tau_o}} \quad (3)$$

where: V_{av} – average velocity of transporting material, m/sec;

d_{av} – average diameter of transporting material, mm.

The most effective vibration effect on the solidifying filling mixture at a speed of its movement is 1.0-1.5 m/s.

The range of feed of the solidifying mixture by the installation can be defined as:

$$l = \frac{\rho_c}{\Delta\rho} \quad (4)$$

where: p_c – the density of solidifying filling mixture, kg/m^3.

The distance between the elastic supports should not exceed l_{o1}:

$$l_{o1} = 0,5 l_o \quad (5)$$

The design diagram of the pipeline section is given in Fig. 3.

 DOI: 10.1051/e3sconf/20172101018

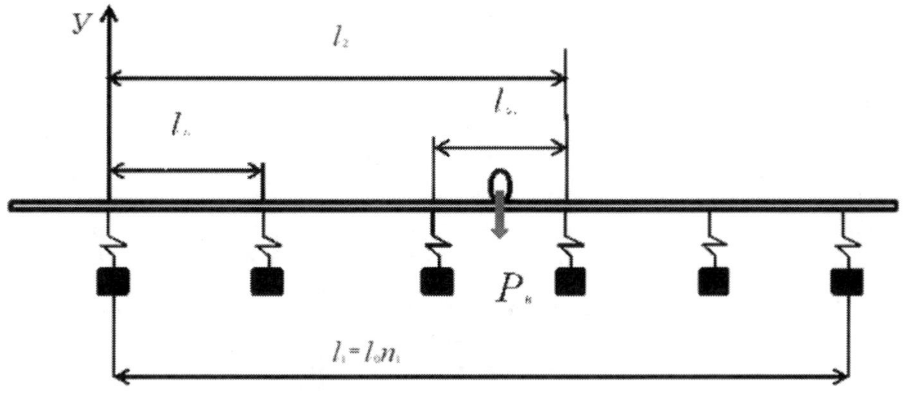

Fig. 3. The design diagram of the pipeline section.

The dynamics of hydraulic losses during transportation of water are shown in Tab. 2.

Table 2. The dynamics of hydraulic losses

Velocity, m/sec	Chezy's velocity factor,	Coefficient of hydraulic friction	Loss of pressure calculated	Pressure loss experienced	Error, %
1	38.5	0.05	5.8	6.6	-11.7
1.5	-	-	12.1	13.0	-6.8
2.0	-	-	21.5	21.7	-0.6
2.5	-	-	33.6	33.5	+0.5

Vibro-transportation increases the length of delivery of the solidifying mixture, which allows solving the problem of their transportation, both for the peripheral sections of the deposits and on the ground surface.

The increase in the activity of the hardening mixture during its transportation makes it possible to involve low-active production waste into recycling, to save the earth's surface from destruction during the extraction of mineral components of the mixture and to reduce the costs of surface and underground mining facilities and related industries objects construction [7].

The results of the study correspond with the specialists in this area of mining opinion [8-9].

4 Conclusions

1. The quality of concrete mixtures increases with the rational use of physical and energy resources during their transportation.

2. Activation in the pipeline improves the quality of the mixture due to a more even distribution of the aggregate, better hydration conditions for the binders.

3. The use of the new technology opens up new possibilities for using resource-saving technologies for underground mining of ore deposits with the achievement of a multi-area effect, including by peripheral areas installing without the construction of additional filling complexes.

Acknowledgement

The article contains the results of researches made under Programme Erasmus + 574061-EPP-1-2016-1-DE-EPPKA2-CBHE-JP "Modernization of geological education in Russian and Vietnamese universities".

References

1. G. Li R. Wang, E. J. M. Carranza, F. Yang. Ore Geol. Rev. **71**, 592–610 (2015)
2. M.E. Jarvie-Eggart, *Responsible Mining: Case Studies in Managing Social & Environmental Risks in the Developed World* (Society for Mining, Metallurgy and Exploration Englewood, 2015)
3. Xiao Li-ping, *Study on Pollution Laws of Coal Gangue Leaching Solution to Ground-water System* (Fuxin : Liaoning Technical University, 2007)
4. Onica, E. Cozma, T. Goldan, AGIR Rev. **3**, 14-27 (2006)
5. J. Broder, *Groundwater Geochemistry. A practical guide to modeling of natural and contaminated aquatic systems* (Springer, New York, 2005)
6. Weijing Wang, Shaofeng Huang, Xiaobo Wu, Qingfei Ma, J. Soft. Engin. Applic. **4**, 329–334 (2011)
7. Zhen-Dong Liu, Gan-Jiang Tao, Qing-Yun Ren, Dong-Sheng Yang, J. Ch. Coal Soc. **36:4**, 450-455 (2011)
8. Haifeng Wang, Yaqun He, Chenlong Duan, Yuemin Zhao, Youjun Tao, Cuiling Ye, *Development of Mineral Processing Engineering Education in China University of Mining and Technology* (Springer-Verlag, Berlin Heidelberg, 2012)
9. V.I. Golik, Yu.I. Razorenov, V.N. Ignatov, Z.M. Khasheva J. Soc.l Sci. **11:15**, 3742-3746 (2016)

Drilling Rig Operation Mode Recognition by an Artificial Neuronet

Fares Abu-Abed[1], *Nikolay* Borisov[2]

[1]Tver State Technical University, A. Nikitin Street, 22, 170026, Tver, Russia
[2]Lobachevsky State University of Nizhny Novgorod, Gagarin Avenue, 23, Nizhny Novgorod, Russia

Abstract. The article proposes a way to develop a drilling rig operation mode classifier specialized to recognize pre-emergency situations appearable in commercial oil-and-gas well drilling. The classifier is based on the theory of image recognition and artificial neuronet taught on real geological and technological information obtained while drilling. To teach the neuronet, a modified backpropagation algorithm that can teach to reach the global extremum of a target function has been proposed. The target function was a relative recognition error to minimize in the teaching. Two approaches to form the drilling rig pre-emergency situation classifier based on a taught neuronet have been considered. The first one involves forming an output classifier of N different signals, each of which corresponds to a single recognizable situation and, and can be formed on the basis of the analysis of M indications, that is using a uniform indication vocabulary for all recognized situations. The second way implements a universal classifier comprising N specialized ones, each of which can recognize a single pre-emergency situation and having a single output.

1 Introduction

While drilling oil-and-gas wells, the drilling rig operator has to quickly solve many technological problems including an early detection of oil-and-gas-and-water showings and mud losses while drilling, drilling optimization depending on the geological tasks, technological operations recognition and timing, choosing and maintaining an efficient drilling mode, pre-emergency situation diagnostics in real time scale, and drilling equipment operation diagnostics [1]. The quality of solving the above problems firstly depends on the skills, and some ambiguous situations can demand some essentially subjective solutions.

At the same time, there exists an image recognition theory [2-3] that can process information as a set of parameters describing the recognized target, in order to conclude what is the class the recognized image (target) belongs to. That theory has been applied in different mining branches [4-6], so it seems prospective to use that apparatus for the automated recognition of the drilling rig recognition in order to minimize the effects of subjective factors on the drilling rig operation.

The principal functions of the drilling rig state classifying stage are 1) real-time raw information processing and 2) classified target state determination based on its indications, using a special algorithm.

The drilling rig state determination uses a ready classifier which should determine the dynamic target state already in the real time. Here, both effective and quick target state determination is crucial.

2 Materials and Methods

The mathematical methods for solving problems in complex technical system diagnostics have been developed for several decades. The researchers proposed numerous ways to check hypotheses and draw conclusions. In 1960s they developed analysis methods which enjoyed some expansion and raised a wave of publications. They had a common feature uniting them – evident decision-making algorithms: the diagnostic algorithm includes a set of rules that determine the raw material processing order in order to determine the diagnosis – that is to classify the system's state [7].

Many years of investigations on most different evident algorithm showed that the implicit geological problems can be solved by the evident methods with such an accuracy and a convenience, that are absolutely insufficient for their wide practical application in particular problems of diagnostics, forecasting and decision-making [8].

Implicit algorithms (that provide automatic experience accumulation and following use in teaching) have been searched for and studied for more than 50 years [9]. However, first serious attempts to form the neural networks were made in 1940-1950s, when Warren S. McCulloch and Walter Pitts proposed the principles of their brain operation theory. With the advent of low-cost computers, there was an abrupt progress in this sphere. In early 1980s, this sphere transformed into an entire brand-new science - neuroinformatics. [10].

The image recognition can be the most popular problem type to solve using neuronets today. The artificial neuron networks are mathematical simulations with their structures and functions formed as in their biological analogs (neural cells or neurons in the brain). The neuronets are formed on the basis of the idea that the neurons can be simulated as quite simple automatic devices (named as artificial neurons), and all the brain complexity, functioning flexibility and other major properties are due to the links among the neurons [11].

The biological neuron is a complex system which mathematical simulation has not been done yet. The ANN theory is based on the assumption that all that complexity is insignificant, and the brain properties can be attributed to the neuron contacts. That is why a simple simulation of a so called formal neuron [12] is used instead of precise mathematical neuron simulations.

A network type can be chosen on the basis of the problem statement and teaching data available. Teaching by a teacher requires an "expert" estimation for every teaching sample element. Sometimes it is impossible to obtain such an estimation for a great data set. In these cases the natural choice is a network taught without a teacher - for example, a Kohonen self-organizing map or Hopfield neuron network [13-14].

The estimation of the number of neurons in the hidden layers of the homogeneous neuron networks can use the formula estimating the necessary number of synaptic weights Lw (in a multilayer network with sigmoid transfer functions):

$$\frac{mN}{1+\log_2 N} \leq Lw \leq m\left(\frac{N}{m}+1\right)(n+m+1)+m \tag{1}$$

where n is the input signal dimension, m is the output signal dimension, N is the number of teaching sample elements.

Estimating the necessary number of weights allows calculating the number of neurons in the hidden layers. For example, the number of neurons in the double-layer network is:

$$L = \frac{Lw}{n+m} \tag{2}$$

Other similar formulas are also known, for example:

$$2\,(L+n+m) \le N \le 10\,(L+n+m) \tag{3}$$

$$\frac{N}{10} - n - m \le L \le \frac{N}{2} - n - m \tag{4}$$

By the same way, the number of neurons in networks having a greater number of layers can be calculated. Using such multilayer neuron networks is sometimes reasonable because they can have smaller dimensions of their synaptic weights of the neurons in their single layer, than double-layer networks implementing the same reflection.

Unfortunately, the above formulae can give only approximate estimations, so obtaining accurate values of the neuronet parameters requires experimental investigations in that particular target sphere.

For the drilling rig pre-emergency situation recognitions, the teaching sample size should be quite great. That is why the network teaching effectiveness is the crucial parameter for choosing the number of layers and neurons in every layer. For that, the time spent on teaching (the number of teaching algorithm iterations) and teaching accuracy (the percentage of situations truly recognized by the taught network).

To teach the neuronets, the backpropagation algorithm is widely used. It is an iterative gradient algorithm applied to minimize the root-mean-square deviation of the current output of the multilayer perceptron and desired output [15].

Major modifications of the backpropagation algorithm involve different error functions, various procedures for increment size and direction evaluation, teaching timing procedures, etc. Integral error functions for the whole set of all teaching examples as well as integral and fractional power functions can be used as error functions.

There exist two teaching modes: a series one and a pack one. While teaching a multilayer perceptron with the backpropagation algorithm, it is repeatedly offered a preset set of teaching examples. A single full cycle of offering a full set of teaching examples is called an epoch. In the series mode, the weights are corrected after feeding each example. This mode is used most often. In the pack mode of teaching, the weights are corrected after feeding all teaching examples of a single epoch to the network. Computingly, the series teaching mode is much faster. This is especially effective when the teaching set is big and highly excessive.

The classifying tasks based on the series teaching of the neuron network with the backpropagation algorithm usually use a method of a random change of the sequence of examples fed to the multilayer perceptron input, from one epoch to another. Ideally, such randomization attributes the effectively processed examples to different classes.

Ideally, all multilayer perceptron's neurons are to be taught at the same rate. However, the last layers usually have higher values of their local gradients than the first layers of the network. So, the teaching rate parameter of the backpropagation algorithm should be set to lesser values for the last layers and greater values for the first layers. To provide approximately the same teaching time for all network neurons, neurons having a greater number of inputs should have a lower value of the teaching parameter than the neurons having a lesser number of inputs. There is an opinion that it could be reasonable to specify the teaching rate parameter value inversely as the square root of the sum of its synaptic contacts – for every neuron.

The above ways to modify the basic teaching algorithm do not solve one of the major problems of this algorithm, namely they do not answer the question: What kind of the target function has been found – a local one or a global one?

 DOI: 10.1051/e3sconf/20172101019

To teach the neuronet to recognize the drilling rig pre-emergency situations, a modification of the backpropagation algorithm was proposed. In contrast to the standard algorithm, it features the following:

1. After reaching the local minimum (completing the teaching), a new start point is randomly generated, and teaching recommences.
2. After reaching (with a preset precision *E*) the same minimum value of a teaching error from different start points *K* times (*K* is an algorithm parameter), parameter *R* – start point generation for teaching – changes (the search radius increases), and the global extremum search continues (the neuronet is taught anew).
3. After parameter *R* reaches *Rmax* (*Rmax* is an algorithm parameter), the neuronet is considered to be taught, and the algorithm stops.

The algorithm flowchart is shown in Figure 1.

For the preset value of parameters *K* and *Rmax*, the algorithm does the first teaching of the neuronet, saving the *Wi* weight values found as a local minimum point. Then, a start value of *R* finds new start values of the neuronet weights in a randomly chosen direction from the minimum found, and the teaching recommences.

If the search from a new start point finds a point with a lower value of the criterion, the attempt counter resets to zero, and new start points will be found from the new-found minimum with the same radius *R*.

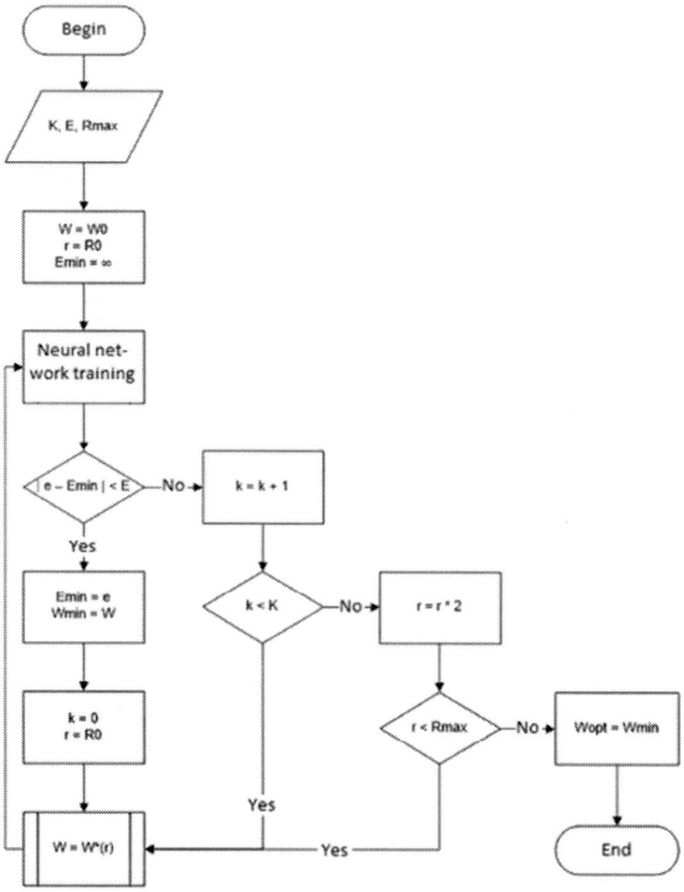

Fig. 1. Neuronet teaching algorithm.

After reaching the same minimum point for K times (with the preset precision), the search radius redoubles, and the attempt counter resets to zero. When the radius reaches the search for the preset value of parameter $Rmax$, the last point is considered to be a global extremum point, and the algorithm completes its operation.

3 Results and Discussions

The pre-emergency situation recognition classifier based on the neuron network can be implemented in two ways.

The first way involves generating N different signals on the output, each of which corresponds to a single recognized situation and is formed on the basis of the analysis of M indications, that is using a common indication vocabulary for all recognized situations.

The alternative way includes implementing a universal classifier comprising N specialized classifiers, each of which can recognize a single pre-emergency situation and has a single output and m_i inputs, with overlapable indication sets used to recognize different pre-emergency situation:

$$\sum_{i=0}^{N} m_i \leq M \tag{5}$$

To establish the structure of the neuronet classifier, experiments were conducted to evaluate the influence of the neuronet parameters on the effectiveness of teaching for the recognition of artificially generated situations.

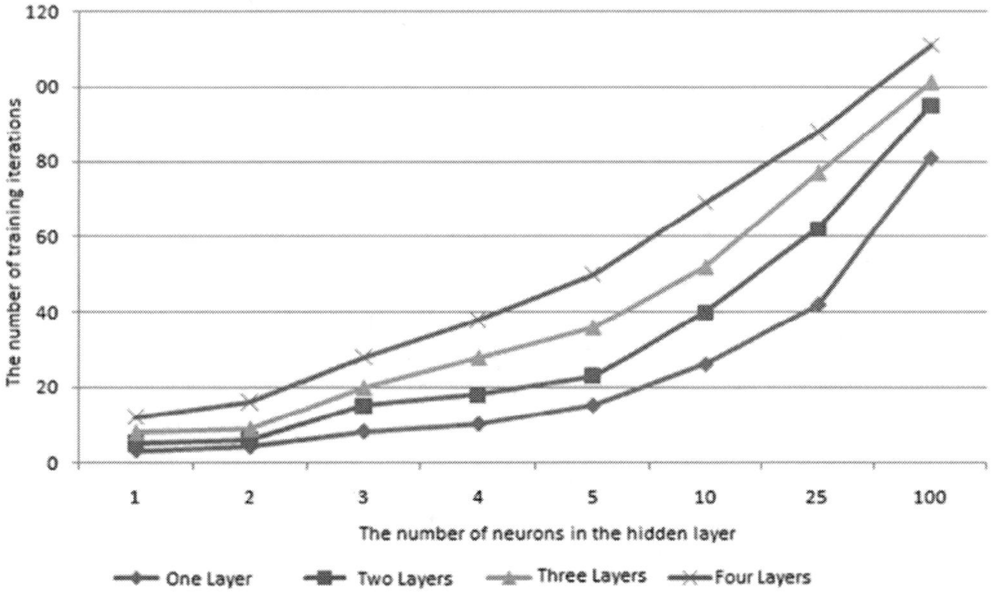

Fig. 2. The numbers of teaching iterations as a function of the number of layers and the number of neurons in each layer.

This Fig. shows that the optimum number of the hidden layer neurons in the case under investigation is four - which is equal to the number of classifier's inputs. Any further increase in the number of neurons is not reasonable as it prolongs the time necessary to teach the neuronet, without giving any increase in the pre-emergency situation recognition precision.

4 Conclusions

The results of the research conducted allowed establishing the neuronet classifier structure. It was found out to comprise a single hidden layer with the number of neurons equal to the number of classifier's inputs.

Consequently, an option of the pre-emergency situation classifier formed on the basis of several specialized classifiers with the above chosen structure will be the target of the further research. Validation of the choice and teaching algorithm parameters evaluation demands research using real geological and technological information in order to make a final decision.

References

1. G. Wittenberger, M. Cehlár, Z. Jurkasová, Acta Montanistica Slovaca, **17:4**, 241-246 (2012)
2. R.O. Duda, P. E. Hart, D. G. Stork. *Pattern Classification* (N.Y., Wiley, 2001)
3. R.C. Gonsalez, R.E Woods, *Digital Image Processing* (N.Y., Pearson Education, 2011)
4. O.E. Baklanova, M.A. Baklanov, ICSI 2016: Advances in Swarm Intelligence, 253-262 (2016)
5. F.N. Abu-Abed, Oil Industry, 100-102 (2014)
6. *Predictive Asset Analytics to Improve Mining Operations. Schneider Electric Software* (2015)
7. J. Korbicz, M. Kowal, *Intelligent Systems in Technical and Medical Diagnostics* (N.Y., Springer, 2014)
8. V.I. Grajfer, A.A. Favorsky, V.A. Shumilov, Oil Industry, **10** (2003)
9. S. I. Gallant, *Neural Network Learning and Expert Systems* (Mass., MIT Press, 1993)
10. M.A. Arbib, *Computing the Brain: A Guide to Neuroinformatics* (Chicago, Academic Press, 2012)
11. C.M. Bishop, Neural Networks for Pattern Recognition (Oxford, Oxford press, 1995)
12. F.N. Abu-Abed, Oil Industry, 107-111 (2015)
13. J. J. Hopfield, Proceed. Nat. Acad. Sci. **79:8**, 2554–2558 (1982)
14. T. Kohonen, *Self-Organizing Maps*, (New York, Logan, 2001)
15. D.E Rumelhart., G.E. Hinton, R.J. Williams, Parallel Distributed Processing, **1**, 318-362 (1986)

Perspectives for application of moulded sorption materials based on peat and mineral compositions

Oleg Misnikov[1]

[1]Tver State Technical University, A. Nikitin Street, 22, 170026, Tver, Russia

Abstract. The paper discusses scientific and technological grounds for using peat, its wastes and processing products as sorption materials for liquidation of pollutions on water and oil basis. A negative effect on the hydro-physical properties of peat by hysteresis phenomena occurring during drying is noted. Use of optimal concentrations of mineral hydrophilic additives in peat matrix allows to increase speed and capacity of sorption. A working hypothesis is proposed that explains the physical and chemical mechanism for increasing quality of composite sorbents. A principal technological scheme for obtaining sorption materials based on peat is developed. It allows to take into account possible transitions between stages of the technological process, which reduce its cost.

1 Introduction

Various types of sorption materials are widely used for purification of liquid and gaseous mixtures, as well as for selective extraction of pollution agents from solutions. Currently, depending on the requirements for the process of sorption, synthetic and natural materials are used, as well as various types of processing wastes. Peat as a natural sorbent and a raw material occupies a special place among them [1-3]. It is due to the variety of raw material resources, as well as its unique physical and chemical properties.

Organic matter of peat has the ability to form bonds between water molecule and the material by means of hydrogen, ion-dipole, dipole-dipole interactions. Weak interactions can be carried out with aromatic sections and sections with unsaturated bonds. This occurs due to unpaired electrons present in them and due to the Van der Waals forces. The micro-structure of native peat associates is responsible for the amount of physico-chemically and osmotically bound water.

Moisture can be absorbed and retained in peat by the following types of bonds:

1. Chemically bound water (crystalized water of mineral part of water). Its binding energy is 40-400 KJ/mol.
2. Physico-chemically bound water. Its binding energy depends on the amount of absorbed water. The water of monosorption is most strongly bound, its amount ranges from 0.15 to 0.2 kg of water per 1 kg of dry matter (hereinafter we will use the dimension - kg / kg for simplicity). The total amount of moisture of this category in peat can be 0.45-0.55 kg / kg.
3. Entropy-bound water (retained by osmotic forces). Its binding energy is up to 2

 DOI: 10.1051/e3sconf/20172101020

KJ/mol. The moisture content of peat in the presence of moisture of this category reaches 1.0-1.4 kg/kg.

4. Water of mechanical retention. This category includes capillary, immobilized, intracellular and structurally trapped water. The main share is accounted for capillary moisture, the binding energy of which is about 1 KJ/mol.

Absorption of moisture by peat and sorbents on its basis is a reverse (in relation to drying) process. The properties of the material to absorb different types of sorbates is determined by its porosity as well as by the nature of its binding centers. The filling mechanism can be different as the pores can be of three types: macropores, mesopores (transitional) and micropores. There are all types of pores in peat [4, 5]. In addition, peat systems have a large number of functional groups.

The main indicator, which makes it possible to assess absorbing capacity of peat is its total moisture capacity (W_t). The total moisture capacity of peat is affected by its current moisture content and decomposition degree (R_p). But extraction and subsequent drying of peat significantly reduce the value of the total moisture capacity. This is due to oxidation processes during drying and air sorption on structural elements [6]. The effect of partial hydrophobization during peat drying is manifested to a greater degree in systems with a low concentration of mineral components (peat ash).

At the end of the drying process, the physico-chemical water bound to the functional groups is removed [7]. In this case, conditions are created for interactions between macromolecules of organic matter due to hydrogen bonds and through polyvalent cations. The change in the structural characteristics of the material leads to a decrease in the immobilization capacity of the associates of macromolecules. Subsequent water treatment of peat partially splits the hydrogen bonds between the macromolecules. Thus, vacancies are formed for its sorption on functional groups.

The effect of dehydration degree on the reversibility of hydro-physical properties increases with increasing content of humic substances in peat [3]. Another important factor in reducing the water absorption of peat is its mechanical processing. Peat dispersion increases its physical and chemical activity. This entails a decrease in the hydrophilic properties of peat processing products.

In addition to the absorption of water-based pollution, hydrophobic sorbents capable of absorbing oil, its products and waste processing products are of great interest. Such sorbents must absorb a large amount of oil products, remain on water surface for a long time, be easy for collection and recycling. Analysis of conducted studies [8] showed that the characteristics of peat sorbents are sufficient to compete with other sorption materials.

Among non-organic sorbents scientists distinguish clays and diatomites due to their comparatively low cost. Quartz sand is also used for liquidation (sanding-up) small oil spills on land. However, non-organic sorbents have a low absorption capacity (about 0.5 kg of oil / kg of sorbent), they do not retain light oils, and when oil spills are liquidated on water they sink together with absorbed oil products.

Natural organic and organo-mineral materials are the most promising type of sorbents for the elimination of oil-based contaminant. The first task to be done on the stage of justification of the technology for obtaining sorbents is the choice of the raw material base. The main indicators that determine the properties of peat raw materials are a deposit type and decomposition degree. Therefore, the initial evaluation of the sorption properties was carried out according to these characteristics. In practice, other approaches are also used: according to the composition of peat-forming plants, the component composition, mineral elements of ash, etc. [9].

Domestic and international experience of production and use of peat sorbents [10-14] demonstrates that the most suitable type of raw material is the upper sphagnum peat of low degree of decomposition. Hence, the primary regulation of the sorption properties of peat is carried out by selecting the material with the largest content of least decomposed plant residues. Technological operations during peat extraction disperses its particles. This has a negative effect on the absorption capacity. Therefore, it is necessary to develop extraction technologies that would minimize mechanical effect on the primary peat structure.

Another advantage of peat raw material is its good plasticity and, accordingly, its ability to mould. The main binding components of peat are humic substances. Therefore, when processing peat of low degree of decomposition, their small content does not allow obtaining high-quality moulded products. To solve this problem, various plasticizing substances (clays, organo-mineral deposits, bitumens, gypsum) and other methods for increasing mouldability of peat are used.

Clay modifiers not only increase plasticity of a formed organo-mineral material, but also significantly reduce the hysteresis phenomena in peat, which appears upon its drying. They do not have a significant effect on the porous structure, but due to physical and chemical interactions of the components, a preliminary modification of the composition occurs. Usually, metal salts and oxides, silicones, halogen compounds of zinc, aluminium, magnesium and iron are used as modifiers of carbon fibers. They serve as structuring agents, and also affect the kinetics of pyrolysis (if it is necessary to conduct it).

2 Method used

There are two basic methods of sorbents moulding: cylindrical extrusion and pelletizing with disk or drum granulators. In the work to obtain sorbents, the pelletizing method was used, which made it possible to obtain sorbents of spherical shape. The process of moulding by the pelletizing on a disk granulator consisted of four main stages:

1. mixing of the particles of source dispersed material with the particles of additives (in our case, mineral clay inclusions) and binding components;
2. formation of granules from fine particles and destruction of large lumps;
3. pelletizing, consolidation and compaction of granules as a result of their movement along the surface of the disk;
4. strengthening of bonds and stabilization of granule structure.

In peat and mineral compositions, water together with clay inclusions served as a binding material at the initial stage of moulding. They helped to accelerate the process of granule formation of mixtures with relatively low initial humidity. In the processes that we consider, granule formation and structural changes during the subsequent drying of the granules occurred due to hydrogen and Van der Waals interactions. It is known that peat contains mainly hydrophilic compounds, but part of its organic matter is composed of hydrophobic components. The increase in the hydrophilicity of organic systems in our case was accomplished by adding clay additives (dispersed kaolin clay).

In the moulded organic-containing mixtures, moisture content of the main component (peat) was $w \approx 65$-75%, depending on R_T. At lower values of w, additional time was required to absorb moisture. Partial hydrophobicity of peat, acquired by it during drying, affected here. During the movement, the granules were compacted by impact-rotational motion along the fixed material layer and along the board (bottom) of the disk of the granulator. At the same time, a less wet loose material adhered to the surface of the appearing pellet. The lumps of material were compacted as a result of repeated impacts and interspersion.

The granules obtained by this method were separated into fractions using sieves and dried at temperature $T = 80°C$ till constant weight. In a number of experiments, the dried pellets were additionally subjected to thermal exposure without access to oxygen at higher temperatures ($T \approx 300°C$).

 DOI: 10.1051/e3sconf/20172101020

3 Results and Discussion

Clay materials are hydrophilic and possess ion exchangeability. They absorb moisture almost immediately compared to peat. This is due to the fact that the main structural hydroxyl groups (OH^- crystalline faces) of minerals are free. Therefore, they have the ability to form hydrogen bonds with water molecules [4]. At the same time, a relatively small water-absorbing ability of clays does not ensure their use as effective sorption materials. The use of peat and mineral compositions allows to strengthen an insufficient hydrophilicity in the peat matrix, as well as sorption capacity in the clay material.

Experiments with sorbents obtained from various types of peat and compositions based on them demonstrated that the additions of clay material (kaolin clay) do not reduce (which was assumed a priori), but rather increase sorption capacity of the granules. At the same time, there is a persistent tendency to intensify the process of moisture absorption by the compositions (Figure 1).

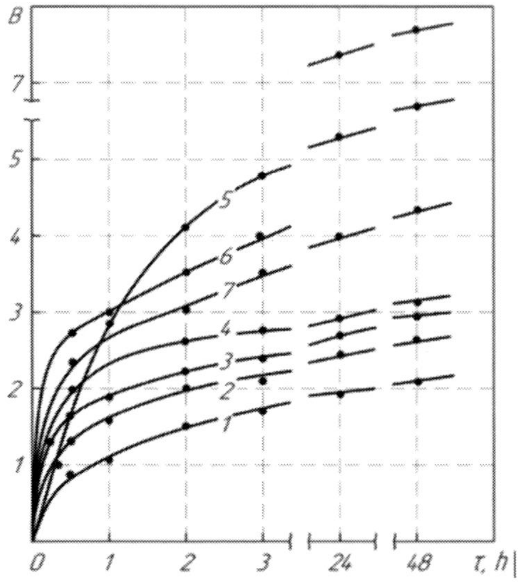

Fig. 1. Kinetics of water absorption *B*, kg/kg of granulated upper cotton grass-sphagnum peat R_p = 25 % (1); peat and mineral compositions: peat kaolin clay in proportions 90:10 (2), 85:15 (3), 75:25 (4); upper magellanicum peat R_p = 5-10 % (5); peat and mineral compositions: peat kaolin clay in proportions 60:40 (6), 50:50 (7)

Studies on the increase in the pelletizability of peat of various degrees of decomposition in the presence of clay components made it possible to establish that introduction of these additives increases the rate of granule formation and improves their shape. The moisture content of moulding is inversely proportional to the content of mineral components, which makes it possible to reduce the amount of moisture, which evaporates during drying the materials.

When studying the swelling of peat composite sorbents obtained under a "soft" drying regime ($T \approx 20°C$), it was found that adding clay allows to obtain an additional volume of absorption as the moisture absorbs. That is, an increased swelling is distinctive for organo-mineral mixtures compared with the original peat. As follows from the analysis of Figure 2, with the increase of clay content the absorption capacity increases as well, which is due to the increased swelling of the composition.

DOI: 10.1051/e3sconf/20172101020

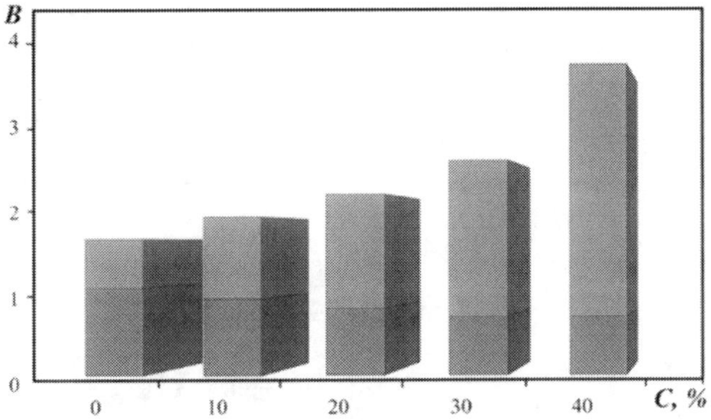

Fig. 2. Initial porosity and increment in pore volume during water absorption by granules (average diameter 5 mm) with different content of *C*,% kaolin clay: initial porosity in the lower part of the diagram, the upper part shows the increment in porosity that occurs when the material swells

Drying is the next step in the preparation of composite moulded sorbents based on peat. The temperature regime of drying should ensure the necessary shrinkage strain (expressed in the density and porosity of the sorbents), as well as the required hydrophilicity or hydrophobicity of the material that determines wettability by various types of liquids. The control of the drying process consists in varying temperature regime, humidity of heat-carrying agent in a drying chamber, processing time, etc. For composite materials (Figure 3), when the drying temperature increases, the moisture absorption decreases, which indicates the initial hydrophobization processes in their structure.

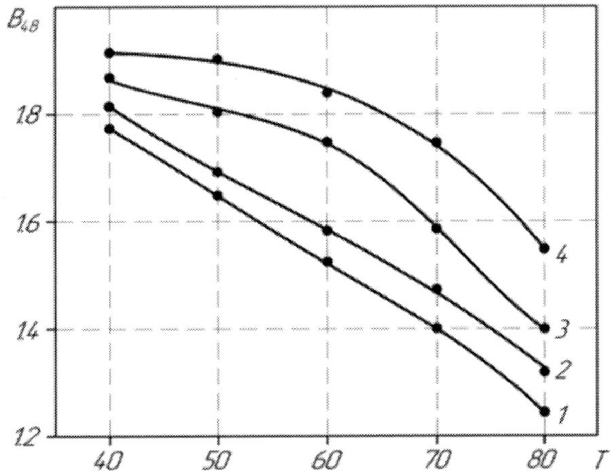

Fig. 3. Dependence of water absorption for 48 hours B_{48}, kg / kg on drying temperature *T*, °C for peat and mineral compositions with kaolin clay content: 1-10%, 2-20%, 3-30%, 4-40%

Thermal modification is the final area of the change (if necessary) of the properties of composite sorbents. For peat and organo-mineral mixtures (peat with kaolin clay) it was established that depending on the conditions of thermal treatment materials with different water absorption parameters can be obtained. If the treatment is carried out with oxygen access, then due to the burning out of combustible components a material with a developed porous structure that retains its hydrophilic components is formed.

A study of the water absorption ability of such sorbents indicates an increased capillary absorption of moisture with a limited but stable swelling of the clay components in the modified peat matrix. Thermal treatment of peat and waste products of its production [15, 16] with insufficient access of oxygen makes it possible to obtain a hydrophobic material.

DOI: 10.1051/e3sconf/20172101020

After selective sorption of liquid hydrocarbons, the material forms a stable floating conglomerate that can be easily removed from the surface and sent for processing or incineration. Many analogues [8] often have lower consumer properties (absorption capacity, porosity, biological stability, retention capacity, etc.). In addition, peat properties allow further modification of the sorbent with oil-oxidizing biological products, which will promote rapid destruction of oil products left in the water.

Analysis of the data, obtained for peat materials subjected to heat treatment without access of oxygen, demonstrates that this type of modification increases the sorption capacity for petroleum products by 45-50% (Table 1).

Table 1. Oil capacity of peat sorbents

Material	Sorbent capacity of a source material	Sorbent capacity of a thermally treated material	Change in sorbent capacity, %
Milled peat	5-5.4	5.8-7.8	Increase by 20-45
Granulated peat	2.4-2.6	3.3-4	Increase by 40-50

The sorbent has a low density, which allows it to be used for collecting oil during spills on water surface (bulk density of granular peat of low decomposition rate reduces from 120 kg/m^3 to 100 kg/m^3).

4 Conclusions

In all cases, modification is possible (thermal, thermochemical, mechanochemical, chemical, electromagnetic, biochemical). It is necessary that attention is paid to the fact that a greater effect is achieved by introducing additional modifying additives that increase wettability, swelling, granulation, porosity, etc. In this case, a transition to composite sorption materials takes place.

Thus, use of the proposed approach makes it possible to comprehensively take into account properties of peat raw materials, extraction technology, possible directions of its processing, requirements to the qualitative characteristics of sorbents, as well as the cost of various options of the final product.

References

1. P. Bel'kevich, L. Chistova, L. Rogach, M. Sobol, Solid Fuel Chemistry, **14:4**, 125-129 (1980)
2. P. Bel'kevich, L. Chistova, L. Rogach, T. Sokolova, Solid Fuel Chemistry, **18:5**, 91-95 (1984)
3. Y. Navosha, S. Prokhorov, V. Strigutskiy, A. Tomson, Eurasian Soil Science, **24:5**, 1-6 (1992)
4. A. Afanas'ev, Colloid Journal of the Russian Academy of Sciences: Kolloidnyi Zhurnal, **53:3**, 425-430 (1991)
5. A. Afanas'ev, A. Boltushkin, Colloid Journal of the Russian Academy of Sciences: Kolloidnyi Zhurnal, **58:2**, 139-144 (1996)
6. O. Misnikov, E. Chertkova, Eurasian Mining, **1**, 63-68 (2014)
7. A. Afanas'ev, S. Gamayunov, O. Misnikov, Colloid Journal, **61:3**, 274-279 (1999)
8. O. Misnikov, A. Timofeev, O. Pukhova, Polymer Science. Series D, **8:1**, 66-74 (2015)

9. N. Gamayunov, B. Maslennikov, Eurasian Soil Science, **24:7**, 32-37 (1992)
10. N. Gamayunov, B. Maslennikov, Y. Shul'man, Eurasian Soil Science, **24:4**, 122-126 (1992)
11. L. Ansone-Bertina, M. Klavins, Open Chemistry, **14:1**, 46-59. (2016)
12. C. Zheng, S. Feng, P. Liu, E. Fries, Q. Wang, Z. Shen, H. Liu, T. Zhang, Clean – Soil, Air, Water, **44:9**, 1163-1173 (2016)
13. A. Robalds, L. Dreijalte, O. Bikovens, M. Klavins, Desalination and Water Treatment, **57:28**, 13285-13294 (2016)
14. E. Bannova, N. Kitaeva, Solid Fuel Chemistry, **49:6**, 377-380 (2015)
15. J. Lee, X. Yang, H. Song, Y. Ok, E. Kwon, Energy, **120**, 929-936 (2017)
16. O. Misnikov, Mires and Peat, **18**, 1-15 (2016)

Technogenic Rock Dumps Physical Properties' Prognosis via Results of the Structure Numerical Modeling

Sergey Markov[1,*], *Victor* Martyanov[1], *Elena* Preis[1], and *Asmelash* Abay[2]

[1] T.F. Gorbachev Kuzbass State Technical University, 650000, 28 Vesennyaya St., Kemerovo, Russia
[2] Department of Earth Science, College of Natural and Computational Sciences, P.O.Box.3066, Mekelle University, Mekelle, Ethiopia

Abstract. Understanding of internal structure of the technogenic rock dumps (gob dumps) is required condition for estimation of using ones as filtration massifs for treatment of mine wastewater. Internal structure of gob piles greatly depends on dumping technology to applying restrictions for use them as filtration massifs. Numerical modelling of gob dumps allows adequately estimate them physical parameters, as a filtration coefficient, density, etc. The gob dumps numerical modelling results given in this article, in particular was examined grain size distribution of determined fractions depend on dump height. Shown, that filtration coefficient is in a nonlinear dependence on amount of several fractions of rock in gob dump. The numerical model adequacy both the gob structure and the dependence of filtration coefficient from gob height acknowledged equality of calculated and real filtration coefficient values. The results of this research can be apply to peripheral dumping technology.

1 Introduction

Using of gob dumps as filtration massifs for waste mine waters treatment is widespread at open pit mines in a major North Eurasia coal region Kuzbass. Nevertheless, frequent used peripheral dumping technology lead to the irregular ("stratisfied") structure of dump, as it shown at Fig. 1.

Therefore, unevenness of physical parameters (density, porosity, filtration coefficient etc.) to evolved from grain sizes segregation in dump height [1].

2 Material and Method

The research methods include photomapping, field observations, statistical analysis, and the discrete elements method (DEM, implemented in the algorithm of particles' packing for numerical counting filtration coefficient) [2-5].

* Corresponding author: markovso@kuzstu.ru

DOI: 10.1051/e3sconf/20172101021

Fig. 1. Uneven structure of gob dump originate from peripheral technology.

3 Results and Discussion

The results of the photomapping analysis of the grain size distribution on dump height are presented in Fig. 2.

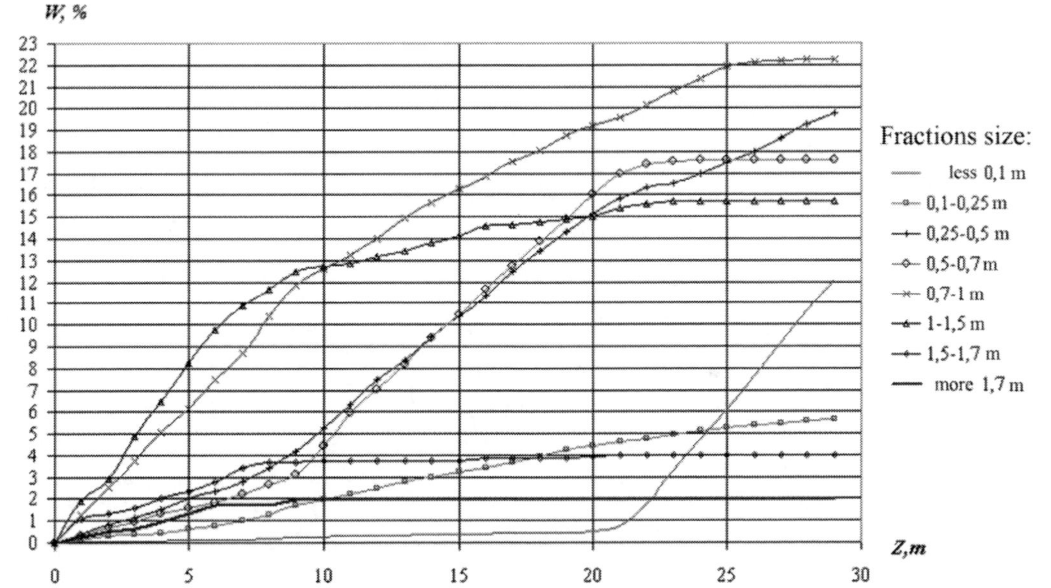

Fig. 2. Integral distribution of grain fractions W (%) depending on distance Z (meters) from base of gob pile.

To identify fractions have the most significant impact to filtration coefficient K_f was applied the folded Plackett-Burman design $2^8 \cdot 3/32$. Design initial data presented in Table 1. Based on the data of the variance analysis folded Plackett-Burman design $2^{8}\cdot 3/32$ (see Table 2), fractions with grain sizes <0.1 m, 0.5-0.25 m and 0.7-1.0 m have a *p*-value less than 0.178, which correspond to it strong influence on filtration coefficient at a confidence level of 82,2%. Volumes of other fractions in gob dump have no significant effect on filtration coefficient.

Table 1. Initial data for design

Distance from bottom of dump, m	Grain size distribution, % of overall volume								Filtration rate K_f, mps
	≤0,1 m	0,1–0,25 m	0,25–0,5 m	0,5–0,7 m	0,7–1,0 m	1,0–1,5 m	1,5–1,7 m	>1,7 m	
0–1	0.033	0.155	0.378	0.311	1.27	1.905	1.049	0.242	0.017
1–2	0.017	0.138	0.469	0.356	1.24	1.021	0.262	0.242	0.017
2–3	0.018	0.076	0.295	0.311	1.24	1.974	0.262	0.121	0.016
3–4	0.018	0.098	0.401	0.341	1.331	1.565	0.437	0.363	0.016
4–5	0.015	0.136	0.454	0.267	1.089	1.769	0.35	0.363	0.015
5–6	0.02	0.174	0.34	0.282	1.301	1.497	0.437	0.363	0.015
6–7	0.026	0.219	0.476	0.341	1.24	1.157	0.612	0	0.014
7–8	0.04	0.242	0.59	0.43	1.664	0.749	0.262	0	0.014
8–9	0.058	0.444	0.771	0.445	1.452	0.817	0	0.242	0.013
9–10	0.018	0.272	1.074	1.334	0.786	0.272	0.087	0	0.012
10–11	0.029	0.261	1.074	1.512	0.605	0.136	0	0	0.008
11–12	0.022	0.28	1.149	1.112	0.756	0.34	0	0	0.008
12–13	0.02	0.263	0.877	1.112	0.968	0.204	0	0	0.008
13–14	0.021	0.212	1.089	1.215	0.696	0.408	0	0	0.008
14–15	0.017	0.236	0.96	1.112	0.635	0.34	0	0	0.008
15–16	0.02	0.225	0.945	1.171	0.605	0.408	0.087	0	0.008
16–17	0.022	0.261	1.112	1.067	0.696	0.068	0	0	0.008
17–18	0.028	0.261	0.975	1.171	0.454	0.136	0	0	0.008
18–19	0.021	0.278	0.915	0.993	0.726	0.136	0	0	0.008
19–20	0.025	0.219	0.817	1.156	0.454	0.136	0.087	0	0.008
20–21	0.253	0.197	0.711	0.963	0.393	0.34	0.087	0	0.0023
21–22	1.04	0.132	0.446	0.415	0.575	0.204	0	0	0.0023
22–23	1.497	0.178	0.242	0.163	0.605	0.136	0	0	0.0023
23–24	1.493	0.163	0.439	0.015	0.544	0	0	0	0.0023
24–25	1.325	0.136	0.491	0	0.575	0	0	0	0.0023
25–26	1.582	0.112	0.484	0	0.242	0	0	0	0.0023
26–27	1.506	0.095	0.658	0	0.03	0	0	0	0.0023
27–28	1.462	0.085	0.628	0	0.06	0	0	0	0.0023
28–29	1.439	0.106	0.507	0	0	0	0	0	0.0023

 DOI: 10.1051/e3sconf/20172101021

Table 2. Data of the variance analysis folded Plackett-Burman design

Source fraction (grain size)	*P*-value	Sum of squares	Df	Mean square	F-ratio
<0.1 m	0.1742	0.0000046957	1	0.0000046957	2.03
0.1-0.25 m	0.5162	0.00000102028	1	0.00000102028	0.44
0.25-0.5 m	0.0963	0.00000726736	1	0.00000726736	3.15
0.5-0.7 m	0.2417	0.0000034287	1	0.0000034287	1.49
0.7-1 m	0.0447	0.0000110785	1	0.0000110785	4.80
1-1.5 m	0.3365	0.00000227591	1	0.00000227591	0.99
1.5-1.7 m	0.4301	0.0000015178	1	0.0000015178	0.66
>1.7 m	0.4569	0.000001346	1	0.000001346	0.58
Total error		0.0000346204		0.00000230803	

If differential percent amount of gob dump fraction with particles diameter less 0.1 m is as W_1 (or $W_{<0.1}$), particles diameter 0.25–0.5 m is as W_2 (or $W_{0.25-0.5}$), and particles diameter 0.7–1 m is as W_3 (or $W_{0.7-1}$), then response function as nonlinear multifactor regression model $K_f = f(W_1, W_2, W_3)$ will be:

$$\begin{aligned}K_f = {} & 0.011 \cdot W_1 \cdot W_3^2 + 0.062 \cdot W_2 \cdot W_3^2 + 4.786 \cdot 10^{-3} \cdot W_3^3 - 0.056 \cdot W_3^2 - \\ & -0.025 \cdot W_1 \cdot W_2 \cdot W_3 + 0.071 \cdot W_2^2 \cdot W_3 - 0.21 \cdot W_2 \cdot W_3 + 0.127 \cdot W_3 + \\ & +3.624 \cdot 10^{-3} \cdot W_1 \cdot W_3 - 0.012 \cdot W_1^2 \cdot W_3 - 0.072 \cdot W_1 \cdot W_2^2 + 0.054 \cdot W_2^3 - \\ & -0.172 \cdot W_2^2 + 0.202 \cdot W_2 + 0.079 \cdot W_1 \cdot W_2 - 0.022 \cdot W_1^2 \cdot W_2 - \\ & -0.052 \cdot W_1 + 0.056 \cdot W_1^2 - 0.014 \cdot W_1^3 - 0.071\end{aligned} \tag{1}$$

Correlation coefficient of this model is 0.98, which confirms the adequacy of the model (1).

4 Conclusions

The proposed high reliability prognosis model establishes a dependence of filtration coefficient from differential amount of gob dump fraction with particles diameter less 0.1 m is as W_1 (or $W_{<0.1}$), particles diameter 0.25–0.5 m is as W_2 (or $W_{0.25-0.5}$), and particles diameter 0.7–1 m is as W_3 (or $W_{0.7-1}$). It can be used for estimate gob dump as filtration massif for purifying of quarry wastewater [6-8].

References

1. S. Markov, M. Tyulenev, O. Litvin, E. Tyuleneva, E3S Web of Conf., **15**, 01011 (2017)
2. J.M. Kemeny, A. Devgan, R. Hagaman, X. Wu, J. Geotech. Engin. **119:7**, 1144-1160 (1993)
3. J.M. Kemeny, Min. Engin, **November,** 1281–1284 (1994)
4. J.C. Russ, *The Image Processing Handbook* (CRC Press, 1995).
5. W.X. Wang, F. Bergholm, O. Stephonsson, *Image analysis of fragment size and shape* (Blasting, Balkema, 1996)

DOI: 10.1051/e3sconf/20172101021

6. M. Tyulenev, E. Garina, A. Khoreshok, O. Litvin, Y. Litvin, E. Maliukhina, IOP Conf. Ser.: Earth Environ. Sci., **50:1**, 012035 (2017)
7. M. Tyulenev, Y. Lesin, , E.Tyuleneva, , E. Murko, E3S Web of Conf., **15,** 02003 (2017)
8. M. Cehlár, P. Varga, Z. Jurkasová, M. Pašková, Acta Montanistica Slovaca, **15:2**, 132-138 (2011)

Modeling of Energy-saving System of Conditioning Mine Air for Shallow Underground Mines

Alexandr Nikolaev[1*], Timur Miftakhov[2], Evgeniya Nikolaeva[3]

[1]Perm National Research Polytechnic University, Department of Mining Electrical Engineering, 29 Komsomolsky prospect, Perm, 614990, Russian Federation
[2]Saint-Petersburg Mining University, Department of Electroenergetics and Electromechanics, 21 Line, 2, St. Petersburg, 199106, Russian Federation
[3]T.F. Gorbachev Kuzbass State Technical University, Department of Mathematics, 650000, 28 Vesennyaya St., Kemerovo, Russia

Abstract. Mines of Verkhnekamsk potassium-magnesium salt deposit in Perm Krai can be subsumed under shallow mines (depth less than 500 meters). At the present moment in shallow underground mines the are problem of condensate formation in large quantities, when ventilation warm seasons of the year. This problem is more actual for salt mine, where during contact between water and potassium-magnesium ore produced electrolyte, which give rise wear of equipment. For prevent/quantity reduction condensate formation in mine used system of conditioning (refrigerating and dehumidifying) mine air (ACS). However, application this system is limited by reason of tremendous costs of electric energy for their work.

1 Introduction

Famously climatic conditions of mine weather in salt mines are of interest from the point of view of: evaluating their effects on the state of health and working capacity of miners; dynamics of dust content in mine atmosphere; works of ventilation structures. In addition climatic conditions of mine weather in salt mines can be of interest from the point of view of effects of moisture precipitation in the warm season in underground mining workings, because of hygroscopicity, solubility adjacent stratas and mineral product. System of conditioning (refrigerating and dehumidifying) mine air (ACS) can be used on mines for improve sanitation and hygiene standards of works by miners and increase security depth of mining in salt mine [1, 2 et al.]. Herewith the supplied air in ACS must be go through the process of dehumidification, because if relative air humidity 80% and more there is an effect of secondary condensate formation (moisture precipitation is not only in field and in-seam workings of mine, but also in air shaft, because decrease temperature therein [3].

The issue about the use of system of conditioning mine air (ACS) is more actual for under shallow salt mines (depth is less than or equal to 500 meters). This is due to the following reasons. Rock temperature in field and in-seam workings of mine is practically unchanged

* Corresponding author: nikolaev0811@mail.ru

 DOI: 10.1051/e3sconf/20172101022

during lifetime of mine. The value of temperature air supplied to mine in warm season is often far higher than the value of rock temperature. For example, rock temperature in Mines of Verkhnekamsk potassium-magnesium salt deposit is 7-11 ^{0}C, at the same time the average temperature of air supplied to mine is 21,5 ^{0}C with relative humidity of 92% to 95% [3]. Moreover, during the motion in air shaft the temperature of air increases by a few degrees (approximately 1 ^{0}C / 100 meters) by reason of rising barometric pressure [2, 4].

Surface system of conditioning mine air is not applicable in under deep mines (deep is more than or equal to 500 meters), because rock temperature is close to or sometimes higher than temperature of air supplied to mine. At such mining companies for mine-air refrigerating in particular areas mainly used spot air conditioners are located in work places of field and in-seam workings of mine [4]. In this case the problem of moisture precipitation is absent because of heating process (incoming cold air from mine rocks) happens at a constant humidity content [2, 4].

If in under shallow mines ACS is absent then during heat exchange process of air moisture begins accumulating in mines. It happens because value of temperature air supplied to mine is more than the value of rock temperature and also because reducing the relative humidity of air in result his refrigeration. Moisture precipitation may lead to consequences such as: water-flooding a mine workings; pillars disintegration; slip of conveyor belt (with the possibility of burning) etc. In salt mines during contact between moisture and potassium-magnesium ore produced electrolyte which give rise wear of equipment.

However, the use of system of conditioning mine air (ACS) is limited by reason of tremendous cost to electric energy for work these systems. According to the data [5] maintenance cost of surface ACS reach 0,76 mln EURO/year for 1 MW the cooling capacity of the installation.

2 Theory

Authors in patent «RU 140553 Ventilation system at under shallow mine» offers method of conditioning mine air for under shallow mines. This method allows increase energy efficiency of air handling process in warm season. In the proposed method air supplied to intake shaft 1 and 2 must be cooled and drained in surface-underground system of conditioning air (Figure 1). Air-cooler (evaporator) 4 of surface ACS located in surface complex 4 of intake shaft №1, but evaporator of underground ACS located in shaft inset 6 of intake shaft №2. Due to work of main mine ventilation fan (MMVF) 10 located in surface complex of MMVF 11 cooled air enters in blind workings 7 (field and in-seam workings without the way to the surface) and removed through main airways 8 on the air shaft 9 to the surface.

The air is cooled in evaporators, because of refrigerant (refrigerant vapour) circulates in heat-exchange tubes. Refrigerant vapour give the cold to air – is heated and transferred to liquid form. After that heated refrigerantis cooled by condensers.

The heated refrigerant vapour in evaporators of surface and underground ACS comes in condensers 14 and 15 through the pipeline 12, which located in breakthrough 13. Refrigerant vapour is cooled in condensers 14 and 15, because cooling ventilators is working. Air door 16 is located in breakthrough 13 for prevent short-circuiting of air stream between intake shaft 2 and air shaft 9. In result the heated air 17 flows from condensers 14 and 15 to main airways 8 and further in air shaft 9. If there are difficulties associated with laying of pipeline with refrigerant 12 from evaporator of surface ACS, condensers of underground ACS 15 will be only installed in main airways 8.

You must consider that thermal drops of ventilation pressure (natural drafts) h_{ni} (Figure 1) operate between mine shafts during the ventilation of mine. Natural drafts occur in results of convection heat transfer, ie when more warm (more easy) air seeks to rise above, while more cold (more heavy) air seeks to descend down [5-8]. Thermal depressions occur between communicated shafts. Thermal depressions is cause of drops in hydrostatical pressure in shafts and prevent or to help motion of air. Combined action of thermal depressions forms resulting natural draft of mine (h_{ni}). The resulting natural draft of mine influences the mode of main mine ventilation fan (MMVF) depending on direction. The resulting natural draft can either increase or decrease the total air volume supplied to mine. The mode of main mine ventilation fan can be switched to area of less pressures if resulting natural draft is positive. In this case, reduced electric power consumption [2]. You need to consider difference in height grades of shaft mouths and surface leakages of air in shaft house when choosing the regulation mode of main mine ventilation fan subject to the resulting natural draft of mine [9].

The location option of surface and underground ACS is shown in Figure 1. Bleeding of warm air issued from condenser of ACS will increase the value of positive resulting natural draft of mine. This, in turn, will contribute to the reduction of electric power costs for work of ACS [2].

3 Results and discussions

The purpose of this paper is derivation of mathematical functions which allow to choose parameters of underground ACS and determine the effect of this method on the energy efficiency of air handling.

Need to determine the refrigerating power ($q_{c.s.}$) of underground ACS, when choosing an evaporator for underground ACS. The refrigerating power depends on the necessary value of cooling temperature of air in mine working T_a. For determine the refrigerating power ($q_{c.s.}$) can then be used a formula: [10, 11]:

$$q_{c.s.} = c_V\left(\frac{\partial\rho_a T_a}{\partial t} + \frac{1}{S_w}\frac{\partial\rho_a T_a Q_a}{\partial z}\right) - S_w\left[\frac{\lambda_{rm}\tau\sqrt{1+0,76\sqrt{\frac{a\tau}{r_w^2}}}}{0,5\sqrt{\pi a\tau}+\frac{2\lambda_{rm}}{\alpha_0\sqrt{\pi}}\sqrt{1+0,76\sqrt{\frac{a\tau}{r_w^2}}}}\right]\frac{(T_r - T_a)}{3600}, J \quad (1)$$

where c_V – specific heat capacity of air at constant volume, J/kg·^{0}C; ρ_a (t, z) – air density in working the necessary value of cooling temperature of air T_a, kg/ cubic meter; $S_в$ – cross-sectional area of working, square meter; Q_a – required capacity of evaporator, cubic meter/sec.; λ_{rm} – thermal conductivity coefficient of rock massif, J/kg·hour·^{0}C; τ – time since the start of the installation, hour; a – thermal diffusivity coefficient of rock massif, square meter/hour; α_0 – heat transfer coefficient (total), J/square meter hour·^{0}C; $r_в$ – equivalent radius of working, the value of which can be found as $r_w = \sqrt{\frac{S_w}{\pi}}$, meter; T_r –temperature of non-refrigerated mine rocks (see Figure 2).

Cool air temperature must be less than or equal to temperature of non-refrigerated mine rocks ($T_a \leq T_r$) in order to ensure that moisture does not fall out on rock massif walls of field and in-seam workings.

Also in this method the possibility of secondary condensate formation is excluded, because of bleeding of warm air issued from condensers of underground and/or surface ACS (heating temperature of air supplied in air shaft is higher than dew point temperature) [2,3].

Thus, the value T_a will be determined depending on the magnitude of temperature of non-refrigerated mine rocks T_r. According to [4] T_a depends from such parameters as: depth of mine (H_m), temperature of neutral zone ($T_{n.z.}$) and stratification depth of neutral zone ($H_{n.z.}$), geothermic gradient (σ):

 DOI: 10.1051/e3sconf/20172101022

$$T_a = T_r = T_{\text{n.z.}} + \sigma\left(H_m - H_{\text{n.z.}}\right) \tag{2}$$

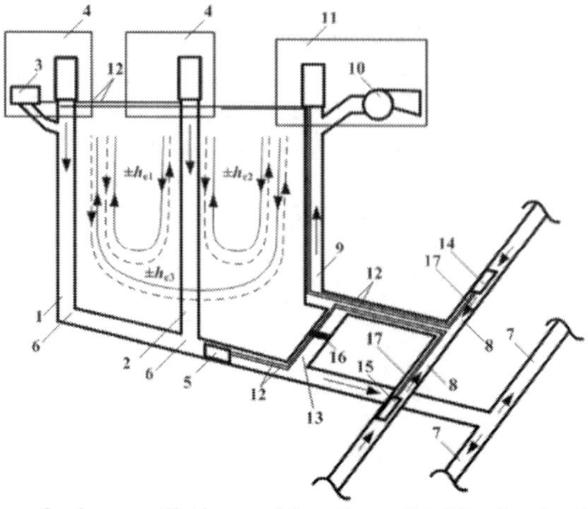

Fig. 1. Simplified scheme of mine ventilation and location of ACS : 1 – intake shaft № 1; 2 – intake shaft № 2; 3 – evaporator (air-cooler) of surface ACS; 4 – surface complexes of intake shafts; air supplied in mine; 5 – evaporator (air-cooler) of underground ACS; 6 – shaft inset of intake shaft; 7 – field and in-seam workings without the way to the surface; 8 – main airways; 9 – airshaft; 10 – main mine ventilation fan (MMVF); 11 – surface complex of MMVF; 12 – pipeline with refrigerant; 13 – breakthrough; 14 – condenser of surface ACS; 15 – condenser of underground ACS;16 – air door; 17 – air heated in condensers 14 and 15

However, keep in mind the fact that in field and in-seam workings without the way to the surface 7 in addition to air cooled in evaporator of underground ACS, air will be flown from intake shaft № 1 and №2. This air will be heated during the descent on shaft. The heated air from intake shaft №1 will be flown in shaft inset of intake shaft 6, in evaporator 5 and workings 7. Part of the air supplied on intake shaft №2 will go on workings without the way to the surface past the evaporator 5 (Figure 1).

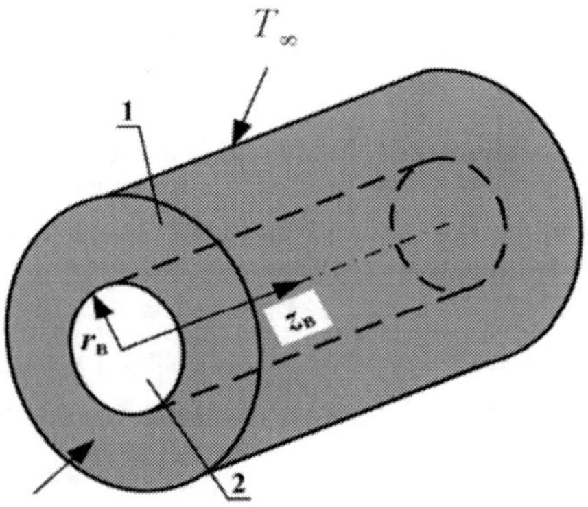

$$T_{\text{п}} = \left(\alpha T_{\text{в}} - \lambda_{\text{гм}} \frac{\partial T_{\infty}}{\partial r_{\text{в}}}\right) \Big/ \alpha$$

Fig. 2. Scheme to determine the amount of heat emitted from rock massif: 1 – rock massif; 2 – mine working with equivalent radius r_w; α – heat transfer coefficient between mine air and rock massif; T_∞ – temperature of mine rocks on removal from the location of rock massif

According to the output from the formulas of paper [2] the required evaporator capacity of underground ACS can be determined by the formula

$$Q_a = \frac{Q_p \rho_p I_p - Q_{s1} \rho_1 I_1 - Q_{s2} \rho_2 I_2}{\rho_a I_a} \tag{3}$$

where Q_p, Q_{s1}, Q_{s2} – air-flow rates by air, which enters accordingly in field and in-seam workings without the way to the surface, intake shafts №1 and №2, cubic meter/sec.; ρ_p, ρ_1, ρ_2 – density of the air, which enters accordingly in accordingly in field and in-seam workings without the way to the surface, in shaft insets of intake shafts №1 and №2, kg/ cubic meter; I_p, I_1, I_2 – specific enthalpy of air, which enters accordingly in accordingly in field and in-seam workings without the way to the surface, in shaft insets of intake shafts №1 and №2, kJ/kg; I_a – specific enthalpy of air on output of ACS, when the value T_a is required

If the values of temperature are known then the density of air in the i-th point can be defined according to the well-known formula:

$$\rho_i = \frac{0{,}465P_i}{273{,}15 + T_i} \qquad (4)$$

According to [9] barometric pressure and temperature of air in shaft insets of intake shafts can be found by the formulas:

$$P_{i.s.} = a_1 \cdot P_a^{n_1} + a_2 \cdot T_{sh}^{n_2} + a_3 \cdot H_{sh}^{n_3} + a_4 \cdot D_{sh}^{n_4} + a_5 \cdot Q_{sh}^{n_5} \qquad (5)$$

$$T_{i.s.} = a'_1 \cdot P_a^{n'_1} + a'_2 \cdot T_{sh}^{n'_2} + a'_3 \cdot H_{sh}^{n'_3} + a'_4 \cdot D_{sh}^{n'_4} + a'_5 \cdot Q_{sh}^{n'_5}, \qquad (6)$$

where a_i, a'_i, b_i and b'_i – regression coefficients [9]; n_i, n'_i, m_i и m'_i – the degree of the parameter [9]; T_{sh}, H_{sh}, D_{sh} and Q_{sh} – accordingly temperature of air, which enters in intake shaft (^{0}C), depth (meter), diameter (meter) of intake shaft and volume of air, which enters in intake shaft, cubic meter/sec.

It should be noted that in formulas (5) and (6) for intake shaft for the value T_{sh} taken air temperature after mixing flows of atmospheric and cooled in evaporator of surface ACS

According to [4] specific enthalpy in the i-th point can be defined by the formula:

$$I_i = 1{,}005T_i + (2500 + 1{,}8T_i)\frac{d_i}{1000} \qquad (7)$$

where d_i – temperature (^{0}C) and moisture content (g/kg) of air in the i-th point.

According to [4] moisture content of air in the i-th point can be found by the formula:

$$d_i = 623\frac{\phi_i P_{t.p.}}{P_i - \phi_i P_{t.p.}} \qquad (8)$$

where φ_i – relative air humidity air in the i-th point; $P_{t.p.}$ – partial pressure of water vapor at the total saturation (millimeter of mercury).

Note that the evaporator of underground ACS will work as installation with ejection: when air stream on output of underground ACS revealed at a certain angle θ and fills all drift size. It's like a work of pneumatic piston (Figure 3).

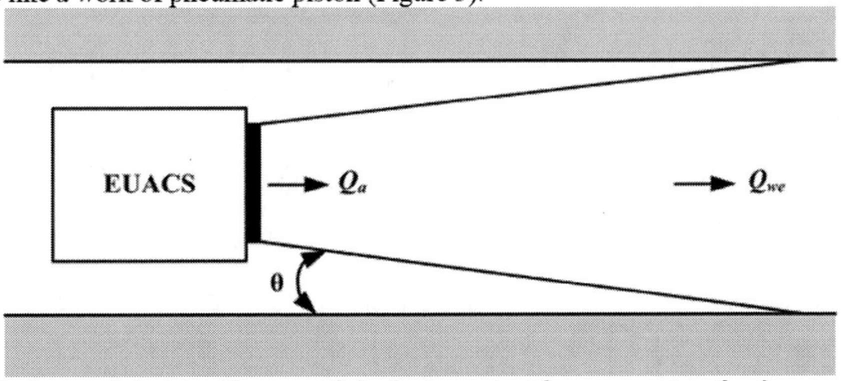

Fig. 3. Disclosure of air stream (process of ejection) on output from evaporator of underground ACS (EUACS)

4 Conclusions

1. The described method of conditioning the air will allow to cool mine air to a temperature of rocks. In result will decrease electric power costs for work of ACS.

2. The bleeding of warm air in the air shaft will allow to exclude possibility of occurrence of secondary condensation in it. Also the process of bleeding warm air in the air shaft will contribute to a observances resulting positive natural draught in mine (thermal depression). As a result in mine will flow more air. In this case, the mode of operation MMVF can be switched to an area of lower pressure. The electric power costs for work of MMVF will be decrease.

3. According to the above order of calculation you can select the required capacity of the evaporator fans and condensers of underground and/or surface of ACS and evolving process of ejection, which increases the total air volume in the working. For work installations also need less electricity.

In the course of calculations it is established that the proposed system will pay off (considering the modern prices for equipment and electricity) for 3.5 – 4 years (on the example of the mine BKPRU-2 of PAO "Uralkali").

Taking the above into account in the implementation of conditioning mine air by the proposed method significantly reduce the cost of electricity for operation of ACS and MMVF. In the mine will stop (significantly reduced) the process moisture precipitation and associated with them the negative and dangerous phenomenon. Held, in the course of mathematical modeling in Solidworks software package flow simulation, research have a high convergence with the results obtained analytically.

References

1. M.J. McPherson, G. Robinson, Mine vent. South Africa, **33**, 145 (1980)
2. A.V. Nikolaev, Gornyi Journal, no. 3, 71 (2017)
3. S. Knothe, B. Nowak, M. Szlazak, Arch. Gorn, **31**, 289 (1986)
4. S.G. Gendler, Eurasian Mining, **2**, 41 (2016).
5. G.B. Lyal'kina, A.V. Nikolaev, Journal of Mining Science, **51**, 342 (2015)
6. W.E. Bruce, *Natural draft: its measurement and modeling in underground mine ventilation systems* (US: Dept. of Labor, Mine Safety and Health Administr., 1986)
7. P.F. Linden, Annual Review of Fluid Mechanics, **31**, 201 (1999)
8. Jianwei Cheng, Yan Wu, Haiming Xu, Jin Liu, Yekang Yang, Huangjun Deng, Yi Wang, Tunneling and Underground Space Technology, **45**, 166 (2015)
9. N.I. Alymenko, A.V. Nikolaev, Journal of Mining Science, **47**, 636 (2011)
10. L.Yu. Levin, M.A. Semin, A.V. Zaitsev, Journal of Mining Science, **50**, 371 (2014)
11. B.P. Kazakov, A.V. Shalimov, M.A. Semin, International Journal of Heat and Mass Transfer, **86**, 288 (2015)
12. A. Shadrina, L. Saruev, International Journal of Mining Science and Technology, **24(2)**, 245 (2014)

Knowledge Assessment Software in Mining Specialist Training

Vladimir Lebedev[1], and *Olga* Puhova[1]

[1]Tver State Technical University, A. Nikitin Street 22, 170026, Tver, Russia

Abstract. The article reviews the knowledge assessment software module of electronic teaching and testing in mining specialist training. To develop the software module integrated programming environment state-of-the-art is used. Its advantage consists in small computer resource consumption, simple editing, and protection against the users' trying to find out the correct answers to test tasks. The software makes it possible to learn the ongoing learning information systematically and consistently as well as to assess the current knowledge in mining. The developed module meets the following requirements: a software module user-friendly interface, the storage of passed test results to be used for subsequent viewing, analyses, and evaluation, fast troubleshooting in case of any troubles with a stable module operation, and further software function extension and upgrading.

1 Introduction

To ensure the competitiveness and employability of young mining specialists of graduate open-cast education the quality of higher education should be improved. The modernization of Federal State Educational Standards calls for some software [1-3] meant for studying technological and industrial material independently [4-10] with the test of the knowledge gained to follow up and the human-generated interference to be eliminated.

When testing in a classical way an examiner is likely [11] to misunderstand the answer or not to take the correct answer in the account due to some possible nuances. As a result the test will not be passed in spite of correct answers. Software is free from such misunderstanding. Another important advantage of the software module is the possibility to calculate test results automatically and store all the information for subsequent analysis and qualitative assessment.

In line of rapid implementation and development of modern information technologies [12-14] the introduction and use of such software has many significant advantages. You can:

- study new materials prepared by a lecturer systematically.
- test and assess students' knowledge distantly.
- learn some information independently.
- reduce the time of current knowledge assessment considerably.

2 Materials and methods

To develop the software of students' knowledge assessment we used the state-of-the-art integrated programming environment such as Microsoft Visual Studio and EmEditor since they consume small computer resources and have reliable protection against trying to crack program source files. Visual Studio provides the development of various applications with a wide variety of functions and GUI functionalities. Text editor EmEditor allows different file formats to be edited and supports a wider range of functions than the standard means of Windows.

To implement the interface of electronic knowledge assessment software we choose the standard interface of Windows applications implemented with Windows Forms. The application advantages are as follows:

- software development ease and convenience
- typical controls
- small memory requirements which provides the opportunity to install the software on different computers.

For convenience the software module has several functional windows:

- a main window of students' training
- a window of users' authorization
- a window of students' knowledge testing
- a window of test result logging.

Such structure reduces the number of controls in each window and allows software functions to be distributed in order to ease and simplify the software learning by the users.

Each application window is made with individual Windows form.

The main controls used in developing the software were chosen as follows: Button, Label, Treeview, TextBox, RichTextBox, NumericUpDown, RadioButton, ListBox.

The authorization window allows a student to enter their personal data (a full name, a student group) so that the application module records the data in the result log after testing to maintain test result reports. The size of the window can be neither changed nor closed with standard ControlBox. The students are signed in only once when launching an application, with the data being kept as long as the program is active.

The window of students' testing allows the level of knowledge after completing the material study to be checked. Each time when finishing the test the program enters the result in the general table.

The testing window provides the user with the following functions:

- To choose one of the several tests proposed.
- To view the general test result table.
- To move to other application windows or to exit the program.

Being user-friendly the window displays the question number and the section title of the test being passed.

When the window of the electronic testing system is downloaded, the code executed initially is the following:

```
public Form2()
{
InitializeComponent();
label1.Visible = false;
button1.Enabled = false;
radioButton1.Visible = false;
radioButton2.Visible = false;
radioButton3.Visible = false;
radioButton4.Visible = false;
radioButton5.Visible = false;
button6.Enabled = false;
```

```
}
private void Form2_Load(object sender, EventArgs e)
{
NewFile = File.ReadAllLines(Application.StartupPath + «\\config\\2.dat»);
}
```

The code prevents visibility or access to some controls unused at the moment. Function Form_Load downloads the data of folders storing the question and answer option files into array NewFile from the file. This allows new tests to be added and available tests to be edited.

The array is a data structure of some number of variables accessed by indices computed. The array variables are its elements and have the same type called a type of array elements. In programming the numeration usually starts with zero while the test numbers start with one. Thus the users were restricted to select the test number not less than one and not more than the total test number is.

The code uploads the titles of main training sections represented in a tree-like structure from the main application window table of contents (Fig. 1). Test titles are on the main (zero) level of a tree structure.

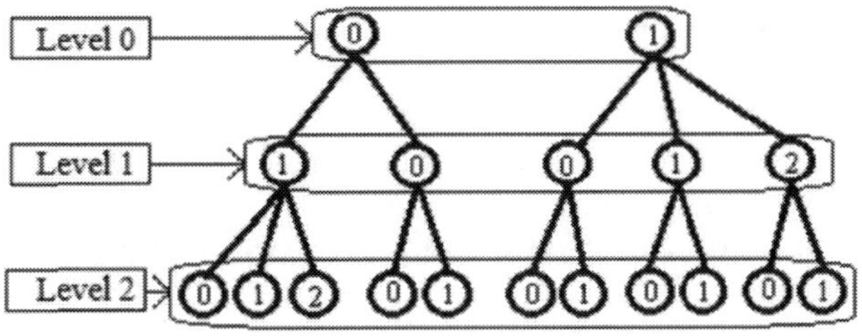

Fig. 1. Tree structure levels

The code downloads the total list of questions and displays one of them. A random sequence of questions is generated with random numbers.

For the user's convenience the window displays a question number and a section title being tested. The test number is selected with left corner buttons. Button Next Question at the bottom of the testing window uploads the next question from the test. загружает следующий вопрос из теста. At the end the program displays the appropriate message and writes the data obtained.

A result logging window displays the total list of all the tests ever completed.

The window is intended for a lecturer conducting testing and shows the test results with a comprehensive set of data related to each one being tested. The window is of a static size and has only one close button.

To log test results class Filestream is used. It represents Stream in a file and supports synchronous and asynchronous read and write operations, file opening and closing operations in a file system, and the change of other operating system descriptors when processing files. To achieve the best performance FileStream buffers I/O. It is important to remember that after the work with them the data streams should be closed to free the memory from their children. This is particularly important given the work with several children. The streams are closed with the following code:

```
srFromFile.Close();
fsFile.Close();
```

When the form is loaded, the code executed is the following:

```
string strText = ««;
FileStream fsFile = new FileStream(Application.StartupPath + «\\config\\res.dat», FileMode.OpenOrCreate, FileAccess.Read);
```

```
StreamReader srFromFile = new StreamReader(fsFile);
while (!srFromFile.EndOfStream)
{
strText = srFromFile.ReadLine();
listBox1.Items.Add(strText);
}
srFromFile.Close();
fsFile.Close();
```

The data read from the file are displayed string by string with element ListBox. The information is written in one string so there is no necessity to organize a big database to write and store test results.

3 Results and discussion

The test includes the tasks of various types, e.g. multiple choice, an answer keyboard input, matching, ordering, classifying. In online testing a tester sees the details on every tested student's progress on their computer. After the test the data are archived so that test results can be viewed and analyzed with the embedded application means. The application provides for test generation as offline executable files which can be distributed for testing offline and without storing the results. The mode is intended, first of all, for self-examination. To begin testing, simply start the output file on any computer with Windows OS, the installation of any applications is not required.

When a test type is selected and an answer for the first question is given, a user should click the button to choose the next question. The code being executed is the following:

```
RadioButton[] RB = { radioButton1, radioButton2, radioButton3, radioButton4,
radioButton5 };
foreach (RadioButton rb in RB)
{
if (rb.Checked && rb.Text == right)
{
res++;
}
}
number++;
if (number < ansnum)
{
count++;
label3.Text = «Question: « + Convert.ToString(count) + « from « + ansnum;
if (count == ansnum)
{
button1.Text = «End Test»;
}
RND = new Random();
int k = RND.Next(0, L.Count);
Find(L[k]);
L.Remove(L[k]);
}
else
{
MessageBox.Show(«No more questions»);
number = 0;
```

```
double allans = ansnum;
double endres = res / allans * 100;
endres = Math.Round(endres, 2);
string test;
string verdikt;
if (endres >= 80)
{
test = «You passed the test»;
verdikt = « (The test is done)»;
}
else
{
test = «You did not pass the test»;
verdikt = « (The test is not done)»;
}
}
```

First a programmer checks if RadioButton has been chosen. RadioButton corresponds to the correct answer and increases the value of the variable meaning the quantity of correct answers. Then a new question is generated with function Find. When all the questions are answered, the application calculates the user's result and displays through MessageBox a message about the percentage and informs if the threshold of 80 % is passed which means that the test is passed.

When completed the application also writes the full data on the test passed into a file which can be viewed in the result table window. The application executes the following code:

```
resultate = «Student: « +Form3.fam + « « + Form3.name + « « + «Grope: « +Form3.group + « «;
resultate = resultate + «Test Number: « + Convert.ToString(numericUpDown1.Value) + « «;
resultate = resultate + «Result: «+ Convert.ToString(endres) + «%» + verdikt +» Date: « + DateTime.Now;
FileStream swFile = new FileStream(Application.StartupPath+»\\config\\res.dat», FileMode.Append, FileAccess.Write);
StreamWriter sw = new StreamWriter(swFile);
sw.WriteLine(resultate);
sw.Close();
```

The information obtained can be viewed after finishing the test by clicking Test Result Table.

4 Conclusions

The developed software module has a simple, convenient, and plain user interface with distributed functionalities. The software provides the possibility of fast editing and has the following functions:

- To show text and graphical information to study systematically.
- To authorize software users.
- To assess students' current knowledge through automated testing.
- To eliminate the human-generated interference when the test is being passed.
- To store test results through logging for subsequent viewing and analyzing.

All the tests and test results are coded which excludes the possibility of fraud and falsification completely. There is also a possibility of determining the ceiling of one test passing number which raises the students' testing standards. The software module does not require large memory which allows it to be installed on any modern computers. There is also a possibility of quick troubleshooting and resolving any issue.

The software module of mining students' knowledge assessment (graduate education program 21.05.04 Mining, Open-cast specialization) has been successfully piloted at the Chair of Geotechnology and Peat Production, Tver State Technical University. The software of electronic testing system has received the Certificate of State Registration.

References

1. N. Nagashree, N.V. Pujari, Comp. Hum. Behav., **65**, 1-131 (2016)
2. E. Sarmiento Leite, E. Almentero, Sotomayor Alzamora, El. Not. Theor. Comp. Sci, **329**, 123-148 (2016)
3. T. Wang, Z. Zhang, X. Jing, L. Zhang, Autom. Soft. Engin., **23:4**, 569-590 (2016)
4. A.E. Afanas′iev, S.N. Gamayunov O.S. Misnikov, Colloid J., **61:3,** 274-279 (1999)
5. O.S. Misnikov, O.V. Dmitriev, V.I. Popov, E.Yu. Chertkova, Polym. Sci. D, **9:1**, 133-139 (2016)
6. O.S. Misnikov, O.V. Dmitriev, E.Yu. Chertkova, Euras. Min. Gornyi Zhurnal, **2:24**, 30-34, (2015)
7. O. Misnikov, Polym. Sci. D, **7:3**, 252-259 (2014)
8. O.S. Misnikov, E.Yu. Chertkova, Euras. Min. Gornyi Zhurnal **1:21,** 63-68 (2014)
9. O. Misnikov, A. Timofeev, O. Pukhova, Polym. Sci. D, **8:1,** 66-74 (2015)
10. O. Misnikov, Mires and Peat, **18:22**, 1-15, (2016)
11. K. Eck, G.P. Alleman, V. Quick, J. Martin-Biggers, N. Hongu, C. Byrd-Bredbenner, J. Com. Health **41:6**, 1187-11951 (2016)
12. F.N. Abu-Abed, D.V. Martynov, A.V. Ivanova, R.V. Dopira, R.Y. Kordyukov ARPN J. Eng. App. Sci., **11:16**, 9636-9645 (2016)
13. B.V Palyukh., V.K. Ivanov, I.A. Egereva, IJAER, **10:24**, 45736-45740 (2015)
14. Y.N. Matveev, B.V. Palyukh, N.A. Stukalova, V.N. Bogatikov, IJAER, **10:24**, 45717-45723 (2015)

Modeling of Three Flat Coal Seams Strata Developing at Open Pit Miming

Tatiana Gvozdkova[1,*], *Sergey* Markov[1,2], *Nuray* Demirel[3], and *Serony* Anyona[4]

[1]Kuzbass State Technical University, Mezhdurechensk branch, Mezhdurechensk, 652881, Stroitelei Avenue, 36, Russia
[2] T.F. Gorbachev Kuzbass State Technical University, 650000, 28 Vesennyaya St., Kemerovo, Russia
[3] Mining Engineering Department, Middle East Technical University, Üniversiteler Mah., Dumlupınar Bulvarı, No: 1, 06800 Çankaya, Ankara, Turkey
[4] Jomo Kenyatta University of Agriculture & Technology, Mining, Materials & Petroleum Engineering Department, P.O. Box 62000 – 00200, Nairobi, Kenya

Abstract. The use of low-cost direct dumpling technology, as is well known, has a relatively limited field of application: flat coal seams, and the higher the dip angle of the seam, the more difficult it is to place the necessary volume of overburden rock in the dumping layers. For this, we have to pour four-tier dumps. In this article, four possible options for piling the dump have been studied and prerequisites have been made for further research aimed at improving the efficiency of the use of direct dumpling technology in the development of flat coal seams.

1 Introduction

At present time economic development of Kuzbass coal industry is the key to the neo-industrial transformation of such mineral resources extracting region. Development of coal mining must accelerate making deep processing of coal and raise the common technological level of regional industry [1-4]. For Kuzbass, the biggest Russian coal basin (located Western Siberia) development of coal mining means expansion of open pit mining.

However when open pit mines increase in numbers it means growth of the area of land occupied by quarry fields, additional load on environment, mainly on air and water sources. This can be reduced by backfilling internal dumps with the change of surface mining technology from transport to direct dumping. The increase in the volume of the rock, moved into the internal dump, will improve the economic indicators of open pit coal mining [5-8]. In addition, because of the flat bedding of the coal seams of open pits of South Kuzbass (Western Siberia, Russia), the height of the rock mass, which is processed by transport technology, is constantly increasing [9-10]. At present, this has led to a decrease in the speed of moving the front of mining work of the transport zone and, as a result, to the inhibition of the development of the transportless [11]. The redistribution of volumes of overburden between zones in the direction of increasing the volume of direct dumpling technol-

[*] Corresponding author: kuzstu@kuzstu.su

ogy will increase the speed of movement of the transport zone (while maintaining the same complex of mining and transport equipment).

Thus, the increase in the volume of the rock moved into the internal dump in these deposits contributes to the improvement of the seam development indices, and the solution of this issue is an actual task. Also the transport technology change to direct dumping can contribute to the improvement of environmental problems caused by surface mining like air and water pollution [12-16].

A particular feature of this study is the analysis of a group of deposits that are complex in terms of mining and geological conditions and are systematized in terms of τ, instead of observing a separate deposit, represented by strata of three flat coal seams. Consequently, the solution of the task based on the justification of the structure of the excavation schemes is not for any separate strata, but for a group of stratas, which implies the multivariance of the solution.

Such research object requires an appropriate methodological approach to the solution of the task.

2 Materials and Methods

The main direction to increase the competitiveness of Kuzbass coals produced by the open pit miming is to reduce overburden costs. On the open pits of Southern Kuzbass, the stratas of flat coal seams are developing with direct dumpling technology for overburden mining.

At present, in scientific research, modeling is used with sufficient confidence to predict the parameters of technological processes.

As is known from theory, modeling is the replacement of the study of an object under observation in nature by a similar study of it on a model. In turn, the model is a system of thoughts fixed on paper in the form of drawings, graphs, formulas, equations, etc., and is characterized by the fact that the unity and integrity of each such system is due to the relationship of properties and interrelations in the object under study.

Any model has some ideal content. At the same time, the spatial properties of the object, the interaction and interconnection of parts, etc. should be transferred to the model. Models can be ideal, not completely reflecting reality, and material - in the form of machines, mechanisms, etc.

Proceeding from these methodological provisions, an ideal grapho-analytical model for the development of a series of three flat coal seams is proposed to solve the scientific task.

3 Results and Discussion

The initial position of the model consists of two elements: the face and the dump sides (Figure 1).

In Fig.1 the following legend is accepted: A – the width of the excavator (dragline) stope, m; R –digging radius, m; R_p – downloading radius, m.

The face side of the model includes options for developing a strata of three flat coal seams. The idealization of this element of the model is as follows. The first (lower) and third (upper) seams are stationary; the distance between them (from the soil of the upper seam to the roof of the lower one) is constant and equals to 52 m, which is accepted according to the statistical analysis of bedding of the considered stratas [17]. The middle seam has a variable position, which is fixed by the value of the exponent τ at a given thickness or lower interbed (H_1), or upper (H_2). The height of the bench marking (H_3) is set in the range from 10 m to $H_{c.max}$, where $H_{c.max}$ is determined by the formula, m:

$$H_{c.\max} = H_{s4} - H_1 - H_2 \tag{1}$$

where: H_{s4} – total height of the developed rock mass, determined by the maximum capacity of the dump, m.

The coal seams' thickness m_i (m_1, m_2, m_3) and their dip angle (φ) are determined from the results of a statistical analysis of the six stratas [18] and are calculated by the formula (2):

$$m_i = \frac{1}{6} \cdot \sum_{n=1}^{n=6} \overline{m}_{i.n} \tag{2}$$

where: the sequence number of the formation from the bottom up (i = 1; 2; 3); n - serial number of the suite according to the developed systematization;

$\overline{m}_{i.n}$ – the average value of the thickness of the i-th seam, obtained during a statistical analysis of bedding parameters, m;

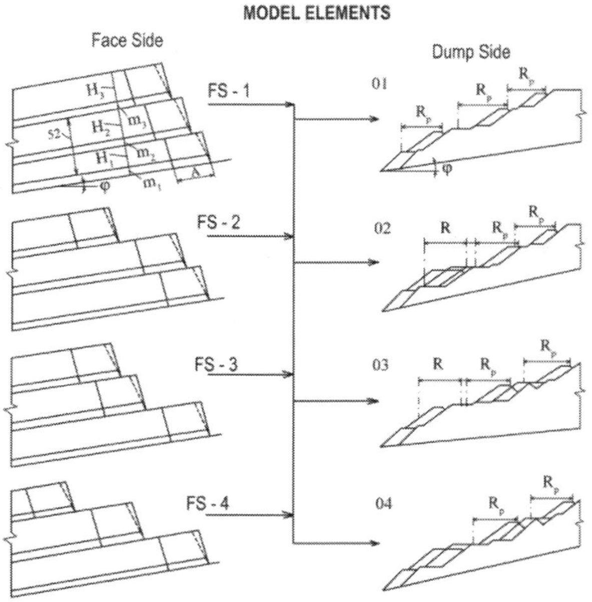

Fig. 1. Structure of the grapho-analytical model for the development of three flat coal seams strata, taking into account the different dumping schemes.

The average angles of bedding are in the range of 6^0-10^0, so the modeling of development of the flat coal seamsof strata is made within these limits.

The dump side of the model includes four schemes of dumping, according to their systematization. The idealization of this part of the model is as follows. The full use of the working parameters of the draglines during the backfilling the dump's tiers is accepted, and the images of the elements of the dump and the dumping tiers are taken by regular polygons.

The proposed structure of the graph-analytical model is subject to the requirement of a methodical approach to solving mining problems of open pit mining for the movement of rock masses in the space.

In this study, the solution to the problem of developing the structures of excavation schemes is connected with the analysis of transshipment of the rock mass by a dragline excavator from three overburden benches to dumping tiers. Such problems belong to the class of problems of dynamic geometry and are solved by dividing the process of transshipment of the rock into separate static stages; each of them is solved by methods of classical geometry.

With this methodological approach, the face and dump sides in the proposed model are the initial and final static stage. The dynamics of moving rock between these stages will depend on the way the side is developed, the bedding parameters of the coal seams, the width of the drilling and blasting stope, the type of the dump formation scheme, the working parameters of the dragline, which will determine the structure of the excavation scheme.

One of the tasks of the excavating scheme selection is to identify typical schemes. In this regard, it is necessary to clarify the notion of a "typical excavation scheme".

The whole process of extracting the overburden rock and moving it to the inner dump is, according to the general theory of open pit mining, an excavation scheme. Schemes of excavation differ in their structure by the number of stages of development and transshipment of the overburden rock, as well as by technological interconnection between them (in different mining and geological conditions or under the same conditions) when the locations of the dragline installation are changed in the profile of the face.

If the number of stages and the technological connection between them are the same, but the geometric shape of the stages and their sizes may differ, then such excavation schemes can be defined as monotype. In the paper, based on the construction of a number of options for excavation schemes, typical schemes are used.

4 Conclusion

According to the proposed structure of the model, each element of the face side (the initial stage of the excavation scheme) can interface with any element of the dumping site (the final stage of the excavation scheme). In addition, by changing the position of the middle seam on the geological structure of the face side, it is possible to study the relationship between the conditions of seams' bedding in the strata and the method of its development, taking into account any of the heaping patterns.

Thus, the model allows assigning any number of variants of excavation schemes for interconnection of the main influencing factors.

References

1. Zhironkin, S.A., Ugol', **6**, 62-63 (2001)
2. A.V. Myaskov, A.S. Ilyin, S.M. Popov, Gornyi Zhurnal, **2**, 51-56 (2017)
3. E. Dotsenko, N. Ezdina, E3S Web of Conf., **15**, 04012 (2017)
4. M.A. Gasanov, K.A. Kolotov, K.A. Demidenko, E.A. Podgornaya, O.V. Kadnikova, IOP Conf. Ser.: Earth Environ. Sci., **50:1**, 012025 (2017)
5. M. Cehlár, L. Mihok, Theor. Emp. Res. Urb. Man., **8:4**, 60-72 (2013)
6. Z. Jurkasová, M. Cehlár, S. Khouri, Int. Conf. on Engin. Sci. and Prod. Man., ESPM, 409-412 (2015)
7. M. Cehlár, J. Janočko, Z. Šimková, T. Pavlik, E3S Web of Conf., **15**, 01019 (2017)
8. M. Cehlár, Z. Šimková, Economics and Innovation Management, **1**, 52-65 (2017) DOI: 10.26730/2587-5574-2017-1-52-65
9. M. Tyulenev, S. Zhironkin, E. Tyuleneva, A. Abay, S. Anyona, M. Hellmer, Coal Int., **265:3**, 30-34 (2017)

10. S. Markov, M. Tyulenev, O. Litvin, E. Tyuleneva, E3S Web of Conf., **15**, 01011 (2017)
11. M.A. Tyulenev, S.A. Zhironkin, O.I. Litvin, E.A. Tyuleneva, O.V. Zhironkina, S.O. Markov, Geotech. Geol. Eng. **35:5**, 2065–2077 (2017)
12. M. Tyulenev, Y. Lesin, E. Tyuleneva, E. Murko, E3S Web of Conf., **15**, 02003 (2017)
13. M. Tyulenev, E. Garina, A. Khoreshok, O. Litvin, Y. Litvin, E. Maliukhina, IOP Conf. Ser.: Earth Environ. Sci., **50:1**, 012035 (2017)
14. Tyulenev, M., Zhironkin, S., Kolotov, K., Garina, E., Poll. Res. **35:2**, 221-227 (2016)
15. M.I. Agienko, E.P. Bondareva, G.V. Chistyakova, O.V. Zhironkina, O.I. Kalinina, IOP Conf. Ser.: Earth Environ. Sci., **50:1**, 012022
16. M.A. Tyulenev, T.N. Gvozdkova, S.A. Zhironkin, E.A. Garina, Geotech. Geol. Eng., **35:1**, 203-212 (2017)
17. T. Gvozdkova, E. Kuznetsov, A. Rudakova, S. Markov, E3S Web of Conf., **15**, 01008 (2017)

Parameters of Transportation of Tailings of Metals Lixiviating

Vladimir Golik[1,2,*], and *Yury* Dmitrak[1]

[1]North-Caucasian State University of Technology, 44 Nikolaev street, Vladikavkaz, 362021, Russia[1]
[2]Vladikavkaz Scientific Center of the Russian Academy of Sciences and the Government Republic of North Ossetia-Alania, 22, Markusa street, Vladikavkaz, 362027, Russia

Abstract. The article shows that the change in the situation in the metals market with a steady increase in production volumes is intensified against the tendency of the transition of mining production from underground mining to underground mining for a certain group of ores. The possibility of a non-waste metals extraction from not only standard, but also from substandard raw materials, is currently provided only by technology with the lixiviating of metals from developing ores. The regular dependences of the magnitude of hydraulic resistances on the hydro-mixture velocity and its density are determined. The correct values of the experimental data convergence with the calculated values of these parameters are obtained. It is shown that the optimization of the transportation parameters of lixiviating tailings allows reducing the level of chemically dangerous pollution of the environment by leachate products. The direction of obtaining the ecological and technological effect from the use of simultaneously environmental and resource-saving technology for the extraction of the disclosed metals is indicated.

1 Introduction

The development of the human society is accompanied by a steady increase in metals producing to meet the need for consumption. For example, over 100 years the consumption of non-ferrous and alloying metals has increased fivefold.

With the development of technological capabilities and the improvement of mining technology, low-grade and complex ores that require pre-concentration are increasingly involved in production. The lag in the possibilities of ore processors from the opportunities of its miners is evidenced by the increase in tailings processing on the surface and the volumes of substandard ore reserves lost in subsoil.

It is generally recognized that the complete elimination of the tailings of heavy metal ores processing is a radical measure of reducing the global risk for the environment of the Earth's biosphere by the toxic components.

* Corresponding author: v.i.golik@mail.ru

The environmental concept for the extraction of metallic raw materials is based on the fact that, since there is no way to prevent damage to flora and fauna, tailings should be disposed properly.

The condition of the metal raw materials base allows us to summarize that Russia's supply of non-ferrous metals is insufficient, and the most of the non-ferrous metals are mined casually, and a significant number of metal ores are exported in the form of raw materials [1-2].

The possibility of wasteless extraction of metals from not only the standard, but also from the substandard raw materials, is currently provided only by the first-step technology with the lixiviating of metals (Fig. 1).

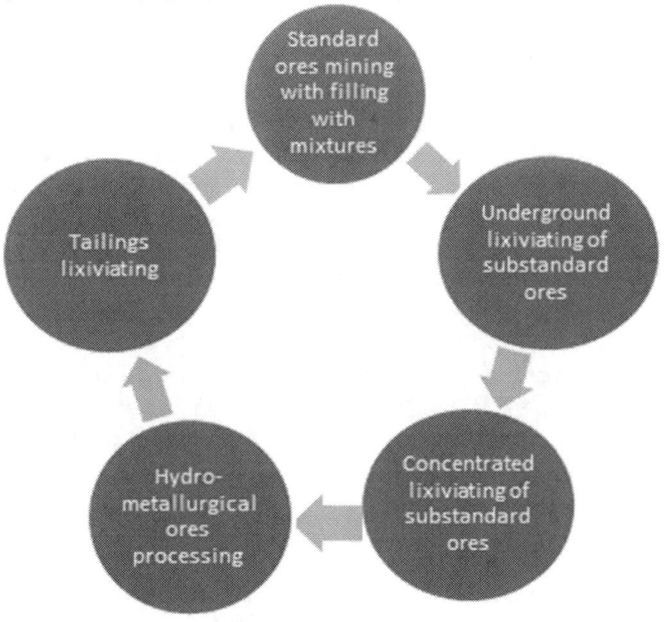

Fig. 1. Options for lixiviating metal from ores.

The concept of obtaining metals by lixiviating includes elements of a single technological cycle, including:

- extraction of relatively standard ores from the subsoil for traditional processing and creation of a compensation space for the removal of other stocks;
- extraction of metals by concentrated lixiviating from mined ores non-conditional for plant processing;
- extraction of metals from current and pre-accumulated tailings of processing and metallurgy with activation of processes by non-traditional methods [3].

Lixiviating produces a significant part of metals, for example, gold, uranium, copper. A large enterprise Priargunsky RPC (Transbaikalia) extracts most of the uranium with different lixiviating options.

The prospect of an increase in metals production by advanced lixiviating technologies forms a new problem handling with the tailings of technological lixiviating.

Technology of delivery and placement of processing tailings includes the hydraulic transport of mineral media at a ratio of solid to liquid (S : L) as 1 : 6, which is primarily done to improve the transportability of the mixture. The wastewater excess is dangerous to the hydrosphere. In mining practice, the density of transported liquid tailings is less than 1.25 t /m3 [4-5].

The mode of operation of the hydro transport system, primarily the flow velocity and the pressure depend on the density of liquid lixiviatings. Empirical methods for determining delivery parameters are imperfect in nature and when solving design problems, they need experimental verification in specific conditions. To achieve the objectives of this study, similar work can be used to optimize the pipeline transport of solidifying filling mixtures [6].

The purpose of research in this area of technology is to increase the reliability of hydro-transportation of condensed pulp to the storage location. The objectives of the study include: studying the composition and properties of lixiviating tailings, determining hydraulic losses and the critical speed of the hydro-mixture and developing recommendations for calculating hydro-transport parameters.

2 Materials and methods

Experimental studies of tailings hydro-transport of lixiviating of gold-bearing raw materials were carried out on a hydro-transport unit including pipelines, a pump, a repulpator, instruments and meters. The researched processes were modeled in a 5.5 m pipeline.

The critical velocity was determined from the flow rate of the hydro-mixture at the time of formation in the pipeline of a fixed layer about 1 mm thick. Three variants of the ratio of the solid component to the liquid component were studied: 1 : 1.0; 1 : 1.5 and 1 : 2.

To determine the base, the values of the parameters of hydro-transport of clean water have been determined. To compare the parameters of hydro-transport of tailings with the process of thickening for S : L = 1 : 6, the possible variants of the ratio of solid to liquid are investigated.

The investigated material is represented by fine sandy (0.1-0.06 mm) and coarse dusty (0.06-0.01 mm) fractions with a plasticity number in the range of 1-7%.

The obtained data were used in mathematical modeling of processes using alternative calculations to develop recommendations in considered case.

3 Results and discussion

3.1 Loss of fall when moving the hydro-mixture

In the course of the studies, the values of fall losses during the motion of hydro-mixtures with different pulp density γ n = 1.11, 1.13, 1.19, 1.22, 1.26, 1.32, 1:40 were obtained.

The method of least squares the dependencies were found (Table 1):

$$100 J_r = a_1 V^2 + a_2 V \quad (1)$$

where: a_1, a_2 – the empirical coefficients;

V – average velocity of the flood, m / sec.

Table 1. Dynamics of hydraulic resistance.

Sample	Density, t/m3	Hydraulic resistance
1	1.0	$100I = 4.334V^2 + 2.158V$
2	1.11	$100I = 4.819V^2 + 2.323V$
3	1.13	$100I = 4.923V^2 + 2.347V$
4	1.19	$100I = 5.199V^2 + 2.385V$
5	1.22	$100I = 5.308V^2 + 2.523$
6	1.26	$100I = 5.577V^2 + 2.384V$
7	1.32	$100I = 5.726 + 2.747V$
8	1.40	$100I = 6.029V^2 + 3.007V$

 DOI: 10.1051/e3sconf/20172101025

The experimental data were compared with the calculated values of the required parameters:

$$J_r = J_b \frac{\gamma_r}{\gamma_b} \quad (2)$$

where: J_b – specific fall loss when moving clean water; γ_r – density of hydro-mixture; γ_b – density of water.

$$J_r = J_b(1 + C_o) \quad (3)$$

The relative density of a solid with Archimedean force (a) can be found as:

$$a = \frac{\gamma_s - \gamma_b}{\gamma_b} \quad (4)$$

where: γ_s – the density of solid particles, t/m^3

The value of hydraulic losses in the transport of clean water by the Darcy method:

$$J_b = \frac{\lambda V}{D^2 g} \quad (5)$$

where: λ – coefficient of hydraulic friction.

The choice of the calculated dependences for λ depends on the ratio of the Reynolds R_e and the limiting values of R_e^1 R_e^2 :

$$R_e = \frac{VD}{\nu} \quad (6)$$

where: ν – coefficient of kinematic viscosity, equal to: 1 x 10^{-6} .

R_e values are in the limits of 75000-150000. The limited values R_e are:

$$R_e^* = \frac{10}{\bar{\Delta}} u R_e^{**} = \frac{500}{\bar{\Delta}} \quad (7)$$

where: $\bar{\Delta}$ – relative roughness (Δ / D); Δ – absolute roughness, (0.4).

Found values of R_e^* and R_e^{**} appeared to be equal to 1250 and 62500 respectively.

For λ calculation the following formula was accepted:

$$\lambda = \frac{8g}{C^2} \quad (8)$$

where: C - Chezy's velocity factor.

For the area of quadratic resistance:

$$C = \frac{R^{\frac{1}{6}}}{n} \quad (9)$$

where: R – hydraulic radius; n – the roughness coefficient of steel pipes (0.0125).

The experimentally determined loss values have a sufficiently high convergence with the calculated values, while the specific loss at a velocity of less than 1.6 m / sec turned out to be higher than the calculated values.

3.2 Critical flow rate of the hydro-mixture

The velocities of movement along the pipeline were determined from the measured flow rate of the hydro-mixture at the time of fixed silt layer formation in the pipeline.

The convergence of the values of obtained critical velocities with the calculated values is given in Table 2.

 DOI: 10.1051/e3sconf/20172101025

Table 2. Dynamics of hydraulic losses during water transportation.

Velocity, m/sec	Chezy's velocity factor,	Coefficient of hydraulic friction	Loss of pressure calculated	Pressure loss experienced	Error, %
1	38.5	0.05	5.8	6.6	-11.7
1.5	-	-	12.1	13.0	-6.8
2.0	-	-	21.5	21.7	-0.6
2.5	-	-	33.6	33.5	+0.5

The experimentally established dependences of pressure losses on the flow velocity are interpreted in Fig. 2.

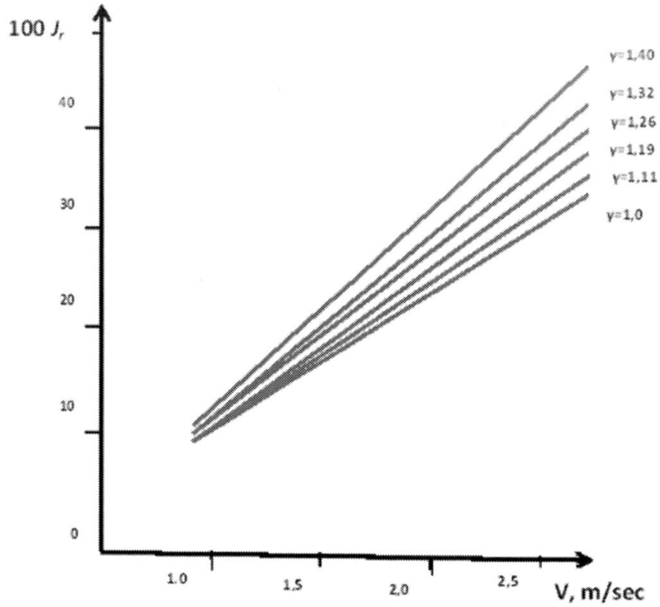

Fig. 2. The graph of the dependence of fall losses on the flow velocity.

The analysis of the results shows that the values of the critical velocities differ from the experimental values by 0-34%. The best convergence of the experimental values of the critical velocities with the calculated values was established using the A.P. Yufin formula.

The established dependencies may be in demand when designing enterprises using options for lixiviating metals from developing ore.

Transportation of tails of non-ferrous metals ores hydrometallurgical processing in the optimum mode of operation of the pipeline expands the field of application of resource-saving technologies for ore deposits development with lixiviating [7-9].

Technology with industrial controlled lixiviating reduces the danger of natural uncontrolled lixiviating, which is carried out spontaneously at mining facilities.

4 Conclusions

1. The danger of chemical contamination of the natural environment by the liquid tailings of lixiviating is reduced by their thickening with increasing density.
2. Optimum parameters of transportation of condensed hydro-mixtures are provided when calculating by the proposed method.

3. Determined dependencies allow optimizing parameters of transportation of lixiviating tailings and to reduce the volume of chemically hazardous lixiviating products with the formation of ecological and technological effect.

Acknowledgement

The article contains the results of researches made under Programme Erasmus + 574061-EPP-1-2016-1-DE-EPPKA2-CBHE-JP "Modernization of geological education in Russian and Vietnamese universities".

References

1. V.I. Golik, Int. Bus. Man. **9:6**, 1209-1216 (2015)
2. G. Wang, R. Li, E.J.M. Carranza, F. Yang, Ore Geol. Rev. **71**, 592–610 (2015)
3. Weijing Wang, Shaofeng Huang, Xiaobo Wu, Qingfei Ma, J. Soft. Eng. App. **4**, 329–334 (2011)
4. Xiao Li-Ping, *Study on Pollution Laws of Coal Gangue Leaching Solution to Ground-water System* (Fuxin: Liaoning Technical University, 2007)
5. I. Onica, E. Cozma, T. Goldan, AGIR Rev. **3**, 14-27 (2006)
6. Zhen-Dong Liu, Gan-Jiang Tao, Qing-Yun Ren, Dong-Sheng Yang, J. Chi. Coal Soc. **35:4**, 450-455 (2011)
7. Haifeng Wang, Yaqun He, Chenlong Duan, Yuemin Zhao, Youjun Tao, Cuiling Ye, *Development of Mineral Processing Engineering Education in China University of Mining and Technology* (Springer-Verlag, Berlin, 2012)
8. T. Chen, C. Lei, B. Yan, X. Xiao, Hydrometallurgy, **147**, 178–182 (2014)
9. C. Zhou, Z. Gong, J. Hu, A. Cao, H. Liang, Waste Manag. **35**, 191-198 (2015)

Ecological and Economic Prerequisites for the Extraction of Solid Minerals from the Bottom of the Arctic Seas

Alexander Myaskov[1,*], and *Alexander* Gonchar[1]

[1]The National University of Science and Technology MISiS 119049, Leninsky avenue, 4, Moscow

Abstract. The world ocean has huge reserves of minerals that are contained directly in the water, as well as on the surface of its bottom and in its subsoils. The deposits of solid minerals on the surface of the bottom of the World Ocean are considered the most promising for industrial extraction. The deposits of ferromanganese nodules, cobalt-manganese crusts and polymetallic sulphides are considered as an object of extracting more often than others. There are the largest deposits of ferromanganese nodules in the central and southern parts of the Pacific Ocean, in the central part of the Indian Ocean, and in the seas of the Arctic Ocean near Russia. The deposits of ferromanganese nodules are a serious alternative to deposits of manganese ore on land. However, there are many factors influencing the efficiency of the development of ferromanganese deposits, the most significant are: the content of the useful component in the ore, the depth of the bottom and the distance from the seaports. It is also necessary to take into account the possible environmental consequences of underwater mining.

1 Introduction

The depletion of certain types of minerals in previously explored and now developing deposits, the growth of population in the world, the increase in demands and living standards, the rapid development of science and technology pose a challenge for mankind to increase the extraction of minerals from new sources. A huge role in the solution of this problem will play the resources hidden under the World Ocean surface.

2 Material and Methods

Available materials on the study of the World Ocean allow us to conclude that the ocean not only can serve as a supplier of food products and hydrocarbon raw materials, but also inexhaustible in the foreseeable future a source of solid mineral raw materials. The world ocean is a giant "pantry" with three "regiments" [1-6]:

* Corresponding author: Myaskov@misis.ru

- sea water with 95 chemical elements dissolved in it, which in total amount to 50,000,000 billion tons (sodium, potassium, lithium, rubidium, iron, manganese, molybdenum, uranium, zinc, chromium, vanadium, copper, gold, etc.);
- solid minerals from the ocean floor: deep-sea polymetallic sulphides (SPS), ferromanganese concretions, cobalt-manganese crusts, placer tin, gold, titanium, zirconium, rare earth elements, metalliferous silts, amber, etc.;
- mineral resources (oil, gas, ferrous and non-ferrous metals, etc.).

The most promising from industrial development point of view are the mineral resources of the seabed – the shelf zone, the continental slope and the ocean floor, in connection with:

- the absence of minerals contained in offshore fields in the range of minerals extracted;
- lower cost of production compared to existing deposits on land;
- higher content of useful components in the ore in comparison with existing continental deposits;
- less environmental damage, compared with the development of existing deposits on land;
- depletion of developed polymetallic ores deposits on land.

3 Results and Discussion

Total reserves of minerals in the ocean floor are difficult to estimate, but the development of explored deposits covers the need for solid minerals by 50-250% for various types of raw materials in upcoming decades.

The major attention of technologically advanced countries of the world is attracted to the iron-manganese formations of the seabed, deep-sea polymetallic sulphides and metalliferous silt, which contain the multicomponent composition (up to 50 elements), as well as the high content of useful components.

Iron-manganese concretions of the bottom of the World Ocean are concentrated as much as possible in several ore fields with uneven distribution. More than 50% of the bottom area is covered by the nodules in some sections. Most concretions are concentrated in the center of the Pacific Ocean between the latitudinal faults of Clarion and Clipperton along the equator for thousands of kilometers. In addition to the Clarion-Clipperton areas with high concentrations of ferromanganese formations are also found in the central Indian Ocean, in the seas of the Arctic Ocean, in the southern and southeastern parts of the Pacific Ocean [7-13].

To understand the scope of possible alternative to terrain deposits of polymetallic ores (in this case manganese ores), it is best to turn to the comparative table of geological and economic indicators of the Clarion-Clipperton field, the Kara Sea, the most studied sea of the Arctic Ocean, and to Usinskoe manganese deposit . 1).

Table 1. Geological and economic indicators of manganese deposits

	Clarion-Clipperton field	**Kara Sea**	**Usinskoe deposit**
The content of the useful component in the ore,%	20.8-21.8	6.8-7.9	16.7-19.6
Ore reserves, million tons	300	34.9	98.5
Value, $ billion	127.8	5.2	35.7
The self-cost of mining 1 ton of ore, $	322	203	672

DOI: 10.1051/e3sconf/20172101026

In addition to the concentration of minerals in the ore, a number of other factors, especially geographic ones such as depth of development and remoteness from land, have a significant impact on the economic efficiency of solid minerals mining from the bottom of the World Ocean.

Depending on the depth of the seabed, the methods of deposits development vary considerably, and even it is possible to modify existing technologies from related industries, for example, using dredges, then at considerable depths there is a need to create and maintain new mechanisms and machines, requiring significant costs (Figure 1).

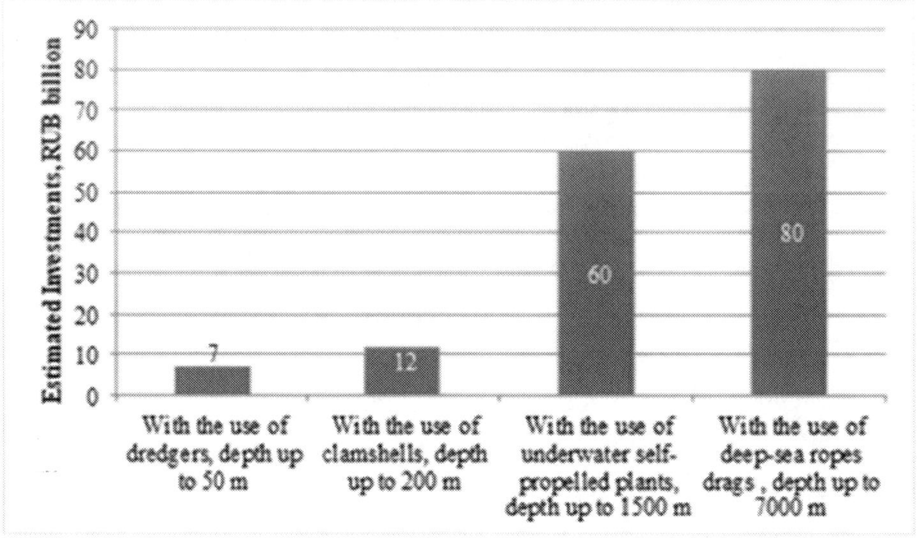

Fig. 1. Estimated value of capital expenditures for the projects designing for development of ferromanganese nodules deposits of the ocean floor

The next important geographical factor affecting the expediency of offshore mineral deposits development is their remoteness from seaports. First of all it affects transport costs, in addition, it leads to increasing the losses of the mineral (Figure 2).

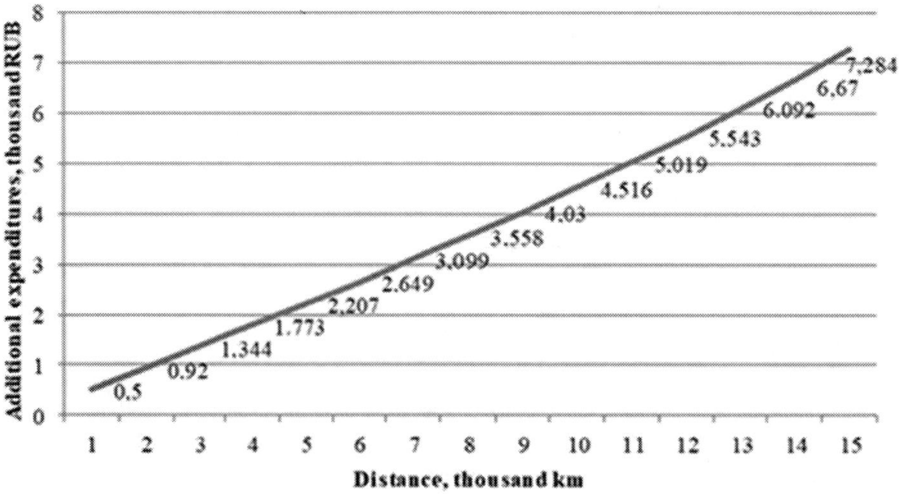

Fig. 2. Additional transportation costs per one ton of minerals, depending on the remoteness of the field from seaports

Due to the small depth of the Arctic seas in the water area of Russian Federation and the short distance to seaports, it can be concluded that the deposits of ferromanganese nodules of the Arctic seas, including the Kara Sea, have some competitive advantages for Russia before other deposits manganese.

In addition to the economic side of the issue, it is also necessary to take into account the consequences of sea deposits development on the surrounding natural environment. In the last century the International Oceanographic Commission of UNESCO adopted the rule that sea pollution is the introduction of substances or energy into the marine environment that leads to harmful or irreversible effects that damage the living resources and health of people, interfere with the activities of marine organisms, deteriorate the quality of the marine environment and disrupt the elements of the "ocean-atmosphere" system [14-18].

The main negative impacts for the marine environment during the seabed field's development will be:

- increase in the concentration of suspended material, leading to the loss of a significant amount of flora and fauna;
- violation of the oxygen regime in the region;
- transformation of the relief in the place of development;
- accumulation of chemicals.

Due to a simpler configuration of the surface of the seabed of the Arctic seas, solid minerals extraction in their water areas is safer because the relief transformations, as well as changes in the concentration of the material weighed in sea water, are minimized. In addition, due to the small depth, there is no serious disturbance of the oxygen regime resulting from the rise of deep waters, which are harmful to the inhabitants of the upper layers of the water column. This allows us to conclude that the development of solid mineral deposits in the Arctic seas has its positive aspects in relation to the development of deposits located in three other oceans.

4 Conclusion

At present, the Russian Federation does not extract manganese ore, so needed by metallurgy. Raw materials for them are imported from China, Kazakhstan and Ukraine. The beginning of national manganese ores deposits development would allow reducing dependence on foreign partners and also lowering expenses for manganese raw materials for domestic enterprises. Given the low production costs, relatively low environmental risks, the deposits of ferromanganese nodules of the Arctic seas should be regarded as a strategic source of manganese. So the further steps towards the beginning of industrial development of these deposits must be taken.

References

1. E. Baker, Y. Beaudoin, *Deep Sea Minerals: Seafloor Massive Sulphides, a Physical, Biological, Environmental, and Technical Review* (London: SPC, 2013)
2. M.L.C.M. Henckens, P.P.J. Driessen, E. Worrell, Resour. Conserv. Recycl. 93, 1–8. (2014)
3. K.V. Ragnarsdottir, Nat. Geosci. Comment. 1, 720–721 (2008)
4. W.J. Rankin, *Minerals, metals and sustainability, Meeting Future Material Needs* (New York, CRC Press, 2011)

5. R. Spickermann, *Rare earth content of manganese nodules in the Lockheed Martin Clarion–Clipperton Zone exploration areas. Offshore Technology Conference Number* (OTC-23084-MS, 2012)
6. A. Thurnherr, C. Young, Front. Mar. Sci. 2 (2015)
7. R.E. Boschen, A.A. Rowden, M.R. Clark, J.P.A. Gardner, Ocean. Coast Man. 84, 54-67 (2013)
8. M.D. Hannington, J. Jamieson, T. Monecke, S. Petersen, Soc. Econ. Geol. Spec. Publ. 15, 317–338 (2010)
9. M. Hannington, J. Jamieson, T. Monecke, S. Petersen, S. Beaulieu, Geol. 39, 1155-1158 (2011)
10. J.R. Hein, T.A. Conrad, R.E. Dunham, Mar. Georesour. Geotechnol. 27, 160–176 (2009)
11. J.R. Hein, T.A. Conrad, H. Staudigel, Oceanography, 23, 184–189 (2010)
12. J.R. Hein, *Prospects for rare earth elements from marine minerals. Briefing Paper 02/12, International Seabed Authority* (2012)
13. A.A. Rowden, J.F. Dower, T.A. Schlacher, M. Consalvey, M.R. Clark, Mar. Ecol. 31, 226–241. 2010
14. K. Bohmann, A. Evans, M.T.P. Gilbert, G.R. Carvalho, S. Creer, M. Knapp, D.W. de Bruyn, Trends Ecol. Evol. 29, 358-367 (2014)
15. P.C. Collins, R. Kennedy, C.L. Van Dover, Mar. Ecol. Prog. Ser. 452, 89-107 (2012)
16. ISA, *International Seabed Authority. Biodiversity, species ranges, and gene flow in the abyssal Pacific nodule province: predicting and managing the impacts of deep seabed mining* (Kingston, Jamaica: ISA Technical Study, 2008)
17. R. Martone, L. Hannah, L. Greig, B. Boutillier, S. Patton, 2015. DFO Can. Sci. Advis. Sec. Res. Doc. 072:66 (2014)
18. M. Prekopová, J. Janočko, Geol. Carp., **60:6**, 485-494 (2009)

 DOI: 10.1051/e3sconf/20172101027

Efficiency of Low-Profile External Dumping at Open Pit Coal Mining in Kemerovo Region

Alexey Selyukov[1], *Vyacheslav* Ermolaev[1], and *Irina* Kostinez[2]

[1] T.F. Gorbachev Kuzbass State Technical University, Department of open pit mining, Vesennyaya St. 28, Kemerovo city, Russian Federation
[2] T.F. Gorbachev Kuzbass State Technical University, Belovo Branch, 652644 Ilyicha st. 32, Inskoy, Kemerovo oblast, Russia

Abstract. Kemerovo region is one of the largest industrial regions of Russia, with a raw material specialization. The rapid growth of the coal industry in recent years has been greatly facilitated by the expansion and development of open pit mining for coal seams extraction, accompanied by an increase in the volumes of overburden and the height of the dumps. There are about 400 objects in the Russian Federation Government Register of Waste Disposal Facilities 80% of which are dumps. Approaches both to external dumping and to the technical stage of reclamation currently contribute to the growth of geomorphic system's instability. Thus, it is proposed to slightly change the approaches to external dumping: the essence consists in the formation of an external dump of overburden, which in future would represent a favorable landscape unit of a flat surface relief used for subsequent differently directed purposes.

1 Introduction

Sustainable development of Kuzbass (Kemerovo region, Western Siberia, Russia) depends both on economic growth and natural environment preservation [1-2]. This problem is quite typical for old mining regions of Europe and Russia [3-6]. Despite implicating new equipment on Kuzbass open pits [7] and technological development of surface mining in the region [8], the practice of open pit coal mining in Kemerovo region shows that the backfilling of overburden dumps by the methods of 1970-1980 sharply reduces the ecological efficiency and raises material and financial costs of coal extraction [9-10]. The area of reclaimed dumps is getting reduced, the area of natural landscape on which the dump is placed is also getting smaller [11, 12, 13].

When analyzing the technical and economic indicators of the low-profile dump backfilling, it is necessary to find the optimal values of its parameters, such as: transportation distance, height, area and, correspondingly, the land capacity. First, it is needed to set the required amount of overburden, which will later contain the dump. Secondly, it is required to find the most convenient location of the future dump with a minimum range of transportation from the quarry to its most remote point. The dump should be located taking into account the surface relief and factors (water resources, geological conditions, etc.) affecting the convenience and the possibility of overburden disposal in the place of its perspective storage.

2 Materials and methods

Taking into account the performance of other models, it is possible to estimate the economic efficiency of designing a low-profile dump. Efficiency lies in the best combination of geomechanical, ecological and operational components of the area dump backfilling compared to the pyramidal technology which is common for Kuzbass. Also, this dumping pattern allows optimizing the process of overburden transportation from the quarry to the dump taking into account the variation in the transportation distance that is, reducing or increasing the transportation distance from the largest to the smaller, depending on the overall technical and economic situation of the mining enterprise. Another reason to consider the low-profile external dumping is increasing dump's stability as an important ecological factor [14-15].

Let's consider the effectiveness of low-profile dumping on the example of the Tersinsky geological and economic area. The projected open pit spreads from the south-west to the northeast. Its area is 510.7 hectares with a length of 4.5 km and a width of 1.5 km, and its maximum depth varies from 260 m to 180 m, depending on the topography of the surface. Considering the volume of rock mass extracted from the quarry (about 800 million m3), and the average stripping ratio (10 tons/m^3), it was calculated that the reserves within the licensed boundaries vary in the range of 60-80 million tons. A quarry with such a number of reserves can be attributed to the average – with a production capacity of 3-5 million tons per year. Thus, taking into account the volume of coal reserves and the possible production capacity, we conclude that the life of the quarry is approximately 20 years, without taking into account development and damping.

The initial height of the dump is formed on the basis of the surface relief, the optimum ratio of its capacity (volume, m^3) and the transportation distance (length, km). For an example of calculations, we take an even surface and a quarry, located next to the dump. Depending on the volume of rock mass extracted from the open pit mime, the area and height of the future dump are calculated. The relief feature of the coal deposit allows us to place the dump considering the difference between the lowest and the highest points on the surface. Backfilling the hollows with overburden will not only allow making the dump more stable, but also creating an absolutely flat horizontal platform with the option of further use for any industrial or civil purposes. The dump placement can be carried out in narrows and hollows at a distance from the quarry not exceeding the economically advantageous transportation length. With the use of AutoCAD CIVIL 3D software, the body of the dumps was projected (Fig. 1a, b) and its capacity and occupancy were calculated (Table 1). The average transportation distance did not exceed 5 km.

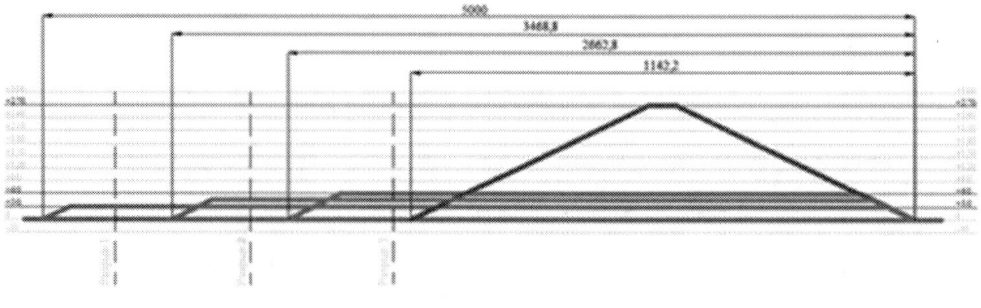

 DOI: 10.1051/e3sconf/20172101027

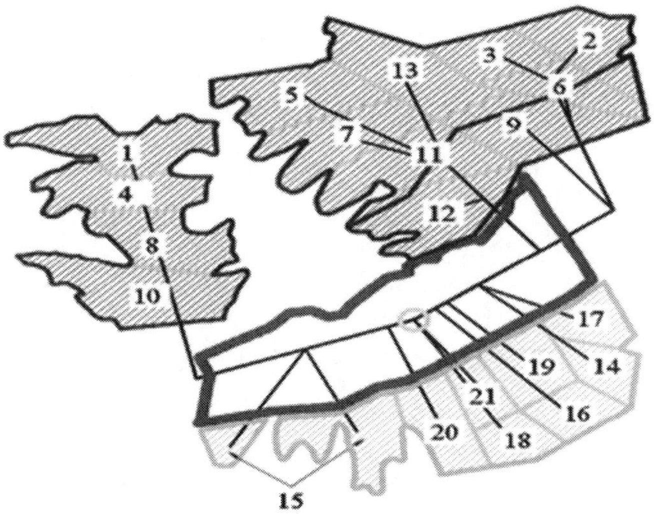

Fig. 1. a) Profiles of projected low-profile and pyramid dumps; b) plan for distributing the parts of the backfill of the reformed low-profile dump by years with tracing to each section

Table 1. Low-profile dump parameters

Year of backfilling	Overburden volume, million m^3			Area of dump's part, hectares			Transportation distance, km		
	Body			Body			Body		
	1	2	3	1	2	3	1	2	3
1	34.9			102.5			5.5		
2		36.6			82.5			5.2	
3		36			58.7			5.1	
4	34.6			86.7			5		
5		37.1			105.1			4.9	
6		35.7			106.1			4.4	
7		35			60.2			4.3	
8	35.4			101.8			4.3		
9		40.2			89.9			4.3	
10	37.2			108.9			4		
11		32.2			134			3.5	
12		36.5			111			3.4	
13		31.1			113.3			3.1	
14			30.4			45.6			2.5
15			27.5			110.5			2.5
16			29			49.3			2.2
17			27.6			48.5			2.1
18			28.2			43.3			1.9
19			29.9			67.7			1.6
20			28.4			71.1			1.6
21			28.9			40.5			1.4
Total	142.1	320.4	229.9	399.9	860.8	476.5	Average = 3.5		
	692.4			1737.2					

3 Results and discussion

From Table 1 it can be seen that, along with a decrease in the average length of transportation, the area of land that is leased under the dump is also decreasing, which will also bring a positive impact on the economic performance of the low-profile dump with the given layout of the dump bodies.

The option of a step-by-step layout of the low-profile dump also gives us the opportunity to start backfilling from the most remote point of the projected dump and to reduce the transportation length as the pit is deepened, thereby maintaining a balance and helping to regulate the technical and economic performance over the quarry life. This cannot be achieved by backfilling a pyramid dump. Also, when planning a low-profile dump, it becomes possible to lease a site for a part of the dump for a short time, and after backfilling, shrinking and reclamation of a dump site, immediately return it for further alternative use. Thus, we economically exploit leased territory and in a short time we return some of the lands environmentally suitable and economically feasible for farming or subsequent development. To reduce overburden transportation distance, it is proposed to transfer the projected areas to the dump from the longest distances to shorter distances. This will reduce both the peak values of each of transportation distances and their average value (Fig. 2).

Graphical analysis and modeling have shown that the most rational option for reducing the transportation distance is the transfer and placement of the dumping rock volume from the dump bodies located from the north of the open pit to the body of the dump projected in the southern part of it. Moving a part of the dump from the north side jf the pit to the southern part, we reduce the transportation distance (Fig. 3).

Re-planning of the dump bodies is carried out depending on the location of a particular site and its distance from the pit from larger to smaller. This makes possible to more flexibly adjust the backfilled overburden into the low-profile dump (reformed) by introducing a new body and adjust it to the required volume. Along with the decrease in the average transportation range, the area of land leased for filling (Table 1) also decreases.

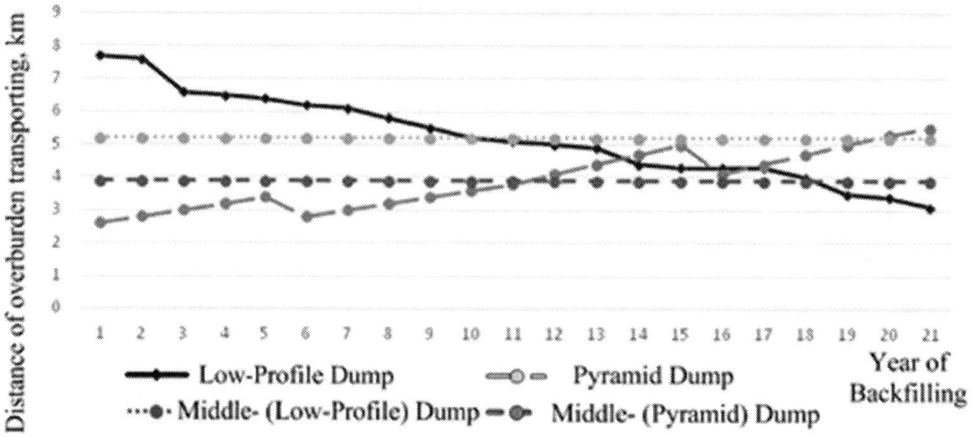

Fig. 2. Comparing the transportation distances for low-profile and pyramid dumps

DOI: 10.1051/e3sconf/20172101027

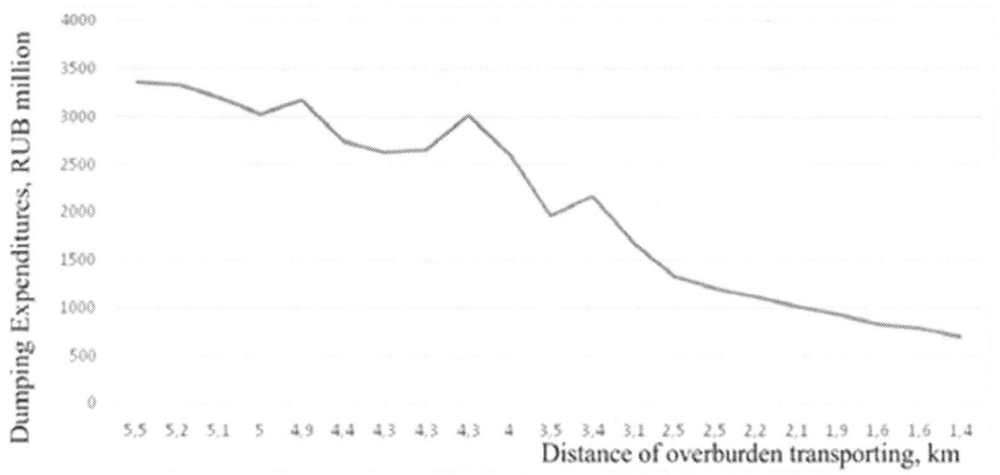

Fig .3. The dependence of transportation expenditures on the transportation distance

Table 2. Summary table of dump backfilling expenditures

	Pyramid dump	Low-profile dump
	RUB million	RUB million
Land leasing	1062.2	868.6
Transportation	68613.48	43560.3
Leveling	4317.1	3994.6
Reclamation	1062.15	868.6
Total	75055	49292

It can be seen from the graph in Fig. 3 and Table 2 that the main part of the costs is connected with overburden transportation and much exceeds other costs: leasing, leveling, reclamation. Therefore, the cost of overburden transporting in the dumps is considered to be top priority and dominant. Reduction of transportation costs is possible in different ways. If the backfilled low-profile dump is placed along the open pit on both sides, it becomes possible to significantly reduce the transportation distance and the associated costs. Low-profile dump makes it possible to regulate the range of transportation, taking into account the deepening of the open pit. The pyramid dump is deprived of the possibility of a thorough planning of the surface taking into account the development of open pit mining operations in an upright direction.

4 Conclusion

The system of step-by-step backfilling allows conducting dump's local reclamation and lease of land for overburden storage for shorter time periods, and reclaimed lands can yield income from alternative land use, giving an advantage to this scheme of dumping in economic and social development. The geomechanical stability of the low-profile dump excludes the possibility of man-made disasters, such as landslides, mudflows, etc., and also avoids additional costs of cumulative risk factor and protects the enterprise from adverse economic and technical consequences.

References

1. V. Trifonov, O. Loyko, D. Nesteruk, S. Zhironkin, E. Strekovtsova, AIP Conf. Proceed., **1800**, 050009 (2017)
2. A.V. Kapitskaya, I.N. Paskar, Economics and Innovation Management, **1**, 39-44 (2017) DOI: 10.26730/2587-5574-2017-1-39-44
3. O. Erdem, T. Güyagüler, N. Demirel, J. Sout. Afr. Inst. Min. Met., **112:5**, 405-412 (2012)
4. M. Cehlár, L. Mihok, Theor. Emp. Res. Urb. Man., **8:4**, 60-72 (2013)
5. M. Erkayaoğlu, N. Demirel, J. Env. Man., **174**, 1-6 (2016)
6. M. Cehlár, J. Janočko, Z. Šimková, T. Pavlik, E3S Web of Conf., **15**, 01019 (2017)
7. M.A. Tyulenev, S.A. Zhironkin, O.I. Litvin, E.A. Tyuleneva, O.V. Zhironkina, S.O. Markov, Geotech. Geol. Eng. **35:5**, 2065–2077 (2017)
8. M. Tyulenev, S. Zhironkin, E. Tyuleneva, A. Abay, S. Anyona, M. Hellmer, Coal Int., **265:3**, 30-34 (2017)
9. M.A. Tyulenev, T.N. Gvozdkova, S.A. Zhironkin, E.A. Garina, Geotech. Geol. Eng., **35:1**, 203-212 (2017)
10. T. Gvozdkova, E. Kuznetsov, A. Rudakova, S. Markov, E3S Web of Conf., **15**, 01008 (2017)
11. A.V. Selyukov, *Advanced technology based on new technological and organizational principles of spatial development of front of mining operations at open pits* (Taishan Academic Forum - Project on Mine Disaster Prevention and Control, 2014)
12. A.V. Selyukov, J. Min. Sci. **51:5**, 879-887 (2015)
13. A.V. Selyukov, Bulletin of the Tomsk Polytechnic University, Geo Assets Engineering, **326:12** (2015)
14. S. Markov, M. Tyulenev, O. Litvin, E. Tyuleneva, E3S Web of Conf., **15**, 01011 (2017)
15. I. Túnyi, D. Vass, S. Karoli, J. Janocko, E. Halásová, A. Zlínská, B. Beláček, Geol. Carp., **56:3**, 273-384 (2005)

 DOI: 10.1051/e3sconf/20172101028

Numerical Simulation of Aerogasdynamics Processes in A Longwall Panel for Estimation of Spontaneous Combustion Hazards

Sergey Meshkov[1], and *Andrey* Sidorenko[2*]

[1]JSC SUEK Kuzbass, 652507, 1, Vasilyev St., Leninsk-Kuznetsky, Russia

[2]Saint Petersburg Mining University, 199106, 21 line, 2, Saint Petersburg

Abstract. The relevance of a solution of the problem of endogenous fire safety in seams liable to self-ignition is shown. The possibilities of numerical methods of researches of gasdynamic processes are considered. The analysis of methodical approaches with the purpose to create models and carry out numerical researches of aerogasdynamic processes in longwall panels of gas mines is made. Parameters of the gob for longwall mining are considered. The significant influence of geological and mining conditions of conducting mining operations on distribution of air streams on longwall panels and effective management of gas emission is shown. The aerogasdynamic model of longwall panels for further research of influence of parameters of ventilation and properties of gob is presented. The results of numerical researches including distribution of air streams, fields of concentration of methane and oxygen at application of various schemes of airing for conditions of perspective mines of the Pechora basin and Kuzbass are given. Recommendations for increase of efficiency of the coal seams mining liable to self-ignition are made. The directions of further researches are defined.

1 Introduction

The endogenous fires in coal mines still represent a danger being the second most important after explosions of methane for all leading coal-mining countries as despite the low frequency of emergence result in significant economic damages and extreme danger to life of miners [1, 4, 15, 17]. Modern researchers even more often point to the endogenous fire, as to the most probable cause of explosion of metane-air mix at gob areas of the seams liable to self-ignition. The main place of emergence of the endogenous fires are the gob areas - both the operating mining panels and those which have been mined earlier [8, 13]. The complexity of studying the process of self-ignition in mine conditions predetermines wide use of laboratory researches [1, 11] and computer modeling [2, 3].The existing modern program complexes, for example, of Ansys Fluent, Ansys CFX, FlowVision and others, allow to model gasdynamic processes in systems of varying complexity that causes expediency of their use for studying aerogasdynamic processes on longwall panels of the gas mines in the course of

* Corresponding author: sidorenkoaa@mail.ru

 DOI: 10.1051/e3sconf/20172101028

mining coal seams liable to self-ignition. The main directions of aerogasdynamic researches with the use of program complexes are connected with studying the distribution of air streams and methane [16] and also with studying the distribution of oxygen on a longwall panel when using various schemes of ventilation with identification of dangerous zones, with the subsequent processing or change of parameters of ventilation and draining of methane [7, 9].The complexity of ensuring adequacy of the models to real conditions used by researchers is connected, first of all, with insufficient study of a gob area and influences of their characteristics (collapse zone height, porosity, permeability, etc.) on aerogasdynamic processes [10]. In this connection studying of influence of the specified characteristics of a gob when using various ways of management of gas in entries of longwall panels on formation of the zones dangerous on self-ignition of coal is of considerable scientific and practical interest [14].

2 Materials and methods

The three-dimensional aerogasdynamic model of the longwall panel including entries, longwall and gob has been developed for carrying out researches. Modeling was carried out in scale 1:1. The gob was modelled as a porous body whose parameters were defined with the methodical approach stated in earlier published of other researchers [2, 3, 5]. The need of allocation for the gob areas of zones with various permeability was emphasized by many researchers [2, 6], however, the impossibility of exact determination of permeability in mine conditions has caused the need of searching for other solutions. In the used model the extinguished developments, the part of the gob which is directly adjoining a longwall, the top and lower parts of the gob on removal from a longwall equal to 6 steps of collapse of the main roof and other gob have various permeability. Now the permeability of the gob when carrying out numerical researches is determined either on the basis of numerical researches with the use of methods of mechanics of the discrete medium which allow to define distribution of rock pressure in the gob, or with use of empirical formulas. So, for example, it is proposed to determine permeability by the formula [2]:

$$K = \frac{K_0}{0.241}\left(\frac{n^3}{(1-n)^2}\right), \tag{1}$$

K_0 – basic permeability, at the maximum porosity of the broken-down rocks; n – porosity.

Methane emission on the longwall panel was set: in the gob - from under- and above-mining seams, from the surface of a coal face and from the rotted coal. Parameters of gas emission, methane drainage and ventilation at the initial stage were set according to the actual mine data for further comparison of results of modeling and correction of model. Carrying out numerical researches of aerogasdynamic processes on longwall panel with use of the program ANSYS complex included the following consecutive actions: a task of geometry of entries and the gob taking into account specific geological and mining conditions, installation of parameters of a spatial grid and splitting model into elements, determination of physical parameters of model and boundary conditions (input-output parameters), carrying out the numerical solution, processing and visualization of results, the analysis of results.

3 Results and discussion

As a result of the conducted numerical researches the assessment of efficiency of various schemes of ventilation and gas drainage was made applied in Kuzbass in the field of ensuring effective gas management and endogenous fire safety at longwall mining. Visualization of results of researches has allowed to reveal the most dangerous zones, to estimate influence various technological (ventilation schemes and gas drainage and their parameters) and geological (extracting seam thickness, parameters of a gob, gas emission) factors and has provided fuller understanding of aerogasdynamic processes on a longwall panel and a gob. As a result of researches the distribution of gases within entries and the gob of longwall panel for various geological and mining conditions is established. As an example of the received results figure 1 shows methane in the working area (in entries, longwall and gob) in the plane passing parallel to bedding at the height of 2.5 m from the coal floor is presented, and fields of concentration of oxygen are given in the figure 2 in the same plane.

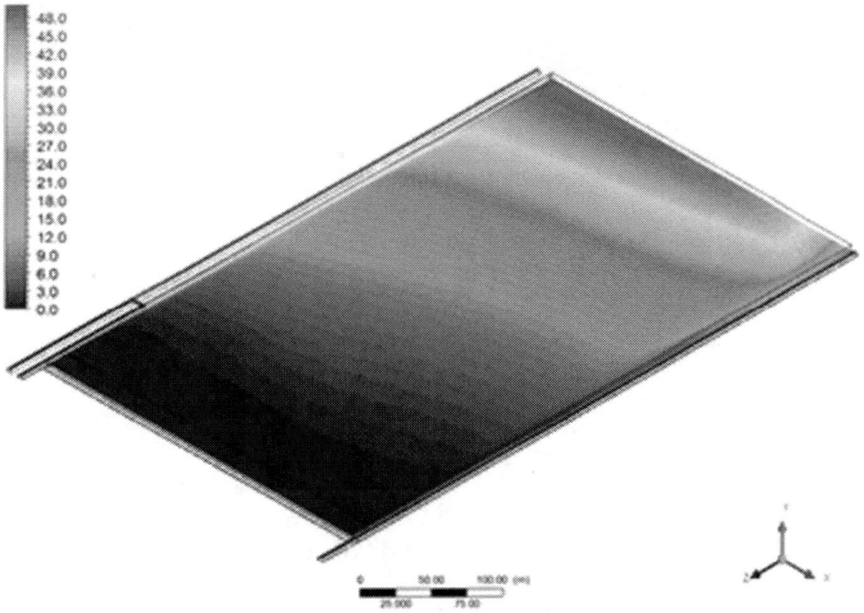

Fig. 1. Concentration of methane in the longwall panel, %.

By researches it is established that concentration of oxygen in the gob remains within fire-dangerous values (10%) on considerable removal (up to 400 m and more) from the longwall. The intensity of ventilation of the gob and possibility of formation of dangerous zones substantially depend on permeability of the broken-down rock and parameters of the applied way of gas management on the longwall panel. Researches into influence of parameters of the combined schemes of ventilation providing maintenance of conveyor entry behind a longwall on the site to the next air connection have shown that significant effect in a longwall the size of residual cross section of the airtaking-away development supported behind a longwall and also timeliness (lag from a longwall) and quality of repayment behind a longwall of the air-giving development for minimization of leak of air in a deadlock part of entry on the headgate (at the descending order of mining).

DOI: 10.1051/e3sconf/20172101028

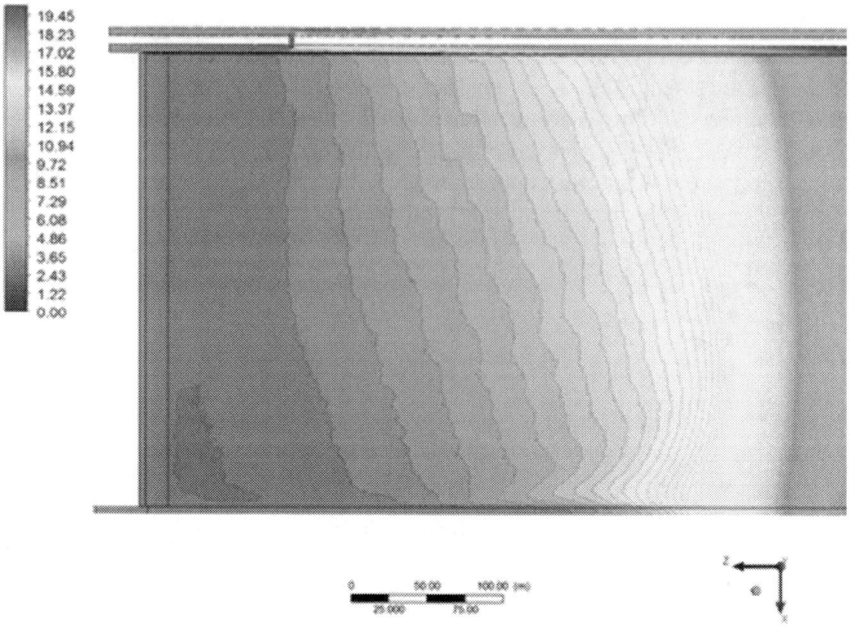

Fig. 2. Oxygen concentration in the working area, %.

Researches have shown that the permeability of the developed spaces is one of the main characteristics defining intensity of processes of ventilation, filtration and movement of gases. The summing up of research results of various authors [16] shows that the minimum permeability of the developed spaces as a result of application of such approaches can be found in the range from 3 x 10^3 to 2.03 x 10^9mD, and maximum from 1.25 x 10^5 to 2.03 x 10^{10}mD, and the values of permeability for conditions of one and the same mine applied by different authors may vary up to 1000 times. At the same time the distinction between maximum and minimum permeability, by results of researches of various authors, is from 10 to 58 times. Such considerable range of the used values of permeability may mean high variability of geological conditions at gob of longwall panels which length reaches 4-5 km and more now and also significant influence of technological parameters (length of a longwall and speed of a coal face) on processes of deformation and the collapse of rocks of a roof. The ratio of the extracting seam thickness and thickness of the immediate roof as well as the sizes of blocks of immediate and main roof have significant effect on permeability of the developed space. It should be noted the permeability of the gob decreases in the process of removal from a longwall and also with time (in connection with development of geomechanical processes). Formation of aerodynamic communication exerts considerable impact on distribution of gases on the gob and has to be considered when carrying out numerical modeling. According to the present research, it is recommended to update the characteristics of the developed space after comparison of results of numerical researches and mine observations which can be carried out on the basis of assessment of distribution of air on the longwall panel and leak of air in the gob.

4 Conclusions

Numerical researches of aerogasdynamic processes are one of the main methods of the researches providing the solution of a problem of distribution of air streams, concentration of methane and oxygen for the wide range of geological and mining conditions that proves high efficiency of the considered method for assessment of endogenous fire safety of the applied technological schemes of mining and schemes of ventilation. The fields of distribution of oxygen received as a result of numerical researches in gob areas allow to predict formations of dangerous zones and to provide decrease in endogenous fire danger of mining of the coal seams liable to self-ignition, by change of parameters of schemes of development of longwall panels, schemes of ventilation and gas drainage or carrying out preventive processing of dangerous zones.

Creation of three-dimensional aerogasdynamic models of the longwall panels which are adequately reflecting the processes happening at gob areas of longwall mining flat coal seams with roof caving requires the correct determination of key parameters of the developed spaces: porosity and permeability. As the most effective method of determination of the specified parameters the empiric-analytical method based on use of the known methodical approaches [2, 5] and adjustment of the received sizes taking into account the actual ratios of change of permeability in mine conditions with carrying out final control of model can be offered.

The roof-caving increment of the main roof and ratio of the extracting seam thickness and thickness of the immediate roof are one of the major factors determining the volume of ventilation loss through the gob and the sizes of intensively ventilation zones. In the conditions of intensive mining the speed of longwall operations leads to the growth of the roof-caving increment in the main roof and increase of permeability of the developed space. The carried out researches have confirmed the need of the complex solution of issues of prevention of the endogenous fires and gas management in entries of longwall panels of the gassy seam liable to self-ignition.

The direction of further researches will be connected with improvement of methodical approaches to determination of parameters of gob areas and validation of the developed theoretical methods in mines.

References

1. A. Arısoy, B.B. Beamish, E. Çetegen, Turkish J. Eng. Env. Sci., **30**, 193 (2006)
2. G.S. Esterhuizen, C.O. Karacan Proceeding of the 40th U.S. Rock Mechanics Symposium. Alexandria, VA: American Rock Mechanics Association., 156-160 (2005)
3. G.S. Esterhuizen, C.O. Karacan, *A methodology for determining gob permeability distribution and its application to reservoir modeling of coal mine longwalls* 07 (Dover, SME 2007)
4. B. Ham, Coal2005: 6th Australasian Coal Operators' Conference, 237 (2005)
5. C.Ö. Karacan, Transport in Porous Media, **82:2**, 413 (2009)
6. C. Zhang, S. Tu, L. Zhang, Q. Bai, Y. Yuan, F. Wang, Journal of Geophysics and Engineering., **13:2**, 181 (2016)
7. Y. Lu, S. Shi, Q. Ye, Z. Tian, Journal of Residuals Science & Technology (JRST)., **14:3**, 373 (2017)
8. W.X. Ren, B.W. Wu, Journal of Mining & Safety Engineering., **26:2**, 198 (2009)
9. A.C. Smith, L. Yuan, Mining Engineering, **60:8**, 61 (2008)
10. G. Xu, K. D. Luxbacher, S. Ragab, S. Schafrik, Tunnelling and underground space technology, **33**, 1 (2013)
11. H. Wen, Xi an University of Science and Technology, 46 (2003)
12. L. Yuan, A.C. Smith, Journal of loss prevention in process industries., **25**, 131 (2011)

13. R. Qin, L. Teng, S. Yuan, L. Shi, AGH Journal of Mining and Geoengineering., **36:3**, 283 (2012)
14. L. Yuan, A.C. Smith, Proceedings of the 12th., 141 (2008)
15. R.V.K. Singh Procedia Engineering., **62**, 78 (2013)
16. D.M. Worrall, *Modeling gas flows in longwall coal mines using Computational Fluid Dynamics, Ph.D. Thesis*, (The Colorado School of Mines, 2012)
17. O.I. Kazanin, Journal of mining institute, **215**, 38 (2015)

The Development of Environmentally Friendly Technologies of Using Coals and Products of Their Enrichment in the Form of Coal Water Slurries

Vasily Murko[1*], and *Veniamin* Hamalainen [1]

[1] T.F. Gorbachev Kuzbass State Technical University, 650000, 28 Vesennyaya St., Kemerovo, Russia

Abstract. The article presents the current state of the technology for production and combustion of fuel coal water slurries in Russia and foreign countries. Experimental and industrial facilities show the technological and economic efficiency of using this technology for disposal of wastes resulting after coal processing and enrichment. The feasibility studies of use of the technology at large Kuzbass thermal power stations are presented. The possibility of solving a serious environmental problem of reducing storage of the most toxic waste of coal enrichment in the location areas of coal washing plants and coal mining enterprises is demonstrated.

1 Introduction

The technology of CWF production is one of the most effective technologies for obtaining environmentally friendly and technological fuel. This technology is successfully used in energy industries of a number of foreign countries, although the main developments were carried out in Russia in the middle of the last century. Today, China's energy industry is using this experience successfully [1-8].

There also is a long and successful experience of operating the Emil Gyushe power plant in France, where a project of direct combustion of coal slimes in the form of a suspension in boilers with a circulating fluidized bed has been realized. A complete disposal of coal slimes using this technology at several thermal power stations in the USA has been carried out. The technology for preparation and combustion of CWF obtained both from coal and slimes became a frequent practice in China.

In recent years, the Scientific and Production Enterprise «Sibecotechnika» (Novokuznetsk) has developed this technology in Russia very successfully. Thus, at the turn of the 20th and 21st centuries, technological complexes were developed and implemented both in Russia and abroad.

The purpose of the work is to show the possibilities of wide introduction of the ecologically friendly technology for preparation and combustion of coal and coal slimes in the form of coal water slurries, while solving the problems of resource saving and improving the environmental situation in coal-mining regions.

* Corresponding author: author@email.org

2 Results and discussion

At present, the formulations of effective plasticizer reagents are worked out, which makes it possible to obtain a statically and dynamically stable CWF with the required thermophysical characteristics for direct combustion. To stabilize technological characteristics of the CWF prepared on the basis of fine-dispersed waste of coal enrichment, a technological scheme (figure) has been developed in which the industrial product of the enrichment factory (EF) plays the role of «stabilizer» of the finished fuel ash contents.

The results of industrial testing of the developed technology on the technological complex of OJSC «Mezhdurechye» are given in the work. This technology can be successfully implemented at the EF, enriching coals of coking grades, and which have some difficulties with sale of industrial products. The fact is that its burning by traditional methods (on a grate, in a fluidized bed, etc.) is ineffective. The organization of construction of a mini TPP with production of own cheap electric and heat energy based on combustion of industrial products and filter cake using CWF technology is economically viable.

A vortex combustion system with organization of adiabatic or similar conditions in the boiler has been developed and is successfully used for reliable combustion of coal water slurry fuel, ballasted with moisture and mineral constituents, prepared on the basis of finely-dispersed waste and other products of coal enrichment. This technology is currently implemented by «ProEnergoMash» (Barnaul) to burn many local ballasted fuels.

The use of the vortex combustion technology made it possible to successfully carry out several projects with the reconstruction of boilers of small and medium capacity (Table 2).

Table 1. Characteristics of the heat generators operation.

Indicator name	**Technological complexes**					
	Boiler house «Zarechnaya» mine, Leninsk-Kuznetsk	**SibUME of the Russian Agricultural Academy SD, Novosibirsk**	**Temirtau, Kazakhstan**	**Mezhdurechensk**	**Hanoi, Vietnam**	**Cherepanovo, Novosibirsk region**
Heating effciency, Gcal /h	0.50÷0.58	0.25	0.3÷0.6	4.05	1.08	0.3÷0.6
Fuel consumption, l/h	120÷130	55	110÷220	Up to 1300	220÷210	110÷220
Temperature in a furnace, 0C	950	950	950÷1050	1050÷1150	1000÷1100	950÷1050

However, taking into account, that the amount of finely-dispersed coal enrichment (filter cake) wastes at modern EF ranges from 0.5% to 10% of the processing of ordinary coal, and their annual output is from 150 to 1 000 tons per year, the use of these wastes in their own boiler houses with a fuel requirement of ~ 5 ÷ 15 thousand tons / year does not solve the problem of their full disposal. Therefore, to use the entire volume of formed coal slimes, it is advisable to burn them in the form of CWF, either at a nearby coal-fired thermal power plant or at the own mini-TPP, to generate electricity and heat. In this case, the industrial product of EF can be effectively used (as shown above). Tables 2-5 provide technical and economic calculations of the use of fine coal enrichment waste at Kuznetsk and Central TPPs of Novokuznetsk, Belovskaya SDPP and mini TPPs with capacities of 1.5 and 10 MW.

Table 2. Technical and economic indicators of the complex for preparation, transportation and combustion of coal water fuel, prepared on the basis of coal enrichment waste of Novokuznetsk EF at Kuznetskaya TPP.

Indicators	Units	CWF
Annual production of CWF	thousand t.	500
Lower calorific value of fuel	kcal / kg	3000
Capital costs, including:	million rubles	225
- CWF preparation workshop (capacity 75 t / h)	million rubles	130
- Transfer of TPP boilers to co-burning of pulverized and coal water fuel	million rub.	95
CWF cost at TPP	rub./ t	600
The cost of fuel for production of 1 Gcal of thermal energy	rub.	184
Cumulative economic effect	million rubles / year	207,5
Payback period of investments	year	< 1,5

Table 3. Technical and economic indicators using coal slurry fuel, prepared on the basis of coal and coal slimes, by LLC «Central TPP».

The name of indicators	Units	Coal	Coal slimes	Coal and coal slimes
Annual capacity of the CWF producing installation	thousand t.	708	383	740
Lower calorific value of fuel	kcal / kg	3600	3300	3444
Capital costs, including:	million rubles	430	285	490
The cost price of slurry fuel	million rubles	1134	738,5	858
The cost of fuel for production of 1 Gcal of energy (70% -CWF + 30% -gas)	million rub.	371	214	440

Table 4. Technical and economic indicators using coal slurry fuel, prepared on the basis of filter cake and coal slimes, by EF «SUEK-Kuzbass» at Belovskaya SDPP (1 block of the PK-40 boiler).

Indicators	Units	CWF
The annual CWF demand	thousand t.	92,4
Share of CWF in the fuel balance of the boiler block	%	20
Lower calorific value of fuel	kcal / kg	3200

Capital costs, including:	million rubles	40,5
CWF cost at SDPP	rub./t	500
The cost of fuel for production of 1 Gcal of thermal energy	rub.	184
Cumulative economic effect	million rubles / year	30,2
Payback period of investments	year	< 1,5

Table 5. Economic indicators of technological complexes (mini TPP).

Indicators	**Electric power, MWh**		
	1,0	5,0	10,0
The cost price of fuel, rubles / t	300	270	250
The cost price of the produced electric energy, rubles / kWh	0,95	0,80	0,70
Cost of electrical energy, rub. / kWh	2,50	2,50	2,50
The economic effect of generating 1 kWh, rubles / kWh	1,55	1,70	1,80
Capital costs, mln. rub.	96	400	640
Payback period, year	5	4	3,5

3 Conclusion

The technology of production and combustion of coal water slurry fuel in boilers of different capacity is being developed in many countries. Experimental and industrial tests showed that the biggest economic efficiency and the essential environmental effect are reached using WCF, produced from coal slimes. In this case the serious problem of disposal of the most toxic finely-dispersed wastes of coal enrichment is solved, as their storage requires large place in the areas of coal mining and conversion.

References

1. V. Murko, A. Djundubaev, M. Baranova, A. Biibosunov, V. Kulagin, H*ydrotransport fuel and energy complexes. The Russian-Kyrgyz scientific and technical cooperation in the field of power engineering*, 250 (Krasnoyarsk, Sib. Feder. University, 2015)
2. V. Biletskyy, P. Sergeyev, O. Krut, Mining of Mineral Deposits, **45**, 105 (2013)
3. S. Mochalov, I. Rybenko, L. Ermakova. World Applied Sciences Journal, **19:1**, 20 (2012)
4. A. Kijo-Kleczkowska,. Archives of Thermodynamics, **32:1**, 45 (2011)
5. A. Kijo-Kleczkowska, Fuel, **90:2**, 865 (2011)
6. E. Karpov. Ecological systems, **1**, (2008)

 DOI: 10.1051/e3sconf/20172101029

7. F. Serant, A. Tsepenuk, Yu. Ovchinnikov, S. Luschenko, E. Karpov, Collection of scholarly articles "Modern science", **1:6**, 95 (2011)
8. O. Nosko, *Coal water fuel - environmentally friendly and efficient*. (News of the Siberian science. IT SB RAS, 2016)

Assessing the effects of underground mining activities on high-voltage overhead power lines

Vladimir Gusev[1], *Alexei* Zhuravlyov[2], *Elena* Maliukhina[1*]
[1]Department of Mine Surveying, Saint-Petersburg Mining University, 2, 21st Line, St Petersburg 199106, Russia
[2]Jet Technology & Construction Company, Dnepropetrovskaya st.,14 ,St. Petersburg 192007, Russia

Abstract. This paper introduces a technique for predictive assessment of changes in the position of power transmission towers and condition of overhead power lines, located in the zone of influence of displacements and deformations of the Earth's surface caused by mining activities. A special approach for monitoring the technical condition of towers and cables is proposed. It is intended to address the issue of controlling the condition of transmission lines that are under the influence of underground mining activities and to checkmate such impact.

1 Introduction

The issue of the effect of underground mining activities on overhead power transmission lines in the major coal basins in Russia is addressed by determining a safe depth for transmission towers [1, 2] situated in the zone of influence of the mining activities. The safe depth is calculated using the formula:

$$H_s = K_6 \cdot m, \quad (1)$$

where m is the mineable seam thickness, K_s is the safety factor, which depends on the type of tower (anchor-angle, intermediate angle and straight, dead-end towers) and the voltage of the power transmitted ($K_s = 60 \div 100$). Layers under transmission towers can be excavated at a depth no less than a safe depth calculated by formula (1). The conditions for such formations are given in [1]. It is necessary to take protective measures when conducting mining activities whose depth goes beyond the safe depth. For transmission towers situated in the zone of influence of subway tunnel construction, the mineable seam thickness in formula (1) can be replaced by the sum of displacements of the tunnel arch crown before tunnel fixing and displacements of the hinge part of the crown after tunnel fixing. We now proceed further as in [1, 2].

2 Research techniques

Since transmission towers situated within the boundaries of the zone of influence of underground mining or construction of underground subway facilities will be affected by displacements and deformations (Fig. 1), there is need to level individual poles and re-stretch

* Corresponding author: elenamaliukhina@gmail.com

the cables and wires to normal state. This is regardless of whether the mining or subway tunnel construction will be carried out above or below the safe depth horizon.

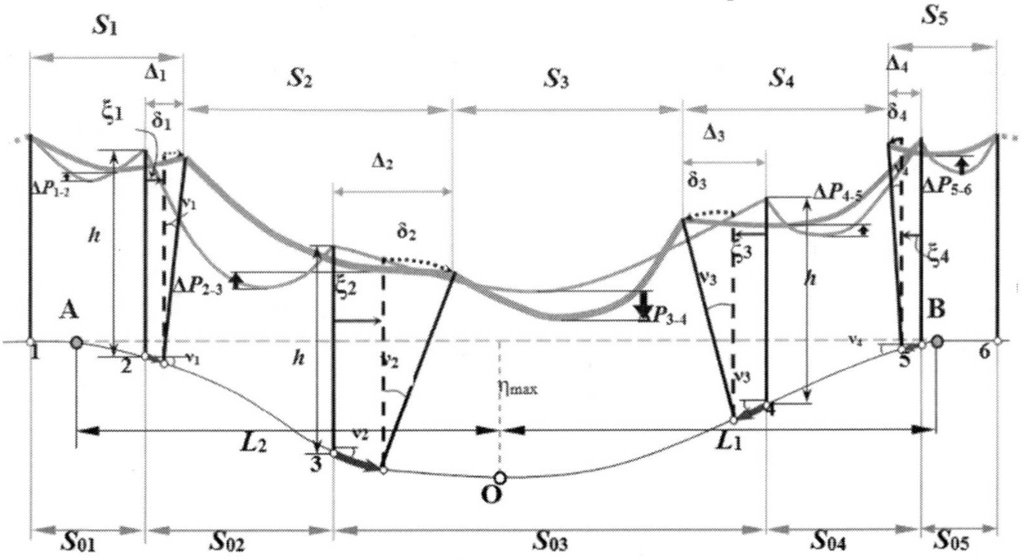

Fig. 1. Change in transmission line sag ($\Delta P_{2\text{-}3}$, $\Delta P_{3\text{-}4}$, $\Delta P_{4\text{-}5}$, $\Delta P_{5\text{-}6}$,) caused by horizontal displacements and slopes at tower installation sites. ξ_1, ξ_2, ξ_3, ξ_4 – horizontal displacement at tower installation sites. ν_1, ν_2, ν_3, ν_4 – tower slope angles caused by the sloping of tower installation sites. δ_1, δ_2, δ_3, δ_4 – tower displacement caused by slopes. Δ_1, Δ_2, Δ_3, Δ_4 – total tower displacement caused by slopes and horizontal displacements. S_{01}, S_{02}, S_{03}, S_{04}, S_{05} – distance between wire suspension points on towers before underground mining activities. S_1, S_2, S_3, S_4, S_5 – distance between wire suspension points on towers after underground mining activities. h – height of wire suspension points from the Earth's surface. L_1, L_2 – semi-troughs across the strike (falling and rising respectively). A, B – subsidence trough boundary. O – point of maximum subsidence (η_{max}).

The boundaries of the zone of influence of underground mining activities or underground construction – subsidence troughs – are established along the length of semi-troughs (L_1, L_2, L_3 in Fig. 1, 2), which in turn are determined by boundary corners [1, 2]. So, boundary corners are defined according to [1] for major coal basins, and [2] for subway tunnel construction in St. Petersburg. The zone of influence of subway tunneling can be established based on modeling of geomechanical processes by finite element method (FEM). Changes in wire tension in areas between transmission towers are due to inclination and horizontal displacements of these towers, caused by slope deformations and horizontal displacements within the subsidence trough boundaries [3-13]. The combined impact of these factors determines how closer or farther away the wire suspension points have moved, and accordingly how the point with maximum wire sag in the span between towers has been lowered or lifted. Lowering of the maximum sag point entails weaker wire tension, while lifting means stronger wire tension. The mechanism for the lifting or lowering of transmission line sag, which is caused by the impact of slopes and horizontal displacements on towers, is shown schematically in Fig. 1.

A shift in the position of towers situated within subsidence trough boundaries (AOB in Fig. 1 and *abcde* in Fig. 2), which is caused by the cumulative effects of horizontal displacements and slopes, are defined by the following formulas:

$$\Delta_1 = \xi_1 + \delta_1,\ \Delta_2 = \xi_2 + \delta_2\ \ \ldots\ \ \Delta_j = \xi_j + \delta_j \tag{2}$$

where Δ_1, Δ_2, …, Δ_j are the total displacements for towers 1, 2, ..., j respectively in the zone of influence of underground mining activities (Fig. 1, 2); ξ_1, ξ_2, …, ξ_j – displacement of towers 1, 2, ..., j respectively, caused by horizontal displacements; δ_1, δ_2, …, δ_j – displacement of towers 1, 2, ..., j respectively, caused by slope deformations (Fig. 1).

Displacements caused by the effect of slope deformations is determined by the following formulas

$$\delta_1 = h \cdot i_1 = h \cdot tg\nu_1,\ \delta_2 = h \cdot i_2 = h \cdot tg\nu_2 \ldots \delta_j = h \cdot i_j = h \cdot tg\nu_j \qquad (3)$$

where h is the height of wire suspension point on tower from the Earth's surface; i_1, i_2, ..., i_j are the slopes at the installation sites of towers 1, 2, ..., j, determined by predictive calculations and equal to the slope of the towers; ν_1, ν_2, ..., ν_j – slope angles of towers 1, 2, ..., j, defined as:

$$\nu_1 = arctg(i_1), \nu_2 = arctg(i_2) \ldots \nu_j = arctg(i_j) \qquad (4)$$

The predicted values for horizontal displacements (ξ_j) and slopes (i_j) can be obtained either by standard curve method [1, 2], or by numerical modeling of deformation processes via finite element method (software package PLAXIS), or by analytical methods .

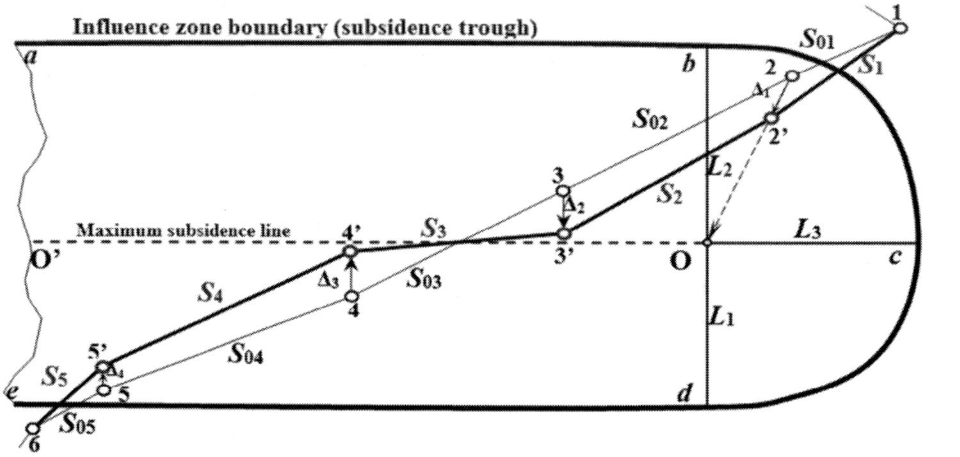

Fig. 2. Changes in distance between transmission towers located diagonally to the subsidence trough boundaries. The designations are the same as in Figure 1.

AutoCAD – a software application for computer-aided design – is used to construct the subsidence trough boundary and mark the position of power transmission towers. Next, the distances S_{01}, S_{02},..., S_{05} between these transmission towers are calculated. They correspond to the distances between wire suspension points on towers 1–2, 2–3, 3-4, 4-5, 5-6 before underground mining activities (Fig. 2). From these transmission towers located within the subsidence trough, tower displacements Δ_j calculated by formula (2) are laid in the direction of horizontal displacement vectors. Within subsidence trough zone *abde*, horizontal displacement vectors are directed towards the maximum subsidence line OO' at an angle of 90° to the line. In subsidence trough zone *bcd*O, these vectors are directed towards point O (Fig. 2). As a result, new positions for transmission towers 1', 2', 3', 4', 5' under the influence of mining activities is determined. From here, we get the distance between transmission towers S_1, S_2, ..., S_5, corresponding to the distance between wire suspension points on towers 1'–2', 2'–3', 3'-4', 4'-5', 5'-6' after underground mining activities (Fig. 2). Hence, the increase or decrease in the distance between wire suspension points as a result of tower displacement caused by underground mining activities is determined as follows

$$\Delta S_1 = S_1 - S_{01}, \Delta S_2 = S_2 - S_{02} \ldots \Delta S_j = S_j - S_{0j}, \qquad (5)$$

If $S_j > S_{0j}$, then $\Delta S_j > 0$, that is, the sag will rise as a result of an increase in the distance between the towers. If $S_j < S_{0j}$, then $\Delta S_j < 0$, that is, the sag will fall as a result of a decrease in distance.

Next, we determine the changes in wire sag (rising and falling) using increment ΔS_j in the distance between wire suspension points. Fundamental calculation of the extent of change in wire sag is shown in Fig. 3 .

In the diagram, S_0 is the distance between towers m and n. If this distance is reduced by ΔS, the wire takes the position of Acm arc with radius r. As a result, maximum sag $p_{m,n}$ is formed relative to mn (Fig. 3), which can be defined as
$p_{m,n} = OC - OD = r - r\cdot\cos\alpha = r\cdot(1 - \cos\alpha)$.

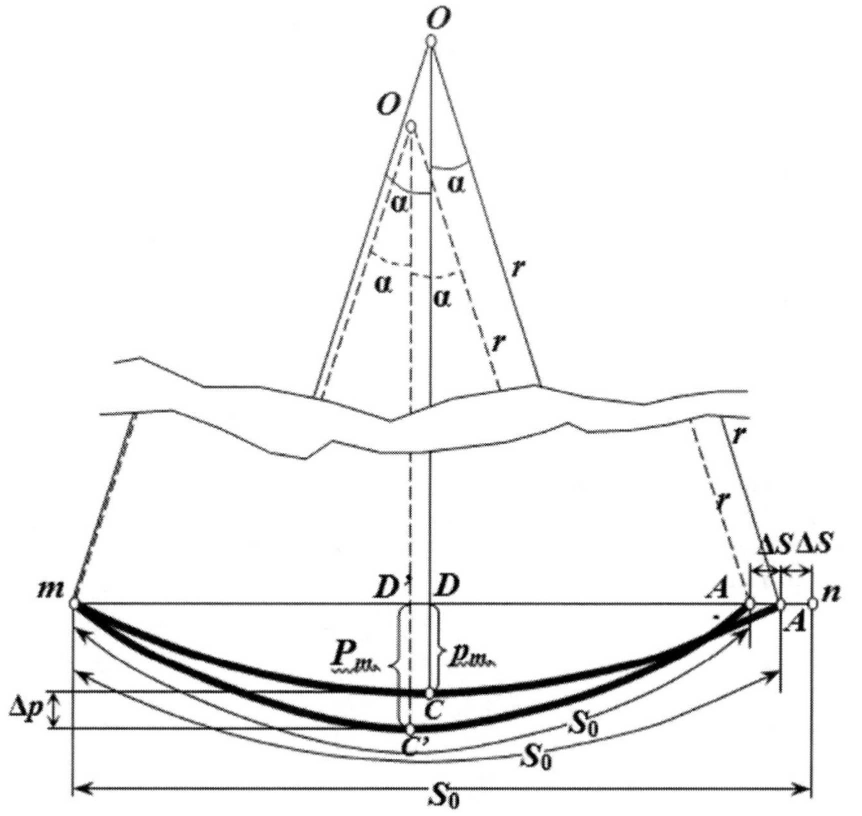

Fig. 3. Calculating the changes in wire sag between transmission towers.

ΔS is calculated as

$$\Delta S = S_0 - Am = 2r\cdot(\alpha - \sin\alpha)$$

Due to the smallness of angle α, using trigonometry formulas and expanding trigonometric functions in series, we get

$$\Delta S = \frac{8p_{m,n}^2}{3S_0}$$

From this we can determine the sag corresponding to ΔS decrease in the distance between towers m and n (Fig. 3):

$$p_{m,n} = \sqrt{\frac{3\cdot|\Delta S|\cdot S_0}{8}}. \quad (6)$$

To predict the extent sag will increase as a result of shorter distance between towers m and n, it is necessary to calculate the sag resulting from a $2\Delta S$ decrease in distance between towers m and n. The sag will be (Fig. 3):

$$P_{m,n} = \sqrt{\frac{3\cdot|(2\cdot\Delta S)|\cdot S_0}{8}}. \quad (7)$$

Then an increment or change in sag, due to shorter distance between towers m and n will be (Fig. 3):

$$\Delta p = P_{m,n} - p_{m,n} \quad (8)$$

So, using (6), we determine the size of wire sag between towers 1, 2, ..., j (Fig. 1, 2) caused by increments ΔS_1, ΔS_2, ..., ΔS_j (see (5)) between wire suspension points on the towers. Then, using formula (7), we determine the sag between the same transmission towers, but at double increment ($2\Delta S_1$, $2\Delta S_2$, ..., $2\Delta S_j$) in distances between their wire suspension points. From here, formula (8) is used to calculate the change in sag between the transmission towers (Δp_j).

If for predictive calculations, we put in (6) and (7) a decrease in distance between wire suspension points (i.e., put increment $\Delta S_j<0$), then the value Δp_j obtained from expression (8) will represent a lowering in the wire sag between the transmission towers. If in (6) and (7), we put an increase in the distance between wire suspension points (i.e., $\Delta S_j>0$), Δp_j obtained from (8) will represent a lift in wire sag between the towers.

As a result, for transmission line sections situated in the zone of influence of underground mining activities or underground construction (tunnels, underground subway stations, reservoirs, etc..), we can highlight the greatest sag changes (lifting or lowering) in the middle of the wire span between transmission towers, which correspond to an increase or decrease in wire tension, and assess the slope angles of the towers. All these are required to ensure that tower leveling and wire re-stretching are effected quickly to prevent the cables from breaking or the middle of the wire span from lowering too much. This would enable increase the distance from the middle of the wire span to the Earth's surface to a safe minimum.

To ensure that tower leveling and wire re-stretching are carried out promptly during the influence of underground mining activities, it is necessary to carry out special monitoring surveillance over the state of transmission lines. Given the peculiarities of the object, such as its reaction to the impact of underground mining activities, the following method for monitoring the technical condition of transmission lines is proposed.

3 Conclusions

The above method enables one to determine the extent of wire elongation between high-voltage transmission towers without direct contact with the high-voltage wires suspended from the towers at high altitude. That is, there is no need to disconnect the power transmission line when checking the technical condition of the cables.

References

1. *Rules for protection of structures and natural objects from harmful effects of underground mining activities at coal fields* (VNIMI, St. Petersburg, 1998)
2. *Manual for measures aimed at protecting buildings and structures from the effects of mining activities in subway construction* (Stroyizdat, Leningrad, 1973)
3. Franzius J. N. PhD thesis (London, 2003)
4. *French Society for Trenchless Technology. Microtunneling and Horizontal Drilling. Recommendations* (FSTT, London, ISTE, 2006)
5. Ghilani, Ch. D. *Adjustment Computations. Spatial Data Analysis* (NJ, John Wiley & sons, 2010)
6. Martos, F., *International Strata Control Congress* (Leipzig, 1958)

 DOI: 10.1051/e3sconf/20172101030

7. Schmidt, B. Ph.D. Thesis (University of Illinois, Urbana, 1969)
8. V.N. Gusev, D.A. Iliukhin, A.E. Zhuravlev, *Proceedings of XV International ISM Congress*, 600, 2013.
9. E. Leca, Tunnelling and Underground Space Technology, **15**, 119 (2006)
10. Puller, M. *Deep Excavations: a practical manual* (Thomas Telford, London, 2003)
11. H. Quick, J. Michael, V. Schöttner, R. Katzenbach, U. Arslan, *Tunnelling and Deep Excavation in Soft Ground*, 1998
12. W. J. Rankin, Engineering geology of underground movements, 20, 79 (1988)
13. B. Schmidt, Ph.D. Thesis (University of Illinois, Urbana, 1969)

Using of Wide Stopes in Coalless Zones Mined by Shovels and Backhoes

Valery Kolesnikov[1,*], *Oleg* Litvin[1], *Juraj* Janočko[2], and *Andrey* Efremenkov[3]

[1]T.F. Gorbachev Kuzbass State Technical University, 28 Vesennyaya st., Kemerovo, Russian Federation, 650000
[2]Technical University of Kosice, Faculty of Mining, Ecology, Process Control and Geotechnologies, Letná 9, 04200 Kosice, Slovak Republic
[3]Yurga Institute of Technology, TPU Affiliate, 26 Leningradskaya st., Yurga 652050, Russian Federation

Abstract. The examples of mining rock panels by different types of equipment are given in this paper. In such conditions, it is most expedient to use wide stopes to improve the productivity of equipment by reducing the time for auxiliary works. Also technological schemes of equipment operation in different conditions for several quarries of the Kuzbass are proposed. The coal-bearing zone is worked “layer by layer” with one or two benches or subbenches. The bottom of the quarry in this case is flat, without leaving "peak" of overburden rocks in the pillars. Cutting of the benches is carried out by trench with a wide bottom on the side of the roof of the coal bed and simultaneously working it with the same excavator or with setting up a additional excavator for coal mining operations. Interbeds are mined out with the use of wide stopes, often to the theirs entire horizontal width. Complex rock-and-coal blocks, including one-three coal beds, are also worked out by wide stopes.

1 Introduction

In the total volume of rock-and-coal mass in quarry fields, the coal-bearing zone consists 72-84% in the deposits of the central Kuzbass and about 70% in the deposits of the north and south of the basin. From the standpoint of the separate excavation of coal and rock, the coal-bearing zone is a complex structure, which is determined by the bedding of coal strata of varying thickness and angles of incidence; the presence of thin rock interpbeds, folding (plicative dislocations), disturbance and hard bedrocks (strength limit 60-80 MPa, less often up to 130 MPa). Uncertainty of capacity and frequent variability of elements of occurrence of separate rock layers within the limits of a quarry field both in length, and in depth are observed. Thus, the development of the coal-bearing zone represents the main difficulty in the open pits, and consequently, the increase in the efficiency of surface mining is directly related to creating of new technological and technical solutions.

* Corresponding author: ormpi@kuzstu.ru

It should be noted that in the classical theory of open mining, the concept of "coal-bearing zone" for bed deposits is absent. Describing the theory of development of open pit mining on inclined and steep deposits, the development of just one layer is considered.

In this case, the width of the bottom of the pit is equal to the horizontal width of the coal seam.

2 Theory

The development of coal-bearing zones by different types of excavation equipment was considered in a number of articles [1-8]. It is recommended to develop of coal-bearing areas (zones) by subbenches (sublayers). In this case, excessive coal losses are excluded, because the approach of mining to each of the strata is always carried out from the side of the hanging side (the roof of the formation).

Experience has shown that when shovels work in a coal-bearing zone, the change in the type of excavator work (the transition from coal extraction to overburden) occurs relatively frequently (up to 3-5 times a month). As a result of such transitions, the use of the excavator in the main work (excavation and loading of the overburden) deteriorates and its productivity decreases by 18-20%.

Along with the complex operating conditions of machines in the face, there are some features of the opening-up and preparation of new horizons (benches). For example, the opening-up of benches is carried out during the whole period of exploitation of the quarry, and the methods of opening-up of individual benches or their groups may be different (external or internal trenches, sliding trenches). Preparation of working benches, as a rule, is carried out simultaneously along several of the thickest beds of the formation.

The notion of a "wide stope" was formulated in [9], where it is noted that its development is characterized by the variable (shuttle) direction of the excavator's movement in the plan. The use of wide stopes is possible with maneuverable quarry trucks, what simplifies the organization of mining operations.

3 Results and discussion

Cutting of the behches by trenches with a wide bottom and working out of the coal-and-rock blocks by wide stopes with the use of shovels is shown in Fig. 1 on the example of the "Bachatsky" open pit.

As it shown on Fig.1, the width of the trench bottom is 53…120 m when cutting a new bench. The trench can be worked out both in interbeds without coal mining operations (Figure 1-a), and with them (Figures 1-b, 2-a, 2-b). On a open pit "Bachatsky" when coal dip angles are 60÷75°, the excavation of coal beds is made to the full height of the bench.

The application of wide stopes on some of Kuzbass open pits for working out by backhoes is shown in Fig. 3. As can be seen from this table, wide stopes are using for interbeds mining. The width of the stope is 40÷65 m. The benches are mined by layers of 3÷5 m, usually 4 m.

Shovels do not always ensure a high efficiency of development of mineral deposits due to their inconsistency with working conditions. It is manifested in the limited choice of the location of the transport setting for loading (only on the level of the excavator standing), at a low height of the formation of the soil (roof) of the formation, in particular, at dip angles of incidence 15-50°, in the inability to provide the specified angle of the slope of the face, in the need to create wide cut trenches. These disadvantages are particularly noticeable in difficult conditions, but they are almost uncharacteristic of backhoes, which has the ability to dig below the level of standing of the excavator.

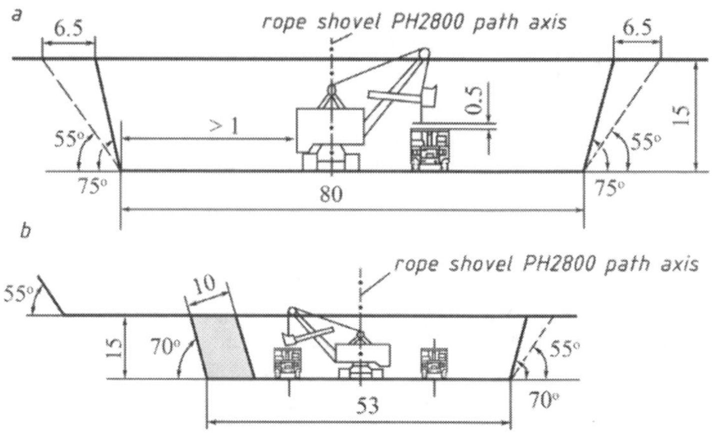

Fig. 1. Using of wide stopes for coal-bearing zone excavation on open pit "Bachatsky": a – cutting of new bench on thick interbed; b – cutting of new bench with simultaneous coal extraction on seam Burned.

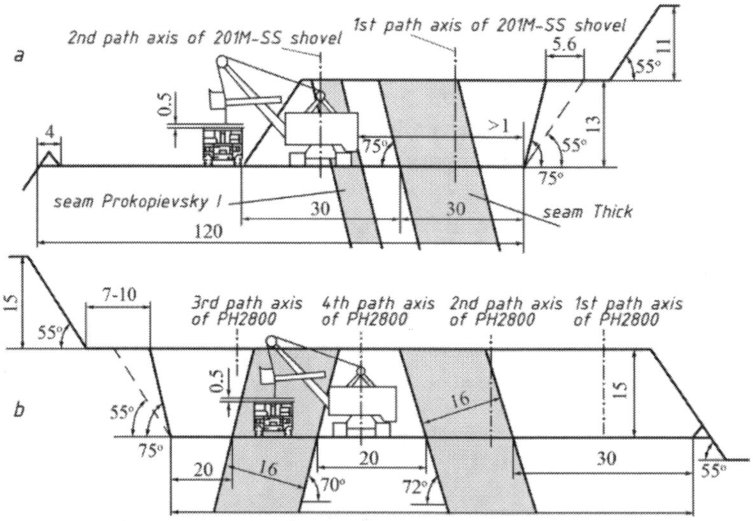

Fig. 2. Selective mining of hard-structural stope including two nearby seams: a – stope includes two close located seams Prokopievsky I and Thick; b – stope included seam Burned in zone of its plicative dislocation (left and right wings of fold).

When using such machines, there is no need to set transport on the excavation horizon. The excavator has a great digging force, changing trajectory of the bucket during all the operations of the cycle; the overall dimensions of the excavators are relatively small, which is very important for the development of complex coal-bearing zones, often with an unsettled occurrence of seams and a variety of forms of their location in the stope.

A feature of the technology with backhoe is the maximum use of shunting capabilities of dump trucks due to their selective setting at the bottom of the bench in conditions of work with the bottom loading. As a result, the minimum value of the angle of rotation of the excavator is ensured, which does not depend on the width of the stope.

The cutting of a new bench with the making of coal mining operations, both for inclined beds and for steep ones is made by wide stopes from 30 to 100 m. The cutting trenches are also excavated by layers (subbenches).

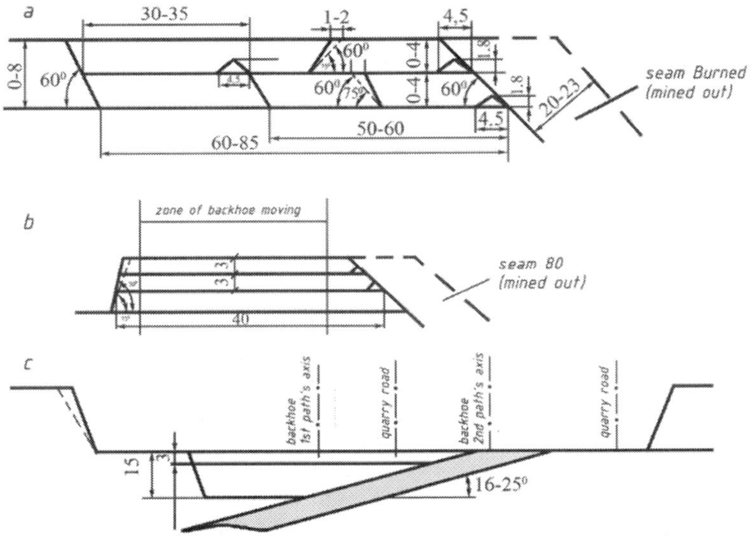

Fig. 3. Excavation of interbed by backhoe: a – open pit mine "Krasnobrodsky" after mining out seam Burned, backhoe Terex RH-200; b – open pit mine "Taldinsky" after mining out seam 80, backhoe Liebherr R9350; c – cutting of new bench on open pit mine "Kedrovsky", backhoe Caterpillar 385C.

4 Conclusion

Based on the analysis we can conclude the following.

1. At present, a wide range of excavator models of the following types is used for the development of coal-bearing zones: rope shovels, hydraulic shovels, backhoes. For the mining of rock interbeds, rope shovels with bucket capacity of 5…30 m^3 and, less often backhoes with a bucket capacity of 3.4…7 m^3 are used.

2. Cutting of the benches is carried out by trenches with a width on the bottom from 53 m to 120 m from the side of the roof of the coal bed and simultaneously working it with the same excavator or with the setting of additional excavator for coal mining operations. Interbeds are worked out with the widespread use of wide stopes with horizontal width up to 100 m. Complex rock-and-coal blocks, including one-three layers, are also worked out by wide stopes.

References

1. E.I. Vasil'ev, V.F. Kolesnikov, Soviet Mining Science **5**, 532 (1969)
2. P.M. Alabuzhev, O.D. Alimov et al., Soviet Mining, **2**, 183 (1966)
3. Yu.I. Belyakov, V.E. Boguslavskii and S.A. Skachkov, Soviet Mining, **21**, 165 (1985)
4. B.N. Lokhanov, Y.A. Zakharov et al., Sov. Min., **3**, 523 (1967)
5. M.M. Bereznyak, A.V. Kalinin and V.G. Pronoza, Soviet Mining 6(6), 638 (1970)
6. I.V. Nazarov, J. Min. Sci., **48**, 55 (2011)
7. N. Demirel, J. Min. Sci., **47**, 441 (2011)
8. I. Vukotic, V. Kecojevic, W. Zhang, Q. Cai, Int. J. Min. Sci. Tech., **23**, 901 (2013)
9. Rzhevsky V.V. *Open pit mining* (Nedra, Moscow, 1985)

E3S Web of Conferences **21**, 01032 (2017) DOI: 10.1051/e3sconf/20172101032
The Second International Innovative Mining Symposium

Intelligent Mining Engineering Systems in the Structure of Industry 4.0

Marina Rylnikova[1*], *Dmitriy* Radchenko[1], and *Dmitriy* Klebanov[2]

[1]IPKON RAS, 111020, 4 Kryukovsky tupik, Moscow, Russia

[2]VIST Group AO, 107078, 1/2 Dokuchaev pereulok, Moscow, Russia

Abstract. The solution of the problem of improving the human environment and working conditions at mines is based on the provision of the rationale of parameters and conditions for the implementation of an environmentally balanced cycle of comprehensive development of mineral deposits on the basis of the design of mining engineering systems characterized by the minimization of the human factor effect in danger zones of mining operations. In this area, robotized technologies are being developed, machinery and mechanisms with the elements of artificial intelligence, and mining and transport system automatic controls are being put into service throughout the world. In the upcoming decades, mining machines and mechanisms will be virtually industrial robots. The article presents the results of zoning of open-pit and underground mine production areas, as well as mining engineering system of combined development depending on the fact and periodicity of human presence in zones of mining processes. As a surface geotechnology case study, the software structure based on a modular concept is described. The performance philosophy of mining and transport equipment with the elements of artificial intelligence is shown when it is put into service in an open pit.

1 Introduction

The first three industrial revolutions have occurred as a result of mechanization, electrification and information technologies. The progress of Internet and satellite navigation have determined a new industrial epoch – the fourth industrial revolution – Industry 4.0 [1]. Global trends of each and all large production corporation entry to Industry 4.0 give clear-cut interpretation of the prospects of mining industry development – transition to intelligent control systems with the human factor role changing in design, operation of mining engineering systems and their condition monitoring [2].

Industry 4.0 is not only and not so much new technologies as a new approach to production and consumption. It is based on the collection of big data, their processing and use for the perfection of coordinated operation and processes without human interference [3]. The base trend of Industry 4.0 development is the assignment of optimization functions to machines and mechanisms and customization of their operation in autonomous mode. A

* Corresponding author: rylnikova@mail.ru

trend is now emerging, according to which the objects of industrial Internet will independently perform the monitoring of their operation, adjust the parameters sending data to the manufacturer for the elimination of machine design shortcomings and estimation of the remaining operation life of some of the machine components, replacement parts, expendable materials, depending on operation in specific geological mining and natural conditions, and climate. It means that one of the most significant aspects of the fourth industrial revolutions is the idea of mining machinery "oriented design service". For instance, for surface mining operations Komatsu has already designed a system of remote prediction of operability and condition monitoring of a diesel engine [4]. The system envisages sending of daily statistics on the variation of shovel engine operation parameters to a remote operator. These parameters include: pressure, consumption of fuel, temperature and rpm of the engine. Komatsu engineers note [4] that this system including the necessary algorithms of data processing, namely, the estimates of the engine current conditions, identification of the source of abnormal behavior and, what is most important, forecast of the remaining operation life of the diesel engine, is a pioneer system. In the upcoming decades, mining machines and mechanisms will be virtually industrial robots [5–8], and their application will change the requirements to mining engineering systems design that in its turn will result in large-scale changes in the effect of mining engineering systems on human environment.

2 Materials and methods

The solution of the problem of human environment and working conditions at mines is based on the provision of the rationale of parameters and conditions of the implementation of an environmentally balanced cycle of comprehensive development of mineral deposits on the basis of the application of mining engineering systems with intelligent control [9]. In this area, robotized technologies, automatic controls for mining and transport systems are being developed throughout the world. For design and perfection of mineral mining and processing processes it is important to find the solutions aimed at the development of resource-saving and resource-reproducing geotechnologies [10], elimination or minimization of human presence in danger zones of mining operations: development headings, faces and stopes, zones of intensive deformation of the rock mass, including those with dynamic rock pressure occurrence, areas with high radioactive background, unfavorable environment – high or low temperature of the rock mass, high gas content, low oxygen content or pressure of the atmosphere. Of significant interest is the establishment of fundamental objective laws of the interoperability of mining engineering systems employing industrial robots with the natural – human environment in the mining engineering system and in the affected area [11,12].

To provide the rationale of an environmentally balanced cycle of comprehensive development of mineral deposits on the basis of the application of mining engineering systems with intelligent control an approach has been proposed aiming at the creation of micro-climate zones characterized by particular, specified and controlled parameters of the environment listed above depending on the fact and periodicity of human presence in the zones of mining processes.

3 Results and discussion

In mining engineering systems, where it is planned to operate the equipment with the elements of artificial intelligence, all zones must be divided into three fundamentally different geotechnological areas, in each of which the specific condition of the human environment are maintained (fig. 1):

I. conditionally manless zones or, as it is referred to in the internationally established terminology, Zero Entry Production Areas (ZEPA) [13];

II. zones of human presence in a mining engineering system to perform main and auxiliary work and maintenance of machines and mechanisms;

III. zones of permanent human presence.

Such zoning dictates the necessity of top-priority solution of the problem of the joint presence of people, mechanized and robotized equipment with the establishment of relevant safety requirements.

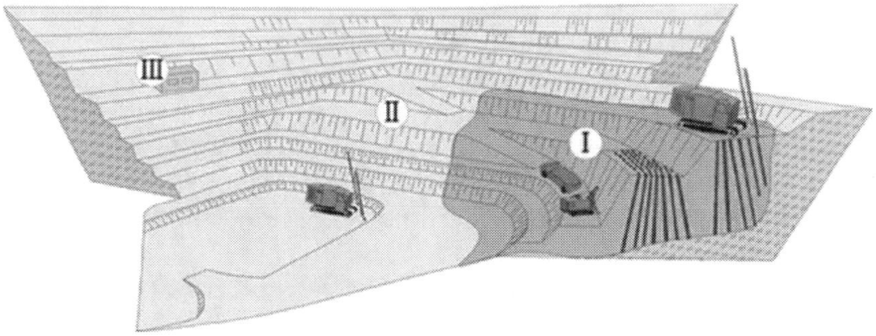

Fig. 1. Principle of zoning of a mining engineering system with the use of industrial mining and transport robots.

In surface mining, zones with "zero entry" will cover open-pit areas specially planned for operation of equipment with elements of artificial intelligence and characterized by smaller width ramps, greater slope angles of benches and pit walls, poorly ventilated and other danger areas. As it is shown in fig.1, equipment operation in zones I-III will be provided by the software including the appropriate modules optimizing the performance indices of an intelligent mining engineering system as a whole. For instance, to control a load-haul complex in the Intelligent Mine system a special module is provided, which combines the functions of loading and hauling control – shovels and loaders, and robotized mine dump trucks. Loading machinery control is performed by an operator from the cab in a manual mode, while dump truck control is performed autonomously without interference of operators.

The main task of the module is in automated control of dump trucks (without interference of operators), while they are on specified routes, in autonomous control mode, including loading and discharge functions. With the help of this module the system transmits data about the route, end point of travel, current task, information about the equipment in the working zone and other necessary data to on-board computers of dump trucks.

The module also includes a dump-truck remote control gear. The control is performed by the supervisor or operator of remotely controlled machinery from the control center throughout the whole route.

For robotized dump trucks the module performs the following base functions:

 DOI: 10.1051/e3sconf/20172101032

- monitoring of simultaneous travel of several automated dump trucks on prescribed routes in autonomous control mode with a feature of remote control from the control center throughout the whole route;
- monitoring of travel and condition of automated dump trucks, haulers, which are not furnished with autonomous/remote control equipment, and auxiliary machinery shown on the map, the required reporting, storage of the history of travel and events on a centralized server;
- providing runs for loading and discharge in autonomous and remote control modes;
- providing interoperability with the personnel for fuelling and maintenance of automated machinery in a dedicated zone;
- stop and overspeed monitoring;
- automatic identification of dump-truck hauls and routes;
- remote monitoring of the condition of on-board equipment for autonomous/remote control and a feature of connectivity with the available dump-truck diagnostic systems;
- keeping a supervisor informed on an accident hazard;
- export and e-mailing of reports;
- import of a mine map from mine surveying-geological and mine surveying systems.

For remotely controlled machinery the following functions are typical:

- remote control mode for machinery control;
- displaying the main parameters at the operator's workplace;
- augmented reality for a remote control operator. Various information helping the performance of control is displayed over the surveillance camera images;
- control of location on the map with the display of auxiliary information: current condition of every unit (speed, coordinates, fuel level, etc.);
- control of equipment condition and recording of machinery misuse;
- keeping an operator informed on emergency situation, and failure to further travel in a remote control mode;
- issue of shift tasks and formation of immediate reporting, shift reports and reports on other periods.

At a time one operator from the remote control point can remotely control one unit of equipment, and if needed, switch the control to some other machinery, thus having an opportunity of serving several units.

For drilling in autonomous mode a separate software module is provided, which performs the autonomous job of a drill rig and monitoring of drilling parameters. A drill rig is directed to boreholes and drills the rock mass without interference of an operator. The module performs the following functions:

- automatic recording of task performance;
- automatic monitoring and control of a drill rig position;
- collection of production and technical information on the equipment condition;
- collection and accumulation of geological data generated in the process of drilling;
- automatic collection of information about actually drilled boreholes (there is no need for mine surveyor to take measurements of actually drilled boreholes).

An autonomous drill rig performs the following processes without interference of an operator:

- direction to planned boreholes;
- platform alignment;
- drilling, including rod handling;
- monitoring of drill rig location on the map with the display of auxiliary information: current condition of each unit (speed, coordinates, fuel level, etc.).

Transmission of accurate information on drilling location to an on-board computer of a drill rig together with the use of a high-accuracy navigation system also helps eliminate the problem of low-quality drilling and blasting processes – great share of oversize.

The efficiency of coordinated operation of separate modules is determined by a module of static optimization. This module is activated by a supervisor from the operating control module, and it displays the results of its operation in a window of an operating control module. It automatically performs the initial distribution of dump trucks at the beginning of a shift by route groups, and redistribution of machines in the course of a shift in case of open pit situation changing (breakdown of a shovel, changes in quality requirements of ore supplied to a processing plant, etc.). The results of this module operation are used as source data for the module of dynamic optimization. The module of dynamic optimization performs the redistribution of robotized dump trucks within each optimized group in the course of a shift in compliance with the specified criteria of optimization. It is automatically activated after each discharge of a dump truck and in automatic mode transmits the instructions to the control system of a dump truck if a change of its route is needed. Change of the route is also displayed in a window of the operating control module. Change of a dump-truck route is also envisaged in a semi-automatic mode, when the system issues a notification to a supervisor on the necessity of changing a route, and a supervisor takes a decision.

One of the most important control elements of the Intelligent Mines system is a module of maintenance and repair planning. This module becomes a logical complement of accounting of all service factors of machinery operation and provides impartial estimation of time to the subsequent maintenance and repair forming the chains of intervals between maintenance and repair for equipment and machinery by motor-hour, mileage or time. Estimated time of maintenance can be adjusted manually for optimal operation of maintenance and repair crews. Maintenance planning and repair mechanism (hereinafter referred to as M and R module) provides the solution for the following tasks:

- keeping the directory of maintenance intervals. Intervals can be specified by time, mileage and motor-hour. Eventually, this information is used in automated calculation of the subsequent maintenance and repair process.
- automatic prediction of the date of the subsequent planned maintenance and repair based on the information about motor-hours, mileage or time, with a feature of manual input and updating of information. This form is intended both for planning and for accounting of actually performed work.
- accounting for mobile object availability in the formation of a shift job order.

For combined surface-underground geotechnology mining engineering systems are zoned depending on geological and mining conditions of deposit development. For instance, with a significant depth of transition from surface to underground mining zero entry zones will include in addition to the above listed the bottom elevations of an open pit (fig. 2), which are poorly ventilated areas, as well as deposit sites of the surface-underground level. For instance, in conditions of combined development of the Trubka Udachnaya deposit open pit dimensions are large in plan and in depth. Steep slope angles, risks of aggregation of mined material due to slumping, consolidation or freezing, formation of non-drainable slurry-like material in an open-pit shell, inrush of gases, pressure water or slurry to mined-out areas make preferable the mineral mining technology in "ore triangles" and open-pit bottom, which does not require human presence in a stoping zone. Without human presence in this zone the ventilation parameters can be any and they

DOI: 10.1051/e3sconf/20172101032

are limited only by the conditions of mining and transport machinery operation. Overlying formations of an open pit, where transport communications are preserved, belong to the second ecological zone – the one of human presence in a mining engineering system with an aim of performance of auxiliary processes and servicing of machinery and equipment. The third type of zone with human presence is characterized by comfortable conditions of work and is aimed at monitoring of automated mining processes control system. This work is performed in office buildings.

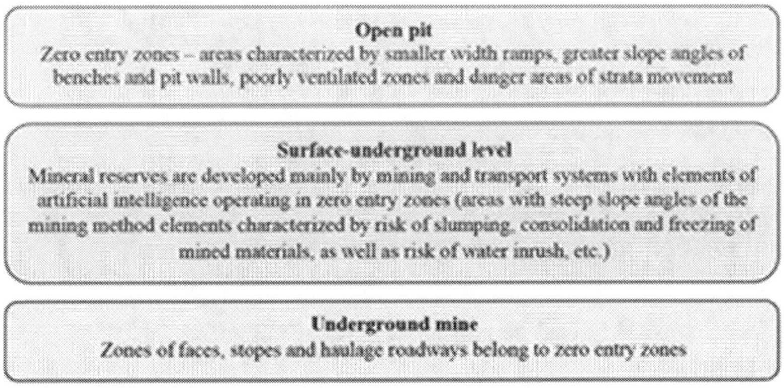

Fig. 2. The principle of identification of conditionally manless zones (zero entry areas) in a mining engineering system.

Underground mining operations are characterized by the zoning based on the above described philosophy. In zero entry zones the microclimate required only for machinery and equipment operation can be created, therefore, planning of underground workings parameters, as well as ventilation system for such conditions are completely different. Underground location with ZEPA is isolated from the mine atmosphere. New opportunities are opened of spot mine air intake and purification via a system of boreholes and pipelines, without integration of contaminated air streams into a common mine network. Human entry to a ZEPA zone, for instance, for the elimination of a possible accident consequences or for robot withdrawal, is allowed only after implementation of special ventilation measures , or without them, but with the use of special personal protection devices. New vision is acquired to industrial safety requirements concerning underground mine ventilation after blasting, ventilation of dead ends and many others. With that it is evident that a system of new requirements of industrial and ecological safety is coming into existence.

4 Conclusions

A mining engineering system, which was earlier designed based on the conditions and limitations concerning human working environment from this time onward, with the development of remote and robotized geotechnologies, autonomous means of the rock mass state monitoring and their combination will assume the grounding of parameters aimed only at the provision of conditions for the presence of robotized and autonomously controlled machinery and equipment in zones of mining operations. Removal of sanitary-hygiene requirements to human presence in these zones, as well as limitations and requirements of industrial safety determines new requirements to the formation of mining engineering systems and grounding of parameters of human environment in a mining engineering system and in the affected area.

 DOI: 10.1051/e3sconf/20172101032

The studies are performed with the support of the Russian Science Foundation (Project No14-37-00050)

References

1. *Recommendations for implementing the strategic initiative Industrie 4.0. Industrie 4.0* (Working Group, 2013)
2. B. Johansson, Int. J. Min. Miner. Eng, **5:4**, 350 (2014)
3. L. Abrahamsson, B. Johansson, J. Johansson, Int. J. Min. Miner. Eng, **1:3**, 304 (2009)
4. K.N. Trubetskoy, Gorn. Zhurnal, **5**, 21 (2016)
5. B. Autonomous, Auton. Min., **1**, 30 (2012)
6. J.J. Green, Int. J. Eng. Adv. Technol., **1:4**, 8 (2012)
7. D.R. Kaplunov, Gorn. Zhurnal, **7**, 49 (2014)
8. K.N. Trubetskoy, J. Min. Sci., **47:3**, 317 (2011)
9. M.V. Rylnikova, D.N. Radchenko, Gorn. Zhurnal., **12** (2014)
10. K.N. Trubetskoy, Y.P. Galchenko, J. Min. Sci., **51:2**, 407 (2015)
11. G. A. Nikolakopoulos, IFAC., 66 (2015)
12. O. Golovina, J. Teizer, P. Nipesh, Autom. Constr. **1** (2016)
13. D.R. Kaplunov, D.N. Radchenko, Gorn. Zhurnal. **5** (2016)

Causes Of Low Efficiency Of Combined Ventilation System In Coal Mines In Resolving The Problem Of Air Leaks (Inflows) Between Levels And Surface

Valeriy Popov[1*], *Yuriy* Filatov[1], *Hee* Lee[2], and *Anatoliy* Golik[2]

[1]JSC «NC VostNII», 650002, 3 Institutskaya Str., Kemerovo, Russia

Abstract. The paper discusses the problem of the underground mining safety control. The long-term air intake to coal accumulations is reviewed as one of the reasons of endogenous fires during mining. The methods of combating air leaks (inflows) in order to prevent endogenous fires are analyzed. The calculations showing the discrepancy between the design calculations for the mine ventilation, disregarding a number of mining-and-geological and mining-engineering factors, and the actual conditions of mining are given. It is proved that the conversion of operating mines to combined (pressure and exhaust) ventilation system in order to reduce the endogenous fire hazard of underground mining is unreasonable due to impossibility of providing an optimal distribution of aerodynamic pressure in mines. The conversion does not exclude the entry of air into potentially hazardous zones of endogenous fires. The essence of the combined application of positive and negative control methods for the distribution of air pressure is revealed. It consists of air doors installation in easily ventilated airways and installation of pressure equalization chambers equipped with auxiliary fans near the stoppings, working sections and in parallel airways.The effectiveness of the combined application of negative and positive control methods for the air pressure distribution in order to reduce endogenous fire hazard of mining operations is proved.

1 Introduction

Currently, the principal direction of the technical policy in the coal industry is the abandonment of the extensive development path. In future, the coal production should be increased mainly through the reconstruction of existing enterprises and their technical re-equipment, concentration and intensification of the mining operations, which will lead to an accelerated dynamics of deep mining and, consequently, deterioration of the geological and mine-engineering factors. These circumstances contribute to the frequency and severity of accidents, and therefore, the problem of ensuring the proper safety of underground mining is of significant importance.

The analysis of accidents in the Kuzbass mines showed that over the last decade more

* Corresponding author: a.zakirova@nc-vostnii.ru

than 30% of such accidents were endogenous fires. In this regard, prevention and control of dustiness and endogenous fire are among the most important issues associated with the overall problem of lowering the accident rate in the coal industry. Thus, many Russian and foreign researchers addressed the issue of dust control in coal mines [1-9], highlighting it as one of the factors affecting industrial safety. Some researchers offered various methods of dust control and dust removal [10-15].

2 Materials and methods

A necessary and virtually sufficient condition for the formation of a spontaneous ignition source in a fire-hazardous seam is long-term (exceeding the incubation period) airflow to accumulations of broken-down, collapsed or loose coal; such air flow contributes to the oxidation processes and leads to the breaking-out and development of endogenous fires.

To exclude such negative consequences, it is necessary either to avoid loss of coal, or to prevent the flow of oxygen to the coal accumulations. Considering that the first condition is nearly impracticable at the modern technical level of underground coal mining development, the main attention should be given to the second solution.

Based on the extensive scientific research and experiments, the following methods for combating air leaks (inflows) have been developed:

- isolation of mined-out working fields, blocks, extraction columns, sections of certain inactive workings,
- ensuring the optimal distribution of aerodynamic pressure along the network of mine workings, which excludes the entry of air into potentially hazardous, fire-prone areas with the use of negative and positive control methods;
- use of the combined ventilation system to maintain the required aerodynamic head pressure in the extraction areas.

The existing means and methods of isolation include the use of rather expensive airtight materials, structures and fixtures. However, in spite of the significant material and labor costs of isolation works, the commonly required conditions of preventing oxidation in isolated workings are not ensured.

Better effect can be achieved by application of aerodynamic means to combat air leaks (inflows). Such means are based on the regulation of the distribution of ventilation pressure and air flows in the network of mine workings, using negative and positive control methods.

The negative control method of regulation consists in the installation of ventilating windows in easily ventilated branches. At that, the resistance of the window determines the amount of flowing air, and the place of installation, the distribution of ventilation pressure in the branch. This makes it possible to control the ventilation pressure by installing ventilation windows simultaneously at both ends of the ventilated branch. The total aerodynamic resistance of these windows provides flow of the required amount of air, and the ratio of their resistances creates the necessary ventilation pressure inside the branch.

The required values of the ventilation pressure at a given air volume distribution can be ensured simultaneously in several different branches with the help of ventilation windows if each branch has only one point in which the required ventilating pressure should be provided, and if no branch with a given pressure is ventilated in sequence with another branch. Such conditions limit the applicability of this method.

The positive control method has a wider use. Here, the air terminal boxes equipped with auxiliary fans are installed at separate stoppings, in active working faces and in parallel branches.

A special place among the aerodynamic methods of combating air leaks (inflows) belongs to the combined method of ventilating the mines. It is typically used to create a zero-pressure zone in the extraction areas and, thus, reduce the amount of air flowing through the mined-out sections to minimum values. Theoretically, this method of ventilation should give the same effect as the use of air terminal boxes. However, in practice, after conversion of the operating mines to combined ventilation, increase in the level of endogenous fire safety of mining operations is, as a rule, not observed. This can be explained by the fact that, due to low compressions of the forced-draught fans, the zero-pressure zone is mainly located near the downcast shafts, while in the extraction areas negative pressure occurs. In such cases, the combined ventilation is equivalent to the exhaust ventilation in terms of fire safety.

Insufficient compression of the forced-draught fans in the combined ventilation system is primarily determined by the fact that a number of factors are not taken into account in the ventilation system design. Thus, the designed rate of surface leakage, equal to 10% (recently, 20%), can be achieved only in special airshafts that are not used for lifting operations. As targeted aerodynamic surveys showed, in the cage shafts leakage is 30-40% in average and in some cases reaches 50%. There are still no reliable methods of sealing the shafts equipped with a lift, so that the actual surface leaks would be equal to the designed ones. With poor sealing of the overhead buildings, the aerodynamic resistance, that the ventilator works for, reduces significantly and, as a result, the compression on the pressure side falls lower than designed.

Low compressions of the forced-draught fans is also pre-determined by the fact that the amount of air supplied to the mine is often greater than the designed due to the intensified production, as compared to the designed scope, and inaccurate gas content forecast. The amount of flowing air can be augmented by improving the capacity of the forced-draught fans or increasing their number. In such cases, the actual individual aerodynamic characteristics of the fans are aligned, which leads to a drop in their compressions.

When choosing forced-draught fans, as a rule, the loss of negative pressure in the air heater and air supply duct is not taken into account; it is assumed that the differential pressure created by the forced-draught fan is used only to overcome the resistance of underground workings, starting from the evasee. In fact, as the negative pressure surveys showed, up to 50% of the differential pressure is consumed by the forced-draught fans on the suction side, which leads to a nearly twofold drop in compression.

It should also be noted that such factor as the change in aerodynamic parameters of the mine ventilation networks in the course of operations is underestimated. In connection with the extraction of the upper horizons and application of the pillarless technology, the worked-out space acquired a special role in setting of ventilation modes. Large areas and volumes of excavations, numerous aerodynamic connections with the operated workings and the soil surface significantly affect the ventilation characteristics of the mines and the distribution of air between the objects of ventilation. The presence of numerous yet neglected parallel air paths in the aerodynamic systems pre-determines the displacement of the fan operation points towards the axis of abscissas, and, as a consequence, the decrease in the aerodynamic pressure heads developed by them.

Eventually, neglect of the above factors leads to the fact that with the combined ventilation, the zero-compression point often appears to be directly under the downcast shaft, instead of the extraction area, and the operation of the forced-draught fan has no effect on the endogenous fire safety of mining operations.

Consider a specific example. Figure 1 shows a design model of the mine ventilation system.

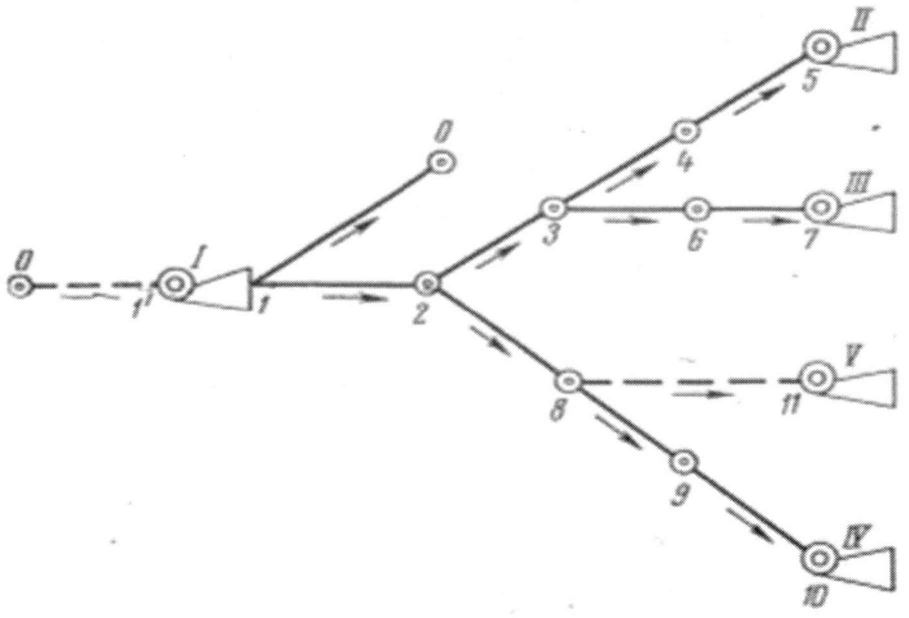

Fig. 1. Design model of the mine ventilation system.

At leakage resistance of 8.4 kμ (branch 1-0) and the operating modes of the forced-draught fans specified in Table 2 (Variant I), the BOD-21 fan (n = 750 rpm, Θ = 30°) will operate in the mode denoted by point *a* in Figure 2 (point *a′* denotes the operating mode of the fan in the mine). Air leakage (10%) corresponds to the designed value. The zero-pressure zone is located in points 4, 6, 9 (Figure 1), which corresponds to the extraction areas.

If now the resistance of leakage paths is reduced from 8.4 kμ to 0.39 kμ, then in order to maintain the amount of air entering the mine at the same level, it is necessary that the suction fans worked in the modes given in Table 1 (Option II).

Table 1. Operating modes of the forced-draught fans

No. of the fan in Figure1	Operating modes of the forced-draught fans							
	Option I		Option II		Option III		Option IV	
	h, daPa	Q, m^3/s	h, daPa	Q, m^3/s	h, daPa	Q, m^3/s	h, daPa	Q, m^3/s
I	210	50	100	61	131	57.5	91.0	62
II	20	15	130	15	169	15.0	274.2	15
III	16	15	126	15	165	15.0	270.2	15
IV	15	15	125	15	164	15.0	358.7	15
V	-	-	-	-	-	-	290.0	12

Then the forced-draught fan will operate in the new mode denoted by the point *b* in Figure 2, and the leakage will be 26% (point *b′* indicates the operating mode of the fan, modified to the mine conditions). Comparison of the two operating modes of the forced-draught fan (points *a* and *b*) shows that the increase in the leakage rate sharply reduces the compression of the VOD-21 fan.

Introduction of an air heater and its duct into the resistance scheme will require the suction fans to be transferred to the operating modes given in Table 2 (Option III) in order to maintain the calculated amount of air. At that, the total differential pressure of the forced-draught fans will increase (refer to point *c* in Figure 2), but the compression on the pressure side will be less than it would be without an air heater (point *c'*).

Thus, the discrepancy between the actual leaks and the designed ones and the neglect of the depression losses on the suction side lead to the fact that the forced-draught fan develops the compression of 61 daPa, instead of the initially adopted compression of 210 daPa. As a result, the negative pressure of 149 daPa occurs in points 4, 6, 9 (Figure 1), instead of the zero pressure.

If now the air intake to the mine is increased from 45 m^3/s to 57 m^3/s by changing the operating mode of the suction fans and connecting an additional suction fan, the operating mode point of the forced-draught fan will move downward along the descending line of pressure characteristics and take the position indicated by the letter *d* in Figure 2, while the compression will be only 10 daPa (point *d'*). At such compression value, the zero pressure zone does not reach beyond the shaft bottom and the application of the combined ventilation method, according to the pressure distribution conditions, substantially loses its meaning.

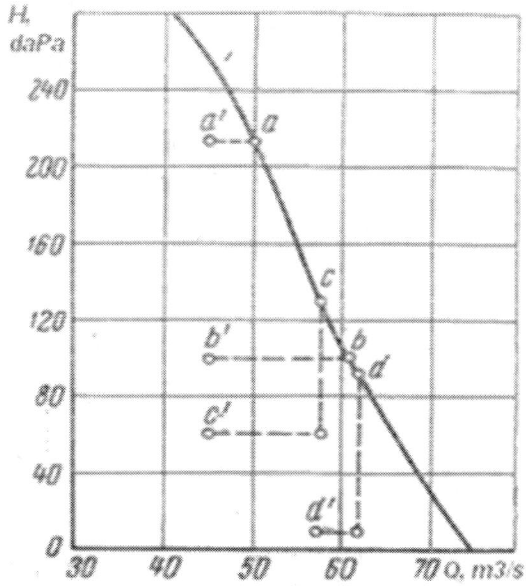

Fig. 2. Changes in operating modes of the forced-draught fan at various leakage values and upon introduction of the air heating duct and the additional suction fan.

3 Conclusions

Thus, it can be concluded:

Firstly, conversion of the operating mines to combined ventilation in order to reduce the endogenous fire hazard of mining operations is impractical, for it is virtually impossible to ensure the optimal distribution of aerodynamic pressure in the existing ventilation networks and exclude the possibility of air cross-flows over the mined-out areas and between the working horizons and the soil surface.

Secondly, on the basis of the aerodynamic principle, the proper prevention of air leaks (inflows) into the potentially hazardous, fire-prone areas can only be achieved by the complex application of negative and positive methods of the ventilation pressure distribution.

References

1. H. Breuer, Glückauf, **108:18**, 806 (1972)
2. A.G. French, Proceedings of the European Economic Communities Conference on Dust Control, 57 (1983)
3. G.V.R. Goodman, T.W. Beck, D.E. Pollock, J.F. Colinet, J.A. Organiscak, Proceedings of the 11th U.S., North American Mine Ventilation Symposium London, 211 (2006)
4. A.B. Cecala, J.A. Organiscak, W.A. Heitbrink, J.A. Zimmer, T. Fisher, R.E. Gresh, J.D. Ashley Transactions of Society for Mining, Metallurgy, and Exploration, **314**, 31 (2003)
5. F.N. Kissell, G.V.R. Goodman, *Continuous miner and roof bolter dust control. In: Handbook for dust control in mining* Pittsburgh, 23 (DHHS, NIOSH Pub., 2003)
6. G.J. Chekan, J.M. Listak, J.F. Colinet Proceedings of the Seventh International Mine Ventilation Congress, 291 (2001)
7. J.M. Listak, G.J. Chekan, J.F. Colinet, Transactions of Society for Mining, Metallurgy, and Exploration, **310**, 155 (2001)
8. T.F. Tomb, Transactions of Society for Mining, Metallurgy, and Exploration, **288**, 1874 (1992)
9. J.A. Organiscak, A.B. Cecala, E.D. Thimons, W.A. Heitbrink, M. Schmitz, E. Ahrenholtz, Yernberg WR, ed. Transactions of Society for Mining, Metallurgy, and Exploration, **314**, 145 (2003)
10. J.F. Colinet, R.A. Jankowski, Transactions of Society for Mining, Metallurgy, and Exploration. **302**, 104 (1996)
11. K.G. Fields, D.J. Atchison, R.A. Haney, Proceedings of the Third Symposium on Respirable Dust in the Mineral Industries, 349 (1991)
12. F.N. Kissell, *Dust control methods in tunnels and underground mines. In: Handbook for dust control in mining*, 3 (DHHS, NIOSH, 2003)
13. G.J. Chekan, J.M. Listak, J.F. Colinet, Yernberg WR, ed. Transactions of Society for Mining, Metallurgy, and Exploration. Littleton, **316**, 186 (2004)
14. J.F. Colinet, E.R. Spencer, R.A. Jankowski, Proceedings of the Sixth International Mine Ventilation Congress, **55**, 345 (1997)
15. E.F. Divers, R.A. Jankowski, J. Kelly, Proceedings of the Third U.S. Mine Ventilation Symposium, 85 (1987)

Problem Of Methane-Air Mixture Explosions In Working Faces Of Coal Mines At Mining Intensification And Ways Of Its Solution

Sergey Novoselov[1*], *Valeriy* Popov[2], *Yuriy* Filatov[1], *Hee* Lee[2], and *Anatoliy* Golik[2]

[1]International Academy of Ecology and Life Protection Sciences, 194021, 3 Institutsky per., Saint-Petersburg, Russia

[2]JSC «NC VostNII», 650002, 3 Institutskaya Str., Kemerovo, Russia

Abstract. The paper presents the problem of methane explosions prevention in the most hazardous mine section — working face, based on the preventive measures to exclude key factors of methane explosion. The measures of the methane explosions prevention in the working faces during the intensification of coal mining are determined. A method for the aggregated calculation of absolute methane emission in the working face at work of a shearer is proposed. An explosive concentration can occur at any time during cutting coal due to irregularity of methane emission and ventilation processes in the face, which makes the risk of explosion uncontrollable. Consequently, the effective preventive drainage of coal seams is primarily required for the intensification of coal mining. A series of new studies on intensity of methane emission will be required for per face output which is about 30,000 tons per day or more to create new regulatory documents.

1 Introduction

The problem of methane explosions has existed since people began to extract coal from the coal seams that contain significant volumes of occluded methane. To solve this problem and prevent methane explosions, it is necessary to exclude the joint effect of two risk factors: the presence of explosive methane-air mixture and the source of ignition (striking of a spark, an arc or methane ignition temperature of 650-750°C). These hazard factors occur under the following circumstances:

Methane explodes at volumetric concentration in the air of 4.9 to 15.4%. The most flammable volume fraction (volume concentration) is 8%. Methane reaches maximum explosive force at volume fraction equal to 9.5%.

Explosiveness of methane-air mixtures at atmospheric pressure is determined by the graph shown in Figure 1.

* Corresponding author: nowosyolow.sergej@yandex.ru

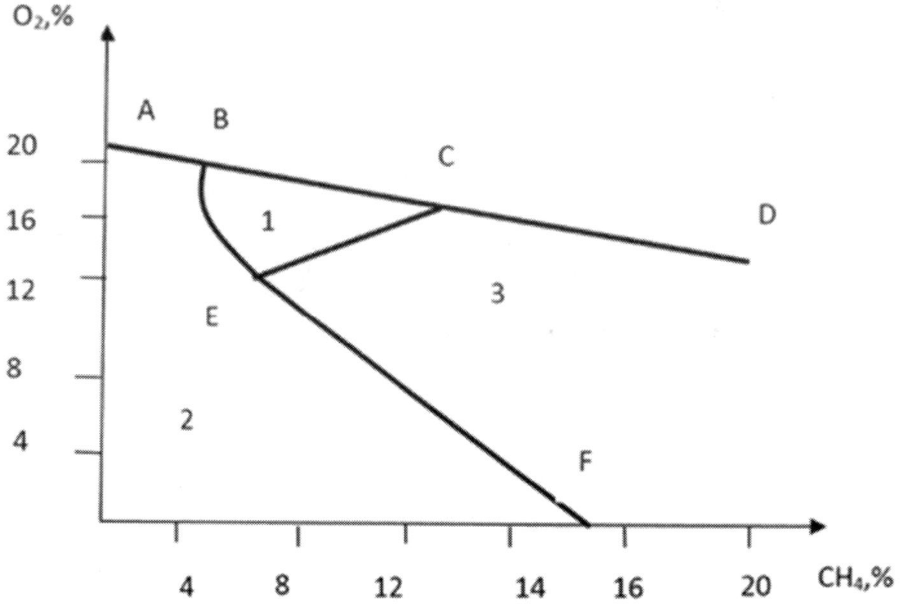

Fig. 1. Explosiveness of methane-air mixtures as a function of concentrations of oxygen and methane in the mine atmosphere, where 1 is explosive mixture; 2 is non-explosive mixture; 3 is mixture that becomes explosive when diluted with oxygen.

Point A characterizes the oxygen content in atmospheric air; points B and C are, respectively, the lower and upper limits of the explosiveness (LEL and UEL) of the methane-air mixture. Point E corresponds to the LEL of the mixture against the oxygen content. All points below the AD line characterize flammable and explosive mixtures of methane with the air oxygen. Interval BE corresponds to LEL of the mixture, interval CE to the UEL. Mixtures in area 1 are explosive, in area 2 flammable in the presence of a heat source, in area 3 flammable and explosive when diluted with oxygen. In area above the AD line, the methane-air mixture cannot be prepared. Triangle CBE is referred to as the explosiveness triangle for methane-air mixtures. For example, at the methane concentration of 8% and the oxygen concentration of 16%, the methane-air mixture is explosive, but at the oxygen concentration of 8%, non-explosive.

The increase in temperature leads to reduction in the LEL of the methane-air mixture. At the temperature of 400°C this limit is 3%. When the methane-air mixture is saturated with water vapour, the explosive limit is reduced. Carbon dioxide also reduces the UEL. As the dust content of the mine atmosphere increases from 5 to 30 g/m^3, the LEL falls from 3 to 0.5%. At that, the ignition energy of the dust-gas mixture increases by a factor of 100-300, compared to the methane-air mixture.

Methane is found in coal seams and the host rock. When conducting mining operations, methane is released into the workings faces from the developed, undermined and overworked coal seams and the host rock and the mined-out areas. The rate of methane release from the coal seams and rock is determined by the gas content in the strata, pressure and temperature, but also by permeability, porosity, gas release capacity, gas retention capacity of the coal or rock, structure of the coal seam and mining technology.

High gas content in the coal seams increases the risk of methane explosions, depends on their depth, metamorphism intensity, tectonic faults, and association with certain geological structures and can reach 40 to 45 m^3 per 1 ton of combustible matter. Russian coal industry, including mining companies, Federal Service of Environmental, Technological and Nuclear Supervision and scientific community, gives paramount consideration to the methane explosion risk assessment and explosion safety of coal mines. Relevance of analysis and assessment of the methane explosion risk factors in the mining industry is acknowledged internationally [1-11].

Based on the present-day practices of coal mining operations and content analysis of both domestic and international published works and regulations, it can be said that simultaneous application of three basic actions strategically prevents the methane explosion hazard: (1) preliminary gas drainage from the coal seams, (2) adequate and efficient ventilation systems, and (3) improved automatic gas monitoring system (GMS).

In addition, it is necessary to eliminate the following factors that disturb the mode of ventilation:

- fault trips or performance degradation of the fans,
- operation of the fans in recirculation mode,
- air losses in the ventilation ducts, through air crossings, stoppings, etc.,
- short-circuiting of ventilation air,
- reduction in the actual cross sectional area of mine openings as compared to the designed area (obstructing, rock spalling, etc.).

To eliminate the sources of explosive gas ignition (static electricity sparks; electric arc; flame; smoking; incandescent particles from electric detonators and parts of mechanisms; frictional impact; metal hammering against metal, pyrites or siliceous rocks; spontaneous inflammation), it is required to monitor the conditions in the mine, follow the safety rules and conduct of efficient preventive maintenance of machinery and equipment. Modern coal shearer-loader are equipped with sensors that control position of the working element, motion speed and the position of the coal-cutting machine [9]; thus, the machine operator is able control the intensity of methane emission by changing the feeding speed.

Moreover, a sufficient number of remote monitoring and alarm systems of various levels and purposes have been developed. Such systems perform the following functions:

- monitoring and operation of the mine technological systems and equipment,
- video surveillance of work stations,
- voice and video wireless communications,
- tracking personnel and vehicles, locating trapped miners,
- recording of working hours and time-keeping,
- safety monitoring of the mine atmosphere, maintenance of its gas composition within the standard limits,
- required air distribution among the ventilated facilities, control of ventilation doors and monitoring of their position,
- control of personnel safety near the moving parts of mining machines and mechanisms [10].

Dispatching control of international mining enterprises appears an operative means of productivity enhancement and a decision-making tool for both personnel and management [11].

However, even close consideration of all the above risk factors of methane explosions in coal mines seems insufficient to radically eliminate such accidents at hazardous production facilities unless a comprehensive approach to the problem is introduced and the human factor that contributes into the occurrence emergency situations is taken into account. The influence of human factor can be minimized by:

- proper selection of professional personnel (implementable yet a difficult task),
- systematic safety training and briefing on compliance with the occupational health and safety rules, safety exams for personnel engaged into the in the production process.

2 Materials and methods

 DOI: 10.1051/e3sconf/20172101034

Recently, two mines of SUEK-Kuzbass JSC — Taldinskaya-Zapadnaya-1 mine and V.D. Yalevsky mine — reached the milestone of monthly coal production from a single working face, equal to 1 mln tons, and in May 2017, V.D. Yalevsky mine established an absolute record of production — 1.407 mln tons per month from a long face. Thus, the rate of stope-face advance was at least 20 m/day and, accordingly, the technology of coal mining can be determined as intensified.

A number of methods for determining the methane content in strata, where the main parameters are the relative rate of methane release from the coal seam, the time factor and production parameters, including broken-down coal, were analyzed. Such analysis demonstrated that, omitting certain empirical dependencies on gas dynamics, the rate of the absolute methane release (J, m^3/min) into the long-face areas can be calculated by proposed formula 1.

$$J_I = m \cdot l \cdot \gamma \cdot q \cdot r \cdot k \quad (1)$$

Consider an aggregated calculation based on actual figures if seam thickness (m) is 5 m, maximum feed rate (l) of a modern coal-cutting machine (for example, shearer-loader SL-900) is 36 m/min, relative methane-bearing capacity (q) is 5, 10 and 15 m^3/t for the I, II, III category mines respectively, average specific gravity of coal (γ) is 1.34 t/m^3, methane desorption factor (k) is 20%, and operating width of the coal-cutting machine (r) is 1000 mm. If the shearer-loader advances in the working face at the rate of 36 m/min, the rate of the absolute methane release (J) shall be:

for the I category mines – $J_I = 5 \cdot 36 \cdot 1.34 \cdot 5 \cdot 1 \cdot 0.2 = 241.2$ m^3/min,

for the II category mines – $J_{II} = 10 \cdot 36 \cdot 1.34 \cdot 5 \cdot 1 \cdot 0.2 = 482.4$ m^3/min,

for the III category mines – $J_{III} = 15 \cdot 36 \cdot 1.34 \cdot 5 \cdot 1 \cdot 0.2 = 723.6$ m^3/min.

Therefore, to ensure the standard content of the mine atmosphere, it is required to intake 25÷75 thousand m^3 of fresh air per minute.

However, at a maximum airflow velocity of 4 m/s (according to the OHS rules) and the neat cross-sectional area of about 10 m^2, the amount of air flowing through the face area cannot be more than 2,400 m^3/min. Insufficient air intake leads to an increased concentration of methane in the working face atmosphere. It should be taken into account that, according to the OHS rules, the CH_4 concentration in the ventilation airflow coming from the extraction area should be less than 1%, and the GMS sensors are adjusted to the concentration of 1.3%. It should also be noted that, when cutting the coal, the CH_4 concentration is even higher under the roof in the zone of the shearer cutting drums.

In such conditions, with the GMS enabled, continuous operation in the long face appears impossible because of regularly occurring emergency power shutdowns due to excessive methane concentrations. That is why the CH_4 control sensors are frequently disengaged deliberately, which creates potential hazardous situations.

These calculations are presented as tentative, for they do not take into account the specifics of the gas dynamics, gas release factors, and the coefficients of variation, which can actually alter the resulting figures, because the range of gas dynamics values for the Kuzbass coal seams is significant.

However, these calculations correlate with the results of research conducted by Zakharov, Zaburdyaev and Artemyev in the Kuzbass mines. According to the presented graph, which characterizes the dependence of the methane release rate on the average daily coal output in the Kuzbass mines, the methane release rate (I, m^3/min) reaches 640 m^3/min at the production rate (A, t/day) of 30,000 t/day.

Taking into account the above calculations, it can be concluded that in very gassy mines, at the relative methane-bearing capacity of 15 m^3/t or more and intensification of coal production to 20,000-30000 t/day per a working face, safety issues related to the presence of methane will definitely occur, unless preliminary gas drainage from the coal seams and actions (2) and (3) proposed above are applied.

3 Conclusion

The concept of 'methane safety' in the Kuzbass coal mines suggests a comprehensive proactive approach to choosing methane explosion prevention methods. This problem cannot stand a one-legged approach, but must be solved in all its sides simultaneously: explosive methane-air mixtures in the faces are to be avoided; the level of explosion protection of the electrical equipment is to be monitored and improved, and personnel's professional attitude is to be encouraged. All of the above can be ensured by systematic monitoring of activities at the hazardous production facility - a coal mine - using automated means of control.

References

1. *The Precautionary Principle. World Commission on the Ethics of Scientific Knowledge and Technology*, 52 (UNESCO, 2005)
2. *AS/NZS ISO 31000.. Risk management* (2009)
3. *CAN-CSA-Q850-97. Risk Management Guidelines for Decision-Makers. National Standard of Canada*
4. *Coal Mining Safety and Health*, (Queensland, Australia, 1999)
5. *Coal Mining Safety and Health Regulations*, (Queensland, Australia, 2001)
6. G. Nugent, *Emergency Mine Entry. Re-entry and Knowledge Management. Queensland andp NSW Mines Rescue Services.*
7. *The RiO Declaration on Environment and Development.* (1992)
8. *QCVN 01: 2011/BCT. Quy chuẩn Kỹ thuật Quốc gia về an toàn trong khai thác hầm lò.* (Vietnamese, 2011)
9. M. Bekker, P. Martin, Mining Report Glückauf, **4**, 10 (2015)
10. M. Bekker, P. Martin, Mining Report Glückauf, **4**, 38 (2015)
11. R. Kipp, M. Junker, Mining Report Glückauf, **2**, 10 (2014)

Coal Squeezing-Out, its Description and Conditions of Development

Svetlana Kostyuk[1*], *Anatoliy* Gegreen[2], *Vladimir* Meljnik[3], *Mihail* Lupeey[4]

[1] T.F. Gorbachev Kuzbass State Technical University, 650000, 28 Vesennyaya St., Kemerovo, Russia

[2] Ltd Geotechnology-Vzryvozaschita

[3] Mining Instiitute NITU MISiS

[4] JSC SUEK-Kuzbass, 652507 1 Vasilieva st., Kemeriovo region, Leninsk-Kuznetsky, Russia

Abstract. The article reviews the need to develop a normative document in Russia on sudden coal squeezing-out based on the analysis of scientific studies. Descriptions of geomechanical, mining, and geotechnical conditions for development of sudden coal squeezing-out are given including the indicators characteristic to sudden coal squeezing-out.

1 Introduction

Information about the first coal squeezing-out dates back to the end of the 40s of the XX century, when gas and dynamic events began to be recorded. In the beginning, these phenomena carried a variety of names: 'coal movement', 'buckling of the coal face', 'face slip'. Later they received the term of 'gas and dynamic events of the 3rd kind', and at present their official term is 'sudden coal squeezing-out (slip).

Seams prone to rock bumps and explosions are being developed in many coal fields of the world: in USA, Australia, India, Poland, China, and other countries.

The earliest studies of rock bumps abroad were carried out by Cook. The overview of current scientific achievements in studying gas and dynamic events was prepared by the Institution of Mining and Metallurgy (London, UK) [2]. In this work, it was noted that the most important thing in combatting, in particular, rock bumps was not the overall reduction of the released energy, but the restriction of stress concentration ahead of the working face. The study details the concept of the effect of coal and rock strata strength in the three-dimensional stress state on the nature of gas and dynamic events development.

2 Materials and methods

Geomechanical conditions of sudden coal squeezing-out development

Geomechanics impact takes place in coal seams grouped in formations when development of one of the seams leads to a change in the stress-strain state of other seams in the formation. This impact varies depending on the interaction of mining operations at adjacent seams.

* Corresponding author: ksg.adm@kuzstu.ru

To determine the type of the geomechanics impact, it is good to review the general scheme of rock strata moving in and deformation during driving of a working (Fig. 1).

A part of the rock strata where the change of the stress state takes place is called the zone affected by the production face 1. Two large geomechanics zones are identified therein – abutment pressure zone 2 and pressure release zone 3. In the abutment pressure zone, normal stresses $K_{\sigma}\gamma H$ are 5 times higher than in the tight rock 6 (K_{σ} – stress concentration factor, its value reaches 5 and higher) due to the fact that a part of weight of undermined rock is transferred to the edges of the coal seam during mining.

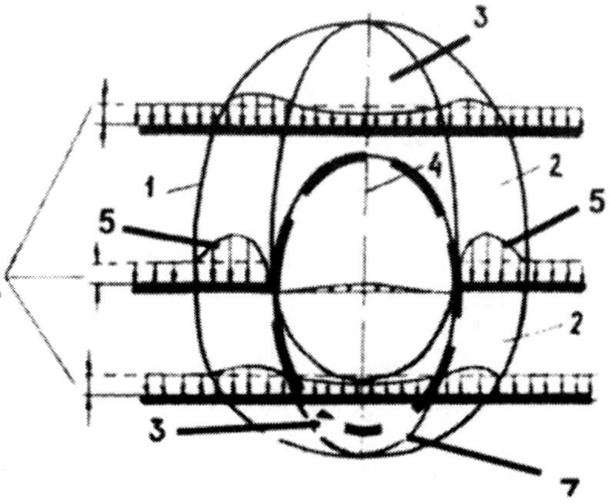

Fig. 1. Geomechanical zones affected by mining.

Zones at the level of over- or underworked seam (rock layer) are called the abutment pressure zones (APZ). Their dimensions decrease when moving away from the seam.

In the release zone, normal stresses are less than the stresses in the tight rock. Within this zone, protected zone 4 is identified, in which normal stresses and stresses along the formation are lower γH_0 (H_0 – critical depth from which gas and dynamic events start to develop) [5, 6].

In the release zone during overworking of the seam, one can distinguish the artificial protected zone 7 which has formed due to degassing of the overworked rock strata [7].

The sizes and parameters of the geomechanical zones depend on the geological and geotechnical mining conditions of development: the depth of work, the thickness of the protective layer, the dip angle of the rocks, the interbed thickness, the composition and properties of the interbed rock, the dimensions of the heading, the method of rock pressure management etc.

In the process of seam underworking (overworking), the coal seam undergoes three main stages of deformation: the contraction stage in the abutment pressure zone; the tensile stage in the release zone; the stage of repeated contraction in the zone of loads restoration.

The abutment pressure zone is characterized by an intense manifestation of rock pressure (increased rock displacements at the contour of headings, loads on the supports, etc.), as well as an increased hazard of gas and dynamic events development. It is not advisable to locate mine workings (including production ones) in these zones. When abutment pressure zones form as a result of production (near the borders of protective pillars) or mining/geological reasons (near the borders of intransgressible geological faults), passing of these zones is accompanied by additional measures aimed at eliminating gas and dynamic events development.

The analysis of distribution of sudden squeezing-out in the abutment pressure zones formed by the junction areas of production faces shows the following:

- The overwhelming majority of the events (99%) occurred when crossing the junctions of production faces in the direction under unmined coal;
- The most dangerous are the abutment pressure zones formed by the junctions of stopped production faces (93% of face slips occurred) as compared to the abutment pressure zones from junctions formed by the face entry;
- No significant distinctions were observed in development of squeezing-out in relation to location of the sources of the abutment pressure zones.

In the overworked zones of abutment pressure with interbed thickness up to 40 m no sudden coal squeezing-out was observed.

One of the main factors influencing the protective effect is thickness of interbed rock strata which is the distance from the protective to the protected seam along the normal. As it was established by the methods of correlation regression analysis, the dependence of the number of sudden coal squeezing-out N on the thickness of the interbed rock strata M during overworking operation has the following character.

$$N = a \cdot \exp(b \cdot M) \tag{1}$$

where a, b – regression coefficients which have been found by the least square method for the linearized form of equation are 0.0526 and 0.0897, respectively.The correlation coefficient of this equation in the linearized form is $K = 0{,}947$.

Mining and geological conditions for development of sudden coal squeezing-out

Let us analyze the dependence of the manifestation of sudden coal squeezing-out on the main mining and geological factors.

With an increase in the depth of development of coal seams, the number of sudden coal squeezing-out increases.

The mathematical model of the dependence of the number of sudden coal squeezing-out N on the depth H is as follows.

$$N = a \cdot \exp(b \cdot H) \tag{2}$$

where a, b – regression coefficients which have been found by the least squares method for the linearized form of equation are 1.2617 and 0.003736, respectively.

The correlation coefficient of this equation in the linearized form is $K = 0{,}765$.

The distribution of sudden coal squeezing-outs by the thickness of the protective seam is characterized by the following data: the range of value variation is 0.42-2.2 m; the average value of 1.25 m, and the standard of 0.37 m. These data were obtained for conditions without taking into account the dip angle, since the differences in this factor turned out to be insignificant.

The sudden coal squeezing-outs were happening within the entire range of dip angles of the coal seams being developed.

Analysis of the composition of the strength properties of the roof rock strata of 33 seams in the Central district of Donbass, the most dangerous by sudden coal squeezing-outs, showed that 70% of the roof rock strata were prone to periodic hang-ups in the mined-out space and to subsidence with dynamic impact on the supports and the face area of the seam.

The analysis of statistical data of the effect of geological disturbances on the occurrence of sudden coal squeezing-out showed that in the areas affected by geological faults there were 88.5% of sudden coal squeezing-outs in flat seams and 15.2% in steep seams.

The histogram of the distribution of the natural gas content of the seams prone to sudden coal squeezing-outs is governed by the lognormal law, which has the following form

DOI: 10.1051/e3sconf/20172101035

$$f(x) = \frac{1}{x\sqrt{2\pi}\sigma}\exp\left(-\frac{(\ln x - \mu)^2}{2\sigma^2}\right) \tag{3}$$

where: μ – the mathematical expectation of the logarithm of a random variable is assumed equal to 2.87 m3 / ton dry ash-free mass; σ - RMS deviation is assumed equal to 0.26 m3 / t dry ash-free mass.

The histogram of the distribution of the relative gas make during sudden coal squeezing-outs is governed by the lognormal distribution law with the probability density (3) and parameters $\mu = 2.84$ and $\sigma = 0.77$.

Geotechnical conditions for development of sudden coal squeezing-outs

The analysis of the mining technical conditions presented in [4] for the development of sudden coal squeezing-outs shows the following:

1. Sudden coal squeezing-outs occur in all types of mine workings, with all technological schemes for the development of coal seams, in all types of mechanization in coal production and development faces, with all the main methods of rock pressure control, however, the level of sudden coal squeezing-outs development in these conditions is different.
2. In flat seams, the vast majority of sudden coal squeezing-outs (90% of their total number) occurred during the implementation of the main production processes of coal mining: mining coal from longwalls and driving development headings. In steep dipping seams, the overwhelming majority of sudden coal squeezing-outs (about 90%) occurred in production faces, and in the shield faces the number of sudden coal squeezing-outs is 3.2 times greater than the number of sudden coal squeezing-outs in stepped faces.
3. By the type of coal mining, the majority of sudden coal squeezing-outs occurred during the most common methods of mining: low dip seams – when mining by shearers and ploughs; steep seams - when mining by conveyor-plough (shield faces) and by a jackhammer (stepped faces).
4. By the rock pressure control method, the majority of sudden coal squeezing-outs, both in low dip and steep seams, occurred at full roof caving; in steep seams, the level of coal slips is also significant when supporting the wall rock by cribs (21%).
5. In the long workings, the maximum number of sudden coal squeezing-outs in flat seams occurs during driving headings by road headers and extraction of coal by jackhammers; the level of sudden coal squeezing-outs is significant (21%) when driving by drill and blast method. In steep-dipping seams, the greatest quantity of sudden coal squeezing-outs (77%) is recorded when mining coal with jackhammers.
6. The analysis of operations preceding the development of sudden coal squeezing-outs allows us to state unambiguously that the majority of them (76% on low-dip and 94% on steep seams) occurred during extraction of coal. It is necessary to pay attention to the high level (12.2% on low-dip seams) of sudden coal squeezing-outs development during implementation of measures aimed at prevention of gas and dynamic events.

The most characteristic features of sudden coal squeezing-outs are as follows:

- the intensity of sudden coal squeezing-outs is much lower than the intensity of sudden coal and gas outbursts and coal falls; the intensity of the latter two types of gas and dynamic events can reach over several hundred tons;
- the average value of the relative gas emission at sudden coal squeezing-outs is 1.2 times higher than the average value of the natural gas content of the seam, while at sudden outbursts of coal and gas this ratio can be several times;
- the shape of the cavity during sudden coal squeezing-outs is mostly pocket-like or dome-shaped;
- in production faces the width of the cavity is, as a rule, bigger than its depth;

- the average value of the slope angle of the squeezed-out coal is approximately equal to the value of its depositional angle;
- the minimum ratio of cavity depth at sudden coal squeezing-outs to the seam thickness *m* is 0.6, while the minimum ratio between the distance to the maximum of the abutment pressure to the thickness of the seam is 2; consequently, a part of these events occurs in the face area of the seam to the maximum of the abutment pressure.

3 Conclusions

In the development of coal seam formations, the zones of abutment pressure are hazardous in relation to sudden coal squeezing-outs. The most hazardous are the abutment pressure zones formed by the junction areas of production faces when they are crossed over towards the coal mass.

The following are the main mining and geological factors affecting the development of sudden coal squeezing-outs, including the depth of occurrence, seam thickness, its gas content, volatile matter yield, strength properties of the wall rock, presence and type of geological disturbances.

Sudden squeezing-out of coal occurs almost in all mining conditions of the development of seams prone to squeezing-out. The highest level of coal squeezing-out development has been recorded: on steep dip seams - when coal is mined by conveyor-ploughs in shield faces and by extraction of coal by jackhammers in stepped faces; on low dip seams - when coal is mined by narrow-web cutter-loader and by ploughs in longwall faces; in development headings - when cutting coal with jackhammers on steep seams and when driving roadways by roadheaders in flat seams; by the method of the rock pressure control - at complete caving of the roof strata.

The characteristic parameters of sudden coal squeezing-outs were analyzed.

References

1. *Rockbursts: prediction and control*, (London: Institution of Mining and Metallurgy, 1984)
2. I.M. Petukhov, A.M. Linkov, International Journal of Rock Mechanics and Mining Sciences and Volume **16:2**, 57 (1979)
3. I.M. Petukhov, Proc. 6th congress International Society for Rock Mechanics, Montreal, **2**, 1207 (1987)
4. I.M. Petukhov, A.M. Linkov, Proceedings - Congress of the International Society for Rock Mechanics, 113 (1983)

 DOI: 10.1051/e3sconf/20172102001

Energy and Resource-Saving Sources of Energy in Small Power Engineering of Siberia

Marina Baranova[1,*]

[1] Institute of Engineering Systems and Energy FSBEI "Krasnoyarsk SAU", Mira Street, 90, Krasnoyarsk, Russia

Abstract. The sustainable development of distant areas of Siberia is associated with the structures of energy demand and supply, the implementation and promotion of the process of environmentally safe restructuring of the energy supply system. It has been established that suspension coal fuels derived from brown coal, coal mining, coal processing wastes can be used as fuel. The results of experimental and industrial boilers on suspension water coal fuel are presented. The designs of vortex combustion chambers of various powers are developed and tested. The possibility of using coal-enrichment wastes and substandard coals for the production of manure-coal fuel briquettes was studied. It is shown that the strength and thermal power characteristics of briquettes depend on the moisture content and degree of metamorphism of the raw materials. The most effective percentage of the solid phase and manure, as a binder, was determined.

1 Introduction

A lot of coal-mining and coal-processing enterprises in the sludgeponds and settling basins accumulate a large amount of coal mined (only in the Kuzbass more than 150 million tons), presented in the form of fine coal slurries, the production of which into a transportable and technologically acceptable fuel will not only improve the ecological situation in the coal mining regions, but also to obtain a significant economic effect [1-3].

The use of highly effective methods of mechanical dewatering allows to eliminate the application of thermal drying, which significantly reduces the first cost of the enrichment process and makes it fire and explosion-proof. A closed water-slime cycle without external sludgeponds and settling basins leads to the fact that up to 7% of processed coal fine waste with a moisture content of 20-45% and an ash content of 18-60% is not produced and piled into dumps [4,5].

As a result, a by-product of enrichment appears – filter cake, high moisture and ash content and small particle size of which do not allow it to be sent to consumers. The utilization of such wastes is a vital task, which can be solved by creating a coal-water fuel from coal-enrichment sludge.

The purpose of this work is to demonstrate the possibility of using locally available and

* Corresponding author: marina60@mail.ru

 DOI: 10.1051/e3sconf/20172102001

inexpensive resources as a source of energy.

2 Results and discussion

Water-coal suspensions. The production of suspension water-coal fuels (WCF) is one of the developed and applied technologies. This technology allows to create efficient and reliable systems for the production of high-quality heat energy in the close proximity to local consumers, taking into account their specific needs. Investment attractiveness is due to the compactness and high environmental friendliness. To burn such fuels effectively, the development of special technologies and technical means is required. In this case, the technology of low-temperature vortex combustion in adiabatic combustion chambers is successfully used, such chambers are either integrated into the combustion chamber of existing or newly developed boilers, or installed next to the operating boiler [6-8]. At the same time, the designs of the combustion chambers must take into account not only the low reactivity of the WCF, but also the fact that when spraying WCF the flame length reaches 2.5 m and more, and the time for the WCF solid phase particles in the combustion chamber is from 1 s to 4 s. The heat generators work experience has shown that with a heat output of more than 2.0 - (3.0) megaW, the combustion chamber is built into the furnace space of the boiler without increasing its dimensions. At a heat output of less than 2.0 - (3.0) megaW, the installation of a stand-alone (remote) combustion chamber is required [7-9].

Below are the results of the experimental and industrial boilers work on suspension water-coal fuel and local brown coals. The CJSC «SPE Sibecotechnika» has developed and tested the design of vortex combustion chambers of various power (Table 1).

Table 1. Characteristics of the operation of heat generators

Indicator name	**Technological Complexes**		
	Technological complex KuzSTU, Kemerovo	**Heat generator in SibEME SD, Krasnoobsk, Novosibirsk region**	**Boiler house in Cherepanovo, Novosibirsk region**
Heating capacity, Gcal / h	0.05	0.25	0.5
Fuel consumption, l / h	120÷130	55	110÷220
Temperature in the furnace, ^{0}C	950÷1050	950	950÷1050
Characteristics of WCF, burned in heat generators			
Moisture, general, %	40÷42	42÷43	38÷43
Ash content (on dry), %	26÷45	8.2÷8.5	30÷45
The yield of volatiles, %	23.0÷41.2	42.3÷43.1	6÷42
Particle size, micron	0÷500	0÷500	0÷500

As a raw material, both coal concentrate (heat generator in SibEME SD of Krasnoobsk, Novosibirsk Region) and coal slurries were used. The results of pilot tests of the developed heat generators have shown that even at low values of the lowest calorific value (2500 kcal / kg) of WCF from different grades of coal, it is possible to obtain the efficiency value more than 80%. At the same time (first cost) 1 Gcal is reduced by 30-300% in comparison with burning of grade coal or liquid petroleum fuel [10].

At the present time in Barnaul a boiler plant has been set up (LLC «ProEnergoMash») for the production of specialized boilers with the «Tornado» vortex combustion system, which allows to burn efficiently local, including ballasted, fuels.

On the territory of the Krasnoyarsk Region there are quite a lot of different scale of brown coal deposits. Brown coals, because of the low degree of metamorphism, lose their energy properties during storage, and also differ in their low net calorific value. To determine the possibility of the boiler KV-1,2-105 ShpVT with a furnace «Tornado» produced by LLC «ProEnergoMash» on the brown coal of mine «Chulymsky» (Krasnoyarsk Region), special tests were carried out. The coal had the following characteristics: humidity - 43.2%; ash content - 11.3%; net calorific value - 2820 kcal / kg; poured density – 868 kg / m3. The test results are shown in Table 2.

Table 2. Test results.

Parameter name	Unit of measure	Numeric value	
		diapason	average
Total amount of heat produced	Gcal	-	0,350
Outgoing gases temperature	°C	41-87	64
The temperature of the hot layer above the grate in the furnace on the grate	°C	1000-1100	1050
The rarefaction in the furnace	mm WG	1,5-3,0	2,4
Composition of flue gases behind the boiler: CO_2 O_2 CO NO SO_2	 %, % mg/m^3 mg/m^3 mg/m^3		 14 13 185 110 320
Heat of fuel combustion	kcal/kg		2820
The control combustion conditions			
Weight of coal supplied	kg		150
Quantity of slag	kg		17,5
Quantity of fly ash	kg		1,5

During the tests in selected calculated stationary modes it was determined that:

- the boiler works steadily on non-calculated fuel, the average temperature of the layer is 1000-1100 ° C;
- possible steady combustion in the combustion chamber at modes from 100% to 25% of the load and manual regulation;
- removal from the combustion chamber is shallow, insignificant, there are no visible smoke emissions from the chimney;
- coefficient of excess air behind the boiler is 1.4-1.6;
- the temperature behind the boiler does not exceed 120 ° C.

Fuel briquettes. The technology for producing coal briquettes using pig manure as binders has been known for a long time. In Siberia and the Krasnoyarsk Region, as in all coal-mining regions, there is a problem of efficient use of substandard coal and coal processing waste. In addition, at the present time in the region for an average of 858 thousand tons of manure per year, the utilization of which is a big problem [10-12].

During the fuel briquettes burning the coefficient of usefulness of coal chemical energy is 70-80%. The advantage is explained by the fact that the flops of unburned coal are practically completely excluded from the grate; decreases physical and chemical underburning due to better conditions for air access to the surface of briquettes; the content of carbon monoxide and soot in flue gases decreases [10]. In the course of the work, the influence of the raw material characteristics on the strength and heat power characteristics of fuel briquettes from sub-standard coals and manure was determined.

As a raw materials were used: pig manure (LLC «Emelyanovskoye»), coals of B2 grade Berezovskoye deposit and D grade Balakhtinskoye deposit, coarse-grained and fine-grained sludge («Polosukhinskaya» mine, Kemerovo region). The characteristics of the starting materials and the grain-size classification are presented in Tables 3 and 4.

During the experimental work, the composition of the mixture was calculated in accordance with GOST 27313-95. Next, the starting components were mixed in a mixer for 7 minutes. The mixture was pressed on a hydraulic press with a pressure of 300 to 500 kgf / cm^2 in a matrix having 20 molds with a diameter of 35 mm each.

Table 3. Characteristics of raw materials.

Component	Humidity, %	Ash content, %
Pig manure	80,2	5,8
Coal of B2 grade	21,9	8,7
Coal of D grade	3,7	12,3
Coarse-grained sludge	14,5	14,7
Fine-grained sludge	34,6	27,8

Table 4. Granulometric composition of starting materials.

Class, mm	Content, %		
	Coal of D grade	Coal of B2 grade	Fine-grained sludge
+3	2,1	-	-
1-3	21,1	-	-
0,63-1	12,0	0,1	0,7
-0,63	64,8	99,9	99,3
Итого	100	100	100

The received briquettes were analyzed for moisture content in accordance with GOST 27314-91, ash content according to GOST 11022-95. Drying of briquettes took place at 100^0 and 150^{0}C. The received briquettes were tested for drop resistance according to GOST 21289-75. The dried briquettes were burned in a laboratory installation.

The research of the influence of the quantitative composition of the initial raw components in briquettes on their quality have been carried out. On the basis of these studies, the optimal ratios of the initial components in the briquette mixture were revealed. Briquettes with a manure ratio were obtained: coal in % - 5:95; 10:90; 15:85; 20:80; 25:75; 30:70. During the experimental work, it was established that briquettes from coal grade D at all ratios do not retain their shape when processed in a press at a pressure of 250 to 500 kgf / cm^2. Briquettes from brown coal and scraps are more stable when processed with a pressure of 300 to 500 kgf / cm^2 and retain their shape after drying and drop resistance testing. The dropping strength was basically 80-90% for all of the above ratios.

The coal compositions of different stages of metamorphism and coal processing waste are studied in the process of obtaining briquettes. It is established that the best compositions, taking into account the strength of the obtained briquettes, are briquettes with a coal ratio of grade D - up to 10%: grade B2 - from 60%: manure with a moisture content of 72% - not more than 30%. It should be noted that the production of briquettes based on sludges, i.e. waste of coal mining, practically does not cause technological and technical problems. Heating the initial mixture up to 50-60 °C leads to an improvement in the strength characteristics of briquettes from all coals.

Analysis of the data on the kinetics of briquette drying showed that at a humidity of 36.8% and a temperature of 100 °C, the drying time is 78 minutes, at 150 °C - 52 minutes, that is, a reduction in drying time of about 27%; at a ratio of 20% of a fine-grained sludge: 80% of a coarse-grained sludge with a moisture content of 15.3% and a temperature of 100 °C, the drying time is 43 minutes; at 150 ^{0}C, 27 minutes, that is, a reduction in drying time of 32.5%. Residual humidity of briquettes from sludges after drying in the air for 14 days amounted to W_t^r= 1.9%; after the air drying for 3 days, the manure-coal briquettes amounted to W_t^r = 36.8%. The weight of briquettes averaged 17.3-18.1 grams.

During the work, calculations were made of the net calorific value. It is established that for briquettes (manure (30%): coal DR (70%) with a humidity of 30%, the lowest calorific value is 3000 kcal / kg; for briquettes after drying, manure (30%): coal B2 (70%) at a humidity of 36,8% - 2676 kcal / kg; for briquettes manure (30%): fine-grained sludge (70%) with a moisture content of 52.1% - 2031 kcal / kg; for briquettes fine-grained sludge (20%): coarse-grained sludge (80%) at a humidity of 10%, the lowest calorific value is 6050 kcal / kg; for briquettes fine-grained sludge (30%): coarse-grained slurry (70%) with a humidity of 10% - 6000 kcal / kg.

References

1. *BP Statistical Review of World Energy* (British Petroleum, 2013)
2. V. G. Lury, A.N. Pankratov, Coal, **11**, 36 (2013)
3. A.V. Papin, A.V. Nevedrov, A.I. Sechin, Polzunovsky vestnik, **3**, 220 (2014)
4. C.H. Kuzmin, R. L. Isyemin, V. V. Konyakhin, E.V. Budkova, A.B. Mikhalev, Coal, **5:937**, 51 (2004)
5. Huang Bo, Zhao Yu-feng, Gong Xu-wen, J. Clean coal technology, **6**, 23 (2011);
6. V. I. Fedyaef, V. I. Murko, H. L. Inetdinof, M. P. Baranova, Clean Coal Technology, **6**, 43 (2011)
7. V. I. Murko, V. N. Delyagin, N. M. Ivanov, V. Ya. Batishchev, V. I. Bocharov, I. Shcheglov, V. I. Fedyaev, V. I. Karpenok, Polzunovsky vestnik, **2:1**, 239 (2011)
8. V.I. Murko, V.I. Fedyaev, H.L. Aynetdinov, M.P. Baranova, The 17th International Coal Preparation Congress, 679 (2013)
9. V. I. Murko, V. N. Delyagin, M.P. Baranova, S. N. Shakhmatov, Vestnik of KRASGAU, **11**, 103 (2015)
10. A.N. Koshka, In: Waste management - a basis of restoration of ecological equilibrium in Kuzbass, 136 (2008)
11. L.A. Antonenko, E.V. Pilipenko, K.I. Domnin, E.P. Volynkina, A.E. Anikin, Science and youth: problems, searches, decisions, 188 (2012)

The Increase of Power Efficiency of Underground Coal Mining by the Forecasting of Electric Power Consumption

Vladimir Efremenko[1*], *Roman* Belyaevsky[1], and *Evgeniya* Skrebneva[1]

[1] T.F. Gorbachev Kuzbass State Technical University, 650000, 28 Vesennyaya St., Kemerovo, Russia

Abstract. In article the analysis of electric power consumption and problems of power saving on coal mines are considered. Nowadays the share of conditionally constant costs of electric power for providing safe working conditions underground on coal mines is big. Therefore, the power efficiency of underground coal mining depends on electric power expense of the main technological processes and size of conditionally constant costs. The important direction of increase of power efficiency of coal mining is forecasting of a power consumption and monitoring of electric power expense. One of the main approaches to reducing of electric power costs is increase in accuracy of the enterprise demand in the wholesale electric power market. It is offered to use artificial neural networks to forecasting of day-ahead power consumption with hourly breakdown. At the same time use of neural and indistinct (hybrid) systems on the principles of fuzzy logic, neural networks and genetic algorithms is more preferable. This model allows to do exact short-term forecasts at a small array of input data. A set of the input parameters characterizing mining-and-geological and technological features of the enterprise is offered.

1 Introduction

The coal branch is one of the most power-intensive in economy of Kuzbass. The consumption of boiler-furnace fuel is more than 9%, heat power is more than 19% and electric power is about 23% of all consumed energy resources by the region.

At coal enterprises in underground coal mining the share of the conditionally constant component of power consumption (ventilation, dewatering, lifting, compressor installations, etc.) is large [1]. In some mines, the share of these consumers accounts for more than 50% of all consumed electricity (Fig. 1).

*Corresponding author: brv.egpp@kuzstu.ru

DOI: 10.1051/e3sconf/20172102002

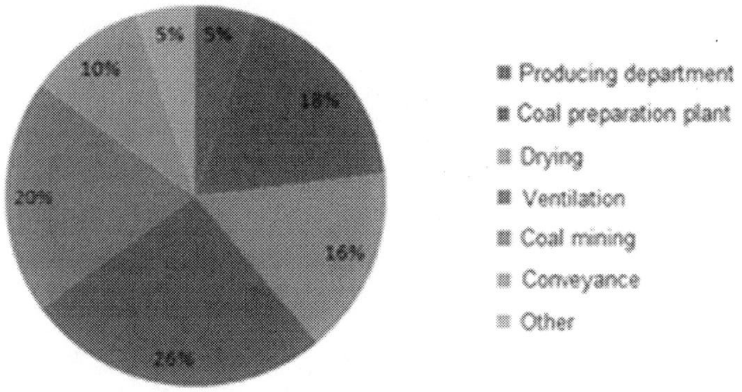

Fig. 1. The electric power consumption on technological processes of the coal mine.

Along with the increase in power consumption for the main technological processes, there is a constant increase in power consumption for drying and ventilation of mine workings (Fig. 2). This is due to the deterioration of mining and geological conditions of underground coal mining.

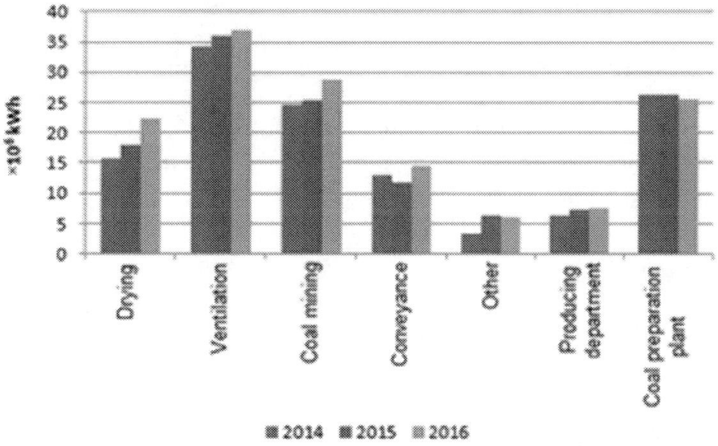

Fig. 2. The dynamics of power consumption in various parts of the coal mine.

The price and tariff policies that have changed in recent years in relation to energy resources and the increase in specific power consumption, have led to a significant increase in the energy component of the cost of production. For some mines, the energy component of the cost of production reaches 18-22%. Analysis of the cost of coal mining shows that the energy component is uneven throughout the year, but has a pronounced seasonal character (Fig. 3). Maximum falls on the autumn and winter period. This is due to increased fuel consumption in own boiler houses and increased power consumption during this period (boiler equipment, heaters, lighting, etc.).

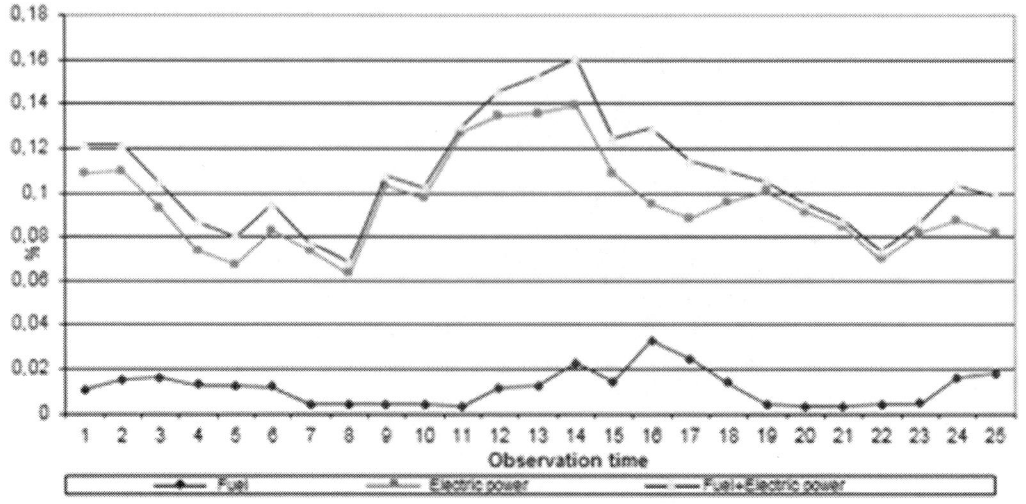

Fig. 3. The change of the energy component of coal production costs.

The long-term program of development of the coal industry of Russia until 2030, along with other purposes, provides for ensuring of competitiveness in the markets, primarily by reducing the cost of coal mining and processing. Experience shows that with the energy component in the production cost of more than 10%, energy-saving measures are required. However, at present power saving in the coal industry is not going to be intensive enough. Nevertheless, the power saving potential is high and amounts to 15-30% of annual consumption in the complex solution of power saving problems [2, 3]. It means that saving fuel fuel and energy resources in coal mines requires serious attention.

2 Materials and Methods

When considering the problems of energy resources saving in coal mines, the following main areas should be singled out:

- boiler-furnace fuel saving;
- thermal power saving;
- electric power saving;
- other measures that increase the power efficiency.

In this article, measures aimed at boiler-furnace fuel saving and thermal power saving, as well as other measures related to the change in the technology of coal mining and processing, are not considered.

Let us dwell on the main directions of electric power saving. In electric grids of coal mines, consisting of branched transmission lines, transformers, electric machines and apparatus, up to 20% of the consumed electric power is lost. This is due to many reasons, including the irrational construction and operation of networks in suboptimal mode. It is caused by the deviation of the actual state of production from the design state. This is expressed in the deviation of the centers of electrical loads from the places of installation of transformers, the overloading of some lines and the underload of others, the lack of optimal reactive power compensation, and the reduction of electric power quality. As a result, in the networks of coal mines there are significant power losses. Experience has shown, only the elimination of the above shortcomings through the holding of different technical measures will reduce the power losses by half.

Another important direction in reducing of electric power costs is the forecasting and monitoring of electric power consumption.

Enterprises through their energy sales organizations participate in power trading on the day-ahead market and the balancing market [4]. Day-ahead market assumes that the price for electric power is determined by the consumer demand and the proposal of suppliers during the auction. The balancing market plays the role of a mechanism that ensures a balance between supply and demand with possible deviations in the production and consumption of electric power from the forecasted parameters the next day. In the event of a deviation of the actual electric power consumption from the application, the electric sales organizations either sell surpluses or purchase the missing electric power in the balancing market. This is unprofitable for enterprises, as the prices of the balancing market take into account the need to cover the costs associated with the redistribution of electric power.

As the installed power of the electric receivers is much higher than 670 kW, for the mines there is a choice from the third to the sixth price category. This requires the provision of an hourly demand for electric power consumption for the next day.

Analysis of costs associated with the acquisition of electric power, shows that the enterprise can actually reduce the cost of electric power by changing the planned electric power consumption and reducing the deviation of actual consumption from the declared. In other words, this is achieved through a more accurate prediction of electric power consumption for the day ahead.

Elimination of power losses in case of deviation of actual consumption from planned consumption is possible due to strict limitation. In this case, the power consumption is limited when approaching the planned value. Also, the elimination of power losses is possible due to the use of forecasting methods based on analysis and consideration of all factors affecting on electric power consumption. The first method in conditions of coal mines is practically difficult to implement, because there are electric receivers providing safe conditions under the ground. Restriction of other energy-intensive consumers can lead to long-term destruction of technological processes. In this case, economic damage can significantly exceed the possible additional costs to acquiring of electric power in a balancing market.

At present, the demand for the electric power consumption is formed empirically. It is based on previous experience, including the experience of the person responsible for this direction. Adoption of optimal solutions in a complex rapidly changing environment requires a high qualification of the person making the decision and the use of modern decision-making methods. It should be noted that the specialists for predicting of electric power consumption do not always have the necessary competences for optimal power management. At the same time, many foreign companies have special power management services. In some countries, such as the USA, Japan, etc., this is determined by law [5].

In general, the problem of optimal control of electric power consumption when working in the day-ahead market is to provide an application for electric power under the influence of various factors. As a criterion for optimization, the minimum of deviations of the actual and declared power consumption (the minimum of the error of the application) is taken.

The forecast system should be a real-time system. It means that electric power consumption has to be formed under the influence of various internal and external factors that change in real time. The high dynamism and uncertainty of the impact make it necessary to quickly solve problems, and sometimes even ahead of schedule. At the same time, the "control clock", that is, the time interval for forecasting is set by the requirements of the wholesale market based on the hourly forecast for the next day.

Factors affecting power consumption can be divided into external and internal. External factors affect electric power consumption, but these factors can not be influenced by the entity (enterprise), for example, gas and water profuseness, depth and thickness of coal seams. Internal factors arise from the activities of the enterprise. These factors change in order to optimize electric power consumption (production volume, structural and technological factors, social and economic factors, etc.).

To solve the problem of forecasting electric power consumption, traditional statistical methods based on specific norms for electric power consumption and models based on expert systems and artificial neural networks can be used [6–9].

Traditional statistical methods have the following disadvantages:

- direct participation of analysts at their application;
- complexity of calculations;
- inadequate forecast accuracy;
- the ability to apply only to a certain type of forecast;
- sensitivity to input parameters [10, 11].

Artificial neural networks are more preferable. In this case, there is no need to build a consumer model and the operability is not lost with incomplete input data. Besides, artificial neural networks have high speed and stability. It is preferable to use the configuration of artificial neural networks of direct propagation (perceptrons) with learning by the back propagation error method, even in spite of the slow convergence. Sometimes training of perceptrons takes a several days. It is important to consider the number of hidden layers and neurons in these layers. Low number of neurons provides poor training and malfunction. A large number of neurons gives a long learning process. Then the network can not be suitable for solving this problem at all. As a rule, the number of neurons in a hidden layer is determined experimentally. For this, it is necessary to put several experiments with different network configurations by the number of neurons in the hidden layer and compare the results of training [11–13].

3 Results and Discussions

The forecasting system based on an artificial neural network allows us to analyze groups of electric power consumers taking into account poorly formalized and nondeterministic factors. It allows to increase the accuracy of the forecast for electric power consumption.

The change in the level of electric power consumption is a nonlinear function and its prediction needs to be made in real time. For these purposes, it is preferable to use neuron-fuzzy systems, which belong to the class of hybrid systems. The fuzzy logic, neural networks and genetic algorithms are the cornerstone of these systems. This model allows you to make accurate short-term forecasts with a small amount of input data.

To increase the level of forecasting, it is necessary to divide all electric power receivers into homogeneous classes in terms of technological parameters and power efficiency. In this case, the architecture and factors of the artificial neural network (neural-fuzzy system) are selected separately for each class.

It is necessary to solve two main tasks:

1. To carry out the analysis of structure of electric power consumption of the research object.
2. To construct a predictive model of electric power consumption based on the data of the analysis [15].

With a known coal production plan for the next period and factors characterizing the mining, geological, and technological conditions of coal mining, a multifactor model based on an artificial neural network (neural-fuzzy system) is adequate and acceptable for planning of electric power consumption for a next day or other period.

To increase the level of forecasting of electric power consumption it is necessary to use individual models characterizing the individual features of each enterprise. It is also possible to carry out preliminary classification of mines by cluster analysis methods taking into account mining, geological, and technological parameters that affect the level of electric power consumption. However, the number of factors should not be large, since this will lead to more complicated analysis and practical use of the results.

For the selection of factors, the Delphi method was used. This method made it possible to select five factors, which in aggregate fully characterize the size of enterprises, their technological, mining and geological features. In addition, the selected factors are easily determined both during the operation of the mine and at the stage of its design:

W – this parameter determines electric power consumption for mining, transport, lifting and processing of coal at existing surface complexes, as well as the required power of the electric drive of mining vehicles and machinery, t/h;

m – average dynamic thickness of the seams determines the choice of equipment for the integrated mechanization and automation of coal mining, m;

Q – water profuseness determines the number of pumping installations of the main dewatering system and the power of electric motors, m^3/h;

V – gas content of layers determines the ventilation system, the number of fan units and the power of their electric motors, m^3/h;

H – depth of the mine determines the power and operation modes of the lifting installations, m.

4 Conclusions

Thus, the important direction of increase of power efficiency of coal mining is forecasting of a power consumption and monitoring of electric power expense. One of the main approaches to reducing of electric power costs is increase in accuracy of the enterprise demand in the wholesale electric power market. To forecasting the day-ahead electric power consumption, it is necessary to use artificial neural networks. When using artificial neural networks (neural-fuzzy systems), along with informational and computational tasks (choice of neural network structure, learning strategies, application of fuzzy inference strategy, etc.), it is necessary to define a set of input parameters that determine the electric power consumption in the mine. These factors should be presented and fully characterize the variety of conditions for underground coal mining.

The order of construction of the economic-mathematical model is the following:

1. To identify a research object (the mine as a whole or a separate technological site).
2. To formulate the purpose of researches as minimizing deviations of actual and planned values of electric power consumption.
3. Within the framework of the research object, to identify the structural and functional elements that most significantly affect the achievement of the stated purpose.
4. To construct a mathematical model of electric power consumption in a coal mine.
5. To carry out calculations for the model and analyze the results. If the results are unsatisfactory, the process of electric power consumption should be analyzed in more detail, taking into account a wider set of factors.

References

1. V. Efremenko, R. Belyaevsky, MINING 2014: Taishan Academic Forum – Project on Mine Disaster Prevention and Control, **190** (2014)
2. J.W. Taylor, R. Buizza, Intern. J. Forecasting, **19**, 57 (2003)
3. T. Mount Decision Support Syst., **30**, 311 (2001)
4. K. Metaxiotis, A. Kagiannas, D. Askounis, J. Psarras, Energy Conversion and Management, **44**, 1525 (2003)
5. F.B. Verona, M. Ceraolo, Neurocomputing, **23**, 135 (1998)
6. A.K. Topalli, I. Erkmen, Neurocomputing, **51**, 495 (2003)

7. S. J. Yao, Y. H. Song, L. Z. Zhang, X. Y. Cheng, Energy Conversion and Management, **41**, 1975 (2000)
8. B. Kermanshahi, H. Iwaminya, Electrical Power and Energy Syst., **24**, 789 (2002)
9. V. Kodogiannis, E. Anagnostakis, Fuzzy Sets and Systems, **128**, 413 (2002)
10. M. Hudson Beale, M. Hagan, H. Demuth, Natick: MathWorks, **22**, 255-271 (2014)
11. J. Nowicka-Zagrajek, R. Weron, Signal Processing, **82**, 1903 (2002)
12. H.C. Huang, R.C. Hwang, J.G. Hsieh, Electrical Power and Energy Syst., **24**, 245 (2002)
13. D. Srinivasan, S.S. Tan, C.S. Chang, E.K. Chan, IEE Proc. Gener. Transm. Distrib, **145:6**, 409 (1998)

Scientific Background for Processing of Aluminum Waste

Olga Kononchuk[1,*], *Alexey* Alekseev[1], *Olga* Zubkova[1] and *Vladimir* Udovitsky[2]

[1]Saint-Petersburg Mining University, 21st line of V. I., 2, 199106, St. Petersburg, Russia.
[2]T.F. Gorbachev Kuzbass State Technical University, 650000, 28 Vesennyaya St., Kemerovo, Russia

Abstract. Changing the source of raw materials for producing aluminum and the emergence of a huge number of secondary alumina waste (foundry slag, sludge, spent catalysts, mineral parts of coal and others that are formed in various industrial enterprises) require the creation of scientific and theoretical foundations for their processing. In this paper, the aluminum alloys (GOST 4784-97) are used as an aluminum raw material component, containing the aluminum component produced as chips in the machine-building enterprises. The aluminum waste is a whole range of metallic aluminum alloys including elements: magnesium, copper, silica, zinc and iron. Analysis of the aluminum waste Al- Zn-Cu-Si-Fe shows that depending on the content of the metal the dissolution process of an aluminum alloy should be treated as the result of the chemical interaction of the metal with an alkaline solution. It is necessary to consider the behavior of the main components of alloys in an alkaline solution as applied to the system Na_2O - Al_2O_3 - SiO_2 - CO_2 - H_2O.

1 Introduction

The urgency of the waste disposal problem is realized by the society, but the methods of processing of many kinds of aluminum waste have not yet been developed or have inefficiently elaborated. According to Rosprirodnadzor, Russia annually produces about 35-40 million tons of solid industrial waste and almost the entire amount of it is placed on landfills, sanctioned and unsanctioned dumps and only 4-5 % are involved in recycling.

2 Materials and methods

The papers [1, 2, 3] present the theoretical and thermodynamic fundamentals of alkaline aluminate solutions using Na_2O–CaO–Al_2O_3–SiO_2–H_2O system obtained at the Russian industrial enterprises processing the Kola nepheline concentrate and other aluminosilicate materials. The process of interaction of aluminum or aluminum oxide α,γ-$Al_2O_{3ТВ}$ and hydroxides: gibbsite γ-Al $(OH)_3$, bayerite α-Al $(OH)_3$, boehmite γ-AlO (OH), diaspore α-AlO (OH) is

* Corresponding author: kononchuk-olga@rambler.ru

carried out according to the reactions (1-6) with a transformation OH^- of alkali solution group to the complex ion $Al(OH)^-_{4\,aq}$ and is shown in Table 1.

On the basis of thermodynamic calculations of the Gibbs energy value ΔG^0_T, there was identified a number of energies of aluminum metal and aluminum hydroxide compounds relative to 1 molar solution of NaOH:

Aluminum [Al] (-435,13 kJ/mol)] → gibbsite [$Al(OH)_3$] (-5,9 kJ/mol) → bayerite [α-Al $(OH)_3$ (-2,75 kJ/mol)] → boehmite [AlOOH] (0,54 kJ/mol) → diaspore [AlOOH (2,28 kJ/mol)] → corundum [α, γ-Al_2O_3 (5,24 kJ/mol)].

Table 1. The theoretical and thermodynamic fundamentals of alkaline aluminate solutions using Na_2O–CaO–Al_2O_3–SiO_2–H_2O system.

No.	Compound name	Chemical reaction
1	Aluminum	$Al_{solids} + OH^- + 3H_2O = Al(OH)_4{}^-_{aq} + 1{,}5H_2$
2	$Al(OH)_{3amorphous}$	$Al(OH)_{3amorphous} + OH^-_{aq} = Al(OH)_4{}^-_{aq}$
3	bayerite Al $(OH)_3$	$Al(OH)_{3solids} + OH^-_{aq} = Al(OH)_4{}^-_{aq}$
4	boehmite AlOOH	$AlOOH + OH^-_{aq} + H_2O = Al(OH)_4{}^-_{aq}$
5	diaspore AlOOH	$AlOOH + OH^-_{aq} + H_2O = Al(OH)_4{}^-_{aq}$
6	corundum α, γ-$Al_2O_{3solids}$	α, γ-$Al_2O_{3solids} + 3H_2O + 2OH^- = 2Al(OH)_4{}^-_{aq}$

The values obtained for the Gibbs energy indicate a high probability of interaction of aluminum, amorphous hydroxide, bayerite with an alkaline solution at a temperature of 298K.

Table 2. The chemical reactions 3-6 are also possible, but they are less likely as they can only execute with increasing temperature, as evidenced by the calculated standard enthalpy of the reactions.

Reaction Nr.	1	2	3	4	5	6
Thermicity, kJ/mol	-409,2	15,79	33,52	12,57	3,21	6,16

Currently, the secondary processing of aluminum-containing raw materials (aluminum alloys) is essential, because they contain a significant amount of very valuable items: Al - Mg - Ca - Sc - Zn - Cu - Sc - Cr - Zr - Fe - Hf.

Production of recycled aluminum requires fewer energy costs and substantially lower emission of toxic substances into the environment than in the production of primary aluminum. According to forecasts, the proportion of recycled aluminum in the total consumption in 2030 could rise to 22 - 24 million tons per year [4].

Study of the problem of aluminum waste recycling for various enterprises shows that aluminum waste containing aluminum are classified according to their properties: nonrigid and cast alloys [5].

3 Results and discussion

Aluminum waste is a whole range of metallic aluminum alloys with the inclusion of significant amounts of elements of the D.I. Mendeleyev Periodic System of Elements: calcium, magnesium, copper, manganese, silica, zinc, iron [GOST 1639-2009].

Changing the technological properties in comparison with the state diagram shown in Fig.1 shows that the alloys containing alloying component less than the solubility limit have the highest ductility and the lowest strength at high temperature. Analysis of the chemical composition of the aluminum waste containing Al-Zn-Cu-Si-Fe shows that depending on the content of the metal the dissolution process of an aluminum alloy should be treated as the result of the chemical interaction of the metal with an alkaline solution containing ionsOH^-.Aluminum alloys should be regarded as a uniform distribution of elements in the crystal lattice of aluminum alloy. From the reactivity point of view, aluminum alloys are local galvanic cells which arise when exposed to water and alkali.

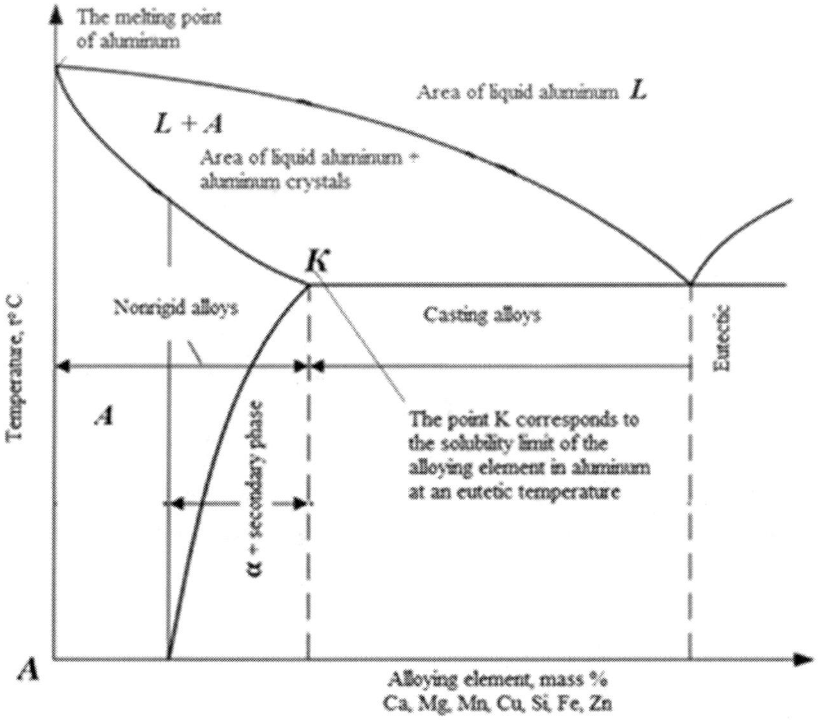

Fig. 1. The circuit diagram of a typical aluminum - alloying element.

The reactions of aluminum alloy components in water and the alkaline solution of the molar concentration (NaOH) used for preparing the aluminate solution are shown in Tab. 3.

Table 3. Gibbs energy and reaction potentials of the dissolution of aluminum waste components in an alkaline solution.

Name of alloying element	Chemical reaction	ΔG^0_{298}, kJ/mol	Standard potentials of metals, V
Beryllium	$Be_{solids} + 2OH^- + 2H_2O = Be(OH)_4^{2-} + H_2$	-309,38	-1,847
Aluminum	$Al_{solids} + OH^- + 3H_2O = Al(OH)_4^{-} + 1{,}5H_2$	-337,77	-1,66
Manga-nese	$Mn_{solids} + OH^- + 2H_2O = Mn(OH)_3^- + H_2$	-97,32	-1,18
Chromium	$Cr_{solids} + OH^- + 2H_2O = Cr(OH)_3^- + H_2$	49,27	-0,852
Zinc	$Zn_{solids} + 2OH^- + 2H_2O = Zn(OH)_4^{2-} + H_2$	-74,05	-0,763
Cadmium	$Cd_{solids} + 2OH^- + 2H_2O = Cd(OH)_4^{2-} + H_2$	46,76	-0,403
Iron	$Fe_{solids} + 2OH^- + 2H_2O = Fe(OH)_4^{2-} + H_2$	35,64	-0,037
Silicon	$Si_{solids} + 2OH^- + 2H_2O = H_2SiO_4^{2-} + 2H_2$	-398,63	-
Copper	$Cu_{solids} + 2OH^- + 2H_2O = Cu(OH)_4^{2-} + H_2$	152,09	+0,337

To evaluate the dissolution of aluminum waste components in an alkaline solution the free Gibbs energy was selected as a criterion ΔG^0_{298}, which is connected to the electromotive force of the element with an equation $\Delta G^0_{298} = n \cdot F \cdot E^o$, where n - the charge of the ions; F - Faraday constant, equal to 96485 C/mol; E° - EMF of the element, V.

Thus, when analyzing the reactivity of the aluminum alloy elements, it must be kept in mind that these compounds release hydrogen from the water molecules as a result of a chemical reaction and the role of alkali is reduced to the dissolution of the corresponding hydroxide. For example, the chemistry of the process for aluminum is carried out as follows:

1. Stage $Al + H_2O = Al(OH)_3 + 3/2\ H_2$;
2. Stage $Al(OH)_3^+ + OH^- = Al(OH)_4$

Therefore, it reacts with water to form aluminum hydroxide which being an amphoteric compound further exhibits acidic properties sufficiently and easily neutralized with alkali (NaOH) to form the aluminate complex anion.

The element silica has similar properties. Other elements such as Mg, Ti, Mn, Ni, Fe and Cu exhibit basic properties and even in the case of their dissolution, they will be presented only by hydroxides. The possibility of their dissolution in an alkaline solution can be evaluated using the solubility data for the corresponding hydroxyl compounds. For example, if an aluminum alloy contains iron and copper, that may form local galvanic element copper - iron - (NaOH solution).

In this case, the iron elements become Fe^{2+} ions and then pass into the solution due to the fact that the formation of $Fe(OH)_2$ on the surface robs free electrons from the copper $2Fe \rightarrow 2Fe^{2+} + 4e$; $2H_2O + O_2 + 4e = 4OH^-$. The presence of oxygen leads to the formation of different versions, for example, $FeO(OH)$.

Thus, a variety of different elements in the aluminum alloy results in the need to approach the thermodynamic analysis of the behavior of all the components when they are dissolved in an alkaline solution.

To develop the technological process of processing of aluminum alloys it is necessary to determine what amount of aluminum oxide can be converted into an alkaline aluminate solution, so the dissolution of the aluminum alloy should be seen as a heterogeneous system Na_2O - Al_2O_3 - SiO_2 - CO_2 - H_2O, which is shown in Fig. 2.

DOI: 10.1051/e3sconf/20172102003

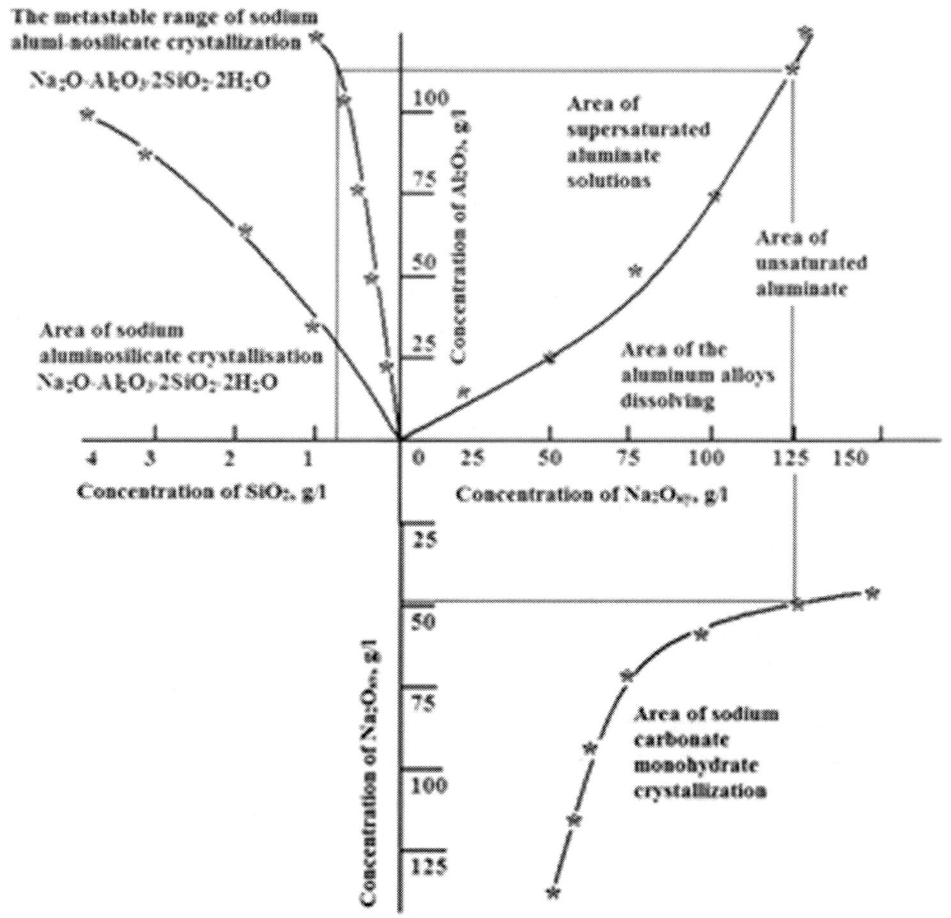

Fig.2. The equilibrium diagram of the system Na_2O - Al_2O_3 - SiO_2 - CO_2 - H_2O in the temperature the interval 80-90 °C.

The dissolution process of aluminum alloy and its components by reaction with an alkaline solution may be carried out only when a free caustic alkali is present in a solution.

As an example, consider the interaction of aluminum $Al_{solids} + OH^- + 3H_2O = Al(OH)_4^- + 1{,}5H_2$ and silicon $Si_{solids} + 2OH^- + 2H_2O = H_2SiO_4^{2-} + 2H_2$ in an alkaline solution and determine the areas of alkaline aluminate solutions, which affect its dissolution.

Since the dissolution of the alloy is performed in reactors with an open surface, the caustic alkali is able to absorb CO_2 from the air. In this case, the neutralization reaction proceeds $2NaOH + CO_2 = Na_2CO_3 + H2_O$ and, consequently, a decrease of free alkali.

These data suggest that the amount of free alkali for the process should be calculated from the formula:

$$Na_2O_{ку} = Na_2O_{total} - Na_2O(NaOH) - Na_2O(Na_2CO_3) - Na_2O_{impurity}, \quad (1)$$

where:

$Na_2O_{total.}$ - the total caustic alkali, $Na_2O_{ку}$ alkali connected in NaOH, $Na_2O_{гасн}$ alkali related to sodium hydro aluminosilicates, $Na_2O(Na_2O{\cdot}Al_2O_3{\cdot}2SiO_2{\cdot}2H_2O)$, $Na_2O(Na_2CO_3)$ - the alkali carbonate is Na_2CO_3, $Na_2O_{impurity}$ - alkali connected to other inorganic compounds.

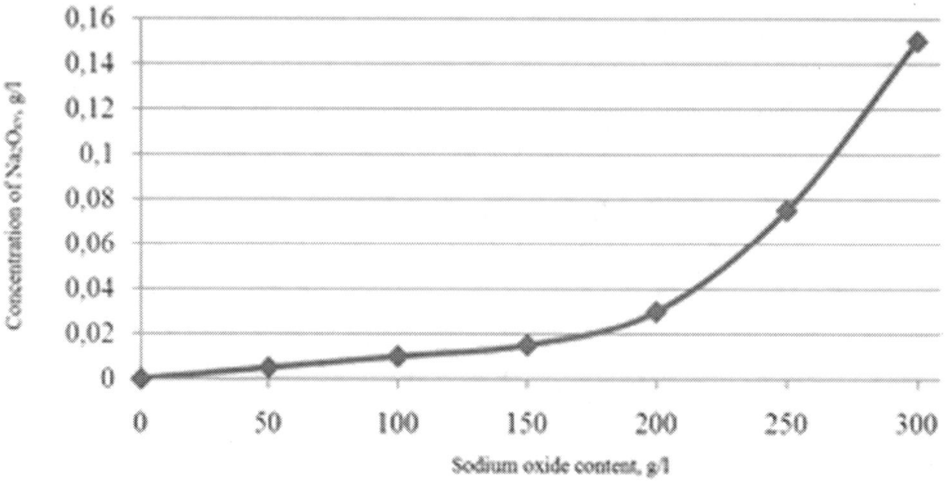

Fig.3. The solubility of iron hydroxide in an alkaline solution of various concentrations of a temperature of 80 – 90°C.

The experimental data on the solubility of iron hydroxide shown in Fig. 3 shows that the alkaline solutions containing Na_2O_{Ky} of 100 - 300 g/l the solubility of iron hydroxide is (0.003 - 0.05) Fe_2O_3 g/l. It should be noted that the solubility of iron hydroxide in alkaline solution at pH = 10 is $2.5 \cdot 10^{-26}$ g/l.

Calcium as an alloying element in aluminum significantly changes its properties, and introduced into a new aluminum gives it special properties and plasticity. When the calcium content is of 5 %, the alloy has the effect of superplasticity [6].It is known that the solubility of the hydroxides of alkaline earth metals depends on an excessive amount of hydroxyl ions in the form of NaOH solution and the temperature rise which causes a decrease in the solubility of calcium hydroxide.

4 Conclusions

1. On the basis of thermodynamic calculations of the Gibbs energy value ΔG^0_{298}, there was identified a number of energies of aluminum metal and aluminum hydroxide compounds relative to 1 molar solution of NaOH:

Aluminum [Al] (-435,13 kJ/mol)] → gibbsite [$Al(OH)_3$] (-5,9 kJ/mol) → bayerite [α-Al $(OH)_3$ (-2,75 kJ/mol)] → boehmite [AlOOH] (0,54 kJ/mol) → diaspore [AlOOH (2,28 kJ/mol)] → corundum [α, γ-Al_2O_3 (5,24 kJ/mol)].

2. Calculated values of ΔG^0_{298} show that only five alloy elements (aluminum, silica, and others) are dissolved during alkaline chemical processing of aluminum alloy. Standard potentials of metals are used to roughly estimate the electrochemical corrosion in alkaline solutions at normal temperatures and to select the contact pairs of dissimilar metals.

3. To determine the dissolution rate of aluminum waste, the experiments were performed with the aluminum alloy specimens of 40x40 mm, which were placed in an alkaline solution of various concentrations (10 - 160 g/l) and temperature (60 - 90 °C) to achieve the same result (4 μm yield on the surface side).

4. The etching equation was obtained based on experimental data that can be written as: $\alpha = k\, C_{NaOH}\, 2^{T-40}\, \tau$, wherein:α - the value by which to determine the number of metal aluminum passed into an alkaline solution, expressed in g/l Al_2O_3, *k* - the speed constant of

1.5 10^{-4} g/s; C_{NaOH} - NaOH concentration, mol; τ - etching time, s. For example, at a temperature of 70 °C the concentration of 120 g/l of NaOH (3 mol) and a time of 5 seconds, α = 1.5 10^{-4} 3 2 $^{70-40}$ 5; concentration α = 0.018 g/l Al_2O_3.

5. On the basis of literature and experimental data considering the separate sections Na_2O - Al_2O_3 - SiO_2 - CO_2 - H_2O we developed the comprehensive chart with the separate sections Na_2O - Al_2O_3 - H_2O (crystallized $Al(OH)_3$; Na_2O - Al_2O_3 - SiO_2 - H_2O (crystallized $Na_2O{\cdot}Al_2O_3{\cdot}2SiO_2{\cdot}2H_2O$, sodium hydro aluminosilicate);Na_2O - Al_2O_3 - CO_2 - H_2O (accumulation of CO_2 in alkaline solution, an aqueous sodium carbonate is crystallized in the form of $Na_2CO_3{\cdot}nH_2O$.

References

1. Tabereaux A., JOM, **2**, 343 (2009)
2. Thonstad J. et al. *Aluminium electrolysis. Fundamentals of the Hall-Heroult Process* (Aluminium Verlag Marketing & Kommunikation GmbH, 2001)
3. Lillebuen B. et al., Light Metals **33**, 389 (2009)
4. Marks J., Light Metals **33**, 251 (2009)
5. Haupin W. E., Kvande H., Light Metals **12**, 257 (1993)
6. Haupin W. E., Kvande H., Proceedings from the International Jomar Thonstad Symposium, 53 (2002)
7. Hyland M.M. et al., Light Metals **16**, 113 (1997)
8. Cochran C.N. et al., Metals **22**, 54 (1970)

Research of Environmental and Economic Interactions of Coke And By-Product Process

Vladimir Mikhailov[1*], *Tamara* Kiseleva[2] , *Svetlana* Bugrova[1], *Alina* Muromtseva[1], and *Yana* Mikhailova[1]

[1] T.F. Gorbachev Kuzbass State Technical University, 650000, 28 Vesennyaya St., Kemerovo, Russia

[2] Siberian State Industrial University, 654007, 42 Kirova st., Novokuznetsk, Russia

Abstract. The issues of showing relations between environmental and economic indicators (further - environmental and economic interactions) of coke and by-product process are considered in the article. The purpose of the study is to reveal the regularities of the functioning of the local environmental and economic system on the basis of revealed spectrum of environmental and economic interactions. A simplified scheme of the environmental and economic system "coke and by-product process - the environment" was developed. The forms of the investigated environmental -economic interactions were visualized and the selective interpretation of the tightness of the established connection was made. The main result of the work is modeling system of environmental and economic interactions that allows increasing the efficiency of local ecological and economic system management and optimizing the "interests" of an industrial enterprise - the source of negative impact on the environment. The results of the survey can be recommended to government authorities and industrial enterprises with a wide range of negative impact forms to support the adoption of effective management decisions aimed at sustainable environmental and economic development of the region or individual municipalities.

1 Introduction

Coke and by-product process is one of the most dangerous sectors of the real economy, which is associated with a wide range of negative effects on all elements of the environment [1, 2]. An additional problem is caused by the historical territorial location of enterprises near residential areas. Toughening national environmental legislation and international requirements motivate coke production plants to use the innovative modern technologies called BAT (best available technologies). Despite the intensification of this activity, and by-product process is a potential source of high environmental and economic risks, which requires greater efficiency and diversification of management methods used. In such conditions, it is advisable to build a local eco-economic system and its main elements (the enterprise - the source of the negative impact, the control center and the environment). To implement the proposed approach to management, it is necessary to identify the most

*Corresponding author: mvg.eohp@kuzstu.ru

significant environmental and economic interactions and to determine the strength of relationship between the correlating variables.

2 Materials and methods

The object of the study is Kemerovo Coke and Chemical Enterprise PJSC "Koks". The subject of the study is a system of environmental and economic interactions between Coke and by-product process and the environment. The study is based on an analysis of domestic and foreign literary sources on this issue. Particular attention is paid to studying the dependencies between the basic technical and economic indicators of production and the level of environmental and economic risk, which is of practical importance for the effective management of the local environmental and economic system. The verification of the adequacy of the theoretical developments is based on the confirmation of the factual data of the enterprise taken from the environmental reporting forms. The elements of system analysis and the results obtained by specialists in the management of environmental and economic systems are also used in the work.

3 Results and discussion

The integration of the specific eco-economic system, where one of the key places is occupied by the coke and chemical enterprise, requires the analysis of domestic and foreign literature sources, according to which the basic model of the environmental and economic system includes three types of participants [3]:

- management bodies ("center");
- economic agents ("enterprise");
- environment.

The management bodies are "interested" in the effective environmental and economic activity of the managed enterprises, which is expressed in the establishment of environmental and economic requirements for the state of the environment and the use of nature (fines for violation of environmental legislation, provision of credit and tax incentives, etc.). On a qualitative level, the task of the management bodies is to ensure the conditions that motivate enterprises to the most beneficial results for the governing bodies.

The main tasks of the management determine the specifics of environmental and economic systems [3]:

- multivariance of the results of the activities of the managed entities and exposure to a variety of uncontrolled, uncertain and random factors;
- multidirectionality of interests of the management bodies and the enterprises;
- significant costs of the regular receipt of reliable and complete information;
- inability to independently "defend" their interests, including the inertial nature of the reaction;
- crucial importance of institutional constraints (regulatory and legal framework) of enterprises' activities and their interactions with the management bodies.

Fig. 1 presents a simplified scheme of the environmental and economic system "coke and by-product process - the environment".

DOI: 10.1051/e3sconf/20172102004

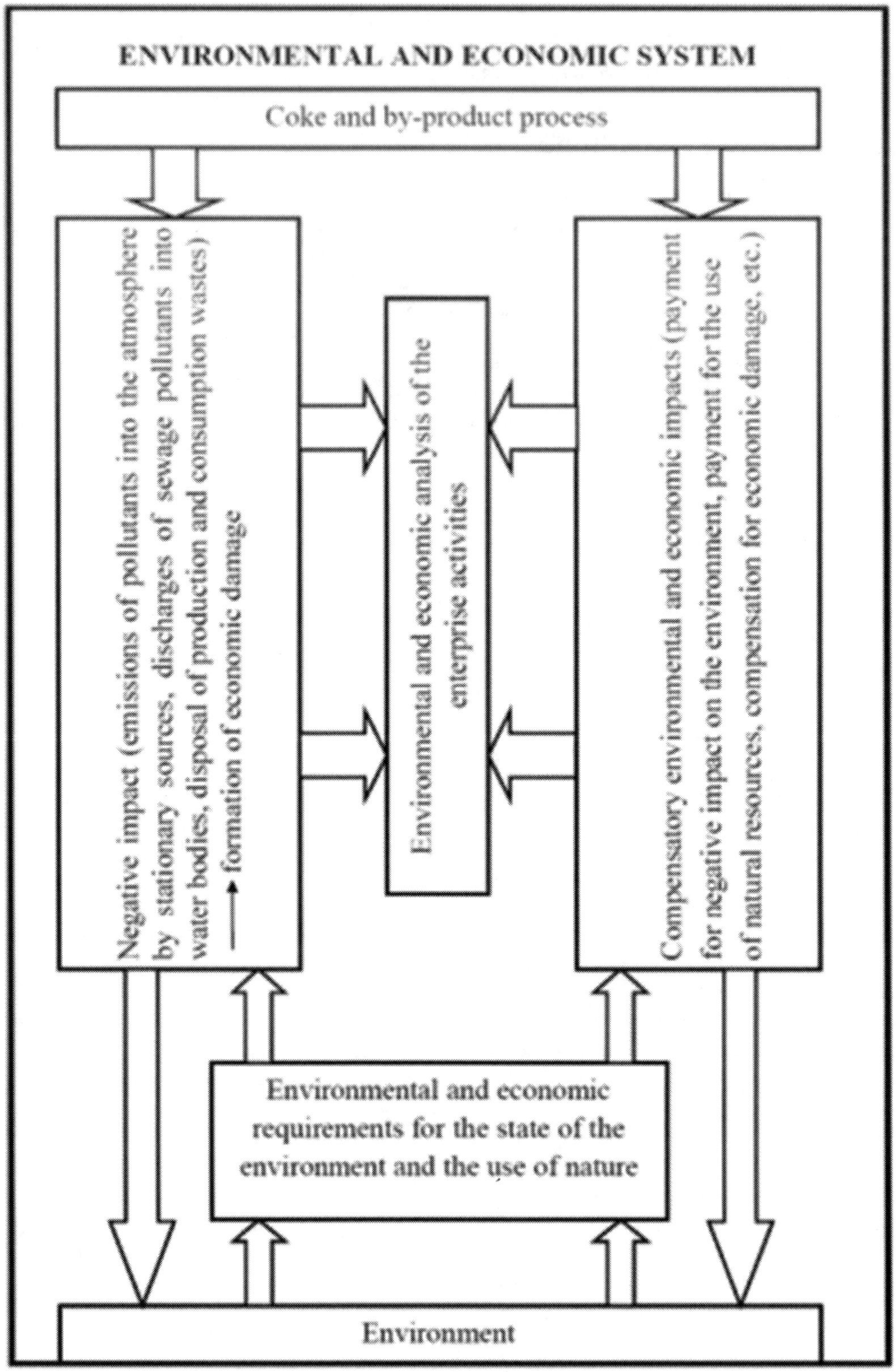

Fig. 1. Simplified scheme of the environmental and economic system "coke and by-product process - the environment".

A number of authors [4] note that the key element in the management of the local ecological and economic system is the nature protection organizational and economic mechanism, the performance of which is assessed by groups of interrelated indicators:

- economic (production volume, profit, cost, technical condition of fixed production assets, level of use of production capacity, etc.);
- environmental indicators, including parameters of negative impact on the environment and other characteristics;
- ecological and economic, characterizing the value of the actual, predicted and damped negative impact on the environment (economic damage, payment for negative environmental impact, fines for violation of environmental legislation, current costs of maintaining environmental fixed assets, etc.).

It is important to clearly structure the organizational and economic mechanism of environmental management consisting of three blocks (analytical, implementation, evaluation of efficiency), which correspond to the following innovative elements [5]:

- reduced integral indicator of the state of the aquatic ecosystem;
- system of enterprises' penalization for providing unreliable information on the level of negative impact on the environment;
- system of universal environmental and economic performance of the enterprise;
- system of environmental and economic indicators of production and consumption wastes;
- diversification of the enterprise's production program according to environmental restrictions.

As a result, a conclusion was made about the maximum belonging of these elements to the analytical block, which additionally reflects its significance.

If we consider the possibility of using these elements to manage the environmental and economic system “coke and by-product process - the environment", then in the conditions of Kemerovo PJSC "Koks" it is possible to use a system of environmental and economic indicators, including those characterizing the movement and efficiency of use of production and consumption waste [6].

One of the most important structural elements of the environmental and economic system presented in Fig. 1, are environmental and economic requirements for the state of the environment [7] and the nature use that are generated by public authorities. The implementation of established requirements is carried out through the implementation of environmental measures at various levels, where BAT is at the center [8-10].

The current problem for the conditions of the unstable economy is the financing of nature protection activities. The analysis carried out in allows us to draw the following conclusions:

- financing of nature protection activities in the constituent entities of the country is carried out on a residual basis;
- significant part of sources of nature protection financing is not available to economic entities;
- indirect assistance in the implementation of environmentally oriented management is not practical.

The theoretical analysis of the features of the functioning of environmental and economic systems made it possible to reveal the main environmental and economic interactions presented in Fig. 2. The "+" sign in the cells means high correlation ratio; "-" - the weak correlation ratio; empty cells show that the study of the correlation ratio between environmental and economic interactions was not carried out.

y \ x	1. Production volume	2. Profit	3. Current costs of environmental protection	4. Payment for negative impact on the environment	5. The amount of current costs and fees for negative impact on the environment	6. Risk level (specific gravity of payment for excess environmental impact)	7. Coefficient of compensation for economic damage
1. Production volume							
2. Profit	-						
3. Current costs of environmental protection							
4. Payment for negative impact on the environment			-				
5. The amount of current costs and fees for negative impact on the environment							
6. Risk level (specific gravity of payment for excess environmental impact)	+	-	+		+		
7. Coefficient of compensation for economic damage	+	-	+		+		

Fig. 2. Matrix of connections between the basic environmental and economic interactions of coke and by-product process.

The conducted approximations and correlation analysis of environmental and economic interactions showed a different level of correlation ratio between the variables x and y. To visualize the result obtained the environmental-economic interactions 1-6 and 3-7, shown in Fig. 3, which have a close statistical dependence and the possibility of their practical management, are chosen.

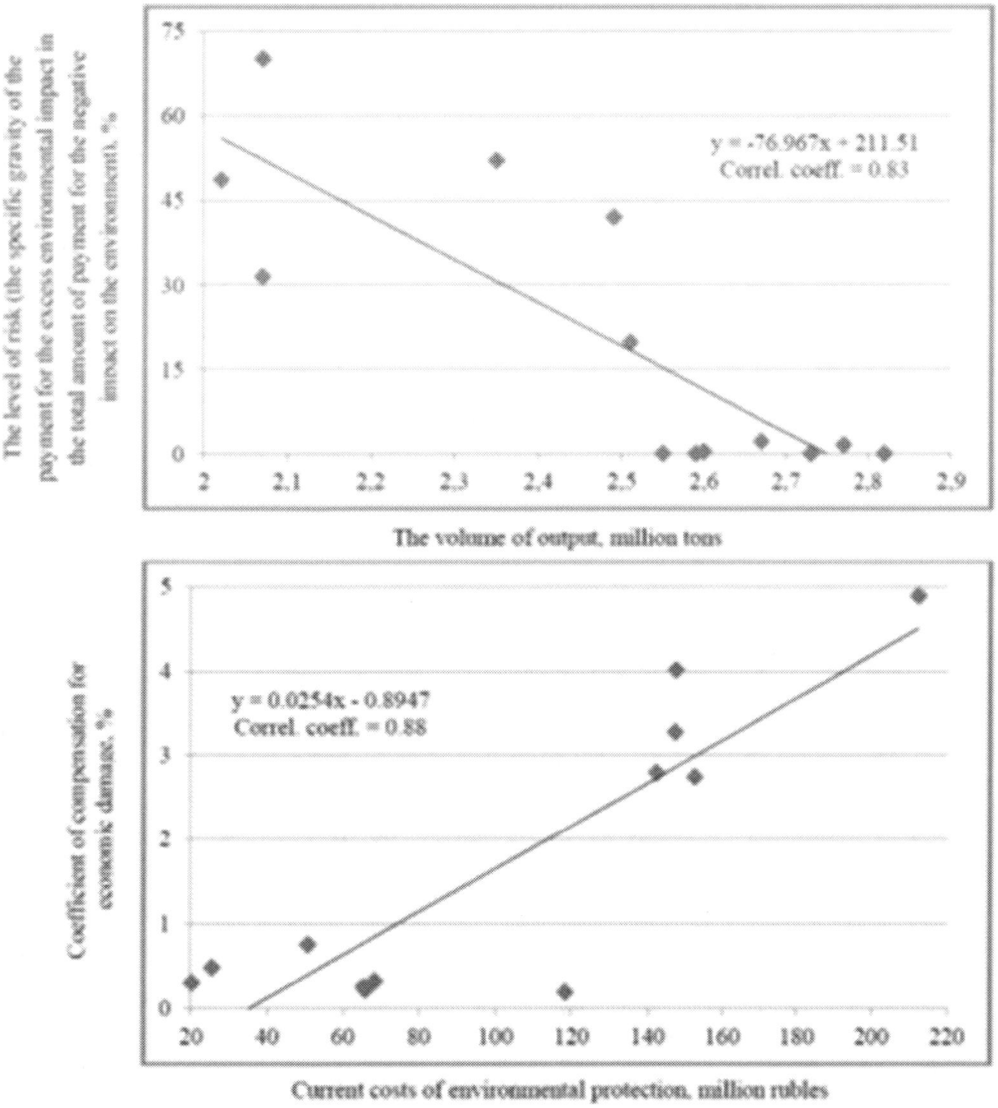

Fig. 3. Correlation fields of the dependencies of some environmental and economic indicators of coke and by-product process.

Environmental-economic interaction 1-6 from Fig. 2, shown in Fig. 3 is satisfactorily described by the linear dependence and reflects the decrease in the level of risk with increasing output, for example, in the range from 2.55 to 2.82 million tons of coke, the level of risk is zero, and when a production volume is of 2.07 million tons, the level of the risk is 70.13 %. Environmental-economic interaction 3-7 is most adequately described by a linear dependence. As a result, the growth of current costs and fees for negative impact leads to an increase in the economic damage compensation ratio. The maximum value of the economic damage compensation coefficient (4.9 %) is observed with the current environmental expenditures amounting to 212.4 million rubles.

4 Conclusions

The conducted research of environmental and economic interactions of coke and by-product process allowed drawing the following conclusions:

- the main environmental and economic indicators, between which an attempt was made to establish statistical links, were identified;
- an analysis of various theoretical approaches to the construction and interpretation of environmental and economic systems was carried out;
- a simplified scheme of the environmental and economic system " coke and by-product process - the environment" was built;
- a matrix of connections between the basic environmental and economic interactions of coke and by-product process was developed;
- a correlation analysis of the dependencies between the most significant indicators of the environmental and economic system was made establishing the correlation ratio between them;
- linear dependencies between the volume of output and the level of risk, as well as between the amount of current costs for environmental protection and the compensation coefficient of economic damage with a high degree of relationship are revealed, which allows managing these indicators to stabilize the state of the environmental and economic system.

References

1. S. V. Gerasimov, N. G. Kolmakov, Coke and Chemistry, **58:5**, 184 (2015)
2. V. N. Burkov, I. V. Burkova, Game Theory and Application. **17**, 17 (2015)
3. T. V. Kiseleva, V. G. Mikhailov, V. A. Karasev IOP Conf. Series: Earth and Environmental Science, **45**, 012013 (2016)
4. V. G. Mikhailov, T. V. Kiseleva, V. A. Karasev, G. S. Mikhailov, V. A. Skukin, IOP Conf. Series: Earth and Environmental Science, **66**, 012008 (2017)
5. V. G. Mikhailov, N. N. Golofastova, T. V. Galanina, T. G. Koroleva, Ja. S. Mikhailova, IOP Conf. Series: Earth and Environmental Science, **50**, 012038 (2017)
6. S. N. Dyakov, A. V. Papin, A. V. Nevedrov, E. V. Zhbyr, Coke and Chemistry, **55:10**, 363 (2012)
7. H. P. Tiwari, R. Sharma, R. Kumar, P. Mishra, A. Roy, S. K. Haldar, Coke and Chemistry, **57:12**, 477 (2014)
8. S. Kordana, E3S Web Conf., **17**, 00042 (2017)
9. N. N. Yashalova, D. A. Ruban, Economic and Social Changes: Facts, Trends, Forecast, **47:5**, 219 (2016)
10. A. Pimonov, E. Raevskaya, T. Sarapulova, E3S Web Conf., **15**, 01021 (2017)

Coal Producer's Rubber Waste Processing Development

Evgeniya Makarevich[1*], *Andrey* Papin[1], *Alexander* Nevedrov[1], *Tatyana* Cherkasova[1], and *Alla* Ignatova[1]

[1]T.F. Gorbachev Kuzbass State Technical University, 650000, 28 Vesennyaya St., Kemerovo, Russia

Abstract. A large amount of rubber-containing waste, the bulk of which are worn automobile tires and conveyor belts, is produced at coal mining and coal processing enterprises using automobile tires, conveyor belts, etc. The volume of waste generated increases every year and reaches enormous proportions. The methods for processing rubber waste can be divided into three categories: grinding, pyrolysis (high and low temperature), and decomposition by means of chemical solvents. One of the known techniques of processing the worn-out tires is their regeneration, aimed at producing the new rubber substitute used in the production of rubber goods. However, the number of worn tires used for the production of regenerate does not exceed 20% of their total quantity. The new method for processing rubber waste through the pyrolysis process is considered in this article. Experimental data on the upgrading of the carbon residue of pyrolysis by the methods of heavy media separation, magnetic and vibroseparation, and thermal processing are presented.

1 Introduction

The decommissioned rubber goods are environmental pollutant. At the same time, the worn automobile tires, conveyor belts, etc., are a source of valuable secondary raw materials, such as rubber, technical carbon, metal cord. Ensuring the extraction and further use of the components of rubber goods will significantly reduce the consumption of some scarce natural resources.

The processing of rubber goods can be roughly divided into three categories: grinding, pyrolysis (high and low temperature), and decomposition by means of chemical solvents.

One of the techniques of processing the worn-out tires is their regeneration, aimed at producing the new rubber substitute used in the production of rubber goods. However, the number of used tires used for the production of regenerate does not exceed 20% of their total quantity [1-9].

In some countries, the worn automobile tires rubber crumbs are used for the production of rubber crumb, which is used in rubber-asphalt mixtures for road construction, for the partial bitumen substitution, for the production of building and technical materials and products, and as a component of polymer mixtures. In many countries, a promising solution

*Corresponding author: evgeniyamakarevich@mail.ru

to the problem is burning tires to produce heat and energy, and also as a fuel in the cement industry. In this way, it is possible to achieve a significant reduction in the amount of worn tires.

However, as a result, burning was not profitable either from an environmental or economic point of view. An alternative to burning is pyrolysis of worn tires. In pyrolysis, the feedstock in the reactor is decomposed at a temperature of about 450 ° C, resulting in the production of intermediates: gas, a liquid fuel fraction, a carbonaceous residue and a metal cord. The advantage of pyrolysis is its environmental safety, as a consequence of the process in the absence of atmospheric air, resulting in pyrolysis gases containing no toxic compounds such as sulfur dioxide, nitrogen oxides and carbon monoxide. During pyrolysis, a solid carbon residue is formed in the form of pieces and particles of broad fractional composition. This residue is of interest as a secondary raw material in certain branches of the chemical industry.

After quenching and cooling the carbonaceous solid residue is subjected to magnetic separation or sieved through a sieve, while the metal cord is separated. Liquid fuel and steel cord are sent to the warehouse for further shipment to the consumer.

Liquid products consisting of a mixture of gasoline, diesel fuel and mazut are used as fuel for boiler houses. Low-quality carbon often cannot find its application and is stored on the industrial site of the enterprise.

The greatest interest of industrialists and researchers is caused by the possibility of using a solid carbon residue of pyrolysis of tires in various spheres of economic activities. However, there are a number of application difficulties caused by the variability in the composition of the pyrolysis products, the poor quality of the products (high ash content and sulfur content), poor state of knowledge of the effects on pyrolysis products in order to improve their quality characteristics. The relevance of these studies is of great importance.

Technical carbon has an ash content of up to 20% by weight, because of the additives in the rubber, is very toxic due to technological process disturbances and the absence of aromatization and polycyclization reactions with the elimination of gaseous products that intensify at temperatures above 550 ^{0}C. It is not suitable for use as a sorbent, nor as a fuel, nor as a raw material for the electrode industry.

Thus, it becomes necessary to upgrade the carbon residue, taking into account its physical and chemical properties and the regularities of the processing processes.

2 Materials and methods

In the course of the work, studies were carried out to improve the solid carbonaceous residue of tire pyrolysis by magnetic separation methods and gravity concentration, heat treatment.

The solid carbon-containing residue of tire pyrolysis by OOO "Ekoshina" (Kemerovo) was taken as an object of study.

At the first stage of the research, the technical analysis of the carbon residue was carried out, when ash, moisture, analytical, sulfur content and heat of combustion were determined.

Then, the carbon residue was upgraded by magnetic and vibroseparation. After grinding the pieces of the carbon residue to the particle size of less than 1 mm, the metal cord particles were observed, which were separated by the magnetic separation method.

The experiments on the upgrading of the carbon residue by the method of heavy media separation were conducted. Heavy media wet separation is based on differences in the densities of the separation media, the recovered product (concentrate) and the waste. gravity concentration processes are characterized by high productivity of concentrating devices, simplicity of the production complex, relative cheapness and high efficiency of separation of mineral mixtures.

Upgrading solid carbonaceous residue of tire pyrolysis was carried out in the experimental installation (Fig. 1), according to the principle of operation based on the method of heavy media separation. The processes of separation of the organic component of the carbon residue of the pyrolysis of tires and mineral particles differing in density occurred, were carried out.

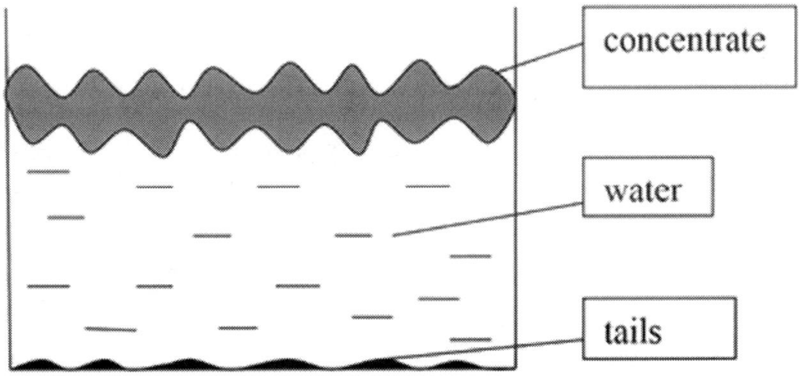

Fig. 1. Scheme of the installation for upgrading the solid carbonaceous residue of tire pyrolysis by the method of heavy media separation

The experiments of concentration of the solid carbonaceous residue of pyrolysis of tires of different grain sizes: > 0,315; 0,315÷0,8; 1,0÷>1,2 were conducted. For the maximum extraction of the organic constituent, the fraction 0,315÷0,8 is optimal, therefore it was taken for the study. Water (ρ=1000 кг/м^3) was used as heavy medium.

Upgrading of the carbon residue of tire pyrolysis by heat treatment was carried out by calcination in the muffle furnace at a temperature of 800±25°C.

3 Results and discussion

The results of the technical analysis of the feed solid carbon residue of tire pyrolysis are presented in Table. 1. From the data presented in Table. 1, it can be seen that the carbon residue has high values of ash content and volatiles yield.

Table 1. Technical analysis of the carbon residue of pyrolysis of worn tires

Test object	**Analyte**	**Analyte content**
Solid carbon residue of tire pyrolysis	Ash content: A^d, % wt.	8.9 – 20.40%
	Volatile content V^{daf}, % wt.	7.50 – 16.55%
	Moisture of analysis sample W_a, % wt.	0.65 -1.35%
	Sulfur content, % wt.	0.50 – 2.00
	Calorific value Q_s^r, kcal/kg	5500 - 6000

After grinding the pieces of the carbon residue to the particle size of less than 1 mm, the metal cord particles were observed, which were separated by the magnetic separation method.

It was possible to isolate up to 5% wt. magnetic inclusions by this method. However, the metal cord residues that do not have magnetic properties were visually observed. For the purpose of their isolation, the method of vibroseparation was applied. Metal inclusions had greater weight compared to the organic part and were easily separated using the vibroseparator. The yield of non-magnetic metal cord residues amounted to 1% by weight.

The experimental data on the concentration of the carbon residue by the method of heavy media separation are presented in Table 2.

the solid carbonaceous residue concentrated by means of this method, has low ash content (Ad), good concentrate yield (more than 80% by weight).

The obtained results showed high selectivity of the heavy media separation process and the possibility of producing the low-ash concentrate. From the data presented in Table 2, the concentration process effectiveness is evident, by producing low ash concentrates.

Table 2. Experimental data the carbon residue processing by the method of heavy media separation

Indicator	**Heavy media separation**
Ash content: A^d, % wt.	5.0-9.3
Volatile content V^{daf}, % wt.	12.0-15.5
Moisture of analysis sample W_a, % wt.	0.6
Sulfur content, % wt.	0.5
Calorific value Q_s^r, kcal/kg	6000-6200

Upgrading the carbon residue of tire pyrolysis by heat treatment was carried out by calcination in the muffle furnace at a temperature of 800±25°C. The mass loss after calcination was from 18.14 to 25.07% by weight.

The quality of the carbon-containing residue of tire pyrolysis was improved (Table 3) after calcinations. The sharp toxic smell disappeared, the pores were cleared and the cylindrical macro-openings became visible to the naked eye (Figure 2), which opens up the prospects for using the produced carbon residue as an adsorbent.

Fig. 2. Carbon residue of tire pyrolysis upgraded by heat treatment

DOI: 10.1051/e3sconf/20172102005

Table 3. Data of the technical analysis of carbon residue of tire pyrolysis upgraded by heat treatment

Test object	Indicator	Value
Upgraded concentrate of solid carbonaceous residue of tire pyrolysis	Ash content: A^d, % wt.	6.5-10.0
	Volatile content V^{daf}, % wt.	0.27-0.38
	Moisture of analysis sample W_a, % wt.	0.20-0.26
	Sulfur content, % wt.	0.25-0.5
	Calorific value Q_s^r, kcal/kg	6250-6850

The solid carbonaceous residue upgraded by heat treatment had lower volatile yield and higher porosity. This makes it possible to use it as an adsorbent, raw material for the production of smokeless fuel, reducing agent for metals, etc.

4 Conclusion

The most promising direction of processing of rubber waste of coal-mining and coal-processing enterprises is pyrolysis. The solid carbon residue produced during pyrolysis is of poor quality, so further processing is required to uprade it.

To remove iron-containing and other mechanical impurities from the carbon residue, it is rational to use magnetic and vibroseparations.

The heat treatment it is most effective to reduce the content of volatile substances and increase the porosity of solid carbon residue.

Processing of rubber waste of coal-mining and coal-processing enterprises by means of pyrolysis will allow processing the significant amount of waste into commercial products and improve the environmental safety of these enterprises.

References

1. C.A.Nau, G.Neal, V.A.Stembridge, Arch. Indust.Health, **17**, 21 (1998)
2. H. John, Converting, Fader American Tire, **3**, 45-56 (2000)
3. Kautschuk. Gummi. Kunststoffe., **48:12**, 909 (1995)
4. Rapra Review Report, **99**, 23 (1997)
5. *Reclaimed Rubber-are our technical abilities at the end by Klaus Knorr/Germany presented at tiie meeting of the Rubber Division, American Chemical Society Cleveland,* **2:1** (Ohio, Rubber and Plastics News, 1996)
6. Scrap tires. TAB special report, **1**, 8 (1989)
7. H. Terrel Ocean Ind, **22:9**,102 (1987)
8. *Tests show powder retrieves dispersing hydrocarbons in spill,* (Ney York, Offshore, 1988)
9. B. Weaver, Pit and Quarry, **11** (1994)

Enhancement of Operating Efficiency Of The Central Coal-Preparation Plant of "MMK – UGOL" Ltd. Under Current Conditions

Maksim Basarygin[1*]

[1] Central coal-preparation plant of JSC «MMK - UGOL», 652607, 27/1 Teleut st, Belovo, Russia

Abstract. In this article the subject of enhancement of operating efficiency of the central coal-preparation plant of OOO «MMK-UGOL» is encompassed. Modern trends in the development of technologies and equipment for coal beneficiation are due to the following requirements: improving competitiveness of coal products, improvement of quality of marketable products, reduction of coal production cost, environmental requirements: polluting emission abatement, prepared coal saving, improvement of the effectiveness of resource conservation; complex mechanization and beneficiation process automation. In the article the contemporary problems of raw coal benefication under current conditions of the increased dilution of withdrawable coals with rock fractions are considered. Comparative analysis of efficiency of application of modern concentrating equipment under the conditions of the CCPP of OOO «MMK-UGOL» is carried out on the basis of research works. Particular attention is paid to dehydration of produced coal concentrate with content of volatile agents of more than 35.0% and content of fine-dispersed particles in flotation concentrate of more than 50.0%. Comparative analysis of the coal concentrate dehydration technologies is conducted.

1 Introduction

Under current mining conditions the coal is re-grinded and the clogging of extracted coal is increased by rock fractions, the thickness of worked layers being reduced. All this determines the need for reequipment and application of progressive processes, effective apparatuses allowing to reduce losses of combustible matter in production wastes, lowering costs of salable production [1-5].

Choice of the optimal technological scheme of beneficiation is the main task of beneficiation specialists. When the raw material base of the coal-preparation plant changes, it becomes necessary to adapt the beneficiation technology. At the same time, under current mining conditions the consumers are placing ever higher demands on the coal concentrate quality and uniformity. When developing new technological schemes, it is always necessary to take into account the environmental requirements for coal-preparation plants.

* Corresponding author: office_cof@mmk-coal.ru

Strong volatility in the coal concentrate market dictates the implementation of projects with minimum payback periods.

Thanks to the use of present-day developments and efficient equipment introduction in coal benefication processes the concentrate yield increases and quality of the marketable product improves.

Work objective: Enhancement of operating efficiency of the central coal-preparation plant of OOO «MMK - UGOL».

Work tasks: ash content reduction and increase in concentrate yield, cost reduction, reduction of losses in the raw coal beneficiation, consideration of the efficiency of use of the most advanced technologies in coal beneficiation under the conditions of the central coal-preparation plant of OOO «MMK - UGOL».

2 Materials and methods

Situation at present. Existing scheme of the raw coal beneficiation at the central coal-preparation plant of OOO «MMK - UGOL» provides for the operations:
1. 13-150 mm grade beneficiation – using the heavy medium separation method on the wheel separator SKVP-20.
2. 0,5 -13 mm grade beneficiation – using the jigging method on the "Batac" jigs.
3. 0-0,5 mm grade beneficiation – using the flotation method on the Wemco L144 flotation machines/MFU-12 mechanical flotation machines 2.

Existing problems:
1. Increased clogging of raw coal by rock fractions, reduced concentrate yield and deterioration of the concentrate quality.
2. Imperfection of the water-slurry circuit: presence of circulating loads, content of fine-dispersed particles in flotation concentrate of more than 50.0%.

Let's consider a solution of the problem of the raw coal clogging by rock fractions, reduction of the concentrate yield and deterioration of the concentrate quality.

For successful achievement of the objective it is necessary to consider the efficiency of the process of the raw coal beneficiation in regards to operations.

For this purpose the research work was carried out by the institute together with the Institute of OOO "Sibniiugleobogashchenie", during which the operation of the technological processes of the coal beneficiation was researched. Data on technological efficiency of the processing equipment during the charge processing with participation of the "Zh" brand are given in the table 1.

Table 1. Operating efficiency of the processing equipment at the central coal-preparation plant of OOO "MMK – UGOL".

Item No	Size grade, mm	Beneficiation process name	Process efficiency ,%
1	13-150	Heavy medium separation	99.2%
2	0.5- 13	Jigging:	
		Rock compartment	89.7%
		Mid-coal compartment	66.2%
3	0-0.5	Floatation	

According to the research data, the least efficient work on the processing equipment in is in the jigging process. The reason for the inefficient operation of the jigging process is a change of the charge washability category from medium to heavy. However, it is commonly known that the jigging process is best suited for easy and average raw coal washability categories, the washability category rise to the heavy one leading to excessive losses of the coal concentrate.

To assess the increase in the raw coal beneficiation, it is necessary to carry out a comparative analysis of the 0.5-13 mm grade beneficiation by the jigging and heavy medium separation method.

Table 2. Calculation of the practical balance of the 0.5-13 mm grade coal beneficiation products in heavy medium cyclones.

Denomination	Before implementation		After implementation	
	Yield, %	**Ash content, %**	**Yield, %**	**Ash content, %**
Concentrate	65.5	9.5	67.74	9.5
Waste	34.,5	77.3	32.26	82.0
Feedstock	100	32.9	100	32.9

According to the research data, the existing 0.5-13 mm grade concentrate yield on the jigs at the 9.5% concentrate ash content is 65.5% of the feedstock of the jigs.

Due to the use of a more efficient beneficiation process in heavy medium cyclones, the increase in the 0.5-13 mm grade concentrate yield will amount to 2.24% of the initial grade, what will amount to 1.52% in terms of the initial raw coal (67.9 % of the yield of the raw coal being supplied to the jig). When the jigs are replaced with heavy medium cyclones, the increase in the concentrate yield will be 1.52% at the E_{pm} separation error index * decrease from 150 kg/m3 to 50 kg/m3.

E_{pm} index (Separation error index) is recommended by the International Organization for Standardization as an international standard for evaluating the efficiency (accuracy) of the coal separation in a suitable separator.

Improvement of the water-slurry circuit.

The plant was the first in Russia in 1994 to install the closed water-slurry circuit.

To install the heavy medium cyclones in replacement of the jigging process, it is necessary to ensure the solid content in the return water and exclude the circulating loads.

To achieve the solid content up to 1 (3) g/l in the return water, it is necessary to obtain a clean filtrate from the flotation concentrate filtration process, the equipment also having to meet the solid high output requirements. At the present moment, the filtrate content less than 3 g/l can be obtained by using the modern equipment of the chamber filter presses and hyperbaric disc filters. Chamber filter presses and hyperbaric disc filters are the fully automated equipment that facilitates operation and reduces the human factor.

As a result of the studies in respect of the conditions of the central coal-preparation plant of OOO «MMK - UGOL» despite the pressure difference of 10 bar between two filters (pressure of the chamber filter presses 16 bars and pressure of the hyperbaric disc filters 6 bars), the lowest residual moisture content is 15-18%. A significant difference between the kinds of equipment is the continuous operation of the hyperbaric filters compared to the cyclic operation of the chamber filter presses. Continuous flotation concentrate filtration operation with due regard to the residual moisture content reduction from 27.0% to 18.0% will improve the values of the thermal drying of the concentrate. It is worth noting the improvement of the technological parameters of the thermal drying, what is especially important because the release of volatile substances as for the "Zh" brand is higher than 35.0%, also allowing to reduce the thermal drying costs.

It is also very important that the filtration will be carried out with the flotation concentrate having the content of fine-dispersed particles of more than 50.0%.

Table 3. Granulometric composition of the flotation concentrate.

Size grade, mm	Content,%	Ash content,%
+1.0	-	-
0.5-1.0	1.3	3.9
0.3-0.5	5.5	3.9
0.2-0.3	11.2	4.1
0.1-0.2	20.8	5.2
0.05-0.1	19.5	7.0
0-0.05	41.7	19.6
Total	100.0	11.3

To ensure the operation of the water-slurry circuit in a closed loop in addition to the equipment already installed, it is required to install a radial thickener with two belt filter presses for the flotation tailings, using the anionic and cationic flocculate in the thickening and filtration processes to intensify the process.

By reducing the circulating loads, concentrate clogging with slurry, the increase in the concentrate yield will be 1.48% (due to the increase in the specific gravities of the separation in the gravitational processes of the beneficiation).

Impact of the project on the environment. Due to the flotation concentrate moisture content reduction from 27.0% to 18.0% during the filtration process, it will allow to reduce the environmental emission by 95 th m3/year during the thermal drying of the concentrate. Increase in the concentrate yield will allow to reduce the amount of beneficiation waste coal stored on the rock dump by 118,437 th t/y.

Cost-benefit analysis of re-equipment of the central coal-preparation plant of OOO «MMK - UGOL».

Table 4. Operating highlights of the re-equipment implementation.

Denomination	Before implementation	After implementation	Deviation +/-
Concentrate yield.%	50.3	53.3	3.0
Concentrate ash content.%	10.8	9.5	-1.3
Environmental emission abatement. t m^3	285	190	95
Reduction of the amount of beneficiation waste coal stored on the rock dump. th t	1922.6	1804.2	118.4

Table 5 List of costs of the re-equipment implementation

Item No	Denomination	Quantity, units	Costs of the reequipment, thousand USD
1	Throw screens	10	
2	Heavy medium cyclone	2	
3	Outlet sieve bends	10	
4	Magnetic separators with permanent magnets	4	9 456
5	Radial thickener	1	
6	Hyperbaric filter	2	
7	Project design work		
8	Construction and installation operations		
9	Costs of technological materials		
10	TOTAL		

Table 6 Calculation of net annual income from project implementation

Item No	Denomination	Quantity
1	Annual volume of raw coal processing. th t	3947.884
2	Expected minimum increase in concentrate yield. %	3.0
3	Cost of concentrate in market (min). USD/t	120
4	Increase in concentrate output will be. th t	118.4
5	Increase in income from output. th USD	14 212
6	Reduced equipment repair costs (reduction in quantity of units of equipment). th USD	240
7	Energy usage reduction (reduction in quantity of units of equipment). th USD	134.9
8	Income from improving the concentrate quality. th USD	6275.6
9	Reduction in expenses for removal of waste. th USD	85.5
10	Reduction of waste storage costs. th USD	95.7
11	Property tax. th USD	- 190.6
12	Profit tax. th USD	- 4134.4
13	TOTAL net annual income from the project	16718.7

Total payback period of the project is 7 months.

3 Conclusion

On the basis of the article it is possible to draw a conclusion on the possibility of increasing the efficiency of coal-preparation plants under current conditions by reequipment: replacement of equipment - jigs with heavy medium cyclones, modernization of the water-slurry circuit (elimination of circulating loads, reduction of the concentrate clogging by slimes), what will allow to increase the concentrate yield, improve the concentrate quality in terms of the ash content value, reduce the equipment operating costs, abate the environmental emissions (by reducing the floatation concentrate moisture content).

References

1. L.A. Antipenko, Ugol', **2**, 87 (2017)
2. A.V. Bauman, XVIII International Coal Preparation Congress, 345-351 (2016)
3. A. Sherritt, G. Jones, A. O'Keeffe, Int. J. Coal Prep. Util. **5**, 67-74 (2014)
4. A.D. Meyers, G. Sherritt, International Coal Preparation Congress, 334-340 (2010)
5. E. Bekker, XVIII International Coal Preparation Congress, 233-240 (2016)

Diagnostics of Oil Pollution Zones by Electro-Physical Method

Sergey Prostov[1*], and *Evgeniy* Shabanov[1]

[1] T.F. Gorbachev Kuzbass State Technical University, 650000, 28 Vesennyaya St., Kemerovo, Russia

Abstract. The article presents the rationale and development of electro physical methods of control of the degree of soils pollution with petroleum products. The ranges of variation for the parameters of the pore space structure and wettability for the Kuzbass conditions are determined by inverse calculations on the basis of experimental data. The theoretical dependences are shown which allow us to assess the degree of soil pollution with oil products according to the results of longitudinal and areal electrical sounding. The study confirms the possibility of rapid prediction of the degree of soil contamination by measuring its electrical resistance. The database has been compiled for interpretation of the results of experimental sounding when monitoring the changes in the degree of contamination. The use of the express forecast of numerical values of the contamination factor was studied.

1 Introduction

One of the most dangerous soil pollutants are petroleum products: lubricants and liquid fuel [1, 2]. The relevance of the problem of cleaning soils from oil pollution in mining industry (coal surface and underground mines, ore mines) due to large amounts of fuel consumption by road and rail transport, and lubricants in the operation of mining equipment, and technical oils in electric power plants. The refined petroleum products contain toxic water-soluble fractions [3, 4], also, gasoline, and especially motor oils contain up to 20% of additives, solvents, heavy metals. One liter of used oil may contaminate one million liters of water, not more than 60% of their volume being regenerated

One of the effective methods for managing the soil properties, especially of low permeability (filtration coefficient $K_f < 10\text{-}8$ m/s), is the method of electrochemical treatment [5, 6], its main impact on the contaminated soil being the electrical destruction of ecotoxicants and the electroosmotic displacement of the diluted pollutant in the area of mechanical removal (pumping). Experimental and theoretical bases of the electrochemical method were developed for the solution of problems of technical reclamation (drainage) and for stabilizing the unstable waterlogged soils. The criteria for the rational use of this method are determined: the ratio of the coefficients of electroosmotic activity and filtration $K_e/K_f > 10^7$ and electrical resistance (SER specific electrical resistivity) $\rho < 8$ Ohm·m.

Direct and indirect methods are applied to control the processes of cleaning of soils from pollution. Direct methods [7-9] (infrared spectrophotometry, ultraviolet luminescence,

* Corresponding author: psm.kem@mail.ru

gas and gas-liquid chromatography) require sophisticated equipment, significant duration of the analysis and extraction of soil samples, the measurement error can reach 40-50%.

2 Materials and methods

Diagnosis and ongoing monitoring of soil condition, properties and physico-chemical processes in the area of electrical treatment can be provided by geophysical methods that are effective supplement to geotechnical investigations. Very promising for the study of contaminated soils are electrical and electromagnetic methods [10] based on the measurement of electric fields, because oil and petroleum products exhibit pronounced dielectric properties and therefore they are electrically contrasting.

To control the degree of soil pollution with oil products, it is advisable to use the dependence of electrical resistivity of the three-phase medium on porosity, moisture and electrical resistivity of the pore filler that is used in electrical prospecting and includes structural-textural empirical parameters [11, 12]:

$$\rho_k = \frac{\alpha \cdot K_n}{m^{\beta} \cdot W^{\gamma}} \cdot \rho_{\text{в}} \tag{1}$$

where m is porosity (pore cavitation); W – coefficient of the saturation space of the pores and cracks; $\rho_{в}$ – electrical resistivity of the solution, filling pore space, Ohm·m; K_n, α, β, γ - are empirical parameters that depend on the structural and textural features of the studied soils; K_n – is the parameter that takes into account the surface conductivity of clay microlayer on the surface of the pores; a – is the parameter depending on the type of geological deposits; β is the parameter determined by the structure of the pore space (mainly by tortuosity of the channels); γ – parameter that depends on the wettability of surface pores by the solution.

The ranges of variation of the parameters of the pore space structure β = 1.3-2.2 and wettability γ = 1.8-3.3 should be taken for conditions of Kuzbass on the results of the inverse calculations based on experimental data; and electrical resistivity of the pore filler is calculated by the formula of harmonic weighted average for the two-component media [13]:

$$\rho_{\text{в}}^{-1} = \sum_{i=1}^{n} V_i \cdot \rho_i^{-1} = V_{э} \cdot \rho_{э}^{-1} + V_{н} \cdot \rho_{н}^{-1}, \tag{2}$$

where $\rho_{в}$, $\rho_{н}$, $\rho_{э}$ – electrical resistivity of, respectively, weighted, oil and electrolyte, Ohm·m; $V_{н}$,$V_{э}$ – volume fraction in the solution, respectively of the oil and electrolyte.

Theoretical dependence (1) and (2) basically allow us to assess the degree of soil pollution with oil products according to the results of single measurements, longitudinal and areal electrical sounding [14].

To further analyze the process of electrochemical treatment of contaminated soils, it is necessary to have information about the electrophysical properties of natural aqueous solutions and liquids containing contaminants or saturated soils. Experimental studies of electrical properties of electrolyte solutions, containing oil-products, with the measurements at permanent and variable electric current.

Figure 1 shows the measurement results of the electrical resistivity $\rho_{э}$ of the NaCl solution depending on the salt concentration in solution and the a. c. frequency, and Figure 2 shows the same dependence of the specific electrical resistance $\rho_{н}$ of motor oil and gasoline. It follows from these data that all petroleum products are electrically quite contrasting, because their electrical resistivity is by at least 3 orders of magnitude higher than electrical resistivity of natural water solutions.

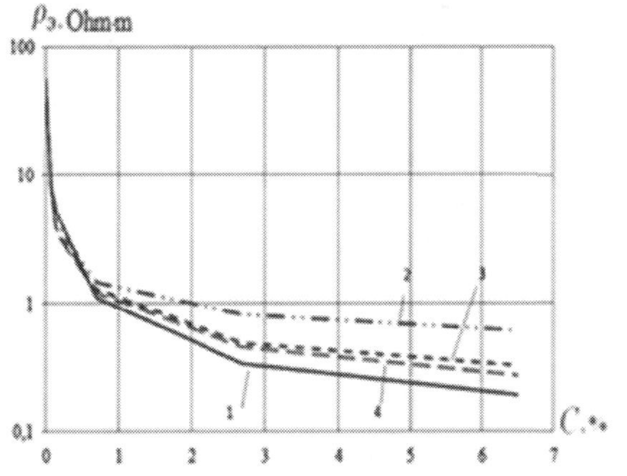

Fig. 1. The dependence of the electrical resistivity of solution $\rho_э$ on the salt concentration C in the solution: 1 – constant current; 2 - alternating current, f=10 Hz; 3 – 200 Hz; 4 - 1000 Hz.

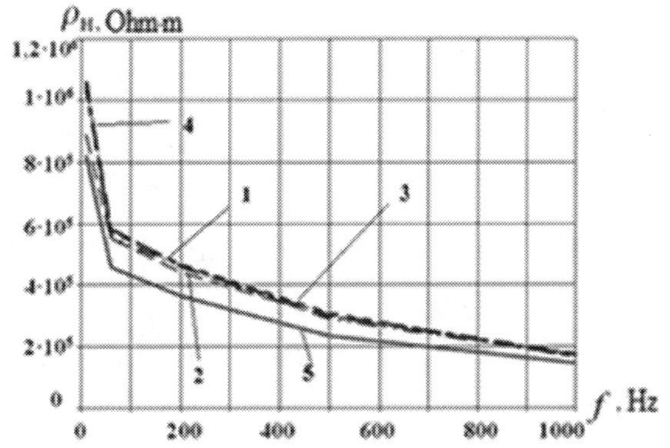

Fig. 2. The dependence of the electrical resistivity $\rho_н$ of motor oil and gasoline on the frequency of the alternating current: 1 - G-Energy F Synth 0W-40; 2 - Mobil Synt S 5W-40; 3 - Shell Helix Ultra (spent); 4 - Shell Helix Ultra; 5 - gasoline AI-92.

By substitution into equation (2) the data of Figure 1 and Figure 2, the dependences of the average values of $\rho_в$ on the relative content of petroleum products $V_н/V_э$ for DC (Fig. 3) and AC (Fig. 4) currents.

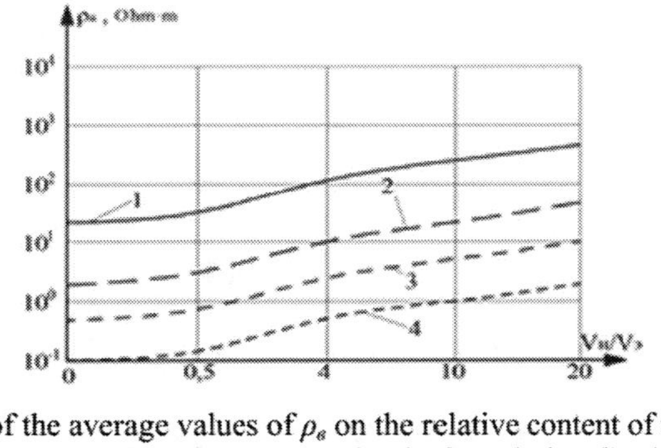

Fig. 3. The dependence of the average values of $\rho_в$ on the relative content of the petroleum products when measured at DC current: 1 - the salt concentration in the solution C = 0.1 g/l; 2 – 0.5 g/l; 3 - 2 g/l; 4 - 10 g/l.

The given dependences allowed us to form the database for interpretation of the results of experimental sounding when monitoring the changes in the relationship $V_н/V_э$, e by $\rho_к$ value [15].

 DOI: 10.1051/e3sconf/20172102007

To quantify the degree of saturation by the pollutant, the coefficient of soil contamination k is introduced that is equal to the ratio of the volume of oil contained in the soil pores to the total volume of the pore fluid. By transformation of equations (1) and (2) it is possible to obtain the following expression:

$$V_{н} = \frac{\rho_{н} - \frac{\alpha \cdot k_n \cdot \rho_{э}}{\rho_k \cdot m^{\beta} \cdot W^{\gamma}}}{\rho_{н} + \rho_{э}} \approx 1 - \frac{\alpha \cdot k_n \cdot \rho_{э}}{\rho_k \cdot m^{\beta} \cdot W^{\gamma}}, \qquad k = m \cdot V_{н} \qquad (3)$$

where $V_{н}$ – the volume fraction of the pores filling with oil.

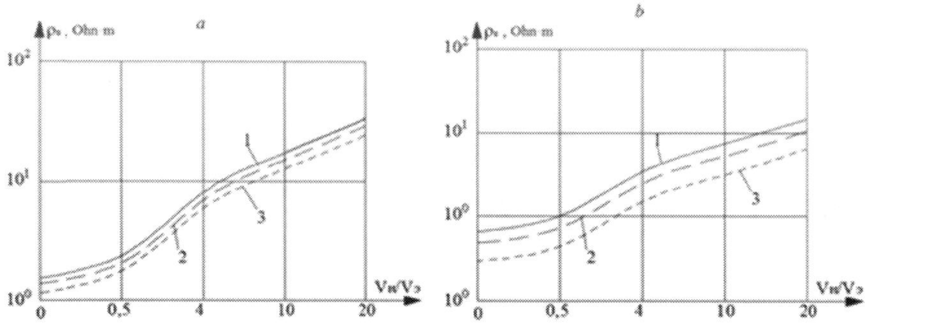

Fig. 4. The dependence of the average values of electrical resistivity on the alternating current frequency and concentration of salt in the electrolyte C = 0.5 g/l (a); 10 g/l (b): 1 - f= 10 Hz; 2 - 200 Hz; 3 - 1000 Hz

The theoretical dependence (3) allows us to estimate the degree of soil pollution with oil products according on the results of single measurements, profile and areal electrical soundings using cyclic algorithms that are implemented in the software package created in the software product Lazarus (environment Free Pascal).

3 Results and discussion

To control the processes of cleaning soil of oil pollution by the electrophysical parameters, it is recommended to use two schemes of sounding from the earth's surface: for partially open facilities and for closed facilities. When using the four-electrode method of sounding from the earth's surface, standard metal feeding electrodes are grounded in the soil beyond the treated area boundaries, using electrodes-injectors as measuring ones. The location of the electrodes-injectors and feeding electrodes for partially open and closed objects is shown in Fig.5.

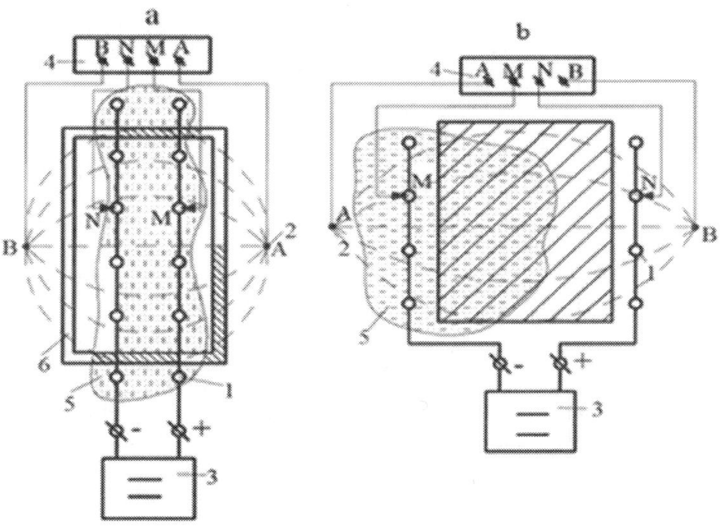

Fig.5. Circuit that control the processes of electrical treatment for partially opened (a) and closed (b) objects: 1 – pollution zone; 2 – feeding electrodes A, B; 3 – electric power unit; 4 – electrical exploration equipment; 5 – area of contamination; 6 – structure

The control results are used to plot the graphs of change in the effective electrical resistivity $\rho_к$ along the main axis of the electrical treatment and depending on the power consumption or time (Fig.6).

The main purpose of monitoring is to identify points of correction of the modes or the termination of electrical treatment, as well as the calculation of the coefficients of pollution before and after treatment.

For Express forecast of the numerical values of the contamination factor, they use four design dependences obtained in the processing of the data received from experimental field studies: linear, logarithmic, parabolic and exponential. The constants a_1-a_4 were determined for a known initial contamination factor k_o = 7 % that is equal to the ratio of the volume of contaminant to volume of contaminated soil. According to the equations given in the table, the contamination factor k was found at each point in time of the measurement of the effective resistivity.

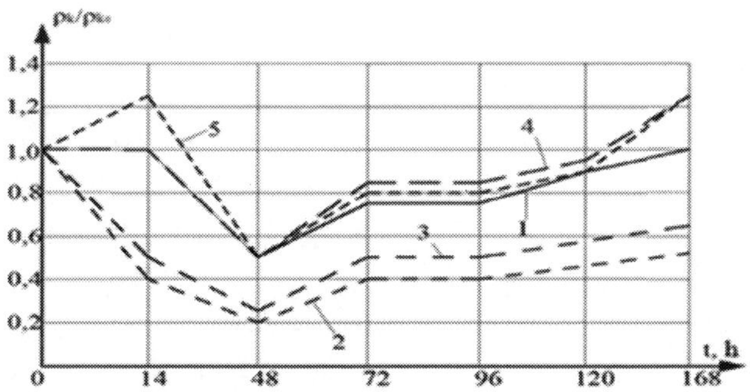

Fig.6. Graph of changes in time t of the relative effective electrical resistance $\rho_к$ at electrical sounding in the process of experimental electrical treatment: 1 – clean clay massif; 2 – contaminated array treated with oil; 3 – array contaminated with gasoline; 4 - array contaminated with used oil and treated with solvent, 5 - array contaminated with gasoline and treated with solvent

Table 1. The dependence of the contamination factor k on the relative change of the effective resistivity of the soil in the treatment area (experimental data)

Equation	The value of the constant a at pollution		The value k		
			initial k_0, %	final $k_к$, %	
	oil	gasoline		oil	gasoline
$k = a_1 \cdot \frac{\rho_к}{\rho_{к0}}$	a_1=2.8	a_1=3.5	7	3.64	4.55
$k = a_2 \cdot \ln(\frac{\rho_к}{\rho_{к0}})$	a_2=7.64	a_2=10.1		2	2.65
$k = a_3 \cdot (\frac{\rho_к}{\rho_{к0}})^2$	a_3=1.12	a_3=1.75		1.9	2.96
$k = a_4 \cdot exp(\frac{\rho_к}{\rho_{к0}})$	a_4=0.575	a_4=0.95		2.11	3.49

Fig.7 presents the change in the coefficient of pollution $k_з$ (concentration of oil) in time. The graph shows that the coefficient of pollution $k_з$ with time decreases until t = 48-72 h, and then stabilizes. The time of stabilization of the calculated value of k is an objective criterion for the termination of electrical treatment as any further exposure of the array does not give any significant positive effect.

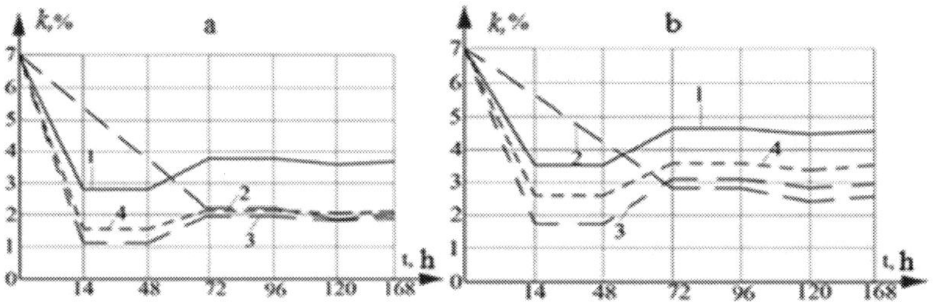

Fig.7. Graphs of change in concentrations of waste oil (a) and gasoline (b) during electrical treatment depending on time t: 1 – linear dependence; 2 – logarithmic dependence; 3 – parabolic dependence; 4 – exponential dependence.

4 Conclusions

The dependence of electrical resistivity (electrical resistivity) of the three-phase medium on porosity, moisture and electrical resistivity of the pore filler, calculated according to the formula of the harmonic weighted average is the physical basis of electrophysical methods of control of the degree of pollution of soils with petroleum products, while for the conditions of Kuzbass the ranges of change in structural parameters of the pore space and wettability are, respectively, β = 1.3-2,2 and γ = 1.8-3.3. On the basis of experimentally established ranges of change of the resistivity of an aqueous solution ($\rho_в$ = 0.2-80 Ohm·m) and major types of petroleum products (fuel, engine oil) ($\rho_н$ = (0.2-1.1)·10^6 Ohm·m), the databases were calculated for dependences of the effective specific electric resistivity $\rho_к$ on the ratio of the volumetric proportion of oil and moisture in the range of $V_н/V_в$ = 0-20.

Express-control of changes in contamination factor k in the process of electrical treatment of the soil is ensured according to the value of the effective electrical resistivity $\rho_к$ for the scheme with the zone-by-zone integral sounding using as electrodes-injectors as the measuring ones. During the experimental trials after treatment for 168 h, the value of k when contaminated with used oil declined from 7 to 2.11 %, and when polluted with gasoline – to 3.49 % (exponential dependence).

Stabilization of the *k* value at a minimum level is a criterion for the transition to a more intensive mode of electrical treatment or, if the capacities of the power electric unit are exhausted – this is a sign that further treatment makes no sense.

References

1. V. P. Seredina, M. E. Sadykov, Contemporary Problems of Ecology, **5**, 457 (2011)
2. V. A. Efremova, E. V. Dabakh, L. V. Kondakova, Contemporary Problems of Ecology, **5**, 561 (2013)
3. S. I. Kolesnikov, M. G. Zharkova, K. Sh. Kazeev, Russian Journal of Ecology, **3**, 157 (2014)
4. T. I. Kukharchyk, V. S. Khomich, Eurasian soil science, **2**, 145 (2013)
5. V. A. Korolev, A. M. Romanyukha, O .V. Abyzova, Journal of Environmental Science and Health, **43:8**, 876 (2008)
6. I. Gratchev, I. Towhata, Soils and foundations, **3**, 469 (2013)
7. C.J. Tsai, C.H. Feng, J.H. LI, Journal of chromatography, **1410**, 60 (2015)
8. Z. Zhou, F. Bing, W. Tao, X. Song, IEEE transactions on dielectrics and electrical insulation, **5**, 1498 (2012)
9. E.Y. Kovalenko, T.A. Sagachenko, E. B. Golushkova, Petroleum chemistry, **2**,101 (2016)
10. S.A. Smernikov, A.I. Pozdnyakov, E.V. Shein, Eurasian soil science, **10**, 1059 (2008)
11. B. Leiss, K. Ullemeyer, K. Weber, Journal of structural geology, **11:12**. 1527 (2000)
12. S. M. Prostov, V. A. Khyamyalyainen, S. P. Bakhaeva Journal of Mining Science, **4**, 349 (2006)
13. G.G. Shtumpf, Journal of Mining Science, **6**, 59 (1993)
14. S. M. Prostov, E. A. Shabanov, Proceedings of the 8th Russian-Chinese Symposium "Coal in the 21st Century: Mining, Processing, Safety". Advances in Engineering Research, 175 (2016).
15. S. M. Prostov, M. V. Gucal, E. A. Shabanov, Proceedings of the Taishan Academic Forum – Project on Mine Disaster Prevention and Control, 433 (2014)

Ensuring the Environmental and Industrial Safety in Solid Mineral Deposit Surface Mining

Kliment Trubetskoy[1], *Marina* Rylnikova[1*], and *Ekaterina* Esina[1]

[1]Federal State-Funded Research Establishment – Academician N.V. Melnikov Institute of Comprehensive Exploitation of Mineral Resources, Russian Academy of Sciences (IPKON RAS) 111020, 4 Kryukovsky tupik, Moscow, Russia

Abstract. The growing environmental pressure of mineral deposit surface mining and severization of industrial safety requirements dictate the necessity of refining the regulatory framework governing safe and efficient development of underground resources. The applicable regulatory documentation governing the procedure of ore open-pit wall and bench stability design for the stage of pit reaching its final boundary was issued several decades ago. Over recent decades, mining and geomechanical conditions have changed significantly in surface mining operations, numerous new software packages and computer developments have appeared, opportunities of experimental methods of source data collection and processing, grounding of the permissible parameters of open pit walls have changed dramatically, and, thus, methods of risk assessment have been perfected [10-13]. IPKON RAS, with the support of the Federal Service for Environmental Supervision, assumed the role of the initiator of the project for the development of Federal norms and regulations of industrial safety "Rules for ensuring the stability of walls and benches of open pits, open-cast mines and spoil banks", which contribute to the improvement of economic efficiency and safety of mineral deposit surface mining and enhancement of the competitiveness of Russian mines at the international level that is very important in the current situation.

1 Introduction

Less favorable geological and mining conditions of mineral deposits being developed and some deposits planned for mining, ever growing ore and fuel demand dictate the necessity of refining the existing regulatory framework with due account for the commercialization of novel geotechnologies to enhance the efficiency of mining and minimize the environmental pressure.

2 Materials and methods

In conditions of mineral reserves mining at deeper horizons a top-priority area of environmental and industrial safety and enhanced efficiency of mining processes involves

* Corresponding author: rylnikova@mail.ru

the development of robotized geotechnologies eliminating the human presence in danger zones of mining operations. For instance, the employment of a self-driving intelligent dump truck operating in an autonomous mode can significantly expand the area of the efficient surface mining application with radical upgrading of an open-pit infrastructure.

In this case, the requirements to open-pit roads are updated, including the provision of the relevant quality of the road surface, as for the application of robotized transport technologies the motor road irregularities, fast turns and other open-pit road quality deviations from standard values may results in the changing of the path of transport machinery travel that will eventually affect the open-pit safety-in-operation. Main criteria of road surface quality assessment are as follows: pavement strength, smoothness, hardness, tires' grip on the road surface.

Application of an autonomous automated dump truck with intelligent control can minimize or completely eliminate the human factor effect on the performance of operational processes. In this case, intelligent functions of mine engineers and managers increase, values of open-pit main parameters, some structural elements of the mining system, first of all the height and slope angle of working benches, change qualitatively, thus significantly enhancing the economic and ecological efficiency of surface mining operations [1-4].

Application of robotized mining and transport machinery provides an opportunity of more complete mining of mineral reserves, which earlier were not included in the planned final pit envelope. In providing the rationale for a new final pit envelope the width of a bench is determined only by the performance capabilities and parameters of the equipment with due account for non-presence of the personnel directly in a face. As robotized equipment is furnished with sensors monitoring the space position of working elements and the machinery itself with due account for the requirements to the minimal width of a bench it is important to proceed from the minimal dimensions and off-contact clearance providing for safe operation of the process equipment.

Studies of IPKON RAS performed in cooperation with VIST Group have proved that application of robotized machinery allows significant reduction of the bench width and thus the increase of the open-pit depth for final extraction of reserves occurring below its planned final boundary [5-8]. With robotized machinery, due to non-presence of the personnel in the face it is possible to minimize the values of some parameters and to eliminate some elements of the open-pit design as a mining structure, which are intended for safety-in-mining, as compared with similar conditions, however envisaging the presence of miners in a working zone.

Furthermore, one of the promising areas of mining efficiency enhancement is the development of inclined and steeply dipping ore and coal beds with the use of high overburden benches. High benches are seen as benches with a common slope plane, which parameters exceed the potential of loading and haulage machinery, and these benches are developed by hole blasting for the full height and mined by separate layers – sub-benches. A sub-bench height is determined by the potential of excavation and loading machinery.

The increase of an overburden bench full height provides a perfect logistic pattern of an ore open pit and allows the minimization of the total number of open-pit transportation machinery, with that, the time of machinery and equipment running, number of working benches, length of transportation routes and number of dumping points are minimized, while the effective depth of surface mining operations increases. All items listed above minimize the negative environmental effect of surface mining.

It has been proved that a high overburden bench is feasibly developed in two layers [9]. As a result of assessment of the efficiency of different high overburden bench mining scenarios it has been found that the most advantageous mining method is the one employing an excavation and loading system consisting of a downward-digging excavator (dragline type) located at the upper site of a bench and excavating the upper layer, with the material loading at the level of its location, and a face shovel excavating the lower layer, with the mined material loading onto the lower transport level (fig. 1).

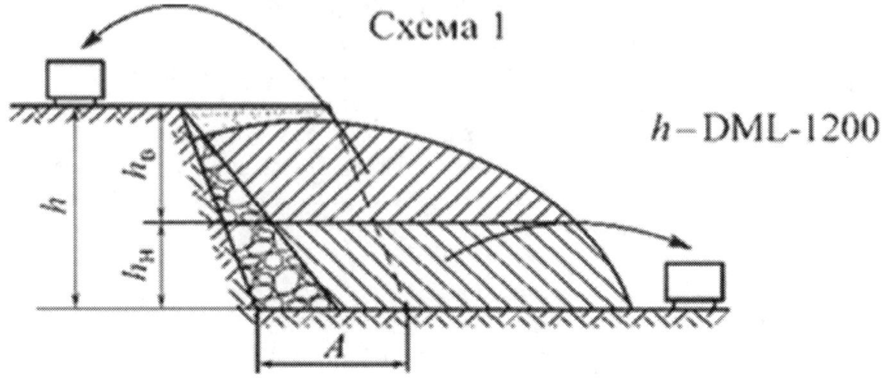

Fig. 1 – High overburden bench development in two layers with ESh-10.70 (ESh-14.50) excavator for the upper sub-bench and EKG-10 face shovel for the lower sub-bench development: $h_Н$ and $h_В$ – height of upper and lower layers accordingly, m; *A* – height of drilling and blasting cut, m.

It has been proved that the highest extension of reserves to be developed due to greater open-pit depth is provided by the transition to high overburden benches at the moment of achieving the equivalence of the current and boundary overburden ratio – at the stage of the maximum development of mining operations. According to the principle of assessment of the final depth of surface mining operations by the equivalence of the boundary and current overburden ratio, pit walls have slope angles equal to those of working benches, and the open-pit width increases until the current overburden ratio reaches the boundary value, i.e., until the moment of the maximum development of surface mining operations. Further development of mining operations with the working bench slope angle being maintained at a base planned level is not economically feasible. In this case the volume of overburden removed at a later stage is minimized.

It is evident that the above described innovations and some other surface geotechnologies, new mining and transport machinery eliminating the need of operator's presence in unfavorable conditions of the open-pit working zone change the requirements to the level of risk and dictate the necessity of refining the regulatory and procedural framework governing the efficient and safe development of solid mineral deposits.

3 Results and discussion

The currently applicable regulatory documentation governing the procedure of ore open-pit wall and bench stability design for the stage of pit reaching its final boundary was issued several decades ago. Moreover, the status of applicable document is granted only to the "Rules for ensuring the stability of slopes at coal opencast mines" of 1998, while outside the coal sector the status of this document is in question.

Over recent decades, mining and geomechanical conditions have changed significantly in surface mining operations, numerous new software packages and computer developments have appeared, opportunities of experimental methods of source data collection and processing, grounding of the permissible parameters of open pit walls have changed dramatically, and, thus, methods of risk assessment have been perfected [10-13]. The necessity is ever growing of implementing the projects meeting the international standards [14-16]. Imperfection, and in some cases, lack of the applicable regulatory framework create unjustified difficulties in the preparation of design documentation for mineral deposit development, as well as its audit and review.

A number of mining companies, design and research establishments, experts of the Federal Service for Environmental, Technological, and Nuclear Supervision have recognized the fact that at the present moment there is a pressing need of elaborating a new regulatory document on geomechanical substantiation and provision of pit wall and bench stability, in which not only the full scope of accumulated Russian and foreign practices of such substantiation will be utilized and the legitimate framework of design and calculation methods will be expanded, but methodological approaches making easier the interaction of mineral developers, license holders, production personnel, researchers, designers and experts will be also determined.

In that context the elaboration of Federal norms and regulations of industrial safety "Rules for ensuring the stability of walls and benches of open pits, open-cast mines and spoil banks" (FNR), facilitating the enhancement of economic efficiency and safety of mineral deposit surface mining, as well as the competitiveness of Russian mines at the international level is very important in the current situation.

The necessity of elaborating the FNR is predetermined by the following factors:
- advent of qualitatively new methods of engineering-geological assessment of the rock mass;
- necessity of giving due account to the existing tectonic stresses and bulk forces caused by the slope curvature in plan;
- expansion of the application area of mineral deposit combined mining method;
- advent of new computational tools and slope stability design methods;
- removal of severe constraints for bench height and safety berm width at the stage of reaching the final pit boundary that allows higher angles of pit wall recovery;
- advent of new surface geotechnologies and novel mining and transport machinery, which operation eliminates the necessity of operator's presence in the working area of an open pit and changes the requirements to the risk level;
- advent of new methods of the rock mass state monitoring and control;
significant difference between Russian and international norms and standards of slope stability design rationale;
- greater dependence of mine economics on the angle of safe slopes and overburden ratio in case of large-scale production;
- severization of the requirements of the General Board of State Expert Review to engineering design freezing in compliance with the country's applicable legislation.

By the decision of the Federal Service for Environmental, Technological, and Nuclear Supervision the elaboration of norms and regulations "Rules for ensuring the stability of walls and benches of open pits, open-cast mines and spoil banks" was assigned to the Academician N.V. Melnikov Institute of Comprehensive Exploitation of Mineral Resources of the Russian Academy of Sciences. It is evident that the elaboration of this large and important regulatory and procedural document can be implemented only on condition of joining and coordination of efforts of all experts in this area. A megaproject "Ensuring the stability of slopes in surface mining" was launched for the preparation of the FNR. The main objective of this project is the elaboration of the regulatory documentation putting into law the rules ensuring pit wall and bench and spoil bank slope stability specified in accordance with the established procedure and giving due account for the accumulated practices and achievements of the mining science oriented towards an opportunity of application of novel developing geotechnologies.

Project tasks:
- customization of norms ensuring the slope stability in line with the current conditions of surface mining, methods of the rock mass state assessment and control;

DOI: 10.1051/e3sconf/20172102008

- synchronization of the applicable Russian norms of ensuring the slope stability with the international practice;
- utmost full utilization of the potential of modern means, methods and technologies for ensuring the slope stability and enhancing the efficiency of surface mining operations;
- removal of the barriers and provision of an opportunity of novel technology implementation in mining projects;
- enhancement of the efficiency of surface mine operation in the Russia's legal framework of mineral resources management;
- taking into account the interests of mineral developers in the enhancement of the efficiency and safety-in-mining.

In the process of the FNR preparation the requirements and recommendations will be classified by type of mined minerals, and due account will be given to an opportunity of developing and updating of all supplements to the FNR in a certain period of time.

4 Conclusions

Application of robotized machinery in solid mineral deposit surface mining provides an opportunity of the maximum improvement of mining and transport equipment performance due to minimization of time expenditure for mining and mined material transportation processes, and elimination of the human factor effect on the performance of operational processes. Application of intelligent machinery, due to the non-presence of the personnel in a face opens an opportunity of minimizing the values of some parameters and eliminating some elements of an open-pit design as a mining structure, in compliance with safe working practices in mining.

Application of high overburden benches in surface mining of inclined and steeply dipping deposits provides higher recovery ratio of deposit reserves with higher efficiency of surface mining and minimized environmental effect.

Commercialization of novel technologies of solid mineral deposit development dictates the necessity of refining the regulatory and procedural framework governing the efficient and safe development of mineral deposits. Basically, the megaproject "Ensuring the stability of slopes in surface mining" is aimed at the solution of all these tasks.

The studies have been performed with the support of the Russian Science Foundation, Project No 14–37–00050

References

1. C. Brown, Autonomous Mining, **1**, 30 (2012)
2. S. Arvidsson, Géosciences, **1**, 64 (2005)
3. E. Nebot Robotics Research, **28**, 268 (2007)
4. J.J. Green, K. Hlophe, J. Dickens, R. Teleka, M. Price International Journal of Engineering and Advanced Technology, **1:4**, 8 (2012)
5. K. N. Trubetskoy, D.R. Kaplunov, M.V. Rylnikova Journal of Mining Science, **48:4** 688 (2012)
6. K.N. Trubetskoy, D.J. Vladimirov, I.A. Pytalev, T.M. Popova, Mining magazine, **5**, 21 (2016)
7. M.V. Rylnikova, D.J. Vladimirov, I.A. Pytalev, T.M. Popova, Journal of Mining Science, **1**, 92 (2017)
8. K. N., Trubetskoy, Mining Journal, **7**, 4 (2013)
9. V.A. Kovalev, V.S. Fedotenko, Journal of Mining Science, **51:5**, 865 (2015)
10. C. O. Brawner, Mining Engineering, **38:8**, 823 (1986)

11. R. Thompson, S. Hahn, S. Pastor, Mining Engineering, **67:9**, 34 (2015)
12. H. Tonderai Chiwaye, *A comparison of the limit equilibrium and numerical modeling approaches to risk analysis for open pit mine slopes: Dissertation for to degree of Master of Science in Engineering* (University of the Witwatersrand, 2010)
13. E. Hoek, C. Caranza-Torres, B. Corcum, Proceedings of the North American Rock Mechanics Society. Mining Innovation and Technology, (2002)
14. J. Read, P. Stacey, *Guidelines for Open pit slope design* (CSIRO 2009)
15. G. Beale and J. Read *Guidelines for Evaluating Water in Pit Slope Stability* (CSIRO, 2013)
16. E. Hormazabal, F. Rovira, El Teniente mine in Chile, **1** (2009)

 DOI: 10.1051/e3sconf/20172102009

Rare and Rare-Earth Metals in Coal Processing Waste

Tatiana Cherkasova[1*], *Elizaveta* Cherkasova[1], *Anastasia* Tikhomirova[1], *Alyona* Bobrovnikova[1] and *Irina* Goryunova[1]

[1] T.F. Gorbachev Kuzbass State Technical University, 650000, 28 Vesennyaya St., Kemerovo, Russia

Abstract. An urgent issue for power plants operating on solid fuels (coal) is the issue of utilization or use of accumulated production waste - ash and slag materials - in the related production. Ash-slag materials are classified as "waste", usually grade 5; tens of millions of tons of them being produced annually in the Kemerovo region, which threatens the ecology of the region. At the same time, ash and slag is a very promising raw material. The use of this material as a base for the final product allows us to significantly expand the possibilities of using coal. The most widespread is the system of ash and slag involving in construction or as a replacement for sand in road construction, or as an additive to building mixtures. However, there are both industrially valuable and environmentally dangerous elements in ash-slag materials. Ash-slag materials can be considered as independent ore deposits located on the surface and requiring the costs of their extraction.

1 Introduction

Kuzbass is a coal region, where a large amount of ash-and-slag waste is produced during the operation of thermal power plants, which "dust", are washed out by atmospheric precipitation, polluting the environment [1-4].

The properties of thermal power plant ash and slag materials depend primarily on the characteristics of the mineral part of the source fuel and the methods of its combustion. In those years, when the boiler units of currently operating Kemerovo state district power station were designed, the main fuel was coal from the Kedrovsky mine and Butovskaya coal mine. It is for the burning of these coals that the equipment of the Kemerovo State District Power Plant and other power plants within the city of Kemerovo, where the boiler units with pulverized fuel combustion are installed, at which the temperature in the zone of active combustion exceeds the melting temperature of the ash, where designed, was projected.

In recent years, the list of coals supplied to thermal power plant from other open pits and deposits has expanded: burning of coals of "Zarechny", "Kotinskaya" and "Taldinskaya" mines began. It should be noted that these coals are close in their technological characteristics to the design ones.

* Corresponding author: ctg.htnv@kuzstu.ru

Ash-and-slag of thermal power plants are solid products of coal combustion formed in the furnace as a result of heat treatment of the initial mineral part of fuels, consisting mainly of rock-forming components, the remaining components are carbon, $CaSO_4$ anhydrite, sometimes iron oxides and calcium aluminosilicates. The 85-95% of mineral part the fuel consists of clay minerals, mudstones, siltstones, i.e. of the same sedimentary rocks that make up the human habitat. The remaining 5-15% are mainly compounds of iron, calcium and trace elements.

For the years of power system operation, 10 million tons of ash and slag have been accumulated at the ash dump of the Kemerovo State District Power Plant. The total annual output of the ash-and-slag materials is about 175 thousand tons.

The composition of slag includes small amounts of rare and rare-earth elements (REE), for the extraction of which highly selective technologies are required, and the extraction of the REE complex, rare and noble metals increases the profitability of coal processing waste [4-15].

And-slag materials are rare and rare-earth elements.

The consumption of REE is growing quite rapidly, since modern high-tech devices and installations, from a mobile phone to a nuclear reactor, cannot exist without them. Currently REE is supplied to the whole world by China, but now these supplies are reoriented to the interior of the country. Thirty years ago, Russia took the third place in production and second place in the reserves of REE, fully meeting the needs of the domestic market, supplying, in addition, products for export. At present, all the needs of Russia are met through imports, that is, Russia is on the "rare earth addict" [12]. There is only one way out: to revive the extraction and production of REE; all the more so as the reserves in the Russian Federation are 30% of the world's, in addition, the ash of natural coal is a source of raw materials for rare earth elements and rare elements.

2 Materials and methods

The objects of research were slag and fly ash from the Kemerovo SDPS.

The preliminary magnetic separation of objects was carried out. It was found that the share of the magnetic fraction in the fly ash is about 5%, the slag was not subjected to separation (there was no magnetization). The lots of slags were crushed to 0.2 mm in size. Analyzes for the content of chemical components were carried out by the method of optoelectronic spectroscopy with inductively coupled plasma on the spectrometer iCAP 6500DIO and consisted of the following stages:

1) preparation for solid sampling;
2) extraction by royal vodka;
3) analysis of compacts;
4) analysis of the extract.

The leaching of the samples was carried out by solutions of mineral acids and alkalis. To win and separate the components, the ion flotation method was used, which is applicable for the extraction of substances from very low concentration solutions, including REE [20,21]. Flotation machine parameters, flotation time, gas or air flow rate are very important for ion flotation, collectors [15] should be selected. The efficiency of the flotation REE winning in a foam product is determined by the pH of the medium, the consumption of the collector and the concentration of REE in the solution.

To carry out experiments on the extraction of rare metals and REE, anionic superficially active substance - surfactant sodium dodecyl sulfate $C_{12}H_{25}SO_4Na$, which is the sodium salt of lauryl sulfuric acid, was used. This non-toxic and inexpensive reagent has both blowing agent and collector properties.

3 Results and discussion

The magnetic fraction of ash contains mainly iron. The non-magnetic fraction contains SiO_2, Al_2O_3, CaO, MgO oxides, which are of commercial interest for the use in building materials. REE and rare elements are distributed between two fractions.

Analytical results on the content of rare elements and REE are presented in Table. 1-3.

Table 1 The content of rare and rare-earth elements in fly ash of Kemerovo State District Power Plant

Elements	**Content**	
	%	g/t
Sr	$1.1 \cdot 10^{-2}$	110.00
Zr	$2.3 \cdot 10^{-4}$	2.30
Nb	$7.0 \cdot 10^{-4}$	7.00
Ga	$9.0 \cdot 10^{-4}$	9.00
Mo	$8.7 \cdot 10^{-4}$	8.70
Au	$1.2 \cdot 10^{-4}$	1.20
Ag	–	–
V	$5.3 \cdot 10^{-3}$	53.00
РЗЭ: Y	$1.4 \cdot 10^{-3}$	14.00
Eu	$6.8 \cdot 10^{-5}$	0.68
La	$1.9 \cdot 10^{-3}$	19.00
Pr	$7.0 \cdot 10^{-4}$	7.00
Sm	$1.5 \cdot 10^{-4}$	15.00

Table 2 The content of rare and rare-earth elements in the slag of Kemerovo State District Power Plant

Elements	**Content**	
	%	g/t
Sr	$1.9 \cdot 10^{-2}$	190.00
Zr	$4.7 \cdot 10^{-3}$	47.00
Nb	$8 \cdot 10^{-5}$	0.80
Ga	$1 \cdot 10^{-4}$	1.00
Mo	–	–
Au	–	–
Ag	$1 \cdot 10^{-5}$	0.10
V	$6.6 \cdot 10^{-4}$	6.60
РЗЭ: Y	$3 \cdot 10^{-4}$	3.00
Eu	$1.5 \cdot 10^{-5}$	0.15
La	$4.7 \cdot 10^{-4}$	4.70
Pr	$1.7 \cdot 10^{-4}$	1.70
Sm	$1.1 \cdot 10^{-4}$	1.10

Table 3 The content of rare and rare-earth elements in concentrates

Elements	**Concentrate 1 content, %/g/t**	**Concentrate 2 content, %/g/t**
Sr	$2{,}7 \cdot 10^{-1}$ / 2700	$1{,}3 \cdot 10^{-1}$ / 1300
Zr	$1 \cdot 10^{-1}$ / 100	$1{,}2 \cdot 10^{-1}$ / 1200
Nb	$9 \cdot 10^{-3}$ / 90	$8{,}5 \cdot 10^{-3}$ / 85
Ga	$5 \cdot 10^{-3}$ / 50	$4{,}8 \cdot 10^{-3}$ / 48
Mo	$3 \cdot 10^{-2}$ / 300	$2 \cdot 10^{-2}$ / 200

DOI: 10.1051/e3sconf/20172102009

V	$2\cdot10^{-1}$ / 2000	$3\cdot10^{-1}$ / 3000
∑РЗЭ	$8{,}5\cdot10^{-2}$ / 850	$7{,}7\cdot10^{-2}$ / 770

The resulting concentrates are suitable for further processing.

4 Conclusions

In Russia over the past 40 years, more than 300 technologies for the use of ash and slag materials in more than 20 areas have been developed. Nevertheless, today we are here at one of the last places in the world. The main reason until recently there was no any state policy in the field of ash and slag waste utilization - in the country and until now there are no regulatory legal acts regulating their use. As a matter of principle, energy companies themselves could be involved in ash and slag waste processing, but when there is no state policy and in the conditions of restricting tariff growth, it is very difficult for them to justify investments in such projects.

The problem of ash and slag is particularly relevant for Siberia. 44 percent of all domestic thermal power engineering is concentrated here, the overwhelming majority of which are coal stations.

The recovered value of rare earth elements and rare metals, as well as the huge need of high-tech industries in them and environmental improvement of the region necessitate the implementation of works concerning the deep processing of ash and slag waste.

References

1. B.F. Nifantov, V.P. Potapov, N.V. Mitina, Processing prospects, 104 (2003)
2. V.V. Krapivintseva, Pacific Geology, **24:1**, 73 (2005)
3. M.D. Skurskiy, *Forecast of rare earth - rare metal - oil-and-gas fields in Kuzbass*, (Kuzbass University Puiblishing, 2005)
4. A. A. Cherepanov, V.T. Kardash, Geology and minerals of the World Ocean, **2**, 98 (2009)
5. V.A. Salikhov, Bulletin of Tomsk State University. Economy, **1:25**, 123 (2014)
6. M. I. Gamov, N.V. Granovskaya, S.V. Levchenko, *Metals in coals*, 45 (Rostov-on-Don : YuFU., 2012.)
7. S.I. Arbuzov, V.V. Yershov, A. A. Potseluev, L. P. Rikhvanov, *Rare elements in Kuznetsk basin coals*, 248 (Kemerovo Printing and Publishing Complex, 1999)
8. S.I. Arbuzov , *Geochemistry of rare elements in Central Siberia coals: thesis for a Doctor's of Science degree in Geology and Mineralogy*, 499 (Tomsk, 2005)
9. S. I. Polkin, *Processing of ores and placers of rare and precious metals*, 428 (Moscow: Nedra, 1987)
10. V. A. Salikhov, Bulletin of KuzSTU, **4**, 255 (2011)
11. S. I. Arbuzov, , Bulletin of TPU, **311:1**, 77 (2007)
12. K. Shunsuke, K. Yuichi, K. Hiroshi, W. Kazuaki, Open Journal of Physical Chemistry **3**, 69 (2013)
13. F. Wojciech, M. Małgorzata M. Wiatros-Motyka, M. Wdowin, Environmental Science and Pollution Research, **22,** 9464 (2015)
14. X. Tran, W. Ping, M. Dustin, Elsevier. ScienceDirect, **163**, 129 (2016)

15. N. Lange, I. Camargo, A. Figueiredo, L.Castro, M. Vasconcellos, R. Ticianelli, Journal of Radioanalytical and Nuclear Chemistry, **311**, 1235 (2017)
16. P. Orlov, *Rare earth addiction* (The Russian Atomic Community, 2005)
17. P.A. Vlasov, DonNTU, **6:125**, 152 (2007)
18. G.L. Pashkov, S.V. Saikova, V.I. Kuzmin, M.V. Panteleeva, A.N. Kokorina, E.V. Linok, Journal of Siberian Federal University. Engineering & Technologies, **5:5** ,520 (2012)
19. O.V. Afanasieva, G.R. Mingaleeva, E.V. Shamsutdinov, S.Yu. Gorbunov, Tatarstan power engineering, **3:39**, 41 (2015)
20. O.V. Afanasieva, G.R. Mingaleeva, A.D. Dobronravov, E.V. Shamsutdinov, Bulletin of higher education institutions. Issues of Power Engineering, **7:8**, 26 (2015)
21. N.V. Fedorova, D.A. Shafrost, Heat power engineering, **1**, 53 (2015)
22. L.N. Adeeva, V.F. Borbat, Bulletin of Omsk University, **2**, 141 (2009)
23. B.S. Ksenofontov, A.S. Kozodaev, R.A. Taranov, M.S. Vinogradov, I. A. Butorova, E.V. Senik, A. A. Voropaeva, Engineering Bulletin, **11**, 121 (2014)
24. Yu. A. Vershakova, *Extraction of rare-earth elements by the method of ion flotation in the case of nitric acid processing of apatite: thesis for a Csandidate's of Technical Science degree*, 180 (Apatites, 2003)
25. D. V. Abryutin, K.A. Streltsov, Bulletin of higher education institutions. Ferrous Metallurgy, **3**, 3(2013)
26. O. A. Lobacheva, N. V.Dzhevaga, National Mineral and Raw University. Mining, 106 (2016)

DOI: 10.1051/e3sconf/20172102010

Effective Processing of the Iron Ores

Vadim Kuskov[1,*], *Yana* Kuskova[1], *Vladimir* Udovitsky[2]

[1]Saint-Petersburg Mining University, 21 Line, 2, St. Petersburg, 199106, Russian Federation
[2]T.F. Gorbachev Kuzbass State Technical University, 650000, 28 Vesennyaya St., Kemerovo, Russia

Abstract. Effective technology for a complex wasteless processing of the iron ores has been designed and includes three main components (plats): comminution plant, briquette plant, pigment plant. The comminution is done per energy effective technology. Using of briquetting for ores clotting enables the costs cut and brings to a higher level of environmental safety of the process. Briquette formation can be done as a regular pressing, as an extrusion. Developed technology allows to produce high quality competitively products for metallurgy industry and red iron oxide pigments. The whole production line impacts the environment in a minimal manner.

1 Introduction

Iron is the most widely adopted metal in industry today. Iron ores are the primary source for iron melting. Almost all of the iron ores are transformed into steel, cast iron and iron based alloys. Cast iron and steel are the fundaments of modern industries. The processing of iron products goes in a few stages. Usually it is mining, mineral processing, agglomeration, casting, depuration. Each step essentially increases the cost of production.

Competitiveness of national iron ore producers can be achieved on the world and domestic market with increase of products quality, decrease of costs, producing with high added value and complex usage of mineral raw materials.

The iron ores are falling into the following classes - high grade ore, ordinary grade ore, and low grade ore. Most of the iron ore deposits in Russian Federation are known for showing comparatively low grade ores and are consequently subjects for a necessity of ore dressing that requires obligatory intense grinding. The result is that such concentrates are not adopted for direct utilization in metallurgic separation and require clotting. Currently two methods of the iron ore clotting are found industrially efficient: agglomeration and pelletizing. Agglomeration is a process of sintering powder products with high temperature of burning fuel. Pelletizing is a process of producing pellets, based on the property of humid fine particles to roll into a spherical grain (pellet). Besides being wide spread these processes have both a sort of disadvantage. Sinter cannot be transported on a long distance and stored for a long period. Pellets can be used only for fine products. For instance, pelletizing and agglomeration need heating of the product to high temperature parameters, then the finished product cooled down and heated again in blast furnace thus leading to inefficient energy allocation. Within these

* Corresponding author: opikvb@mail.ru

processes a lot of fine particles form that have negative side for blast furnace effectiveness. That is why these particles are sieved for the later removal and thus an abundance of by product arises. An alternative to agglomeration and pelletizing is the briquetting process. Briquetting is a process of mechanical and thermo mechanical briquette making that can be used for various raw materials, e.g. pieces with regular shape and size.

Besides producing various ferrous alloys, iron ore can be used for other industries, for example, pigments production. Pigments is a fine compounds, insoluble in water, organic solvents, firm forming agents and etc, with special optical, mechanical and sorbing properties. Pigments can be organic and non-organic. It can be natural and synthetic. Non-organic pigments includes iron oxide pigments, which also can be natural and synthetic. Iron oxide pigments are the most frequently used among colour pigments. The colour of pigments is determined by iron oxides: FeO, Fe_2O_3, Fe_3O_4. Iron oxide pigments have wide colour palette due to chemical compounds diversity.

2 Subject of investigation

Besides relatively low grade iron ores Russian Federation has deposits of what is mostly known as high grade ores. These are usually result products of natural quartz iron rock dressings formed within long period under weathering or metamorphosis impacts. An example of such ore are ones found in Yakovlev deposits. The deposit is unique, it is mined by Yakovlev underground mine. The ores of the deposit have miscellaneous mineral composition. Follow general mineralogical types of ores were determined: micaceous iron oxide and martite; martite-hydrohematite; hydrohematite-hydrogothite, etc. As a mean content of 60.5% iron along deposit. Harmful contaminants content (S, P, etc.) is minimal. The most common ores in the deposit are micaceous iron oxide and martite ores with the highest iron content [1, 2].

As the investigations showed special sort of martite-hydrohematite ores is an excellent base for manufacture of red iron oxide pigments.

3 Substantiation and developing of complex iron ore processing

The investigations showed the ore can be an excellent primary product for metallurgical industry due to low harmful contaminants (production of high quality cast, steel, ferrous alloys). One of its types (pigment type) is excellent base for manufacture of red iron oxide pigments.

But these are mouldy on the contrary and are requiring agglomeration. In present the ore is sintered. An alternative to agglomeration may be presented by briquetting especially by the so called cold briquetting when the final pellets are not heated to a high C°. From the point of technology and economy cold briquetting has a range of advantages to compare with other methods of agglomeration. First, this process is cheaper than common agglomeration and it has much better environmental characteristic as compared to the traditional ones. Second, rather simple technological process and possibility to use flux with wide PSD (from 0 to 10 mm and more). Third, the addictives can be easily put in the flux. Fourth, briquettes have higher density, strength and transportability to compare with sinter. Fifth, ore contains the minimum of harmful contaminants; it can be agglomerated with unroasted method, in briquettes the composition and properties will remain. Sixth, briquetting is more environmental friendly than sintering and pelletizing [3-5]. Red iron oxide pigments price is much higher than price of iron ore concentrate. The share of Russian Federation in a global iron oxide pigment manufacture is below 1%. Meaning that the country imports almost all of its consumed pigment. Currently the main product to produce in Russian Federation is synthetic red iron oxide pigment. There is no industrially developed production of natural iron oxide pigment in Russia nowadays. Cost of natural iron oxide pigment is two times less to compare with synthetic pigments. Taking into consideration the characteristic the natural iron oxide pigment has no analogues as being cheap in comparison to the others and as per crucial quality standard matches the best counterparts of synthetic iron oxide pigment. As per some characteristic the pigment even surpasses its synthetically revised counterparts. It almost doesn't contain sulfur and chlorine ions, has optimal particle size distribution. Pigment may be applied for paint and coating industries, anti corrosion primers, liners, dry compounds, wall plasters on cement, colour source for the concrete items (roof tiles, colorful floor tiles, pavement tiles), for dying caoutchouc compound, rubbers, leathers, paint manufactures adopted in marine industry, glass, glass wares, ceramics, polishing liquid, food industry.

The research of briquettes production has been established for briquettes for metallurgical industry and pigments for paint production. Within the research carried the technology of complex wasteless processing of high grade iron ores was evaluated. The general physical and metallurgical properties of briquettes are known: fineness, coverage rate, oil adsorption, etc.

It should be mentioned that during the researches the prospect of processing iron ore with iron content 69% and more ("superconcentrate") has been evaluated. It can be a raw material for direct reduced iron. The concentrates can be produced with environment friendly (without flotation) methods, for example, sensor-based sorting, fine and ultrafine classification, special methods of magnetic separation and flocculation, etc. Also the opportunity to create ultrafine paint raw material has been shown for special purpose pigments. Other type of ores can be used as sorbents. The data is not included in the paper. The technology of complex processing high grade iron ores unites three main components (plants) - comminution plant, briquette plant, pigment plant. Selective ore mining enable separate feed of metallurgy class and pigment class for re-fabrication.

In comminution plant the feed material is crushed and screened. Within the stages of crushing and screening of metallurgy class we do receive large particles that is further lead to metallurgy treatment thus enabling for the energy efficiency as per the principle "should not we crush what we should not crush". (The size of particles crushing and screening can change in collision with the granulometry content of the initial compound). The products of small mesh size later transported to the briquette plant into the preservation bunkers.

Metallurgic class of the ore is subject to briquetting as well as sands of initial classifications and magnet fractions from the pigment plant that are the byproduct of pigment. This enables efficient raw compound utilization. These product is mixed with a binder agent and the bonded material formed in briquettes. Further the briquettes meets screening to throw non conditional in terms of size briquettes that are returned into mixing stage. Briquettes that meet requirements of size pass to dryer zone. The dryer temperature depends on the binder type. For some types the drying process with an additional temperature is quite not needed. Several types of agents are applied for briquettes production during evaluation procedure- CMC, molasses, modified vinyl acetates, concrete, bentonite, liguid glass and other ones. Dry briquettes are screened and after we have: non-conditional briquettes are returned to the process and final product is stored and transported.

As it is known, the roll presses are frequently used for briquettes producing with binder compounds. The general advantages are versatility, rather simple construction, high capacity, absence of dynamic loads, small dimensions, reliability in intense mechanical and thermal loads. But a few disadvantages should be mentioned. For example, the asymmetric briquettes are formed due to a slight cell displacement. The strength of the briquettes is low, sometimes the briquettes are torn apart. The surface of cells should be ideal smooth, without any roughness. The bond material should have less adhesion properties to compare with the slurry.

Otherwise the roughness brings to higher adhesion and sticking during the exploitation and as a result – destruction of the wet briquettes. Either the irregularity of roll bondage wearing (the central part wears faster than the sides) and inconstant pressure due to the irregular fill of cells.

That is why the screw press (extruder) was tested with binders. The idea of the method is to form the material of different length with pressure in die with a special profile. The main advantages of extrusion formation are: high strength of the briquettes right after the die, low capital costs and exploitation, rather low energy consumption for forming, slightly lower binder agent loss, chance to have the finished product of various sizes and cross section types on the same machines as well as option to process highly sticky (up to 20%) product effectively. Extrusion technology is quite similar to press one. The extruder for formation is used instead of roll press.

The developed briquetting technology is quite flexible and setup due to customer requirements. E.g. the shape and size of briquettes and extrudats can be easily modified with forming zones of the machine. The strength, waterproof and other properties can be changed with pressure, drying settings, adding other type of binder. The technology is energy efficient.

As the investigation unveiled, the briquettes and extrudats are adaptable for metallurgy industry utilization. «Pigment» ore type is located in different fields of the deposit. To let plant work without any technologic interruptions is thanks to warehouse close to the pigment plant. Selective «pigment» ores type after crushing is accumulated in the consolidation warehouse.

To the pigment finalization the crushed paint grade of row material (pigment type) fed onto the mill from warehouse, mill fills the magnetic separator with low magnetic field. Magnetic fraction clarified on the separator is composed of mainly magnetite relict that are harmful agent for pigments. This fraction goes to briquette plant. Non magnetic fraction falls into final classification steps (in spiral classifier area). Sands after the classifier also passed to briquette plant. Classifier passes to cleaning and control classification stages in the hydro-cyclone of different diameters. The classifier overflow to the cleaning hydro-cycloning first and after to the ultrasound clarification. Underflow of cleaning operations pass to I control hydro-cycloning leading also to the ultrasound clarification via the output frames and the sands itself to II control hydro-cycloning. An underflow of II control hydro-cycloning back onto cleaning step and underflow to control classification. Ultrasound clarification is applied if increased raw material sulfide contents. After the ultrasound clarification the material thickens and dries. The dried pigment transported for packing and delivery to client or to the finished product storage. Produced pigment is suitable for paint and coating industries and others. The technique of concentration raw ore on shaking tables as alternative to the magnet separations has also been adopted. The utilization of tables enabled complete exclusions of the ultrasound clarification of products as sulfide containing mineral can almost completely be evolved to the corresponding product. The scheme of pigment processing is technologically flexible, i.e. it is easy to modify and automate. It is coming from that classifiers (hydrocyclones) are simply adapted. The fineness of the pigment can be easily changed. The amount of hydrocyclones, assemble place, type of hydrocyclones is also modifying because it is moveable and rather cheap equipment. The processing is environmental friendly with no chemical and flotation processes involved.

4 Future prospects

Positive aspects for developing the metallurgical briquettes production are: almost unlimited mineral raw resources; rather low capital costs (lower than sinter and pelletizing processes); lower costs of production to compare with sinter and pelletizing; high ecological clearness of the process to compare with sinter and pelletizing, the process can be established close to the mine; good transportability of the product; convenient location that minimizes transport costs [6-10]. Positive aspects for developing the red iron oxide pigments production from enriched iron ores are: low capital costs and low producing costs with the resulted high quality product, it is connected with simple and flexible technology and unlimited access to raw materials; import replacement claim and lack of national products in international sanctions situation; lack of competitors on domestic and global market; possibility for expansion of market place due to stable demand; absence of high quality product on the national market with moderate price; environmental friendliness of the processing and the product; end of the economical crisis and market growth.

Follow prospects are: the increase of the capacity; production of briquettes for direct reduced iron. Addiction to the pigments market, replacement in a prospect of a certain synthetic red iron oxide pigment brands due to high quality and low costs of the production. Production of ultrafine pigments for special cases.

5 Conclusion

Thus, a technology of rational wasteless processing of the high grade iron ore has been promoted granting release of a high quality products for metallurgy industries equaling the existing industrial analogues and red iron oxide pigment quite demanded as well applied in paint and coating industries and in many other. The technology examined here has minimal effect on ecological issues.

References

1. V.P. Orlov et al., *Iron ores KMA* (Moscow, Publisher Geoinformark, 2001)
2. I. Túnyi, D. Vass, S. Karoli, J. Janocko, E. Halásová, A. Zlínská, B. Beláček, Geol. Carp., **56:3**, 273-384 (2005)
3. V.B. Kuskov, J.V. Kuskova, A.V. Kornev, V.L. Trushko. Preparation of iron ore for metallurgical processing. CIS Iron and Steel Review, 2011, p. 4 – 5.
4. V.B. Kuskov, A.V. Kornev. Briquettes for metallurgical industry. CIS Iron and Steel Review, 2011, p. 6 – 7.
5. V.L. Trushko, R.E Dashko, V.B. Kuskov, A.V. Kornev, Mining university notes, **194**, 120 (2011)
6. V.B. Kuskov, Y.V. Kuskova, A.I. Mikheyev, I.S. Rogova, CIS Iron and Steel Review **7** (2009)
7. V.B. Kuskov, Y.V. Kuskova, Metallugist, **3**, 70 (2010)
8. V.B. Kuskov, Y.V. Kuskova, Metallurgist, **54**, 192 (2010)
9. A.S. Klyamko, V.L. Trushko, V.B. Kuskov, Mining university notes, **197**, 245 (2012)
10. V.B. Kuskov, Y.V. Kuskova. V.L. Trushko, Processing of ores **1**, 39 (2014)

 DOI: 10.1051/e3sconf/20172102011

Influence of Coal Industry Enterprises on Biodiversity (on the Example of Formicidae)

Svetlana Blinova[1,*], and *Tatiana* Dobrydina[2]

[1]Kemerovo State University, Department of Ecology and Nature Management, 650000, 6 Krasnaya st., Kemerovo, Russian Federation
[2]Kemerovo State University, Department of Foreign Languages in Professional Communication, 650000, 6 Krasnaya st., Kemerovo, Russian Federation

Abstract. The fauna, the population density and the types of ants' nests on the territory affected by the coal industry have been studied. It has been found that the level of environmental contamination influences all the indicators: the minimum indices of density, 1-2 species of ants and only underground nests near the pollution sources. On the contrary, in the areas with a weak influence of coal enterprises, 9 species with a maximum density of up to 15.2 nests / m^2 have been observed with predominance of nests in the form of earthen mounds. Lasius niger are the most resistant to pollution, while the Myrmica do not stand such an impact.

1 Introduction

The constantly increasing anthropogenic influence on natural ecosystems which reduces their biodiversity, and, therefore, their stability, is one of the current problems. The study of the green areas of industrial cities, including parks, squares and lawns is becoming particularly relevant. One of the indicators of the stability of such sites is insect-herpetobionts which react sensitively to the changes in environmental conditions. In the opinion of a number of authors [1-2], ants are an informative component of the biocenosis and able to serve as indicators of certain habitat conditions because of their abundance, high species diversity, complex structure of the formed communities, permanent habitation on the same site. According to the observations [3-4], industrial pollution leads to limiting the number of species, reducing the density of nests and the number of individuals in them. Intensive industrial development leads to reducing the number of anthills Formica rufa by 66.5 % over 5 years [5].

In the Czech Republic, in places where brown coal is extracted, the same species of ants as in natural cenoses have been observed. But the mountain species Formica lemani and Manica rubida have been found only on dumps and not in natural cenoses. In contrast, the nests of Myrmica rubra have been registered only in natural cenoses [6]. In Italy, the quantity of cold-loving and forest species and the representation of the individual genera Formicidae change in the species richness depending on the age of the dumps [7].

* Corresponding author: sv_blinova@mail.ru

However, urban conditions are more severe for the life of ants: only some species settle and survive there [8]. Therefore, ant complexes in urban areas are characterized by poor species composition and the presence of 1-2 dominant species [8-11].

It is shown that ants' communities are sensitive to the environmental consequences of land use in Brazil's mangrove systems [12]. In Japan, to assess the changes in soil in the cities, it is proposed to use Messor aciculatus [11]. The example of Italy shows that the indicators of species richness and abundance of ants can act as bioindicators of the degree of restoration/disturbance of various biotopes [13]. The other authors suggest using a species composition to control the degree of development of agrocenoses. At the same time, it is pointed out that there should be its own set of indicator species for each locality [14].

Some species of ants can act as indicators of land restoration after mining [7]. Our studies [15] of Formicidae have shown the possibility of judging the degree of influence of industrial production by the reaction of ants. Therefore, the purpose of this work is to analyze the impact of coal industry enterprises on ants on grass-cereals meadows located at different distances from the sources of impact.

2 Materials and Methods

The studies were conducted from 1996 to 2013 in Prokopyevsk (Pr) located on the territory of the Kuznetsk coal basin (Kemerovo Region, Russia). Coal mining is the main economic activity in the city with 79 % deep mining and 21% surface mining. During the study, the total release and emission of pollutants into the city environment averages about 61.127 ± 12.75 thousand tons. The coal industry and municipal coal-fired boiler houses make the main contribution to the city air pollution. Dust, soot, carbon monoxide, phenols, and sulfur oxide (IV) have been found in the air within a radius of 2 km from the mines. Their concentration decreases in some father distance from the emission sources but the average concentration in the 2 km area is 1.5-2 times lower than in the 500-m radius area.

At the beginning of the study, within the city, there operated 16 coal enterprises (not taking into account coal-fired boiler houses); since 2010, the number of mines has reduced gradually. Currently, there is one mine there.

The collection of all the material was carried out using standard myrmecological methods: 5 areas were allocated at different distances from the coal enterprises, 5-10 test sites (5 * 5 m^2 in size) were laid on the area (table 1). An area with a grass-cereals meadow, located opposite the wind rose, 30 km from the city, was used for monitoring.

Table 1. Characteristics of the studied areas.

Area name	Distance from source of emission, km	Pollution level	Projective grass cover, %	Notes
Pr 1	0-0.5	high	40	50% of the territory covered with a layer of 5 cm of coal dust
Pr 2	0.5-1	high	50	Technological dumps formed in 1960-1980
Pr 3	1.5	average	50-70	
Pr 4	5-5.5	average	85	
Pr 5	10	low	95	
Pr M	45	low	90-95	Monitoring

The areas were inspected; if necessary, the soil was scrabbled. When finding a nest, no work was conducted in diameter 0.2-0.25 m: this is the conventional territory occupied by one family. All quantitative material was processed by statistical methods in the Excel program.

3 Results and discussion

Within the investigated territory, 10 species of ants belonging to 5 genera of two subfamilies have been found: Myrmica rubra L., M. scabrinodis Nyl., M. schencki Emery, Tetramorium caespitum L., Formica fusca L., F. cunicularia Latr., F. polyctena Forst., F. rufa L., Lasius flavus Fabr., L. niger L. At the same time, the distribution of the detected species and genera within the investigated territory is uneven. So, in the area with a high pollution level, nests of ants of the genus Lasius predominate. With distance from the coal mining enterprises, the number of nests of the genus Myrmica and Formica increases.

3.1 Species composition and density of nests

In general, within the investigated territory, ants are distributed unevenly: the smallest number of species and the lowest density of nests have been observed in close vicinity to the coal dust sources. With distance, these indicators increase. However, in the monitoring area, in the city, they are somewhat lower than in the areas with a low degree of the coal industry impact (Fig. 1).

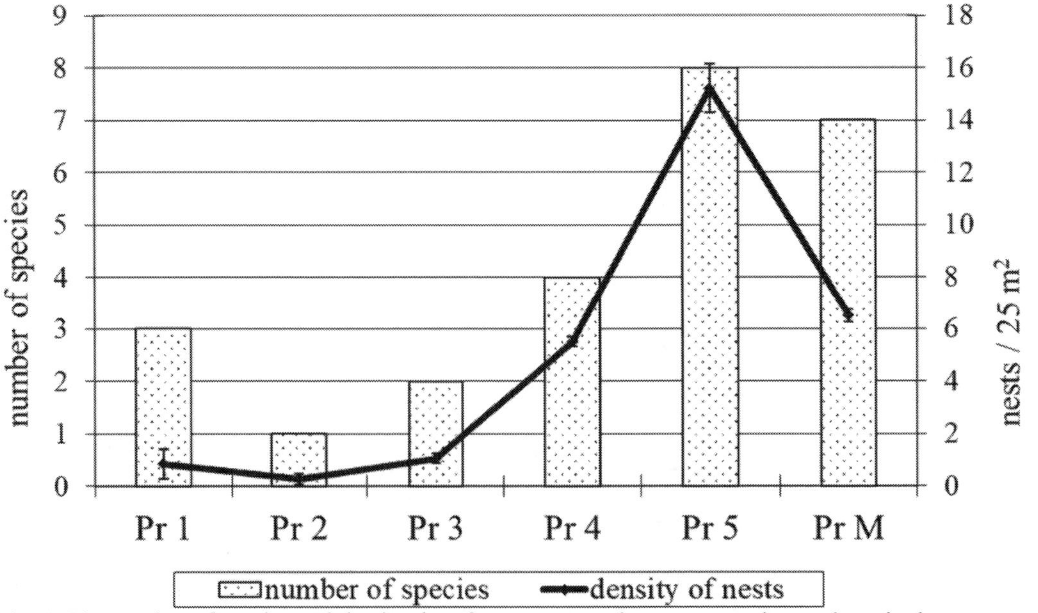

Fig. 1. The number of species and the density of ants nests on the grass-cereals meadows in the area affected by the coal industry (monitoring and average). The explanation of the areas is in Table 1.

In the years 2006-2010, amid the coal production breakdown, enlarging species composition and increasing density of nests were recorded, primarily in the mine areas (up to 3 species and up to 4 nests / 25 m^2 in some areas). However, in the most polluted areas with coal dust and crumbs, there was not a single nest.

After the subsequent reclamation in 10-15 years, a mono-species settlement of L. niger is observed in the vicinity of mines, mine workings and concentrating mills.

There are no nests of ants at the 25-30 years old mine dumps which have never had reclamations (Pr 2). Ants make settlements on the dumps only 35-40 years after the dumps' formation. In this case, only the mono-species settlements of L. niger were recorded for the first 5 years. Thus, on the dumps formed 35-40 years ago, the density of the settlement reached 0.3 nests / 25 m^2, at 45-50 years old dumps there were 0.4 nests / 25 m^2.

Ants of the genus Lasius reach the greatest species diversity and density of settlement in the area with high pollution. In this case, the number of nests increases (from 0.1 to 1.5 nests / 25 m^2) with a large number of small stones (about 15 cm in diameter) or a distance from shafts (on average, 200 m), excavations (for example, near administrative buildings). The distribution of species of the genus Lasius appeared to correlate with the volumes of industrial emissions: the number of species increases with the decrease in the coal industry impact ($r = -0.51$ at $p < 0.05$). Thus, nests of L. niger have been found in all the areas. With the decrease in the volume of emissions near the mines, the nests of L. flavus have also been recorded.

In contrast, the number of species of the genera Myrmica, Tetramorium and Formica increases with the distance from coal mining enterprises; the quantity of these genera nests increases, reaching the maximum in areas with a low level of exposure. For example, there are 37.2 ± 0.7 % in area Pr 5, and 42.9 ± 0.9 % Formica nests in the monitoring area Pr M. The representatives of the genus Formica are not found near the industrial enterprises (within a radius of 4.5 km). The exception is F. rufa, an anthill of which has been found near the shafts of the administrative buildings of one of the investigated mines. At the same time, only the nests of the subgenus Serviformica – F. fusca and F. cunicularia have been recorded in areas with an average level of pollution.

Ants of the Myrmica family have been registered only in the area with a low level of impact of coal industry enterprises. However, the nests of M. rubra and M. schencki are found only in the city. Such observations coincide with the results of our similar work done in the areas influenced by the cement and lead-zinc industries [15].

3.2 Characteristics of the nest-building

The study has detected four main types of nests: underground, with an earthen mound, with a plant dome and under a stone. The analysis of the dependence of the nest types on the distance to the source of emissions has showen that the quantity of nests located under the stones depends only on the presence of small stones on the area and does not relate to the contamination. On the contrary, the quantity of nests of the other types varies depending on the distance to the coal industry enterprises. So, in the area with the greatest pollution, ants settle only in underground nests without land structures. At a distance from the sources of emissions, the number of capsule nests (ground mounds and mounds with a plant dome) increases. An increase in the linear dimensions of the nesting mounds of L. niger ants has also been observed (Fig. 2). It has been noted that the height of the nesting hillock is directly proportional to the distance from the mine faces and descents into the mine ($r = 0.67$ with $p < 0.05$). In the immediate vicinity of the coal mining enterprises, only underground nests have been found. At a distance of 1.5 km (area Pr 3), their quantity decreases to 90.3 ± 11.5 %. The minimum of underground nests has been registered in the monitoring area: 32.3 ± 12.6 %. In contrast, about 65% of all the nests with ground domes have been observed in the monitoring area and in the area with a weak level of influence.

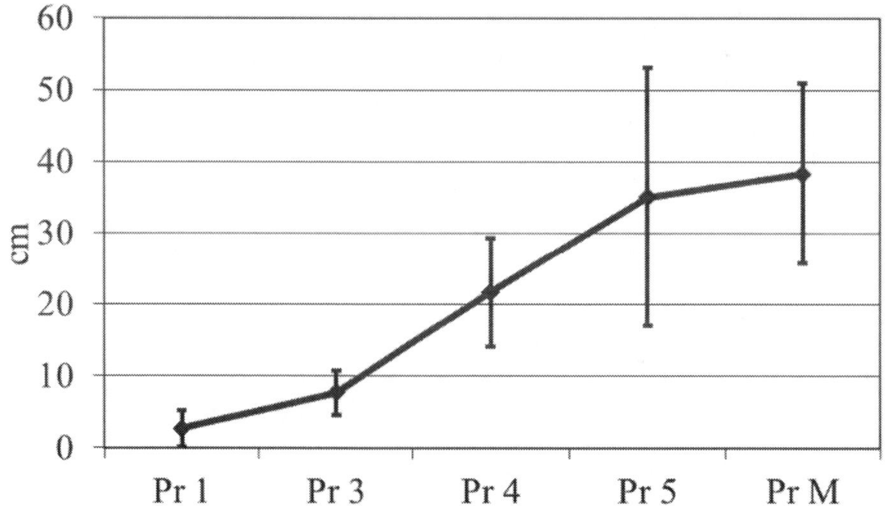

Fig. 2. The average height of the earthen mound depending on the distance to the coal dust sources. The explanation of the areas is in Table 1.

According to our observations, L. niger can use coal crumb as a breeding material only in mixture with the soil. The only nest of T. caespitum, located directly in the coal crumb, was observed. However, in the following years, even with a decrease in the impact, such nests have not been found.

4 Conclusion

Thus, ants exhibit a quasi-adaptive type of reaction to the pollution by coal enterprises: a reduction in the species composition and in density of the ants' nests has been observed near the coal enterprises. With distance from the pollution sources, all the parameters under the study increase first, but then again decrease to average values. The reduction of the total volume of mine emissions that occurred due to the closure of several mines may have influenced the increase in the species and density of the ants' settlement in the investigated areas.

The ants of the genus Lasius, occurring at any level of pollution by industrial emissions, are the most resistant to anthropogenic impact; they numerically dominate in all the areas studied in the city.

Ants of the genus Myrmica do not withstand the influence of solid industrial emissions: they are found only in a relatively clean area.

In the area of coal dust impact on the biotopes, underground nests without ground structures have been observed mainly. Most of the nests found in the monitoring area are in the form of earthen mounds or hillocks of plant residues.

The number of species, the density of nests and the numerical ratios of various types of ants' nests can act as indicators of the coal industry impact on the environment, allowing the rapid identification of a decrease in biodiversity and, consequently, a decrease in the stability of the ecosystem as a whole.

References

1. T. Puszkar, Mem. zool. **29**, 129-142 (1978).

2. J. D. Majer, Env. Man. **7:4**, 375-383 (1983).
3. J.M. Petal, Mem. zool. **29**, 99-108 (1978).
4. J. M. Petal, Pol. Ecol. Stud. (PRL) **6:4**, 665-672 (1980).
5. W. T. Szczepanski, T. Podkowka, Sylwan. **127:5**, 35-44 (1983).
6. M. Holec, J. Frouz, Pedobiologia. **49:4**, 345-357 (2005).
7. L. Ottonetti, L. Tucci, G.. Santini Restoration Ecology. **14**, № 1, 60-66 (2006).
8. K. Vepsalainen, B. Pisarski, *Animals in urban environment. Proceedings of the Symp. on the occasion of the 60–th anniversary of the Institute Zool. of the Polish Academy of Sciences*. 155-168 (1982).
9. B. Pisarski, W. Czechowski, Mem. Zool. **29**, 109-128 (1978).
10. J. D. Majer, K. R. Brown, J. R .Soc. West. Aust. **69**, 13-17 (1987).
11. T. Yamaguchi, Entomol. Sci. **8:1**, 17-25 (2005).
12. J. H. Delabie, V. R. Paim, I. C. Do Nascimento, S. Campiolo, C. S. Neotrop. Entomol. **35:5**, 602-615 (2006).
13. L. Ottonetti, L. Tucci, G., Redia **LXXXVII**, 199-202 (2004)
14. C. Castracani, A. Piotti, D. A. Grasso Redia. **XC**, 67-70 (2007).
15. S.V. Blinova, Russian Journal of Ecology. **39:2**, 148-150(2008).

Land Resource Management as the Ground for Mining Area Sustainable Development

Aleksander Solovitskiy [1], *Olga* Brel[1*], *Nikolai* Nikulin [1], *Ekaterina* Nastavko [1], and *Tatayna* Meser[1]

[1] Kemerovo State University, Department of Geology and Geography, 650000, Kemerovo, Krasnaya Street, 6, Russia

Abstract. It is established that the problem of sustainable development of Kuzbass cities is their being tied to a single production and income from other sources is not considered. Therefore, their economy is underdeveloped, depends entirely on one city-forming enterprise (single-industry city), which causes response to the slightest changes in the economic situation. In Kuzbass, all cities, except Kemerovo, are mono-dependent, including Kiselevsk, which economy mainly consists of coal mining enterprises. In the circumstances, there is a need to develop a set of measures for management the urban land, primarily aimed at ensuring the sustainable development of Kiselevsk city. The development of principles and management mechanism of the urban territory land fund determines its effectiveness. Establishing the dependence of rational use of land resources and sustainable development characterizes a new level of information interaction between sciences (land management and economy). Practical use of this theory is to overcome the mono-urban development of mining cities, taking into account effective subsoil management.

1 Introduction

In the typical Russian mining area – old industrial coal town Kiselevsk (Kemerovo region, Western Siberia, Russia) – the urgency and priority in solving the issues of urban land use are related to the specificity of settlement land that is characterized by multifunction and concentrated industrial and residential buildings. The mechanism of management and disposal of the lands of mining regions in Russia is based on the expansion of the authority of local government, the introduction of payment for land use, close connection of land management and urban development [1-2]. The absence of such mechanism at the end of the last century [3-5] in the city led to the negative consequences: the closure of six mines, the mine construction trust, three ore-processing plants, two mine-building departments; decreasing the volume of production of machine-building, chemical, food and timber manufacturing industries by 30%-40%. The special negative impact of mining on the neighbor towns comes from water resources pollution by mines and open pits [6-8]. Due to the lack of product competitiveness at a number of machine-building and consumer goods industry enterprises, as well as the reduction of consumer demand, whole production was suspended

* Corresponding author: brel_o_a@mail.ru

 DOI: 10.1051/e3sconf/20172102012

(Kuzbassobuv, Gormash plant). This period was characterized by a significant dominance of the coal industry share in the overall structure of industrial production, the value of which was up to 75% (Figure 1).

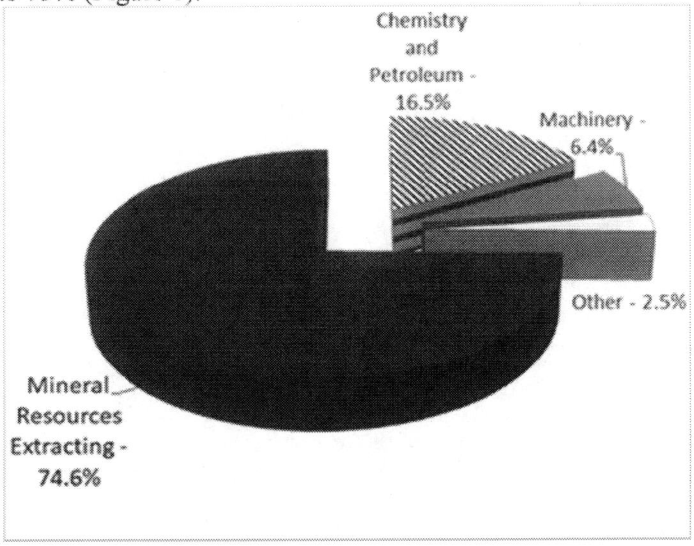

Fig. 1. The structure of production in Kiselevsk economy by industries in 2009.

The problem of unemployment is increasing in the city. Due to the released labor force (miners) there is an increase in unemployment. Only 1.5 thousand out of 4.9 thousand workers after the closure of mines and ore-processing plants were employed. The dynamics of the unemployment rate is shown in Figure 2.

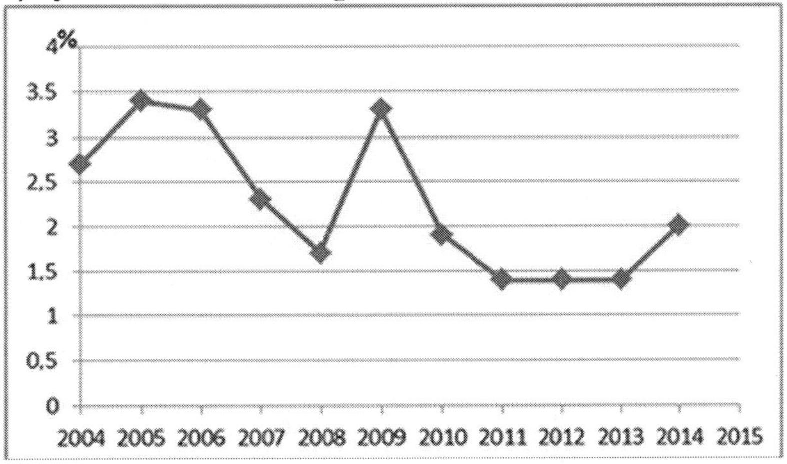

Fig. 2. The dynamics of unemployment rate (% of the working population).

Mineral resource extracting in Kiselevsk resulted in a significant amount of disturbed land, which is about 4888 hectares (43.4% of the total area), including 2060 hectares in open pit mining, 1550 hectares occupied by external dumps. The soil layer and underlying rocks are completely destroyed on 40% of the territory. The technogenic consequences of these enterprises operations are the following:

- air pollution caused by emissions from mining equipment, industrial and municipal boiler facilities, burning rock dumps;
- pollution of water bodies with mine, quarry, industrial and domestic wastewater;

significant disruption of the terrain;

DOI: 10.1051/e3sconf/20172102012

- exemption from land use, pollution and disturbance of land.

In Kiselevsk one of the most significant environmental problems is the work of the 4th sewerage basin. All the treated wastewater from this facility is discharged into the Akchurla River. At present, the facility's treatment equipment can not cope with the actual load, as more than 45,000 m / day of sewage flows, and the design capacity of the facility is less than 28 thousand m / day, which exceeds the norm by 1.6 times.

The foregoing indicates that keeping the dependence of the city's economy only on subsoil management does not ensure the sustainable development of Kiselevsk.

2 Materials and Methods

For the comprehensive modernization of urban land use in the Kiselevsky urban district within the framework of sustainable development and in order to overcome the mono-urban development [9-11], we propose the following basic principles for managing the land fund:
1. Preservation of the total area of the land fund.
2. Redistribution of the land balance to ensure revenues to budgets.
3. Improving urban area.
4. Improving the ecological situation of the territories.

In addition, we proposed a criterion for this overcoming as the initial stage of the transition to sustainable development

$$\frac{B_c}{B_a} \leq 1 \qquad (1)$$

where: B_c – receipts to the budgets from coal enterprises;

B_a – receipts to budgets from alternatively developed enterprises of other industries or businesses.

The indicated revenues to the budgets B_i (i means c for coal enterprises or a for alternative) are proportional to the area of the land fund occupied by them P_i and are the following:

$$B_i = k_i \times C_i \times P_i \qquad (2)$$

where: k_i – coefficients; C_i – tax rates.

Proceeding from the principle of equal influence n of the revenues from (1) and (2), it is easy to obtain a mathematical description of the management of the area of such an alternatively developed enterprise from other industries or business, depending on budget revenues

$$P_i \geq \frac{B_a}{n} \qquad (3)$$

Based on the mentioned above, the management of the land fund includes the following main areas:
1. Changes in the type of coal mining technology on the land of the industrial zone.
2. Development of mechanical engineering and other types of industry on the lands of the industrial zone.
3. Land allotment for the development of trade and small business.
4. Land allotment for the service sector.
5. Changes in the legislation of urban land.

3 Results and Discussion

 DOI: 10.1051/e3sconf/20172102012

The management of land fund of coal enterprises is based on a change in the method of deposits development. For the effectiveness of subsoil management it is planned to develop open pit mining where the cost of a ton of produced coal is in half less than that produced by underground mining. Due to this approach the profit increases and it should be used for the own needs: technical support for the elimination of underground mining, modernization of enterprises and implementation of social programs. This management ensured the taxes revenues from mining operations to the city budget as compared to 2015 by 1.2% higher (2015 - 2 433 713 rubles, 2016 - 2 463 976 rubles).

The management of the land fund of non-coal enterprises has its own peculiarities for overcoming the mono-urban development. The following is suggested:

• the use of land fund of existing enterprises (for the machine-building complex, two enterprises (LLC «Gormash», LLC «United Machine-Building Technologies») which specialize in the production of equipment for coal and chemical industry enterprises;

• the use of adjoined territories.

For example, it is planned to build a plant for the production of mining equipment and metal structures on the territory of LLC «United Machine-Building Technologies». The effectiveness of the project includes:

• social impact (313 new jobs);

• economic effect (more than 3.5 million rubles of additional revenues to the city budget per year, an annual increase in the volume of industrial production by 3.5-4%);

• overcoming mono-urban development (reduction of the coal industry share by 3-4%).

To implement the project it is proposed to use the existing infrastructure on an industrial site with an area of 12 hectares, where the allocation of this project is appropriate. This site has a good location (in the prospectively developing district «Red Stone») and is not far from the nearest freight railway station (3 km), and from the nearest Kemerovo-Novokuznetsk highway (30 km). In addition, it is planned to use an adjoined land plot of 29.2 hectares. It is characterized by the ability to connect to the entire complex of engineering networks at a sufficiently close distance, and has near railway access roads. Thus, the planned task of managing the land fund of the machine building complex is a real opportunity to attract investments for sustainable development of the city. The mechanism for its solution is the rational use of the land fund. The management of the land fund is aimed at expanding production and increasing the industrial output of the chemical industry, which is also a factor in overcoming the mono-urban development. The chemical industry of the urban district is represented by OJSC «Znamya», which employs more than a thousand people. The enterprise produces industrial explosives and is one of the main city taxpayers. In order to develop the city's chemical industry it is necessary to provide the manufacturing of new products, as well as the modernization of the production technology of industrial explosives. This management provides:

• economic effect (expansion of existing chemical production, creation of new jobs, additional revenues to the city budget of more than 3.0 million rubles a year, an increase in the output of industrial products);

• social effect (creation of new jobs);

• overcoming of mono-urban development (reduction of coal industry share by 4.5-5.5%).

Tax revenues from the wholesale and retail trade to the municipal budget increased by 9% compared to 2015 (217,925 rubles in 2015 and 238,098 rubles in 2016). It is noted that the production of building materials in the city is not developed enough. Therefore, the task of the city authorities is to create the necessary land fund for the development of this type of production [11]. In the short term, it is planned to organize such production, as well as the development of output of related products. In connection with the opening of the ore-processing plant, it is possible to build a factory in the city to produce silicate bricks and

other construction materials based on ash and slag waste. This will contribute to the solution of:
• environmental problems (utilization of ash and slag waste, improvement of the ecological situation);
• economic problems (providing the Kemerovo region with affordable building materials, additional revenues to the budget of the urban district of more than 0.7 million rubles a year, an increase in industrial production by 1.5-2.5% annually);
• social problems (creating new jobs);
• overcoming the mono-urban development (reduction of the share of the coal industry by 1.5-2.5%).

Today small business provides a significant range of products. So, for example, for 2014 entrepreneurs of the urban district developed 16 investment projects (50% in production and 50% in services). 52 jobs were created. Increasing of small and medium-sized businesses is necessary to overcome the mono-urban development. The share of tax revenues in the city budget from this type is constantly growing and reaches 16-18%.

The main problems of small business and local industry development are not only related to the land fund, but also:
• project financing;
• orientation of industries;
• lack of sponsors and a narrow focus of investment activities.

The main areas of development of the land fund of small business include:
• food industry enterprises;
• consumer goods manufacturing.

Land management based on an integrated approach to sustainable development requires the creation of a transport infrastructure. Within this territory it is necessary to build new and reconstruct existing roads as well as railways. For example, it is necessary to improve the municipal bypass road in the village Uskat, as well as to coat with asphalt the constructed bypass road with the exit to the Leninsk-Kuznetsk-Novokuznetsk highway.

The tasks of road construction have been developed:
• improving the quality of the city's transport infrastructure;
• increasing the quantitative indicators of traffic;
• increasing the quality of passenger service;
• satisfaction of new transport needs;
• increasing social impact.

As part of an integrated development approach, the creation of a railway project is also proposed. On the field "Karagailinsky 1-2" it is necessary to provide the construction of private railway tracks with an adjoining to the station Uskat with a length of 1050 meters, which will allow to expand the production of this project, to optimize the costs for freight traffic.

When designing, it is necessary to ensure:
• operation stability of the railway;
• economic effect;
• obtaining maximum profit;
• improving the environment;
• quality of service.

This project requires taking into account the terrain, types of permitted use of the territory, calculation and establishment of a security zone, which is designed to ensure the safety, strength and stability of railway transport facilities. The requirements for such zones are defined by the regulatory documents (Order of the Ministry of Transport of the Russian Federation of August 6, 2008 N 126 "On approving the norms for allocation of land required for the railway rights-of-way, as well as the norms for calculating the security zone

DOI: 10.1051/e3sconf/20172102012

of the railways"). When designing the road the existing railway line from Uskat station to mine no. 12 can be used. After that, the projected road will be developed to the north to Cherepanovo village, then along the previously existing field road through the river Pryamoy Uskat and to LLC "Shahtoupravlenie Karagaylinskoye".

The world experience shows that it is possible to create parks not only on specially designated places, but also on objects that seem not only inappropriate for this, but also dangerous. There are 5 small recreation areas on the territory of Kiselevsk, but they all have a narrow focus. Multifunctional parks are a relatively new type of Public Park, which includes mass cultural and recreational, physical, and children's recreation zones, as well as a walking and economic zone. For example, it is proposed to build a new park on the territory of the former industrial enterprise the Kiselevskaya mine located in the city.

4 Conclusions

Based on the study results the authors developed a theory that determines the basis of managing the land fund of mining towns, and the following conclusions were made:

1. It is established that the creation of cities in the Kemerovo region on the basis of the city-forming coal enterprises does not contribute to the sustainable development of the territories.
2. It is justified that one of the first stages of sustainable development of the mining city Kiselevsk is to overcome the mono-urban development.
3. An economic mechanism for managing the land fund of a mining city has been developed to overcome the mono-urban development.
4. It is shown that a multidimensional program has been developed to ensure sustainable development. Its implementation is effective and shows positive results. The planned measures will allow reducing the coal industry share in the overall structure of the city economy to 46%, which will allow eliminating the majority of risks, first of all, unmanageable transfer into managed ones to ensure further sustainable development of Kiselevsk.

References

1. M.A. Tyulenev, T.N. Gvozdkova, S.A. Zhironkin, E.A. Garina, Geotech. Geol. Eng., **35:1**, 203-212 (2017)
2. T. Gvozdkova, E. Kuznetsov, A. Rudakova, S. Markov, E3S Web of Conf., **15**, 01008 (2017)
3. H. Gans, *Urbanism and suburbanism as ways of life: A reevaluation of definitions* (Columbia University Press and Russel Sage Foundation, New York, 1991)
4. H. Molotch, Amer. J. Sociol. **82:2**, 309-332 (1976)
5. R. Park, *Human communities; the city and human ecology* (Free Press, Glencoe, 1952)
6. M. Tyulenev, Y. Lesin, , E.Tyuleneva, , E. Murko, E3S Web of Conf., **15**, 02003 (2017)
7. M. Tyulenev, E. Garina, A. Khoreshok, O. Litvin, Y. Litvin, E. Maliukhina, IOP Conf. Ser.: Earth Environ. Sci., **50:1**, 012035 (2017)
8. M. Cehlár, P. Varga, Z. Jurkasová, M. Pašková, Acta Montanistica Slovaca, **15:2**, 132-138 (2011)
9. Shrinking cities, Urb. Des. Int. **18:1** (2013)
10. Eur. Plan. Stud. **23:1** (2015)
11. A. Haase, D. Rink, K. Grossmann, M. Bernt, V. Mykhnenko, Env. Plan. **46**, 1519-1534 (2014)

 DOI: 10.1051/e3sconf/20172102013

Adaptive Adjustment in Taraxacum Officinale Wigg. in the Conditions of Overburden Dump

Olga Legoshchina[1], *Irina* Egorova[1,*], and *Olga* Neverova[2]

[1]Institute of Human Ecology of the SB RAS, Leningradskiy avenue, 10, Kemerovo, 650065, Russia
[2] Kemerovo State University (KemSU), 73, Sovietskiy pr., Kemerovo, 650000, Russia

Abstract. Morphological and anatomical features of the leaves and roots of *Taraxacum officinale* Wigg., growing under the conditions of the rocky dump of the Kedrovsky coal mine of the Kemerovo region, were studied. It was revealed that the specific environmental conditions of the dump cause morphological and anatomical changes in the leaves and roots of the dandelion. At the level of morphology, a decrease in the average leaf area, a thickening of leaf blades, a tendency to decrease the number of leaves in the rosette, a significant decrease in the mass and length of the roots. At the level of the anatomical structure of the leaves, there is a significant increase in the thickness of the mesophyll, a tendency to decrease the thickness of the tissues of the upper and lower epidermis, a decrease in the number of cells in 1 mm^2 and an increase in the size of stomata in the tissues of the lower and upper epidermis, a decrease in the number of stomata by 1 mm^2 and a stomatal index on the upper epidermis. At the level of the anatomical structure of the roots, the radius of the root decreases, the radius of the cortex and phloem, the diameter of the xylem.

1 Introduction

As a result of coal mining, the area of disturbed land in Kuzbass is more than 100 thousand hectares. As a result of the open method of coal mining, the emerging rock dumps are characterized by specific ecological conditions – a lack of productive moisture, failed water permeability, a contrasting temperature regime on different relief elements; low potential fertility of embriozems, therefore they are a classical object for studying the adaptive mechanisms of plants planted on them. A high degree of resistance of some species of higher plants is provided by a combination of morphological, physiological, anatomical adaptive reactions, which makes these plants potentially capable of improving the quality of the natural environment – urban areas, industrial zones and technogenically disturbed lands. Adaptations of plants are associated, first of all, with the rearrangements of their assimilating organs, which, in connection with their gas exchange function, are highly sensitive to external influences [1]. It is known that morphological and anatomical rearrangements change the physiological status of plants in extreme conditions of existence for more efficient use of environmental resources [2]. The aim of the work is to study adaptive rearrangements at the morphological and anatomical level in *Taraxacum officinale* Wigg., which grows in conditions of the rocky dump of the Kedrovsky coal mine.

* Corresponding author: nir_kem@mail.ru

2 Materials and Methods

Studies were carried out on the rock dump "Southern" of the Kedrovsky coal mine.The dump has a flat-inclined relief with a height of 58 m, an area of 599.3 hectares. Dump rocks are represented by sandstone (60%), siltstones (20%), mudstones (15%), loams and clays (5%). The experimental site is located at the top of the dump (the location coordinates of the site are 55°30'39" NL. 86°04'00" EL.). The life of the biocenosis lasts 7-10 years. The technical and biological stages of recultivation (landings of *Pinus sylvestris* L. and *Onobrychis arenaria* (Kit.)), the initial stages of the formation of meadow communities are observed; and there are separate sections of stony scatterings of up to 3 m^2 in size from various rock formations completely devoid of vegetation. Projective coverage is up to 50%, on rocky areas - below 10%. There is no turfness. In the herbage prevail *Onobrychis arenaria*, *Taraxacum officinale* Wigg, *Melilotus officinalis* (L.) Pall., *Artemisia sericea* Web. Ex Stechm., *Picris hieracioides* L. Embriozems are characterized by an alkaline reaction (pH of water extract 7.1-7.7), medium availability of humus (3.5%), low availability of mobile forms of phosphorus (10-50 mg / kg) and nitrogen (3.6-6.0 mg / kg), high potassium (100-140 mg / kg), a lack of productive moisture (19-22 mm in the 0 ... 20 cm layer).

The control plot was located 15 km north-west of the dumps at the border of the birch forest. A large area (about 80%) of the site is occupied by gramineous herb meadow. Projective coverage is 95-100%, the soil is not turfed. The meadow communities are dominated by *Centaurea scabiosa* L., *Hieracium umbellatum* L., *Lathyrus pratensis* L., *Achillea Asiatica* Serg., *Taraxacum officinale Web.*

The object of research was common dandelion (*Taraxacum officinale* Wigg.), as the most widespread type of perennial herbaceous plants growing on dumps. Preparation of raw materials was carried out in dry sunny weather, according to generally accepted rules during the maximum development of the vegetative mass (end of May 2016). Collected raw materials with no visible signs of damage from 15 sites with a size of 1 m^2 from the experimental and control sites. For morphological studies, the sample of plants from each registration site was 10 plants, total from the experimental and control sites –150 plants. The area of the leaves (S) was determined by scanning them and then processing the image using the computer program Image Tools. For anatomic research, plant material was fixed in a 60% solution of ethyl alcohol. From the middle part of the leaf blade and root, transverse sections were made by a microtome and placed into glycerol. Measurements of the anatomical parameters of the leaves (the thickness of the leaf and its tissues, the number of cells in 1 mm^2 of the lower and upper epidermis, stomata sizes) and roots (root radius, cortex, phloem, diameter of the xylem) were carried out with the Axioscope-2 + microscope, model ZEISSN HBO103 and N XBO75 (Germany) with eyepiece micrometer, camera and software. Calculation of the stomatal index (*SI*) was carried out according to the formula:

$$SI = N_{st} *100\% / (N_{st} + N_{ke}).$$

Where N_{st} is the number of stomata per 1 mm^2, and N_{ke} is the number of the major epidermal cells of the leaf per 1 mm^2. Mathematical processing and interpretation of the experimental results were carried out using the Statistika 8.0 application software package, descriptive statistics.

3 Results and Discussion

A comparative analysis of the morphological parameters of a dandelion growing in various ecological conditions has shown that in the waste dumps in plants with a high degree of reliability, the green leaf mass decreases from one plant (by 20%), the average leaf area (by 39%), the mass and length of the root (by 46 and 35%, respectively), there is a tendency to decrease the number of leaves in the rosette, the leaf blade thickens (by 15%) (Table 1, Figure 1).

Table 1. Morphological characteristics of leaves and roots *Taraxacum officinale* Wigg.

Variants	Characteristics					
	Green weight from one plant, gm	**Length root, cm**	**Mass of a root, gm**	**Average area of a leaf, cm^2**	**Number of leaves rosette, pcs.**	**Thickness of leaf blade, µm**
Control:						
Leaves –	20.52±1.97	—	—	40.62±3.83	5.80±0.45	118.10±3.22
Roots –	5.58±0.54	12.87±1.12	5.58±0.54	—	—	—
Experience:						
Leaves –	16.40±1.32	—	—	24.74±1.62	4.92±0.42	136.25±6.22
Roots –	3.01±0.25	8.38±0.63	3.01±0.25	—	—	—

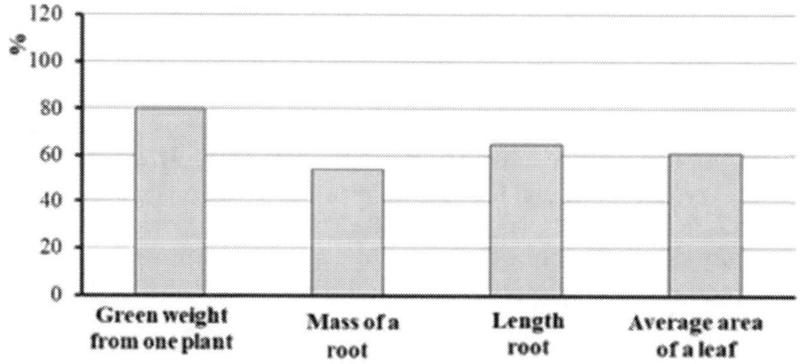

Fig. 1. Changes of morphometric indexes of *Taraxacum officinale* Web. in the conditions of a dump (% of control)

It is known that growth is one of the largest consumers of energy therefore his shutdown expands the power opportunities necessary for adaptive reorganizations. The values of morphometric indexes of leaves and roots of the dandelion growing on pedigree dumps received by us confirm the aforesaid. It is known that at growth in droughty conditions and big insolation grassy plants have an increase in thickness of a plate of a leaf and assimilatory tissue [3]. P. É. Lauri et al. [4] established that the lack of moisture leads to decrease of the area of a leaf blabe and its weight in absolute values that can be considered as adaptive reaction to adverse factors of the environment.

The revealed morphometric changes at a dandelion are manifestation of a kserofitization – the adaptive reaction at the level of the whole plant directed to survival in adverse conditions of a dump. Analysis of the anatomical structure of the leaves showed that under the conditions of the rock dump in the dandelion, the thickness of the mesophyll (by 55%) increases reliably (Fig. 2), there is a tendency to decrease the thickness of the tissues of the upper and lower epidermis (Table 2); with a high degree of reliability, the anatomical indices of the roots change: in the conditions of a pit heap, the root radius decreases by 21.5%, the radius of the cortex and phloem is reduced by 19.6 and 28.7%, respectively, the diameter of xylem decreases by 30% (Table. 2, Fig. 3).

Table 2. Anatomo-morphological characteristics of leaves and roots *Taraxacum officinale* Wigg.

Characteristics	Control	Experience
Area of a leaf, cm2	48.02±4.82	36.43 ± 3.46
Number of leaves per plant, pcs.	5.80±0.45	4.92±0.42
Thickness of a leaf blabe, µm	1180.10±3.22	136.25±6.22
Tissue thickness, µm		
- upper epidermis	19.61±1.24	18.66±1.19
- lower epidermis	16.81±0,65	15.37±0.42
- mesophyll	84.00±2.12	103.11±3.24
Number of mesophyll rows, pcs.	4–6	4–6
Radius of root, µm	2945±23.14	2310±18.69
Radius of cortex, µm	2170±35.12	1743±12.24
Radius of phloem, µm	1708±14.17	1218±10.09
Diameter of xylem, µm	1750±14.26	1225±11.24

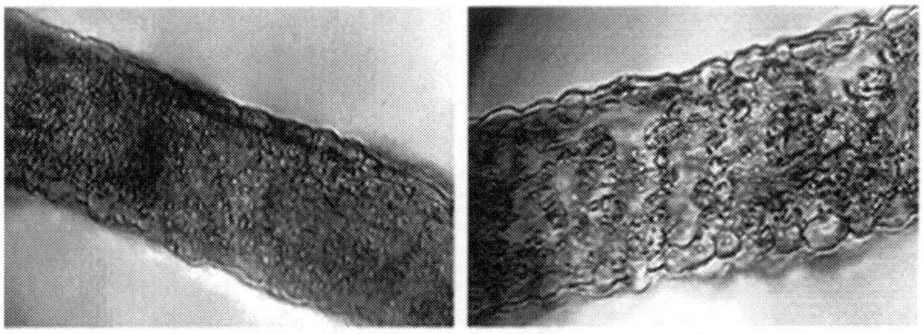

Fig.2. Cross-section of the leaf *Taraxacum officinale* Wigg. (**A** - control, **B** - experience)

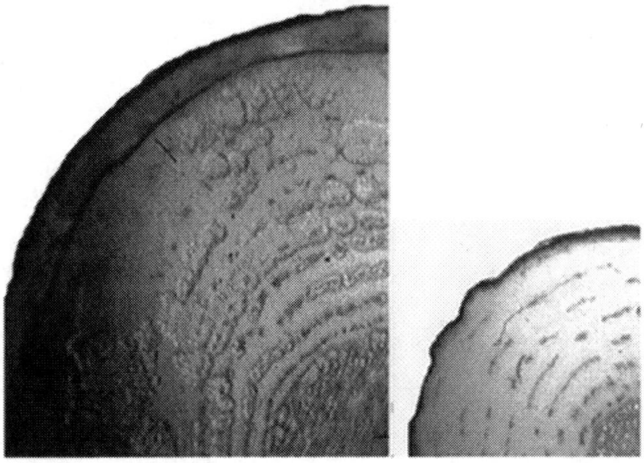

Fig. 3. Cross-section of the root of *Taraxacum officinale* Wigg. (**A** - control, **B** - experience)

It is known that adaptation of a plant to adverse conditions of the environment, as a rule, is followed by change of growth rate of roots and bines.

At action of adverse factors of the environment most visually responses of plants are shown in structure of an epidermis of a leaf. The obtained experimental datas confirm differences of anatomic signs of an epidermis of a leaf of a dandelion, growing in the conditions of the rock dump (Table. 3).

Table 3. Anatomical characteristics of the epidermis of the leaf *Taraxacum officinale* Wigg.

Characteristics	Lower epidermis	Upper epidermis

	Control	Experience	Control	Experience
Number of cells in 1mm^2	978.80±24.32	924.45±16.23	1042.37±25.62	940.44±20.08
Number of stomata in 1mm^2	142.02±1.29	125.01±1.34	81.01±1.14	60.02±0.82
Length of stomata, µm	24.36±0.47	26.12±0.58	23.65±0.52	24.28±0.46
Width of stomata, µm	19.32±0.41	20.54±0.67	17.93±0.36	18.44±0.92
Stomatal index,%	12.67±0.25	11.87±0.11	7.21±0.28	6.21±0.36

In particular, in the common dandelion plants growing on dumps, there is a tendency to decrease the number of cells in 1 mm^2 and to increase the size of stomata in the tissues of the lower and upper epidermis, the number of stomata is significantly reduced by 1 mm^2 (by 12 and 26%, respectively); on the upper epidermis there is a significant decrease in the stomatal index (by 14%). Given the low supply of productive embriozems in the waste heaps, a decrease in the stomatal index is a natural reaction of plants aimed at survival.

There is information in the literature that the low frequency of stomata in combination with their large size contributes to a more effective control of water exchange [5, 6]. The data obtained by us also agree with the results of the studies of I. I. Krokhmal [7] who established that in *C. sibirica* at the level of rosette leaves adaptation to the harsh conditions of growth is achieved by reducing the number of stomata and increasing their size.

The use of correlation analysis showed that the common dandelion growing in conditions of the rock dump increases the number of reliable correlations between the anatomical parameters of the root and the provision of soil with microelements and productive moisture. In particular, in the dandelion, close positive correlations radius of root, radius of cortex and diameter of xylem with phosphorus, nitrogen and productive moisture of embriozems appear in dandelion; the radius of phloem is positively correlated with the nitrogen content and the productive moisture in the soil (Table. 4).

Table 4. Correlation links between anatomical characteristics root *Taraxacum officinale* Wigg. and agrochemical indicators embryozems

Dependence of the studied features	**Reliable correlations** (r), n=150, $p<0.05$	
	Control	**Experience**
Radius of root – phosphorus content	–	+0.85
Radius of root – nitrogen content	+0.83	+0.81
Radius of root – productive moisture content	–	+0.87
Radius of cortex – phosphorus content	–	+0.85
Radius of cortex – nitrogen content	+0.80	+0.81
Radius of cortex – productive moisture content	–	+0.83
Radius of phloem – phosphorus content	+0.81	–
Radius of phloem – nitrogen content	+0.89	+0.81
Radius of phloem – productive moisture content	–	+0.84
Diameter of xylem – phosphorus content	–	+0.80
Diameter of xylem – nitrogen content	+0.80	+0.82
Diameter of xylem – productive moisture content	–	+0.85
Total correlations:	**5**	**11**

It should be noted that an increase in the total number of correlation links occurs due to the formation of positive correlations, which, as is known, control the development of biosystems and arise in critical periods for the body. The coordinated increase in the number of correlation links between the anatomical characteristics of the common dandelion root and agrochemical indices of embryosemes can be considered as a specific reaction aimed at survival and stable existence in conditions of the rock dump.

4 Conclusion

Specific ecological conditions of the rock dump (low availability of mobile forms of phosphorus and nitrogen, lack of productive moisture) of the Kedrovsky coal mine cause morphological and anatomical changes in the leaves and roots of *Taraxacum officinale* Wigg. It has been established that adaptive reactions permitting the survival of *Taraxacum officinale* Wigg. in conditions of a rock dump are:
- at the level of plant morphology – decrease in leaf mass from one plant, average leaf area, weight and root length, formation of a tendency to decrease the number of leaves in the rosette;
- at the level of the anatomical structure – a significant increase in the thickness of the mesophyll, a tendency to decrease the thickness of the tissues of the upper and lower epidermis, a decrease in the number of cells in 1 mm^2 and an increase in the size of stomata in the tissues of the lower and upper epidermis, a decrease in the number of stomata by 1 mm^2, epidermis, a decrease in the radius of the root, the radius of the cortex and phloem, the diameter of the xylem. Identified morphological and anatomical changes in *Taraxacum officinale* Wigg. are a manifestation of xerophytization – an adaptive reaction at the level of the whole plant, aimed at survival in adverse conditions of the heap.

In the conditions of the rock dump, an increase in the number of positive reliable correlations between the anatomical parameters of the root *Taraxacum officinale* Wigg. and the provision of embryozems with microelements and productive moisture has been established. Positive correlations, as we know, control the development of biosystems and arise in critical periods for the body. The coordinated increase in the number of correlation links between the anatomical characteristics of the dandelion root and agrochemical indicators of embryosemes can be considered as a specific reaction aimed at survival and stable existence in conditions of the rock dump.

References

1. O.A. Neverova, O.M. Legoshchina, A.A. Bykov, Middle-East J. Sci. Res. **17:3** (2013)
2. M.Yu. Menshakova, V.K. Zhirov, A.Kh. Khaitbaev, R. I. Gainanova, *The variability of the photosynthetic apparatus of plants: boreal and subarctic ecosystems* (Nauka, Moscow, 2008)
3. L.A. Ivanova, Rus. J. Ecol. **45:2** (2014)
4. P.-É. Lauri, A. Marceron, F. Normand, et al., J. Plant Hydr. **1** (2014)
5. R.Ceulemans, J. Jmpens, R. Lemeur, et al., Oecol. Plant. **13**, 1 (1978)
6. D.R Bissing, Amer. J. Bot., **69**, 6 (1982)
7. I.I. Krokhmal, Ecol. and Noosph. **26**, 1-2 (2015).

 DOI: 10.1051/e3sconf/20172102014

Formation of Mesoherpetobionts Communities on a Reclamated Coal Open Pit Dump

Sergey Luzyanin[1,*], and *Natalya* Eremeeva[1]

[1]Kemerovo State University, 650000, 6 Krasnaya st., Kemerovo, Russian Federation

Abstract. The structure of the mesoherpetobionts arthropod communities of the reclamated dump of the Krasnobrodsky coal pit (Kemerovo region, Russia) has been studied. It was established that the pioneer grouping of mesoherpetobionts arthropod represented by classes of Chilopoda, Arachnida and Insecta-Ectognatha has been formed on the dump for two years after the soil deposition. From the Arachnida, the species of the order Aranei are the most active in the stocking of the dumps. From the class Chilopoda, the species of Lithobiomorpha appear the first on the dump. Insects from the following three orders, Heteroptera, Hymenoptera and especially Coleoptera take the main part in the expansion of dumps and the formation of primary communities. Among the Coleoptera, the beetles of the family of Carabidae (44 species, dynamic density 22.9 specimens/10 trapped per day) dominate. From them, small or medium-sized species are mainly involved in stocking the dumps. There are significant differences in the complexes of ground dump carabid beetles in comparison with the control group, differing in species composition of dominant species, species richness and species diversity parameters.

1 Introduction

Kemerovo region (Russia), in which one of the world's largest coal basins is located, Kuznetsk (Kuzbass), is among the leaders in hard coal excavation. Recently, there has been an annual increase noticed in production volumes. So, according to the results of 2016, 227.4 million tons of coal were produced, which is 5.4% more than in 2015.

The basic technology of excavating minerals is an open (career) method, which accounts for more than 64% of the total production. Such anthropogenic impact inevitably leads to negative consequences for the environment: disturbance of the soil cover and hydrological regime, pollution of the atmosphere with erosion products, dust and gas emissions, etc. In addition, the consequence of the activities of coal mines is the formation of a complex man-made relief, the elements of which are quarries and mined bulk dumps.

An important direction in reducing the impact of coal pits on the environment is the reclamation of disturbed lands. Succession shifts in such territories are accompanied first by the formation of primary microbes and phytocoenoses. Practically simultaneously technogenic substrata are populated by animals, as a rule, first of all microarhropods, and then larger invertebrates (mesoherpetobionts) [1]. This group of animals plays a huge role in the

* Corresponding author: sl_luzyanin@mail.ru

circuit of substances of natural and anthropogenically transformed ecosystems. Biological and ecological diversity, the abundance of mesoherpetobionts arthropods are clear diagnostic signs of the ecosystem condition and the direction of the recovery process.

Studies of the organic communities formation patterns in conditions of technogenesis are becoming increasingly important to scientists. It should be noted that a significant part of the work is aimed at studying the ecological and faunistic features of certain groups of mesoherpetobionts only, such as ground beetles [2-6] and ants [7, 8]. At the same time, there are no data on the structure of the entire complex of mesoherpetobionts arthropods formed on young reclaimed dumps. Therefore, the purpose of our study was to study the structure of the population of mesoherpetobionts arthropods on young reclamated coal pit dumps.

2 Materials and Methods

Studies were conducted from May to August, 2016, at the Krasnobrodsky coal pit (Kemerovo region, Russia). The experimental site was started on a rocky dump (54° 14'48"N, 86° 45'91"E), represented by loose quaternary deposits (brown clays and loess-like loams) with fragments of massive sedimentary rocks of coal-bearing suites (sandstones, siltstones, argillites, limestones). In 2014, it was reclamated, the technical stage of which provided for the smoothing of the dump slope and the application of potentially fertile rocks (PFR) with limited physical and chemical properties good for plant growth. The biological reclamation consisted of planting 5-6-year-old *Pinus sylvestris* L. seedlings. Nowadays, as a result of the natural distribution of herbaceous plants on the dump by ruderal and weed vegetation dominates represented by *Convolvulus arvensis* L., *Carduus crispus* L., *Sonchus arvensis* L., etc. Projective cover degree is not more than 5-7%. There is no turfing.

As a control, a site unspoilt by the activity of the coal pit is selected, located 1 km west of the dump, which is a birchwood surrounded by grassy-cereal meadows.

The collection of material was carried out by the method of soil traps, which were ranged in a line of 10 pcs. with an interval of 10 m. The line of traps was located along the top of the dump. The traps were inspected every 7-10 days. A 10% solution of acetic acid was used as the fixing liquid.

The dynamic density of arthropods was expressed in the number of specimens per 10 trap-days (specimens / 10 trapped per day (tpd)). In total, more than 80 samples were processed, over 5,500 individuals were counted.

Quantitative characteristics of the assessment of species diversity of the communities studied were carried out using traditional indices [9]. The following classes of ground beetle abundance are distinguished [10]: more than 5% of the collections are dominant species, 1.1-5 are subdominant, 0.3-1 are rare, and less than 0.3 are very rare.

When characterizing mesoherpetobia of model sites, the obtained structural and numerical indices of individual groups of arthropods were used, the correlation analysis (Spearman correlation) [9]. Mathematical data processing was carried out using Excel and Statistica 6.0.

3 Results and Discussion

The conducted research made it possible to state that during the first two years after the completion of dumping, along with the formation of phytocoenosis in the reclamated areas, a pioneer grouping of herpetobiontic arthropods is being formed. The arthropod stocking is a spontaneous way – by the migration of native species from the nearest biocenoses. At the earliest stages of the dump succession, the population of mesoherpetobionts is formed,

 DOI: 10.1051/e3sconf/20172102014

consisting of only three groups of arthropods: arachnids, centipedes and insects (Figure 1), while in the control zone, in addition to the aforementioned groups of arthropods, Crustacea (cranberry), as well as a few representatives of the Mollusca type, were also found. The total dynamic density of arthropods in young dumps (44.29 ± 6.35 specimens/ 10 tpd) was significantly lower in comparison with the control site (181.99 ± 10.90 specimens / 10 tpd) (for $p < 0.05$).

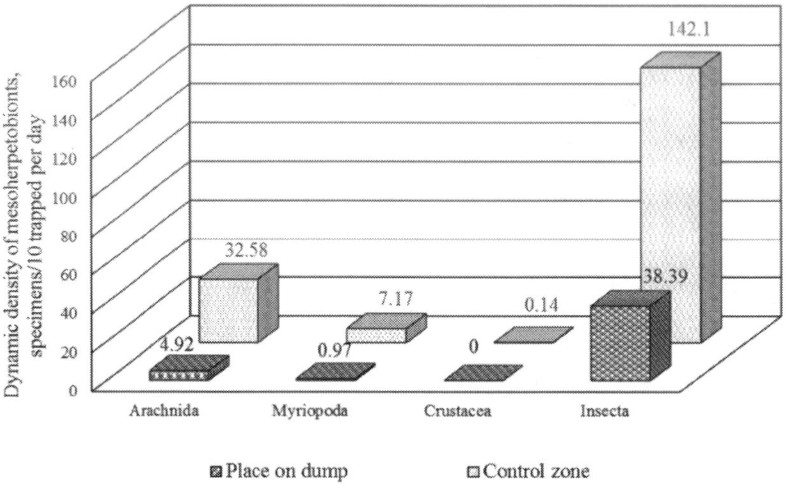

Fig. 1. Dynamic density of different groups of arthropods on the reclamated dump and in the control zone, specimens / 10 tpd.

From Figure 1, it can be seen that very low values of the dynamic density of all the detected groups of arthropods (millipedes – 7.4 times lower, arachnids – 6.6 times, insects – 3.7 times) are noted in the dump in comparison with the control. Moisture-loving arthropods, crustaceans from the order Isopoda and wood lice, are very few in the control, and are not found on the dump.

Insects take an active part in the primary succession. They account for 86.7% of the total collection of all arthropods. This group of arthropods is also numerous in the control zone (78.1%).

The density of the population of the second largest group, arachnids, was 7.8 times lower in the dump than that of insects (in the control 4.4 times less compared to insects). From the class Arachnida, representatives of three orders are noted on the dump: Acariformes, Opiliones and Aranei. Of these, spiders take the largest part in the development of new biotopes. The dynamic density of their population on the dump is 3.46 ± 0.76 specimens / 10 trapped per day (70.4% of the total collection of all arachnids), in the control area – 21.10 ± 2.58 specimens / 10 trapped per day (64.8% of the total collections of all arachnids). Daddy-long-legs and acariform mites on the dump are less common than in the control zone. For example, a very low population density of acariform mites was recorded on the dump – 0.98 ± 0.71 specimens / 10 trapped per day (19.9% of the total collections of all arachnids) compared to the control – 10.40 ± 1.86 (31.9%). The most conservative group of arachnids is the daddy-long-legs, the average dynamic density of which on the dump is 0.48 ± 0.20 specimens / 10 tpd (9.7% of the total collections of all arachnids), in control – 1.07 ± 0.20 specimens / 10 tpd (3.3%).

The dynamic population density of Myriopoda was even more significantly different from that of the insects. On the dump, the density of their population was 39.6 times less compared to the insects, and 19.8 times in the control. The most mobile representatives of centipedes Chilopoda participate in young dumps stocking, while in the control, there are also representatives of another class – Diplopoda. From the Diplopoda class, the representatives of the order Julida were found on the model sites and the Lithobiomorpha order from the Chilopoda class.

The dominant group of mesoherpetobia, insects, is formed on the dump only by species from the class of Insecta-Ectognatha, while in the control there are representatives of two classes – Insecta-Entognatha and Insecta-Ectognatha. From the Insecta-Entognatha class, the species of the Collembola are marked in the control zone (Table 1).

Table 1. Dynamic density of orders of insects on the reclamated dump and in the control zone, specimens / 10 tpd.

Detachment	Place on dumps	Control place
	Insecta-Entognatha	
Collembola	–	3.37 ± 0.76
	Insecta-Ectognatha	
Heteroptera	6.42 ± 1.26	2.59 ± 0.57
Coleoptera	30.16 ± 4.26	66.79 ± 4.67
Hymenoptera	1.82 ± 0.86	69.35 ± 7.48
Total insects	38.39 ± 5.42	142.10 ± 9.71

In succession changes on the reclamated dump, insects of three orders from the class of Insecta-Ectognatha: Heteroptera, Coleoptera and Hymenoptera (Table 1) take part. From these, the basis of the pioneer grouping of mesoherpetobia is Coleoptera, accounting for 78.6% of all insect collections on the dump. In the control zone, the role of two orders - Coleoptera and Hymenoptera – is significant. The number of these detachments is not significantly different; they account for 95.8% of total insect collections.

The Hymenoptera order on the dump is mainly represented by various species of ants from the family Formicidae: the genera Lasius, Myrmica, Formica and Tetramorium. The ants of the genus Camponotus are noted only in control. The most active in the settlement of the dump are ants of the genus Tetramorium, for which arid and highly heated areas of reclaimed dumps are optimal for life. The ants of this genus account for 62.4% of the total gathering of ants on the dump, while on the control plot it is only 0.2%. The ants of the Myrmica family dominate in the control (87.6%).

The pioneer grouping of the Coleoptera order on the dump is represented by the families Carabidae, Silphidae, Staphylinidae, Elateridae, Histeridae, Byrrhidae, Curculionidae (genus *Otiorhynchus* Germ.). In the control, a greater variety of Coleoptera was noted; there are beetles of three more families – Scarabaeidae, Dermestidae, Tenebrionidae. The dynamic density of Coleoptera on the heap is 30.16 ± 4.26 specimens / 10 tpd, which is 2.2 times less than the values in the control section.

In the first stages of succession changes in the dump and in the control zone, two families dominate in dynamic density: Carabidae and Staphylinidae (Figure 2), accounting for 90% of total beetle collections in the dump and 86.1% in the control.

The most significant contribution to the stocking of the reclamated dump is made by predators, ground beetles, one of the dominant groups of invertebrate animals of natural and anthropogenically transformed terrestrial ecosystems. Due to their biological and ecological characteristics, as well as a good ability to react to changes in the ecological situation, they have long been used as bioindicators of the condition of man-made territories.

 DOI: 10.1051/e3sconf/20172102014

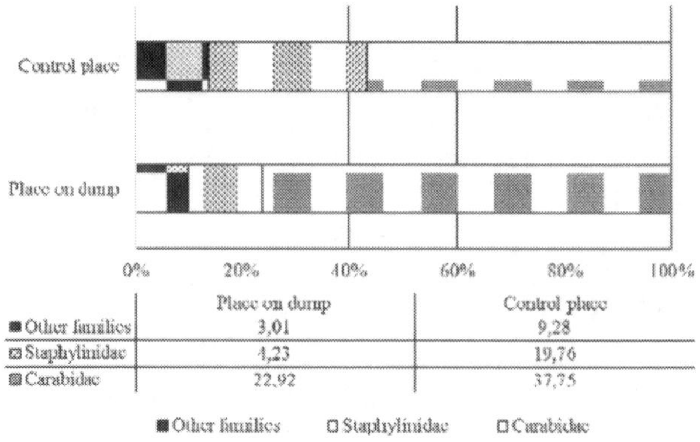

Fig. 2. The ratio of the dynamic density of the dominant families of coleopterans on the dump of the Krasnobrodsky coal pit and in the control zone, %.

Note. The data table shows the dynamic density of Coleoptera, specimens / 10 tpd

The carried out investigations made it possible to identify 68 species of ground beetles of 24 genera, 16 tribes, 10 subfamilies. The structure of carabidocomplexes of model sites is dominated by rare species, a significantly smaller number of species of medium abundance and few dominant species. The detailed quantitative characteristics of the species structure of ground beetles are given in Table 2. As it can be seen from Table 2, the area of the young dump compared to the control zone differs significantly in species diversity and abundance of carabids; the value of the Chekanovsky-Sorensen index is only 0.13 (i.e., the similarity of the species composition of these plots is 13%).

Table 2. Some indicators of the structure of the population of ground beetles in the territory studied.

Indexes	**Designation**	**Place on dumps**	**Control place**
Total number, (specimens)	N	1353	1379
Wealth of species	S	44	33
Margalef diversity index	K	0.52	0.48
Margalef richness index	D_{Mg}	5.96	4.42
Shannon diversity index	H	2.72 ± 0.11	1.65 ± 0.07
Pielou's evenness index	e	0.71	0.47
Simpson's dominance index	C	0.1 ± 0.002	0.32 ± 0.014
Chekanovsky-Sorensen index	I_{CS}	0.13	

Analyzing the species composition of ground beetles in the areas under consideration, it can be seen that small or medium-sized species are mainly involved in the expansion of dumps. From the tribe of Bembidiini, it is *Bembidion quadrimaculatum quadrimaculatum* (L.) (dynamic density 5.27 ± 1.66 specimens/ 10 tpd), Zabrini – *Amara ingenua* (Duft.) (3.14 ± 1.77), Harpalini – *Anisodactylus signatus* (Pz.) (1.97 ± 0.92), *Harpalus rufipes* (1.59 ± 0.62) (Deg.), *Harpalus distinguendus distinguendus* (Duft.) (1,3 ± 0,67).

Very interesting is the discovery on the dump of a new type of *Calathus ambiguus ambiguus* Pk., for Kemerovo region, whose dynamic density was 0.015 ± 0.015 specimens/ 10 tpd. It is a subarid western Palearctic species. Previously, in Western Siberia, the species was known only from the vicinity of Tyumen and the village of Verkh-Tula (Novosibirsk region). In addition, a fairly high dynamic density of the steppe species *Harpalus nigrans* (Mor.), very rare for Western Siberia, is recorded here (dynamic density 0.41 ± 0.13 specimens / 10 tpd).

On the control place, in contrast, the collection is dominated by ground beetles of large size characteristics, preferring forest biotopes. From the tribe Carabini – *Carabus aeruginosus* Fisch. (13.43 ± 4.16 specimens / 10 tpd), *C. regalis regalis* Fisch. (3.85 ± 0.99). In addition to these species, *Pterostihus oblongopunctatus* (F.) (0.39 ± 0.15), *P. magus magus* Man, (3.19 ± 0.48), Zabrini – *Amara communis* (Pz.) (0.63 ± 0.25), Sphodrini – *Synuchus vivalis vivalis* Ill. (0.25 ± 0.14), Platynini – *Agonum gracilipes* (Duft.) (1.22 ± 0.37) can be classified among the dominants of the tribe Pterostichini.

The analysis of the landscape-biotopic distribution of carabides in the territory of the Kemerovo region in order to identify the most preferred biotopes, and also based on literature data, made it possible to combine the species of ground beetles marked in the studied area into six ecological groups (Figure. 3).

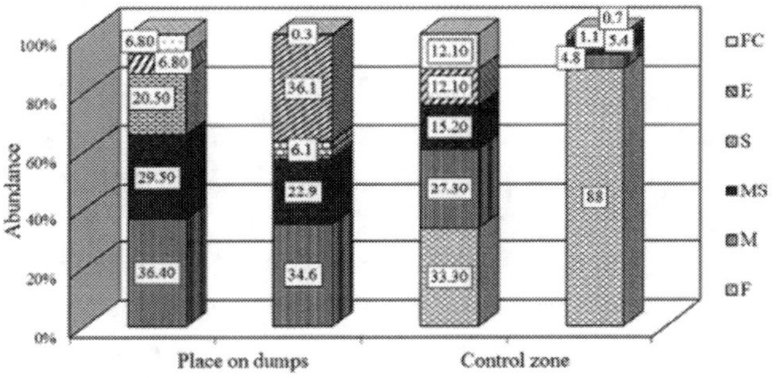

Fig. 3. Species (column 1) and numerical (column 2) abundance of ecological groups of ground beetles in the areas studied.

Note. FC – floodplain-coastal species, E – forest species, S – steppe species, MS – meadow and steppe species, M – meadow species, E – eurytopic species

As it can be seen from Figure 3 there is a clear tendency for the predominance of meadow, meadow-steppe and steppe groups, both in species and in numbers, on the dump, and there are no species of forest group on them. This is quite natural because on the dump there is scant vegetation, it is under prolonged exposure to insolation, which creates extremely arid conditions on its surface.

It is interesting to find three species of the floodplain-coastal group on the dump (*Bembidion obliquum* Sturm, *Acupalpus meridianus* (L.), *Apristus striatus* (Motsch.)), which prefer strongly or medium-humid areas, such as banks of water bodies, small rivers, streams etc. The dynamic density of these species is quite low – 0.025 ± 0.02 specimens / 10 tpd. Their presence is probably related to the periodic migration of ecosystems adjacent to the dump characterized by the presence of small streams.

Forest species, mainly representatives of the genus *Carabus*, were only encountered in the control. Remarkably, the species of eurytopic biotopic preferences are met at each site. An interesting fact is that there is a decrease in the relative abundance of representatives of this group as succession processes occur. From Figure. 3 shows that the maximum abundance of eurytopic species is noted on the young dump, which is primarily due to the large numerical abundance of the two species – *Bembidion quadrimaculatum quadrimaculatum* and *Harpalus affinis* (Schrank).

4 Conclusion

Thus, the obtained results of the study show that carrying out remediation measures on the dumps of coal enterprises contributes to the rapid formation of the pioneer communities of herpetobionts, which take an active part in the further restoration of biocenoses disturbed.

Subsequent comprehensive studies of the succession processes of mesoherpetobia on the areas transformed by industrial activities will allow us to understand better the processes of formation and functioning of man-made landscapes, assess the suitability of new environmental conditions for coenoses formed, and develop recommendations for remediation work.

Acknowledgement

The reported study was funded by RFBR and Kemerovo Region according to the research project № 16-44-420211.

References

1. I.V. Zenkova, V.V. Kalmykova, A.A. Liskovaya, Eurasian Soil Sci., **42**, 8 (2009)
2. I.Vowinkel, Hess. faun. Briefe, **8**, 4 (1988)
3. C. Haag, M. Depenbusch, Mitt. Dtsch. Ges. allg. und angew. Entomol., **9**, 4-6, (1995)
4. W. Durka, M. Brandle, M. Altmoos, Mitt. Dtsch. Ges. allg. und angew. Entomol., **11**, 1-6 (1997)
5. E. Sigmund, W. Waitzbauer, Verh. Zool.-Bot. Ges. Osterr., **144** (2007)
6. S.L. Luzyanin, R.Yu. Dudko, A.N. Bespalov, N.I. Eremeeva, Euroasian Entomol. J., **14**, 5 (2015)
7. J. Frouz, Sb. Okres. muz. Moste. R. prirodoved., **18** (1996)
8. L. Ottonetti, L. Tucci, G. Santini, Restor Ecol., **14**, 1 (2006)
9. A.E. Magurran, *Ecological Diversity and It's Measurement*, (1988)
10. D. Mossakowski, Z. Wiss. Zool., **181**, 3/4 (1970)

E3S Web of Conferences **21**, 02015 (2017) DOI: 10.1051/e3sconf/20172102015
The Second International Innovative Mining Symposium

Resource-Saving Cleaning Technologies for Power Plant Waste-Water Cooling Ponds

Lyudmila Zakonnova [1,*], *Igor* Nikishkin[2], and *Alexandr* Rostovzev[3]

[1] T.F. Gorbachev Kuzbass State Technical University, Belovo Branch, 652644 Ilyicha st. 32, Inskoy, Kemerovo oblast, Russia
[2] Belovo Fish Center Ltd., 652644 Technologichesky Dist., 3, Inskoy, Kemerovo oblast, Russia
[3] FSUE "Gosrybcentr", Novosibirsk branch, 630091 Novosibirsk, 1 Pisareva st. Russia

Abstract. One of the frequently encountered problems of power plant small cooling ponds is rapid eutrophication and related intensified development of phytoplankton ("hyperflow") and overgrowing of ponds by higher aquatic vegetation. As a result of hyper-flowering, an enormous amount of detritus settles on the condenser tubes, reducing the efficiency of the power plant operation. The development of higher aquatic vegetation contributes to the appearing of the shoals. As a result the volume, area and other characteristics of the cooling ponds are getting changed. The article describes the environmental problems of small man-made ponds of power plants and coal mines in mining regions. Two approaches to the problem of eutrophication are considered: technological and ecological. The negative effects of herbicides application to aquatic organisms are experimentally proved. An ecological approach to solving the problem by fish-land reclamation method is shown.

1 Introduction

Today the correlation between different factors of sustainable development – socio-economic and ecological development of old industrial regions is in focus of many Russian researches [1-4]. A special role in improving of ecological situation in mineral resources extracting areas is played by wastewater purification and water sources preservation [5-9]. Modern monitoring tools ensure the use of remote methods instead of expensive direct measurements, which allows more accurate and more dynamic predicting of aquatic eco-systems development, both of natural and anthropogenic origin [10-15]. At the same time, methods of direct monitoring and experimental approbation of their purification are still relevant to the study of ecological problems of small man-made water ponds of coal-mining and coal-processing enterprises.

We observe two approaches to the power and coal plants waste-water cooling ponds purifying problem solvation: technological and environmental. The technological approach does not take into account the interests of existing community of organisms; its supporters still believe that eutrophication of the pond is most expedient with the help of herbicides, in particular, Cu^{2+} salts. An ecological approach to this problem solvation takes into account

* Corresponding author: nir_belovo@mail.ru

the fact that during the existence of the cooling pond, a co-society of organisms was formed in it, and the pond itself became a part of the ecosystem of the district and a recreation zone for the citizens.

2 Materials and Methods

Among the known measures to fight herbal forms, the most environmentally friendly and economically acceptable is the method for pond cleaning with the help of fish meliorators - microphytophagous carp and macrophytophage white amur. Initially, this method was used for biological cleaning of irrigation systems in Central Asia, but recently it is getting increasingly used in Western Siberia, in particular, in the reservoirs of Belovskaya power plant and Berezovskaya-1 power plant, belonging to the Ob River basin.

With the undoubted advantage of such a method, it has weak points, for example, long-term consequences for aboriginal ichthyocenosis in connection with introduction of competing fish. Microphytophagous carp in the Ob River basin has few food competitors, so it will not reduce the productivity of native fish species. Macrophytophage white amur, with a lack of preferred plant species, extends the nutritional spectrum and becomes a benthophagus, undermining the fodder base of tuvod fish. Therefore, the introduction of white cupid into cooling ponds should be carried out taking into account the conditions of each specific pond.

We considered the situation that developed in the cooling pond of Belovskaya power plant (Belovskoe reservoir). Belovskaya power plant works on a circulating system of technical water supply. A water cooling pond is a reservoir formed by regulating the flow of the Inya River. The project area of the reservoir is 1300 hectares, at present it is 1430 hectares, the volume of water is 47.9×10^6 m^3.

Water to cool the condensers of the power plant is taken from the dam of the pond and getting discharged along a diverting channel about 6 km long into the upper reach. The average flow of circulating water in the discharge channel is 35 m^3 / sec in summer.

A characteristic feature of the Belovskaya power plant is the constancy of the load. The operation of the power plant in a constant mode provides not only a constant flow of circulating water, but also the absence of sharp fluctuations in its temperature in the daily and weekly periods.

At the initial stage of the formation of the hydroecosystem, the cooling pond at the Belovskaya power plant had features of an oligotrophic type of pond with insignificant tendencies towards the mesotrophic state. At present some indicators (chlorophyll content “a”, zoobenthos biomass, oxygen content in bottom water, transparency) have eutrophic values. The biomass of phytoplankton increased in 3.1 times, primary production increased in 2.6 times, chlorophyll a content – in 7.1 times, biomass of zooplankton – in 2.6, zoobenthos – in 3.9 times.

Studies of a number of scientists, including V.V. Kirillov et al. allowed to assess the level of production and destruction processes in the cooling pond of Belovskaya power plant, to estimaye the current state of zoobenthos and zooperifiton and higher aquatic vegetation of the pond and compare the obtained data with the results of other authors [16-21].

Thus, as a result of more than forty years of operation of the cooling pond of the Belovskaya power plant, the eutrophication of the reservoir occurred quite naturally under the existing conditions. The biomass of phyto- and zooplankton, higher aquatic vegetation significantly increased, and in this connection there was a danger that detritus, depositing on condenser tubes of power station, would reduce the degree of water cooling. In 1998-1999 years summer "flowering" of the reservoir was particularly high, therefore, since 2000, works have begun to clean the Belovo Reservoir. According to V.A. Smirnov and co-authors, the index of the species diversity of the Belovo Reservoir reaches 3.2-3.5 (with a maximum possible value of 6), the hydrobiocenosis is represented by 418 organisms, including 17 species of tuvodia and 7 introduced species of fish.

3 Results and Discussion

All the data given indicate that a mesotrophic-eutrophic ecosystem has developed in the Belovskaya power plant cooling pond. In this case, the use of herbicides can lead to the death of many species of hydrobionts and the degradation of the reservoir. Therefore, the fight against further eutrophication of the reservoir should be carried out by soft ecological methods, among which are the use of herbivorous fish - silver carp and white amur - to combat plankton and higher aquatic vegetation. In 2000, we developed the "Fish and Biological Substantiation for Additional Filling of the Belovskaya power plant cooling ponds with herbivorous fish", according to which fat carp had been produced in the reservoir.

Table 1. Fish-breeding indicators, which can be obtained in 4-8 years after stocking the cooling pond of the Belovskaya power plant with silver carp

Timing	**Average piece weight, kg**	**Total weight, tons**	**Survival for the period, %**	**The number of individuals, thousand pieces**	**The increase in the ichthyomass for the period, tons**	**Feed Ratio**
4 years after stocking	0.15	5		33370		
8 years after stocking	4.0	80	60	20022	75	15-60

At the same time, with the undoubted advantages of ecological methods for controlling the overgrowing of water bodies, unauthorized use of herbicides can not be denied, such as copper sulfate, especially since such cases occurred in the late 1970s.

Therefore, in 2000, under the conditions of the Belovo Fisheries, an investigation was made of the possible consequences of the use of Cu^{2+} salts for the ichthyofauna of the Belovo Reservoir by the "fish test" method, whose tasks included the following: the detection of toxic and safe carp endothelial cells and larvae; the concentration of copper penta-sulphate; determination of the effect of copper sulfate solutions of different concentrations on fertilization, embryogenesis and early postembryogenesis of; identification of the long-term consequences of the influence of Cu2 + ions for carp fry.

It is known that the most sensitive to the action of copper are embryos and larvae of fish [22-26]. In connection with this material, developing embryos, pre-larvae, larvae and carp fry were used in research (Cyprinus carpio, L). Young carp were obtained in the production conditions of the Belovo fish farm using standard technology. For the purity of the experiment, the descendants of one female fish were used in all variants.

The mother liquor of copper penta-sulphate ($CuSO_4*5H_2O$) was prepared in distilled water, working solutions of different concentrations were made using the technological water of the Belovskoye Reservoir, since it is known that the toxic effect of copper ions depends on a number of reasons: hardness, oxygen content, organic compounds , suspensions [13]. Solutions containing from 0.02 to 10 mg $CuSO_4 * 5H_2O$ / liter were used (Tables 2, 3).

Copper sulfate solutions of different concentrations were used to activate sperm and the process of eggs developing. To determine the effect on embryogenesis, fertilized caviar was placed in solutions of copper sulphate, for the activation of the sperm and the swelling of the eggs, ordinary technological water was used. In order to detect the effect of copper ions on pre-larvae, intact hatching embryos were taken and placed to stand in solutions of copper sulfate of various concentrations. To determine the long-term consequences of the action of copper ions on developing carp fry, all the larvae surviving in previous experiments were grown during the usual technology for fish farming.

The criteria for the quality of embryos and postembryons were survival, growth rate, and the presence of individuals with various malformations. Each experimental group consisted of 100 individuals.

As a result of the studies, it was found that the maximum toxic effect of copper ions on embryos is manifested when using solutions of copper sulfate directly during the fertilization of eggs, as when swelling solutions enter the eggs. Survival of embryos in a concentration solution of 0.08 mg / liter was 25.6%, in a solution with a concentration of 0.02 mg / liter - 39.2%, in control samples - 86.7%.

When the eggs were incubated, which was fertilized with technological water, the survival of embryos was found to depend on the copper sulfate concentration in solutions (Table 2).

Table 2. Survival of carp embryos during incubation in solutions of copper sulfate of various concentrations

Concentration $CuSO_4 * 5H_2O$, mg/liter	**10**	**5**	**1**	**0.2**	**0.1**	**0.08**	**0.06**	**0.04**	**0.02**	**Control**
Survival, %	9.2	15.0	76.8	75.0	76.9	78.0	76.6	77.7	80.3	81.0

The most toxic were concentrations above 5 mg / liter, concentrations from 00.4 to 1 mg / liter caused a 3-6% decrease in survival rate, a concentration of 0.02 mg / liter practically did not reduce the percentage of survival in embryogenesis. The delay in embryonic growth in comparison with the control samples, and then the death of embryos were observed at the beginning (fouling stage) and at the end of embryonic development during the period of the circulation system activation, as a result of which toxic substances penetrated into the embryo can more effectively approach sensitive organs.

When breeding preliminaries took the long time before switching to external nutrition, it turned out that in solutions with a concentration of 0.6 mg / liter and lower, the survival was closed to 100% (like in the control group). Higher concentrations of copper sulphate caused the death of some individuals (Table 3).

Table 3. Survival of carp larvae with holding in solutions of copper penta-sulfate of various concentrations

Concentration CuSO4 * 5H2O, mg / liter	**3**	**2**	**1**	**0,8**	**0,02-0,6**	**Control**
Survival rate,%	40.0	55.0	80.0	85.0	100	100

The long-term consequences of the action of copper ions in early ontogeny were revealed only in those embryos that were obtained using solutions of copper sulfate during fertilization. Among them, about 30% of individuals with various malformations were found, significantly lacked behind in growth from the rest of the group.

4 Conclusions

 DOI: 10.1051/e3sconf/20172102015

Thus, the advantage of the biological method of cooling ponds cleaning with the help of fish-meliorators can be considered proven.

References

1. V. Trifonov, O. Loyko, D. Nesteruk, S. Zhironkin, E. Strekovtsova, AIP Conf. Proceed., **1800**, 050009 (2017)
2. M.I. Agienko, E.P. Bondareva, G.V. Chistyakova, O.V. Zhironkina, O.I. Kalinina, IOP Conf. Ser.: Earth Environ. Sci., **50:1**, 012022
3. M. Erkayaoğlu, N. Demirel, J. Env. Man., **174**, 1-6 (2016)
4. M. Cehlár, J. Janočko, Z. Šimková, T. Pavlik, E3S Web of Conf., **15**, 01019 (2017)
5. M. Tyulenev, S. Zhironkin, K. Kolotov, E. Garina, Poll. Res. **35:2**, 221-227 (2016)
6. M. Tyulenev, E. Garina, A. Khoreshok, O. Litvin, Y. Litvin, E. Maliukhina, IOP Conf. Ser.: Earth Environ. Sci., **50:1**, 012035 (2017)
7. M. Tyulenev, Y. Lesin, E.Tyuleneva, E. Murko, E3S Web of Conf., **15**, 02003 (2017)
8. M. Cehlár, P. Varga, Z. Jurkasová, M. Pašková, Acta Montanistica Slovaca, **15:2**, 132-138 (2011)
9. M. Prekopová, J. Janočko, Geol. Carp., **60:6**, 485-494 (2009)
10. R.P. Bukata, J.H. Jerome, K.Ya. Kondratyev, D.V. Pozdnyakov, *Optical Properties and Remote Sensing of Inland and Coastal Waters* (Boca Raton, Florida: CRC Press, Inc.; 1995)
11. A. Collin, J.L. Hench, Rem. Sen. **4:12**, 125-144 (2012)
12. M. Deidda, G. Sanna, *Bathymetric extraction using WorldView-2 high resolution images. International Archives of the Photogrammetry, Remote Sensing and Spatial Information Sciences* (Melbourne, Australia, Proceedings of the XXII ISPRS Congress, 2012)
13. A. Minghelli-Roman, A. Goreac, S. Mathieu, M. Spigai, P. Gouton, Int. J. Rem. Sens. **30:2**, 5737-5750 (2009)
14. S. Sathyendranath, *Remote sensing of ocean colour in coastal, and other optically-complex, waters* (Dartmouth, Canada: MacNab Print, 2000)
15. R. Uma, Maheswari Int. J. Mar. Sci. **3:1**, 91-97 (2013)
16. A. Akiyama, Bul. Jap. Soc. Scient. Fish. **36**, 563-570 (1970)
17. N.J. Cazzaniga, Biocell. **26**, 71-81 (2002)
18. C.R. Hazel, S.J. Meith, Water Proj. Branch. Lab. Rep. **16**, 1-8. (1969)
19. A. Hutorowicz, J. Hutorowicz, Ecol. Ques. **9**, 79-86 (2008)
20. A.L. Estebenet, P.R. Martin, Biocell. **26:1**, 83-89 (2002)
21. H.R. Gordon, Applied Optics. **14:2**, 413-416 (1975)
22. C.H. Bronson, Tech. Bul. **3**, 2-4 (2002)
23. A. Kanno, Y. Koibuchi, M. Isobe, IEEE Geosci. Rem. Sens. Let. **8:1**, 64-67 (2011)
24. C. Korschgen, W. Green, *American wildcelery (Vallisneria americana): Ecological considerations for restoration* (Wildlife : Fish and Wildlife Technical, 1988)
25. K. Matsukura, H. Tsumuki, Y. Izumi, T. Wada, J. Exp. Biol. **212**, 246-255 (2009)
26. J.M. McKim, D.A. Benoit,. J. Fish. Res. Board Can. **28**, 655-662 (1971)

Innovative Production of Polyvinychloride on the Basis of Vertical Integration of Business and Cluster Organisation

Irina Kudryashova[1], *Natalia* Zakharova[2], and *Evgeniy* Kharlampenkov[3]

[1]Kemerovo institute (branch) of the Russian Plekhanov Economic University, Administration, 650992 Kemerovo, 39 Kuznetskiy av., Russia
[2]Plekhanov Russian University of Economics, Department of the World Economy, 117997, Moscow,36 Stremyanny lane, Russia
[3]Kemerovo institute (branch) of the Russian Plekhanov Economic University, Department of Commercial Affair, 650992 Kemerovo, 39 Kuznetskiy av., Russia

Abstract. The paper presents the results of the research of the world and Russian market of polyvinyl chloride (PVC), a comparative assessment is made and the similarities and differences of these segments at the present have been identified. The theoretical possibility of considerable prospects of the development of domestic production of PVC is justified, ensure that it can achieve import substitution and export substitution of this product in Russia. Based on the theory of value chains and cluster organization of the enterprises of coal and chemical industries, ferrous metallurgy and other enterprises of Kuzbass have been proposed new technological solutions for the production of PVC, taking into account regional features of natural-resource base and capacity of national and world markets. It is established that the organization of production of PVC, using coal chemistry technology can be a promising growth point for the business of the Kemerovo region, will help to diversify the coal industry, to expand its product structure and export potential of the region, increase its competitiveness.

1 Introduction

Kuzbass at the turn of XX – XXI centuries moved into the category of old coal regions experiencing pressure from newly developing coal basins located in Yakutia and Tuva, having rich deposits of coking and energy coal. In this case the logistics component of delivery of coal to the countries of South-East Asia and Japan for these fields, given the proximity of BAM and the ports of Primorye, significantly reduces the competitiveness of Kuznetsk coal. In this case the logistics component of delivery of coal to the countries of South-East Asia and Japan for these fields, given the proximity of BAM and the ports of Primorye, significantly reduces the competitiveness of Kuznetsk coal. Despite the fact that the share of Kuzbass in the coal export of Russia at the moment remains high at 77%, and coking grades of 80%, the pressure of new coal basins continues to increase.. It is worth noting that coal is transforming from the fastest growing energy source (increase in world consumption in the period 2000-2013 amounted to, on average, 3.8% per year) to the fuel with the lowest development index of consumption (about 0.8 % per year in the period 2015-2035.). It makes Kuzbass enterprises to search new markets and to solve the issues for creation of manufactures of processing of coal in place, allowing to obtain a product with a higher added value, using the cluster approach, cooperation of industrial relations, integrative approaches of interaction. Significant technological capabilities for the further development of coal in the region and in Russia are prompted thanks to the conversion of coal in non-fuel products (chemical intermediates, compounds, carbon materials), the creation and production of which determines the level of technological sovereignty of the country and readiness for potential changes in the global energy balance and structure of consumption of hydrocarbon resources. Coal, as the experience of a number of countries, primarily China, shows, can be quite competitive raw material for the production of many chemical products. The higher the degree of processing of coal and the level of high-tech product produced from coal, the lower component of the logistics (mainly transport) and the more efficient becomes the process of coal production. If coke tariff rate per 1 ton is 29% of its value, then PVC – 9%. Taking into account the capacity of coal-chemical cluster of the Kemerovo region, the project of PVC production is quite promising, feasible and cost-effective. While the values of non-fuel products, conversion of coal - chemical intermediates and precursors that are in demand in industrial markets, an order of magnitude higher than those of raw coal. Very promising forms of diversification of the resource economies there may be models of vertical integration and cluster organization of the regional economy.

2 Material and Methods

The World market for PVC is one of the most dynamic in the global economy. The main areas of consumption of PVC in the world is the production of rigid and flexible pipes and fittings. About 150 companies in 50 countries today produce PVC. About 70% of global PVC consumption accounts for the construction industry, where the polymer is an essential component in the process of making different designs. The demand for PVC is in direct proportion to the dynamics of the construction industry.

The highest rates of growth of consumption volume are demonstrated by such countries as China, India, Russia, Middle East countries. The next five years they will have a key influence on the global market [1]. The share of Russia in world volume of production of PVC reaches a 1.6%, share in global consumption - 2.7%. Currently, the PVC market in Russia is totally import-dependent, and the main supply comes from the United States and China. Since 2006, Russia is the main consumer of Chinese PVC in the world. n the structure of export of Chinese PVC Russia plays a key role: in 2014 - 2015, its share was about 50%, and in 2016-2017 the trend has not changed. To reverse this situation is possible only through the creation of the own production, on the basis of carbide and ethylene technology. However, given the experience of the enterprises of the coal and chemical industries of the Kuznetsk basin, coal companies, taking into account the presence of scientific bases in the region, this task at the present stage can be completely solved.

Up to the present time, 74% of global capacity is represented by the countries of Asia, including China (67%) and India (2%). Europe accounts for 11%,Middle East — 6%, other countries — 9%. The Eastern and Western Europe together account for about 26% of world production of PVC [2]. Since the beginning of the century the centre of production and consumption of PVC has moved to China, where, according to some sources, in the period 2005-2010 production capacity grew annually by almost 10%, and according to the other (China Chlorine-Alkali Industry Ass.), over the last decade the production of PVC in China grew by 3-4 times. Really, when you consider that in 2004, China produced just over 5 million tonnes (in 2006 – 11 million tons), and now 2/3 of the capacities are concentrated in China, we can say that the production of PVC in this country has taken very serious steps. Today, Chinese PVC production is estimated to be 15-20 million tons. It should be noted that the basis for PVC production in China is carbide technology, according to which currently China produces 70-80% of PVC [3]. This is due to the cheap cost of raw coal and electricity produced in the North-Western provinces. Chinese companies follow the way of formation of clusters consisting of enterprises in the coal industry, energy production, chemicals, processing of raw materials into finished products with high added value.

In the Russian market of PVC are clearly visible the conditions and limits that stimulate and restrict the localization of production in Russia - the process that better describes the growth of the industry than the so called "import substitution". The capacity of the domestic market is enough to invest in basic raw materials at the scale of a large plant, but, however, is inadequate for special grades of PVC.

3 Results and Discussion

The rapid growth of PVC production in Russia in physical terms in the period from 2012 to 2015 is confirmed by 18% growth in 2015 compared to the previous 2014 and 1,5 times growth compared to the basic 2012 This fact proves the rapid growth of this sub-sector compared to the whole plastics processing industry, where the annual development rate reached 11.8% in 2015.

In continuation of this study highly rational and innovative, in our opinion, is the placement of the PVC production based on vertical integration of the enterprises of the coal industry, enterprises of non-metallic materials, metallurgy and other enterprises of Kuzbass, in particular with leading producers of chemical products in the region, such as JSC "Khimprom" - a potential supplier of chlorine for the production of vinyl chloride monomer.

The production of acetylene, given the resource potential of the Kuzbass and the largest enterprise JSC "Cox", included in industrial metallurgical holding, is possible with the use of three technologies:

-production of acetylene from calcium carbide (carbide),

- production of acetylene from methane extracted from coal seams, the method of electrocracking or oxidative pyrolysis

- production of acetylene from coke oven gas obtained in the process of main production at JSC "Cox" and containing 27% methane and 55% of hydrogen.

Carbide acetylene can compete with ethylene as a raw material for the production of vinyl chloride, if its value does not exceed the cost of ethylene by more than 40%. If the cost of the acetylene produced from methane is quite low, this method will be competitive [4].

In practice, the most appropriate method of production of acetylene is the combined method on the basis of coke oven gas and methane derived from coal beds and delivered in liquefied form to the company. Currently PJSC "Gazprom" together with LLC "Gazprom Mining Kuznetsk" develop a project for production of liquefied methane extraction sites. Thus, this part of the process is quite realizable in practice.

On the basis of the method of obtaining vinyl chloride by acetylene hydrochloride lies the catalytic reaction occurring with high heat. Hydrogen chloride has to come from the Kemerovo enterprise JSC "Khimprom". This method is characterized by the simple technological design process, low investments, high selectivity to vinyl chloride, but the method has not found wide industrial application due to the high cost of acetylene. Hydrochloride process of acetylene is usually carried out in the presence of mercuric chloride, applied in the amount of 10-15% on activated carbon, in a stationary layer of catalyst at 425-535 K and a pressure of 0.2-1.5 MPa. The degree of conversion of acetylene is 98.5% with a selectivity to vinyl chloride of 98% [4].

Polymerization occurs in the droplets of vinyl chloride. In the end it is possible to obtain porous microspheres. Production is fully automated, the polymerization is carried out in reactors whose volume exceeds 200 cubic meters [5].

After completion of the polymerization degree of vinyl chloride, which by this time has reacted, reaches up to 90 percent. Items that were not used are removed, while PVC itself is sifted and dried with hot air, then there is packaging. Polyvinyl chloride in the last step is converted to vinyl or plastic [6]. Feature of plasticity is the presence of stabilizers.

On their nature and amount largely depends the workability (processability) of the compositions of PVC and lifetime of the final product. Indirect characteristics of the selection of these additives is such characteristics of the plastic as the thermal stability, allowing to carry out the technological process of production of the final product in a wider temperature range [6].

Main additives in PVC upon receiving the plastic forms are calcium stearate and flame retardants, manufactured at the Russian enterprises [5]. To reduce the flammability of PVC, the introduction of the plasticized hydroxides of aluminium or magnesium is possible [6]. Suspension PVC is produced in the form of a homogeneous white powder with a bulk density 450-700 kg/m3. An important element of the process is the production of powder PVC products having a higher commodity value, and used in various industries. The most suitable is the production of sheets and films of PVC pipes and profiles [6].

Production of PVC is possible on the basis of industrial cooperation and vertical integration of enterprises of Kuzbass within the frames of the functioning of the coal chemical cluster. In the production chain should be involved, given the characteristics of the production the following companies: coal companies, coal mining coke coal comprising the process enriching enterprises, JSC "Koks", OJSC Khimprom, Kemerovo state district power station and other power generating companies that are part of the Siberian generating company, LLC "Gazprom Mining Kuznetsk." To the chain can be aligned the enterprises of small and average business manufacturing the finished product from the powder (Fig. 1).

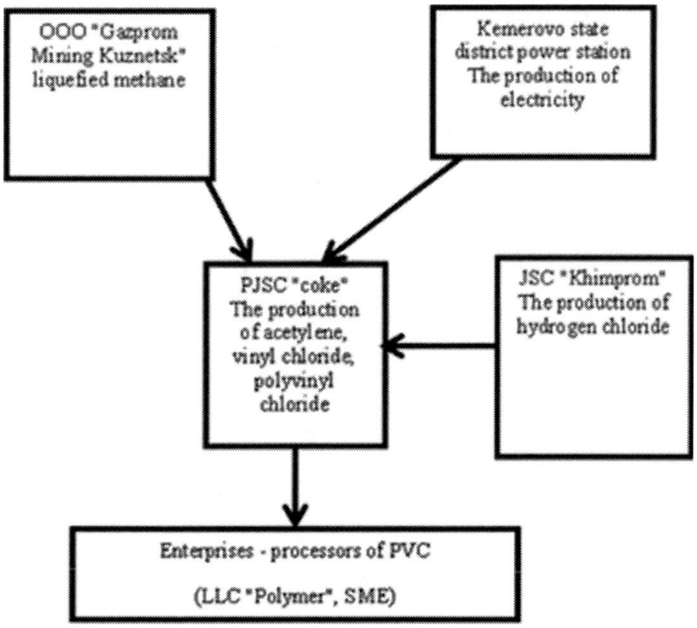

Fig. 1. The generating process PVC production based on vertical integration of enterprises of the Kemerovo region.

These companies will be involved in the process of creating value added chain for the entire process chain, from extraction process to the finished product (services) on the market. The production of such complex product, like PVC, will allow to create downstream value chains that appear around the production and export of innovative, high-tech final goods and services produced by the enterprises of the Kuznetsk basin [7].

The development of coal chemical cluster engaged in extraction and processing of raw materials will generate value-added chain in the Kuzbass, which may be a catalyst of process of economic diversification of the economy of the old coal region and individual industries of the Kemerovo region [8-10].

4 Conclusions

It is established that the organization of production of polyvinyl chloride on the basis of production of PJSC "coke," coal enriching plants using coal chemical technology can be a promising growth point for the economy of Kuzbass, part of a data bank for promising investment projects for potential and existing market area businesses. This will provide some sectors of the economy of Kuzbass with such important materials as PVC, to reduce its import dependence, diversify the production of coal, chemical and metallurgical enterprises, broaden the product structure with the introduction of the polymer and the export opportunities of the region and thus to improve the economic performance of the business, to increase tax revenues to the budgets of different levels. Rational and optimal choice of aggregate technology value chains will allow ensuring the sustainable development of the region's economy to ensure its competitiveness relative to the new coal Russian Far Eastern regions on the principles of rational use of the subsoil

References

1. M. Schiller, *PVC Additives: Performance, Chemistry, Developments & sustainability* (London: Hansen Gardner Publications, 2015)
2. M. Biron, *Industrial applications of renewable Plastics: Environmental, Technological & Economic advances* (Plastics Design Library) (Chicago: William Andrews, 2016)
3. Ye Liping, *Life cycle assessment of polyvinyl chloride production & it's recyclability in China* (Beijing, HDGT, 2010)
4. S. A Pazur, Plast. Eng. **34:1**, 62 (1978)
5. C. Chauffoureaux, C. *Dehennau, Flow and thermal stability of Rigid PVC* (New York, November, 1977)
6. P. Rice, U. Adam, *Developments in PVC Production and Processing* (London, Chem,1977)
7. T. J. Sturgeon, IDS Bull, **32:3**, 9-18 (2001)
8. S. Zhironkin, M. Gasanov, G. Barysheva, K. Kolotov, O. Zhironkina, E3S Web of Conf., **15**, 03012 (2017)
9. F. Agafonov, A. Genin, O. Kalinina, O. Brel, O. Zhironkina, E3S Web of Conf., **15**, 04011 (2017)
10. M. Erkayaoğlu, N. Demirel, J. Env. Man., **174**, 1-6 (2016)

The Environmental Impacts Of The Coal Industry

Sergey Burtsev[1*], *Viktor* Efimov[1], and *Tatyana* Korchagina[2]

[1] AO HK "SDS-Ugol", 650066, 7/2 Pritomsky ave., Kemerovo, Russia
[2] OOO Mining Engineering Institute of Siberia, 650066, 7/2 Pritomsky ave., Kemerovo, Russia

Abstract. The article analyzes the main indicators of the Russian coal industry's impact on the environment: the atmosphere, hydrosphere and lithosphere for the year 2016 compared with 2015. It also identifies actual values of the environmental safety targets of the coal industry and the main investments in the nature protection. For prevention, restriction and minimization the negative impact on the environment, the coal industry has proposed the packages of managerial and technical measures.

1 Introduction

Hundreds of thousands tons of polluting substance are emitted into the atmosphere [3, 4], hundreds of millions cubic meters of polluted manufacturing water are discharged into water bodies [5], thousands of hectares of agricultural land are seized [6] by the coal-mining enterprises of Russia every year.

Coal production directly damages the Earth's interior. This damage is closely related with geological exploration, construction, mining and other works that lead to a change in the vegetation cover, hydrogeological regime of the terrain, the formation of anthropogenic relief and other qualitative changes in the land [7, 8].

Thanks to the analysis of literature sources, official reports about the state and protection of the environment, etc. [9-14] the authors could highlight the main environmental problems in the coal industry and formulate measures to reduce its negative environmental impact.

2 Materials and methods

Effects on the Atmosphere

In 2016 the industrial enterprises spewed into the atmosphere 1007.8 thousand tons of pollutants, including 879.5 thousand tons of hydrocarbons (mine methane). Compared to 2015 the emissions increased by 0.3%. With coal production rising by 3.3% in 2016, specific emissions were reduced by 3.3%. Reduction of specific pollutant emissions into the atmosphere was achieved through the implementation of the following environmental measures at the industrial enterprises (Table 1).

* Corresponding author: s.burtsev@sds-ugol.ru

Effects on the Hydrosphere

437.6 million m3 of sewage were discharged into surface water bodies in 2016, of which 333.2 million m3 were contaminated (exceeding regulatory requirements). The spillover of polluted wastewater increased by 7.8% compared to 2015, the specific spillover per ton of coal production increased by 3.6%. A significant spillover of polluted manufacturing water is caused by the absence of sewage treatment facilities on part and the ineffective operation of existing mechanical treatment facilities. A positive result is a reduction in the spillover of polluted manufacturing water without preliminary cleaning by 13.2 million m3 or by 9.6%. The achieved result was obtained thanks to the construction of new and modernization of existing treatment facilities based on the using of modern technologies in the mines to the A.D. Rubana, "Kotinskaya", "Taldinskaya-Zapadnaya-1" (AO "SUEK-Kuzbass"), AO "Ugolnaya kompaniya Yushnaya", AO "Meshdurechye" etc. [1,2].

Effects on the Lithosphere. 5161.4 hectares of land were destroyed by coal industry in 2016, and 964.2 hectares were remediated. Compared to 2015 the area of disturbed land decreased by 1148.6 hectares or by 18.2%. The area of the remediated lands increased by 141.9 hectares or by 17.3%. The increase in the area of reclaimed lands in 2016 compared to 2015 led to an increase in the level of reclamation from 13.0% to 18.7%. Disturbed and discharged lands that can be found in most enterprises and that are unusable now enable to increase the scale of remediation (Table 1).

The coal industry generated 3236.6 million tons of waste in 2016. 1690.1 million tons of it were used and 1526.9 million tons were disposed in external waste dumps for the various production purposes. The amount of generated waste compared to 2015 increased insignificantly - by 1%, the specific indicator of waste generation per ton of coal production decreased by 2.3%. The volume of used waste increased by 191.2 million tons or by 12.8%. The amount of waste disposed in external waste dumps decreased by 452.2 million tons (23%) (Table 1).

Table 1. Indicators of the negative environmental impact of the coal industry over 2016 compared to 2015

Sequence number	Indicators	Units	Report	
			2015	2016
1	Volume of coal mining	million tons	373.3	385.7
2	Pollutants released, total including:	thousand tons	1004.5	1007.8
	solids	thousand tons	54.9	58.4
	sulfur dioxide	thousand tons	9.3	10.4
	carbon monoxide	thousand tons	35.1	36.0
	nitrogen oxides (in terms of NO2)	thousand tons	14.7	18.8
	volatile organic	thousand tons	2.4	3.7
	hydrocarbons	thousand tons	887.3	879.5
3	Water reclaimed	million m³	36.2	35.3
4	Water used, total	million m³	104.5	99.6
	including fresh water	million m³	38.1	40.1
5	Water recycling	million m³	329.1	305.0
6	Water reuse	million m³	10.3	11.5
7	Wastewater discharged, total	million m³	425.3	437.6

 DOI: 10.1051/e3sconf/20172102017

8	Total volume of wastewater requiring treatment	million m³	397.4	410.5
9	Volume of organized discharge of pollutants into surface water bodies	t	177.852.1	129.033.3
10	Polluted wastewater discharged (without treatment)	million m³	137.4	124.2
11	Polluted wastewater discharged (insufficiently purified)	million m³	171.6	209.0
12	Regulatory clean water discharged (without treatment)	million m³	27.9	27.1
13	Regulatory purified water on the purification facilities was discharged:	million m³	88.4	77.3
	on biological treatment facilities	million m³	0.5	0.8
	on physical and chemical cleaning facilities	million m³	26.4	36.1
	on mechanical cleaning facilities	million m³	61.5	40.4
14	Capacity of treatment facilities before discharge into surface water bodies	million m³	691.6	773.5
15	Waste			
	at the beginning of the year, total	million tons	20.241.8	21.667.8
	at the end of the year, total	million tons	21.667.8	21.029.1
16	Waste generated per year	million tons	3.204.8	3.326.6
17	Waste production disposed in external mine dumps	million tons	1.979.1	1.526.9
18	Waste used, total	million tons	1.498.8	1690.1
19	Waste at the enterprise neutralized, total	t	1.034.9	427.4
20	Waste transferred to third-parties, total including:	thousand tons	35.472.9	365.233.0
	for waste neutralization	t	42.0	58.2
	for use	thousand tons	33.292.0	13.996.6
21	Area of disturbed land:			
	at the beginning of the year	ha	101.800	107.500
	at the end of the year	ha	107.500	111.200
22	The area of disturbed land per year *	ha	6.310.0	5.161.4
	Area of remediated land per year	ha	822.3	964.2
23	Environmental protection capital investments , total including:	thousand rubles	1.095.730.2	902.497.8
	in collection and treatment of wastewater	thousand rubles	780.523.2	511.623.0
	in air protection and prevention climate changes	thousand rubles	2.677.7	10.545.8
	in land protection from production and consumption of wastes	thousand rubles	10.0	37.535.0
	in protection and rehabilitation of land	thousand rubles	3.282.6	18.563.0

 DOI: 10.1051/e3sconf/20172102017

24	Current expenditure on environmental protection, total	thousand rubles	1.229.071.9	1.165.003.0
	including: collection and treatment of wastewater	thousand rubles	462.405.5	535.337.4
	air protection and prevention climate changes	thousand rubles	432.494.2	359.943.0
	land protection from production and consumption of wastes	thousand rubles	148.576.6	154.164.7
	protection and rehabilitation of land	thousand rubles	147.076.2	82.775.2

* The area of disturbed land over a year does not correspond to the difference in the area of disturbed land at the end of the year and at the beginning of the year, due to the inventory of land that occurs during a year at some enterprises or due to the transfer to other enterprises.

The actual values of the environmental safety targets of the coal industry are reflected in table 2.

Table 2. Actual values of the environmental safety targets of the coal industry

Target	For the program 2015	Fact 2015	For the program 2016-2020	Fact 2016
Specific discharge of polluted sewage into water bodies, m^3 / t of mining	1,1 - 1,2	0,8	0,7-0,8	0,86
The level of polluted sewage discharge to the total discharge volume, %	80 - 85	72,7	70-60	76,1
Coefficient of water cycle	0,73	0,77	0,8-0,85	0,78
Specific emission of pollutants into the atmosphere, kg /t of mining	2,6 - 2,7	2,7	2,4-2,5	2,6
The emission's level of pollutants into the atmosphere in the total volume of educated, %	60 - 65	53,5	45-55	58,5
The level of land remediation compared with annual damage, %	40	13,0	65-70	18,7
Specific volume of waste generation, t / t of mining	5,9 - 6	8,6	5,8-5,9	8,4
Level of waste production disposal in external mine dumps, % of generated	42 - 43	61,8	35-40	47,2
Multipurpose use of coal production's waste, %	10	46,8	13	52,2

Investments for the nature protection. In 2016 27 industrial enterprises invested in the environmental protection from their own resources in the total amount of 902,497.8 thousand rubles, 511 623.0 thousand rubles of which were meant for collection and treatment of wastewater, 37 535.0 thousand rubles for land protection from production and consumption wastes, 18 563.0 thousand rubles for protection and rehabilitation of lands, and 10 545,8 thousand rubles for protection of atmospheric air and prevention of climate change. Compared to 2015, investments fell by 18%.

3 Conclusions

In order to prevent, limit and minimize the negative impact on the environment, two main sets of organizational and technical measures are recommended.

A complex of organizational and technical measures to improve the state of the environment and its protection quality:

• introduction of BAT, technological processes and mining and transportation equipment designed to reduce emissions and discharges of pollutants, waste generation volumes and mining capacity of mining operations;

• construction of wastewater treatment facilities and equipping stationary sources of emissions into the atmosphere at new mines, sections and concentrating mills based on new modern technologies;

• equipping sources of emissions and discharges without cleaning facilities, efficient gas treatment facilities and treatment facilities;

• reconstruction of inefficient wastewater treatment plants, gas treatment plants and dust suppression facilities, and other environmental facilities at operating enterprises;

• regular examination of existing environmental facilities and equipment with an assessment of their technical condition, the achievable efficiency, the development and implementation of measures to improve their work;

• implementation of permanent technological control over the technical condition and efficiency of wastewater treatment plants, gas purification plants, waste dumps and other environmental protection facilities, timely taking measures when technological regimes of their operation are violated;

• advanced training of maintenance personnel of environmental facilities, managers and specialists of environmental services of coal companies and enterprises on the basis of scientific centers.

A complex of industrial and technical measures to improve the state of the environment and its protection quality.

For the air protection:

• transition from single-stage and two-stage systems of cleaning off-gas from coal-fired boiler plants and drying plants of concentrating mills to two-stage and three-stage systems respectively;

• replacement of dry cleaning devices for gases (cyclones, battery cyclones) for more efficient wet dust collectors, fabric filters and electrostatic precipitators;

• choosing the ways to reduce the formation and release of dust into the atmosphere during technological processes of mining production, and the parameters of dust suppression equipment operation, taking into account the dust-forming ability of coal;

• use of surfactant's solutions, dusting, dust-binding compounds and foam in the water systems instead of water;

• economic incentives and benefits for enterprises engaged in degassing and utilization of methane (concessional lending, cancellation of MET, etc.);

• development of methods, technologies and technical means for extracting and using methane from vent streams during ventilation of gas mines

For the water conservation:

• adopting effective cleaning methods in the projects of construction and reconstruction of mine and quarry water treatment plants: thin settling, clarification in a layer of suspended sediment and pressure filtration with the use of chemical reagents, electrocoagulation-flotation, filtration, disinfection with ultraviolet radiation, sludge dewatering to tape and chamber filters presses;

• adopting methods of dry enrichment with the use of spiral separators and pneumatic float machines, enrichment technology with thickening of slimes and deep clarification of slime waters using hydrocyclones and radial thickeners, and dehydration of flotation waste by filtering it using filter presses at the undergoing and reconstructed concentrating plants;

• increase in the use of treated mine and quarry water instead of drinking water for production needs of own and related enterprises, including replenishment of recycling water supply systems;
• development of economically acceptable methods and technologies for purification of acidic (pH 2-5) and mineralized (with salt content over 2-3 g / dm3) mine waters with utilization of formed precipitation and brines.

For the land conservation:
• development of remediation projects based on complete initial data on mineralogical, granulometric, chemical, microbiological composition, agrophysical and agrochemical properties of regenerated soils and rocks, taking into account the landscape approach and natural and climatic conditions;
• taking measures to prevent spontaneous combustion of rock dumps prone to spontaneous ignition during remediation projects;
• development on commercial scale of new methods and technologies for the restoration of biological productivity of disturbed, degraded and polluted lands, in particular the microbiological method of accelerated remediation and the method of biological remediation of rock dumps by the Institute of Agrarian Problems of Khakassia;
• proceeding with the development and industrial testing of new methods and technologies for technical and biological remediation, special machines and equipment for removal and deposition of PSP and PPP, hydroseeding on slope surfaces, mechanized planting of trees and shrubs.

For the waste management:

• a detailed study of the current production waste and the rock mass accumulated in the dumps, with the definition of composition and properties, the content of associated minerals and valuable elements, the determination of their reserves and state registration;
• increase in the use of coal mining and preparation waste in construction, in building materials production, for the extraction of coal and valuable elements, in the production of fuel briquettes;
• improvement of existing and development of new efficient methods and technologies for processing waste products into commodity products with obtaining energy fuel and ash residue;
• reducing the volume of waste generation by improving mining technology, including rational cutting of mine fields and the sequence of their mining, optimizing the technological parameters of overburden and mining operations, maximum placement of overburden and enclosing rocks in the worked out area of cuts and mines;
• application of fire-safe technologies for their formation on existing and planned waste rock dumps;
• monitoring of the thermal state of rock dumps in order to identify hotbeds of self-heating and spontaneous combustion and the adoption of operational measures to eliminate them;
• extinguishing burning rock dumps by transforming conical dumps into flat dumps, carrying out preventive measures and subsequent remediation.

References

1. *Official report "About the state and protection of the environment of the Russian Federation in 2015"*, 639 (Ministry of Natural Resources of Russia, NIA-Priroda, M, 2016)

2. N.M. Kachurin, L.L. Rybak, V.I. Efimov, S.A. Vorobiev, Industrial safety, **2**, 24 (2015)
3. N.M. Kachurin, L.L. Rybak, V.I. Efimov, S.A. Vorob'ev, Industrial safety, **3**, 36 (2015)
4. V.I. Efimov, R.R. Minibaev, T.B. Korchagin, S.A. Svinarenko, Coal, **6**, 62 (2017)
5. V.I. Efimov, L.V. Fisherman, *Industry and the environment* 336 (MGGU, M, 2012)
6. V.G. Gridin, V.I. Efimov, *40 environmental questions*, 168 (MGGU, Moscow, 2007)
7. V.G. Gridin, V.I. Efimov, *Industry and the environment: lectures: " Industry and the environment"*, (MGGU, Moscow, 2007)
8. V.I. Efimov, *The Fundamentals of Nature Management: tutorial*, (MGGU, Moscow 2004)
9. V.I. Efimov, R.R. Minibaev, T.B. Korchagin, Ya.A. Novikova, Coal, **1**, 66 (2017)
10. V.I. Efimov, R.R. Minibaev, T.B. Korchagin, Ya.A. Novikova, Mining information-analytical bulletin, **1**, 85 (2017)
11. VI. Efimov, S.M. Popov, P.M. Fedyaev, N.V. Efimova, *Methodological bases of the ecological and economic evaluation of ways of involving coal waste into economic activities of the regions*, 121 (Tula State University, Tula, 2016)
12. V.I. Efimov, RV Sidorov, T.V. Korchagin, Coal, **12**, 90 (2014)
13. V.I. Efimov, T.V. Korchagin, *Ecological and economic analysis of the impact of coal enterprises on the environment*, 173 (MGGU, Moscow, 2006)

Stimulation Of The Methane Production With The Use Of Changing Of The Rock Massif Physical Conditions

Mikhail Baev[1*], *Veniamin* Khyamyalyaynen[1], and *Aleksandr* Shevtsov[1]

[1] T.F. Gorbachev Kuzbass State Technical University, 650000, 28 Vesennyaya St., Kemerovo, Russia

Abstract. The commercial coalbed methane production success is majorly defined by the effectiveness of the use of special gas inflow stimulation methods. The necessity of using of such methods is subject to the aspects of searching and displacement of methane within the coal compound. The analysis of the ways of methane production stimulation from virgin coal formations is given. The description of the process of hydraulic fracturing (fracturing) as the most common stimulation method during the commercial coalbed methane production as well as its major advantages are presented. The present work provides data about the initiated laboratory research of sands collected from Kemerovo region deposits for the purpose of finding of the most prospective samples by means of anchoring of fractures. The prospectivity and ability to implement the hydraulic fracturing with the use of locally available sands acting as proppants are shown. The influence of the strain-stress state of the rock massif on the alteration of permeability and the necessity of its extension study with respect to different technological features of hydraulic fracturing is shown.

1 Introduction

Today's world economy is characterized with the high level of energy consumption. The development of fuel and energy complex and metallurgy, as well as oil processing and chemical industries is based on the wide using of different types of mineral resource produced from the earth interior. In the capacity of fuel and energy material, oil, natural gas and coal are the most widespread. Growing human demand on the abovementioned resource leads to the continuously increasing of its production. Gradual depletion or immense decreasing of reserves in accessible deposits culminated in higher complexity and the cost of production. Due to that, the one of prior tasks is the development of unconventional hydrocarbon resources, such as: heavy oil and oil sands (bituminous sand), coalbed methane, shale gas and oil, tight gas, gas hydrates and water dissolved gases. Coalbed methane (CBM) shows the number of differences from other unconventional resources.

At the present moment, the unique for Russia project for commercial methane production from the virgin coal seams is being implemented in Kemerovo region. The major issue of the commercial methane production, unlike the conventional natural gas production,

* Corresponding author: bma.gdk@gmail.com

centers around the features of the methane storage within the coal bed. According to existing concepts, methane in coal deposits presents in different states. Upon that, up to 80-95% of the total volume consists of the tied methane, and its major volume is presented as the solution within the hard matter (sorbed methane). Along with this, in the natural conditions of occurrence of coal seams, the process of slow diffusion-induced transfer of methane through coal matter predominates, i.e. the gas transfer from the formation is extremely slow [1, 2]. These particular features of bedding and transfer of methane within the coal matter predetermine the necessity of using of special methods, such as: hydraulic fracturing (the fracking), caving, directional drilling, pneumatic and hydrodynamic impact, well's reactant treatment, wave exposure on the seam, injection into coal seams of helium, nitrogen and carbon dioxide, plasma impulse excitation, foam-nitrogen hydraulic fracturing and so on.

2 Materials and methods

For the effective implementation of stimulation methods the understanding of physical processes occurring within the rock massif is primarily required. Aside from that, the Kuzbass coalbed methane fields development process showed the necessity of methodological adaptation with respect to the features of exact deposits as well as optimizing of economical expenses for their implementation.

In the world's practice of wells performance stimulation, both conventional oil and gas deposits and commercial coalbed methane production, hydraulic fracturing (fracking) is widely used [1-4]. During the process of fracking the special technological fluid is injected into the seam under high pressure, enough for fractures to occur. At the next fracking stage grainy materials – proppants (propping agents or fixing materials) are added to the fracturing fluid. The proppant allocates in fractures and holds them open (wedging them) after the process is finished. As the result of fracking, the created system of fractures deeply penetrating into the seam allows, first of all, to enlarge the seam's surface area connected to the well (well's drainage area), among other things due to the creation of the connection with the natural fractures system not touched with the well. Second, wedged highly-permeable fractures act as the passage ways helping the migration of gas from the seam to the well. Almost in all the cases, the major portion of the production comes from the seam into fractures and after that into the well. And third, due to the increasing permeability conditions for effective underground water drainage are created, which allows to get the maximal possible depression on the coal seam with the aim of enabling the process of gas desorbing.

Consequently, the well performance is connected with the fractures geometry and their residual (effective) permeability. In the work [5] general factors affecting the geometry of fractures created after the fracking are described. Main factors affecting the residual permeability of fractures are subsurface conditions (rock pressure, mechanical properties of rocks, temperature) and proppant's features.

Subsurface conditions are best characterized with the stress-strain state of the examined rock massif. During the long period of geological time the rocks lying within the earth's crust, are exposed to many natural forces and factors. As the result of such impacts within the massif the natural stress-strain state is created. Any human impact (like construction of the well, fracking, well production or extraction of coal) leads to the alteration of the natural stress-strain state. The control of the stress-strain state of the rock massif or the formation properties as the result of the fracking is a quite complicated task. However, during the implementation of the fracking understanding of the stress-strain state allows to find the most prospective intervals for the fracking, as well as to control the fractures growth and to forecast their orientation. Even after the fracking and during the whole period of the well's operation the stress plays an important role for the underground fluid production affecting the reservoir properties. It is proved that the porosity and permeability of the rock massif are decreasing with the growth of the massif stress [6], which works greatly for coal.

The important task during the fracturing designing is the proper proppant selection, which can allow the necessary permeability of the wedged fracture. During the selection of the exact propping material it is necessary to take into account its major physical properties: the strength, grain size and granulometric texture, the quality (the presence of compounds, acid solubility), grain shape (spherical shape factor, roundness) and the consistency. The special emphasis is to be paid on the prevention of the closure of the fixed fractures and providing of the required permeability. As the proppant artificial ceramic materials and natural quartz sands are used. Most of the frackings, including the coalbed fracking, are done with the standard quartz sand sorted by size [1-4, 7-10]. This is largely due to its availability, relatively low price and applicability at different underground conditions. The sand shows quite high strength to stand the closure stress during the fracking of seams bedding at the depths up to 2400 – 2500 m. Coal seams currently used for coalbed methane production in Kuzbass are bedding at the depths up to 1350 m, thus the using of the sand during the fracking is a quite practical solution. Upon that, in order to decrease the proppant cost, the using of local sands is obviously practical. For that purpose it is necessary to prove the feasibility of its using for wedging the fractures during the methane production from virgin coal seams.

3 Results and discussion

Kemerovo region shows the rich material base of sands. There are nine explored foundry sand deposits (reserves are estimated 214 million tons), as well as three glass sand ones (144 million tons), six masonry sand ones (35 million m^3) and thirty deposits of the sand-and-gravel material (189 million m^3). Along that, there is a number of deposits not accounted for by the state balance sheet. Thus, the purpose of the present work is searching of natural sand deposits of Kemerovo region, which can be used as propping agents during the fracturing of coal seams. For research samples of twenty two sand and sand-gravel deposits of Kemerovo region, as well as three deposits of Tomsk region and two Altai Krai deposits have been collected.

As it was mentioned before, the permeability of fractures is directly connected with wells performance. Whereas there is a link between the permeability of the grain material and its porosity (cavernosity) – as a general rule, the permeability is higher with the higher porosity. One of the factors affecting the porosity is the homogeneity or the sorting of grains by size [11]. If the small-sized particles are mixed with bigger sand grains, the effective porosity (represented by the pores connected to each other) will gradually decrease. Thus, for wedging of fractures proppants of defined grades are used, i.e. properly sorted material. Along with that, the permeability of the wedged fracture is greatly influenced with the grains breaking down within the fracture caused by the compression strain. The breaking down intensity relies upon the grain size, its shape, invisible defects, temperature, chemical compound and so on. Grains breaking down leads to the alteration of granulometric texture of the proppant within the fracture and, consequently, to the decreasing of its porosity and permeability. Thus, during the selection of proppant it is necessary to take into account the ability of its grains to withstand the destruction.

The compound and physical properties of untreated sand have been investigated, which included the analysis of the presence of flour and clay particles, as well as granulometric texture and the modulus of size, density and cavernosity of the unwiped and wiped samples. Also, research of the treated sand based on the ISO 13503-12 standard [12] is done for defining of the density and cavernosity, as well as the crush resistance [13].

As for the permeability measuring of the wedged fracture, the following aspects should be taken into account. The residual (or effective) permeability of the wedged fracture within the coal seam can gradually differ from the initial proppant's permeability factor, provided by the supplier. This difference can be explained with the number of causes, such as physical-mechanical properties of coal and bedding properties of the coal seams. These causes are not fully taken into account in the standard method for measuring of the proppant's conductivity in accordance with ISO 13503-5 [14], while the information about its influence on the filterability of wedged fractures at coal seams is insufficient.

By that reason during the research of sands for measuring of permeability it is suggested to use the original laboratory unit developed in KuzSTU. Its major feature is working with samples manufactured from the material comparable to coal which are used instead of the standard laboratory samples of the sandstones [5]. First of all, research for measuring of the fracture's permeability factor wedged with the proppant of given permeability (i.e. measured by the standard unit) has been implemented. Such material is the ceramic aluminosilicate proppant Borprop of the fraction 20/40. The data of its permeability is given at the manufacturer's website. It allowed to compare the given official information with data obtained at the assembled laboratory unit with the use of samples prepared from the equivalent material. The experiment results are shown on the drawing 1.

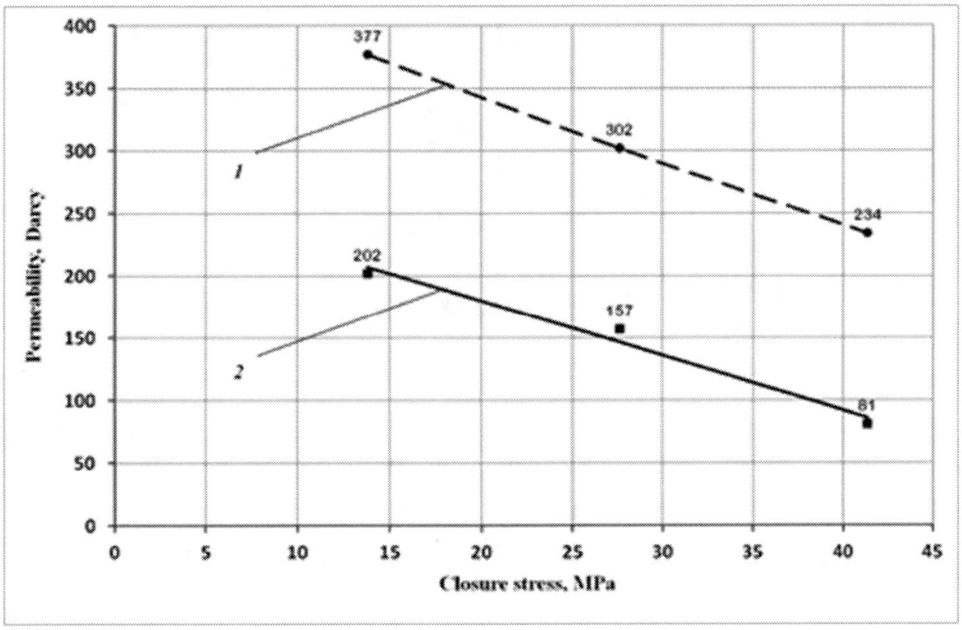

Fig. 1. Dependence of the permeability of the fracture wedged with Borprop 20/40 proppant on the closure stress: 1 – initial permeability; 2 – data obtained at the assembled laboratory unit.

The drawing shows that permeability of the wedged fracture in the coal seam is 2-3 times lower than the initial permeability. This fact shows the influence of coal seam properties on the residual permeability of the wedged fracture.

The world's practice of the strain-stress state's research is widely using the geomechanical modelling. The geomechanical model can be analytical, numerical or the physical one depending on the type of processing of data. The most handling are numerical models which are more accurate and more complex than analytical ones while the physical models are even more acute, however demanding more time and expenses. The key components of numerical models are: vertical stress, minimal horizontal stress, maximal horizontal stress and its direction, as well as the porous pressure and mechanical properties of the rock mas-

DOI: 10.1051/e3sconf/20172102018

sif. Among the existing methods of measuring of the strain-stress state (finite element method, finite difference, boundary element method and so on) the finite-element technique and the finite difference technique can be accentuated, which are used in the majority of today's software suites for engineering measurements. Upon that, the leading role is given to the finite element method, which is explained with the ability of creation of the geometry of any complexity, and the resulting granularity degree [15]. Thus, during the modelling of coal seams with fractures (the opening of which can reach the value of several millimeters) the achievement of results properly describing the actual stress distribution is impossible without getting of the proper degree of detalization.

Results of the geomechanical modelling of the rock massif on the example of Yerunakovsky geological and economic region of Kuzbass in form of the stress patterns are given on the drawing 2.

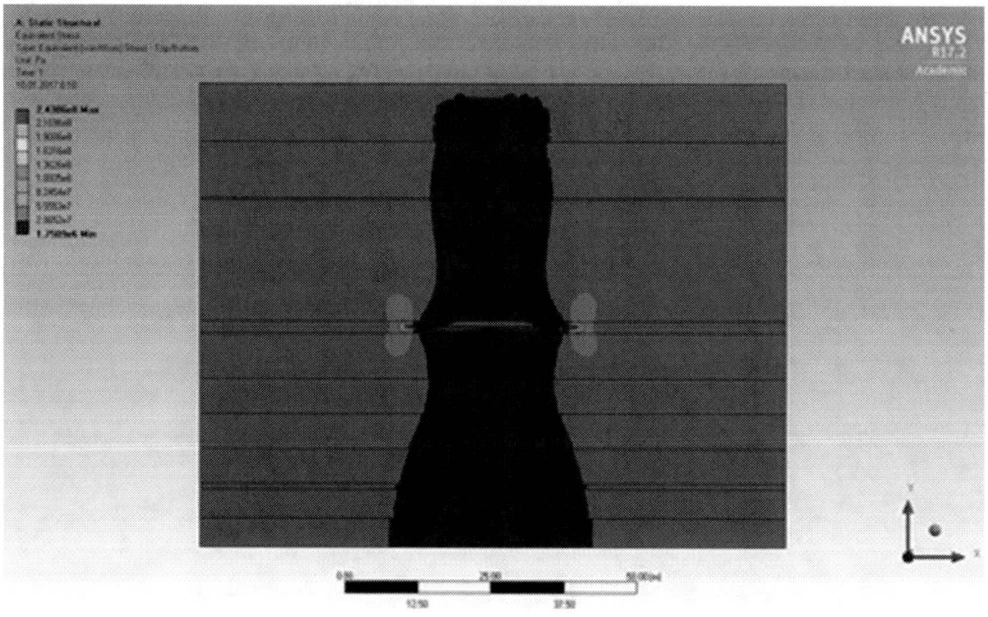

Fig. 2. Distribution of equivalent stresses around the coal seam with the horizontal hydraulic fracturing's resulting fracture of 30 m length and 10 cm opening.

Results show that fractures of the same geometry at different reservoir properties will lead to the different alteration of the natural strain-stress state. Consequently, the hydraulic fracturing requires the individual approach in each particular case.

4 Conclusions

Results of the abovementioned analysis of abstract theorems and research of the possibility of stimulation of coalbed methane production from the virgin coal seams prove the prospectivity of using and improving of the hydraulic fracturing technique with respect to the alteration of strain-stress state of the rock massif around the newly-created fractures. The using of local sand in Kuzbass as the proppant for hydraulic fracturing's resulting fractures appears to be prospective as well.

References

1. R. E. Rogers, K. Ramurthy, G. Rodvelt, M. Mullen, *Coal Bed Methane: Principles and Practices. 2nd ed*, 504 (Starkville, MS: Oktibbeha Publishing Co., 2007)
2. S. Schatzel, K. Aminian, *Coal Bed Methane: From Prospect to Pipeline. 1st ed*, 440 (San Diego, CA, Elsevier Inc., 2014)
3. M. J. Economides, R. E. Oligney, P. Valkó, *Unified Fracture Design: Bridging the Gap Between Theory and Practice*, 262 (Alvin, TX, Orsa Press, 2002)
4. M. J. Economides, T. Martin, *Modern Fracturing: Enhancing Natural Gas Production*, 509 (Houston, TX, Energy Tribune Publishing Inc., 2007)
5. M. A. Baev, Proceedings of the Taishan Academic Forum – Project on Mine Disaster Prevention and Control, 361(2014)
6. I. Palmer, J. Mansoori, SPE Reservoir Evaluation & Engineering, **1:6**, 539 (1998)
7. V. A. Hollub, P. S. Schafer. *A Guide to Coalbed Methane Operations*, 376 (Chicago, IL, Published by Gas Research Institute, 1992)
8. M. J. Economides, A. D. Hill, *Christine Ehlig-Economides. Petroleum Production Systems. 1st ed*, 624 (Upper Saddle River, NJ, Prentice Hall PTR, 1994)
9. F. Gray, *Petroleum Production in Nontechnical Language. 2nd ed*, 288 (Tulsa, OK, Penn Well Publishing Company, 1995)
10. M. J. Economides, K. G. Nolte, *Reservoir Stimulation. 3rd ed*, 856 (New York, NY, John Wiley & Sons Inc., 2000)
11. E. C. Donaldson. *Petrophysics: Theory and Practice of Measuring Reservoir Rock and Fluid Transport Properties. 3rd ed*, 976 (Waltham, MA, Gulf Professional Publishing, Elsevier Inc., 2012)
12. *ISO 13503-2:2006. Petroleum and natural gas industries – Completion fluids and materials – Part 2: Measurement of properties of proppants used in hydraulic fracturing and gravel-packing operations*, 28 (2006)
13. M. A. Baev, V. A. Khyamyalyaynen, Proceedings of the 8th Russian-Chinese Symposium , 330 (2016)
14. *ISO 13503-5:2006. Petroleum and Natural Gas Industries – Completion Fluids and Material.* **5**, 25 (2006)
15. M. D. Zoback, *Reservoir Geomechanics. 1st ed*, 461 (Cambridge, Cambridge University Press, 2010)

Increasing the Reliability of the Work of Artificial Filtering Arrays for the Purification of Quarry Waste Water

Maxim Tyulenev[1,*], *Yury* Lesin[1], *Oleg* Litvin[1], *Elena* Maliukhina[2], and *Asmelash* Abay[3]

[1] T.F. Gorbachev Kuzbass State Technical University, 28 Vesennyaya st., Kemerovo, Russian Federation, 650000
[2] National Mineral Resources University (University of Mines), Faculty of Mining, Department of Surveying, 21 Line 2, St. Petersburg, Russian Federation, 199106
[3] Department of Earth Science, College of Natural and Computational Sciences, P.O.Box.3066, Mekelle University, Mekelle, Ethiopia

Abstract. Features of geological structure of the Kuznetsk coal basin stipulate the application of a low-cost open technique of coal mining, which is more advantageous both from the economic standpoint, and by safety criteria of mining. However, open mining affects significantly the water resources of region. Intensive pollution of reservoirs and water courses, exhaustion of the underground water-bearing layers, violation of a hydrographic network, etc. be-long to the main disadvantages of an open technique of coal mining. Besides, the volume of the water coming into the mining producers exceeds signi-ficantly the needed quantity. According to the data of annual reports of ecology and natural resources department, 348.277 million m3 of water were ta-ken away during production of soft coal, brown coal and lignum fossil from waters of Kemerovo region in 2013 (mostly from underground water objects (96,5%) when draining of mine openings). At the same time, only 87.018 million m3 of water (25%) has been used within a year.

1 Introduction

Industrial complexes concentrated in the most developed part of the biggest Russian coal basin – Kuzbass (Kemerovo Region, Western Siberia) – have the decisive influence on the ecological condition of environmental objects [1-2]. The decrease of production from the mid-1990s in the coal-mining areas of Kemerovo region has not led to the improvement of its environment. The negative effect made by coal-mining enterprises on the natural environment has complex origination. We observe an intensive pollution as a result of the influence of coal deposits open-pit mining on the atmosphere, water resources and landscape complexes. Complex processes of environmental anthropogenic changes caused by open pits' operations have brought to light the problem of surface water pollution near large mining segments. Currently, the volume of coal production in Kemerovo region amount to

* Corresponding author: tma.geolog@kuzstu.ru

more than 210 million tons per year. According to the report of supervisory environmental authorities in 2015, coal-mining enterprises of Kemerovo region disposed 1.239 billion m3 of wastewater, including inadequately treated – 228.8 million m3.

The main source of pollutants' transportation into the surface waters near the territories of open-pit mines is quarry water, resulting from massive extraction of groundwater of developed coal seams for draining purpose.

The issue of wastewater treatment in open pit mining of mineral deposits causes great ecological problems [3-5]. Due to the constant tightening of the requirements to the quality of discharged wastewater their treatment in settling basins and purifying ponds does not provide the parameters of maximum permissible concentration (MPC). Moreover these treatment facilities occupy large areas of land. Also it should be noted that usually there are several distanced weir wastes, which location is changed during development of mining. For example, JSC "KuzbassRazrezUgol" (Kuzbass, Western Siberia, Russia) has 32 places of water use, taking into consideration that the structure of this coal company consists of 7 large open pits.

2 Materials and methods

As an alternative coal enterprises may use expensive wastewater purifying equipment, which include different devices like hydroclassifiers, electrocoagulators, modules of water clarifying and sludge thickening, modules of sludge dewatering, finalizers, aerators and so on. Complex of such equipment for purifying highly voluminous quarry wastewater may totally cost up to several million dollars. So because of these large expenses at present time even prosperous Kuzbass coal open pits conduct wastewater treatment in artificial filter arrays made of waste coal. But the quality of purifying this way is quite low and does not match MPC.

Taking into account the upward trend in open pit mining of coal in Kuzbass, and as a consequence inevitable growth of the volume of discharged water by the coal enterprises, further study of improvement the environmental safety of mining operations is needed.

3 Results and Discussion

The technical solution to the problem of wastewater treatment from suspended solids in the filters of coarse rocks can be searched in two ways. The first of them is to use the discrete rocks arrays existing at enterprises as filters, after a preliminary examination. Such arrays include overburden rock arrays, refuse heaps and various technological dumps.

The second way is the purification of water in specially constructed filtering arrays of rocks and semi-rocks, which are, as a rule, the mining waste.

The filters' design depends on the terrain and the properties of the upper layers of the ground of the underlying surface, and the parameters are determined by the volume of water supply, its pollution density and filtration characteristics of the filtering material.

The main filter elements are the unit for supplying polluted water, filtering array, filter body, a device for collecting and draining the purified water. If there are any natural or artificial hollows (ravines, logs, river beds of dried-up rivers, trenches, ditches, abandoned mines and others) near the water disposal and low permeable rocks that lie at their sides and bottom, the latter can be the body of the filter, fig. 1. In this case, water may be supplied to the filter array by gravity on a surface or pipe and by pumping. The filter array is piled from rocks that meet the relevant requirements using trucks and bulldozers.

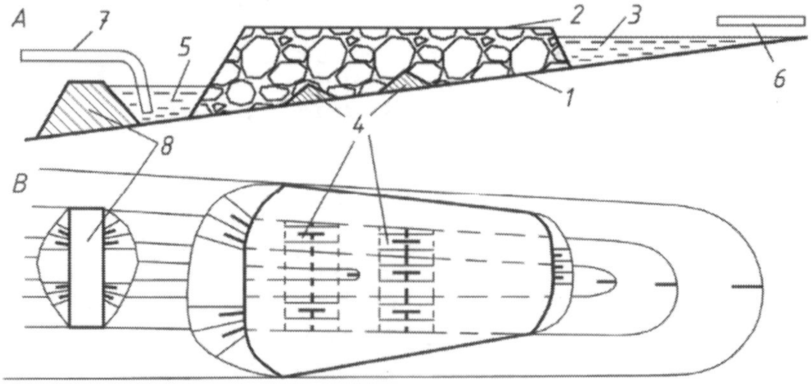

Fig. 1. Construction of the filtering array with waterproof stopping: A – longitudinal section; B – plan view; 1 – the bottom of the filter; 2 – filtering array; 3 – receiver for contaminated water; 4 – waterproof stopping; 5 – lodgment of purified water; 6 – conduit for feeding for contaminated water; 7 – conduit for purified water removing; 8 – water retention levee.

A pond appeared as the result of water infiltration becomes a device for supplying and routing water in the filter. The retaining dam for collection and turning out the purified water must be constructed of the rocks with low hydraulic permeability lower filter array. The disposal of pure water from water reservoir is produced by gravity in chutes or pipes in order to prevent the erosion of soil and new water pollution. When water discharge by gravity is impossible or there is a necessity of its subsequent use for enterprise needs pumping station should be installed.

If rocks of the sides and bottom of the hollows are characterized by high hydraulic permeability, to prevent infiltration of water into the ground and possible pollution of aquifers bottom filter forming is necessary. The bottom is arranged by dumping the low permeable rocks, followed by their layout and thickening by bulldozers. To prevent losses the water for cleaning must be supplied by chutes or pipeline.

Another method of supplying contaminated water to the filtering array is also proposed in this paper. In this case, water is supplied to the additional receiver for contaminated water 9 that created directly in the filtering array 2.

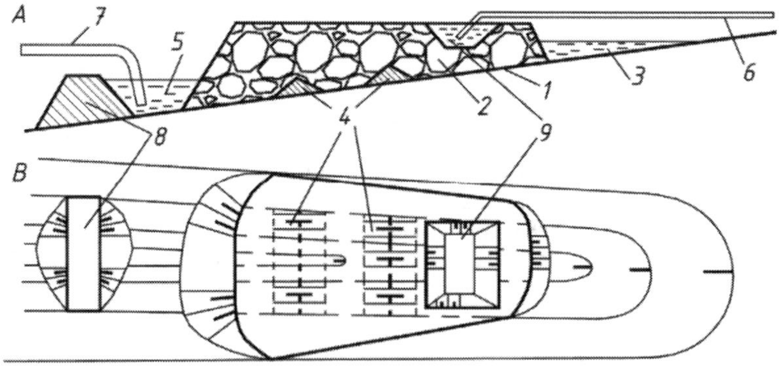

Fig. 2. Construction of the filtering array with waterproof stopping and additional receiver: A – longitudinal section; B – plan view; 1 – the bottom of the filter; 2 – filtering array; 3 – auxiliary receiver for contaminated water; 4 – waterproof stopping; 5 – lodgment of purified water; 6 – conduit for feeding for contaminated water; 7 – conduit for purified water removing; 8 – water retention levee, 9 – additional receiver for contaminated water.

The technology of such massive construction at the initial stage is no different from the previous method, but after the work is completed and the filter body is sealed, the backhoe with a small bucket capacity (and minimum mass, respectively) creates a recess at a safe distance from the top edge of the filter.

The depth and width of this additional receiver are determined in each case individually.

This method has the following advantages:

1) the water entering the treatment is filtered through the upper small-lump layer of the massif, which increases the degree of its purification;

2) the water supplied to the purification is distributed more evenly through the array, which reduces the probability of its untimely siltation.

However, this method has limited application due to the fact that it is not always possible, firstly, to place equipment on the upper platform, and secondly, because of the peculiarities of the terrain that make it difficult to manage and maintain the pipeline of contaminated water.

As an example, Fig. 3 depicts a real array with a potential receiver shown on it.

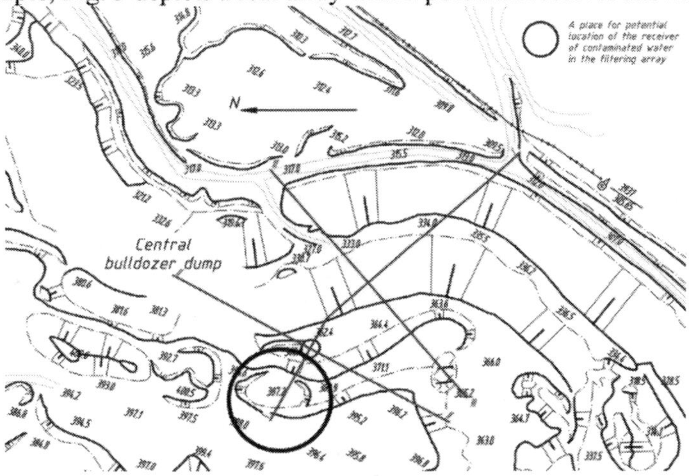

Fig. 3. Scheme of real filtering aeeay with potential place for additional receiver.

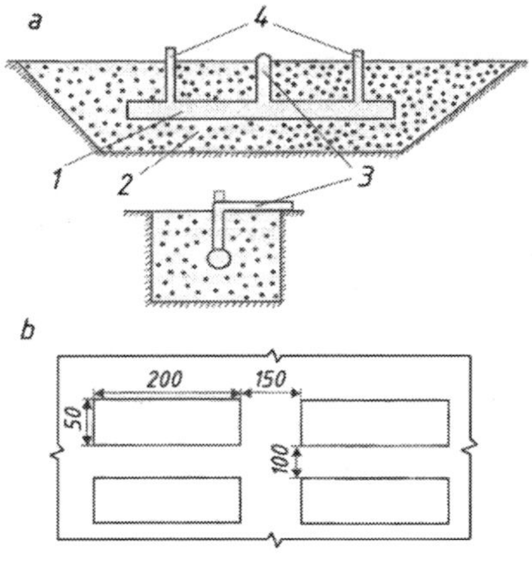

Fig. 4. Constructive sscheme of filtering array.

There is also a method of supplying water to an array similar to the previous one, but without the construction of an additional receiver. In this case, water is supplied through pipes reinforced directly in the body of the massif (Fig. 4). Pipes are equipped with perforation for an even distribution of water, the depth of their laying is usually up to 5 meters.

Such a method guarantees continuous operation of the filter throughout the year and excludes the possibility of freezing the water intake point.

4 Conclusions

Development of effective quarry water purifying technology requires cooperation of technical universities, research organizations and mining enterprises for saving the environment in coal basins. Technology of forming and using filter arrays allows substituting expensive water treatment equipment what was proved by conducted research. In general high perspectives of the new technologies of quarry water treatment implementation at surface mining are expected to expand the limits of using mining machines.

References

1. T.V. Kiseleva, V.G. Mikhailov, V.A. Karasev, IOP Conf. Ser.: Earth Environ. Sci., **45:1**, 012013 (2016)
2. T.V. Galanina, M.I. Baumgarten, V.G. Mikhailov, T.G. Koroleva, G.S. Mikhailov, IOP Conf. Ser.: Earth Environ. Sci., **50:1**, 012030 (2017)
3. M. Tyulenev, Y. Lesin, E.Tyuleneva, E. Murko, E3S Web of Conf., **15**, 02003 (2017)
4. M. Cehlár, P. Varga, Z. Jurkasová, M. Pašková, Acta Montanistica Slovaca, **15:2**, 132-138 (2011)
5. M. Tyulenev, E. Garina, A. Khoreshok, O. Litvin, Y. Litvin, E. Maliukhina, IOP Conf. Ser.: Earth Environ. Sci., **50:1**, 012035 (2017)

Organizational-Legal and Technological Aspects of Ensuring Environmental Safety of Mining Enterprises: Perspective Analysis in the Context of the General Enhancement of Environmental Problem

Elena Vorontsova[1], *Andrey* Vorontsov[1], and *Yuriy* Drozdenko[2,*]

[1] Southwest State University (SWSU)

[2] T.F. Gorbachev Kuzbass State Technical University, 650000, 28 Vesennyaya St., Kemerovo, Russia

Abstract. The article is devoted to the analysis of problems of maintenance of ecological safety of the mining enterprises. The aim of the work was the formulation of proposals, the implementation of which, in the opinion of the authors, is capable of raising the level of environmental safety of the mining industry and ultimately ensuring the environmentally oriented growth of the Russian economy.

1 Introduction

The study is of an interdisciplinary nature. This approach allowed us to assess the problems of the negative impact of the mining industry on the environment in the context of the global environmental crisis and the need to ensure Russia's environmental security.

The authors note the awareness of the global nature of environmental changes by humanity, while pointing to the irreversibility of many natural processes, which confronts mankind as a whole and Russian society in particular with the task of preserving the environment and taking care of natural resources.

For the Russian mining industry, the solution of this task is complicated by the need to intensify production processes, which is due to the needs of our country's economic development and the place of the mining industry in the overall structure of the Russian economy.

The authors draw attention to the limitations of traditional approaches to solving this problem with a comprehensive assessment of the prospective possibilities to reduce the load exerted by mining enterprises on the environment. In their opinion, the improvement of environmental safety of mining enterprises is possible not only through technological innovations (although this is the main direction), but also through a system of organizational, legal and economic measures. As such measures they indicate: the creation of an effective system of environmental monitoring; strengthening the activities of control and supervisory and licensing authorities; introduction of economic and legal mechanisms

* Corresponding author: haa.omit@kuzstu.ru

 DOI: 10.1051/e3sconf/20172102020

that induce owners of industrial enterprises to modernize technological equipment; legislative prohibition of the practice of "temporarily permitted emissions of pollutants".

The realization of global changes that occurred in the natural environment under the influence of man-made impact came to humanity not immediately.For a long time man has been destroying the natural foundations of his existence, carrying out his interaction with nature within the framework of the so-called anthropocentric approach, that is, from the standpoint of a being that is the apex of evolution, and therefore having the absolute moral right to use natural resources exclusively at his own discretion, while forgetting about that that he is only part of a single chain of biospheric processes that occur on Earth.

This "forgetfulness" has led to the fact that, for the last 100 years, the concentration of carbon dioxide, one of the main greenhouse gases, has increased by 12-18% in the atmosphere; the dustiness of the atmosphere increased by 10-20%; the illumination of the Earth decreased by 7-10%; The ozone layer, which protects people from the deadly radiation of the Sun, decreased by 2%. The average annual temperature over the past 30 years has grown by 0.6 degrees; if the temperature increase does not stop - scientists predict that by 2020 the temperature will grow by 2-5 degrees - this will lead to the melting of glaciers, a rise in sea level of 20 centimeters, and by 2100 a rise of 1 meter, and the flooding of huge areas. Already, there is an intensive melting of the ice sheet in Greenland, ice shelves in Antarctica, mountain glaciers in the Alps, the Andes, the Himalayas, which leads to a reduction in fresh water supplies, drying up of mountain rivers that feed on meltwater. The area of tropical forests only in the last 10 years has decreased by 20-30%; forests of Siberia, Canada, the Amazon delta are annually reduced by 20 million hectares; arable lands lose 24 billion tons of fertile layer (especially intensively - India, China, CIS countries due to outdated methods of processing); 70-80% of the world's pastures have already turned into a desert. The World Ocean annually receives 3.5 million tons of oil and petroleum products. About 1.3 billion people use only contaminated water. The Earth's atmosphere annually receives 60 million tons of solid particles; Because of the emissions of sulfur dioxide, acid rains fall out [14].

In the Russian Federation, by the end of the 2000s, natural ecosystems were destroyed in 35% of the territory, 75% of water resources were not suitable for drinking, 56% of agricultural lands were subject to soil degradation [11].

The environmental safety of Russia, like any other country, is determined by the degree of its readiness to prevent threats and risks related to the state of the environment and its impact on the population. This includes factors of chemical and biological safety, because it is associated with the potentially dangerous effects of chemical and biological factors on the environment and the population of the country [7]. In a number of ways, the load on nature has reached critical values. As a result, the annual economic damage reaches 6% of GDP, and taking into account the health consequences of people - up to 15%.

The Yale Center for Environmental Law and Policy published the results of a global study of the world in terms of environmental performance in 2016 (The Environmental Performance Index 2016). Russia ranks 32 out of 180 in the ranking - between Azerbaijan and Bulgaria.

Degradation of the environment poses a significant threat to the population. The list of diseases directly related to ecology, including congenital malformations in infants, is increasing. In terms of the number of such cases, the Russian Federation has come close to the threatening 5% line, beyond which social and biological degradation is possible across the country [11].

The environmental safety in Russia is negatively affected by the imperfection of the system of environmental legislation: many normative acts essentially duplicate each other. In addition, there are contradictions in matters of environmental protection and environmental security, even at the constitutional level. [14].

Is it possible then to call the modern "homo sapiens" really a reasonable person? Let's leave this question open, but to date it is obvious that we have moved beyond the economic capacity of the biosphere, as a result of which the natural processes are destabilizing, creating a threat to the life-support of present and future generations[5]. Thus, ensuring environmental security is a common task for humanity. At the same time, the success of solving environmental safety, preserving a favorable environment and preventing the negative impact of economic and other activities on nature depends on the extent to which states take into account mutual interests in the process of developing natural resources and the level of their ecological communication in the global information space [12].

2 Materials and methods

The materials for the study were statistical data of international and state organizations, scientific publications of Russian and foreign authors, official statements of the leaders of states, normative legal acts, and the authors' own empirical materials.

The conceptual position of the authors is that the problem of environmental safety is complex and includes aspects of various branches of knowledge: technological, legal, philosophical, economic, etc. This necessitates an interdisciplinary approach to its study that allows to consider environmental safety and the environment as a single a global system with a large number of relationships.

As the main methods, based on the interdisciplinary nature of the study, the following were used:

- a statistical method used to process data on the state of the environment;
- a systematic method used as a basis for studying the complex interrelationships between the extraction of solid minerals and indicators of the ecological background;
- The method of economic analysis, which made it possible to draw a conclusion about the economic feasibility of using certain technologies;
- a method of socio-psychological analysis, which made it possible to identify the characteristics of human perception and reaction to changes in the surrounding natural environment;
- a formal legal method used in the analysis of regulatory legal acts;
- futurological (prognostic) method, which made it possible to assess the prospects of organizational and legal and technological measures designed to ensure environmental safety of mining enterprises.

All research work of the authors was based on the principles of scientific, objectivity, correctness of the material presented.

3 Results And Discussion

3.1 Necessity of ensuring environmental safety: scientific understanding of the problem.

The situation described above has become the subject of active attention in scientific research since the 1970s. XX century. And quickly enough environmental problems began to be considered in the context of security. In the English-language scientific literature an expression of «environmental security», this means «safe environment» or «environmental safety». [4]. The same concept began to occur in Russian scientific research [2].

Discussion of environmental degradation problems has shifted from scientific studies to an official international level. The Stockholm Declaration, the Convention on Environmental Impact Assessment in a Transboundary Context, the Rio Declaration on Environment and Development, the Convention on Access to Information, Public Participation in Decision-making and Access to Justice in Environmental Matters and others were accepted. The analysis of these documents allows us to speak about the unconditional awareness by the international community of the real threats that man's ill-considered activities in the environment contain, and attempts to formulate conceptual approaches to solving environmental problems. A clear understanding of the global nature of these problems makes them consider them only as problems that threaten the security of all mankind. In turn, assigning an environmental problem to the category of security problems means giving it the highest international priority [13].

In the context of the foregoing, the most depressing is the fact that mankind has sounded the alarm only in the conditions of the global ecological crisis, when the biosphere of our planet has largely lost its ability to support the natural biochemical cycles of self-healing. Perhaps such a belated response to a change in the environment is due to the inability of a person as a biological species to directly perceive dangerous changes in the quality characteristics of the environment in real time. Long-term research of specialists The Institute of Socio-Political Studies of the Russian Academy of Sciences [1] has shown that human communities tend to assess environmental changes solely from the point of view of their real impact on the level and quality of human life. This is the indirectness of our perception of environmental problems: the reaction is caused by their social consequences, and not by problems as such. Given the irreversibility of many natural processes, such an approach is fraught with disastrous consequences for people, since in solving environmental problems, the time factor that can be missed very often plays a decisive role.

In the case of Russia, the situation described above is exacerbated by the fact that the technological processes of most of the resource-extracting industries that form the backbone of our economy were initially built without due regard for their environmental consequences. Throughout the twentieth century, the dilemma of "economics or ecology" was clearly resolved in favor of the former. Moreover, for many decades this issue has not been raised in our country at all. At the same time, the urgent need for the present day in the development of industrial technologies in the direction of increasing their environmental friendliness does not at all cancel the considerations of the economic profitability of these types of industries, which is dictated by the conditions of a market economy. Thus, it is extremely urgent to develop such models of industrial production that would optimally combine the economic and environmental interests of society. To this we are inspired by the consciousness of social responsibility for future generations in the consumption of material and energy resources.

3.2. Ensuring Environmental Safety of Mining Enterprises: Opportunities and Prospects

Everything that has been said before is fully applicable to the mining industry. For this industry is characterized by a significant load on the environment, which manifests itself in the change in the natural landscape and other negative consequences. Carrying out seismic exploration and blasting operations, mechanical destruction of rocks by cutting is accompanied by noise, shaking, the formation of a large amount of dust and the emission of gases. In addition, some technological processes involve the use of large volumes of water, which significantly reduces the reserves of this critical resource. All these activities are associated with high consumption of materials and energy, the production of which, in turn, also harms nature. In the end, mining processes often lead to the destruction of natural ecosystems, or to negative changes in the habitat.

Soil erosion, disturbance of physical and chemical properties of soil, the same changes in water, air, drainage of lakes and marshes, reduction of biodiversity (both at the level of an individual species and at the level of the ecosystem as a whole), the destruction of biocenoses, their loss of self-regulation and self-recovery , harm to human health - this is an incomplete list of environmental damage that may be caused by mining enterprises. In this regard, we note that in the near future the impact of the mining industry on the environment is likely to increase, which is due to the ongoing intensification of production processes [10]. Of course, this cannot but affect the overall state of environmental security in Russia.

It seems to the authors of the article that the need to solve the above problems requires the rapid development and introduction of alternative technologies into the ore mining and enrichment process, as well as the use of new schemes for the development of deposits that allow to combine economic and environmental interests with maximum effectiveness. For example, experts have long been paying attention to the need to sort the ore to be enriched, depending on the mechanical characteristics of the latter. Of course, this complicates the production process a little, because it forces us to consider, in addition to the mineralogical composition of the rock, its texture and structure, the granulometric composition of the rock and its strength. However, taking into account the given qualities of the rock when it is loaded into the crusher can reduce energy costs and loss of enrichment (provided that the optimal ore dressing option is selected). At the same time, a certain reduction in the harmful impact on the environment is achieved. Despite the obvious advantages of this innovation, this and similar technologies are still not widely used in the mining industry.

An important direction in optimizing the mining industry both from the point of view of profitability and from the point of view of ecological safety is the development of technologies that make it possible to use the full potential of the field being developed. This involves the inclusion in the production process in addition to the main product - ores with a high content of useful component, associated and by-products, that is, ores with a low content of the useful component, as well as other materials of the depositAt present, huge areas are occupied by dumps of poor and off-balance ores, as well as by storages of other breeds that are not included in the technological chain of mining and processing production and that are harmful to the environment, because they are sources of constant dusting of the atmosphere, are located on territories that could be used in agriculture and forestry, etc. Reducing the burden on the environment in this situation could be facilitated by the more active use of low-waste technologies, the introduction of which, apart from the environmental effect, would undoubtedly have an economic effect. Moreover, in a number of cases, both of these effects are fully interrelated - it is enough to recall the use of mining waste to improve the quality of land during reclamation, as well as the possibility of using as mineral fertilizers slurry enrichment of phosphoresce ores, which in addition to phosphorus contain a long list of necessary in agriculture of chemical elements.

It should be noted that significant shortcomings for the environment result in shortcomings in the planning and organization of development systems for mining deposits, as well as inattention to indicators of environmental safety of production materials. All this causes an urgent need to change traditional approaches to the organization of development of deposits and development of alternative production management schemes in mining enterprises, which allow to reduce the negative load on the environment. As an example, we note that even a simple reduction in production losses can significantly improve the state of the environment. And the speech in this case is not only about the savings that can be used for the subsequent reclamation of the deposit. The maximum full utilization of the potential of the developed deposit postpones the beginning of the operation of a new field in time, which, taking into account the environmental consequences of such operations, and taking into account the limited natural resources, reduces the overall negative environmental load.

Of course, innovative technological solutions based on a different perception of the surrounding nature and human place in it play the main role in the processes of increasing the "environmental friendliness" of mining enterprises (like any other). However, it is already impossible to solve problems related to the state of the environment, relying only on the development of engineering thought [6]. A comprehensive approach is needed, based not only on improving industrial technologies, but also on a wide range of economic, legal, ideological, organizational and managerial means.

Some main lines of activities designed to improve the state of the environment and ensure environmental safety of Russia's economic development are named in the Concept of Long-Term Social and Economic Development of the Russian Federation for the period up to 2020. Such areas are: 1. Establishment of a developed waste recycling industry and 2. Expanding the use of renewable energy sources. However, these areas of activity are more closely related to the technological side of the issue. As for other measures, first of all organizational and legal, then, in our opinion, they should assume the following:

1. creation of a modern system of environmental monitoring. The absence of such a system in Russia significantly complicates the possibility of implementing a single state environmental policy stipulated in the Environmental Doctrine of the Russian Federation No. 1225-r of August 31, 2002 and the Basics of State Policy in the Field of Environmental Development of the Russian Federation for the period up to 2030 from 30.04.2012, The lack of a complete and objective, and most importantly - constantly incoming information on the ecological condition of the territories does not allow for the effective planning of environmental scale of the country. It seems that there is no need to say that given the characteristics of the biosphere, these activities should be carried out within the framework of a single plan;

2. Strengthening the activities of control and supervisory and licensing bodies. Particular attention should be paid to the mandatory environmental impact assessment of production projects, environmental auditing, certification and licensing of industrial activities. These measures are largely capable of preventing environmental damage, and it is always better than compensating for this damage;

3. Introduction of economic and legal mechanisms that induce owners of industrial enterprises to modernize technological equipment. As the simplest and most effective measures could be even more significant fines for environmental pollution and tax incentives in the transition of the company to clean technologies;

4. Legislative prohibition of the practice of "temporarily permitted emissions of pollutants".

It seems to the authors of this article that a skilful combination of the above measures and promising technological solutions will ensure a higher level of environmental safety not only for the mining industry, but also for other industries of Russian industry.

4 Conclusion

The conducted research allowed the authors to draw the following conclusions:

1. The global environmental crisis that hit the planet did not bypass Russia. This complicates for our country the task of economic growth and encourages society to develop a new strategy for its interaction with nature based on saving resources and taking care of the environment;

2. Improving the environmental safety of mining enterprises is part of the overall goal of preserving the environment that is conducive to human life and health. This implies a comprehensive approach to solving the problem, which includes not only technical modernization, but also systemic changes in the entire process of development and mining of deposits;

3. The mining industry is one of the most significant factors of the negative impact on the environment. This is due to significant changes in the natural landscape, as well as other negative consequences of this type of industrial activity, which are both direct and indirect. At the same time, the mining industry is one of the foundations of Russia's economic development, which implies an inevitable intensification of the processes of mining of solid

 DOI: 10.1051/e3sconf/20172102020

minerals. Given the severity of the environmental problems that arise and the threat of significant increase while maintaining traditional approaches to the development of ore deposits, the urgent need of today is the improvement of mining processes based on the use of innovative technologies that reduce the burden on the environment;

4. The organizational and legal measures that can ensure the environmental safety of mining enterprises are: the creation of an effective system of environmental monitoring; strengthening the activities of control and supervisory and licensing authorities; introduction of economic and legal mechanisms that induce owners of industrial enterprises to modernize technological equipment; legislative prohibition of the practice of "temporarily permitted emissions of pollutants".

The implementation of these measures in combination with technological developments creates the prospect of an environmentally oriented growth of the Russian economy.

References

1. S.M. Alekseev, I.A. Sosunova, D.A. Boriskin, *Ecology, economy, society: state, trends, prospects*. 302 (NIA-Nature, REFIA, Moscow, 2002)
2. M.M. Brinchuk, State and Law, **9**,30 (2008)
3. J.M. Broadus, R.V. Vartanov *The Oceans and Environmental Security: Shared U.S. and Russian Perspectives. Marine Policy Center.* (Island Press, Washington DC, 1994)
4. S. Dalby, *Environmental security*, 239 (Univ. of Minnesota press, Minneapolis, London, 2002)
5. V.I. Izmalkov, A.V. Izmalkov, *Technogenic and environmental safety and risk management*, 482 (NICEB RAS, 1998)
6. S.V. Kamashev, Philosophy of Education, **3:42**, 119 (2012)
7. B.A. Morgunov, A.M. Bagin, M.L. Kozeltsev, A.A. Terentyev, Human ecology, **4**, 3 (2017)
8. S.N. Rusin, Ecological law, **5**, 12 (2010)
9. A.I. Semyachkov, Yu.O. Slavikovskaya, K. Drebenstedt, *Ecological and economic aspects of the activity of mining enterprises*, 252 (Ekaterinburg: Institute of Economics, UrB RAS, 2013)
10. V.A. Smyshlyaev, *Environmental safety: global contours and the Russian aspect (political monitoring)* 537 (Voronezh, The scientific book, 2009)
11. T.V. Vilova, N.A. Chertova, M.G. Vilova, M.S. Permilovsky, Human Ecology, **5**, 21 (2017)
12. E.V. Vorontsova, A.L. Vorontsov, T.N. Balashova, *17 International Multidisciplinary Scientific GeoConference* (SGEM, Albena, (2017)
13. A.S. Timoshenko, Soviet Yearbook of International Law, **23**, 456 (1989)
14. L.M. Yao, *Social ecology. Tutorial* (Kazan Publishing House. state. technol. University, 2007)

Belt Aligning Revisited

Vadim Yurchenko[1*]

[1] T.F. Gorbachev Kuzbass State Technical University, 650000, 28 Vesennyaya St., Kemerovo, Russia

Abstract. The misalignment causes the greatest damage to the conveyor belt. As a result of the interaction of the moving belt with the stationary parts of the conveyor, the sides of the belt wear intensively. This results in reducing the life of the belt. The reasons for this phenomenon are well investigated, but the difficulty lies in the fact that they all act simultaneously. The belt misalignment prevention can be carried out in two ways: by minimizing the effect of causes and by aligning the belt. The construction of aligning devices and errors encountered in practice are considered in this paper. Self-aligning roller supports rotational in plan view are recommended as a means of combating the belt misalignment.

1 Introduction

The misalignment of conveyor belts contributes to their active wear. According to the place where the causes of wear are concentrated, the conveyor belts are represented by the following elements: the traction core, the conveyor-belt edge surface, the carrying side cover and the bottom cover . The greatest influence on the strength properties of the belts and their serviceability is provided by the causes associated with the wear of the conveyor-belt edge surface (abrasion and breaking). The complexity of the problem lies in the fact that a number of causes of belt misalignment act simultaneously [1, 2, 3, 4]. The most hidden causes are the belt camper and the uneven distribution of tension along the width of the traction core (manufacturing defect). Other reasons can be called man-made, depending on the technical culture of the staff. For example: the incorrect joining of belt segments; skew of segments in plan view or joining of segments (the width of the segments has tolerance ±) not along the axis, but along the conveyor-belt edge surface; installation of the linear flight with a deviation in plan view from a straight line or the installation of flight sections not perpendicularly to the sill (i.e., with the inclination of the flight cross-section); installation of roller supports, even on a straight line flight, not perpendicularly in plan view to the belt movement direction; installation of any drums with a skew in any plane; the increase in the resistance to rotation of the rollers (or failure) along one edge surface of the belt will lead to the redistribution of the tension along its width; the installation of the loading device that does not provide the symmetrical arrangement of the load on the receiving belt – all these errors will contribute to the belt misalignment.

*Corresponding author: yvm@kuzstu.ru

DOI: 10.1051/e3sconf/20172102021

2 Materials and methods

The belt misalignment can be prevented by minimizing the effect of the causes and by aligning the belt. Within the framework of the article, we will consider the aligning devices.

In practice, is the belt aligning by rotating the roller supports to a certain angle in plan view is used the most often [5] (Fig. 1, a) or by inclining in the vertical plane (Fig. 1, b). Simplicity wins. However, in order to avoid errors that are often encountered in practice, it should be noted that the force contributing to the belt misalignment is variable. It depends on the belt tension, i.e. from the amount of the load on the belt and from its varying position along the conveyor length. Therefore, belt aligning by rotating the roller supports to a certain angle in plan view (Fig. 1a) or inclining in the vertical plane (Figure 1, b) can lead to the opposite effect, to the misalignment in the opposite direction.

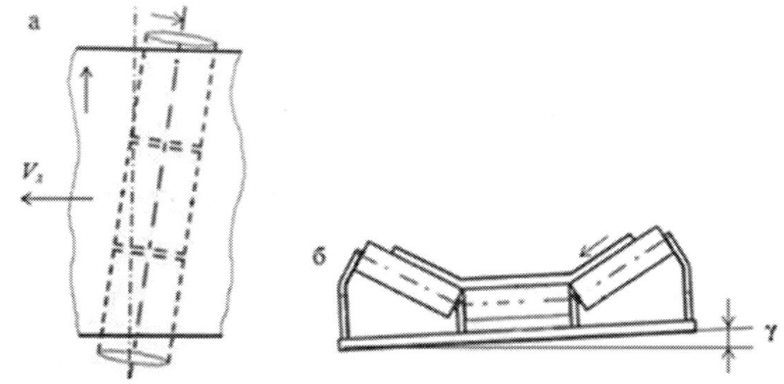

Fig.1. Ali gning the tape by rotating the support roller: a - in plan view, b - in the vertical plane.

In addition, the inclination of the roller supports at an angle γ=4-5° leads to the decrease in the receiving capacity by 10% . The results are obtained by modeling the cross-section of the load on the belt; the inclination angle of the supporting roller being changed in vertical plane (Fig. 2).

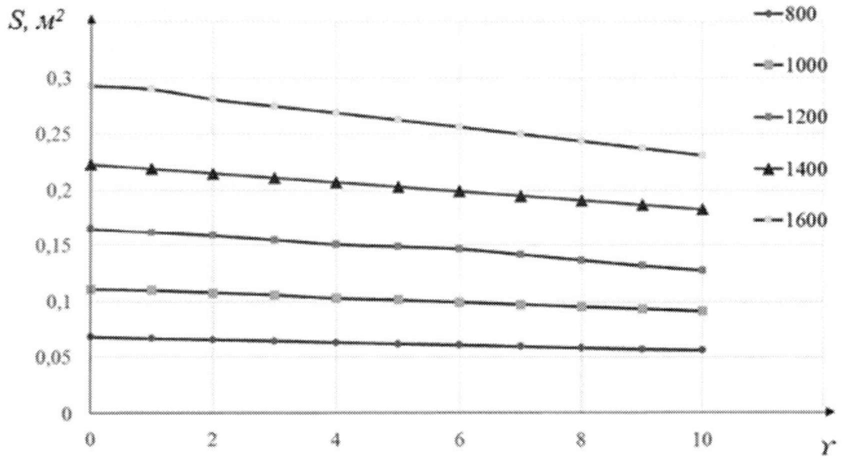

Fig. 2. Dependence of the sectional area of the load on the belt on the angle of inclination of the roller.

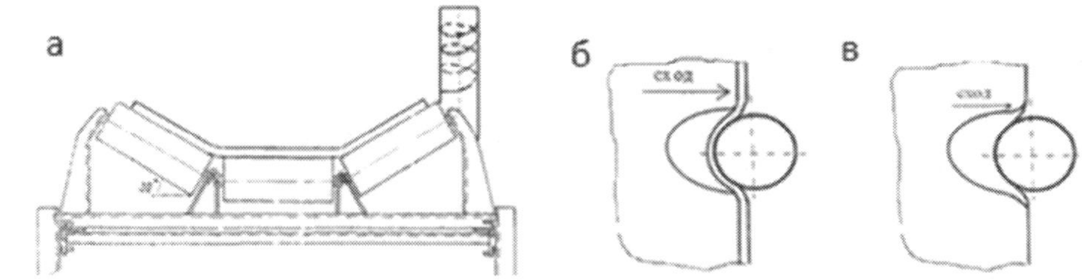

Fig. 3. Belt edge surface interaction with bump roller and deflector roller: а – bump roller, б – being pressed the belt edge surface move upwards, в – being pressed the belt edge surface move downward

Another common mistake is the use of bump rollers to counteract the belt misalignment (Fig. 3, a). Since the conveyor-belt edge surface rigidity is less than the roller shell rigidity, the belt will be pressed into the roller and can bend (move) upwards (Fig. 3, b). When moving upwards, the effect of "screwing" appears (Fig. 3, a), i.e. the belt moves up the bump roller up to the turning point. The process is accompanied by the spillage of the transported material. When pressed, the belt edge surface can be bent (shifted) downward relative to the bump roller (Fig. 3, c). In this case, the belt is directed into the space limited by the bump roller and fixed brackets of supporting rollers. This leads to increased wear of the belt edge surface by friction, up to the detachment of the pieces.

When the bump roller is installed incorrectly, i. e. the roller axis is inclined in the direction of the belt movement (or vice versa), the roller axis is not perpendicular to the plane of the belt interacting with it, and the processes described above are amplified. Stresses resulting from the interaction of the belt edge surface with the bump rollers, contribute to the intensive wear and the stratification of the side of the belt edge surface.

Self-aligning roller supports, which are rotational in plan view, which are represented by two types, are the most effective among the known means of belt aligning.

The first type is a self-aligning roller supports, rotatable in plan view, due to deflector roller levers (Fig. 4, a) [3]. To achieve the aligning effect, it is necessary that the lever arms and the gap between the belt edge surfaces and the deflector rollers allow the roller to rotate through the angle of 3-6°.

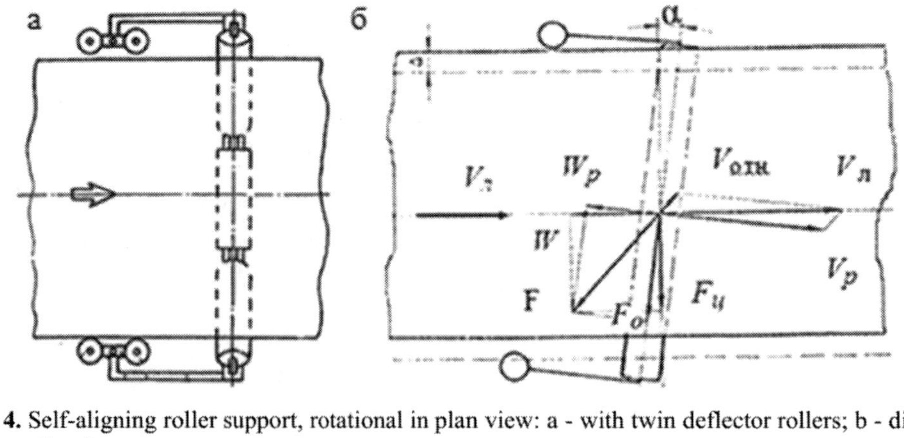

Fig. 4. Self-aligning roller support, rotational in plan view: a - with twin deflector rollers; b - diagram of operating forces.

The advantage of such devices lies in the fact that the value of the aligning force varies depending on the value of the emerging force of the misalignment. When the belt moves along the roller support rotated by a certain angle, the vectors of their linear velocities V_p and $V_л$ are mismatched, as a result of which the relative velocity of the belt $V_{отн}$ (Fig. 4, b)

 DOI: 10.1051/e3sconf/20172102021

appears. As a result, the friction force F acts on the belt in the opposite direction. The projections of this force on the longitudinal axis which is perpendicular to the roller are its components: W_p is the resistance to the belt movement; F_o is the axial friction force. The projection of the axial friction force on the axis perpendicular to the longitudinal axis of the belt represents the aligning force opposing the misalignment of the tape.

The magnitude of the aligning force that occurs when turning the roller support in plan view depends on the force of the normal pressure on it from the side of the belt and the load and the clutch friction coefficient and the angle of rotation. For the effective operation of the self-aligning roller support, as well as the conveyor as a whole, it is necessary to ensure the groove of the belt (Fig. 5). That is, to fulfill the ratio

$$0{,}5B \le (a+b+c) \le B \tag{1}$$

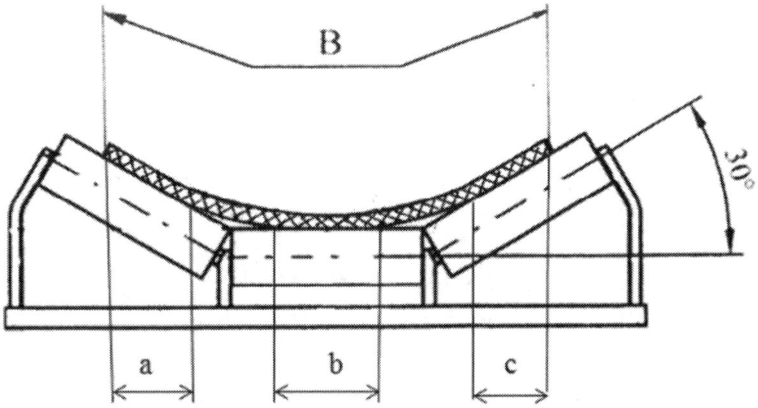

Fig. 5. The groove of the conveyer belt.

This is achieved by the fact that during the assembly of the conveyor the level of the surfaces of the tail and unloading drums should be at the level of the surface of the middle rollers of the supporting grooved rollers. The more contact, the more efficient will be the self-centering roller.

The disadvantage of such self-aligning roller supports is that the contact of the belt edge surfaces with the deflector roller, although with less effort compared to the bump rollers, occurs. And this contributes to the wear and the stratification of the side of the belt edge surface.

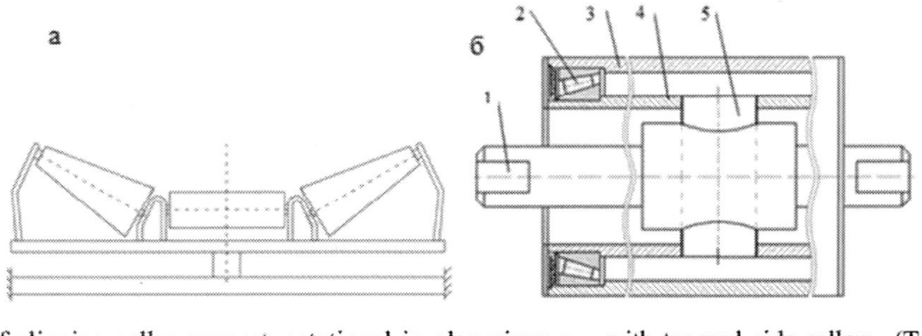

Fig. 6. Self-aligning roller support, rotational in plan view: a – with tapered side rollers (Tru-Trac Tapered Trough Tracker), б – with twin rotating shells (Tru-Trac Flat Return Tracker, Tru-Trac Trough Tracker).

3 Conclusions

The second type is the self-aligning roller support rotational in plan view by changing the position of the belt relative to the vertical turning node and creating greater force of interaction between the belt edge surface and the roller (Fig. 6, a) [6].

The most original solution is the self-aligning roller support with double rotational shells (Fig. 6, б) [7, 8]. The roller consists of an axis 1, in the central part of which there is a pin 5. The roller 4 is hinged on its inner shell. Tapered rollers 2, on which the outer rotating shell 3 rests, are mounted at the ends of this shell. The radial clearance between the axis 1 and the inner shell 4 allows the roller to rotate relative to the pin 5 at the angle of 3 to 6°.

Advantages of self-aligning rollers of the firm "Tru-Trac":

- work with any direction of the belt movement,
- eliminate the wear and stratification of the belt edge surface.

Analysis of the examined devices shows that effective aligning of conveyor belts should be carried out only by self-aligning belt supports, rotational in plan view.

References

1. M. Noncic, Bulk Solids Handling, **32:1**, 34 (2012)
2. F.L. Naples, *Belt Conveyors for Bulk Materials, Sixth Edition published by Conveyor Equipment Manufacturers Association* (Adelaida, CEMA, 2005).
3. R. Todd, L. J. Goldbeck, A. D. Marti, *The Practical Resource for Total Dust & Material Control*, (Detroit, ASSG Found., 2002)
4. N. Schmidgall, Pit and Quarry, **95:12**, 22 (2003)
5. F. Han, F. Xiong, R. He, Applied Mechanics and Materials, **66**, 356 (2011)
6. G. Gerbert, Design Engineering Division **88**, 443 (1996)
7. H. Cheng, S. Yanabe, Y. Iwata, H .Sato, T. Komatsuzaki, K. Sato, Transactions of the Japan Society of Mechanical Engineers, **68:10**, 2911 (2002)

 DOI: 10.1051/e3sconf/20172102022

Spectral Study of Modified Humic Acids from Lignite

Sergey Zherebtsov[1, *], *Natalya* Malyshenko[1], *Ludmila* Bryukhovetskaya[1], and *Zinfer* Ismagilov[1]

[1] Federal Research Center of Coal and Coal Chemistry, 650000 18 Sovetskii av., Kemerovo, Russia

Abstract. The IR-Fourier, ESR and solid-state ^{13}C NMR analysis are used for investigation of unmodified and modified humic acids obtained from Tisul lignite (the Kansko-Achinsk Basin). Treatment with Hydrogen peroxide used for modification of humic acids and it changes the functional-group composition of the humic acids and increases the sorptional capacity.

1 Introduction

The structure of humic acids reflects the material from which they are derived [1–4]. The reactivity of humic acids depends on the presence of different functional groups and it reflects the content of aromatic conjugates and degree of condensation. Chemical modification may change the functional composition of the humic acids.

Humic acids derived from alkylated and debitumenized lignite and peat are characterized by an elevated content of aromatic and phenolic components, as noted in [5–7]. It is important to establish the relation between the composition, structure, and proper- ties of the humic acids, in order to understand their behavior in nature and in technological processes. That, in turn, will permit their more effective use.

The applications of humic acids depend on their chemical composition and properties. One major application is sorbent production [8, 9]. The presence of ion-exchange carboxyl groups and phenolic hydroxyls, as well as electron-donor functional groups, facilitates ion exchange in complex formation and creates prospects for the separation and extraction of metal cations from various media and for the purification of wastewater and industrial water.

Purposeful chemical modification may increase the content of the desired functional groups and thereby improve the sorptional characteristics of the humic acids. The sorption of zinc cations by humic acids derived from lignite and naturally oxidized coal from the Tisul field is due to ion exchange, as shown in [10]. Modification by hydrogen peroxide increases the content of carboxyl groups, whose protons participate in ion exchange, and therefore expands the sorptional capacity of the humic acids [10].

In the present work, we investigate the functional group composition of humic acids derived from lignite after modification by hydrogen peroxide.

[*] Corresponding author: sizh@yandex.ru

 DOI: 10.1051/e3sconf/20172102022

2 Experimental Method

We prepared the humic acids from Tisul lignite (from the Kansko-Achinsk Basin), by the method in [11]. That method is based on treating lignite with potassium hydroxide to produce solutions of potassium humates, to which 10% hydrochloric acid is added. The humic acid deposits are filtered out, washed in distilled water to specified pH, and dried at 70°C. Table 1. presents the characteristics of the initial samples

Table 1. Technical and elementary analysis of the samples, %.

Sample	W^a	A^d	V^{daf}	C^{daf}	H^{daf}	$(O + N + S)^{daf}$ (as a difference)	$(HA)_t^{daf}$
Coal	8.04	6.11	48.14	64.34	4.69	30.97	26.88
Humic acids	4.97	4.01	–	60.84	4.18	36.07	–

Here *daf* denotes a dry ash-free state. $(HA)_t^{daf}$ is the yield of humic acids.

The humic acids are modified by means of hydrogen peroxide (concentration 32.5%) to obtain constant volume of potassium-humate solutions of equal concentration, with mixing for a specified time, and then adding hydrochloric acid. The deposits are filtered out, washed in distilled water to specified pH, and dried at 70°C to constant weight.

High-resolution solid-state ^{13}C NMR spectra are recorded on a Bruker Avance III 300W instrument at 75 MHz, with rotation of the samples at 5 kHz.

IR spectra are recorded on an Infralyum-FT 801 Fourier-transformation spectrometer, by means of tablets with KBr.

The ESR samples are recorded at room temperature on a Bruker EMX-m40X ESR spectrometer at 9.86 GHz. The ESR spectra of all the samples are recorded in identical conditions: power 1.8–1.9 mW; modulation frequency100 kHz. To determine the quantity of organic paramagnetic centers, Mn^{2+} ions in magnesium oxide MgO are chosen as the standard. The characteristics of the ESR spectra are calculated by means of Bruker Win EPR software.

3 Results and Discussion

Humic acids are characterized by typical absorption bands in the IR spectra; that is consistent with the results in [12, 13]. In Fig. 1, we show the IR spectrum of unmodified humic acids and humic acids modified with hydrogen peroxide.

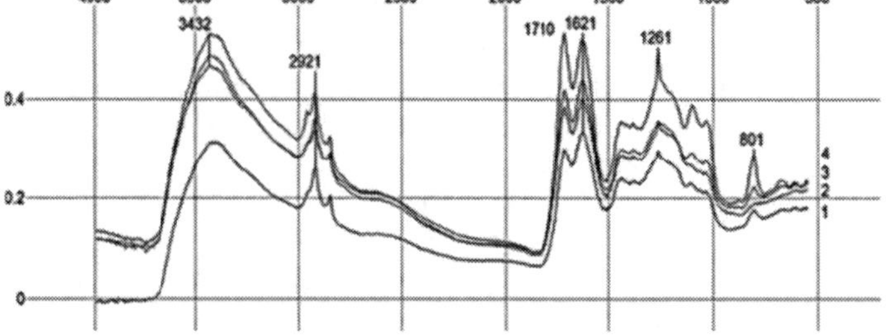

DOI: 10.1051/e3sconf/20172102022

Fig. 1. IR spectra: unmodified humic acids; (*2–4*) humic acids modified with 5, 10, and 15 mL of hydrogen peroxide, respectively.

For all the humic acids obtained, bands of different intensity appear in the range 1720–1710 cm^{-1}, corresponding to valence vibrations of the C=O bond of the humic acids' carboxyl groups. We see that modification by hydrogen peroxide increases the intensity of the bands corresponding to the C=O bond of the humic acids' carbonyl groups (1710 cm^{-1}); the carbonyl groups of the primary amides and the conjugate C=C bonds (1620 cm^{-1}); the C–O bonds of carboxylic acids complex esters, the O–H bonds of phenols and carboxyl groups (1260 cm^{-1}); the C–O bonds of alcohols and esters (1100–1030 cm^{-1}); and hydrogen-bonded OH groups (3400 cm^{-1}). Table 2 presents the ^{13}C NMR and ESR data: the integral intensities of the spectral regions and the para-magnetic centers of the humic-acid samples.

Table 2. Integral intensities of the spectral regions in the ^{13}C NMR spectra (%) and paramagnetic centers in the ESR spectra of the samples.

Sample*	Chemical shift, ppm							PMC
	220–187 C=O	187–165 COOH	165–145 C_{ar}–O	145–108 C_{ar}	108–90 $C_{O\text{-}alk\text{-}O}$	90–48 C_{alk}–O	48–5 C_{alk}	$\times10^{-18}$, spin/g
Humic acids	1.04	5.83	10.93	19.22	0	9.08	53.87	8.92
Humic acids (5)	1.10	6.12	10.01	18.90	0	11.68	52.19	7.18
Humic acids (10)	1.16	6.27	10.00	18.63	0	11.97	51.97	6.42
Humic acids (15)	1.24	6.35	9.98	18.86	0	11.96	51.61	5.13

* The figures in parentheses in the sample notation indicate the quantity of H_2O_2 (mL) used in modification; PMC, paramagnetic centers.

We see that modification by hydrogen peroxide increases the content of carbonyl C=O, carboxyl COOH groups, and groups containing oxygen with an alkyl carbon atom C_{alk}–O. Correspondingly, the content of C_{ar}, C_{ar}–O, and C_{alk} is reduced. Comparing the IR and NMR data, we find that they are generally consistent.

PMC 10^{18}, spin/g

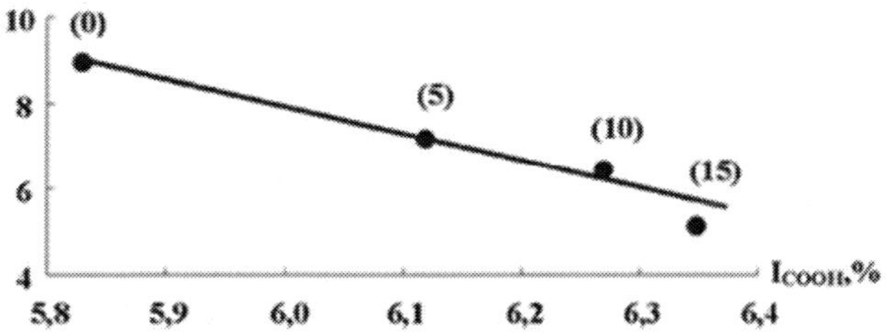

 DOI: 10.1051/e3sconf/20172102022

Fig. 2. Dependence of the concentration of paramagnetic centers (PMC) in the humic acids on the intensity of the carboxyl groups with modification by different quantities of hydrogen peroxide (shown on the curves, ml).

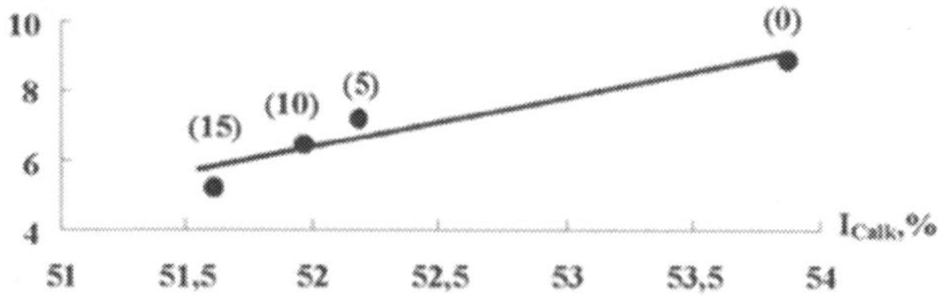

Fig. 3. Dependence of the concentration of paramagnetic centers (PMC) in the humic acids on the intensity of the Calk groups with modification by different quantities of hydrogen peroxide (shown on the curves, ml).

The ESR spectra are recorded with g = 2.003-2.004. That corresponds to the radicals of organic compounds [14]. The signals from the samples are symmetric or slightly asymmetric; this is typical of coal and solid polyaromatic compounds of quinoid type [15]. Modified and unmodified humic acids are characterized by narrow lines corresponding to organic paramagnetic centers. That may indicate a high content of polyconjugate aromatic structures.

Increasing the hydrogen-peroxide concentration on modification reduces the content of organic paramagnetic centers (Table 2), to an extent that depends on the functional-group composition of the humic acids. In Fig. 2, we see that, with increase in the con- tent of carboxyl groups (according to NMR data) during modification, the concentration of paramagnetic centers declines. Inverse proportionality is also seen for the carbonyl groups C=O and the C_{alk}–O groups.

At the same time, the decrease in the C_{ar}–O, C_{ar}, and C_{alk} concentrations during modification by hydro- gen peroxide leads to proportional decrease in the concentration of organic paramagnetic centers. In Fig. 3, as an example, we show the concentration of paramagnetic centers as a function of the C_{alk} content.

The presence of paramagnetic centers in the humic acids may be due to free radicals. In modification by hydrogen peroxide, the decrease in aromatic content (Table 2) lowers the content of free radicals [16]. In addition, hydrogen peroxide is a source of OH radicals and oxygen and hence may lead to the destruction of organic radicals and may facilitate the appearance of new functional groups [17- 20]. The change in composition of the humic acids, with increase in the content of oxygen-bearing groups, may be attributed to oxidative reactions with the participation of hydrogen peroxide. We see from Table 2 that reducing the content of some functional groups increases the content of others. For example, the C_{alk} content declines, while the C_{alk}–O content increases.

4 Conclusion

Modification of the humic acids derived from lignite and naturally oxidized coal by hydrogen peroxide increases sorptional capacity in the sorption of zinc cations by a factor of 2–4 in comparison with unmodified humic acids.

By increasing of the content of various oxygen containing functional groups the expansion of sorptional capacity of humic acids can be achieved with respect to various cations, by both ion exchange and formation of complexes.

References

1. I.V. Aleksandrova, Pochvov. **12**, 47–51 (1993)
2. T.A. Kukharenko, Lignite and Coal Oxidized in the Earth Layers (Moscow, Nedra, 1972)
3. I.D. Komissarov, I.N. Strel'tsova, *Influence of the extraction method of humic acids from the raw material on chemical composition of obtained preparations, in Guminovye preparaty (Humic Preparations)* (Tyumen, Tyumen. S-kh. Inst.)
4. N.V. Yudina, V.I. Tikhova, Khim. Rastit. Syr'ya, **1**, 93–96 (2003)
5. L.G. Sivakova, S.I. Zherebtsov, O.V. Smotrina, Solid Fuel Chem. **39:5**, 20-26 (2005)
6. S.I. Zherebtsov, Yu.V. Musin, A.I. Moiseev, Khim. Rastit. Syr'ya, **2**, 125–130 (2009)
7. S.I. Zherebtsov, Z.R. Ismagilov, Solid Fuel Chem. **46:6**, 339–351 (2012)
8. J.C. Lobartini, K.H. Tan, J.A. Rema, Sci. Total Environ. **113:1**, 1 (1992)
9. N.V. Malyshenko, S.I. Zherebtsov, O.V. Smotrina, L.V. Bryukhovetskaya, Z.R. Ismagilov, Khim. Int. Ust. Razvit. 23:4, 451-457 (2015)
10. E.M. Taits, Methods of Analysis and Tests of Coal, (Moscow, Nedra, 1983)

Substantiation of the Necessity for Design of Geohod Control System

Vladimir Aksenov [1,3*], *Ivan* Chicherin [2], *Irina* Kostinez[4], *Anton* Kazantsev[3], and *Andrey* Efremenkov[3]

[1] The Federal Research Center of Coal and Coal-Chemistry of Siberian Branch of the Russian Academy of Sciences, 650610, 10 Leningradski av., Kemerovo, Russia.
[2] T.F. Gorbachev Kuzbass State Technical University, 650000, 28 Vesennyaya St., Kemerovo, Russia
[3] Yurga Technological Institute branch of Tomsk Polytechnic University, 652052, 26 Leningradskay str., Yurga, Russia
[4] T.F. Gorbachev Kuzbass State Technical University, Belovo Branch, 652644 Ilyicha st. 32, Inskoy, Kemerovo oblast, Russia

Abstract. The article proves the necessity of creating a control system for the geohod. The scope of application of the geohod is described. The general scheme of the geohod with the main structural elements - cutter drum, head and tail sections, external mover - is described. Enlarged tasks that need to be addressed when creating a control system for the geohod are identified. The tasks of motion control of the geohod in the geomedium, improving the efficiency of the geohod, positioning the geohod in space, control of the support systems, diagnostics of the condition of the geohod, interaction of the control system with the operator are described. The directions of further research are identified: to formulate the requirements for the geohod control system; to analyze tasks solved by the geohod control system; develop the structure, mathematical models and algorithms for the operation of subsystems of the geohod control system; to develop software to simulate the interaction of subsystems of the geohod control system with each other, with the geomedium and the operator.

1 Introduction

Geohod technology is a promising method of mining [1,2,3]. A technical tool that implements this method is the geohod. A geohod is a machine that uses the geomedium to move in an underground space. Geohod fields of application are very wide [4,5,6,7]:

- mine working penetration of various locations in space;
- building of underground facilities for various purposes;
- laying of the city's sewer line;

* Corresponding author: 55vva42@mail.ru

 DOI: 10.1051/e3sconf/20172103001

- construction of the main metro tunnels;
- construction of underground warehouses, storage facilities, underpasses, garages;
- conducting emergency rescue operations in the rubble;
- construction of field fortifications.

The construction of the geohod is made on a modular basis (figure 1). The basic components of the geohod are the executive body, which destroys coalface (cutter drums), head and tail sections, the external mover [8,9,10,11].

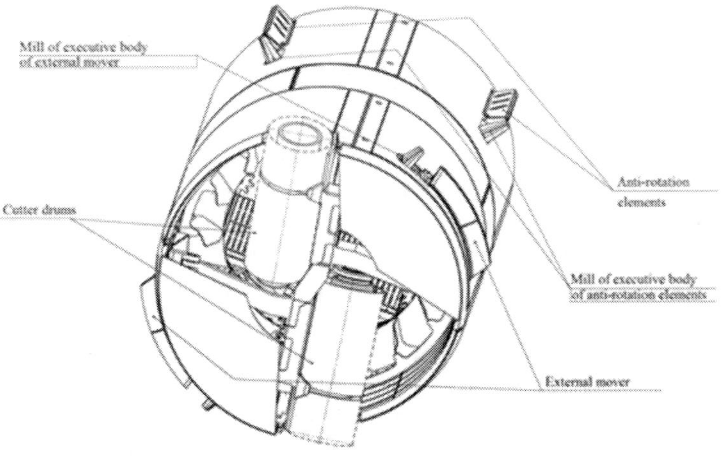

Fig. 1. General view of the geohod.

For effective work of the geohod, a clear interaction of all systems of the geohod is required, in particular, taking into account changing conditions in the external geomedium. To ensure such interaction a control system is required that will allow to control the particular systems and the geohod in general.

Tasks solved by the geohod control system. To ensure effective work of the geohod, it is necessary to create a control system that implements many tasks. The list of these tasks depends on technical characteristics of a particular geohod model. However, all of these tasks can be divided into the following classes:

1) motion control of the geohod in the geomedium;
2) improving the efficiency of the geohod;
3) positioning of the geohod in space;
4) control of auxiliary systems;
5) diagnostics of the state of the geohod;
6) interaction of the geohod control system with the operator.

The task of the geohod motion control, in the first approximation, is the task of controlling executive mechanisms, the basis of which are hydraulic cylinders and hydraulic motors. The hydraulic cylinders and hydraulic motors drive the executive bodies, as a result of which the geohod moves in the geomedium [12, 13]. This task can be carried out both in manual and automatic modes. In manual mode, an operator with the help of controls turns on the executive mechanisms and changes the parameters of their work for maneuvering and changing the speed of geohod movement. In automatic mode, the operator specifies the direction and the speed of movement, and the control system calculates the operation parameters of the executive mechanisms and applies the appropriate control actions.

The task of improving the geohod efficiency is divided into two subtasks:

- improving the geohod efficiency based on optimization criteria;
- improving the geohod efficiency based on the use of adaptive algorithms that take into account the heterogeneity of the geomedium and dynamic parameters of the geohod.

To solve the optimization problem, it is necessary to develop mathematical models and intelligent algorithms [14] for the operation of the geohod control system, taking into account the interrelation of all its elements. These models and algorithms will allow to apply to the executive mechanisms such control actions, in which the work of the geohod will satisfy the specified optimization criterion. The optimization criteria can be the geohod energy consumption per unit path, the accuracy of the trajectory.

Mathematical models for the adaptive control should take into account the heterogeneity of the parameters of geological medium (Protodyakonov scale of hardness, etc.) [12] and dynamic parameters of the geohod, which will depend on the speed, acceleration, direction (right, up, down, left, right) of the geohod.

Currently, the task of orientation of ground objects is effectively solved with the help of global positioning systems (GLONASS, GPS). However, for the geohod and other mining machines working underground, the use of such systems is impossible because the satellite signal does not reach them. At present, positioning systems of miners and transports have become widespread in mines. The work of such systems is based on the detection of radio tags that are embedded in miners' cap lamps and mounted on the transport. Radio tags are detected with the help of readers which are placed in special wells drilled along the route of movement of miners and transport. Readers through the communication channels transmit information about the registered marks to the data collection server. It is obvious that the use of such systems for mining machines, including the geohod, is impossible, since readers and communication channels cannot be set in advance on the route of the machines. Thus, the development of a system of a geohod positioning in space is the subject for further research. The operation of such systems should be based on a combination of laser positioning technology, positioning by radio tags, orientation with the help of gyroscopes and computer modeling of the underground space.

Auxiliary systems of the geohod are systems ensuring the main process – the movement of the geohod in the geomedium. Such systems include: an energy-power device, a transport module and a lighting system. The task of control of these fundamentally different systems should also be solved by the geohod control system.

Difficult operating conditions of the geohod and a high level of dynamic loads predetermine special requirements to its reliability. One of the important factors of increasing the reliability in operation is technical diagnosis. The geohod control system should include a diagnostic subsystem that monitors the technical condition by measuring and indicating the parameters of the geohod elements, automation devices, electrical and hydraulic systems, lubrication systems, bearing assemblies, etc.

The main technical means for interaction of the geohod control system with the operator are the control panel and display. With the help of the control panel, the operator applies control actions to the subsystems, executive mechanisms and geohod devices working in manual, semi-automatic or automatic modes. It is necessary to solve the problem of a clear separation of control functions between the operator and the control device in semi-automatic mode. The display allows the operator to track in real time and retrospectively the status of processes and elements, emergency situations, the results of the system diagnostic. It is also possible to implement an additional sound and light operator alert, for example, to alert emergency situations. The task of the control system is to provide the operator with the necessary information.

2 Results and discussion

Taking into account the complexity and diversity of the formulated tasks concerning the geohod control, and taking into account the complex structure of the geohod, which consists of many interrelated systems, the general structure of the multilevel geohod control system, which is shown in figure 2., was developed.

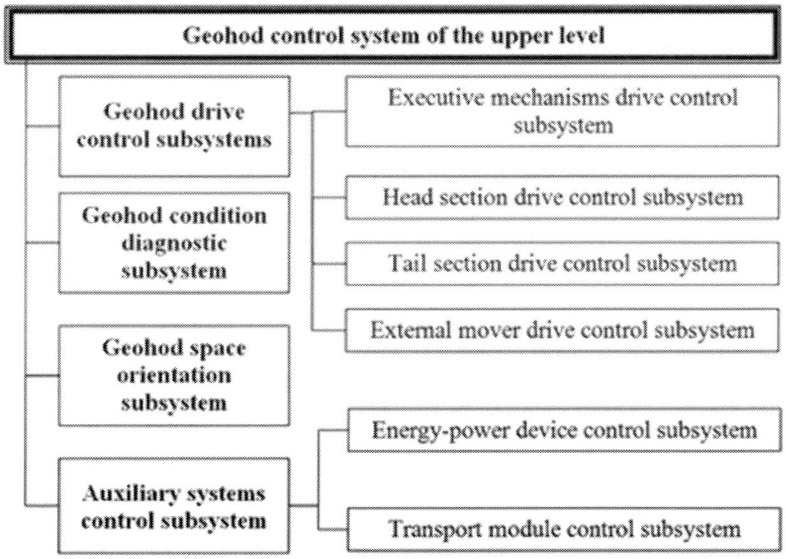

Fig. 2. General structure of the geohod control system.

The upper level control system is an information control system which solves the tasks of coordinating the work of subordinate subsystems, the exchange of information between these subsystems and the interaction of the geohod control system with the operator.

The middle level of the control system is represented by subsystems controlling drives, auxiliary systems, geohod diagnostics and its orientation in space.

Drive control subsystems solve the task of automatic control of executive mechanisms of geohod systems, ensuring the geohod movement in a changing geo-medium, taking into account the specified optimization criteria. The description of the tasks of other subsystems of the middle level is given above.

The lower level is represented by the subsystems controlling specific drives of the main (executive body, head and tail sections, external mover) and auxiliary (energy-power device, transport module) systems.

3 Conclusion

The above description of the tasks and the general structure of the control system make it possible to conclude that the tasks solved by the geohod control system are complex and diverse. However, at present there are no scientific approaches to develop mathematical models and algorithms of the geohod control systems. This inhibits the creation of the geohod with fundamentally new characteristics. Therefore, researches aimed at the solution of these tasks are actual.

To continue the research it is necessary:

- to formulate the requirements for the geohod control system;
- to analyze tasks solved by the geohod control system;
- develop the structure, mathematical models and algorithms for the operation of subsystems of the geohod control system;
- to develop software to simulate the interaction of subsystems of the geohod control system with each other, with the geomedium and the operator.

DOI: 10.1051/e3sconf/20172103001

References

1. R.V. Chernukhin, A.A. Dronov, M.Y. Blashchuk, IOP Conf. Ser.: Mater. Sci. Eng., **91**, 012086 (2015)
2. A.B. Efremenkov, 6th International Forum on Strategic Technology, **2:1**, 348 (2011).
3. V.Y. Begljakov, Timofeev V.Y., Dokhnenko M.V., Applied Mechanics and Materials, **682**, 282 (2014).
4. G. Brierley, Y. Smith, World Tunnel. And subsurface Excav., **9** (1998)
5. J. Carmody, R. Sterling, *Underground space design*, (New York, 1993)
6. T. Wighman, ENR: News – Rec, **4** (1998)
7. S. Nishi, T. Seiki, Mem. Sch. Eng. Nagoya Univ., **1** (1997)
8. M.Y. Blashchuk, A.A. Kazantsev, R.V. Chernukhin, Applied Mechanics and Materials, **682**, 418 (2014)
9. V. Aksenov, V. Sadovets, E. Rezanova, D. Pashkov E3S Web of Conferences, **15**, 03015 (2017)
10. V.V. Aksenov, A.B. Efremenkov, V.Y. Beglyakov, Applied Mechanics and Materials, **379**, 16 (2013).
11. M.Y. Blaschuk, A.A. Dronov, S.S. Ganovichev, IOP Conf. Ser. Mater. Sci. Eng., **127**, 012033 (2016).
12. R.V. Chernukhin, M.Y. Blaschuk, V.Yu. Blumenstein, P.A. Chazov, IOP Conf. Ser. Mater. Sci. Eng., **142**, 012003 (2016).
13. V.Y. Sadovets, V.Y. Beglyakov, A.B. Efremenkov, Applied Mechanics and Materials, **770**, 384 (2015)
14. W.C. Zhong, J. Liu, M.Z. Xue, L.C. Jiao. IEEE Trans. On Systems, Man and Cybernetics, **34:2**, 1128 (2004)

Dependence of Reliability and Resource of the Elements of the Design of Quarry Automatics with the Degrees of their Downloads

Dmitriy Stenin[1*], and *Natalya* Stenina[1]

[1] T.F. Gorbachev Kuzbass State Technical University, 650000, 28 Vesennyaya St., Kemerovo, Russia

Abstract. The article discusses the impact of operational factors on the reliability of reducers for motor-wheels of mine dump trucks, the consequences of oil overheating in the reducer are determined. As a result of the research, the degree of influence of various factors on the oil temperature and the dependence of the oil temperature on the coefficient of load capacity have been determined. And also dependence of expenses on repair of reducers of motor-wheels from loading of a dump truck.

1 Introduction

Coal continues to play a crucial role in the global economy. Coal takes the second place after oil. The growth in Russian coal production 1.5 times from 2000 to 2016 is mainly due to a more economical open-cut mining method, the share of which has increased over this period from 64 to 70,6 %. [1]

The global crisis and anti-Russian sanctions have exacerbated the internal and external problems of the domestic coal industry, have formed a range of risks that limit the development of coal-mining regions.

Russia is one of the world's leading producers of coal. There are one third of the world's coal resources and one-fifth of the explored reserves are 193.3 billion tons in its subsoil. [2]

The diagram of coal production in Russia and Kuzbass for 2006-2016 is shown in Figure 1.

The coal production in Russia and the Kuzbass increased at the last 10 years by 75.7 and 53.9 million tons, respectively.

The dump trucks with the carrying capacity of 130-220 tons are the most common at present time. [3]

* Corresponding author: sdv.ea@kuzstu.ru

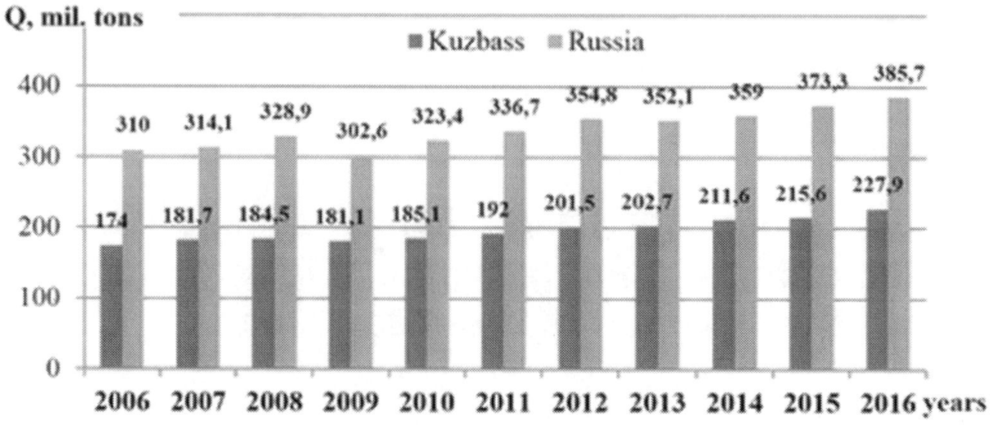

Fig. 1. The coal production in Russia and Kuzbass in 2006-2016.

2 Materials and methods

The most unreliable element of the electro-mechanical transmission of mining dump trucks is the motor-wheel reducer (MWR). The idle-times of the dump trucks by reasons of failures of the carrier part are approximately 28% of the total idle-time and the failures of MWR account for 37% of they [3]. The temperature mode of the MWR depends on its technical condition. The change of the technical condition of the MWR depends on their operating modes which characterized by operating conditions. In work [3] the temperature mode means the mode of changing the temperature of the MWR oil in time during various operating conditions of mining dump trucks.

The influence of the all variety of operational factors on the reliability of the MWR can be estimated through the indicators of their thermal state. Therefore, for the determine of the thermal state it is necessary to establish the dependence of operating conditions from their technical state [4].

Scientific research began with the formulation of the problem and the experimental researches carried out at the area of the branch of "Taldinskiy coal mine" "Mining Company"Kuzbassrazrezugol".

The purpose of the experimental researches was the determination of the various temperature modes of the mining trucks MWR in different operating conditions at different times of the year.

The research tasks included the study of the temperature of MWR oil through of measuring instruments.

For the experiment were chosen the following variables indicators:
- temperature of the oil of the motor-wheel reducer;
- ambient air temperature;
- longitudinal slope of the road;
- length of the route;
- coefficient of load the vehicle.

The all measuring instruments were with a high accuracy and small error. The temperature of the oil was measured at least once a week for several hours.

The main models of dump trucks with electric transmission at the "Taldinskiy coal mine" are the BelAZ-75131, -75302, -75306. Over 90 dump trucks of these models were observed.

The object of the study is MWR (temperature of MWR oil).

The most actual issues related to the development of modern technology, its reliability and durability are the increasing the technical level and quality of mining machine building products. Intensification of mining operations, increasing the productivity of machines and aggregates, intensive growth of mining production is impossible without the increasing the reliability of technical means of mining enterprises. The experience of mining enterprises shows the reliability of mining equipment is still insufficient and depends on mining, organizational, natural and climatic and operational conditions. [5]

The dependencies of reliability indicators from operating conditions were established in work [6]. The obtained regularities allow controlling the technical state of machines during their operation.

The reliability of dump trucks and other mining equipment depends on their constructive features, manufacturing quality and on factors characterizing operating conditions, such as mining technical factors, the quality of road surfaces, the organization of technical operation processes, modes and quality of maintenance. The issue of increasing the reliability of RMK dump trucks is of great importance.

3 Results and discussion

Oil is the most effective, flexible, variable and controlled element. The oil's temperature has the great value in the operation of MWR, because the transmission oil loses its lubricating properties at high temperatures. Because of this, there is an increased wear of gears and bearings in reducers. In practice, the heating of the gearbox is determined by touch, and this often does not provide reliable information.

Studies by various authors show:

1. The intensity of decline in the technical readiness of mine dump trucks is 1,8 ÷ 3,0% per year [7];

2. The TAP-15V gear oil is used in MWR of the dump trucks with the carrying capacity of 80 ÷ 220 tons at an ambient temperature of -25 ° C [7].

3. A decrease the reliability of aggregates of mining equipment with a closed lubrication system is due to a worsening the technical condition of the aggregates themselves as a result of wear, a change of the properties of the operating oil. It is possible to determine the violation of the adjustment of the bearings or the change in the oil level in the reducer by the degree of heating. The oil's condition and its parameters change faster than the failure of the technique. This is justified by the fact that before the failure of MWR in oil the content of wear products increases and the oil's temperature increases [8].

4. The operating oil contains the complex information that allows to predict and diagnose the technical condition of the object and to describe the ongoing processes. It makes the opportunity to identify the reasons of the decrease the reliability of the MWR and to estimate them quantitatively [9].

5. The minimum value of the temperature of the MWR oil is $t_o = 95°$ C (BelAZ-75306), and the average oil temperature – $t_o = 129°$ C and the maximum temperature - $t_o = 161°$ C (BelAZ-75302) exceed the critical already at an ambient temperature more than +15° C [3].

6. The actual technical condition of the motor-wheels reducers is estimated by the temperature of the operating oil and the critical value of the oil operation is 120°C. [8-12]

7. The quantity of failures is reduced by 25% at deviation the oil temperature from the critical value (120°C) to the minimum (95°C), and increases by 9 times with increasing to a maximum value (161°C).

As a result of data processing, the cost ratio for repairing the MWR and the mean temperature of the oil in the reducer was established (Fig. 2).

DOI: 10.1051/e3sconf/20172103002

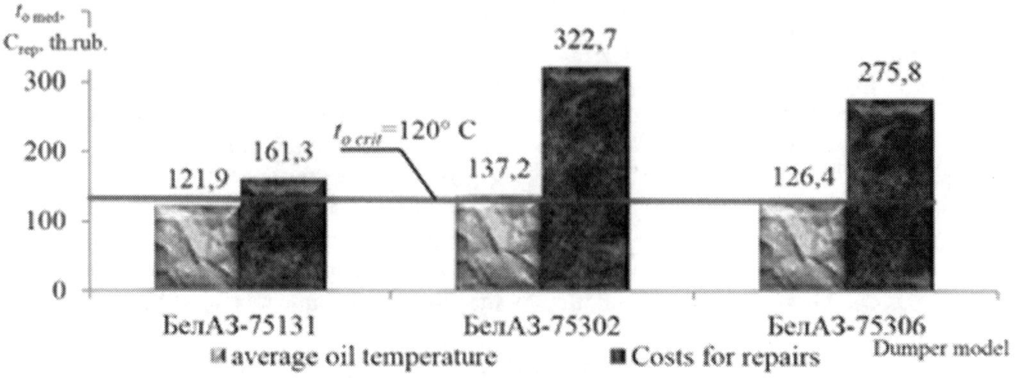

Fig. 2. The value of the average oil temperature and repair costs per one dump truck.

The average temperature of the MWR oil has the maximum values for the BelAZ-75302 quarry dump truck (137,⁰C), the repair costs of the MWR are the highest for the same model of dump truck.

It is established the temperature regime of motor-wheels reducers objectively evaluates their technical condition through reliability indicators: the number of failures, the availability factor and the operating cost factor.

The most indicative index for determining the rational load-carrying capacity coefficient (LCC) is the unit cost of repairs, which allow the influence of the LCC at the efficiency of dump trucks and on the cost of repairing the reducer.

The procedure for determining the rational loading of mining dump trucks with considering the temperature regime of the MWR at the minimum unit cost of repair is as follows:

The unit cost of repairs costs are determined by the formula:

$$C_{un} = \frac{C_{rep}}{Q_y}\text{, rub/t,} \quad (1)$$

where C_{un} - annual costs for repair of RMK, rubles / year .; Q_y - annual efficiency, t/year.

The temperature of MWR oil for all models of dump trucks is determined by the formula:

$$t_o = 70{,}83 + 27{,}09 \cdot \gamma_l + 1{,}04 \cdot t_{am} + 0{,}38 \cdot L_{cc} + 1{,}43 \cdot i, {}^\circ\text{C} \quad (2)$$

where γ_l – load-carrying capacity coefficient.

Annual efficiency (Q_y) is determined by the formula [10]:

$$Q_y = Q_h \cdot T_w \quad (3)$$

where Q_h – is the hourly efficiency of the truck, t/hour; T_w – working time per year, hours.

The hourly efficiency of the truck is determined by the formula:

$$Q_h = \frac{q \cdot \gamma_l \cdot V_t \cdot \beta}{L_{cc} + \text{T}_{it} \cdot V_t \cdot \beta}\text{, т/ч} \quad (4)$$

where q - the carrying capacity of the truck, t; γ_l - load-carrying capacity coefficient; V_t - technical speed, km / h; L_{cc} - length of the carriage with a load, km; T_{it} - idle time of a dump truck for one flight, h; β - the mileage utilization factor.

The coefficient of readiness is determined by the formula [5]:

$$k_r = \frac{D - Д_{rep}}{D}, \quad (5)$$

The coefficient of operating costs Koc is determined by the formula [5]:

$$k_{oc} = \frac{C_{rep}}{C_{MWR}},\tag{6}$$

where C_{rep} - the cost of maintenance before the repair which consist of the costs of repair on maintenance and costs on the fault rectification; C_{MWR} - the cost of MWR.

The relation of unit costs on the load-carrying capacity coefficient is shown in the graph in Fig. 3.

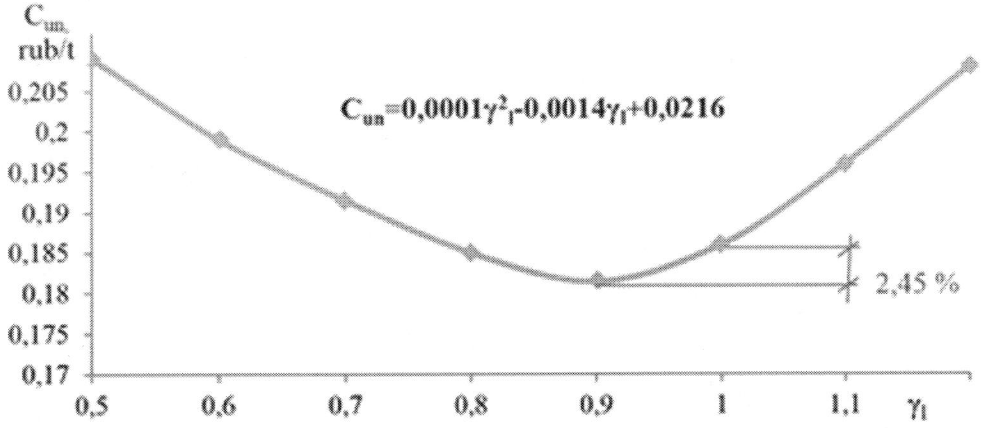

Fig. 3. relation of unit costs (C_{un}) on the load-carrying capacity coefficient (γ_l).

Thus, the rational load-carrying capacity factor during operation dump trucks under these conditions is 0,9. Wherein, the oil temperature increased by 3%, the duration time in the repair for the year did not change; the coefficient of readiness decreased by 0,1%, but thanks to the growth of the efficiency unit costs decreased by 2,45%.

It should be noted that in each case the value of the rational load-carrying capacity coefficient is a different and it depends on the operating conditions.

4 Conclusion

We can to make the following conclusions based on the foregoing:

1. The relations of unit costs, number of failures and days in the repair of oil temperature in the MWR of dump trucks are exponential $n_f = 0{,}0034 \cdot e\,\hat{}\,(0{,}055 \cdot t_o)$, $D_{rep} = 0{,}0093 \cdot e\,\hat{}\,(0{,}0555 \cdot t_o)$; $C_{rep} = 967{,}29 \cdot e\,\hat{}\,(0{,}043 \cdot t_o)$
2. At oil's temperature deviation from the critical value of 120° to the minimum of 95°, the number of failures is reduced by 25% and at increasing to a maximum value 161° - increases by 9 times.
3. The relation of unit costs on the load-carrying capacity coefficient is parabolic, which has a minimum value for the rational coefficient and is described by the equation: $C_{un}=a \cdot \gamma^2_l - b \cdot \gamma_l + c$.
4. The rational value of the the load-carrying capacity coefficient with considering the temperature mode of the MWR and operating conditions is selected at the minimum of unit costs which taking into account the productivity and reliability of mine dump trucks. The number of failures is reduced by 25,7%, the cost of repairing the MWR by 20,8%, the temperature of the MWR oil by 3,21% (from 140,4 ° C to 135 ° C) when decreasing the load-carrying capacity coefficient to 0,8 [3-16].
5. The operating conditions of vehicles reduce performance, reliability and, as a consequence, increase the cost of the obtained final product and makes the increased demands on the fuels and lubricants used in operation.

6. On the one hand, an increase a LCC leads to an increase in productivity. But only hourly, shift or daily, i.e. for a short period of time, when the reliability of the MWR is not yet affected by the performance value. On the other hand, an increase of the load-carrying capacity coefficient (and, in some cases, the overload of dump trucks) leads to an increase of the oil temperature in the MWR. And as a result, their reliability decreases, the number of failures, repair costs and coefficient of readiness increases.

7. The criterion for choosing a rational load-carrying capacity coefficient is as follows: $\frac{C_{rep}}{Q_y} = min \rightarrow \gamma_l^{rat}$.

8. If we know the relation of the idle time of the dump trucks in the repair D_{rep} (because of the failure of the reducers) from the oil temperature in the MWR t_o we can to obtain the relation $D_{rep} = f(\gamma_l)$ with the help of the presented equations. And, thus, we can to determine the most rational load-carrying capacity coefficient in which the duration of downtime in repairs and repair costs will be minimal and the productivity of dump trucks - the maximum.

Thus, the smaller the loading rate of the truck, the better the listed performance indicators. However, with a significant underload of dump trucks, their productivity is significantly reduced.

Therefore, the determination of the degree of loading in this sequence according to the proposed procedure is appropriate.

References

1. A.I. Kopytov, Ugol', **3**, 72 (2017)
2. I.G. Tarazanov, Coal, **3**, 36 (2017)
3. N.A. Stenina, *The influence of operating conditions on the temperature regime of motor-wheel reducer of mining dump trucks. Scientific theses* (Kemerovo, KuzSTU, 2013)
4. S.A. Bogdanov, *The development of a method for determining the change in the technical state of vehicle transmission aggregates in terms of their thermal state: abstract of scientific theses*, (Kharkov, UIAN, 1987)
5. A.A. Khoreshok, G.D. Buyalich, E.V. Preis, M. Yu. Blashchuk, *The reliability of mining machines and equipment: Schoolbook*, (Tomsk: TPU Publishing House, 2008)
6. N.V. Zyryanov, *Determination of the influence of dynamic loads on the resource of mining dump trucks: abstract of scientific theses*, (St. Petersburg, SPbMU, 1995)
7. V.A. Ametov, *The development of ways to improve the reliability of vehicle transmission units in terms of operating oil parameters: Scientific theses*, (Kemerovo, KuzSTU, 1987)
8. A.V. Kudrevatykh, *Justification of methods and parameters for diagnosing reducers of excavator-car complexes: Scientific theses*, 187 (Kemerovo, KuzSTU, 2010)
9. N.V. Ababkov, KuzSTU Bul. **1**, 129 (2016)
10. D.V. Stenin, *Justification of the influence of the resource of bearing systems and the degree of loading on the performance of mining dump trucks: Scientific theses*, (Kemerovo, KuzSTU, 2008)
11. D.V. Stenin, N.A. Stenina, A.P. Stolyarova, Sci. Alm. **9:23**, 501 (2016)
12. D.V. Stenin, N.A. Stenina, Autom. Ind., **10**, 26 (2012)
13. A.S. Furman, D.V. Stenin, V.E. Ashikhmin, KuzSTU Bul. **5**, 92 (2007)
14. D. Basset, M. Herman, J. M. Martin, ASLE Transactions, **27:4**, 380 (1984)

DOI: 10.1051/e3sconf/20172103002

15. Pearse Geoff, Min. Mag., 128 (1987)
16. Q.Sunqing, D. Junxiu, Ch. Guoxu, Lubr. Sci. **11:3**, 165 (1999)

DOI: 10.1051/e3sconf/20172103003

Functional Quality Criterion of Rock Handling Mechanization at Open-pit Mines

Yuri Voronov[1*], and Artyom Voronov[1]

[1] T.F. Gorbachev Kuzbass State Technical University, 650000, 28 Vesennyaya St., Kemerovo, Russia

Abstract. Overburden and mining operations at open-pit mines are performed mainly by powerful shovel-truck systems (STSs). One of the main problems of the STSs is a rather low level of their operating quality, mainly due to unjustified over-trucking. In this article, a functional criterion for assessing the quality of the STS operation at open-pit mines is formulated, derived and analyzed. We introduce the rationale and general principles for the functional criterion formation, its general form, as well as variations for various STS structures: a mixed truck fleet and a homogeneous shovel fleet, a mixed shovel fleet and a homogeneous truck fleet, mixed truck and shovel fleets. The possibility of assessing the quality of the STS operation is of great importance for identifying the main directions for improving their operational performance and operating quality, optimizing the main performance indicators by the quality criterion, and, as a result, for possible saving of material and technical resources for open-pit mining. Improvement of the quality of the STS operation also allows increasing the mining safety and decreasing the atmosphere pollution – by means of possible reducing of the number of the operating trucks.

1 Introduction

Since the quality of the shovel-truck system operation can only be manifested in the course of fulfilling its function in accordance with its purpose (timely excavation, loading and removal of rock) under certain conditions, the work of the STS as an unified system within the mine should be viewed as the result of interaction of its elements (mining shovels and trucks) under restrictions imposed by the external mining environment. The main purpose of assessing the quality of the interaction between shovels and trucks in the STS is to determine how effectively each of them performs its function under specific conditions. For such an assessment, it is necessary to have a functional criterion for joint work of shovels and trucks in the STS structure, taking into account constraints imposed by the external environment – including in terms of its protection and mining safety.

2 Materials and methods

* Corresponding author: ya.voronov454@yandex.ru

A lot of scientific papers are devoted to the problem of determining the rational structure of the STS, for example [1-7]. In them we can single out one common point. From the point of joint work of mining shovels and trucks, the main purpose of shovels, which are leading machines of the system, is to ensure high productivity of the STS; the main purpose of trucks is the ability to ensure that this productivity is achieved, that is, the work of shovel and truck fleets should be balanced as much as possible.

It is possible to balance the total production of N_s available shovels with the help of the necessary number of trucks N_t by comparing the actual production of shovels, which are leading machines in the STS, with the potential production of the truck fleet.

The total shift operational performance (production) of the shovel fleet of the STS can be determined by the formula (m³/shift):

$$P_{\Sigma}^{s} = \frac{E_b^s}{t_c^s} \cdot N_s \cdot T_{sh} \cdot k_{OP}^{s}, \tag{1}$$

where E_b^s – shovel bucket capacity, m³; t_c^s – duration of the shovel cycle (time to load one bucket), h; N_s – number of shovels in the STS; T_{sh} – duration of the working shift, h; k_{OP}^s – operational performance ratio of the shovel fleet in the STS.

The operational performance ratio k_{OP}^s is the ratio of the time of pure work of mining shovels to the duration of the shift, taking into account all regulated and unplanned downtimes. After transformations, it can be represented as:

$$k_{OP}^{s} = 1 - \frac{\sum_{1}^{N_s} t_{DT}^{s}}{T_{sh} \cdot N_s}, \tag{2}$$

where $\sum_{1}^{N_s} t_{DT}^{s}$ – total downtime of N_s shovels during the shift, h.

The total theoretical productivity of the truck fleet (N_t units) is (m³/shift):

$$P_{\Sigma}^{t} = \frac{E_b^t}{t_c^t} \cdot N_t \cdot T_{sh}, \tag{3}$$

where E_b^t – truck body capacity, m³; t_c^t – truck cycle time from the end of previous load to the beginning of the next one, h; N_t – number of trucks in the STS.

The ratio of these productivities (P_{Σ}^{t} and P_{Σ}^{s}) characterizes the balance of the joint work of shovel and truck fleets in the greatest degree, and, consequently, can serve as a functional criterion for assessing the quality of the STS operation. We have

$$\lambda = \frac{P_{\Sigma}^{t}}{P_{\Sigma}^{s}} = \frac{E_b^t \cdot N_t \cdot T_{sh} \cdot t_c^s}{E_b^s \cdot N_s \cdot T_{sh} \cdot t_c^t \cdot k_{OP}^s}. \tag{4}$$

Considering that $E_b^t / E_b^s = n_b$ is the number of buckets loaded into the truck body, and $t_c^s \cdot n_b = t_l^s$ is the time to load the truck by this shovel, after transformations we obtain:

$$\lambda = \frac{N_t \cdot t_l^s}{\left(N_s - \frac{\sum_1^{N_s} t_{DT}^s}{T_{sh}}\right) \cdot t_c^t}. \quad (5)$$

In dependence (5), the value $\sum_1^{N_s} t_{DT}^s / T_{sh}$ is a part of the shovel fleet, which is not used during the shift due to downtimes. It can be designated as N_s^{DT}. Then we have

$$N_s^{DT} = \frac{\sum_1^{N_s} t_{DT}^s}{T_{sh}} \quad (6)$$

and

$$\lambda = \frac{N_t \cdot t_l^s}{\left(N_s - N_s^{DT}\right) \cdot t_c^t}. \quad (7)$$

Without taking into account the shovel downtime, dependence (7) degenerates into $N_t \cdot t_c^s / N_s \cdot t_c^t$, investigated in [8-12] and called "the match factor".

Actually, the functional criterion λ (7) is the ratio of the rate of arrival of trucks for loading λ_t and the rate of service them by shovels λ_s. If $\lambda = 1$, then there is a certain equilibrium point, when trucks arrive to shovels with the same intensity with which they are serviced. If $\lambda > 1$, trucks arrive faster than they are serviced, so we should expect the queues to appear. $\lambda < 1$ shows that shovels operate faster than trucks arrive, and then downtime of shovels waiting for the truck arrival is likely to occur.

3 Results and discussion

Dependence (7) assumes that the arrival of trucks for loading is carried out evenly, that is, each subsequent truck arrives for loading exactly at the moment when the previous one leaves. However, it is almost impossible to achieve an even movement of trucks because of the large number of unforeseen factors. As a result, even if the number of trucks in the STS is quite enough to ensure the removal of all rock, which is loaded by available shovels during the shift, downtime of shovels and trucks will still occur because of the uneven arrival of trucks for loading. Great influence is exerted by the used strategy of truck allocation between shovels (with fixation for particular shovels, without fixation at all or according to a combined scheme). Besides the problem of improving truck loading is absent on the segments with direct dumping [13]. But this technology with internal dumping can be used on the quarry field with flat coal seams. And for inclined and steep seams using the excavator-and-truck complex is necessary. So there are many other aspects that are specific to a particular mine, for example: possibility of unforeseen interaction of trucks at the intersection points of mine roads, impossibility of overtaking, different quality of different road sections, poor crushing of rock during drilling and blasting (affects the loading time), different truck weariness and driver skills (affects the truck speed characteristics).

Dependence (7) also assumes that loading and transporting mining equipment fleets are homogeneous, that is, only one type (size) of trucks and shovels is used. However, in practice, there are mixed equipment fleets everywhere. Therefore, it makes sense to consider three options for joint operation of shovels and trucks in the STS: 1) a mixed fleet of trucks and a homogeneous fleet of shovels; 2) a mixed fleet of shovels and a homogeneous fleet of trucks; 3) both fleets are mixed.

1. Intensity of arrival of trucks for loading (auto/h):

$$\lambda_t = \frac{N_t}{\overline{t_c^t}}, \tag{8}$$

where $\overline{t_c^t}$ – averaged cycle time (including idles) of all trucks during the whole shift, h.

Obviously, loading time of a truck varies depending on the size of this truck. The rate of truck service by shovels λ_{s_i} should reflect the time for servicing each *i*-th truck:

$$\lambda_{s_i} = \frac{N_s - N_s^{DT}}{\overline{t_l^s}}, \tag{9}$$

where $\overline{t_l^s}$ – time to load the weighted average truck, h.

Since the truck fleet is mixed, the value of $\overline{t_l^s}$ will be:

$$\overline{t_l^s} = \frac{\sum_1^{N_t}\left(t_{l_i}^s \cdot N_{t_i}\right)}{N_t}, \tag{10}$$

where $t_{l_i}^s$ – time to load the truck of type *i*, h; N_{t_i} – number of trucks of type *i* in the STS.

From dependences (9) and (10) we obtain:

$$\lambda_{s_i} = \frac{\left(N_s - N_s^{DT}\right) \cdot N_t}{\sum_1^{N_t}\left(t_{l_i}^s \cdot N_{t_i}\right)}. \tag{11}$$

Since the functional criterion λ is the ratio of the intensity of truck arrival to the intensity of shovel service, then dividing expression (8) by expression (11), we obtain:

$$\lambda = \frac{\sum_1^{N_t}\left(t_{l_i}^s \cdot N_{t_i}\right)}{\left(N_s - N_s^{DT}\right) \cdot \overline{t_c^t}}. \tag{12}$$

Despite the fact that the shovel fleet is homogeneous, the variance in the time to load different trucks by the same shovel leads to significant variances in the duration of truck cycles. Therefore, it makes sense to use unique values of cycle times for calculation of the functional criterion. This means that in formula (12), instead of the average cycle for all trucks, the weighted average value of unique truck cycles should be used. We have

$$\overline{t_c^t} = \frac{\sum_1^{N_t}\left(t_{c_i}^t \cdot N_{t_i}\right) + \sum_1^{N_t} t_{DT_i}^t}{N_t} = \frac{\sum_1^{N_t}\left[\left(t_{c_i}^t + t_{DT_i}^t\right) \cdot N_{t_i}\right]}{N_t}, \tag{13}$$

where $t^t_{c_i}$ – cycle time of the *i*-th truck without downtimes, h.

Substituting dependence (13) into expression (12), we obtain:

$$\lambda = \frac{N_t \cdot \sum_1^{N_t} \left(t^s_{l_i} \cdot N_{t_i} \right)}{\left(N_s - N_s^{DT} \right) \cdot \sum_1^{N_t} \left[\left(t^t_{c_i} + t^t_{DT_i} \right) \cdot N_{t_i} \right]}, \tag{14}$$

where N_s^{DT} is determined by dependence (6).

2. When using shovels of different sizes, the time taken to load identical trucks will vary. The exact definition of shovel service rate is achieved by introducing the least common multiple (LCM) for unique values of the loading time in each "truck-shovel" pair. Calculating how many loading cycles is contained in the LCM, we can find the number of trucks serviced in a given period of time. Then the intensity of service of trucks by shovels of different sizes will be determined as

$$\lambda_{s_j} = \frac{\sum_1^{N_s} \left[\left(N_{s_j} - N_{s_j}^{DT} \right) \cdot \frac{Z_{t_l}}{t^s_{l_j}} \right]}{Z_{t_l}}, \tag{15}$$

where N_{s_j} – number of shovels of type *j* in the STS; $N_{s_j}^{DT}$ – number of idling shovels of type *j*; Z_{t_l} – LCM for individual loading times for all *j*-s; $t^s_{l_j}$ – time for shovel of type *j* to load the truck of the only type, h.

Considering that the functional criterion λ is the ratio of the intensity of arrival of trucks (8) to the intensity of their service (15), we obtain:

$$\lambda = \frac{N_t \cdot Z_{t_l}}{\overline{t^t_c} \cdot \sum_1^{N_s} \left[\left(N_{s_j} - N_{s_j}^{DT} \right) \cdot \frac{Z_{t_l}}{t^s_{l_j}} \right]}. \tag{16}$$

If there are several dumps, we can extend equation (16) for different truck cycles:

$$\lambda = \frac{N_t^2 \cdot Z_{t_l}}{\sum_1^{N_s} \left[\left(N_{s_j} - N_{s_j}^{DT} \right) \cdot \frac{Z_{t_l}}{t^s_{l_j}} \right] \cdot \sum_1^{N_t} \left[\left(t^t_{c_i} + t^t_{DT_i} \right) \cdot N_{t_i} \right]}, \tag{17}$$

where $N_{s_j}^{DT}$ is determined by dependence (6).

3. In the case of mixed truck and shovel fleets, it is necessary to first determine the time required for each shovel to service the available fleet of trucks. The loading and idle time (t_{L_j}) for each *j*-th type of shovels is defined as

$$t_{L_j} = \sum_1^{N_t} \left(t^s_{l_{ij}} \cdot N_{t_i} \right), \tag{18}$$

DOI: 10.1051/e3sconf/20172103003

где $t^s_{l_{ij}}$ – time for shovel of type j to load the truck of type i, h.

Then the intensity of service of all trucks by all shovels is

$$\lambda_s = \sum_1^{N_s} \left(\frac{N_{s_j} - N^{DT}_{s_j}}{t_{L_j}} \right) \cdot N_t . \qquad (19)$$

If the truck cycle time is taken as the average over the entire fleet for a given period, then the functional criterion is defined as

$$\lambda = \frac{1}{\sum_1^{N_s} \left(\frac{N_{s_j} - N^{DT}_{s_j}}{t_{L_j}} \right) \cdot \overline{t^t_c}} . \qquad (20)$$

In the case of unique truck cycles, this formula can easily be extended to this:

$$\lambda = \frac{N_t}{\sum_1^{N_s} \left[\frac{N_{s_j} - N^{DT}_{s_j}}{\sum_1^{N_t} \left(t^s_{l_{ij}} \cdot N_{t_i} \right)} \right] \cdot \sum_1^{N_t} \left[\left(t^t_{c_i} + t^t_{DT_i} \right) \cdot N_{t_i} \right]} . \qquad (21)$$

In the dependence (21), four unknown values are fundamental: $t^s_{l_{ij}}$, $t^t_{c_i}$ as well as $\sum_1^{N_t} t^t_{DT_i}$ and $\sum_1^{N_s} t^s_{DT_j}$ (downtime of trucks and shovels waiting for the work, respectively).

The values of $t^s_{l_{ij}}$ and $t^t_{c_i}$ can be determined analytically or taken according to the factual data of the mine. And if it is possible to determine $t^s_{l_{ij}}$ analytically quite accurately, then the truck cycle time $t^t_{c_i}$ is difficult. Therefore, in order to avoid incorrect results, it is desirable to use actual values, especially since automated fleet management systems operating at most mines are able to record the time of all operations to within a second. Analytical methods should be used if such a system is not installed on mining trucks.

Total downtimes $\sum_1^{N_t} t^t_{DT_i}$ and $\sum_1^{N_s} t^s_{DT_j}$ are purely stochastic values, therefore their analytical determination is impossible. They should be taken according to the mine data.

4 Conclusion

A distinctive feature of the functional criterion for a system of machines (STS) is the provision of rational interaction of its component mining shovels and trucks. Shovels as the leading machines of the system should provide the highest possible productivity of the STS, the task of the trucks is to ensure the achievement of this productivity, that is, the work of the shovel and truck fleets should be balanced as much as possible.

The degree of balance of the total productivity of the shovel fleet at a particular mine with the help of a truck fleet, which can be used for this, can be estimated by comparing the intensity of arrival of trucks for loading and servicing them by shovels. Three variants of the STS structure are considered, for each of which specific formulas for calculating the functional criterion were obtained.

As practice shows, the number of simultaneously operating trucks at existing mines is usually overestimated. Formation of the STS with the help of the proposed functional criterion will allow to reduce the number of trucks at the mine. This will reduce the probability of collisions, as well as the level of harmful emissions, thus increasing the safety and environmental friendliness of mining operations.

References

1. G.H. Blackwell, CIM Bulletin, **92:1028**, 143 (1999)
2. C.H. Ta, J.V. Kresta, J.F. Forbes, H.J. Marquez, International Journal of Surface Mining, Reclamation and Environment, **19:3**, 162 (2005)
3. A. Krause, C. Musingwini, The Journal of The Southern African Institute of Mining and Metallurgy, **10**, 469 (2007)
4. S.G. Ercelebi, A. Bascetin, The Journal of The Southern African Institute of Mining and Metallurgy, **109**, 433 (2009)
5. M.A. May, *Applications of queuing theory for open-pit truck-shovel haulage systems*, 79 (MSc thesis, Virginia Polytechnic Institute and State University, USA, 2012)
6. C. Burt, L. Caccetta, Interfaces, **44:2**, 143 (2014)
7. S. A. Dindarloo, M. Osanloo, S. Frimpong, The Journal of The Southern African Institute of Mining and Metallurgy, **115**, 209 (2015)
8. J. Douglas, *Prediction of shovel-truck production: a reconciliation of computer and conventional estimates* (Department of Civil Engineering, Stanford University, USA, 1964)
9. W. Morgan, L. Peterson, Mining Engineering, 76 (1968)
10. C. Burt, *An optimisation approach to materials handling in surface mines*, 168 (Curtin University of Technology, Australia, 2008)
11. C. Burt, L. Caccetta, P. Welgama, L. Fouche, Journal of the Operational Research Society, **62:8**, 1498 (2011)
12. S. Nel, M.S. Kizil, P. Knights, 35th APCOM Symposium, Wollongong NSW, 381 (2011)
13. M.A. Tyulenev, T.N. Gvozdkova, S.A. Zhironkin, E.A. Garina, Geotech. Geol. Eng., **35:1**, 203-212 (2017)

Definition of Static Voltage Characteristics of the Motor Load for the Purpose of Increase in Energy Efficiency of Coal Mines of Kuzbass

Fedor Nepsha[1*], and *Vladimir* Efremenko[1]

[1] T.F. Gorbachev Kuzbass State Technical University, 650000, 28 Vesennyaya St., Kemerovo, Russia

Abstract. The task of determining the static load characteristics is one of the most important tasks, the solution of which is necessary for the correct development of measures to increase the energy efficiency of the Kuzbass coal mines. At present, the influence of electric receivers on the level of consumption of active and reactive power is not taken into account, therefore, the proposed measures to increase the energy efficiency are not optimal. The article analyzes the L-shaped and T-shaped circuit for the replacement of an asynchronous motor (AM), according to the results of which it is determined that the T-shaped replacement scheme is the most accurate for determination of static load characteristics. The authors proposed and implemented in the MATLAB Simulink environment an algorithm for determining the static voltage characteristics of the motor load.

1 Introduction

Electric receivers of the coal mine differ in permissible voltage deviations, depending on the operating mode and their individual properties. The change in the power consumed by the electric receiver when the voltage is changed is usually characterized by a regulating effect. It's expressed as a percentage of the load change when the voltage is changed by 1%.

In the center of the shaft supply and load nodes there is a different combination of different electric receivers, differing in their static load characteristics, which determine the dependence of the power (energy) consumed by the receivers on the voltage, and accordingly the values of the regulating effect. The overall regulating effect of the load depends on the static characteristics of individual electric receivers, the share of each electric receiver, the configuration and parameters of the power supply system of the coal mine.

In the conditions of coal mines, this effect is primarily due to changes in losses in asynchronous electric motors and the power they develop, as well as losses in transformers and networks. To a lesser extent, it is determined by the change in lighting and electric heaters, as well as welding transformers.

It should be noted that the practical results in the thesis of Professor Konyukhova E.V. proves the need to take into account the static voltage load characteristics in the control of normal regimes. The proposed method for determining the rational voltage levels and the degree of reactive power compensation, taking into account the voltage reference voltage,

* Corresponding author: nepshafs@gmail.com

was introduced at the enterprise "AMO ZIL". As a result the active maximum load in the power supply system of the mechanical assembly shop was reduced by 18%.

In addition, the effectiveness of voltage regulation at mining enterprises to manage the level of electricity consumption is noted in the dissertations written under the supervision of Professor Abramovich B.N.

The research of individual and general static load characteristics on the voltage of electric receivers is relevant in the context of solving the problem of increasing the energy efficiency of the Kuzbass coal mines. Currently, all activities aimed at improving energy efficiency are developed without taking into account the specific features of the topology of power supply system, electric receivers to the level of consumption of active and reactive power. Also, to a proper extent, voltage regulation devices are not involved. Because of it the algorithms of their work do not have interrelations with each other.

The motor load on a coal mine has a significant share and is about 60-70% therefore, it is decisive in the development of measures for voltage regulation.

However, at present there are no studies in the field of determining the static voltage characteristics of the engines load used in the coal mines of Kuzbass. In connection with this, the development of a technique for determining the static characteristics of the motor load of a coal mine is an actual scientific and technical task.

The voltage deviation at the AM clips affects the power consumption in a rather complicated way which is confirmed by literary sources and experimental data. These sources show that the consumption of reactive power varies unambiguously: as the voltage decreases, the reactive power consumption decreases. The consumption of active power is characterized by a more complex dependence. With a weak load of AM with a decrease in voltage, the consumption of active power decreases. With the increase in the load factor, the graph approaches a straight line, and as the nominal load is approached, the consumed active power remains practically unchanged.

Nevertheless, there is no generally accepted technique for determining the static characteristics for the motor load, and the question of the mathematical model that can be used for this purpose has not been resolved.

2 Materials and methods

In practical calculations the L-shaped equivalent scheme is used to simplify the calculations (fig. 2). This simplification is made on the basis of an analysis of the relationship between the parameters of the T-shaped and the L-shaped equivalent scheme. In the transition from the T-shaped equivalent scheme to the L-shaped, the magnetization branch is transferred to the place of application of the supply voltage using the conversion coefficient. In this case, when the load changes, the magnetizing current does not change, therefore, in the nominal mode, the magnetizing current is overestimated. However, the most complete characteristics of AM can only be transmitted using a T-shaped circuit for replacing an asynchronous machine (fig. 1). In the article of Plankov A. A. it is noted that the transition to a simplified L-shaped equivalent scheme leads to an error of 3-5% in determining the rotor current strength. Comparison of the results of calculation of the parameters of AM and its characteristics with the use of three equivalent schemes (T-shaped, L-shaped and simplified L-shaped) shows that in the transition from the T-shaped scheme for replacing the AM to the L-shaped scheme in current calculations of AM there is an error of up to 5.8% and up to 7.6% when calculating the overflows of reactive power. In thesis of Professor Konyukhova E.V. the absence of data on resistance of a rotor and a magnetizing branch is resulted as the reason of application of the L-shaped equivalent scheme. Nevertheless, the formulas for calculating the parameters of the T-shaped replacement circuit presented in the literary sources allow us to determine the rotor and stator resistance from the catalog data. Data on the parameters of the magnetizing branch can be obtained from the magnetization curve of the steel used to make the rotor or using the idling characteristic of the motor. Thus, when studying the static characteristics of the voltage of asynchronous motors it is advisable to consider the T-shaped equivalent scheme.

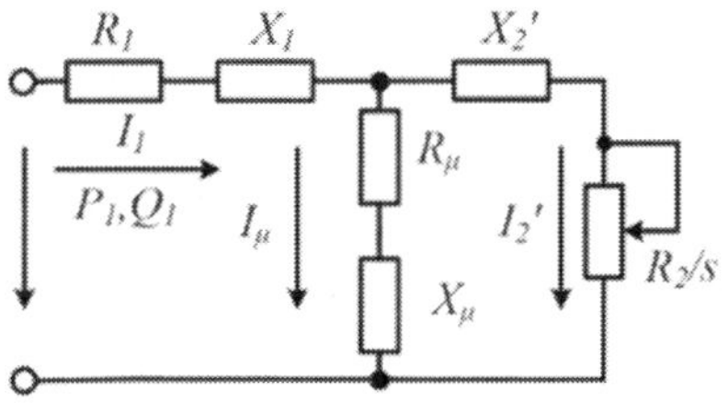

Fig. 1. T-shaped equivalent scheme for AM

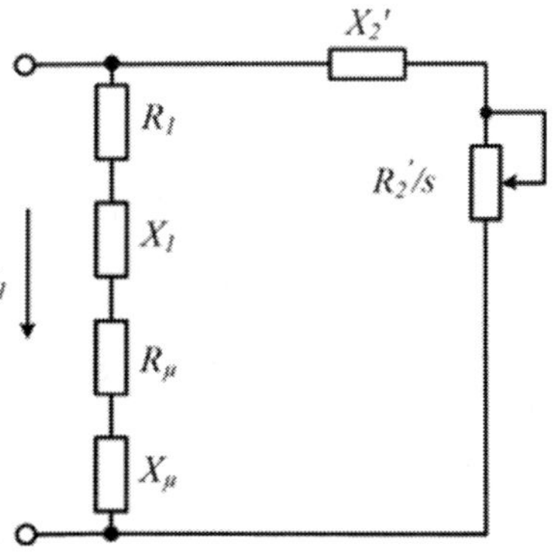

Fig. 2. L-shaped equivalent scheme for AM

It is important to note that a sufficient amount of time has passed since the writing of the thesis of Professor Konyukhova E.V. and new software tools for modeling asynchronous motors have appeared. In particular the authors consider it necessary to single out MATLAB Simulink. In this software complex it is possible to sufficiently accurately simulate an asynchronous motor taking into account the magnetization curve.

The above analysis made it possible to compile an algorithm for determining the static characteristics of the motor load voltage using MATLAB Simulink (fig. 3). This algorithm allows to determine the static voltage characteristics for a number of asynchronous motors.

The algorithm for determining the static characteristics of the motor load voltage is as follows:

2.1 Input of initial data

The initial data for calculating the parameters of AM are the following: V_{nom} - the nominal line voltage [V]; P - rated power [W]; f_1 - main frequency [Hz]; p - number of pole pairs; n - rated speed [rpm]; η - the coefficient of efficiency; $cos\varphi$ - power factor; I_{nom} - rated stator current [A]; k_I - multiplicity of the starting current [pu units]; M_{st} - the multiplicity of the starting torque; M_{max} - the multiplicity of the maximum moment; J - the moment of inertia [kg · m^2].

Initial data of the moment of resistance of the mechanism: M_{fr} - initial moment of friction (s = 1) [N · m], β - load factor of AM; α - an exponent that determines the nature of the load on the shaft.

In the presence of initial data within the program, saturation of AM steel can be taken into account. The parameters of the magnetization curve of the AM are entered in the parameter block of the AM parameters "[i; v] (pu)" in the MATLAB Simulink environment. The magnetization curves $V = f(I_\mu)$ are determined from the current-voltage characteristic of the motor or reference data.

2. 2 Calculation of the parameters of AM equivalent scheme

The process of determining the exact value of the calculated coefficient C is realized by solving the following equation:

$$0 = 1 + \frac{\dfrac{\sqrt{\left[\left(V_{nom}/\sqrt{3}\right)/\left(k_I \cdot I_{nom}\right)\right]^2 - \left(\dfrac{1}{2}\dfrac{V_{nom}^2 \cdot (1 - s_{nom})}{C(1 + C/s_k)m_{\max}(P_{nom} + \Delta P_{mec})} + R_r\right)^2}}{4pf_1}}{L_s - \dfrac{\sqrt{\left[\left(V_{nom}/\sqrt{3}\right)/\left(k_I \cdot I_{nom}\right)\right]^2 - \left(\dfrac{1}{2}\dfrac{V_{nom}^2 \cdot (1 - s_{nom})}{C(1 + C/s_k)m_{\max}(P_{nom} + \Delta P_{mec})} + R_r\right)^2}}{4pf_1}} - C, \qquad (1)$$

where s_k - critical slip; ΔP_{mec} - mechanical losses [W]; R_r - rotor resistance [Ohm]; L_s - stator inductance [H].

The equation is solved using the fzero command built into MATLAB to find the zeros of a function of one variable.

At the end of the calculation, the parameters of the equivalent scheme for each engine under investigation are displayed in the form of a matrix in the working environment of the program.

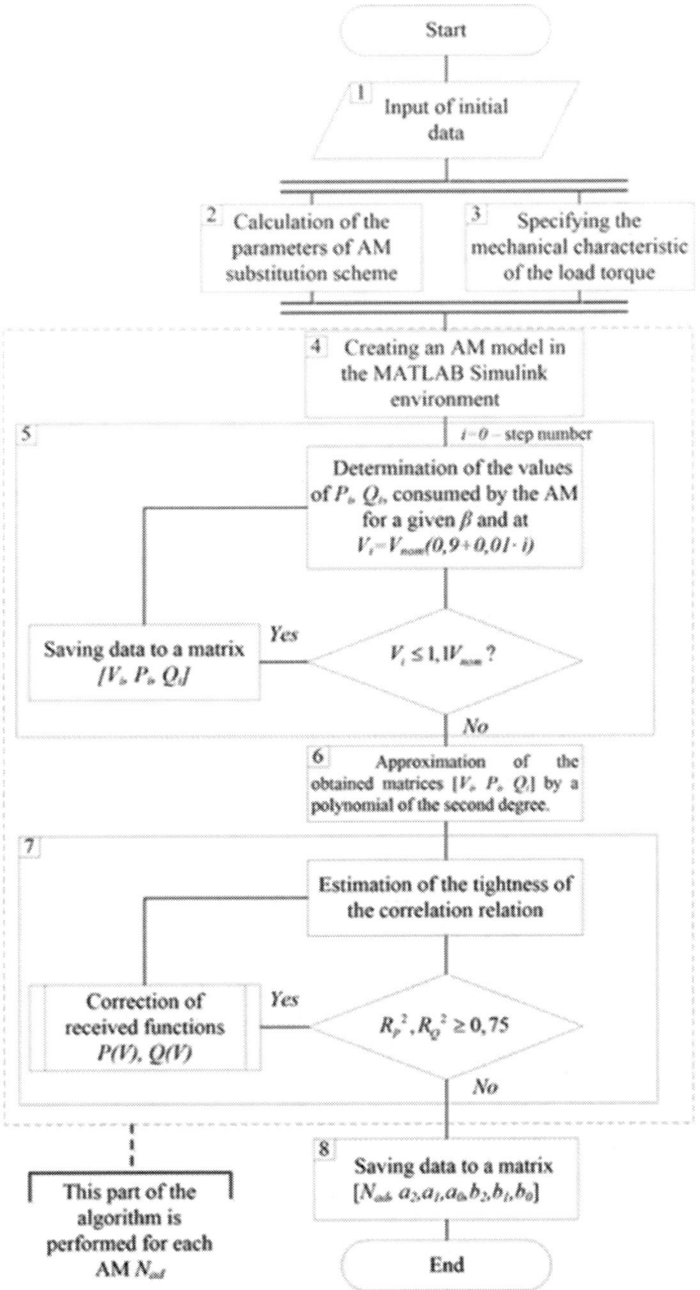

Fig.3. Algorithm for determining the parameters of static voltage characteristics for the motor load voltage using MATLAB Simulink

2.3 Specifying the mechanical characteristic of the load torque

The mechanical characteristic of the moment of load resistance is given by the formula:

$$M_{st} = \beta M_{nom}\left(M_{fr} + (1 - M_{fr})\cdot\left(\frac{1-s}{1-s_{nom}}\right)^{\alpha}\right), \tag{2}$$

DOI: 10.1051/e3sconf/20172103004

The exponent α which takes into account the change in the speed of rotation of the engine is determined by the type of mechanism.

2.4 Creating An AM Model In The MATLAB Simulink Environment

The AM model is created using the MATLAB Simulink toolkit. The parameters defined in items 2-3 are entered in the model parameters. The appearance of the model is shown in Fig.4.

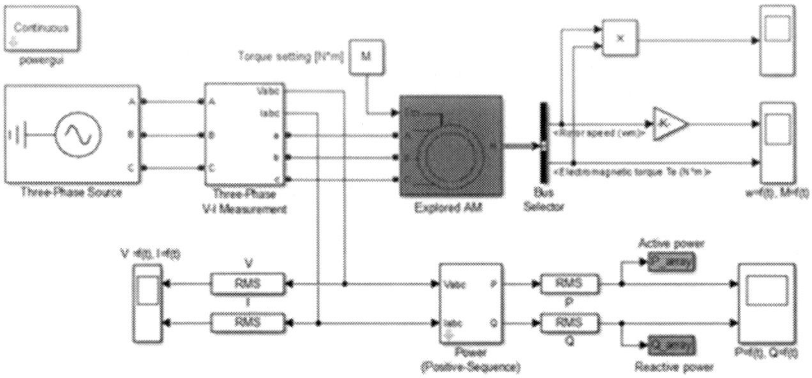

Fig. 4. Model for determining the parameters of static voltage characteristics of motor load

The model can be adjusted depending on the type of motor and the mode of operation of the neutral of the simulated network.

2.5 Determination of the values of active and reactive powers consumed by AM at a different voltage level.

For each engine under investigation, the electric regime is calculated until the steady state is established. The calculation is performed for a voltage range of 0.9-1.1V_{nom} (the calculation is performed 20 times with a resolution of 0.01V_{nom}). The upper and lower limits of the simulation are determined by the permissible voltage level deviations at the motor terminals determined by GOST 52776-2007. The obtained data on the change in the consumption of active and reactive power (blocks P_array, Q_array) are stored in the MATLAB working environment. Also the matrices [V_i, P_i, Q_i] are formed for each simulated AM.

2.6 Approximation of the obtained matrices [V_i, P_i, Q_i] by a polynomial of the second degree.

Approximation of the obtained dependencies is performed using the polyfit function built into MATLAB. As a result for each AM a second-degree polynomial is formed in the form:

$$P(V) = P_{nom} P_*(V) = P_{nom} \left[a_0 + a_1 \cdot \left(\frac{V}{V_{nom}} \right) + a_2 \cdot \left(\frac{V}{V_{nom}} \right)^2 \right], \quad (3)$$

$$Q(U) = Q_{nom} Q_*(U) = Q_{nom}\left[b_0 + b_1 \cdot \left(\frac{V}{V_{nom}}\right) + b_2 \cdot \left(\frac{V}{V_{nom}}\right)^2\right], \quad (4)$$

Where P_{nom} and Q_{nom} are the active and reactive power of the AM at the nominal voltage; $P^*(V)$ and $Q^*(V)$ are the static voltage characteristics of AM in relative units; V_{nom} - the nominal voltage of the AM or network; a_0, a_1, a_2, b_0, b_1, b_2 are the coefficients of the polynomial of degree 2.

2.7 Estimation of the tightness of the correlation relation

To estimate the reliability of the obtained polynomials of the second degree (3,4), the determination of the determination coefficient is carried out:

$$R_P^2 = 1 - \frac{\sum_{i=1}^{20}(P_i - P_{scli})}{\sum_{i=1}^{20}(P_i - \bar{P})}, \quad (5)$$

$$R_Q^2 = 1 - \frac{\sum_{i=1}^{20}(Q_i - Q_{scli})}{\sum_{i=1}^{20}(Q_i - \bar{Q})}, \quad (6)$$

where P_i, Q_i are the active and reactive power values obtained as a result of simulation; P_{scli}, Q_{scli} - the values of active and reactive powers, obtained from (3,4), $\bar{P}, \bar{Q}$ - average values of active and reactive power.

If $R_P^2, R_Q^2 \geq 0,75$ the obtained functions (3,4) have a strong correlation with the data obtained in the MATLAB Simulink. Otherwise, the received functions are corrected (3,4).

2.8 Obtaining the final values of the coefficient of the polynomial of the second degree.

The data is output in the form of a form matrix $[N_{ad}, P_2, P_1, P_0, Q_2, Q_1, Q_0]$ (N_{ad} -the number of AM) and can be used to generate a database for setting the motor load when calculating the electrical regimes of a coal mine power supply systems or for the subsequent determination of the generalized static characteristics of mine loads.

3 Results and discussion

On the basis of the developed algorithm, the authors wrote a program in the MATLAB environment which allows automatic calculation of the static characteristics parameters for the voltage of asynchronous motors.

With the use of the program, the static characteristics of the voltage of the pumping station motors of different power for various load factors have been determined. The obtained static characteristics of the motor load are shown in fig. 5.

DOI: 10.1051/e3sconf/20172103004

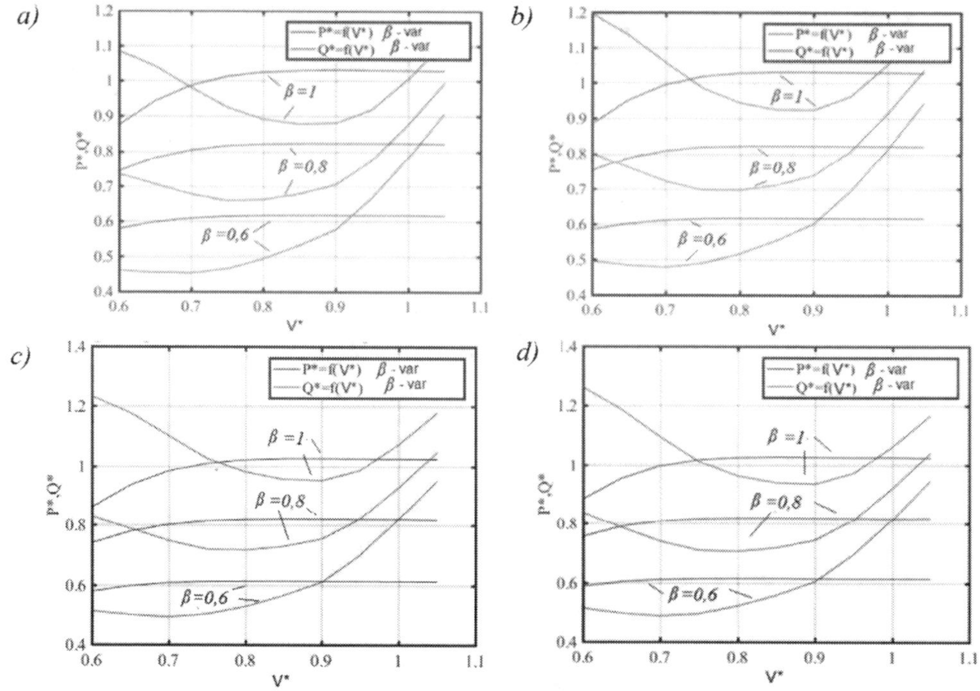

Fig. 5. Static voltage characteristics of the pump station motors: *a* - VAO2-280S4 (132 kW), *b* - VAO2-280L4 (200 kW), *c* - VAO2-315M4 (250 kW), *d* - VAO2-315L4 (315 kW).

The resulting static voltage characteristics of the motors were approximated. As a result the coefficients of the polynomial of the second degree are obtained which are given in the table.

Table 1. The obtained values of the coefficients of the polynomial of the second degree for the engines of the pump station of various powers

Brand AM	P_{nom}, **kW**	β	**Coefficients of a polynomial of degree 2**					
			by active power			**by reactive power**		
			a_2	a_1	a_0	b_2	b_1	b_0
VAO2-280S4	132	0,6	-0,3783	0,6847	0,3089	3,5638	-4,9391	2,1541
		0,8	-0,7927	1,4423	0,1730	4,0222	-6,1293	2,9819
		1,0	-1,5265	2,7930	-0,2350	4,3627	-7,2569	3,8993
VAO2-280L4	200	0,6	-0,3123	0,5633	0,3645	3,8119	-5,3456	2,3436
		0,8	-0,7184	1,2980	0,2428	4,4146	-6,8276	3,3250
		1,0	-1,4689	2,6715	-0,1711	4,9666	-8,3894	4,4728
VAO2-315M4	250	0,6	-0,3577	0,6442	0,3265	3,8225	-5,3867	2,3813
		0,8	-0,7999	1,4509	0,1686	4,4239	-6,8826	3,3840
		1,0	-1,6213	2,9629	-0,3137	4,8723	-8,2884	4,4880
VAO2-315L4	315	0,6	-0,2652	0,4755	0,4030	3,9123	-5,5402	2,4401
		0,8	-0,6545	1,1785	0,2953	4,6627	-7,2989	3,5536
		1,0	-1,4536	2,6339	-0,1530	5,3910	-9,2094	4,8770

From the analysis of the obtained data it follows that the voltage change at the terminals of the pump station motors practically does not lead to a change in the consumed active power regardless of the load factor of the engine and its rated power. When the voltage level rises by 1%, the reactive power consumption increases by 2-4%, depending on the load factor of the engine and its rated power. At the same time with a decrease in the rated power and the load factor of the engine, the efficiency of reducing the reactive power consumption increases.

The use of the developed algorithm is also possible to determine the static characteristics of the voltage of synchronous motors under the condition of replacing the model AM in the MATLAB Simulink model with the SM model.

4 Conclusions

When determining the static voltage characteristics of AM, it is necessary to use the T-shaped equivalent scheme which is much more accurate than the L-shaped equivalent scheme. Under these conditions to determine the static voltage characteristics, it is most expedient to use the AM models presented in the MATLAB Simulink library.

The proposed algorithm makes it possible to obtain reliable data on the static voltage characteristics of the motor load with respect to voltage taking into account the change in the voltage and the moment of resistance.

The data obtained can be used to optimize the operation mode of the coal mine power supply system by voltage according to the criterion of minimum consumption of active power as well as in the development of measures to increase the energy efficiency of power supply systems of coal mines.

References

1. P. Krause, O. Wasynczuk, S. D. Sudhoff, Steven Pekarek, *Analysis of Electric Machinery and Drive Systems*, (IEEE Press, 2013)
2. C.D. Pitis, M.W. Zeller, EPEC, **1:3**, (2008)
3. H. Auinger, *Energy efficiency Improvements in Electric Motors and Drives*, 284 (Springer-Verlag, 1997)
4. *Simulink Design Optimization. User's Guide. User's Guide*, (The Math Works, Inc. New Hampshire, 2015)

Disk Rock Cutting Tool for the Implementation of Resource-Saving Technologies of Mining of Solid Minerals

Leonid Mametyev[1], *Aleksey* Khoreshok[1], *Alexander* Tsekhin[1], and *Andrey* Borisov[1*]

[1] T.F. Gorbachev Kuzbass State Technical University, 650000, 28 Vesennyaya St., Kemerovo, Russia

Abstract. The directions of a resource and energy saving when creating a boom-type effectors of roadheaders of selective action with disc rock cutting tools on a multi-faceted prisms for the destruction of formation of minerals and rocks pricemax are presented. Justified reversing the modes of the crowns and booms to improve the efficiency of mining works. Parameters of destruction of coal and rock faces by the disk tool of a biconical design with the unified fastening knots to many-sided prisms on effectors of extraction mining machines are determined. Parameters of tension of the interfaced elements of knots of fastening of the disk tool at static interaction with the destroyed face of rocks are set. The technical solutions containing the constructive and kinematic communications realizing counter and reverse mode of rotation of two radial crowns with the disk tool on trihedral prisms and cases of booms with the disk tool on tetrahedral prisms in internal space between two axial crowns with the cutter are proposed. Reserves of expansion of the front of loading outside a table of a feeder of the roadheader of selective action, including side zones in which loading corridors by blades of trihedral prisms in internal space between two radial crowns are created are revealed.

1 Introduction

In leading mining countries, the main means of mechanization of mining operations are roadheaders, shearers, plough system and drilling machines. Improvement of the effectors of roadheaders boom type by rational combination and the allocation of cutter and disk tool for the implementation of the principle of the destruction of coal and strong rocks major cleavage to resource and energy conservation is a topical scientific and technical problem.

2 Materials and methods

Modern mining machines is implemented by mechanical destruction method of coal and rock faces using the broad structural spectrum of mining tools. It is known that the destruction of rocks by cutting is an energetically favorable process [1]. However, in the abrasive rocks of medium and above medium strength, this method is not practically acceptable, in connection

* Corresponding author: borisovau@kuzstu.ru

with the intensive wear of the working tool, due to the inability to transfer a large power unit is required for destruction of the face. Therefore, the area of effective application of the destruction of rocks by cutting is limited to breed with a little abrasiveness (less than 15 mg) with a limit of compressive strength up to 60 MPa.

Studies in KuzSTU justified by the prospects of using the disk tools on the effectors of mining machines.

Working conditions mining tool for boom-type roadheaders necessitates the identification of functional possibilities of using the disk tools on the crown effectors.

The complex of requirements to the construction of the crowns boom-type roadheaders selective action is based on the interdependence of kinematic, power and design parameters of the attachment disc tool with the physico-mechanical properties and fracture parameters of faces and the provision of installation and dismantling operations.

To enhance the functionality of the effectors of roadheaders selective action via combining the processes of destruction, loading and crushing of rock mass proposed disk tools with the attachment to the triangular prisms in the construction of two reversible radial longitudinal axis of the crowns (fig. 1) [2].

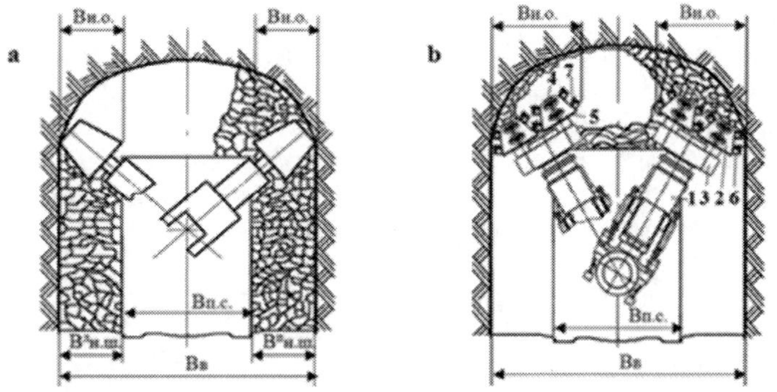

Fig. 1. Work boom-type effector with the radial crowns: a – prior to modernization; b – after modernization.

The effector of roadheader (fig. 1, b) contains an boom 1, which has two destructive loading of the crowns 2, kinematically connected through the transfer case gear 3. Each crown 2 consists of truncated cones or prisms with a small base 4 and a big base 5, a hub which is attached to the output shaft of the gearbox 3. To the outer surfaces of crowns attached triangular prism 6 with disc tools 7. Gear kinematic connection allows the triangular prisms 6 crowns 2 during rotation to create a spatial mazes and corridors, providing combination of the processes of destruction, crushing and loading within the project of the width $B_В$ of the driven working. The process of loading zones in the cut off parts of tunnel production is characterized by the following parameters: $B_{н.о.}$ – working width of the effector along the sides; $B^{л}_{н.ш.}$ и $B^{п}_{н.ш.}$ – width is not shipped stacks of destruction products from the left and right sides; $B_{в}$ – width of project development; $B_{п.с.}$ – the width of the delivery table of the feeder of the loading device, characterizing the width of the zone of the front loading.

The department of mining machines and complexes KuzSTU in cooperation with the department of mining equipment of YUTI (branch) of Tomsk Polytechnic University conducted studies of the stress-strain state of various designs of the cantilever attachment disk tool by the finite element method. One of the results of the study of the stress state of the mounting disk tool to a triangular prism with the nut shown in fig. 2 [2].

 DOI: 10.1051/e3sconf/20172103005

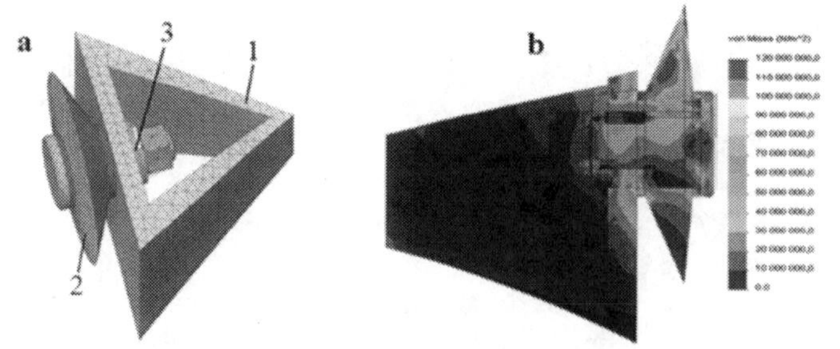

Fig. 2. Finite element model (a) and stress state (b) option of the mount of the disk tool are: 1 – triangular prism; 2 – disc; 3 – nut.

In three constructive variants of the attachment (strap, screw and nut) used biconical and conical disk tools. As an example, in fig. 2, b shows the distributions of the equivalent stress according to Mises for disk tool diameter D = 0,16 m.

The console variants of the attachment of the disk tool on the triangular prisms taking into account the results of modeling the stress-strain state in the destruction of coal and rock faces are recommended to equip their working bodies roadheaders, shearers and drilling machines.

The main disadvantage of many of the effectors of roadheaders cross-axial design is the low productivity of the process of the cutting to the presence of an indestructible block coal and rock faces in the space between the crowns (fig. 3) [3].

The department of mining machines and complexes KuzSTU to improve the efficiency of the process of the cutting of the effector roadheader of selective action with cross-axial crowns separated from each other by the casing of the distributing reducer proposed in the space between the crowns to place the disk tools on tetrahedral prisms.

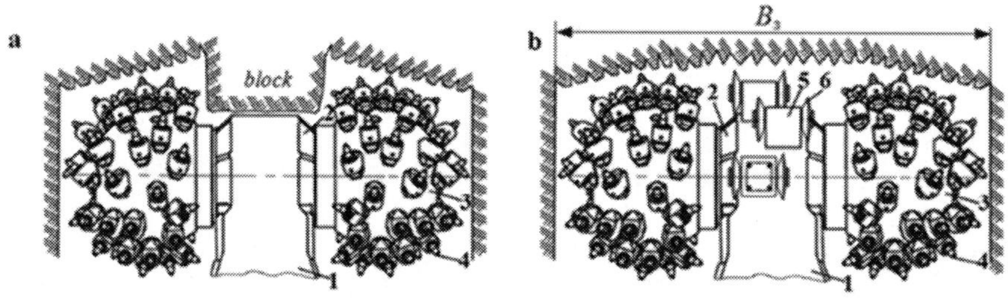

Fig. 3. Cutting effectors: a – analog; b – upgraded

In the process of the cutting (fig. 3) boom 1 the effector with the distributing gearbox 2, axial crowns 3, cutters 4, tetrahedral prisms 5 with disc tools 6 make a vertically rotatable planing motion with simultaneous telescopic effect on the face. In this block (fig. 3, a), which is formed by the standard of the effectors-analogues is destroyed in the space between the crowns of disk tools 6 on the tetrahedral prism 5 (fig. 3, b). Part of the body of the distributing gearbox 2 boom 1 turned to face and made in the form of a cylindrical sector, on the outer surface of which are mounted in a staggered tetrahedral prism 5 with the disk tools 6, fitting into the space gap between lines of cut formed by the outermost cutters 4 of side large grounds axial crowns 3. Thus the radial departure of V-cutting edges of the tool disk 6 does not exceed the radial departure of the extreme cutters 4 on large grounds axial crowns 3. It allows to be effectively cut to the width B_3 (fig. 3, b) due to movement of the boom 1 during the cutting in the vertical plane with a telescopic effect. In the case of step processing slaughtering axial crowns 3 is carried out the turning movement of the boom 1 in a horizontal plane to the right or left sides of the excavation.

Similarly, a procedure for the construction of finite element models and calculation efforts loadings P_z, P_y, P_x, disc tool with the attachment on tetrahedral prisms. In fig. 4, a finite element model of a tetrahedral prism with disc tools creating an additional fracture zone in the space between the crowns effector of roadheader. For the initial conditions to calculate the parameters of the stress state was chosen ranges of the characteristics of the destructible faces: coal ($\sigma_{сж}$ = 12,4; 13,5; 14.8 MPa) and the rock ($\sigma_{сж}$ = 51; 60,6; 78.9 MPa). In the simulation used four possible designs of disk tool diameter D = 0,16 m (three biconical with pointed angles: $\varphi = \varphi_1+\varphi_2$ = 25°+5° = 30°; 20°+10° = 30°; 15°+15° = 30° and one conic φ = 30°) [3].

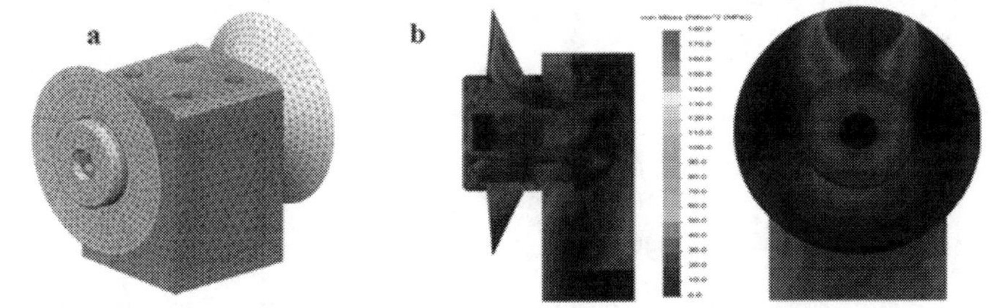

Fig. 4. Finite element model (a) and stress state (b) option of the mount separate disks (φ = 25°+5° = 30°) to the tetrahedral prism in the destruction of the rock mass $\sigma_{сж}$ = 78.9 MPa.

The calculation was performed in SolidWorks Simulation. Meshing was used parabolic finite elements in the form of tetrahedra. The size of the finite elements was chosen so that further increase in mesh density had no significant effect on the results of the calculations. Material details 35HGSA, steel 45. The cutting force P_z, penetration P_y and lateral force P_x on the disc tools are subject to design, operating parameters and characteristics of the destroyed face $\sigma_{сж}$. Computational effort of loading P_z, P_y, P_x was applied to finite element models (fig. 4, a) disk tools the attachment in tetrahedral prisms to produce pictures of the stress-strain state for biconical and conical disc tools (fig. 4, b).

On the basis of the analysis of the basic constructive solutions [4-13] proposed the original technical solution, which includes tetrahedral prisms with dual disc tools (fig. 5), made as a single prefabricated structural unit with the joint possibility of free rotation relative to the fixed axle-bushings. Its implementation provides for joint rotation of the two disc tools on each tetrahedral prism with a decrease in the process of jamming and wear in the cut of the boom-type effector of roadheader in the massif [4].

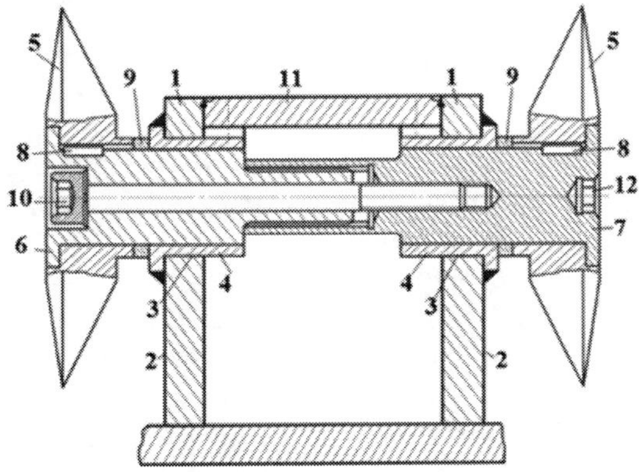

Fig. 5. Design tetrahedral prism with dual disk tools.

Tetrahedral prism 1 (fig. 5) consists of two working faces 2 with a through cylindrical holes 3 and two side faces. Inside each of the through cylindrical holes 3 is placed the axle-bushing 4. While paired disk tools 5 made as a single dismounted and mounted structural unit with the possibility of a joint free rotation about fixed axle-bushing 4. Each axle-bushing 4 consists of an end thrust collar rigidly attached to the outer surface of the working faces 2 tetrahedral prism 1 and the cylindrical surface of the console placed in a through cylindrical hole 3 of the working face 2 in the internal space of the tetrahedral prism 1.

A single dismounted and mounted constructional unit made in the form rigidly attached to each other two axes 6, 7 from the with thrust ribs, one of which contains a slotted shank, and the other contains the spline hub. Axis 6, 7 have the zones providing connection of the disk tool 5 by means of dowel-pawl 8. Remote face of the washer 9 is placed between the end faces of each of the two disc tools 6 and the end faces of the thrust ribs of each of the two axle-bushings 4. The axle 6 is formed with a stepped through hole, inside which is placed remote the screw 10 for fastening a single dismounted and mounted structural unit. For protection of internal space of a tetrahedral prism 1 use a tetrahedral cover 11. From the free end of the axle 7 is made hexagonal recess 12 under the key for the implementation of installation and dismantling operations.

3 Results and discussion

Found that equivalent stress according to Mises all versions of the attachment disk tool radial crowns roadheaders is substantially below the yield strength for steel 35HGSA (σ_T = 490 MPa). With the transition from asymmetry to symmetry biconical disk tools can be traced to the reduction zone settings equivalent stresses in the attachment points with the general increase in maximum stress with the increase of rock strength in a wide range $\sigma_{сж}$ = 50-120 MPa.

It is revealed that, disk tools conical (φ = 30°) and biconical performances (φ = 25°+5° = 30°) implement the process of breaking larger areas of maximum equivalent stresses and displacements than alternatives biconical version (φ = 20°+10° = 30° and φ = 15°+15° = 30°), and the minimum sizes of zones of equivalent stresses and displacements observed for the biconical design (φ = 15°+15° = 30°).

Decrease in the sizes of zones of the maximum equivalent tension and movements on a bottomhole side of the trihedral prism turned to a face for option of knot of fastening of the disk tool by a nut, in comparison with fastening by the screw is noted that characterizes higher rigidity of fastening by a nut.

Technical solutions and test results were obtained within the base part of Ministry of Education and Science of Russia state order, project 632 "Investigation of technologies and techniques parameters for selecting and developing innovative designs to improve operating efficiency of multipurpose mining machines in Kuzbass".

4 Conclusion

1. One of the directions of energy saving in underground coal mining or mineral raw materials is to reduce the downtime of roadheaders and to increase the rate of penetration of preparatory workings by the extension of front loading before the cut off parts zones using the original boom-type effectors with two radial crowns and disk tool on the triangular prisms.

2. Proposed to reduce the energy intensity and time of cutting boom-type by the roadheaders of selective action using the disk tool in tetrahedral prisms in the inner space between the axial cutting units crowns.

3. Additional factors of resource-saving are to a significant increase in durability and service life of disc rock-destructive tools compared to cutters for reducing the cost of maintenance and repair in the face space.

References

1. A.A. Khoreshok, L.E. Mametev, A.Yu. Borisov, A.V. Vorobev, Appl. Mech. Mater. **770**, 429 (2015)

2. A.A. Khoreshok, L.E. Mametev, A.Yu. Borisov, A.V. Vorobev, Applied Mechanics and Materials, **770**, 434 (2015)
3. A.A. Khoreshok, L.E. Mametyev, A.Yu Borisov 2015 I Miner's Week – Reports Of The Xxiii International Scientific Symposium, 482 (2015)
4. B.L. Gerike, P.B. Gerike, V.I. Klishin, A.P. Filatov, J. Min. Sci. **44:5**, 497 (2008)
5. B.L. Gerike, P.B. Gerike, A.P. Filatov, V.I. Klishin, J. Min. Sci. **42:6**, 610 (2006)
6. P.B. Gerike, M.A. Belikov, J. Min. Sci. **39:2**, 162 (2003)
7. P.D. Krestovozdvizhensky, V.I. Klishin, S.M. Nikitenko, P.B. Gerike, J. Min. Sci. **51:2**, 323 (2015)
8. Y Zhang, X W Wang, H F Liu, Appl. Mech. Mater. **487**, 513 (2014)
9. K. Yang, Y. M. Xia, Y. S. Wu, Appl. Mech. Mater. **508**, 159 (2014)
10. A. N. Zhang, Z. F. Zhu, F. Zhu, Adv. Mater. Res.**791:793**, 742 (2013)
11. X H Li, W Du, Z L Huang, W L. Fu, Appl. Mech. Mater. **127**, 385 (2011)
12. G. Wittenberger, M. Cehlár, Z. Jurkasová, Acta Montanistica Slovaca, **17:4**, 241-246 (2012)

Technical Diagnostics of Ventilation Units for Energy Efficiency and Safety of Operation

Evgeny Kuzin*[1], *Vitaliy* Shahmanov [1], and *Dmitriy* Dubinkin [1]

[1]Prokopievsk branch of T. F. Gorbachev Kuzbass State Technical University, 653039, 19a Nogradskaya str., Prokopievsk, Russia

Abstract. The article considers the questions of application of technical diagnostics fan installations methods for providing safe operation, the system of the technical maintenance improvement and repair. Due to the fact that one of the most important aspects in fan operation in mining is energy efficiency and energy saving, the lack of the data in the control of the level in vibration of stationary sensors is shown. The necessity of taking into account the geometric parameters of the intake channel has been shown, and also the necessity of creation of the reference masks for the assessment of technical condition and energy efficiency when operating fan installations in mining. The results of technical diagnostics of the main fans using the methods of vibration diagnostics are provided. Aspects of vibration at characteristic points are shown. The necessity for further accumulation of data characterizing vibration for adjustment of the reference masks and more accurate detection of defects and deviations from the energy-efficient mode of operation of the fan installations is given.

1 Introduction

Fan installation in mining operations provide ventilation of the main and auxiliary mine workings, and also used for preliminary degassing of coal seams. Primarily, personnel safety depends on the safe operation of installations of ventilation and degassing of negative consequences caused by the release of methane in underground mines.

All ventilation systems of coal mines are reserved and supplied with stationary vibration temperature and performance sensors. This controls the working of the ventilation system, but not its energy efficiency.

Due to the continuous growth of energy tariffs, as well as in the framework of the «A policy framework for climate and energy in the period from 2020 to 2030» and the International standard ISO 50001 [1, 2], it is very important not only to reduce energy losses, but also to provide efficient service for the maintenance and repair of equipment of mining enterprises. Energy saving is more beneficial than generation capacity, because the costs of saving 1 kW of power costs four to five times cheaper than the cost of a newly commissioned 1 kW of power, depending on the remoteness of the enterprise from generating plants.

* Corresponding author: kuzinevgen@gmail.com

 DOI: 10.1051/e3sconf/20172103006

2 Materials and methods

Based on the results of the measurements only the overall level of vibration (for example, in a wide range of 2-1000 Hz) changes in the condition of rotating machinery to a condition in which increasing the amplitude of certain harmonics will reach the level of the component with the largest amplitude (usually at a frequency of rotation of the rotor) can not be detected [3]. This condition can be evaluated as satisfactory (exp limited time), in this case the energy saving in the operation of this equipment is not taken into account [4].

For detailing the defects the limits on vibration level for individual spectral components (reference masks) that is associated with different levels of risks of emergency failure of equipment from various defects are introduced. The method of using the reference masks allows you to not only to evaluate, but also to predict the change in the condition of the equipment [5, 6]. In addition, for each fan size, it is advisable to create your own reference mask, as it influences not only frequency of rotation of the rotor, but the number of vanes (blades) of the fan. It is proposed to use the method of reference masks to assess the efficiency of the fan. Defects such as imbalance and misalignment leads to increasing mechanical losses, therefore to a waste of energy.

Much attention is paid to the energy saving in the adjustment of performance, it is also known that the efficiency of axial fans is influenced by the geometric dimensions of the duct, the distance between the impeller and diffuser and other aspects [7, 8].

Thus, in this work the selected method of construction of the reference masks, to ensure safe operation and evaluation of energy efficiency when the fan has been chosen.

3 Results and Discussion

During the operation of the fan 2VDK-32/2 stationary system of vibration diagnostics the repeated exceeding of the maximum allowable level of vibrations at 11.2 mm/c has been noticed. It has decided to conduct the diagnostics in the manual mode. Point measurements of vibration are shown in Fig. 1. and the total vibration is shown in table. 1

Fig. 1. The scheme of measurements of vibration on the fan 2VDK-32/2

Table 1. The overall level of vibration in the range 2-1000 Hz

 DOI: 10.1051/e3sconf/20172103006

Measuring point	RMS, mm/s	Measuring point	RMS, mm/s	Measuring point	RMS, mm/s
1V	1,53	3V	2,80	5V	1,68
1H	4,16	3H	5,85	5H	4,30
1A	2,06	3A	2,98	5A	1,31
2V	3,75	4V	2,04	6V	2,49
2H	5,69	4H	3,68	6H	4,25
2A	7,35	4A	1,95	6A	2,03

Analysis of the data in the table shows that at the point 2A overall vibration level reaches a warning. The results of detailed analysis show the following: current frequency of the rotor

$$f_r = \frac{n}{60} = \frac{750}{60} = 12{,}5 \text{ Hz}; \tag{1}$$

where n=750 rpm, the rotational speed of the rotor of the motor.

Noticed some imbalance of the fan impeller in the General level within zone A (see Fig. 2)

The frequency of the flow on the blades of stage 1 (the number of blades of stage 1)

$$f_{l1} = f_r \cdot z_1 = 12{,}5 \cdot 15 = 187{,}5 \text{ Hz}; \tag{2}$$

The frequency of the flow on the blades 2 of the degree (the number of blades 2 stages)

$$f_{l2} = f_r \cdot z_2 = 12{,}5 \cdot 11 = 137{,}5 \text{ Hz}. \tag{3}$$

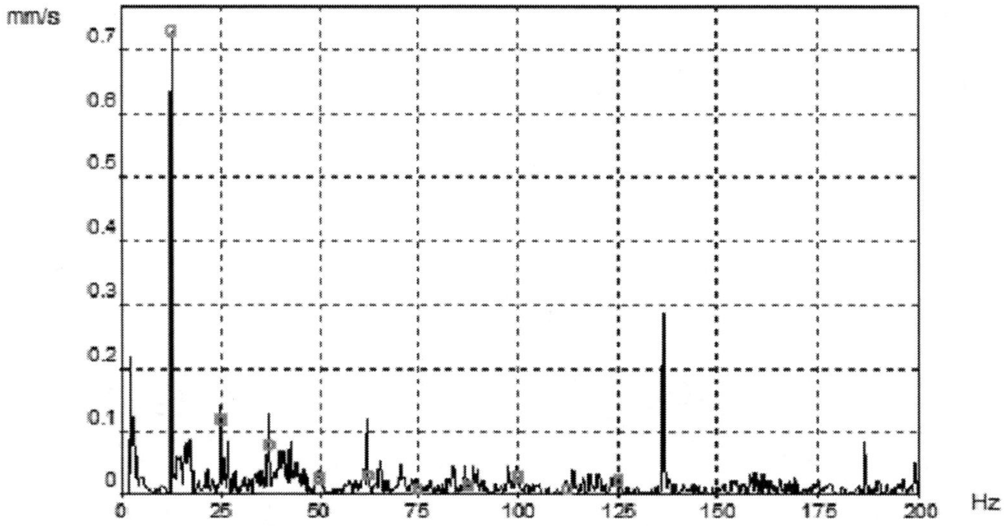

Fig. 2. A vibration spectrum at point 1V stands still from 23.06.2017

This fan has some features, such as the second fan stage 2VDK-32/2 is mounted coaxially of the first stage with the technological gap between the end planes of the runners. In order to avoid the use of guide vanes and thereby to increase the efficiency of the fan rotation direction of impellers they are made opposite [9, 10].

The frequency of collision of flows in the transition to the second stage (taken from a real frequency in the spectra of Fig. 3)

DOI: 10.1051/e3sconf/20172103006

$$f_{l1-2} = f_{l1} + f_{l2} = 186{,}2 + 136{,}2 = 322 \text{ Hz.} \quad (4)$$

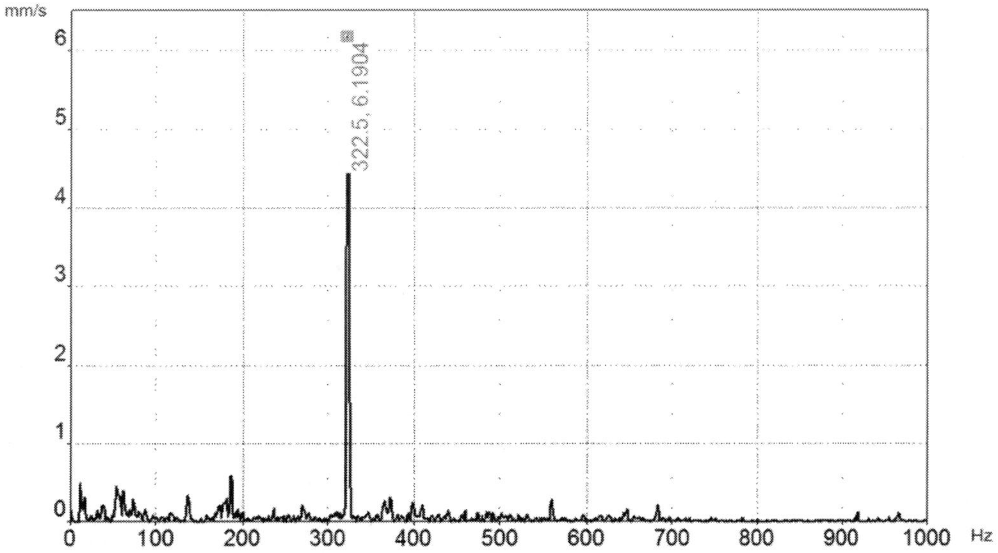

Fig. 3. A vibration spectrum at the point 2A stands still 23.06.2017

To improve the energy efficiency by reducing energy losses on the guide apparatus between the motor and the impeller counter-rotating wheels direct the flow of air from one to another, is confirmed by studies [10 - 13]. For stabilization of the flow in a ventilation pipe a mounted stabilizer has been assembled.

The reference spectra are compiled for each of the characteristic frequencies of the fan, according to the diagnostics carried out after running-in (about 1000 – 1500 hours) for the characteristic points. Compared to the reference spectrum reference masks are generated.

The reference masks designed for this fan, (shown in Fig. 4 to point 2) have warning and alarm levels.

Thus, a relatively high level of vibration is caused by the stall of the vanes of the first wheel when the transition to the second one and does not affect the normal operation of the fan and the vibration of the bearing supports. However, the exceeding of the warning level for the frequency of threads collision means the reducing of the fan efficiency. It is advisable to check the air channel to change the geometric parameters, seals, air suction, etc.

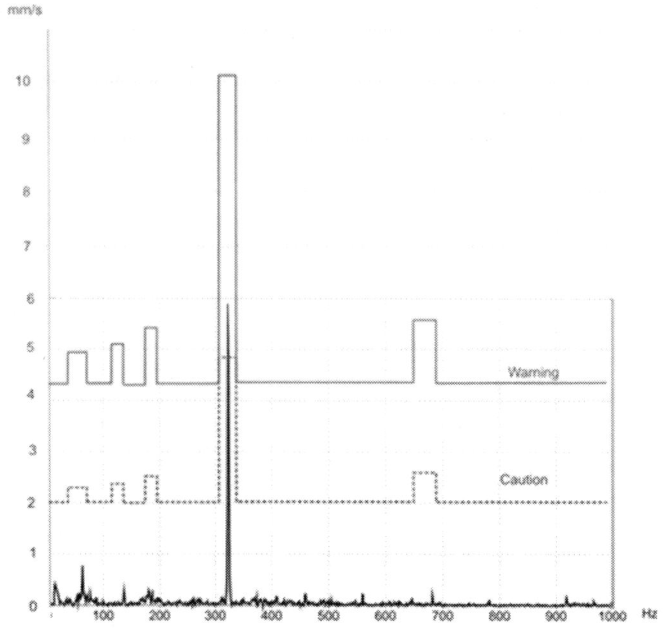

Fig. 4. The reference mask to a point 2A

4 Conclusion

When carrying out the technical diagnostics after comparing the current spectra with the reference it is possible to evaluate the technical condition of the fan to the area of warning as efficient, in excess of the individual frequencies as limited functional, but not energetically efficient. In case of exceeding the level of anxiety, the spectra of vibration must be analyzed. A detailed analysis allows to evaluate the condition of motor bearings and impeller, the blades of the first and second stages, as well as the occurrence of the flow separation due to the irregularities of the aerodynamic characteristics of the duct.

The use of a sufficient number of spectral masks requires a long process of accumulation of experimental data. It makes the method time consuming. However, thanks to the subsequent use of masks the diagnosis will be performed quickly and reliably.

References

1. *Energy management systems - Requirements with guidance for use. ISO 50001* (2011)
2. *A policy framework for climate and energy in the period from 2020 to 2030*
3. *ISO 10816-1: Mechanical vibration - Evaluation of machine vibration by measurements on non-rotating parts - Part 1*
4. B. Gerike, I. Panachev, E. Kuzin, E3S Web of Conferences, **15**, 03008 (2017)
5. XiuJun Yang, Procedia Engineering, **166**, 32 (2016)
6. *Air Pollution and Turbulence. Modeling and Applications*, 324 (CRC Press. Boca Ration. London. NewYork, 2010)
7. Xiaohang Jinl, W. M. Mal Eden, W. S. Chowl, M. Pechtl, Prognostics & System Health Management, **4**, 984 (2012)

8. Shi Lei, Liu Bo, Na Zhenzhe, Wu Xiaoxiong, Lu Xiaofeng. Chinese Journal of Aeronautics, **28:4**, 1044 (2015)
9. B. Dobrzynski, H. Saathoff, G. Kosyna, Proceedings of ASME Turbo Expo 2008: power for land, sea and air, 176-182 (2008)
10. Hao Ming L, Ibrahim Hassan. Heat Transfer Engineering, **36:16** (2015)
11. Howard A. Gaberson, Sound & Vibration, **11** (2013)
12. C. El Ayoubi, WS Ghaly IG Hassan, Journal of Propulsion and Power, **31:2**, 543 (2015)

DOI: 10.1051/e3sconf/20172103007

Load Cases Relevant for Proof of Competence of Fast Running Hoists

Stefan Vöth[1,*]

[1]Technische Hochschule Georg Agricola, Centre for Drive and Lifting Technology ZAFT, Bochum, Germany

Abstract. Dynamically operated hoist systems are exposed to different load cases, which may induce high loads to the drivetrain. The paper shows the load cases and additional equipment, to restrict the loads out of these load cases to the hoist. Especially in focus are optional measures for the exceptional load cases emergency stop snag load. The load cases are listed including affiliated equipment and aspects to be considered during design phase.

1 Technical Development

Cranes in general and hoists especially comprise in transhipment of bulk or containers increasing operational speeds. A special example with regard to this are ship-to-shore container cranes. Today they cover lifting heights up to 72 m with lifting speeds of up to 180 m/min. Planned new ship generations even indicate possible further growth in crane and hoist performance.

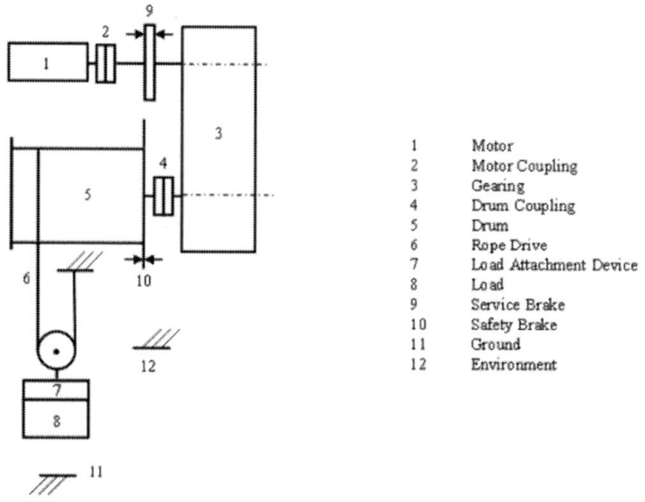

Fig. 1. Hoist Components.

* Corresponding author: stefan.voeth@thga.de

Due to high operational speeds the lifting systems are on high levels of kinetic energy. This will affect system design in order to ensure safety for all operational situations, especially in situations in which the system has to be braked down. Safety comprises for example aspects as observance of intended operation, structural integrity [1], limited braking distances and availability of equipment. Relevant load cases may be correlated with the load case grouping according to [2]: Regular loads, non-regular loads and exceptional loads.

2 Regular Loads

Regular loads appear continuously during operation of the hoist. As they are no exceptional events, they have to be covered by fairly every hoist. Regular loads, neglecting deadweight, are acceleration and deceleration, pick up and drop off of load, slack rope, remaining rope windings and two blocking.

2.1 Acceleration and deceleration

Acceleration will occur at the beginning and the end of every hoist movement.

Acceleration processes typically are not supported by additional equipment to reduce loads to the hoist. Aspects to be considered: Preset motor accelerations, braking torques, dead times and regulative and contractual requirements.

2.2 Pick up and drop off of load

Pick up or drop off of load will occur during every working cycle of the crane.

Pick up or drop off processes typically are not supported by additional equipment to reduce loads to the hoist. Aspects to be considered: Hoisting speeds, lowering speeds, releasing characteristics of load attachment devices as magnets, grabs etc., releasing characteristics of load attachment devices as magnets during unintended release and regulative and contractual requirements.

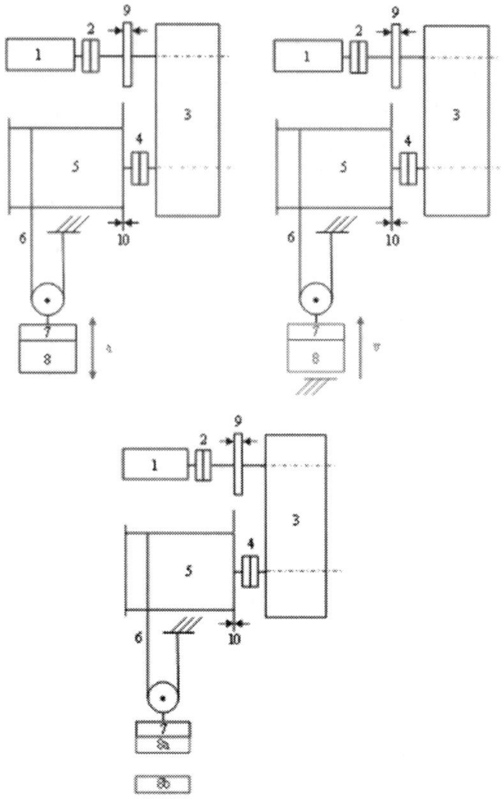

Fig. 2. Load cases acceleration/deceleration, Pick up of load and drop off of load.

2.3 Slack Rope

Slack rope will occur, when a lowering movement continues after putting a load to the ground. System design options:

Lower Limit Switch: With regard to the ground position too much slack rope may be prevented by a lower limit switch. It has to be considered, that this switch can only be adapted to the "lowest ground". If there are different ground levels, the upper levels are not covered by the switch. For these ground levels other strategies may be considered (alternative ground recognition system, slack rope tolerant hoisting system). Aspects to be considered: Operational lowering speeds, rope lengthening, variety of geometric load properties, infrastructural changes, regulative and contractual requirements.

Underload Switch Off: An option is the measurement of the load by a load measuring device and to initiate a braking procedure as soon as a load limit is exceeded. Depending on the load, the switching characteristics and the lifting speed a corresponding slack rope length has to be expected. Aspects to be considered: Relation of maximum lifting height, rope weight and hook block weight (maybe an exclusion condition), measuring frequency, data processing and triggering level, regulative and contractual requirements.

2.4 Remaining Rope Windings

Remaining rope windings on the rope drum have to be ensured, when during lowering operation no ground is reached. System design options:

Lower Limit Switch: Remaining rope windings on the rope drum may be ensured by a lower limit switch at the hoist drivetrain. Aspects to be considered: Operational lowering speeds, regulative and contractual requirements.

2.5 Two Blocking

Two blocking will occur, when a lifting movement continues after reaching the highest theoretical position of the hook block. System design options:

Upper Limit Switches: A feasible method is the positioning of a limit switch to allow a switch off of the hoisting movement on time. Aspects to be considered: Operational hoisting speeds, regulative and contractual requirements.

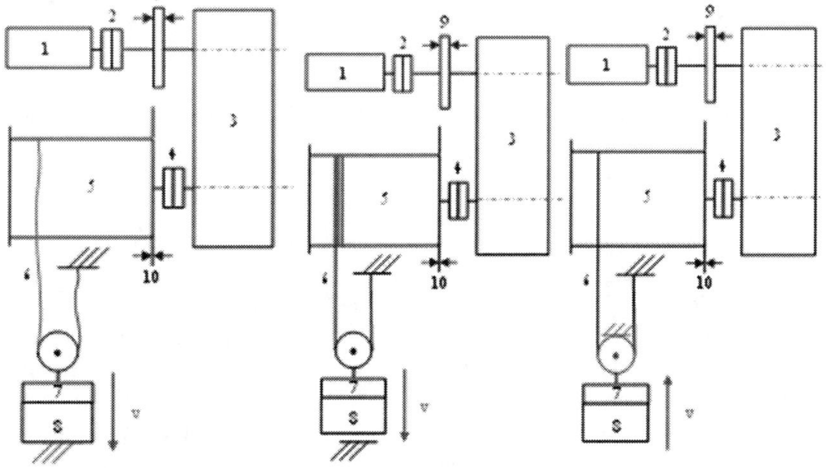

Fig. 3. Load cases slack rope, remaining rope windings, two blocking.

3 Non Regular Loads

Neglecting loads due to snow and ice non-regular loads according to [2] do not appear to hoists in drive direction.

4 Exceptional Loads

Exceptional loads are not planned to appear during operation of the hoist. Nevertheless they can be introduced by exceptional conditions. These exceptional conditions more or less most time have an accidental character. Exceptional loads may be special events which have to be covered only by certain hoists. Exceptional loads are overload, drivetrain failure, service brake failure, emergency stop and snag load.

DOI: 10.1051/e3sconf/20172103007

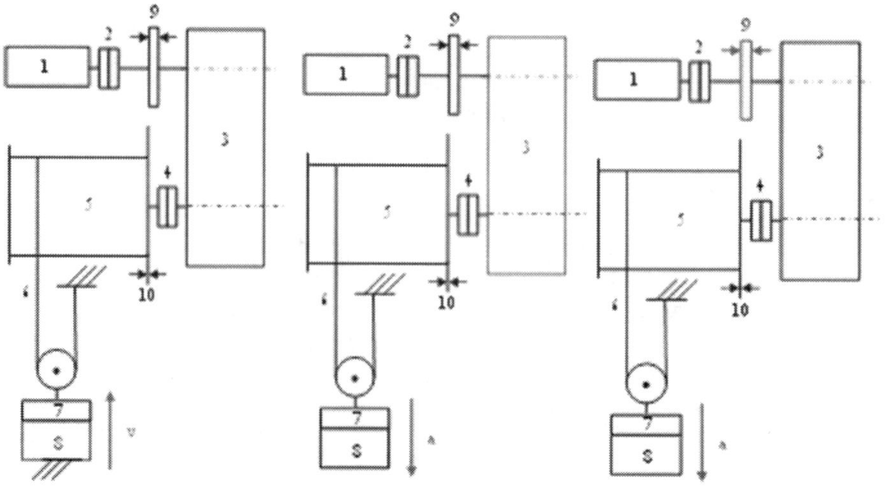

Fig. 4. Load cases overload, drivetrain failure, service brake failure.

4.1 Overload

Overload may happen, when the crane driver tries to lift a load greater than the rated capacity. System design options:

Overload Switch Off: An option is the measurement of the load by a load measuring device and to initiate a braking procedure as soon as a load limit is exceeded. Depending on the load, the switching characteristics and the lifting speed a corresponding maximum load to the system has to be expected. Aspects to be considered: Measuring accuracy, measuring frequency, data processing and triggering level, regulative and contractual requirements.

Frictional clutch: Another option is to locate a frictional clutch in the drivetrain. Depending on the switching torque a corresponding maximum load has to be expected. Aspects to be considered: Switching tolerance, regulative and contractual requirements.

4.2 Drivetrain Failure

Most probable drivetrain failure according to practice is a gearing failure. This failure may occur, when not permissible maximum torques are transferred via the gearing. System design options:

Angle Difference Sensor: The not any more given constant gearing ratio can be realized by a measurement of the input rotating angle and output rotating angle and the check of the relation of both to each other. The detected change in ratio gives the trigger for an automatic emergency stop [9]. Aspects to be considered: Triggering speed difference, brake lining speed, brake lining performance, regulative and contractual requirements.

Speed Sensor: An occurring overspeed of the load side can be realized by a speed sensing device. The detected overspeed gives the trigger for an automatic emergency stop [9, 10]. Aspects to be considered: Triggering speed, brake lining speed, brake lining performance, regulative and contractual requirements.

4.3 Service Brake Failure

A conceivable failure may be a service brake failure. This failure may occur due to lining wear or misadjusting. System design options:

Additional brake: A measure is to place an additional brake on the fast running motor axis. This additional brake may come into action parallel to the service brake. Optional the additional brake comes into action with a certain delay against the service brake [10, 11]. Aspects to be considered: Total braking torque in case of simultaneous brake activity, triggering sequence of brakes, regulative and contractual requirements.

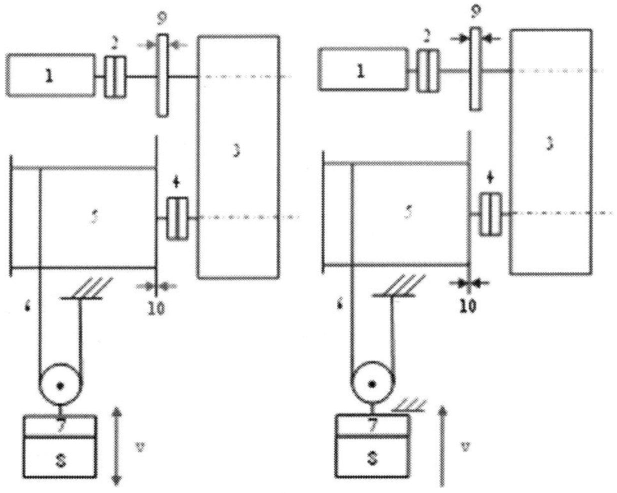

Fig. 5. Load cases emergency stop, snag.

4.4 Emergency Stop

A (manual) emergency stop will occur when introduced by the crane driver. Typically this is a result of a presumed critical operational condition. The following system behaviour depends on the position and the characteristics of the acting brake(s). Mechanical or electrical brakes acting on the fast running motor axis are of interest with regard to the braking distances, because they have to slow down the full rotational mass of the drivetrain. Mechanical brakes acting on the slow running drum axis are, beside the braking distances, of interest with regard to the drivetrain loading. Out of lifting operation about the sum of load torque and braking torque has to be transferred via the gearing. Out of lowering operation in connection with gearing and coupling clearances shock torques have to be transferred via the gearing. System design options:

Safety Coupling: A safety coupling located between motor and gearing is a means to limit the maximum gearing torques transferred by the gearing in case of a safety brake acting on the slow running axis. This especially applies for lifting operation. Aspects to be considered: Conflict between switch off at low load level and unfounded switch off, additional masses to the system, process of recommissioning of the safety coupling after burst, prevention of slack rope conditions, regulative and contractual requirements.

Freewheel Coupling: A freewheel coupling located between motor and gearing is a means to diminish the maximum gearing torques transferred by the gearing close to zero in case of a safety brake acting on the slow running axis. This especially applies for lowering operation [3]. Aspects to be considered: Additional masses to the system, regulative and contractual requirements.

Fast acting service brake: This system basically consists of a fast acting service brake. With a fast acting service brake both brakes come into action synchronous. As the motor masses are braked directly the torque transferred by the gearing is reduced significantly. Furthermore the two brakes realize the fastest braking for the whole system mass possible [4,

5, 6, 7, 8]. Aspects to be considered: Synchronization of brakes, regulative and contractual requirements.

4.5 Snag Load

A snag load occurs, when a load gets stuck during a lifting operation. The immediate fixation of the load causes high loads on the drive train especially at high speeds. System design options:

Hydraulic Protection System: This system is acting to the rope drive of the hoist. Either the rope end or a pulley is adapted to a hydraulic cylinder rod. With an arising overload an according hydraulic pressure in the cylinder is opening a pressure valve. The opened valve allows the cylinder rod to move out and the rope end or the pulley to move with the pending load. This gives a release to the rope drive and the whole drivetrain. A hydraulic protection system probably is combined with a fast acting safety brake, triggered by the hydraulic system condition. Aspects to be considered: Environmental protection, recommissioning after burst, prevention of slack rope conditions, regulative and contractual requirements.

Mechanical Protection System: This system basically consists of a safety clutch between motor and gearing. The occurring overload due to snag forces the clutch to open. This uncouples the dominant motor mass from the drivetrain. The mechanical protection system probably is combined with a fast acting safety brake, triggered by the mechanical protection system condition. The safety brake is enabled to brake down the drum very fast, as the system masses are quite reduced. The motor mass is uncoupled and the service brake disc may consist out of a material of low density. Aspects to be considered: Conflict between switch off at low load level and unfounded switch off, additional masses to the system, process of recommissioning of the safety coupling, after burst, prevention of slack rope conditions, regulative and contractual requirements.

Fast acting service brake: This system basically consists of a fast acting service brake. With a fast acting service brake both brakes come into action synchronous. As the motor masses are braked directly the torque transferred by the gearing is reduced significantly. Furthermore the two brakes realize the fastest braking for the whole system mass possible. [4, 5, 6, 7, 8] Aspects to be considered: Synchronization of brakes, regulative and contractual requirements.

Sensoric Protection System: This system bases on measuring conditional changes at the load. By means of acceleration sensors and inclination sensors the corresponding physical conditions are measured and evaluated. As soon as the parameters indicate a snag case the braking process is triggered. A sensoric protection system probably is combined with a fast acting safety brake. Aspects to be considered: Prevention of unfounded bursts, regulative and contractual requirements.

5 Outlook

For the very basic load cases acceleration and pick up or drop off of load overloads are prevented by suitable design of the working cycle.

For certain load cases measures for preventing overloads to the hoist system are clear. The application of upper and lower limit switches is a quite common application to prevent slack rope, under ranging minimum rope windings on the rope drum and two blocking. This also applies for overload switches / frictional clutches to prevent overloads.

For other load cases, measures in some cases are applied. This applies for angle difference sensors comparing the gearing input and output rotation. Those sensors are just applied for non-serial winches in operations with higher safety requirements.

For emergency stops and snag load cases different equipment options, beside the option of no additional equipment, are given. Whether an additional equipment and which additional equipment is of use, must be determined by thorough transient analysis of the dynamic system for relevant load cases. This requires the treatment of the whole system rather than single components. For experimental evaluation of the system behaviour as a result of the application of optional equipment an adequate testing rig s built up at ZAFT of THGA [12].

References

1. F. Schmeink, *Dynamische Beanspruchung von Hubwerksgetrieben ...* (Tagungsband der 22. Internationalen Kranfachtagung, Magdeburg 2014)
2. DIN EN 13001-2 Kransicherheit, Konstruktion allgemein, Teil 2: Lasteinwirkungen.
3. S. Vöth, STS-Crane-Hoist-Structure with Freewheel Coupling (to be published)
4. S. Vöth, M. Tyulenev, KuzSTU Bul., **1**, 88-96 (2016)
5. S. Vöth, Hebezeuge und Fördermittel, **3** (2015)
6. S. Vöth, Hebezeuge und Fördermittel, **4** (2015)
7. S. Vöth, Hebezeuge und Fördermittel, **5** (2015)
8. A. Khoreshok, M. Tyulenev, S. Vöth, *Conditions for minimum dynamic loading of multi brake hoists, 8th Russian-Chinese Symposium, Coal in the 21st century: Mining, Processing, Safety* (Kuzbass State Technological University, Kemerovo, 2016)
9. RWTÜV-Schriftenreihe, Heft 8, Krane, Bemessung und Sicherheit, 1981
10. N. Schilling, G, Schneider, S. Vöth, Hebezeuge und Fördermittel, **7-8** (2017)
11. S. Vöth, *Loading of hoist gearings due to fast running motors and additional brakes,* (ITI, Dresden, 2016)
12. S. Vöth, E2S Web of Conferences, **15**, 03007 (2017)

The Influence of Parameters on the Generatrix of the Helicoid Form Guide of Geokhod Bar Working Body

Vladimir Aksenov [1,3*] *Vladimir* Sadovets [2,3] , and *Dmitriy* Pashkov [1,3]

[1]Institute of Coal of the Federal Research Center Coal and Coal Chemistry Siberian Branch of the RAS, 650610, Leningradsky av, Kemerovo, Russia

[2] T.F. Gorbachev Kuzbass State Technical University, 650000, 28 Vesennyaya St., Kemerovo, Russia

[3]Yurga Technological Institute branch of Tomsk Polytechnic University, 652052 Leningradskay Str. 26, Yurga, Russia.

Abstract. Influence of geometrical parameters of generatrix of helicoid on a guide of geokhod bar working body is proved in article. The relevance of the conducted research is considered and proved. General characteristics of the geokhod are presented. Features of geokhod working body, in particular formation of irregular shape of a surface of a face and working body are formulated and also it is told that at screw movement of geokhod working body of a face, points of working body will be formed a helicoid (screw) surface of a face. For establishing of the goals and objectives of research general geometrical parameters of generatrix is marked and justified which treat length of generatrix, width of generatrix of helicoid, pitch of helicoid and it's form. Forms of guides of geokhod bar working body based on basis parameters of geokhod and accepted general geometrical parameters of geokhod working body are received and presented. In virtue of the conducted research the dependence of a form of a guide of bar on general geometrical parameters of helicoid is defined and also basis parameters of helicoid in influencing a form of guide of working body.

1 Introduction

Because of increase of the number of vehicles and intensive growth of the population in the cities the critical shortage of territories is observed. In this regard in many large settlements of the world experts actively master underground space. However formation of a cavity in underground space with use of the tunnelling equipment causes need of the solution of questions of safety of carrying out mining operations, increases speed of a driving and labor productivity, decreases in prime cost [1,2,3,4].

The perspective direction which is solved this question is using of geokhod technology for formation of a cavity in underground space where basic element is the geokhod [5,6,7].

The geokhods belong to the new class of the driving equipment. Distinctive feature of the geokhods is screwing of container to the rock mass. So power crowding are created as a

* Corresponding author:55vva42@mail.ru

 DOI: 10.1051/e3sconf/20172103008

result of introduction of elements of the propulsor to the peripheric massif and uses of the normal reaction arising at interaction of elements of the propulsor of the geokhod with the massif of rocks. At present, a prototype of geokhod is manufactured with the cutting body for destruction of rocks of medium hardness (Figure 1) [8,9,10,11].

Fig. 1. Geokhod prototype with drum cutting body (CB).

At the existing development stage of elements of geokhod technology is particularly acute need in development of constructive and technical solutions of working body which is capable to carry out formation of a cavity in underground space on grounds with a hardness up to 1 on M.M. Protodyakonov's scale [12].

2 Materials and methods

Features of geokhod working body. A peculiar features of movement of the geokhod of a face causes formation of irregular shape of a surface not only the face, but also the working body (WB). A face surface, at destruction of his WB of the geokhod, has an appearance of several helicoid surfaces with ledges.

Any point of WB located at distance x (figure 2) from pivot of the geokhod moves to a face at an angle[7]:

$$\beta_x = arctg\frac{h_B}{2\pi x}. \quad (1)$$

Points of WB which are closer to pivot of the geokhod move to a face at an bigger angle, than the points which are on the periphery. Thus, at screw movement of WB of the geokhod of a face of a point of WB forms a helicoid (screw) surface. Therefore, the site of a surface of a face in the sector between adjacent beams after their pass takes a form of a screw surface.

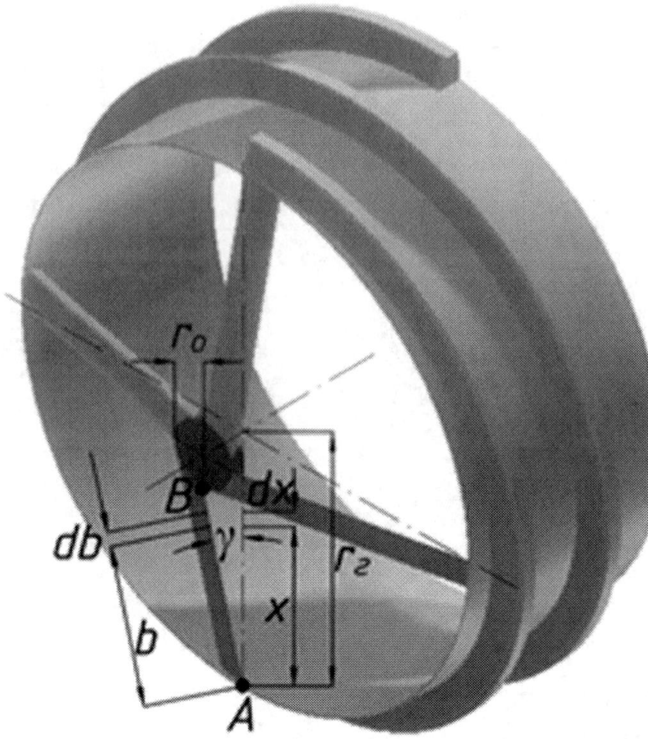

Fig. 2. Scheme of geokhod working body.

It agrees developing technical solutions, on the basis of a structural portrait of the geokhod, beams of WB can be at an some angle to pivot of the geokhod. The corner y (figure 1) can be: equal 90 angle degree, less than 90 angle degree – WB takes a form of a right cone, more than 90 angle degree – WB will be in the form of the back taper [7].

On the basis above stated the problem of the presented research is formulated in the following: to develop options of forms of geokhod working body with geologic environment depending on parameters of generatrix of helicoid.

For development of the forms WB of the geokhod was accepted geometrical parameters of a helicoid into account the sizes of the created geokhod with roller working body for destruction of grounds of average hardness [13] $D_{г}=3{,}2$ м, $\gamma=78^{0}$, $h_{в}=0{,}8$ м and width of generatrix is accepted equal 0,15 m. Besides beams on which the directing WB settle down have been located at an some angle to pivot of the geokhod. The corner y has been accepted: equal 90 angle degree – a direct form, y = 78 angle degree – WB takes a form of a right cone, y =102 angle degree – WB will be in the form of the back taper.

From the existing WB using for destruction of grounds with a hardness up to 1 on M.M. Protodyakonov's scale bar WB are the most suitable [12]. The received forms the helicoid surfaces of WB of the geokhod will be guides for the movement of cutter chain.

3 Results and discussion

Considering the requirements formulated in works [7,10,13] and also conditions given above, we have received 21 forms of a guide of geokhod bar working body. Some options are presented in table 1. The form of a guide of bar is allocated with the red line.

Table 1 - forms of a guide of geokhod bar working body

 DOI: 10.1051/e3sconf/20172103008

№	Geometrical parameters of generatrix	Formation of a face	As seen from a face	Axonometric projection
1		tabulated		
2	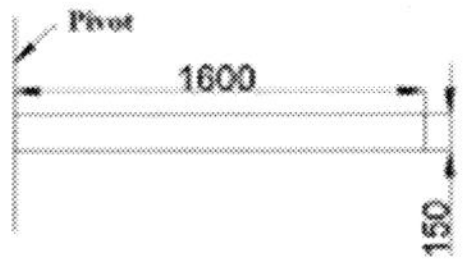	right cone		
3		back taper	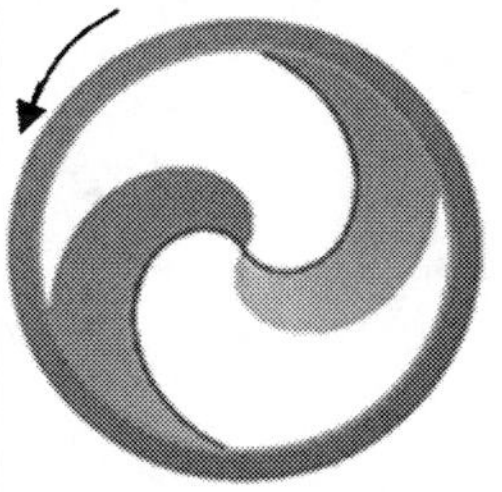	
4	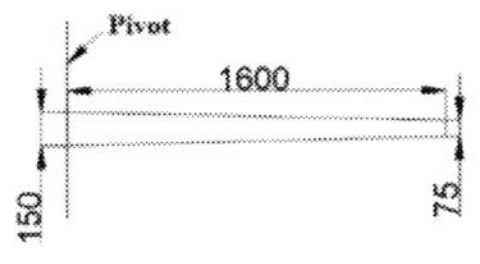	tabulated	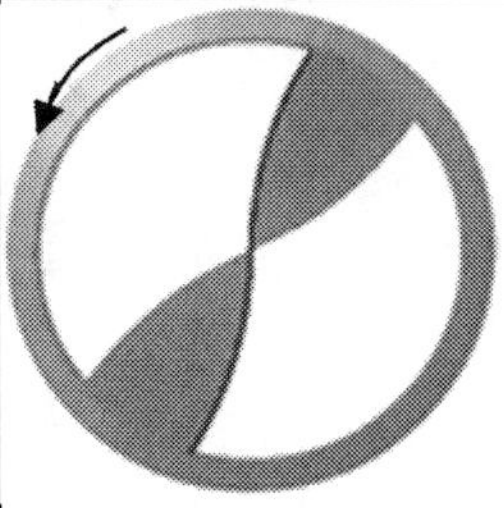	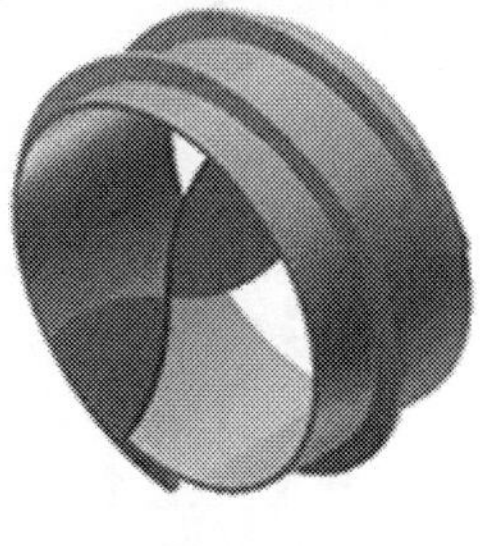
5		right cone		

№	Geometrical parameters of generatrix	Format ion of a face	As seen from a face	Axonometric projection
6		back taper		
7		tabulated	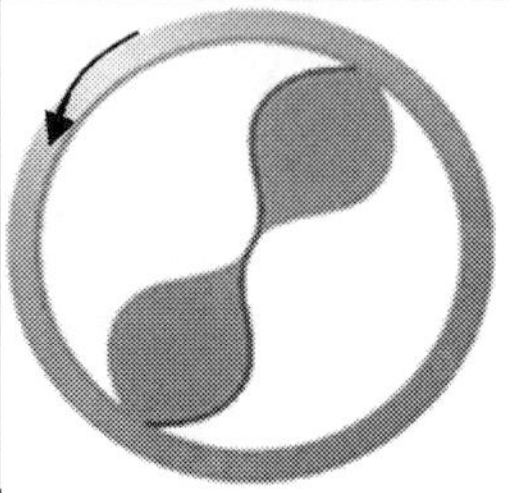	
8	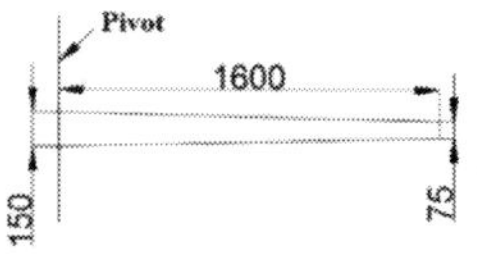	right cone		
9		back taper	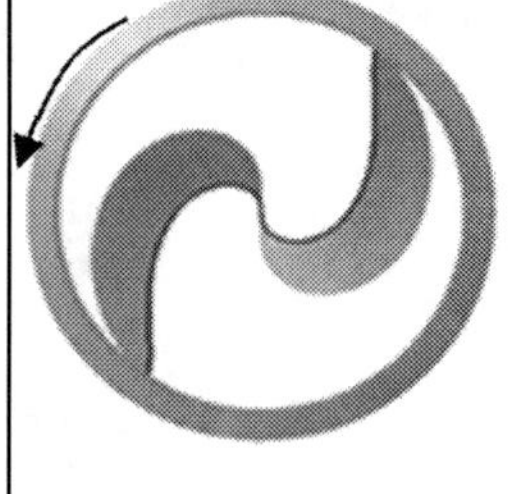	

4 Conclusions

In table it is visible that on the form of a guide influences not only geometrical parameters generatrix of helicoid, but also angle of beam tilt WB to the frontal wall of a face (a face form). For reduction of forces of cutting and a possibility of self-tightening of the tunnelling equipment to the massif of grounds, a particular interest present forms of a guide of bar working body in the form of the back taper.

It is necessary for carrying out further researches:

- to formulate requirements to geokhod bar working bodies for destruction of grounds with a hardness up to 1 on M.M. Protodyakonov's scale;
- to prove, using a finite element method, a rational form of a guide of bar working body.
- to receive dependences of a form of a guide of bar working body on diameter and a step of the screw blade of the geokhod.

References

1. G. Brierley, World Tunnel. And subsurface Excav, **9** (1998)
2. J. Carmody, R. Sterling. *Underground space design* (Reinhold, New York, 1993)
3. T. Wighman, ENR: News–Rec, **4,** (1998)
4. S. Nishi, T. Seiki Mem. Sch. Eng. Nagoya Univ., **1** (1997)
5. R.V. Chernukhin, A.A. Dronov, M.Y. Blashchuk, IOP Conf. Ser.: Mater. Sci. Eng., **91**, 012086 (2015)
6. A.B. Efremenkov, 6th International Forum on Strategic Technology (IFOST - 2011) **2:1**, 348 (2011).
7. V.Y. Begljakov, V.Y. Timofeev, M.V. Dokhnenko, Appl. Mech. Mater. **682**, 282 (2014).
8. M.Y. Blashchuk, A.A. Kazantsev, R.V. Chernukhin, Appl. Mech. Mater. **682**, 418 (2014)
9. V. Aksenov, V. Sadovets, E. Rezanova, D. Pashkov, E3S Web Conf., **15** 03015 (2017)
10. V.V. Aksenov, A.B. Efremenkov, V.Y. Beglyakov, Appl. Mech. Mater. **379**, 16 (2013).
11. M.Y. Blaschuk, A.A. Dronov, S.S. Ganovichev, IOP Conf. Ser. Mater. Sci. Eng., **127**, 012033 (2016).
12. R.V. Chernukhin, M.Y. Blaschuk, V.Yu. Blumenstein, P.A. Chazov, IOP Conf. Ser. Mater. Sci. Eng., **142**, 012003 (2016).
13. V.Y. Sadovets, V.Y. Beglyakov, A.B. Efremenkov, Appl. Mech. Mater. **770**, 384 (2015).

Application of Mathematical and Three-Dimensional Computer Modeling Tools in the Planning of Processes of Fuel and Energy Complexes

Olesya Aksenova[1*], *Evgenia* Nikolaeva[1], and *Michal* Cehlár[2]

[1] T.F. Gorbachev Kuzbass State Technical University, 650000, 28 Vesennyaya St., Kemerovo, Russia
[2]Technical University of Kosice, Faculty of Mining, Ecology, Process Control and Geotechnologies, Letná 9, 04200 Kosice, Slovak republic

Abstract. This work aims to investigate the effectiveness of mathematical and three-dimensional computer modeling tools in the planning of processes of fuel and energy complexes at the planning and design phase of a thermal power plant (TPP). A solution for purification of gas emissions at the design development phase of waste treatment systems is proposed employing mathematical and three-dimensional computer modeling - using the E-nets apparatus and the development of a 3D model of the future gas emission purification system. Which allows to visualize the designed result, to select and scientifically prove economically feasible technology, as well as to ensure the high environmental and social effect of the developed waste treatment system. The authors present results of a treatment of planned technological processes and the system for purifying gas emissions in terms of E-nets, using mathematical modeling in the Simulink application. What allowed to create a model of a device from the library of standard blocks and to perform calculations. A three-dimensional model of a system for purifying gas emissions has been constructed. It allows to visualize technological processes and compare them with the theoretical calculations at the design phase of a TPP and, if necessary, make adjustments.

1 Introduction

At the present day, the fuel industry is the basis for the development of the Russian economy, an instrument for carrying out domestic and foreign policy. An accurate modeling processes of the fuel and energy complex is considered to be very important. In addition, the operation of such facilities due to the products of coal combustion processing entails massive emissions of gases into the environment. Thus, violating the ecological and industrial safety of the world around us. Therefore, when planning and modeling processes of fuel and energy complexes, it is necessary to take into account the availability of waste treatment systems. Namely a system for purifying gas emissions and a system for the full cycle

* Corresponding author: olesya_aksenova42@mail.ru

DOI: 10.1051/e3sconf/20172103009

of processing ash and slag wastes that will not only ensure environmental and industrial safety, but will also be cost-effective for enterprises and the country as a whole.

2 Materials and Methods

At the stage of planning the processes of fuel and energy complexes and the development of technical documentation, a modern range of computer programs allows to present an operation analysis of a future TPP from different perspectives. Mathematical modeling [9] makes it possible to verify the model correctness and adjust it at the stage of construction, while 3D computer modeling allows to obtain a realistic image based on available technical documentation, identify technical errors, and correct, during the planning stage, individual technological processes and a TPP as a whole.

When performing a mathematical analysis, it is advisable to use the E-nets apparatus [2]. As an extension of Petri nets, E-nets are a graphical and mathematical tool used to model systems of different types.

An abstracted model of a TPP, shown in Fig. 1, with the possibility of expansion is considered in this work.

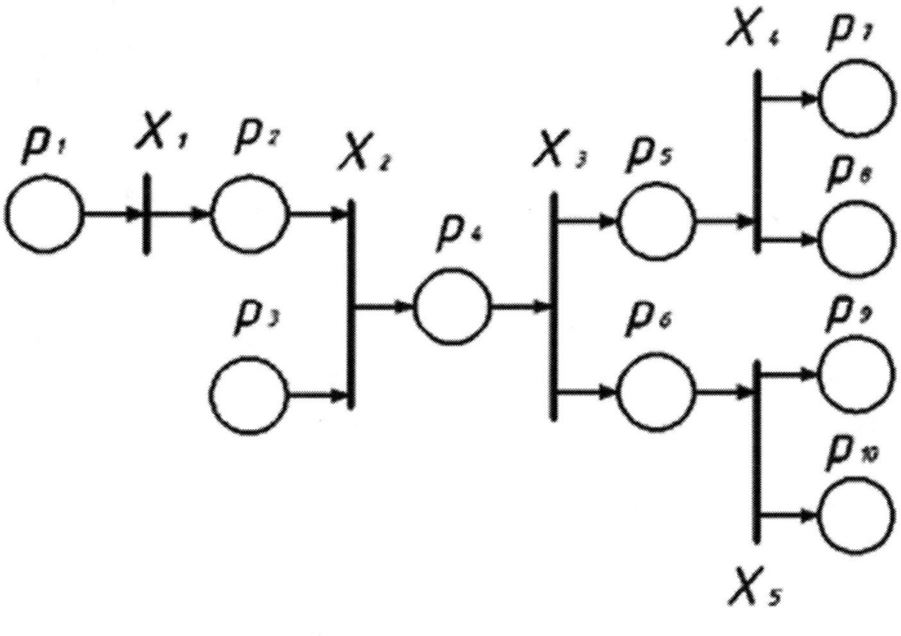

Fig. 1. A TPP model in terms of E-nets.

Where p_1 - a new portion of coal entered a crushing plant; p_2 - coal is processed; p_3 - condensate and replenished water losses entered the storage facility; p_4 - the raw material was supplied to the combustion chamber; p_5 - combustion products are obtained; p_6 - steam is fed to the turbine; p_7 - has and slag waste are transferred to the processing system; p_8 - gas-like wastes enter the purification system; p_9 - electric power is allocated to external consumers; p_{10} - steam from the turbine selections goes to a TPP; $X_1 = (T, 0, \varphi_1)$ - coal processing; $\varphi_1 = L(X_1) = \{p_1X_1|X_1 \rightarrow p_1p_2\} = \{p_1p_2\}$, $X_2 = (I, 0, \varphi_2)$ - loading of raw materials into the combustion chamber; $\varphi_2 = L(X_2) = \{p_2p_3X_2|X_2 \rightarrow p_2p_3p_4\} =>$ $L(X_2) = \{p_2p_3p_4\}$, $X_3 = (F, 0, \varphi_3)$ - raw materials processing; $\varphi_3 = L(X_3) =$

$\{p_4X_3|X_3 \rightarrow p_4p_5p_6\} = \{p_4p_5p_6\}$, $X_4 = (F, 0, \varphi_4)$ - waste disposal; $\varphi_4 = L(X_4) = \{p_5X_4|X_4 \rightarrow p_5p_7p_8\} = \{p_5p_7p_8\}$, $X_5 = (F, 0, \varphi_5)$ - power generation and steam extraction; $\varphi_5 = L(X_5) = \{p_6X_5|X_5 \rightarrow p_6p_9p_{10}\} = \{p_6p_9p_{10}\}$, the initial marking is represented by the vector: (1,0,1,0,0,0,0,0,0,0).

Software packages for numerical computations MatLab [1,3] and Simulink [4] were used to implement the developed E-nets [5] model on a computer.

3 Results and discussion

Modeling using Simulink implements the principle of visual programming, according to which, the user create a model of a device on a screen using the library of standard blocks and performs calculations. At the same time, in contrast to classical ways of modeling, the user does not need to thoroughly study the programming language and numerical methods. Basic computer working skills and, of course, knowledge of the subject area are sufficient.

The model created in Simulink, shown in Fig. 2, is a balance model of material flows [6], in which the following technological phases were realized:

1. The loss of coal in its preparation for burning,
2. Electric power generation,
3. Formation and processing of ash and slag wastes,
4. Formation and purification of gas emissions.

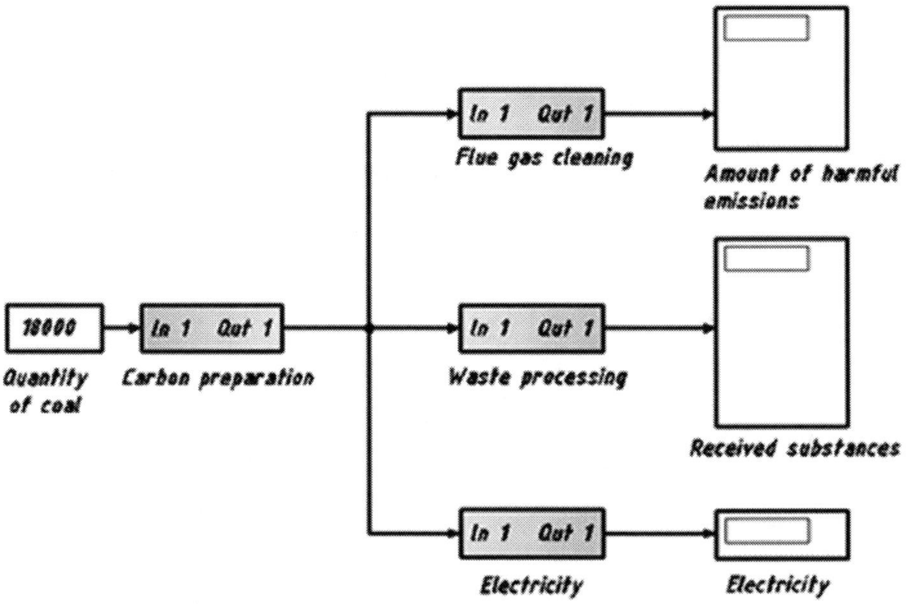

Fig. 2. The balance model of material flows of a TPP.

In order to make the model more understandable and easier to expand, the processing and purification blocks were implemented as subsystems. Consider a subsystem for purification of gas emissions of a TPP, which in turn also consists of several modules.

The first in the sequence of these modules is an electrostatic precipitator. The following formula was used for its efficiency calculation:

$$M_T = A \cdot m \cdot \chi \cdot \left(1 - \frac{\eta_T}{100}\right), \quad (1)$$

 DOI: 10.1051/e3sconf/20172103009

where M_T – is the gross emission of solid particles in flue gases; A is the ash content of fuel, in %; m is the amount of fuel consumed per year, ton; χ is a dimensionless coefficient characterizing the fraction of ash carried away with flue gases, depends on furnace and fuel types; $\eta_т$ is the efficiency of ash collectors, in %.

The next module is a system for purification from nitrogen oxides. To determine its characteristics, the following formula was used:

$$M_{NO_2} = m \cdot Q^н \cdot K_{NO_2} \cdot (1 - \beta) \cdot 10^{-3}, \quad (2)$$

where M_{NO_2} is the gross release of nitrogen oxide in terms of nitrogen dioxide; $Q^н$ is the lowest heat of combustion of natural fuel, MJ/kg; K_{NO_2} is a parameter characterizing the amount of nitrogen oxides generated per GJ of heat, kg/GJ; β is a coefficient which depends on the degree of reduction of nitrogen oxide emissions resulting from the application of technical solutions.

Desulphurization system. When calculating the performance of this module, the following formula was used:

$$M_{SO_2} = 0{,}02 \cdot m \cdot S \cdot (1 - \eta^1_{SO_2}) \cdot (1 - \eta^2_{SO_2}), \quad (3)$$

where M_{SO_2} is the total emission of sulfur oxides in terms of sulfur dioxide; S is the sulfur content, in %; $\eta^1_{SO_2}$ is the proportion of sulfur oxides bound by the flying ash of the fuel; $\eta^2_{SO_2}$ is the proportion of sulfur oxides trapped in the ash collector.

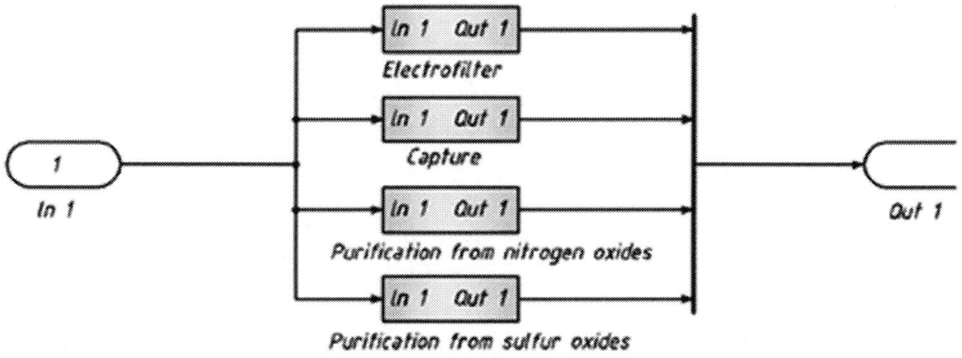

Fig. 3. The subsystem purifying gas emissions.

As an example of mathematical and three-dimensional computer modeling, we consider in more detail the subsystem of carbon dioxide capture.

Carbon dioxide is one of the greenhouse gases, the excess of which in the atmosphere leads to global climate change. Throughout the world, numerous attempts are being made to reduce greenhouse gas emissions and one of these methods - capture and storage of carbon dioxide (Carbon Capture & Storage - CCS) - is becoming increasingly widespread.

The system of carbon dioxide capture and storage is divided into three stages: capture, transportation, and storage. Consider the first stage shown in Fig. 4.

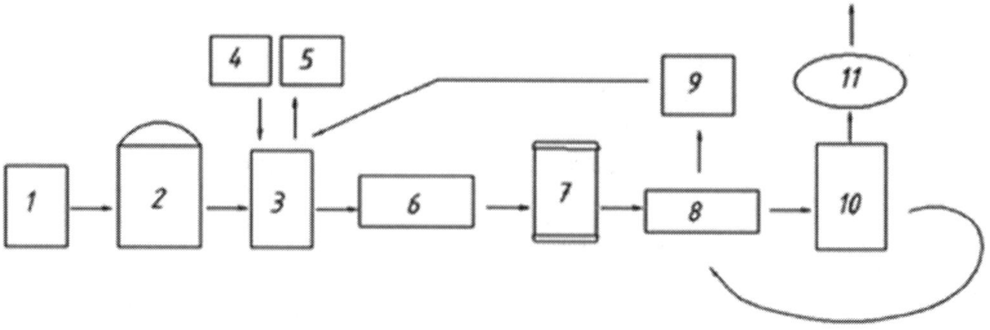

Fig. 4. Flow diagram of the carbon dioxide capture process.

The technology of carbon dioxide capture and concentration is realized according to the following scheme: the flue gases (1) are fed by the induced draft fan to the centrifugal apparatus (2) to remove suspended particles and reduce the temperature of the gas stream by passing the gas mixture through the water layer. The cooled gas stream is fed to the packed absorber (3), which also receives the regenerated absorbent solution (4). The gas purified in the absorber is extracted from the unit, and the solution saturated with an acid gas is removed from the built into the absorber vessel by the pump (6) through the filter (7) and the recuperative heat exchanger (8) and fed to regenerate into the packed desorber (10) heated from below by the heaters. The acid gas from the top of the desorber is fed into the compressor (11), then it is removed from the apparatus. The regenerated solution from the bottom of the desorber is moved through the recuperative heat exchanger (8) and the refrigerator (9) into the absorber for further operation.

The model of the carbon dioxide capture and storage system in terms of E-nets has the following form (Fig. 5): p_1 - flue gases entered the centrifugal apparatus; p_2 - a cooled gas is fed to a packed absorber; p_3 - the gas purified in the absorber is removed from the unit; p_4 - the solution saturated with the acid gas passes through the filter; p_5 - the solution passes through the recuperative heat exchanger; p_6 - the solution is recovered in a packed desorber; p_7 - the acid gas at the top of the desorber is removed from the unit; p_8 - the solution at the bottom of the desorber is regenerated; p_9 - the regenerated solution enters the refrigerator.

$X_1 = (Y, 0, \varphi_1)$- process of regeneration in desorber;

$$\varphi_1 = L(X_1) = \begin{cases} p_1 \to X_1, p_9 \to X_1, r_1 \to X_1, X_1 \to p_2 \leq> r_1 = 0 \vee \\ p_1 \to X_1, p_9 \to X_1, r_1 \to X_1, X_1 \to p_2 <=> r_1 = 1 \vee \end{cases}$$

$$\left.\begin{matrix} \vee\,(p_1 = 1, p_9 = 0, p_2 = 0,) \to p_1p_2 \\ \vee\,(p_1 = 1, p_9 = 1, p_2 = 0,) \to p_1p_9p_2 \end{matrix}\right\} => L(X_1) = \{p_1p_2 \wedge p_1p_9p_2\}$$

$X_2 = (F, 0, \varphi_2)$- the solution is fed for regeneration into the desorber;

$$\varphi_2 = L(X_2) = \{p_2X_2|X_2 \to p_2p_3p_4\} = \{p_2p_3p_4\},$$

$X_3 = (Y, 0, \varphi_3)$- the solution is fed for regeneration into the desorber;

$$\varphi_3 = L(X_3) = \begin{cases} p_4 \to X_3, p_8 \to X_3, r_2 \to X_3, X_3 \to p_5 \leq> r_2 = 0 \vee \\ p_4 \to X_3, p_8 \to X_3, r_2 \to X_3, X_3 \to p_5 <=> r_2 = 1 \vee \end{cases}$$

$$\left.\begin{matrix} \vee\,(p_4 = 1, p_8 = 0, p_5 = 0,) \to p_4p_5 \\ \vee\,(p_4 = 1, p_8 = 1, p_5 = 0,) \to p_4p_8p_5 \end{matrix}\right\} => L(X_3) = \{p_4p_5 \wedge p_4p_8p_5\}$$

$X_4 = (F, 0, \varphi_4)$ – the solution is fed for regeneration into the desorber;

$$\varphi_4 = L(X_4) = \{p_5X_4|X_4 \to p_5p_6p_9\} = \{p_5p_6p_9\},$$

$X_5 = (F, 0, \varphi_5)$ – regeneration in the desorber;

 DOI: 10.1051/e3sconf/20172103009

$$\varphi_5 = L(X_5) = \{p_6X_5|X_5 \rightarrow p_6p_7p_8\} = \{p_6p_7p_8\},$$

r_1 – verifies that the regenerated solution is present; r_2 - verifies that the regenerated solution from the bottom part of the desorber is present.

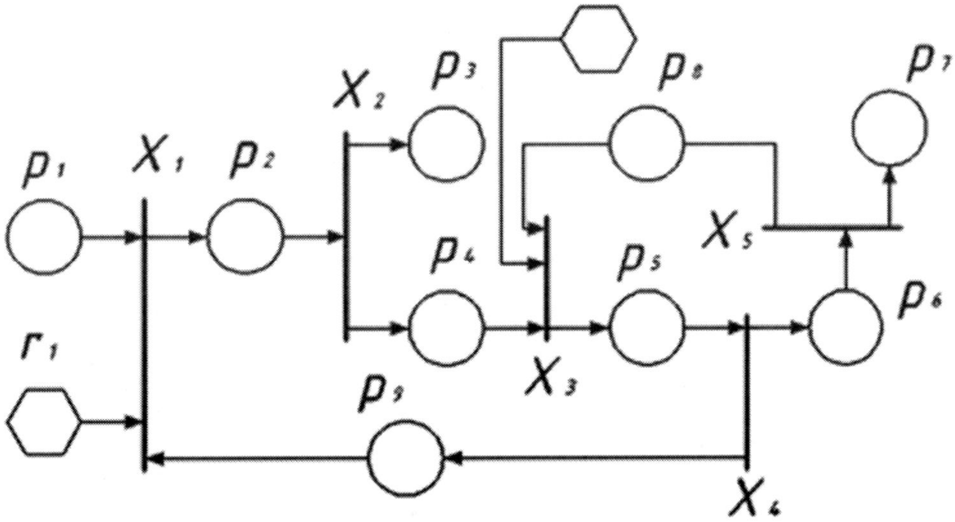

Fig. 5. Model of the carbon dioxide capture and storage system in terms of E-nets

For a realistic representation of the system for purification of gas emission from carbon dioxide, solid models of equipment and pipelines are built in several graphic editors based on the available technical documentation. Then, taking into account the planned design area and arrangement of the equipment, a three-dimensional image of the future operational line and technological process is created as shown in Fig. 6. Such an image allows to visually see the technical flaws in the theoretical planning phase and eliminate them.

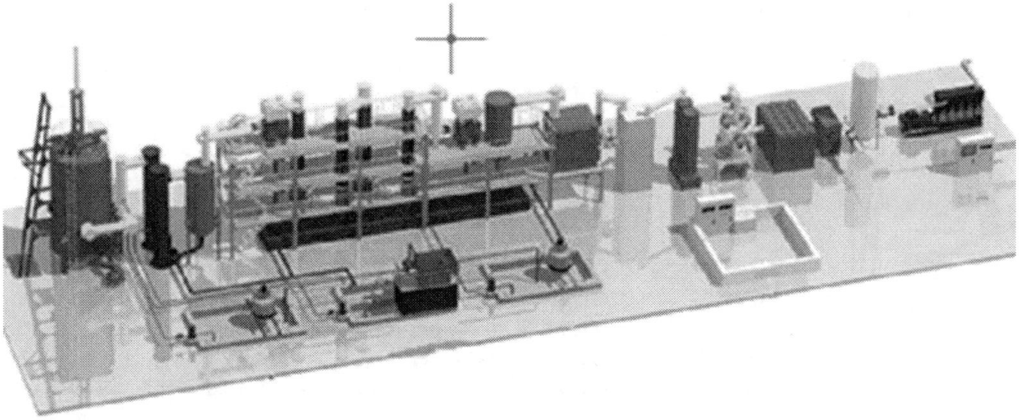

Fig. 6. Three-dimensional model of carbon dioxide capture and storage system.

Similarly, all modules of the subsystem for purification of gas emissions from a TPP can be analyzed. In particular, to calculate parameters of the module of the carbon dioxide capture system, the formula given below is used:

$$M_{CO_2} = 0{,}01 \cdot m \cdot 3{,}664 \cdot C^p \cdot (1 - 0{,}01 \cdot q), \quad (4)$$

where M_{CO_2} is the gross emission of carbon dioxide; C^p is the carbon content in fuel, in %; q is the loss of heat due to mechanical incompleteness of combustion of solid or liquid fuel, in %.

The amount of materials (in tons) formed during the processing of 18,000 tons of coal: solid particles - 0.5837; Carbon dioxide - 105.4; Nitric oxide - 0.01684; Sulfur oxides - 1.782; gold and silver (<10%) - 0.00001818; Coagulant of aluminum sulfate - 2764; Uranium - 6.48; Thorium - 0.1163, Tantalum, Niobium, Cerium, Yttrium - 32.4; Titanium oxide - 170.5; Concentrate of Zirconium with Hafnium - 1494; Gallium - 0.22479; liquid glass - 3161; Electricity (in Mwh) - 112540.06. The results of the model are shown in Fig. 7.

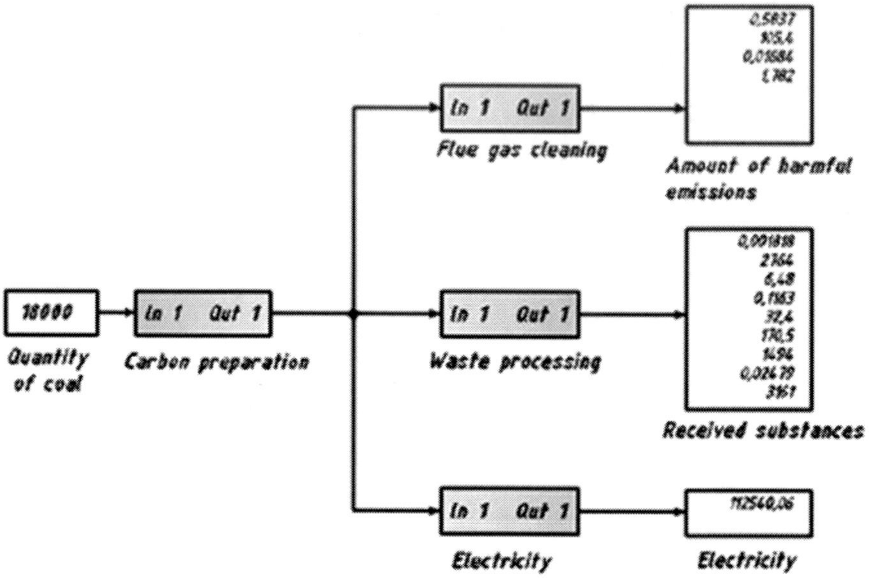

Fig. 7. The emission purification system of a TPP.

In conclusion, it should be noted that the resulting model can be further extended by adding necessary modules to its structure.

4 Conclusions

The problem of the application of mathematical and three-dimensional computer modeling tools for developing processes of fuel and energy complexes at the planning and design stage of a TPP is considered.

The solution of the problem of purification of gas emissions by means of mathematical and three-dimensional computer modeling is proposed. It allows to estimate the planned result, to choose the production technology, and also to ensure a high environmental effect of the planned waste treatment system. The results of processing the planned technological processes and the system for purifying gas emissions in terms of E-nets using mathematical modeling with the Simulink application are presented. A three-dimensional model of the gas emission purification system is constructed. This model allows to visualize the technological process and to compare it to theoretical calculations at the design stage of a TPP and, if needed, to make adjustments.

References

1. Morad Behandish, Horea T., Computer-Aided Design, **70**, 100 (2016)
2. Wenlan Ba, Ning Ren, Graphical Models, **82**, 1 (2015)
3. N.G. Oliveira, J.L. Sotomayor, R.P. Torchelsen, C.T. Silva, L.D. João, Computers & Graphics, **60**, 119 (2016)
4. S.P. Bakhaeva, V.A. Gogolin, I.A. Ermakova, J. Min. Sci. **52:3**, 454 (2016)
5. G.A. Kazunina, D.V. Alekseev, Adv. Mater. Res., **933**, 62 (2014)
6. Hao Yang, Hui Zhang, Computers & Graphics, **65**, 45 (2017)
7. V.M. Zolotukhin, V.A. Gogolin, M.Yu. Yazevich, M.I. Baumgarten, A.V. Dyagileva, IOP Conf. Ser.: Earth Envir. Sci. **50:1**, 012027 (2017)
8. I.A. Ermakova, N.N. Pirieva, Coal in the 21st Century: Mining, Processing and Safety, **90** (2016)
9. M. Tyulenev, A. Khoreshok, E. Garina, O. Litvin, Y. Litvin, E. A. Maliukhina, E3S Web Conf., 012035 (2017)
10. A. Khoreshok, L. Kantovich, V. Kuznetsov, E. Preis, D. Kuziev, E3S Web of Conferences, **15**, 03004 (2017)

Justification of the Shape of a Non-Circular Cross-Section for Drilling With a Roller Cutter

Gennady Buyalich[1*], and *Mikhail* Husnutdinov[1]

[1] T.F. Gorbachev Kuzbass State Technical University, 650000, 28 Vesennyaya St., Kemerovo, Russia

Abstract. The parameters of the shape of non-circular cross-section affect not only the process of blasting, but also the design of the tool and the process of drilling as well. In the conditions of open-pit mining, it is reasonable to use a roller cutter to produce a non-circular cross-section of blasting holes. With regard to the roller cutter, the impact of the cross-section shape on the oscillations of the axial force arising upon its rotation is determined. It is determined that a polygonal shape with rounded corners of the borehole walls connections and their convex shape, which ensures a smaller range of the total axial force and the torque deflecting the bit from the axis of its rotation is the rational form of the non-circular cross-section of the borehole in terms of bit design. It has been shown that the ratio of the number of cutters to the number of borehole corners must be taken into account when justifying the shape of the cross-section, both from the point of view of the effectiveness of the explosion action and from the point of view of the rational design of the bit..

1 Introduction

The non-circular shape of the cross-section, as shown by a number of studies [1-11], affects the process of explosive crushing of rocks. To create an elongated cavity in a rock with a non-circular cross-section, appropriate drilling tools are created [12]. For drilling bore-holes in open-pit mining operations, it is suggested to use roller cutters in order to form a non-circular cross-section, and the effect of the cross-sectional shape on both the effectiveness of the explosion action and on the design of the roller cutter and the drilling process must be considered.

2 Materials and methods

The non-circular shape of the cross-section with stress raisers is presented in variants with notches on the walls of the bore-hole, or as a polygonal cross-section whose angles are also stress raisers. In the latter case, the area of the lateral surface, which is related to the cross-sectional area, changes, which causes an additional effect. Proceeding from this, the criteria for the cross-sectional shape are formulated from the point of view of the effectiveness of the explosion: the relative range of the radii of the cross-section, the connecting angle of the

* Corresponding author: gdb@kuzstu.ru

 DOI: 10.1051/e3sconf/20172103010

walls of the cavity, the increase in the area of the lateral surface of the well with a non-circular cross-section.

The relative range of the radii tends to the maximum with increasing concavity of the walls of the well, while the connecting angle of the walls of the borehole decreases and the area of their lateral surface increases. According to these criteria, the best, in terms of the effectiveness of the explosion, shapes of the cross-section are those with a smaller number of angles and a concave wall shape.

Therefore, biangular, triangular and quadrangular shapes of cross-section with rectilinear and curvilinear shape of the walls of the borehole are considered below (Fig. 1).

The possibility of forming a cross-section in the form of a regular figure by single-cone roller cutters is proved. The number of angles is determined by the gear ratio of the roller bit *i*, and by the shape of the roller cutter (Fig. 2). It is determined that with the help of bits having one, two, three cutters, it is possible to produce blast holes with a concave shape of their face and a regular shape of the cross-section having rectilinear, convex and concave sections of the sides, except for the shape in the form of a regular triangle that can only be obtained by one roller cutter. For rectilinear sides of a polygonal cross-section or with the same curvature, the rational shape of the cross-section from the point of view of the constructive perfection of the tool is the shape of the cross-section with rounded angles, and a larger curvature at which the maximal permissible cone angle of the cutter increases by the condition of its free rotation inside the well.

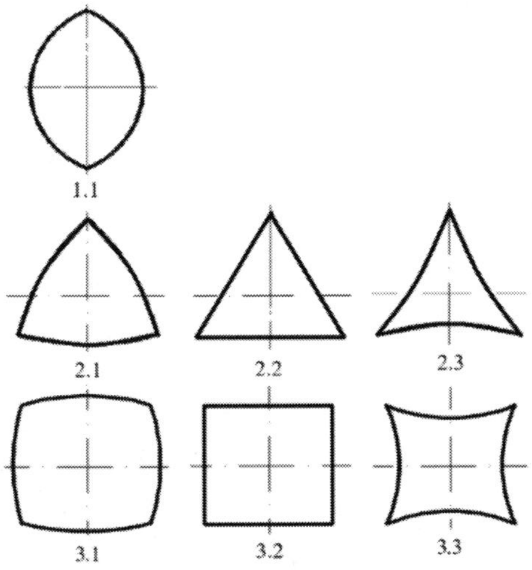

Fig. 1. Variants of shapes of the bore-hole cross-section: 1.1 – biangular; 2.1 – triangular with convex sides; 2.2 – triangular with rectilinear sides; 2.3 – triangular with concave sides; 3.1 – quadrangular with convex sides; 3.2 – quadrangular with rectilinear sides; 3.3 – quadrangular with concave sides.

To obtain a non-circular cross-section, it is proposed to form the base of the circular cone of cutters by planes intersecting along the line of its rotation diameter and passing at an acute angle to its axis (Fig. 3). In this case, with the concave shape of the bottom, the cone angle of the cutter is not limited by the condition of its free rotation inside the well and at an angle greater than or equal to 60 degrees, rolling the cutter along the bottom reproduces the convex sides of the walls of the well of different curvature, and if the angle is less than 60 degrees, the concave elements of the walls of the well of different curvature (Fig. 4).

 DOI: 10.1051/e3sconf/20172103010

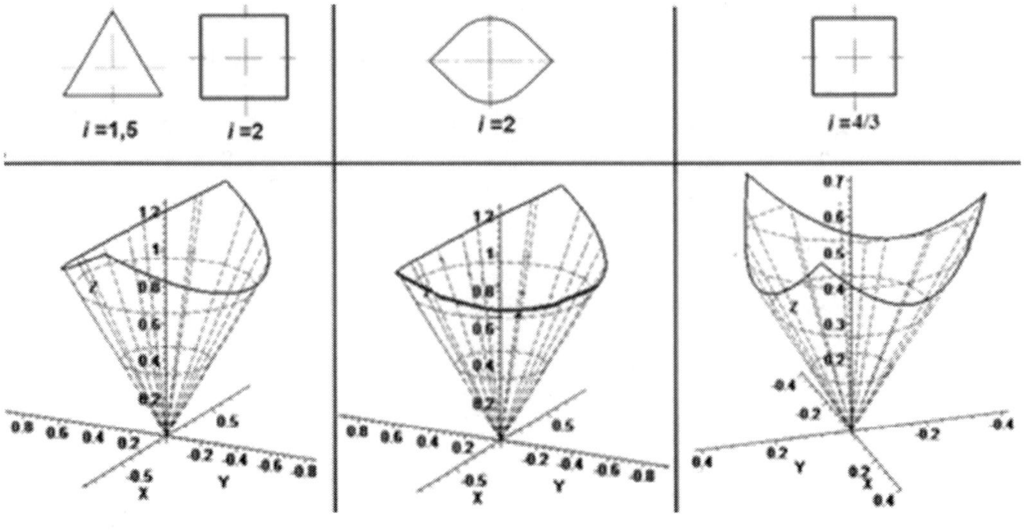

Fig. 2. Concordance of cross-section shapes, gear ratio of the bit *i* and the roller cutter shape

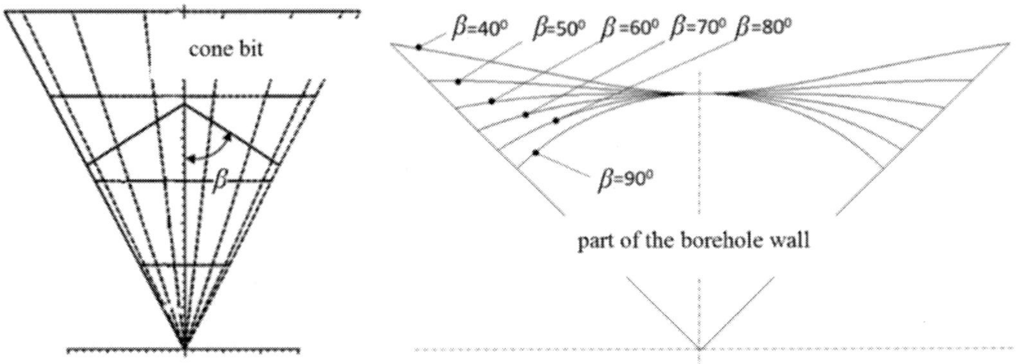

Fig. 3. Forming convex and concave sections of the bore hole wall depending on angle β between the axle of the cutter cone and the planes of its shear.

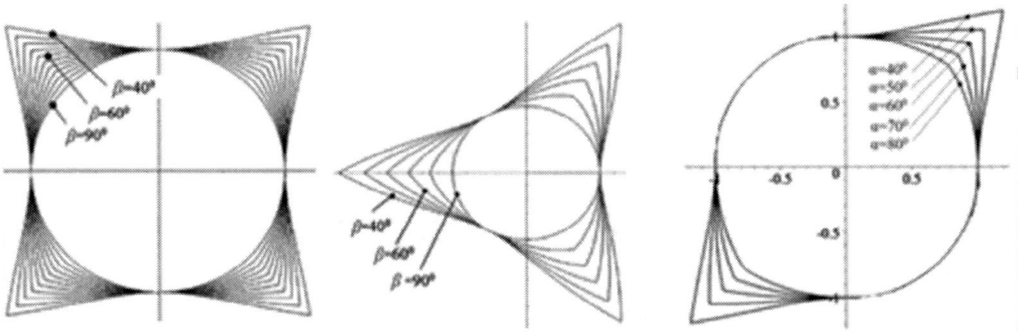

Fig. 4. Impact of angle β on shaping various cross-sections.

To assess the impact of the cross-section shape on the force parameters of the roller cutter drilling process, it is assumed that the cutters present a smooth cone that rolls along a smooth, non-deformable bottom of the hole. Such an assumption helps to assess the change in axial forces during tool rotation caused by a non-circular cross-sectional shape without determining the quantitative parameters of the rock reaction. In this case, the reaction forces of the rock are proportional to the contact length of the cutters (Fig. 5).

The relative range of the total length of the cutter contact with the bore-hole bottom, due to the non-circular cross-sectional shape, decreases with the convex shape of the bore-hole and a greater rounding of the angles of their connection, the fractional ratio of the number of cutters (more than one) to the number of bore-hole corners and the number of cutters greater than the number of bore-hole corners (Fig. 6).

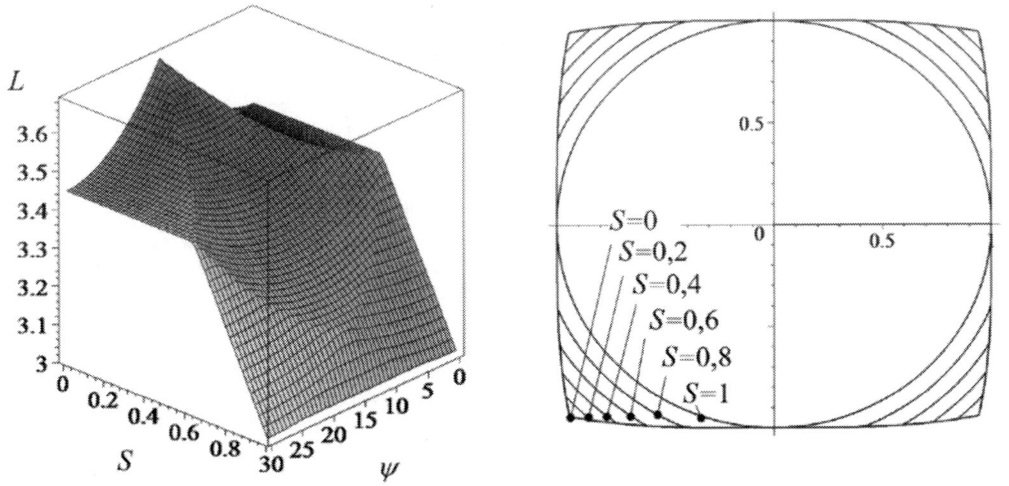

a b

Fig. 5. The impact of the degree of rounding the angles (S) of the cross-section of a quadrangular bore-hole at a shear angle $\alpha = 60^0$ and a unit radius of inscribed circumferential $r = 1$ on: a - the total length of the contact line (L) of three cutters with the bore-hole bottom depending on the angle of the tool rotation (ψ); b – the shape of the bore-hole cross-section.

It is determined that in the case of fractional multiplicity of the number of several cutters and the angles of the bore-hole cross-section, the non-circular shape of the cross-section stipulates the appearance of the torque deflecting from the axis of cutter rotation (Figure 7). The magnitude and frequency of the change in this deflecting torque correlates with the magnitude and frequency of the change in the range of the axial force.

3 Results and discussion

Thus, the parameters of the bore-hole with the non-circular cross-section formed by cutters, including the range of the radii, the angle of the walls connection, and the increase in the area of the lateral surface, are determined by the design parameters of the tool - the gear ratio, the number of cutters, their cone angle and the shape of the gage edge. The total axial force from the bore-hole bottom reactions proportional to the length of the contact line between the cutters and the bottom, and the modulus of the main torque arising from their action that deflects the bit from the axis of its rotation are determined by the parameters of the shape of the non-circular cross-section of the bore-hole and the design parameters of the tool; having the convex form of the bore-hole and a greater rounding of the angles of their connection, a fractional ratio of the number of cutters (more than one) to the number of corners of the well and with the number of cutters greater than the number of corners of the bore-hole, their range stipulated by the non-circular shape of the cross-section decreases.

DOI: 10.1051/e3sconf/20172103010

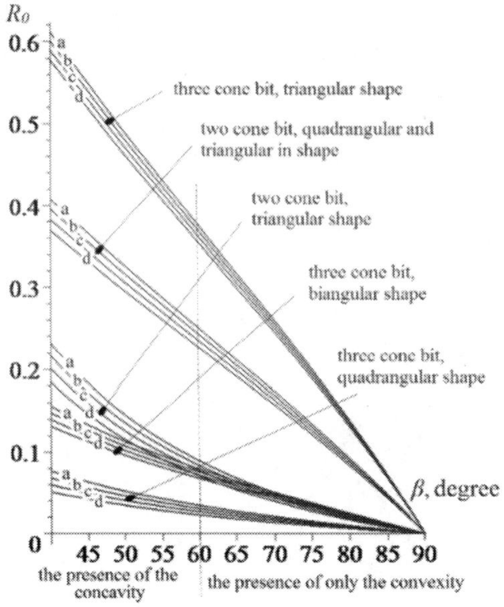

Fig. 6. Relative range R_0 of the total length of the cutters contact with the bottom depending on the shape of the bore-hole, angle β at S: a) 0; b) 0.05; c) 0.1; d) 0.15.

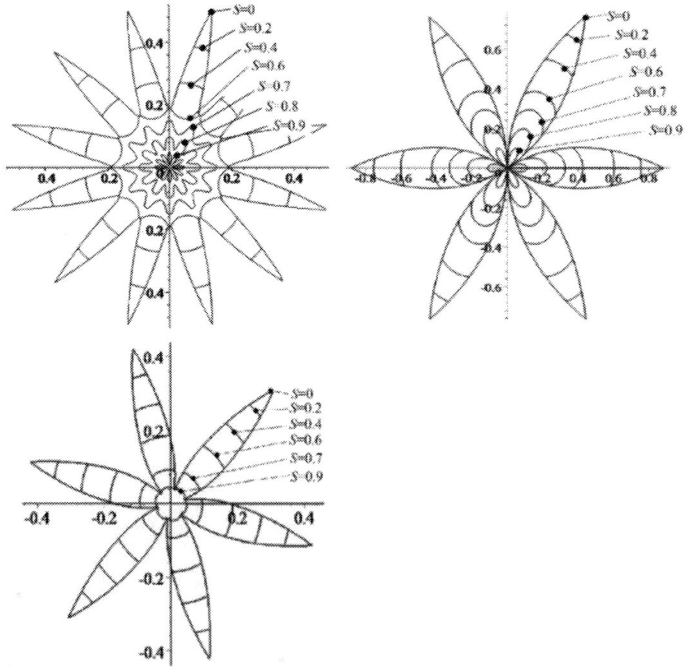

Fig. 7. Change of the modulus of the main torque (Н·м) deflecting the bit from its rotation axis due to the degree of rounding of S at β=60^0, r=1, angle of inclination of the concave bottom 8 degrees for: a) a three-cone drill bit and quadrangular cross-section; б) a two-cone drill bit and triangular cross-section; в) a three-cone drill bit and biangular cross-section.

4 Conclusions

Considering that the range of the axial force and the torque deflecting the bit from its rotation axis are the source of the oscillation processes in the "drill bit – drill pipes – drilling rig" system, the preferred shape of the non-circular cross-section of the borehole from the point of view of the rational design of the bit is the polygonal shape with rounded corners of the borehole wall connection and their convex shape, this ensures a smaller range of the total axial force and the torque deflecting the bit from the axis of its rotation. In the integrated justification of the shape of the bore-hole cross-section both from the view point of the blasting action efficiency and from the view point of the rational bit design, consideration should be given to the ratio of the number of cutters to the number of the bore-hole corners. In this case, the fractional ratio of the number of cutters (more than one) to the number of corners of the bore-hole leads to a decrease in the range of the axial force, and their whole ratio leads to exclusion of torque deflecting the bit from the axis of its rotation.

The proposed criteria for the shape of the cross-section can be used for a multifactor analysis of the design of a drilling tool, taking into account the shape of the cross-section of an explosive cavity created by it for fracturing of rock, and also for consideration of the influence on the duty and power parameters of the drilling process, which are interrelated with the oscillatory processes occurring both in the rock [13] and in the structure of the drilling tool and machine tool [14], which in turn affects the strength of the elements of their design [15].

References

1. S. Bhandari, Proceedings of Rock Fragmentation by Blasting, **511** (2013)
2. A. José, P. Sanchidrián, International Journal for Blasting and Fragmentation, **4**, 1 (2000)
3. S.H. Cho, Y. Nakamura, B. Mohanty, H.S Yang, K. Kaneko, Engineering Fracture Mechanics, **75**, 3966 (2008)
4. Finn Ouchterlony, Mats Olsson, S. O. Båvik, International Journal for Blasting and Fragmentation, **4**, 55 (2000)
5. Zong-Xian Zhang, Rock Fracture and Blasting, **6**, 217 (2016)
6. Wang Yan-mei, Li Cheng-fang, Chongqing architecture, **11**, 51 (2005)
7. Ding Dexing, Zhv. Chenghang, Trans Nonferrous Metals Soc China, **1**, 188 (1999)
8. C. E. Tsoutrelis, N. Gikas, P. Nomikos, G. Exadaktylos, International Journal for Blasting and Fragmentation, **1**, 445 (1997)
9. Yue Zhongwen, Guo Yang, WangXu, Chinese Journal of Rock Mechanics and Engineering, **10** (2015)
10. Yu Musong, Yang Yongqi, Yang Renshu, Zhang Qi, Explosion and shock waves, **2** (1997)
11. A. Zhukov, L. T. Dvornikov and S. M. Nikitenko, IOP Conference Series: Materials Science and Engineering, **124**, 012171 (2016)
12. G. D. Buyalich, K. G. Buyalich, V. Yu. Umrikhina, IOP Conf. Series: Materials Science and Engineering, **142**, 012120 (2016)
13. E.N. Pashkov, G.R. Zijakaev, M.V. Tsigankova, Applied Mechanics and Materials, **379**, 91 (2013)
14. G. D. Buyalich, A. V. Anuchin, K. P. Serikov, IOP Conf. Series: Materials Science and Engineering, **127**, 012012 (2016)

Improving the Repair Planning System for Mining Equipment on the Basis of Non-destructive Evaluation Data

Michael Drygin[1*], and *Nicholas* Kuryshkin[1]

[1] T. F. Gorbachev Kuzbass State Technical University, 28 Vesenija str., Kemerovo 650000, Russia

Abstract. The article tells about forming a new concept of scheduled preventive repair system of the equipment at coal mining enterprises, based on the use of modern non-destructive evaluation methods. The approach to the solution for this task is based on the system-oriented analysis of the regulatory documentation, non-destructive evaluation methods and means, experimental studies with compilation of statistics and subsequent grapho-analytical analysis. The main result of the work is a feasible explanation of using non-destructive evaluation methods within the current scheduled preventive repair system, their high efficiency and the potential of gradual transition to condition-based maintenance. In practice wide use of non-destructive evaluation means will allow to reduce significantly the number of equipment failures and to repair only the nodes in pre-accident condition. Considering the import phase-out policy, the solution for this task will allow to adapt the SPR system to Russian market economy conditions and give the opportunity of commercial move by reducing the expenses for maintenance of Russian-made and imported equipment.

1 Introduction

At present Russian coal mining enterprises use the system of maintaining the technological equipment in the proper working order (SPR). The analysis of the SPR system and the underlying regulatory documentation allows to assert that the SPR system has not yet outlived its usefulness until now, but requires a new approach [1, 2].

The main guideline document that shall be subject to compulsory implementation for all coal mining enterprises of the USSR and then Russian Federation is the Regulation on scheduled preventive repair of the equipment used at open-cast mining works [2]. It was enacted in 1983, and remains valid through present and is applicable to: shovel excavators, drill rigs, transport and overburden bridges, stackers and overloaders, extensible belt conveyers and dredge pumps.

The matter of the SPR system is that after a certain number of machine hours and/or accomplished scope of work scheduled repair of different types is performed, and the sequence of repair must be defined due to component life and operating condition of the equipment. At the same time the Regulation presupposes improvement for SPR system by development and application of automated control systems of maintenance and repair, im-

* Corresponding author: mike.drygin@gmail.com

proving repair technology and use of specific monitoring and metering instruments and means [2].

With the reference to the above mentioned it can be affirmed that the SPR system is based not only on the planned replacement of the equipment components after a certain number of running hours, but on the constant control of the technical state during the shift changeover and recording it to the "Shift turnover register", and planning of medium, normal and annual maintenance. Thus defect list which is an integral part of the SPR system is one of the most important documents, and it should be drafted correctly for the balance of the whole system.

Again, judging by the terminology of the Regulation, scheduled preventive repair is nothing but a system that requires long-term and short-term planning based on planned maintenance works with the schedule made up due to monitoring the technical condition of the equipment at any time in accordance with the defect list. The defect list is drafted by a qualified specialist, and the list of repair parts demanding age replacement is nothing short of long-term planning (including national scale) and partly error insurance for the specialist who drafts the defect list.

At the same time it is important to note that within the last ten years staff competence has dropped dramatically as well as the number of young and qualified workers among engineering control and top management. The result of intensification of production, globalization of economy and the above mentioned processes can be regarded as the factor that has determined inappropriate technical condition of coal mining equipment and excavators in particular.

2 Materials and Methods

Within the period since 2009 to 2013, the authors had performed more than three hundred complex diagnostic tests of the technical state of the excavators at Kuzbass open-cast coal mines. The extended diagram of these tests' result is presented on the Fig. 1. The diagram shows that only 14% of the excavating equipment is in good technical condition, and 27% is in unsatisfactory but still working condition.

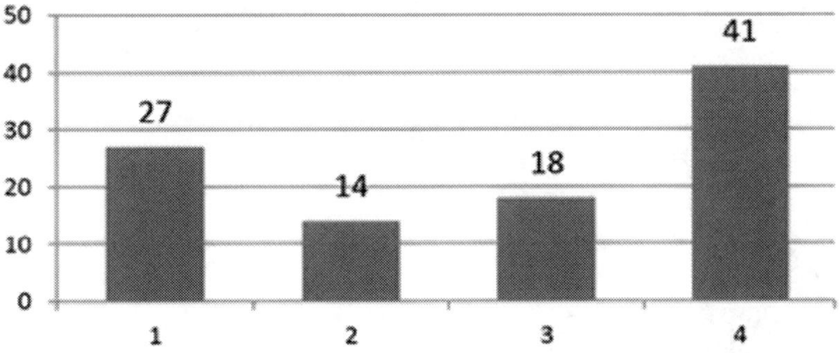

Fig. 1. Real situation of the excavating equipment at Kuzbass open-cast coal mines according to the results of diagnostic studies.

Methods used during the evaluation, such as vibration diagnostics, heat monitoring, visual and dimensional inspection, ultrasonic control, metal magnetic memory method, give a true and fair view of flaws distribution. When adequately interpreted these methods reveal a host of presented defects and an opportunity of system-oriented analysis of cause-and-effect relationship in their development [4, 5, 6].

According to the analysis of the obtained results it can be asserted that the most important problem of retaining excavator fleet survivability is keeping constructional steelwork in good technical condition and avoiding its destruction [7]. According to the above mentioned Regulation on SPR basic methods of constructional steelwork control are visual and dimensional inspection and ultrasonic testing. Time allowance for ultrasonic testing of welded joints are also given in the Regulation (Table 1).

Table 1. Time allowance for ultrasonic flaw detection in welded metallic structures of excavators.

Excavator type	Monitoring time, days	Number of welded seams subject to inspection, items.
EKG-4,6b; EKG-5a	4	26
EKG-8i; EKG-6,3us; EKG-4u	12	12
EKG-12,5; EKG-6,3u	15	8
ESh-6/45	13	16
ESh-10/70	15	26
ESh-13/50	15	–
ESh-15/90	15	33
ESh-20/90	16	–

Even without considering the basics of control process technology but only according to the data presented in Table 1 a conclusion can be made about significant efforts that ultrasonic control requires. For instance, the extension of 26 welding seams of Esh-10/70 mining shovel boom subject to ultrasonic control is 245,60 mm. More than 294,000 mm of the whole boom are strength weld. Fig. 2 shows general view of the Esh-10/70 drag-line boom and allows to evaluate the amount of constructional steelwork that needs control. Constructional steelwork with shop weld that is subject to ultrasonic flaw detection. It is assumed that the rest of the welded seams will not fracture until the moment of disposal, taking into account all performed repairs including flaw detection (for instance, during the capital repair). Repair welded seams are also not subject to control and therefore remain "dangerous" [8].

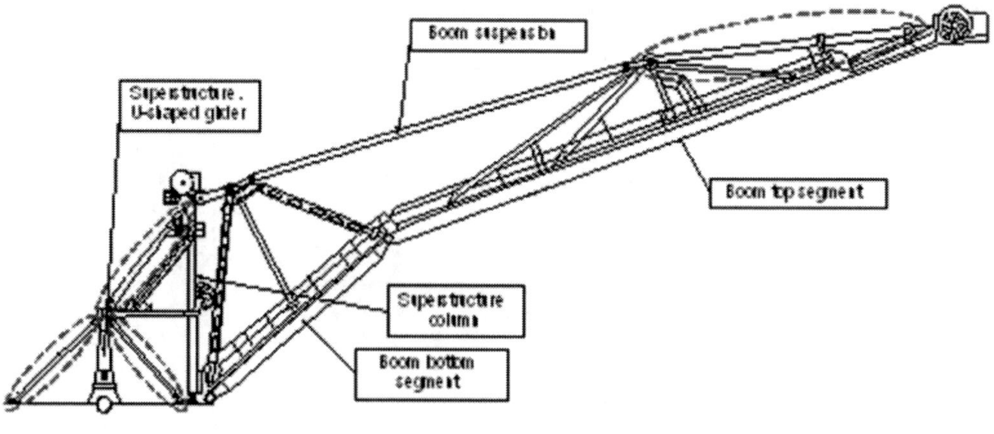

Fig. 2. General arrangement of a mining shovel Esh-10/70 boom.

Fatigue cracks developing from technological repair defects that were not detected during the control procedures performed after repair are the main reason of metallic structures' destruction. Fig. 3 shows a fracture of a mining shovel boom metallic structure along a basic metal crack that developed cyclically. Intolerable slag inclusions of a welded seam which is not subject to control in accordance with the Regulation, served as a concentrator [9, 10, 11, 12].

Fig. 3. Fracture of a mining shovel boom metallic structure.

After analyzing more than fifty similar metallic structures followed a conclusion that 100% welded seams should be subject to control, but that would require other, more highly-efficient control methods.

3 Results and Discussion

According to performed analytical and practical work a scheme of performing non-destructive testing procedures for metallic structures that had reached the end of their service life was worked out [13, 14]. The first accepted method was visual and dimensional control. As one of basic control methods it allows to detect the most part of surface defects. Compared with other methods of non-destructive control visual and dimensional testing is highly-efficient and not expensive and allows the specialist to detect the most part of defects with minimum opening about 10 μm [15]. However the main disadvantage of this method is human factor (fatigability, site preparation quality, physical well-being of the engineer performing the inspection) and absence of possibility to detect developing surface and all subsurface defects. The table of some non-destructive control methods' sensitivity evaluation data is given below (Table 2).

The second accepted method of 100% welded seams control was metal magnetic memory method. It is based on recording and analysis of diffusion self-magnetic fields that appear on the equipment metallic surface in stress concentration zones [16].

In comparison with other non-destructive control methods metal magnetic memory's main advantage is that it does not require any preparation of metal surface before inspection. Besides flaw detection metal magnetic memory method gives information about real state of the object and allows to find reasons of flaw development [17].

Table 2. Evaluation data of sensitivity of some non-destructive testing methods.

Method	Minimum size of detected discontinuity flaws, μm		
	opening	extension	depth

Visual and optical control	5...10	100	–
Dye-penetrant test	1...2	100...300	10...30
Fluorescent-penetrant inspection	1...2	100	10...30
Magnetic particle method	1	30	10...50
Eddy current method	0.5...1	6	150...2000
Ultrasonic method	0.01...1	10	–
Radiographic method	100...500	–	1...1.5

In support of this conclusion Fig. 4 shows the oscillogram of self-magnetic field stress of a weak weld of a Esh-10/70 drag-line boom head. The criteria of detected stress concentration zones' danger level is magnetic field intensity. Thus the level of 27.8 (A/m)/mm is unacceptable – maximum point on the oscillogram below. This is confirmed by calculation of real magnetic deformation capacity index:

$$m_r = \sqrt{\frac{k_{\max}}{k_{\mathrm{m}}}} = \sqrt{\frac{27.8}{3.7}} = 2.74, \tag{1}$$

where $k_{\max}$ is self-magnetic field maximum gradient; k_{m} is mean value of magnetic field gradient.

Extreme magnetic index characterizing metal deformation capacity:

$$m_{ext} = \left(\frac{\sigma_{\text{В}}}{\sigma_{\text{Т}}}\right)^2 = \left(\frac{490}{350}\right)^2 = 1.96, \tag{2}$$

where $\sigma_{\text{В}}$ is metal endurance strength, MPa; $\sigma_{\text{Т}}$ is metal yield strength, MPa.

Deformation index must not excess extreme index. In this case deformation index is 1.4 times more than extreme index. Therefore the section of a welded seam with a stress concentration zone is unfit for operation.

The form of the oscillogram is the third criterion supporting this conclusion. Specific distribution of magnetic field intensity is shown distinctly on the Fig. 4. In this case given the lateral (across the supposed defect) direction of the probe movement on the oscillogram specific distribution with fading-in, spike in and symmetrical descent of magnetic field gradient is formed.

This defect was confirmed during referee ultrasonic control as a subsurface discontinuity flaw with depth of location 2…12 mm, and this allowed to prevent a breakdown with boom head fracture because this welded seam was not mentioned in the "Reference Guide on Ultrasonic Flaw Detection of Single-bucket Excavators" as subject to traditional control.

Therefore complex approach to flaw detection of all welded seams in metallic structures allows to detect up to 100% of defects unacceptable according to regulatory documentation. The generalized result of such approach to the control of a metallic structure that was more than 20 years in operation with total length of welded seams up to 100 meters was the following:

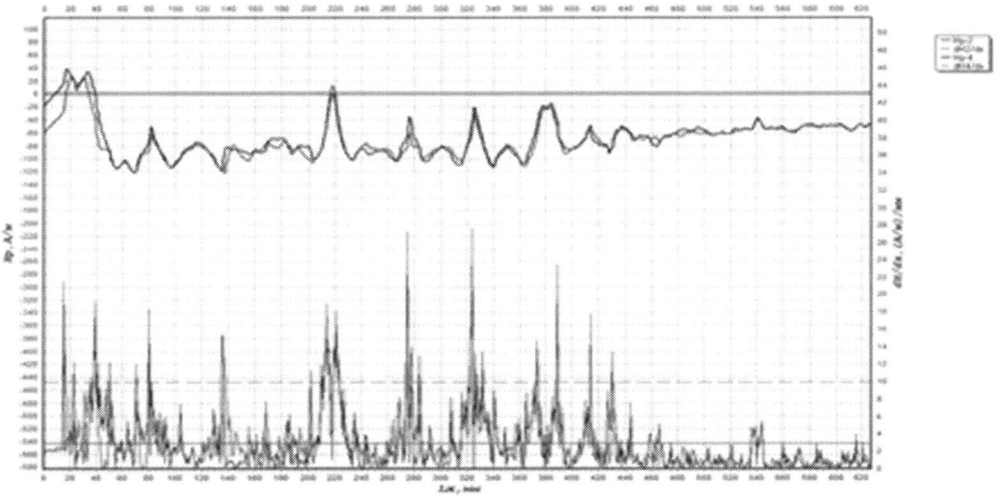

Fig. 4. Oscillogram of a weak weld magnetic field intensity.

- visual and dimensional testing allowed to detect about 40,000 mm of welded seams that are not in compliance with the regulatory documentation;
- metal magnetic memory method allowed to detect about 60,000 мм of stress concentration zones where 12,800 mm of welded seams became critical and demanded subsequent operational supervision;
- Ultrasonic testing allowed to detect 4,500 mm of defects demanding immediate removal.

4 Conclusion

1. Improving efficiency and quality of mining equipment repair based on the system of scheduled preventive repair with wide use of non-destructive control methods and means allows in the short run to achieve good results characterized by reducing emergency downtime and proper repair.
2. The opportunity of using a host of non-destructive control methods and means within the SPR system is included into the regulatory documentation. Therefore implementation of one modern non-destructive control methods – metal magnetic memory – allows to solve one of the most important problems – flaw detection in significant amounts of welded joints of metallic structures within a short time.
3. Using the metal magnetic memory method allows to describe with high level of confidence the type of the defect and the level of its danger that will further allow to remove only dangerous defects, and this will provide efficiency of metallic structures over extended periods.

References

1. V. A. Kovalev, A. A. Khoreshok, O. I. Litvin, The 8th Russian-Chinese Symposium. Coal in the 21st Century: Mining, Processing and Safety, **287** (2016)
2. A. Morshedlou, H. Dehghani, S. H. Hoseinie, JME, **5:2**, 113 (2014)
3. S. Namata, *Performance analysis of heavy earth moving machineries (hemm) in opencast coal mines. Department of Mining Engineering*, (Rourkela National Institute of Technology, 2015).

4. E. Todorov, R. Spencer and M. Lozev, M. E., **72:6**, 812 (2014)
5. L. De Chiffre, S. Carmignato, R. Kruth, J. P. Schmitt, Manuf. Technol., **63:2**, pp. 655 (2014)
6. *ISO 10042 Welding – Arc-welding joints in aluminium and its alloys – Quality levels for imperfections* (2005)
7. *ISO 3834-1 Quality requirements for fusion welding of metallic materials – Part 1: Criteria for the selection of the appropriate level of quality requirements* (2005)
8. *ISO 3834-2 Quality requirements for fusion welding of metallic materials – Part 2: Comprehensive quality requirement* (2005)
9. *ISO 3834-3 Quality requirements for fusion welding of metallic materials – Part 3: Standard quality requirements* (2005)
10. *ISO 3834-4 Quality requirements for fusion welding of metallic materials – Part 4: Elementary quality requirements* (2005)
11. R. Talathi, *Corporate planning: features, process and analysis / management. Your Article Library.*
12. R. Gedrey, *Stress-life fatigue testing basics.* (Quality Magazine)
13. H. M. Kroening, D. Sednev & Y. A. Salchak, Advanced Materials Research, **1040** (2014)
14. *ISO 24497-3, Non-destructive testing – Metal magnetic memory – Part 3: Inspection ot welded joints* (2009)
15. T. Nelligan, *Ultrasonic flaw detection as a quality tool*, (Berlin: Quality Magazine)
16. A. A. Dubov, Al. A. Dubov, and S. M. Kolokolnikov, *Method of metal magnetic memory and inspection devices. Training handbook.* 320 (Moscow: Tisso Co., 2003)
17. *ISO 24497-2: Non-destructive testing – Metal magnetic memory – Part 2. General requirements* (2007)

The Raising Influence of Information Technologies on Professional Training in the Sphere of Automated Driving When Transporting Mined Rock

Andrey Kosolapov [1*], and *Sergey* Krysin[1]

[1]T.F. Gorbachev Kuzbass State Technical University, 650000, Vesennyaya 28 st., Kemerovo, Russian

Abstract. Revolutionary changes in the area of production, holding and exploitation of the automobile as a transport vehicle are analyzed in the article. Current state of the issue is described and the development stages of new approach to driving without human participation are predicted, taking into consideration the usage of automobiles for transportation of mined rock in Kuzbass. The influence of modern information technologies on the development of new sector of automobile industry and on the process of professional and further training of the specialists in the sphere of automobile driving is considered.

1 Introduction

The transformation of automobile industry is coming before our eyes. The transition "from the automobile to automobility" is becoming a common trend. Many traditional automobile producers are gradually changing into "suppliers of mobile services", accumulating digital solutions and additional services. At the same time, more and more IT- giants are appearing in the field. And, Google and Apple are no more the brightest examples here.

On one hand, a human operator (the driver) has more experience in driving so far. They are resourceful and quick-witted in managing force majeure situations on the road. On the other hand, major cases of car accidents are caused by a human factor. A robot, in its turn, is capable of following the algorithm accurately, but it is still to be learnt to deal with nonstandard circumstances. That's why, as many automobile engineers claim, the automobile is going to be driven by a human for a long time, up to 2030 as minimum.

To "exclude" a human from the system of control many IT developers are placing more and more new products in the constantly expanding market. Examples can be seen in [1-6].

Thus, BlackBerry is developing new driver assistants for Ford. They are based on operational system QNX, which has already been used in Ford's onboard media system Sync 3.

IT-giant IBM, which is run by Mark Fields, CEO of Ford, stakes on artificial intelligence system Watson, designed to assist drivers with advanced voice commands. Four "online" prototypes BMW i8 were developed in cooperation with BMW. System Watson will be

* Corresponding author: kosolapovav@kuzstu.ru

adopted by Opel and GM to improve their media systems. IBM cooperates with PSA concern and Ford, working on big data, cloud technologies and "smart city".

Transnational corporation Intel is staking on the automobile: its new subdivision Automated driving group has been developing new technologies for transport since December 2016. Along with Delphi, Intel is working on the solutions to automated driving, which are planned to be presented finished to automobile producers.

Having signed numerous partnership agreements with cloud service Azur (also used by TomTom, Renault-Nissan and Toyota), Microsoft set up a massive campaign on application of its software to automobiles: Skype for Business (is already prepared to be installed in Volvo 90 series) and personal assistant Cortana.

Nokia's chart subdivision Here, designed by American Navteq and known as Ovi, was bought by Audi-BMW-Mercedes in 2015. Nowadays the company competes with Apple and Google, designing digital and multilayered maps, necessary for automated driving.

Nvidia, Santa-Clare, takes revolutionary decisions in the area of automated driving and computerized vision. Its processors and supercomputers with neural networks and self-learning technologies were chosen by Tesla, Volvo, Audi and Mercedes-Benz.

Automobile producers became interested in digitization and new services, staking on the future, when automobiles are autonomous, electrical and distributed shared.

2016 was the year of major changes in automobile industry. It became clear that automobile production would develop new business directions, controlled by special subdivisions and affiliates [6-8]. Thus, Smart Mobility (Ford), Moia (Volkswagen concern), Free2Move (PSA), Maven (GM car sharing), Mercedes-Benz Energy and Accumotive (firms controlled by Daimler and responsible for batteries) were established. Volvo company opened sharing subdivision Sunfleet, BMW expanded activities of DriveNow (in Europe), ReachNow (in the USA). Nissan brand tested systems V2G (vehicle-to-grid) to integrate its electric car Leaf in electricity network of the city. FCA concern made an agreement with Waymo, which continued Google Car project. Perhaps, it is a revolution and it has two directions. On one hand, automobile producers go for high technologies. On the other hand, the producers of electronics began opening automobile subdivisions, concluding bargains of billions, increasing the number of departments, working on software for automated driving. Apart from Google and Apple with Android Auto and CarPlay, respectively, there are a lot of other famous names that are ready to take automotive industry as their major business. Alexander Kotouc, the Head of Product Management BMWi, stated "Automobile industry, in its usual meaning, is already dead. Automobility is our future" [6].

Thus, it's time to speak about new level of opportunities for passenger traffic – mobility 2.0. It means that new requirements appear for specialists, providing the transportation of passengers and freights in various spheres, including those where the exclusion of human participation will significantly raise the level of industrial security and effectiveness. The approach to specialist training ought to be changed adequately.

2 Materials and methods

Specialists, working in the area of transportation, already need to know the principles of modern artificial intelligence technologies in terms of their application to transportation sector. Such technologies ought to be based on interaction of rational intellectual agents in common environment (fig. 1) [9, 10].

Artificial intelligence in terms of passenger and freight transportation is expected to include such segments as "autonomous" vehicles, "intelligent" infrastructure and the systems of control. At the same time, the technologies used for autonomous vehicle driving, will be based on [5, 11]:

- numerous sensors and intercorporate communication of types V2V, V2I and V2C;

 DOI: 10.1051/e3sconf/20172103012

- artificial neural networks for big data processing;
- making decisions according to constantly qualifying rules and logics.

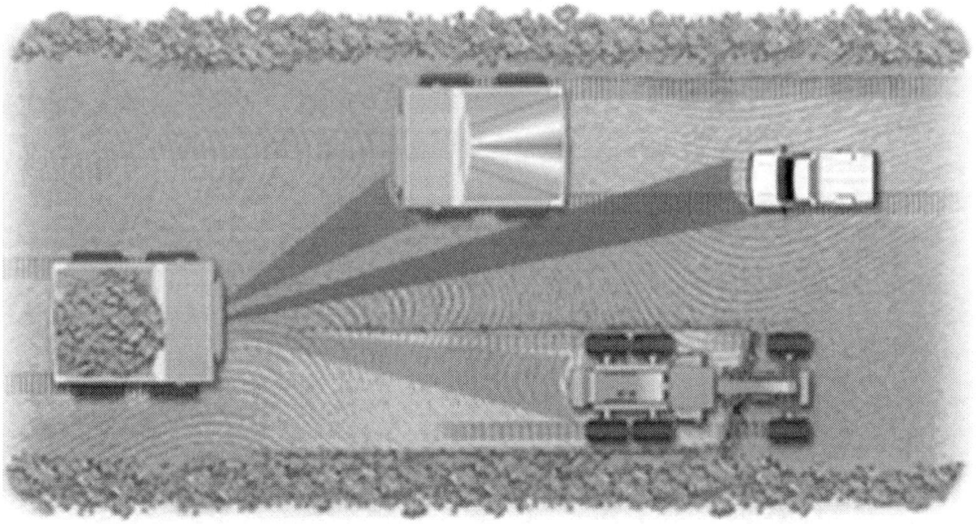

Fig. 1. Representation of traffic as interaction of separate agents in the common information environment.

Road infrastructure is to be integrated in the system of traffic control in a balanced manner.

The current tasks on introduction of artificial intelligence methods into transportation sector are the following:

- the intellectual analysis of big data,
- the developed system of decision making for the control of complicated objects and processes.

The development of information technologies along with the progress in automobile components designing enabled to classify the levels of automated driving. The Society of Automotive Engineers, USA (SAE), defines 6 levels of automated driving, from first assistants, now available even with low-budget economy cars to almost autopilot vehicles [12]. Each level differs by progressive increase of tasks and responsibility transferred from the driver to vehicle electronic systems.

Level 0 represents the systems of forward collision or even of automatic braking, when the vehicle actively interferes with driving. The driver is in complete control of the driving (fig. 2).

DOI: 10.1051/e3sconf/20172103012

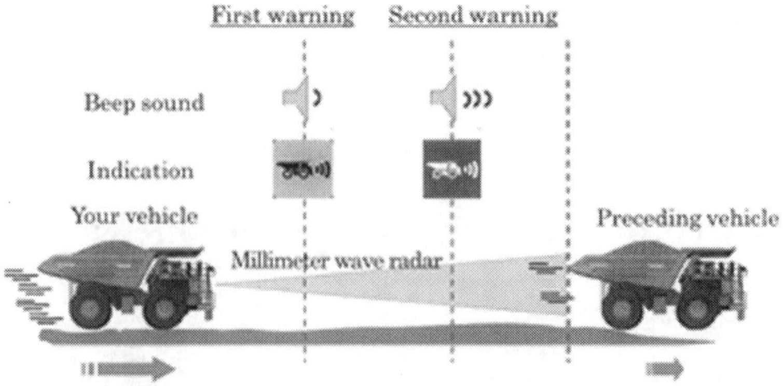

Fig. 2. Level 0. No automation. The driver is in complete control of the vehicle.

Automobile begins cooperating with a human on level 1. Limited number of tasks can be committed to electronics, for instance, active cruise control. However, a human has total control over driving (fig. 3).

Fig. 3. Level 1. Driving with the assistance: electronics only assists.

Level 2 represents modern stage of automated driving development. In certain conditions (for example, while driving in the countryside or if there are clear road markings) the automobile is able not only to keep distance, accelerating and decelerating, depending on the speed of the cars in front, but also to keep the track, staying in the traffic. The driver ought to keep his hands on the steering wheel to provide safety. The human remains the only one responsible for motion and any strategic decision taken.

A sudden leap to automated driving is expected on level 3. The artificial intelligence of the automobile fulfills the task of monitoring the road situation and in certain conditions makes a strategic choice, for example, to take over or to turn, to enter primary roads and etc. The human can take his hands off the steering wheel and temporarily look away but he ought to be ready to take control if situation occurs to be too complicated for electronics. Time horizon is 2018.

In certain conditions the automobile is able to control all driving processes on level 4. The human can drive by himself, but in case he fails or makes a mistake, the system corrects them with the help of the correct decision system. The driver is allowed to take his hands off the steering wheel, look away from the road and relax. However, the driver presence is obligatory as there are still some traffic or weather conditions, when system fails. Time horizon is 2020-2025 (fig. 4).

 DOI: 10.1051/e3sconf/20172103012

Fig. 4. Level 4. High automation: almost autonomous vehicle (e).

Finally, level 5. It is assumed that the automobile will manage all phases of a trip by itself. The driver won't play any role. The driver presence as well as a steering wheel is no longer needed. Mainly, this level is oriented at transportation systems. Time horizon is 2025-2030 (fig. 5).

Fig. 5. Level 5. Full automation: driver presence is no longer needed.

Certainly, "big data" processing will play one of the leading roles. "Big data" is usually defined as a huge amount of data in terms of its volume, diversity, transfer rate and degree of reliability that requires special technologies and methods of analysis to extract valuable information.

Autonomous automobiles are required to have various tools of reality perception: short-range and long-range radars, medium-range lidars, ultrasonic sensors and cameras [12-13]. As a result, the artificial intelligence of the automobile is capable of producing such virtual images, where automobiles, cyclists, motorcyclists and pedestrians are shown as "cubes" of different colours. Virtual bars on the carriageway indicate deceleration zone and braking (fig. 6).

 DOI: 10.1051/e3sconf/20172103012

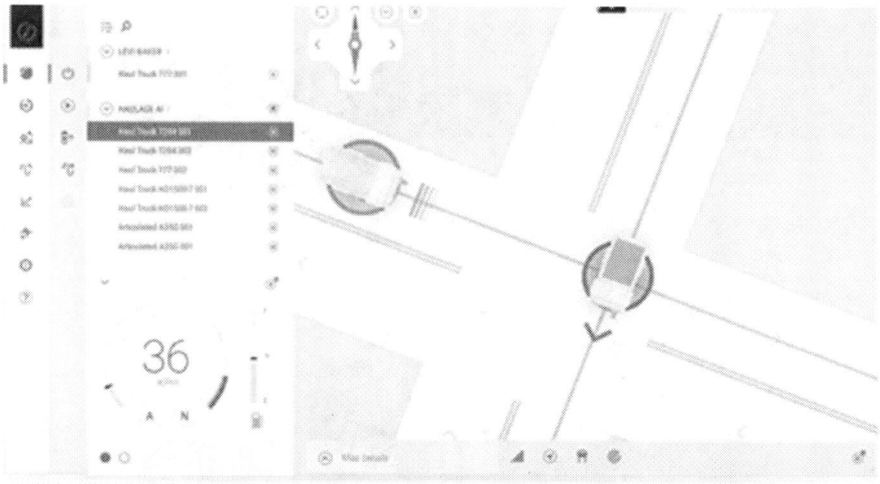

Fig. 6. Virtual presentation of real crossroad.

There are some examples of attempts to introduce autopilot vehicles. Autonomous truck Volvo was put into operation in a Sweden mine in September 2016. According to JSC "KAMAZ", robotized trucks are to be used to fulfill the tasks of mining and other industries in our country in 2020. The representatives of the system designing company stated, that the operation of robotized dump trucks is economically sound now and is possible only for the fulfillment of special tasks in certain conditions, where the presence of humans is limited and the operation of machinery is conditional on safety requirements and economic feasibility. One of such industries is mining.

One more domestic example is launching a pilot root of "MatrЁshka" bus, which is to run round the campus of Far Eastern Federal University on Russkiy island. "MatrЁshka" is a totally autonomous vehicle, designed by Scolkovo residents. It allows transferring passengers and freights and can be used as municipal machinery. The designer ensures that autopilots are safer than traditional buses as a human factor, causing car accidents, is excluded, and the artificial intelligence monitors the technical state of the vehicle and reports all malfunctions to the owners.

3 Results and discussion

It is assumed that in Kuzbass as a major region known for its open cut mining, automated driving can be used for mined rock transportation. The motion of the dump truck is influenced by fewer factors comparing to its motion in rural conditions. The stability of transportation root, the exclusion of conflicts with other participants of the traffic due to their absence, comparatively low speed – all these factors are to simplify the development of driving algorithms.

However, it requires changes in the list of subjects and the curriculum, taught in T.F. Gorbachev Kuzbass State Technical University. It is obvious that the number of subjects and systems of knowledge, learnt by specialists in traffic organization, ought to include not only information technologies but also such disciplines as "big data" management, neural networks and artificial intelligence. The disciplines that ought to be paid special attention may include "Automatics and telemechanics in traffic", "Computer engineering and networks in the industry", "Informatics", "Information technologies in transportation", "Computer technologies in traffic organization" and "Applied programming in traffic".

4 Conclusion

Modern development of automobile industry, forced not only by consumers' needs but also by the change of the approach to exploitation, holding and even driving the automobile, gets a new vector of development, aiming at the exclusion of the man as an operator. One of the gravest revolutions in freight and passenger transportation, comparing to the revolution in information transfer in the society, will happen in the nearest future.

References

1. N. Fleming, New Scientist, **206:2754**, 34 (2010)
2. J. Weyer, R. D. Fink, F. Adelt, Safety Science, **72**, 199 (2015)
3. D. J. Fagnant, K. M. Kockelman, Transportation Research, **77**, 167 (2015)
4. M. Schellekens, Computer Law & Security Review, **31:4**, 506 (2015)
5. C. Katrakazas, M. Quddus, W. Chen, L. Deka, Transportation Research Part C: Emerging Technologies, **60**, 416 (2015)
6. W. Payre, J. Cestac, P. Delhomme, Transportation Research Part F: Traffic Psychology and Behaviour, **27**, 252 (2014)
7. A. Alessandrini, A. Campagna, P. Delle Site, F. Filippi, L. Persia, Transportation Research Procedia, **5**, 145 (2015)
8. H. Martínez-Barberá, D. Herrero-Pérez, Transportation Research Part C: Emerging Technologies, **39**, 94 (2014)
9. D. J. Fagnant, K. M. Kockelman, Transportation Research Part C: Emerging Technologies, **40**, 1 (2014)
10. L. C. Fernandes, J. R. Souza, G. Pessin, P. Y. Shinzato, D. Sales, C. Mendes, M. Prado, R. Klaser, A. C. Magalhães, A. Hata, D. Pigatto, K. C. Branco, Jr. V. Grassi, F. S. Osorio, D. F. Wolf , Journal of Systems Architecture, **60:4**, 372 (2014)
11. *Quattroruote*, 26 (2017)
12. J. Alonso, V. Milanés, J. Pérez, E. Onieva, C. González, T. Pedro, Transportation Research Part C: Emerging Technologies, **19:6**, 1095 (2011)
13. S. Le Vine, A. Zolfaghari, J. Polak, Transportation Research Part C: Emerging Technologies, **52**, 1 (2015)

Estimation of Energy Efficiency of Means of Transport According to the Results of Technical Diagnostics

Anton Shalkov [1], and *Mariya* Mamaeva [1]

[1]Prokopievsk branch of T. F. Gorbachev Kuzbass State Technical University, 653039, 19a Nogradskaya St., Prokopievsk, Russia

Abstract. The article considers the questions of application of nondestructive methods control of reducers of conveyor belts as a means of transport. Particular attention is paid to such types of diagnostics of technical condition as thermal control and analysis of the state of lubricants. The urgency of carrying out types of nondestructive testing presented in the article is determined by the increase of energy efficiency of transport systems of coal and mining enterprises, in particular, reducers of belt conveyors. Periodic in-depth spectral-emission diagnostics and monitoring of a temperature mode of operation oil in the operation of the control equipment and its technical condition and prevent the MTBF allows the monitoring of the actual technical condition of the gearbox of a belt conveyor. In turn, the thermal imaging diagnostics reveals defects at the earliest stage of their formation and development, which allows planning the volumes and terms of equipment repair. Presents diagnostics of the technical condition will allow monitoring in time the technical condition of the equipment and avoiding its premature failure. Thereby it will increase the energy efficiency of both the transport system and the enterprise as a whole, and also avoid unreasonable increases in operating and maintenance costs.

1 Introduction

In the Russian Federation, as in many other developed countries, transport is one of the largest sectors of the economy, and an important part of the production infrastructure. The transport system connects all regions of the country, which is a necessary condition of its territorial completeness. In turn, the wide application of transport communications across the country and across the enterprise makes us think about energy efficiency in the transport system as a whole in view of annual growth of energy consumption, negative impact on the environment and emissions of harmful substances, unjustified increase of costs of operation and maintenance of the equipment [1].

One of the most common means of moving bulk materials is continuous transport system, including belt conveyors. Particularly widespread they last received in coal and mining industry as an intermediate and main conveyor systems high performance, providing transportation of goods over long distances.

The problem of increasing energy efficiency and achieving energy savings is a priority for the country's economy in general and for the mining sector specifically. A new research direction is being currently developing, which develops the theoretical foundations of energy conservation and efficiency, and accumulates experience of their practical implementation [4, 9, 10]. One of the methods of diagnosing the technical condition of the transport unit (e.g. a belt conveyor or dump truck) as a means of transport is a method of nondestructive testing [5]. One of the important elements of nondestructive control is a thermal control and analysis of lubricant condition.

2 Materials and methods

Operating temperature of lubricants that meets the standards of GOST28549.9-90 is of great importance for efficient operation of the transport unit. Raising appropriate temperature standards, lubricants degrade their lubricating property which leads to increased wear of the working friction surfaces and bearing units. In practice, the heating of the reducer is determined by the "manual", and often it does not give reliable information. For this reason, it is advisable to use an integrated approach: continuous monitoring of a temperature mode and periodic in-depth spectral emission diagnostics of the working oil. Spectral analysis of the impurities contained in the transmission fluid determines the concentration of wear products. While diagnosing oil the following parameters are determined: kinematic viscosity, flash point in open crucible, the water content and the products of wear [1, 2]. Particles of metal, indicating the technical condition of friction work surfaces and bearing units are iron, copper, chromium, Nickel and silicon. At a constant excess of any of these metallic elements in the lubricant you need to perform diagnostics of gears, splines and bearings. Only silicon content excess in oil says that the oil is worn out and needs to be fully replaced. Oil change should be made in the case of the gradual accumulation therein of metal particles, which concentration is close to or exceeds the limit value. The presence of small amounts of copper particles in the oil is normal; it is due to the use of bearings with brass cages.

The increase in the concentration of the content of the copper particles in the lubricant indicates the beginning of an intense wear of the cage, and when exceeding the limit values of the content of these pieces of brass can be visually observed in the oil, which is poured in a glass test tube, in the form of a golden glitter. In such cases the oil must be replaced. To perform diagnostics of bearing units is also necessary and in need also will be replaced [2, 3, 8]. This service allows reducing repair costs, preventing emergency downtime and increasing energy efficiency as a transport unit in particular and the enterprise in general [9-10].

3 Results and discussion.

In practice, when periodic analysis of the technical condition of the gear oil in gear boxes of belt conveyors for mining-revealed their incompatibilities with normative values (GOST 23652-79) of the main physical-chemical parameters were revealed. The obtained results are presented in tables 1 – 4.

Table. 1 Organoleptic studies

Settings	**Sample 15.05.15**	**Sample 18.09.15**	**Sample 4.12.15**	**Sample 17.06.16**	**Sample 20.10.16**
Appearance	Dark brown	Dark brown	Dark brown	Dark brown	Dark brown

The presenceof water	Itisnotrevealed	Itisnotrevealed	Itisnotrevealed	Itisnotrevealed	Itisnotrevealed
The presence of insoluble particles on filter paper	Include	Include	Include	Itisnotrevealed	Itisnotrevealed

Table. 2 Kinematic viscosity

Temperature, °C	**Norm (DIN EN ISO 3104)**	**Sample 15.05.15**	**Sample 18.09.15**	**Sample 4.12.15**	**Sample 17.06.16**	**Sample 20.10.16**
	KV, mm²/s	**KV, mm²/s**	**KV, mm²/s**	**KV, mm²/s**	**KV, mm²/s**	**KV, mm²/s**
40	320	346,50	348,92	371,61	373,85	359,46
100	24,1	26,17	21,17	23,17	25,33	25,50

Table. 3 wear Products, g/t

Elements	**tolerancel imits**	**Sample 15.05.15**	**Sample 18.09.15**	**Sample 4.12.15**	**Sample 17.06.16**	**Sample 20.10.16**
Fe	126-200	132,47	29,20	63,39	136,00	299,16
Si	21-30	35,05	35,68	39,81	42,24	51,24
Cu	100-150	3,09	6,25	1,09	7,81	7,13
Al	4-7	1,19	1,29	1,71	2,15	2,95
Cr	2-5	0,83	1,19	1,43	1,39	1,69
Pb	-	3,73	1,33	3,49	4,44	4,03
Sn	-	7,00	4,89	7,54	7,32	12,11

Table. 4 flash point in open crucible

Settings	Norm (DIN ISO 2592)	Sample 15.05.15	Sample 18.09.15	Sample 4.12.15	Sample 17.06.16	Sample 20.10.16
Temperature, °C	255	238	236	228	216	235

According to the data given in tables 1 – 4, it can be noticed that the condition of the reducer is critical, there is a buildup of mechanical impurities in the oil to the limit values and exceeding the standards on the content of silicon (Si) and iron (Fe). There is also a slight deviation of viscosity from normative values, as well as changing the flash point in open crucible oil. According to the obtained results of the oil analysis it is necessary to replace oil, to pay attention to the heating mode of operation and to avoid the entrainment of coal dust in the gearbox.

In this mode of service an individual approach to each piece of equipment is not applied. The reason is that in the process the oil undergoes a number of changes, some of which can reduce the reliability and durability of the movement. To avoid this, the manufacturer position maintenance regulation list of applicable lubricants and frequency of their replacement. However, it does not guarantee reducing performance oils, because the change in each mechanism occurs individually. Moreover, deterioration of the working oil is because of inappropriate quality of service transport unit.

Hence the need is to monitor the temperature regime of the oil in the operation of the control equipment and its technical condition and to prevent failures. The application of

 DOI: 10.1051/e3sconf/20172103013

temperature as a diagnostic parameter of the working oil allows monitoring of the actual technical condition of the gearbox of a belt conveyor.

One of the most progressive methods of diagnostics of the temperature regime of today is recognized as thermal imaging diagnostics [4, 6]. Its application is based on the fact that the presence of almost all kinds of defects in the equipment causes change of temperature of defective elements and, as a consequence, the change in the intensity of infrared radiation registering thermal imaging devices. The defect is detected by comparing temperatures of similar (neighboring) areas of the surface of parts and units working under the same conditions of heating and cooling. Thermal imaging detects defects at the earliest stage of their appearance and development that allows planning the scope and timing of repair. Decommissioning hardware is defective according to the results of its diagnosis, to its total rejection, greatly improves the reliability and safety of transport communications, and reduces significantly the loss of energy. The results of thermal monitoring of a gearbox of a belt conveyor are presented in table 5 and figure 1.

Table. 5 Results of thermal imaging of the gearbox of a belt conveyor

The name of the site	**Permitted temperature, °C**	**The actual surface temperature, °C**	**Actual temperature of the lubricant, °C**
Reducer D3RST82XO	106	79,7	97,7

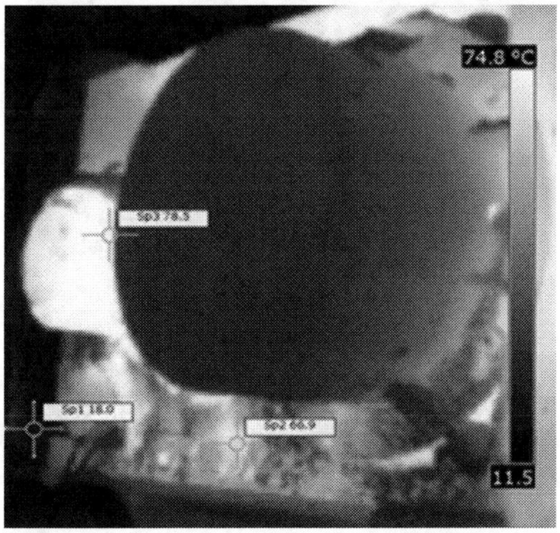

Fig. 1. Results of thermal imaging of the gearbox of a belt conveyor

The results of thermal imaging of the rollers of the gear shown in figure 2 (a, b, c, d).

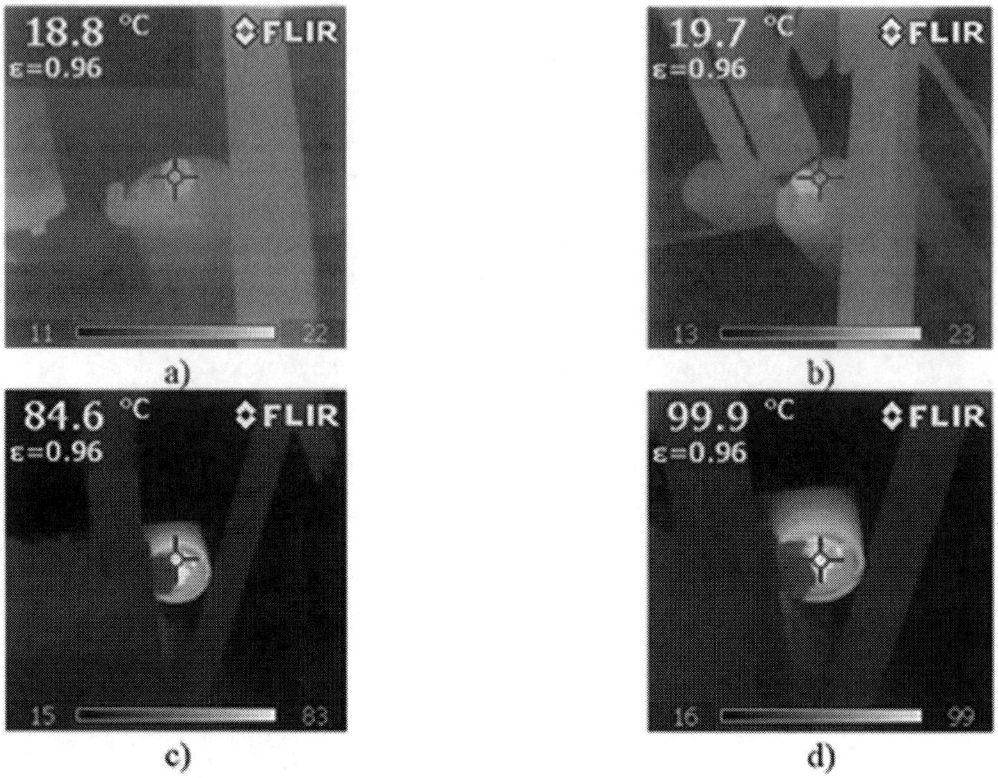

Fig. 2. Results of thermal imaging rollers reducer

Evaluation criteria the results of thermal control are presented in table 6.

Table. 6 evaluation criteria the results of thermal control

Temperature anomaly, ΔT, ℃	Actual temperature of the lubricant, ℃	Evaluation criteria
0-5	Not more than 80	CategoryA – good condition
5-10	80 - 90	Category B – limited working capacity requires additional diagnosis
10-15	90 - 100	Category C – critical condition, surgical intervention is required, additional diagnostics
More than 15	More then 100 - 110	Category D – an unacceptable condition, the possible damage, and surgical intervention is required, possible repair.

The other rotating element of the equipment (e.g. a roller of belt conveyor) also affects the energy saving in haulage. The durability of the belt conveyor and its core components [8] depends on how easily rotates the roller. The more the non-rotating (jammed) rollers on the line, the higher the friction force between the defective roller and a tape sheet (which accelerates wear). This increases the load on the drive stations and the repeater drum, which also entails an increased wear of these elements and increase in energy consumption.

4 Conclusions

Diagnostic performance characteristics of lubricant and temperature of the elements of equipment allows monitoring in time the technical condition of equipment and avoid premature failure, improves the efficiency of the transport system, and the enterprise as a whole.

References

1. Z. Sroka, Open Access Library Journal, **1**, 1100816. (2014)
2. H.C. Thakur, International Journal of Scientific & Engineering Research, **7:12**, (2016)
3. Ying-Chieh Chou, Leehter Yao, *Automatic Diagnosis System of Electrical Equipment using Infrared Thermography* (Taipei University of Technology Taipei, Taiwan, R.O.C., 2010.)
4. A. Tawfik Saleh, Spectral Analysis Review **2:1**, 43252,2 (2014)
5. L.I. Efimenko, M.P. Tichanskiy, KTU Bul. **25**, 163 (2010)
6. R. Zimroz, W. Bartelmus, T. Barszcz, J. Urbanek, Mechanical Systems and Signal Processing, (2013)
7. R. Zimroz, R. Krol, M. Hardygora, J. Gorniak-Zimroz, W. Bartelmus, L. Gladysiewicz, S. Biernat, 22nd World Mining Congress & Expo, **1**, 433 (2011)
8. W. J. Bartelmus, Phys.: Conf. Ser., **305:1**, 012029 (2011)
9. W. Bartelmus, F. Chaari, R. Zimroz, M. Haddar, European Journal of Mechanics., **29:4**, 637 (2010)
10. A. Bellino, A. Fasana, L. Garibaldi, S. Marchesiello, Mechanical Systems and Signal Processing, **24:7**, 2250 (2010)

 DOI: 10.1051/e3sconf/20172103014

Innovations of Engineering Company and Competitiveness in the Mining Equipment Market

Vladislav Pogrebnoi[1], *Lyudmila* Samorodova[2], *Larisa* Shut'ko[2] , *Yulia* Yakunina[2] , and *Oleg* Lyubimov[3,*]

[1]"Transport-Electroprivod" Ltd., Directorate, 6500000, Kemerovo, Russia
[2]KuzSTU, Department of economics, 650000 28 Vesennya st., Kemerovo, Russia
[3]KuzSTU, Department of information and automated manufacturing systems, 650000 Kemerovo, 28 Vesennya st., Russia

Abstract. The article deals with issues related to the development of innovative projects by engineering companies and effective marketing policy as a factor of increasing their competitiveness in the mining equipment market. The paper presents the results of the development of innovative technology in the segment of extensible belt conveyors. The necessity of marketing the innovative project is proved by the example of the development of technology for the production of the motor-drum of a belt conveyor with an adjustable drive on continuous current magnets by the engineering company "Transport-Electroprivod" (LLC). The authors consider the effective marketing strategy of an engineering company as an attractor of increasing demand for innovation and competitiveness in the mining equipment market. The methods of marketing promotion and promotion of innovations are considered within the framework of the ecosystem concept of J. Moore.

1 Introduction

In modern conditions, engineering companies are actively developing in the Kuzbass market, promoting innovative technologies in the production of mining equipment. An example of successful promotion of innovative projects in the mining equipment market is the activity of the group of companies "Transport-Electroprivod". The own production of the group of companies "Transport-Electroprivod" is provided by the plants in cities Kemerovo and Topki, which produce conveyors and conveyor systems of all types and sizes, as well as building metal structures and various tanks, including high pressure.

Management of innovative projects of the engineering company involves solving a whole range of issues related to the feasibility study of the proposed technological solutions and ensuring competitiveness. Along with technical solutions, one of the important problems of the competitiveness of engineering companies is the marketing promotion of

* Corresponding author: oleg_lyubimov@mail.ru

technological innovations in the mining equipment market in conditions of insufficient demand for innovations in the field of coal mining [4-8].

2 Materials and Methods

One of the latest innovative projects of LLC "Transport-Elektroprivod" was a project related to the development of a motor-drum model of a belt conveyor with an adjustable drive on continuous current magnets, in introduction of which the developers face the inadequate demand for an innovative product.

The development of this project is aimed at improving the conveyor systems in terms of:

- improving ergonomics (by reducing the size of the drive station), in mine conditions this is one of the most important requirements;
- Increasing the operational reliability of the conveyors (exception of additional nodes and mechanisms in the chain "transmission of driving torque from the motor", which in turn may fail, removing from the operating state the entire drive group, namely: reducer, underframe or lantern, reactive traction, turbo coupling, elastic clutch, delivered brake, low-speed half couplings and electric motor);
- reducing the costs of the enterprise as a result of the exclusion of the above components from the drive unit to the reserve funds for the organization of consignment warehouses (spare parts) and service from five, six suppliers of this equipment. Nowadays, at least five or six suppliers of components with which it is necessary to sign contracts are involved in the structure of the drive units (with each one, if possible, to sign service agreements, perform an audit, etc.), which is quite costly for the enterprise -producer;
- increasing the labor safety of miners (by reducing the rotation elements in drive unit);
- reducing the time and costs of installing the drive unit;
- the absence of special requirements for the surface on which the drive unit is installed, because the only supporting structure is the frame on which the motor-drum is located;
- the opportunities, in combination with management system, based on a frequency converter to save electricity by automatically reducing the speed at the time of absence or reduction of the volume of the rock mass on the conveyor, as well as transporting people at a safe speed of 3.15 m / s according to the Safety Rules.

3 Results and Discussion

Together with foreign and Russian colleagues, the element base of the product and the performance specification were developed. The technical characteristics of the motor-drum are presented in Tab.1.

Table 1. The technical characteristics of the motor-drum.

Nominal power of electric motor, kW	Belt speed, m/s	Drum rotation frequency, rpm	Torque rating, kN / m
Diameter of the drum 800 mm			
***P=250* kW (*n = 1500 rpm*)**	1.25	30	80
	1.6	37.5	62.2
	2.0	47.6	50
	2.5	60	40
	3.15	75	31.75
	4.0	94	25

It is planned to install this equipment in a belt conveyor L = 1250 meters. The rated belt speed is 4.0 m / s. The frequency-controlled drive must provide a change in speed from the nominal in the range "+" or "-" 20%.

The carried out development and designed work revealed the competitive advantages of the developed innovative motor-drum model in comparison with the available models in terms of technical characteristics and price. The technical characteristics of the pre-calculated motor-drums with liquid cooling are given in Tab. 2.

Table 2. The technical characteristics of the proposed motor drums.

Specifications and parameters	**Options**		
	1	2	3
Power on shaft, kW.	785	314	160
Frequency of rotation, rpm.	60	60	73
Напряжение сети, 50 Гц, кВ Voltage, 50 Hz, kV	6	6	6
Amplitude of phase voltage pulse, kV	1.0	1.0	1.0
Frequency of phase voltage, Hz	60	40	36.5
Rated torque, kNm	125	50	21
Efficiency is not lower	0.97	0.97	0.96
Multiplicity of the starting torque	2.0	2.0	2.0
Diameter of motor-drum, mm	1250	1000	650
Length of motor-drum, mm	1600	1400	1250
Number of phases	6	6	6
Number of packages	2	2	2
Number of parallel branches	2	2	1
Number of poles	18	12	9
Number of stator / rotor teeth	54/60	36/40	27/30
Specific force, N/cm2	4.2	4.0	4.8
Degree of protection of motor-drum	IP54	IP54	IP54
Weight of electrical steel, kg	6000	3800	1800
Weight of electrical copper, kg	560	350	270
Total weight of electrical steel and copper, kg	6560	4150	2070

The method of expert assessments revealed that in result of excluding the above components, the engine mass could be reduced by about 10%, and calculations showed a reduction in drive unit price.

As a result of the correlation analysis performed with the help of software package, it was revealed the following:

- practically proportional dependence in price structure of the share of used electric motor on the diameter of the drum (and, accordingly, its price share);
- close correlation in price structure of the shares of the selected gearbox size and the adopted control system;
- reduction in the share of couplings and brake devices in price structure with an increase in the diameter of the drum;
- insignificant share of metal structures, supports and fasteners in price structure.

The analysis showed convincing superiority in price of the composed motor-drum in comparison with the standard drive unit, which consists of many costly components.

The third stage - perspective – the development of an innovative project - preparation of production and power ascension, i.e. manufacturing of products, which includes arrangements for a new product output. To implement this project, "Transport-Elektroprivod" LLC cooperates with Russian enterprises that have government orders for prototypes of motor-drums, and work under the orders of the Defense Ministry of the Russian Federation.

The fourth stage of the innovative project is the production of the products created in accordance with the orders' portfolio. In this connection, the concept of the ecosystem by J. Moore is of interest. "For the implementation of any innovation, we need partner-consumers and partner-suppliers. And the more radical (and often more valuable) innovation is, the more deeper and wider should be other players, especially consumers involved" [9]. Thus, to promote innovation successfully, cooperation with potential consumers is needed at the stage of development an innovation system.

In addition, it is at this stage that market research is needed, identifying potential consumer companies and promoting the product to the market. Preliminary monitoring of the Russian market of motor-drums for underground conveyors allowed to develop three variants of the forecasted market situation and to obtain the following results regarding potential demand, which is described in Tab. 3.

Table 3. The main technical and economic characteristics of motor-drums market for the underground conveyors in Russia.

Indicators	Units	Options		
		1	2	3
Conveyor Belt Width	mm	1000	1200	1400/1600
Average number of conveyors	Units Annually	25	15	3
Typical power of the drive unit	kW	160	315	500
Rated drum speed	rpm	75	75	60
Drum Diameter	mm	630	800	1250
Total number of drives	Units	75	45	9
The expected percentage of the market share of motor-drums of the total number of drives	Percent	35	60	70
Expected volume of motor-drums market	Units Annually	20	40	30

At the first stage the investments in this area (production of a pilot set of products) will be carried out entirely by LLC «Transport-Electroprivod». The cooperation with partner companies will allow LLC «Transport-Electroprivod» to purchase semi-finished products that will undergo further refinement and screwdriving assembly and certify the finished product.

The fifth stage – the operational period of the project – requires an effective marketing strategy, because the demand for this innovative product must be formed.

At the moment, LLC «Transport-Electroprivod» does not have an established system of marketing innovation in the mining equipment market. It is necessary to generate demand for innovations, because there is a certain inertia and stereotypes of engineering thinking of specialists in coal companies, restraining their introduction into production, due to the increased danger of mining operations. Overcoming this barrier, «Transport-Electroprivod» will have the opportunity to engage in further market research, look for potential consumers, etc.

Within the framework of the chasm theory by G. Moore, the most important in modern conditions is not the existence of two "cracks" in the life-cycle curve of technology: one between innovators and early followers, the other between the early and late majority, but the "deep chasm that divides early followers and an early majority. Today, this is the most significant and complex transition in the life cycle of technology adoption, and it is becoming increasingly dangerous, since it is usually not recognized "[10].

 DOI: 10.1051/e3sconf/20172103014

Of particular importance is a trusted image of an innovative product in the eyes of consumers, and it is necessary to accumulate confidence in product from the consumer's part. In addition, the innovator enterprise should determine the marketing channel for an innovative product, choosing between its own distribution network and marketing through intermediaries. In this case, it is advisable for «Transport-Elektroprivod» not only to use direct contracts, but also to combine sales channels, as the enterprise is included in the integrated group of companies, represented in the domestic and foreign markets of mining equipment. Undoubted competitive advantage of the company in the promotion of innovations will be the system of "smart service", which allows providing service of innovative equipment from the moment of delivery and installation of equipment up to its disposal. In Kuzbass there is an experience of successful use of "smart service" as an innovative service system for mining companies [11].

4 Conclusion

Along with traditional marketing methods (specialized exhibitions, expositions, demonstrations, transfer of equipment for trial operation, personal work with potential clients), we suggest methods of promotion based on the ideas of the chasm theory by G. Moore, his technique "test in the elevator", the use of an innovative system "smart service", the combination of sales channels. These methods will contribute to the formation of a trusted image of innovative goods in the eyes of consumers, and stimulate the demand for innovation.

References

1. *"Transport-Elektroprivod" Ltd. has partner relations with Russian engineering companies,* URL: https://piter-expo.ru/ru/engineering/
2. M.A. Tyulenev, S.A. Zhironkin, O.I. Litvin, E.A. Tyuleneva, O.V. Zhironkina, S.O. Markov, Geotech. Geol. Eng. **35:5**, 2065–2077 (2017)
3. M. Tyulenev, S. Zhironkin, E. Tyuleneva, A. Abay, S. Anyona, M. Hellmer, Coal Int., **265:3**, 30-34 (2017)
4. A. Amosha, B. Birenberg, A. Kabanov, *Methods of managing innovative processes in Ukranian coal industry* (Donetsk, Mining, 1999)
5. S. Valdaytsev, *Technological innovations management* (Saint-Petersburg, 2004)
6. Yu. Fridman, G. Rechko, Ye. Loginova, Vestnik KuzSTU, **6**, 139 (2013)
7. I. Gurkov, Ye. Avraamova, V. Tubalov, Econ. Iss., **2**, 40 (2005)
8. C. Christensen, M. Raynor, *The innovator's solution creating and sustaining successful growth* (Harvard, 2004)
9. J. Moore, *The Death of Competition* (NY, 1996)
10. J. Moore, *Crossing the Chasm: Marketing and Selling Technology Products to Mainstream Customers* (NY, 1998)
11. L. Samorodova, L. Shut'ko, Yu. Yakunina, O. Lyubimov, E3S Web of Conf., **15**, 03013 (2017)

Increasing the technical level of mining haul trucks

Yuri Voronov[1], *Artyom* Voronov[1], *Sergey* Grishin[1], and *Alexey* Bujankin[1]

[1]T.F. Gorbachev Kuzbass State Technical University, Kemerovo, Russian Federation

Abstract. Theoretical and methodological fundamentals of mining haul trucks optimal design are articulated. Methods based on the systems approach to integrated assessment of truck technical level and methods for optimization of truck parameters depending on performance standards are provided. The results of using these methods are given. The developed method allows not only assessing the truck technical levels but also choosing the most promising models and providing quantitative evaluations of the decisions to be made at the design stage. These areas are closely connected with the problem of improvement in the industrial output quality, which, being a part of the widely spread in Western world "total quality control" ideology, is one of the major issues for the Russian economy.

1 Introduction

Technical re-equipment of open-pit mining facilities and further development and upgrades of automotive mining machinery are inextricably linked. Efficiency in execution of hauling operations determines the technical and economic performance indicators of coal production on the whole, as transportation costs make up more than 50% of overall mining costs.

So far skyrocketing payload capacity as the main tendency for mining automotive machinery development has not been accompanied by a corresponding gain in performance and increase in other merits. However, it should be noted that the engineering policy meant to increase the payload and aimed at further mining truck redesign without major quality changes. Open-pit mining transport enhancement can be achieved, in the first place, by optimal tuning of all the structural components, units and systems of a truck not only among themselves, but with the facility environment as well. So when designed, a mining haul truck must be regarded as a complex technical aggregate being in its turn a part of the total surface mining transport system.

Vast amounts of mined bulk to be transported by haul trucks, influence of hauling operations upon the speed of open-pit mining and their high labour intensity stimulate the constant search for large-scale improvements to the mining transport. M.V. Vassiliev, A.A. Kuleshov, M.G. Potapov, K.N. Trubetskoy, S. Alarie, M. Gamache, C.N. Burt, L. Caccetta, R.A. Carter, V.A. Temeng and others contributed greatly to the theory and practice of automotive mining machinery usage at open-pit facilities. In their papers they determined the sphere of haul truck efficient use in open-pit mining [1, 3, 5, 8-10], summed up the experience of its operation [2, 4, 7, 11-14], highlighted the issues of techno-economic studies devoted to automotive mining machinery usage efficiency [1-7, 15-17].

 DOI: 10.1051/e3sconf/20172103015

Belarusian Automobile Plant (OJSC "BelAZ", Republic of Belarus) is a major world manufacturer of heavy dump trucks used in mining and construction. Its production mainly consists of mining dump trucks with a payload capacity of 30 to 360 t. During all the years (since 1948 when the company was founded) "BelAZ" specialists have been doing their best to enhance the truck design. These design implementations contributed to providing mining companies with high-performance equipment. However in recent years "BelAZ" designing engineers have had to face the following problem: increased payload did not lead to the proportional rise in truck performance. The power-to-weight ratio is still insufficient, the payload ratio and life mileage are too low as well as the truck technical level in general and only production and maintenance costs go up with every passing year.

This situation can be due to the fact that the theory of mining machinery optimal design (including mining dump trucks) has not been developed sufficiently yet. Mining truck design framework still needs the procedures for ultimate solutions' synthesis.

2 Results and discussion

In order to estimate adequately the technical level and quality of the mining machinery (including automotive machinery) the most agreeable alternative is the expert assessment method, which allows assessing functionally homogeneous machines of various types, models and modifications. It was proposed by G. Solod. For this purpose by the technical level the truck quality at the stage of the design development is meant.

Another important issue of the assessment is the grounded selection of the parameters determining the technical level of haul trucks to have the essential and sufficient nomenclature. The parameters that determine the technical level to the most are selected from the initial nomenclature after a thorough analysis of the operational processes followed by a further verification of their matching (direct correlation) and representativity (the degree of influence upon the technical level depending, in its turn, on the purpose of its assessment). Eventually the following system of the parameters was obtained (Fig. 1).

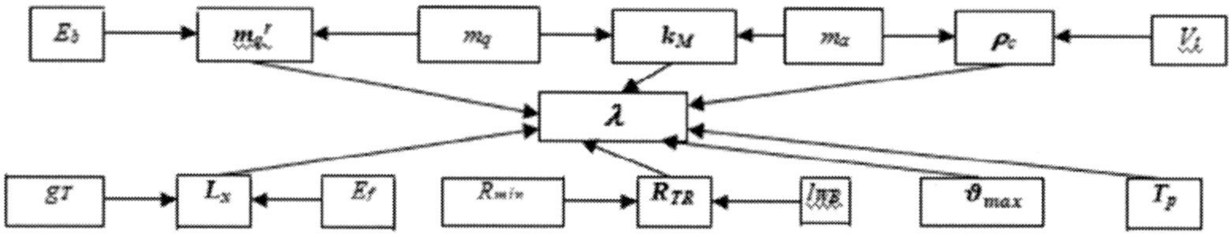

Fig. 1. Complex of haul trucks technical level parameters.

In this system: λ – functional criterion, kW/t; $m_q^r = m_q/E_b$ – truck specific payload, t/m³; $k_M = M/m_w$ – truck weight coefficient (comparative materials consumption); $\rho_c = m_a/V_t$ – component density, t/m³; $L_x = 100E_f/g_T$ – distance to empty, km; $R_{TR} = R_{min}/l_{WB}$ – specific turn radius; ϑ_{max} – maximum speed of the truck, km/h; T_p – predetermined service life (truck mileage), km; m_q – truck payload capacity, t; E_b – body capacity, m³; m_a – empty truck weight, t; V_t – total truck volume (product of length by width and height), m³; E_f – fuel tanks capacity, l; g_T – linear fuel consumption, l/100 km; R_{min} – minimum turn radius, m; l_{WB} – truck wheelbase, m; $M = m_a + m_q$ – gross truck weight, t; $m_w = (0{,}212N_e - 8{,}4)$ – truck weighted average mass (corresponding to the installed engine capacity N_e).

The given parameters can be determined at the earliest stages of truck design. The integrated assessment of the technical level is carried out by means of unique (q_{ij}) and generalized (k_i) parameters according to the formulae of the Solod's method.

In the paper according to the selected parameters a comparative assessment of 22 trucks was made, 14 of which are "BelAZ" products, 4 – of "Caterpillar", 1 of "Komatsu" and 1 of "Unit Rig". This group comprises the whole range of truck payload capacities (30 to 320 t). The trucks are compared in similar conditions.

After the generalized parameter of the truck technical level was calculated and analyzed it can be concluded that for all 22 trucks this parameter exceeds 0,5 (Fig. 2), therefore, this result proves that the technical level of all the trucks is quite high.

The average technical level for all the machines is 0,695. Also it should be noted that the level of "BelAZ" trucks and that of the trucks produced by foreign companies are virtually the same. The average value of the technical level generalized parameter for the trucks considered is 0,698 and that is only 0,7% higher than for "BelAZ" trucks, which means that as far as the vehicle structure perfection and new design ideas and solutions concerned Belarusian Plant products show very little difference.

On the basis of the analysis it can be concluded that power-to-weight ratio of new vehicles is to be increased, empty weight is to be lowered while reliability and service period are to be raised to a higher level; all truck parameters are to be balanced and reviewed as interrelated and interdependent. All this shows that the system approach to new wheeled open-pit mining machinery design is essential.

At the present moment there is a distinctive deficiency of guidance materials and norms and specifications that could standardize products optimal design as an important stage of quality improvement. System approach is to be regarded fundamental for optimal design as it allows perceiving the vehicle as a whole while its individual elements are being designed. The system approach to the vehicle parameters' determination consists of the following: 1) every parameter must correlate optimally with the others; 2) output operational performance established on the basis of these parameters is to comply with the specified values. It is suggested that the mathematical model of truck parameters optimization should be presented as a set of regression analysis equations directly connecting output operational performance and the parameters which are to be optimized.

Haul truck as a system is characterized by its own structure and parameters. The truck structure predetermines its elements and their interrelationships; it is to ensure the truck optimal functioning and for this purpose the parameters characterizing the truck are to be optimized. It is reasonable to use generalized evaluations of the technical level as optimization criteria and its unique indicators as optimized parameters. Such approach enables to reduce the problem dimensions, to simplify its solution and also to ensure objective assessments of the suggested solutions.

After carrying out the analysis of the output operational performance (such as payload ratio k_T and 90% predetermined service life (truck mileage) T_p) specified in the regulations (including interstate GOST 30537-97) and specific truck performance P_{T_y} as the main technical and operational parameter for mining haul trucks and taking into account their interrelationships (dependence (3)) the essential and sufficient nomenclature of the optimized parameters was compiled. It is stated in Fig. 3.

The problem of finding out how much each of these parameters influences the technical level generalized parameter chosen as the optimization criterion and subsequent generating an objective function model on this basis is a problem of approximation type and can be solved with the help of the theory of experiment design. Having been carried out, the correlation analysis of the interrelations between the parameters that are to be optimized determined the degree of their independence and allowed diminishing their number from 8 (Fig. 3) to 3 (N_e^r, k_T and T_p).

DOI: 10.1051/e3sconf/20172103015

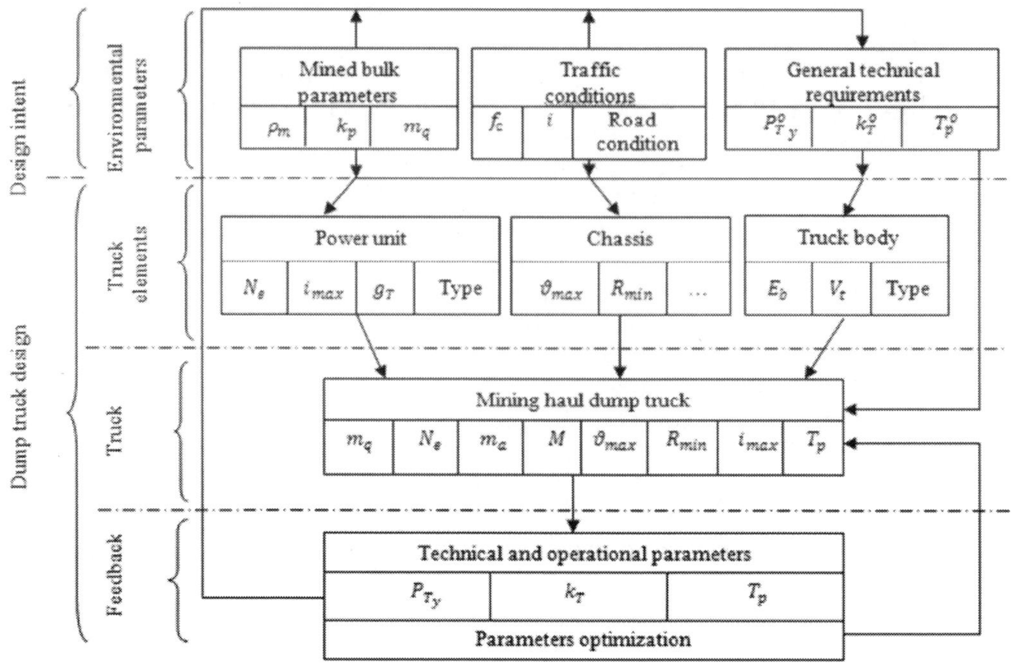

Fig. 2. Flow chart for haul truck design and truck parameters optimization ($\boldsymbol{\rho_m}$ – mined bulk density; $\boldsymbol{k_p}$ – mined bulk loosening factor).

The evolved method is based upon the technical requirements regulated by GOST 30537-97 but it can be used for the purpose of optimization of haul truck parameters and other external conditions established by the customers. The method suggests a complete rethinking of the establishment of critical parameters for truck specifications.

The experiment confirmed linearity of the objective function models and all the constraints. The linearity is conditioned by the fact that all interrelations between the generalized and unique parameters of the technical level and between the truck parameters can be viewed as linearity with a reasonable degree of accuracy as suggested by regression parameters high values and correlation indices reliability.

3 Conclusion

The analysis of conditions and practices in mining machinery operation enabled to ascertain that the structure of "BelAZ" truck fleet is of relatively low quality. At present the number of guidance papers and materials for mining trucks adequate assessment and recommendations aimed at their technical level increase is not sufficient.

Integrated comparative assessment of the technical level of mining machinery produced by both domestic and foreign manufacturers highlighted the prospect of the technology advances in mining machinery and their implementation as early as the design stage. Increase in the technical level of newly developed mining automotive machinery can be achieved through the usage of theoretically grounded methods of optimal design which allow balancing all parameters.

Optimal design is based upon the optimization mathematical modeling of machinery operations and parameters. The objective function and constraints of the optimization mathematical model are viewed as a set of regression analysis equations which connect corresponding figures with the optimized parameters directly. It is suggested to use the technical level generalized parameters as an optimization criteria and its unique indicators as optimized parameters.

The implementation of scientific methodological and technical recommendations will allow bringing newly developed mining machinery up to the level that matches world standards.

References

1. S. Alarie, M. Gamache, Int. J. Surf. Min. Reclam. Env. **16**, 59 (2002)
2. C.N. Burt, L. Caccetta, Int. J. Surf. Min. Reclam. Env. **21**, 262 (2007)
3. C.N. Burt, L. Caccetta, Int. J. Surf. Min. Reclam. Env. **22**, 84 (2008)
4. R.A. Carter, Eng. Min. J. **213**, 28 (2012)
5. R.A. Carter, Eng. Min. J. **221**, 30 (2010)
6. V.A. Temeng, F.O. Otunoye, J.O. Frendewey, Int. J. Surf. Min. Reclam. Env. **11**, 203 (1997)
7. V.A. Temeng, F.O. Otunoye, J.O. Frendewey, Miner. Resour. Eng. **7**, 59 (1998)
8. G. Lumley, *Trends in performance of open cut mining equipment* (GBI Mining Intelligence white paper, 2012)
9. C.H. Ta, J.V. Kresta, J.F. Forbes, H.J. Marquez, Int. J. Surf. Min. Reclam. Env. **19**, 162 (2005)

A. Krause, C. Musingwini, J. S. Afr. I. Min. Metall. **107**, 469 (2007)

10. S.V. Mkhatshwa, J. S. Afr. I. Min. Metall. **109**, 223 (2009)
11. S. Tan, *Production planning and operational control; new algorithms with application to the mining industry* (PhD Thesis, Pennsylvania State University, USA, 1992)
12. Kolonja, N. Vasiljevic, *Mine Planning and Equipment Selection* (Balkema, Rotterdam, 2000)
13. J.W. White, J.P. Olson, S.I. Vohnout, CIM Bull. **86** (973), 43 (1993)
14. S.G. Ercelebi, A. Bascetin, J. S. Afr. I. Min. Metall. **109**, 433 (2009)

B. Brown, Eng. Min. J. **213**, 30 (2012)

15. Y. Kuo, *Highway earthwork and pavement production rates for construction time estimation* (PhD Thesis, University of Texas, USA, 2004)

Forecasting of a Thermal Condition of Pneumatic Tires of Dump Trucks

Anna Kvasova[1,*], *Boris* Gerike[1], *Elena* Murko[1] and *Dmitriy* Skudarnov[1,2]

[1] T.F. Gorbachev Kuzbass State Technical University, 650000 Kemerovo, Russia
[2] VIST GRUPP, 650002 Kemerovo, Russia

Abstract. Over the last 10 years the world consumption of coal has grown almost by 50%. Coal is one of the main energy resources capable to satisfy basic energy demands of increasing population and developing world economy. On January 24, 2012 the long-term Coal Industry Development Program for the period till 2030 was approved in Russia. According to this Program coal mining in Kuzbass in 2030 will make 260 million tonns of coal per year. Development of the coal industry is impossible without upgrade of coal production by avoiding inefficient technological, organizational and economic solutions. Off the road (OTR) tires play an important role in ensuring effective, continuous and safe work of mining motor transport.

1 Introduction

At present time open pit mining is the growing segment of coal industry of Kuzbass – region with 210 million tons of coal mined annually [1-2]. So ecological and economic problems here is increase simultaneously with growth of open pot coal mining [3-5].

In cost value of one tonne of the transported products using an open-cut method of exploitation of mineral deposits 65% are occupied by costs for motor transport 18% of which are costs for tires. Not by coincidence purchase and service of OTR tires are among main expenditure items of the open coal mining enterprises. Considering a major role of OTR tires both in the expenditure pattern, and in the technical support of mining operations, the leading manufacturers and consumers of OTR tires are interested in maximum effective management of tire supply.

JSC "Razrez Berezovskiy" of the group of LLC "Stroyservis" enterprises use heavy payload dumper trucks: BelAZ-75306 (loading capacity of 220 t) and Komatsu HD830E (loading capacity of 231 t). These dump trucks are equipped with tires of standard sizes: 40.00-57 and 46/90-57.

According to JSC "Razrez Berezovskiy", in 2017 in operation there are 258 tires of standard sizes 40.00-57 and 46/90-57, what is 72 pieces more in comparison with 2016. It is connected with commissioning in 2017 of four new dump trucks: BelAZ-75306 and nine dump trucks Komatsu HD830E. From the analysis of tire accidental failures, it is revealed that the number of tires written-off because of the natural wear in January-April, 2017 is 11

* Corresponding author: kvasovaaa@kuzstu.ru

pieces more than for the similar period in 2016, and because of the accidental failure it is 70 pieces more.

Table 1. Number of tires for BelAZ-75306 dump trucks (35 dump trucks in operation).

Tire model	Number of tires in use, pcs.	Standard operational kilometers
40.00 R57 Michelin XDR	27	105 000
40.00 R57 Bridgestone VELS E-4	24	110 000
40.00 R57 LUAN HA-368	1	65 000
40.00 R57 Yokohama	15	60 000
40.00-57 Goodyear	10	100 000
46/90-57 Goodyear RM-4B+**4SL	60	100 000
46.90 R57 LUAN HA-569	5	65 000
46/90 R57 Bridgestone VRPS E-4	4	110 000
46/90-57 Belshina Bel-160D	43	47 000
40.00 R57 LUAN HA-368	1	65 000

Table 2. Number of tires for Komatsu HD830E dump trucks (19 dump trucks in operation)

Tire model	Number of tires in use, pcs.	Standard operational kilometers
40.00 R57 Michelin XDR	71	105 000
40.00 R57 Bridgestone VELS E-4	12	110 000
40.00-57 Goodyear	6	100 000
46/90-57 Goodyear RM-4B+**4SL	24	100 000
46/90-57 Belshina Bel-160D	1	47 000

Table 3. Data on tires life and write-off in 2016 and 2017 (January-April)

Tire brand	Average actual life in 2016, %	Number of tires written-off in 2016, pcs.		Average actual life in 2017, %	Number of tires written-off in 2017, pcs.	
		Natural wear	Accidental failure		Natural wear	Accidental failure
Luan	-	-	-	49.9	-	4
Bridgestone	80.3	6	5	66.35	5	16
Michelin	68.8	0	5	63.5	4	27
Yokohama	83.8	2	0	50.1	3	18
Goodyear	82.2	3	5	68.8	12	9

DOI: 10.1051/e3sconf/20172103016

Belshina	101.1	11	1	66.1	9	12
TOTAL	85.94	22	16	62.56	33	86

2 Experimental research

Losses of the enterprise in 2016 due to reduction of operational kilometers made 116.8 million rubles.

As the main materials and adhesives used for production of tires are sensitive to high temperatures, OTR tire life substantially depends on its thermal condition. Quite often the generation of temperatures in OTR tires used on dump trucks reaches the maximum, that is caused by their low cooling rate. The reason of it is the feature of tire design that has big thickness. Especially it is easy to see in summer, when the average level of OTP tire life considerably decreases because of the tires which failed as a result of thermal destructions (up to 70% of all failures). When the internal temperature of a tire reaches extreme values (110° C), the emergence of the phenomenon of rubber pyrolysis is probable, that leads to release of combustion gases (methane, hydrogen), sharp increase in internal pressure in the tire which in the presence of oxygen can lead to ignition and blow-out of the tire.

Fig. 1. The tires failed as a result of thermal destructions

It is possible to reduce failures of tires because of thermal destructions. In order to prevent increase of heat generation, it is necessary, following the operation manual of OTR tires, prevent exceeding of the working capacity indicator "tonnes-kilometers per hour" (TKPH), as the internal temperature of the tire directly depends on this indicator (fig. 2).

During the work of dump trucks there is alternately a warming up and a cooling of tires because of stops for loading and unloading, lunch breaks and shift turnarounds.

Working capacity TKPH can be calculated by the formula:

$$TKPH = Qav.*Vav.w. \qquad (1)$$

where: Qav - average load of a tire, t; Vav.w. - average working speed of a dump truck, km/h..

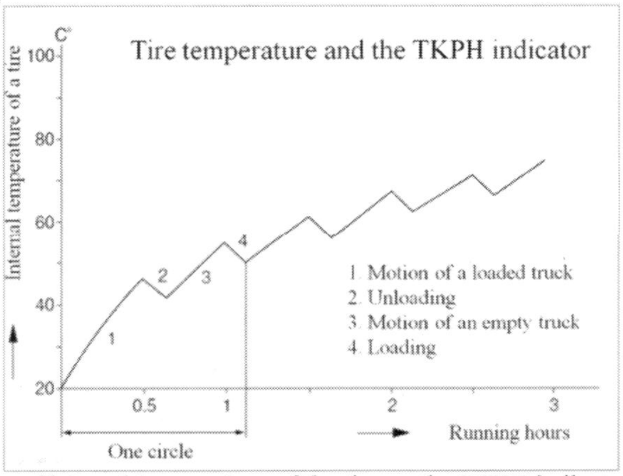

Fig. 2. Dependence of internal temperature of the tire on the TKPH indicator

3 Conclusions

The JSC "Razrez Berezovskiy" enterprise refused use of OTR tires 46/90-57 Belshina Bel-160D and 40.00 R57 Yokohama on Komatsu HD830E dump trucks because of frequent accidental failures, but, due to the lack of tires of other brands in the material warehouse, they are anyway installed periodically on these dump trucks by production necessity. TKPH values for these tires are 680 and 598 respectively. The average distance of transportation by Komatsu HD830E dump trucks is 4.5 km, average working speed is 14.5 km/h. As per calculations, in order to prevent exceeding of the admissible TKPH indicator, these dump trucks should make no more than 14-16 runs per shift, in fact they perform 22-25 runs per shift that corresponds to the working capacity indicator of 913-1038 TKPH. Working speed of dump trucks BelAZ-75306 is 12.6 km/h, therefore they perform less runs per shift and the tires 46/90-57 Belshina Bel-160D and 40.00 R57 Yokohama on these dump trucks work better.

Also, low pressure in a tire, partial transfer of heat from brakes and gears, slipping, overload, exceeding of a speed limit can be the cause of thermal destructions of OTR tires, however, a lot of attention is paid to control these factors. As for working capacity, now the enterprises perform calculations of the TKPH indicator only after failures of OTR tires for identification of failure causes, however, forecasting and control of the TKPH indicator of working capacity in modern conditions is an important task of the mining enterprises. If the predicted TKPH indicator is higher than the standard one, it can be corrected by changing of the load of a tire, dump truck speed, transportation distance, and idle times.

References

1. M.A. Tyulenev, T.N. Gvozdkova, S.A. Zhironkin, E.A. Garina, Geotech. Geol. Eng., **35:1**, 203-212 (2017)
2. M. Tyulenev, S. Zhironkin, E. Tyuleneva, A. Abay, S. Anyona, M. Hellmer, Coal Int., 265:3, **30-34** (2017)
3. S.A. Prokopenko, V.S. Gornyi Zhurnal, **1**, 47-49 (2014)
4. T.V. Galanina, M.I. Baumgarten, V.G. Mikhailov, T.G. Koroleva, G.S. Mikhailov, IOP Conf. Ser.: Earth Environ. Sci., 50:1, 012030 (2017)
5. T.V. Kiseleva, V.G. Mikhailov, V.A. Karasev, IOP Conf. Ser.: Earth Environ. Sci., **45:1**, 012013 (2016)

Perfection Of Methods Of Mathematical Analysis For Increasing The Completeness Of Subsoil Development

Mariya Fokina[1*]

[1]South Ural State University, 454080, 76 Lenina ave., Chelyabinsk, Russia

Abstract. The economy of Russia is based around the mineral-raw material complex to the highest degree. The mining industry is a prioritized and important area. Given the high competitiveness of businesses in this sector, increasing the efficiency of completed work and manufactured products will become a central issue. Improvement of planning and management in this sector should be based on multivariant study and the optimization of planning decisions, the appraisal of their immediate and long-term results, taking the dynamic of economic development into account. All of this requires the use of economic mathematic models and methodsApplying an economic-mathematic model to determine optimal ore mine production capacity, we receive a figure of 4,712,000 tons. The production capacity of the Uchalinsky ore mine is 1560 thousand tons, and the Uzelginsky ore mine – 3650 thousand. Conducting a corresponding analysis of the production of OAO "Uchalinsky Gok", an optimal production plan was received: the optimal production of copper – 77961,4 rubles; the optimal production of zinc – 17975.66 rubles. The residual production volume of the two main ore mines of OAO "UGOK" is 160 million tons of ore.

1 Introduction

Russia is the richest country in terms of natural resources. Mining industry occupies an important place in the country's economy, providing 10% of the GDP of the Russian Federation. But, according to actual data, the level of natural resources usage is on average 17% of the technologically possible one. It is also necessary to increase an investment in this industry, since they represent only 1% of the revenues that this industry provides. Thus, it becomes important to increase the pace of domestic mining industry development to the level of developed countries. The tasks set by the country's metallurgical industry require further improvement of the planning and management systems in the industry. In the conditions of high competitiveness of enterprises in this industry, the issue of improving the efficiency of the work and products becomes very important. The growth of industrial production has led to the accumulation of more than 100 billion tons of industrial waste on the country's territory and constant growth of their quantity continues nowadays. At the same time, about 35% of the total waste generated during the year is used and rendered harmless.

* Corresponding author: gva.pm@kuzstu.ru

The need for recultivation of lands, disturbed by mining operations, and the lack of territories for the construction of landfills for waste disposal leads to the use of quarries for keeping there industrial waste of the mining industry. The location of industrial wastes of various industries in the quarry capacity allows expanding the exploration of the Earth's interior, as well as gaining additional profit from the implementation of such projects.

2 Materials and methods

The existing scientific and methodological base for the design and formation of mining structures does not take into account the purposeful usage of the worked out space for the disposal of industrial wastes, which hinders the increase of integrated development efficiency of subsurface resources.

It is necessary to fulfill the basic requirements for the rational use and protection of mineral resources, which are as follows:

1) observance of the procedure for granting subsurface resources for usage and prohibiting unauthorized usage of subsoil;

2) ensuring the completeness of geological study, rational integrated usage and conservation of mineral resources;

3) conducting advanced geological study of subsoil, providing a reliable estimate of mineral reserves or properties of the subsoil plot granted for usage for purposes not related to the extraction of minerals;

4) carrying out state expertise and state accounting of mineral resources, as well as subsoil plots used for purposes not related to the extraction of minerals;

5) ensuring the most complete extraction from the depths of reserves of basic and jointly occurring minerals and associated components;

6) true recording of recoverable and retained in the bowels of reserves of basic and jointly occurring minerals and associated components in the development of mineral deposits;

7) protection of mineral deposits from flooding, fires and other factors that reduce the quality of minerals and the industrial value of deposits or complicate their development;

8) prevention of subsoil contamination during subsoil usage, especially in the underground storage of oil, gas or other substances and materials, disposal of waste of I-V hazard classes, wastewater discharges, placement of associated waters and waters in rock formations used by users of subsoil for own production and technological needs;

9) observance of the established order of conservation and liquidation of enterprises for the extraction of minerals and underground structures not related to the extraction of minerals;

10) prevention of unauthorized development of mineral deposits and compliance with the established procedure for the use of these areas for other purposes;

11) prevention of the disposal of production and consumption wastes in the catchment areas of underground water bodies and in groundwater areas that are used for the purposes of drinking water supply or technological water supply to industrial facilities or agricultural facilities or which reservation is carried out as sources of drinking water supply.

Thus, the purpose of this research is to improve the efficiency of mining enterprises using an improved mathematical model for the improvement of a developed section of the Earth's interior.

To formulate the strategy for further development of JSC UGOK, it is necessary to solve the following tasks:

- to prove that UGOK is a self-sufficient enterprise and does not receive financing from the main Holding of the "Ural Mining and Metallurgical Company"
- to determine the optimal production plan for the given enterprise;

- to form conclusions and proposals for further development of JSC "UGOK".

Before modeling and analyzing the activities of JSC UGOK, and before forming a strategy for further development, we will determine how high the financial stability of this enterprise is.

In order to prove that UGOK is a self-sufficient enterprise and does not receive financing from the main Holding, we need to consider the model of the optimal allocation of investments.

Let's write down the mathematical model of the task. We will find the optimal set of investment allocation:

$$X^* = (x_1^*, x_2^*, \dots, x_k^*, \dots, x_n^*) \tag{1}$$

satisfying the following conditions:

$$\begin{cases} x_i \geq 0; i = 1, \dots n, \\ \sum_{i=1}^{n} x_i = B. \end{cases} \tag{2}$$

Then we will divide the optimization process into n steps and at each k - th step we will optimize the investment of not all enterprises, but only enterprises with the k -th to the n -th. At the same time, it is natural to assume that funds are invested in other enterprises (from the first to (k -1) -e), and therefore not all funds remain for investment of enterprises from k -th to n-th, but some smaller amount $C_k \leq B$. This value will be the state variable of the system.

The control variable at the k -th step is the amount x_k of funds invested in the k -e enterprise. As a function of Bellman $F_k(C_k)$ at the k -th step, we can choose the maximum possible income that can be obtained from enterprises from k -th to n-th, provided that their investment remains C_k of funds. It is obvious that investing x_k of the funds in k -th the enterprise will provide profit $g_k(x_k)$, and the system will have moved move to the state $(k + 1)$ to the state $S_k + 1$ and, consequently, the investment of enterprises from the $(k+1)$ - th to the next n will remain $C_k + 1 = (C_k - x_k)$ of funds [3].

Thus, at the first step of conditional optimization $k = n$, the Bellman function represents profit only from the enterprise from the n. At the same time, the amount of money remaining to invest in it $C_n, 0 \leq C_n \leq B$.. To get the maximum profit from this enterprise, you can invest in it all these funds, i.e., $F_n(C_n) = g_n(C_n)$ and $x_n = C_n$.

At each subsequent step to calculate the Bellman function, it is necessary to use the results of the previous step .

Let it be at the k -th step for investing enterprises from the k -th to n-th left money C_k ($0 \leq C_k \leq B$). Then, from the investment in k -e the enterprise x_k of the funds will receive a profit $g_k(C_k)$, and the investment of the remaining enterprises will remain $C_k + 1 = (C_k - x_k)$ of the funds. The maximum possible income that can be received from enterprises will be:

$$F_k(C_k) = \max_{x_k \leq C_k}\{g_k(x_k) + F_{k+1}(C_k - x_k)\}; k = 1, \dots n. \tag{3}$$

The value of the Bellman function $F_1(C_1)$ is the maximum possible income from all enterprises, and the value x_1^* at which the maximum income is achieved is the optimal amount of funds invested in the first enterprise. Then, at the stage of unconditional optimization for all subsequent steps, the value $C_k = (C_{k-1} - x_{k-1})$ of the optimal control at the k -th step is calculated to be the value x_k that provides the maximum revenue with the corresponding state of the system S_k .

To solve the second problem, consider a multi-product model of production management. Considering that c_1 is the cost of production per unit of production per unit time, q is the volume of output, t_s is the time between recharges, and $n = T / t_s = R / q$ is the number of replenishment periods, the expression for the total cost of production in this case has the following form [4]:

$$D = \left(\frac{c_1 q t_s}{2} + C_s\right) \cdot n = \frac{c_1 q T}{2} + \frac{C_s R}{q} \tag{4}$$

It is assumed that the demand for products is constant and within the time interval T, R units will be sold. The optimum value for the volume of the replenishment product is determined from the condition that the next derivative $dD / dq = 0$.

Solving this equation, we determine the value of the optimal replenishment of reserves q :

$$q^* = \sqrt{\frac{2C_s R}{c_1 T}} \tag{5}$$

In the presence of many products, the use of this formula becomes difficult, because for each product will determine its optimal value of the purchased product and it's time to resume delivery.

To overcome such difficulties, the idea arises to produce products of several types simultaneously so that by the beginning of the next stage of production the previous batch of products would become zero.

The costs of production during the period between replenishments will be:

$$\bar{D} = \sum_{i=1}^{L} \frac{c_i q_i t_s}{2} \tag{6}$$

where L - number of product types in a group, q_i - number of units of the i -th product, t_s - time interval, c_i - the cost of producing a unit of product i -type in a unit of time.

Since the demand r_i for the i -th product is constant, the volume of ordered products should be equal to the quantity of output that will be consumed within a time t_s , i.e $q_i = r_i t_s$.

Based on the assumptions made, the total costs for storage and delivery organization for a group of products take the form:

$$D(q) = (\tilde{D} + C_s) \cdot n = \left(\sum_{i=1}^{L} \frac{c_i q_i t_s}{2} + C_s\right) \cdot n, \quad (7)$$

where $n = T / t_s$, $q_i = r_i t_s$.

Substituting the expressions for n and q_i in this relation:

$$D(t_s) = \left(\sum_{i=1}^{L} \frac{c_i \cdot r_i \cdot t_s^2}{2} + C_i \right) \cdot \frac{T}{t_s} = \sum_{i=1}^{L} \frac{c_i \cdot r_i \cdot t_s \cdot T}{2} + C_i \cdot \frac{T}{t_s} \quad (8)$$

The function $D(t_s)$ is a continuous function t_s, and its value tends to $+\infty$ as $t_s \rightarrow 0$.

Therefore, the minimum value of the cost function and, accordingly, the optimal value t_s is reached when $dD(t_s) / dt_s = 0$, i.e when

$$\frac{dD(t_s)}{dt_s} = \sum_{i=1}^{L} \frac{c_i \cdot r_i \cdot T}{2} - \frac{C_s \cdot T}{t_s^2} = 0 \quad (9)$$

From these conditions, and determines the optimal time between adjacent replenishments of the warehouse t_s^*:

$$t_s^* = \sqrt{\frac{C_s \cdot 2}{\sum_{i=1}^{n} C_i \cdot r_i}} \quad (10)$$

Then, in accordance with the expressions determining the quantities q_i, we get:

$$q_i^* = r_i \cdot t_s^* = r_i \cdot \sqrt{\frac{C_s \cdot 2}{\sum_{i=1}^{n} C_i \cdot r_i}} \quad (11)$$

3 Conclusions

Applying the economic-mathematical model for determining the optimal production capacity of the mine, an indicator equal to 4712000 tons was obtained. The production capacity of the Uchalinsky mine is 1560 thousand tons, and the Uzelginsky mine is 3,650 thousand tons. Having carried out the corresponding analysis of the production of JSC Uchalinsky GOK, the optimal production plan was obtained: the optimal copper production - 77961.4 rubles; the optimal production of zinc is 17975, 66 rubles. The residual volume of production of the two main mines of JSC UGOK is 160 million tons of ore.

References

1. H.N. Gizatullin, *Mathematical modeling of the development of mining and processing plants*, 125 (Academy of Sciences of the USSR. The Ural Scientific Center, 1983)
2. L. Kagramanyan, *Modeling and management of mining enterprises*. 360 (M., Nedra, 1989)
3. A.U. Kazak, *Financial policy in the corporate governance system: a textbook* (Ekaterinburg: AMB Publishing, 2004)

4. N.M. Kalinin, *Models of management multicommodity stocks*, 2 (Moscow, Institute for System Analysis, Russian Academy of Sciences)
5. M.S. Krass, *Mathematical methods and models for undergraduates of economy: Text-book*, 496 (St. Petersburg: Peter, 2010)
6. E.N. Lomkova, *Economic-mathematical models of production management: a text-book*, (Volgograd: RPK Polytechnic, 2005)
7. M.S. Fokina, A.V. Panyukov, Mining Information and Analytical Bulletin (Scientific and Technical Journal), **5** (2013)
8. G. Aggarwal, Computer Science Department, Stanford University, **94305:12**, 3 (2013)
9. J. Chapuis, Revue Juridique polynesienne, **14:13**, 6 (2007)
10. G. Gallego, Operations Research, **45:1**, 5 (1997)
11. A. Halanay, *Differential equations, discrete systems and control. Economic models*, (Kluwer Academic Publishers, 2010)
12. M. P. Moffatt, Electronic Thesis and Dissertation Repository, **176**, 155(2012)

Factors Determining the Size of Sealing Clearance in Hydraulic Legs of Powered Supports

Gennady Buyalich[1*], *Konstantin* Buyalich[1], *Maxim* Byakov[1]

[1]T.F. Gorbachev Kuzbass State Technical University, 650000, 28 Vesennyaya St., Kemerovo, Russia

Abstract. The factors that directly influence the formation of a sealing clearance between the piston and the working cylinder in hydraulic legs of powered supports are considered in this article, the size thereof has a direct effect on the tightness of the legs and, as a result, on the safety of work at the production face. A detailed description of these factors is given, which is supported by the dependencies obtained from the results of finite element modeling of various types of legs under various strength and geometric parameters, external load types and locations in the support section. The problems of formation of radial deformations of a working cylinder loaded with working fluid pressure are considered, and their dependence on the level of this pressure and extension. Based on the simulation results, the mechanisms for the formation of additional clearances due to rod and cylinder misalignments are described, and the conditions and causes of the resonant phenomena development are given. The classification of these factors according to the degree of generalization and the functional interaction between each other is proposed.

1 Introduction

The main trend in underground coal mining in recent decades is reduction of the total number of integrated mechanized faces with a simultaneous sharp increase in their efficiency. At the same time, this situation is accompanied by deterioration of the geological and mining conditions for functioning of such sets of equipment. Thus, in July 2017 the mine named after V. D. Yalevsky of the coal mining company SUEK-Kuzbass mined 1.5 million tons from one face setting a record of monthly production. It all poses higher requirements to the reliability of operation of integrated mechanized faces.

The main elements of the powered support, directly taking up rock pressure and ensuring reliable and safe working conditions for labour and mechanisms at the production face, are hydraulic legs, their functioning determines the operation of the set of mining equipment as a whole [1-5].

2 Materials and methods

* Corresponding author: gdb@kuzstu.ru

Radial deformations of the wall (dR) of the working cylinder are determined by the design and geometric dimensions of the hydraulic legs, the main of which are wall thickness, length, and extension [6-7].

As shown by numerous studies carried out by the authors using finite element methods on hydraulic legs of various designs [8,9] installed in powered supports of different manufacturers, the shape of a typical curve of radial deformations of the working cylinder wall is shown in Fig. 1. This curve has 5 characteristic key points. Points 1 and 5 (p.1 and p.5 in Figure 1) are located in the places of the largest deformations of the cylinder, from the piston and the bottom sides of the cylinder, respectively.

Key point 2 (p.2) corresponds to the deformations in the main part of the cylinder which are slightly smaller than deformations in pp. 1 and 5.

Point 4 (p. 4) is located at the position of the first seal on the side of the piston cavity of the hydraulic leg. Radial deformations at this point directly form a clearance to be sealed in order to prevent the flow of working fluid from the high pressure zone into the rod cavity, i.e. directly affect the tightness of the leg.

Key point 3 (p.3) is located on the deformation curve having the minimum value with a (-) sign. If this point is located within the limits of the piston, the excess of deformations in it (dR3) of the total tolerance zone for the manufacture of the piston and cylinder will lead to damage to the working cylinder mirror in the form of longitudinal scratches through which leakages of the hydraulic fluid will occur. This will make it impossible for the leg to perform its main functions in supporting the roof and controlling the rock pressure.

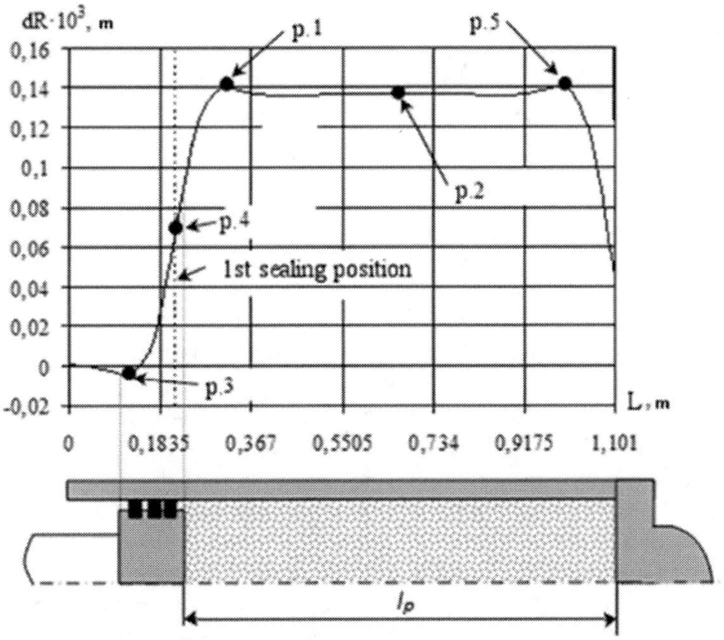

Fig. 1. Example of radial deformations (dR) of the working cylinder of M130 support hydraulic leg along the length of the working cylinder (L) at its full extension *(lp)*, working liquid pressure 50 MPa, and location of key points (p.1-p.5)

The value of radial deformations of the working cylinder in the area of the sealing gap (dR_4) is also determined by the pressure of the working fluid in the piston cavity of the hydraulic leg and its expansion. These dependences are linear in nature and they are shown for hydraulic legs of different types of powered supports in Fig. 2 and Fig.3.

 DOI: 10.1051/e3sconf/20172103018

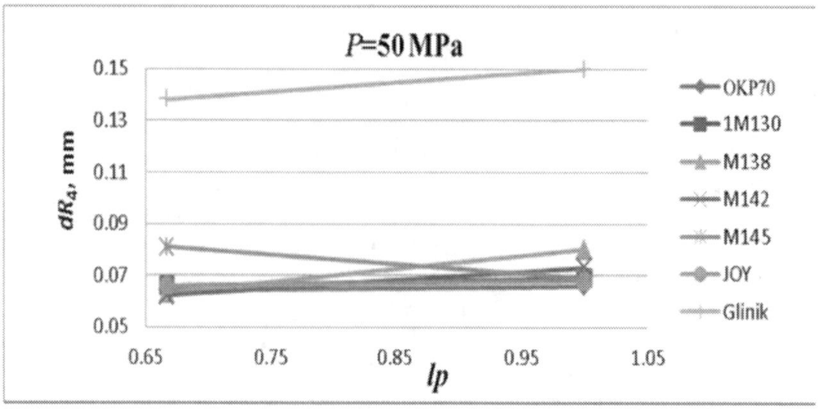

Fig. 2. Dependences of radial deformations of the working cylinder in the area of sealing (dR_4) on pressure of the working liquid in the piston cavity of the leg (P) at its full extension (*lp)*

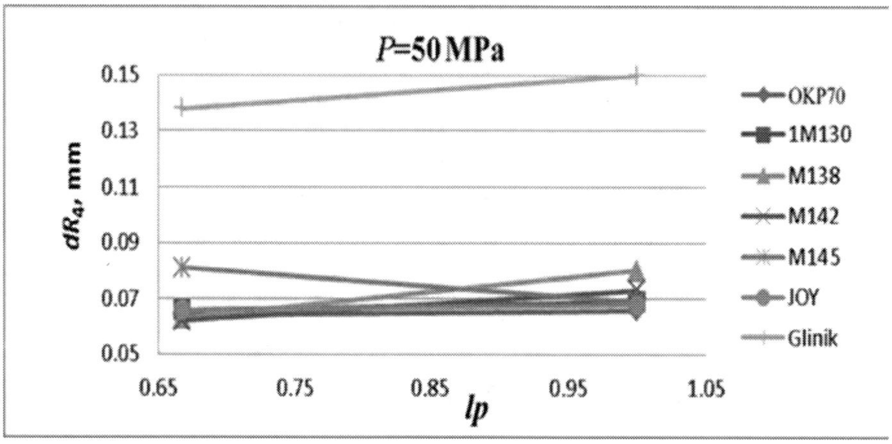

Fig. 3. Dependences of radial deformations of the working cylinder in the area of sealing (dR_4) on extension of the hydraulic leg (*lp)*

In order to increase the passages for the movement of miners, hydraulic legs are installed at angle α to the support elements of the support section (Fig. 4). In this case, the load on the rod and the bottom of the cylinder is applied eccentrically, that is, the direction of the forces from these loads do not coincide with the longitudinal axis of the leg. In this case, additional forces Fa and Fb arise in the leg, under the action of which the axis of the rod deviates from the axis of the cylinder by angle γ, the value of which is determined by the angle of inclination of the hydraulic leg to the support surfaces in the support section, by the leg extension, and also by the clearances in the joints between the rod and the sleeve, and clearances in the joints between the piston and the working cylinder.

The location of the hydraulic leg in the section of the powered support, the scheme of forces application, and the formation of the clearance are shown in Fig. 4.

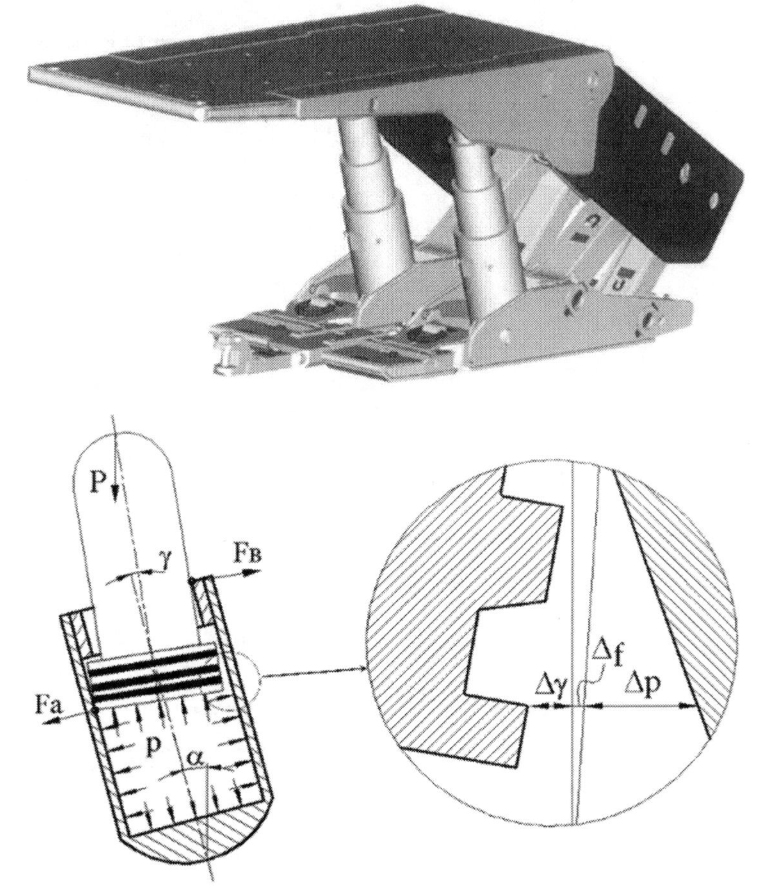

Fig. 4. Location of the hydraulic leg in the section of DBT powered support крепи DBT and the scheme of a sealing clearance formation when rod misalignment occurs

Due to its kinematic connections, the angle of installation of the hydraulic leg in the support section (α) also depends on the extracting seam thickness, i.e. in fact, on the leg extension.

To study the effect of the parameters of this phenomenon, a special finite element model was developed that takes into account the geometric dimensions of the cylinder, piston, rod, bush, leg extension, the direction of the external load from the rod and cylinder supports, the pressure of the working fluid in the piston cavity, the clearances between the piston and the cylinder, as well as between the rod and the bushing, determined by their tolerance zones, and the properties of the cylinder material.

The solution was carried out in two stages. First, the deformations of the cylinder from the action of the working fluid pressure (Δp) were identified, the identified values were used to define forces Fa and Fb (see Figure 4), arising in the joints due to the misalignment of the rod and the cylinder. After this, the final calculation is performed and the total deformations from all loads ($\Delta p+\Delta f$) are determined.

The total largest clearance to be sealed is

$$\Delta = \Delta\gamma + \Delta p + \Delta f,$$

where $\Delta\gamma$ – the clearance determined by the dimension tolerance zones and mutual location of the piston in relation to the cylinder;

Δp – radial deformations caused by working fluid pressure;

Δf – radial deformations caused by the additional forces Fa and Fb resulting from misalignment of the rod in relation to the cylinder.

Fig. 5 shows the results of the calculations made using the model described above applied to displacement (dR) of the internal walls of the working cylinder (1 and 3) of the hydraulic leg (M130 powered support) along its length (L) at the nominal operating resistance, zero installation angle, maximum extension and sealing of the cylinder support. The symmetry axis displacement is shown as a dotted line (2).

Analysis of calculation results showed that the initial maximum possible clearance in the compaction area determined by the tolerance zones of the cylinder and piston (280 μm) increased under load up to 490 μm (on the right side in Figure 4), which could result in its insufficient overlapping by the piston seals.

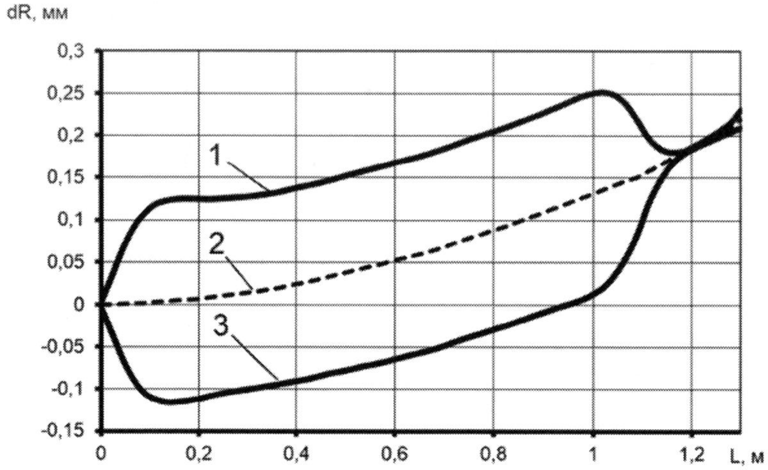

Fig. 5. Displacements of M130 leg cylinder walls dR under load taking into account rod misalignment

For double- and triple-acting legs, this effect is enhanced, because the total misalignment value is determined by the total clearances at all stages of the hydraulic leg in the joints between pistons and working cylinders and joints between bushings and rods.

When the external loads, which are dynamic or periodic in nature, influence the powered support, the frequency of these influences can coincide with the natural oscillation frequency of both the hydraulic leg and the system as a whole "support section - roof - coal seam" [10-15]. In this case, a phenomenon of resonance may occur, in which the radial deformations are further increased reducing the tightness of the legs.

Such loads are possible during secondary settling of the main roof when periodic fluctuations of fragile rock blocks occur. In addition, the frequency component of the load on the support can be transferred from the closely located rotating cutting body of the shearer crushing the coal mass of the seam.

As shown by theoretical studies, the frequencies of natural oscillations of the "powered support-roof-seam" system are close to the possible frequencies of external loads, and, consequently, the development of resonant phenomena has a very high probability.

3 Results and discussions

From all said above, it is possible to highlight main factors affecting the radial deformations of the working cylinders of hydraulic legs and formation of clearances that are to be reliably sealed to prevent loss of legs tightness and to maintain the working capacity of the complete set of mining equipment.

Fig. 6 shows the main factors that form the size of sealed clearances on hydraulic legs of the powered supports. Arrows indicate the direction of the subordination of factors and their increasing impact. For example, the frequency of natural oscillations (2) is determined by the geometric parameters (4) and the pressure of the working fluid (8).

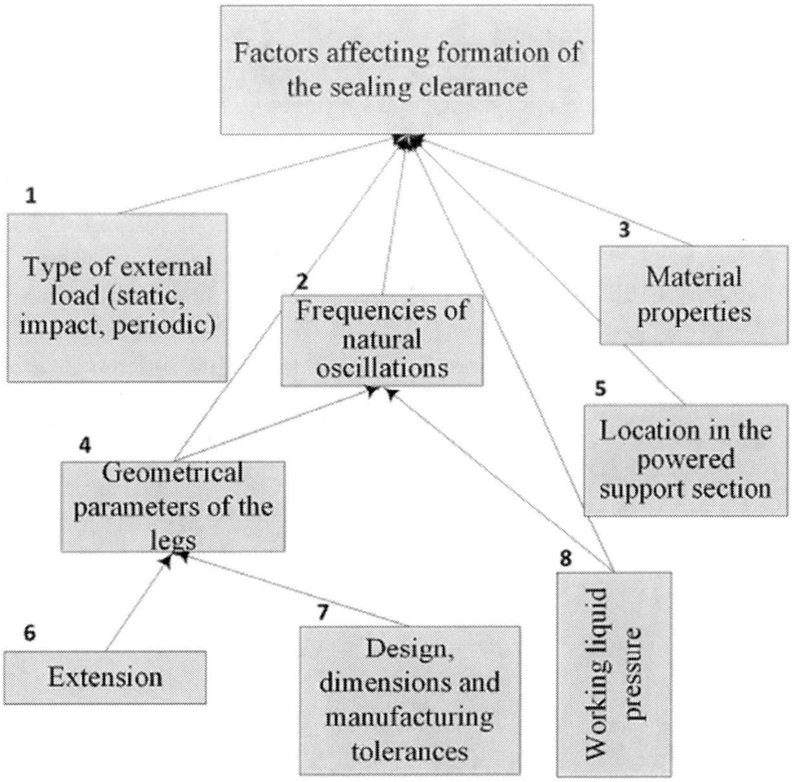

Fig. 6. Classification of factors affecting formation of the sealing clearance of the hydraulic leg

4 Conclusions

The size of the sealing clearance in the hydraulic legs of the powered supports is determined by their strength and geometric parameters, their location in the support section, properties of the materials and the type of external load. The integrated record of these parameters for the specific operating conditions of the mining set of equipment will help to avoid loss of tightness of the main support assembly and increase reliability and safety of mining operations.

References

1. V.I. Klishin, Y.V. Matviets, Fiziko-Tekhnicheskie Problemy Razrabotki Poleznykh Iskopaemykh, **2**, 23 (1993)
2. S. V. Klishin, V. I. Klishin and G. Yu. Opruk, Journal of Mining Science, **49:6**, 932 (2013)

3. V. I. Klishin, Yu. S. Fokin, Yu. M. Lekontsev and D. I. Kokoulin, Gornyi Zhurnal**, 12,** 112 (2005)
4. V. I. Klishin, Journal of Mining Science, **30**:**4**, 390 (1995)
5. V. I. Klishin and T. M. Tarasik, Journal of Mining Science, 37:1, 77 (2001)
6. D. A. Chinakhov, J. Applied Mechanics and Materials, **52:54**, 442 (2011)
7. D. A. Chinakhov, J. Applied Mechanics and Materials, **762**, 717 (2013)
8. G. D. Buyalich, A. V. Anuchin, K. P. Serikov, IOP Conf. Series: Materials Science and Engineering, **27**, 012012 (2016)
9. G. D. Buyalich, K. G. Buyalich, V. V. Voevodin, Source of the Document IOP Conference Series: Materials Science and Engineering, **127:1**, 012034 (2016)
10. A. Putnic, Archives of Mining Sciences, **60:2,** 595 (2015)
11. S.V. Uvakin, Ugol, **7**, 53 (2016)
12. G. D. Buyalich, K. G. Buyalich, V. Yu. Umrikhina, IOP Conf. Series: Materials Science and Engineering, **142**, 012120 (2016)
13. E.N. Pashkov, G.R. Zijakaev, M.V. Tsigankova, Applied Mechanics and Materials, **379**, 91 (2013)
14. M. V. Novoseltseva, I. A. Masson, E. N. Pashkov, IOP Conference Series: Materials Science and Engineering, **127:1**, 012007 (2016)
15. A. V. Krainov, E. N. Pashkov, R. E. Lushnikov, V. A. Arkhipov, Key Engineering Materials, **685**, 280 (2016)

Individual Learning Route as a Way of Highly Qualified Specialists Training for Extraction of Solid Commercial Minerals Enterprises

Elena Oschepkova [1*], *Irina* Vasinskaya [2], and *Irina* Sockoluck [2]

[1]T.F. Gorbachev Kuzbass State Technical University, street Vesennyaya, 28, 650000, Kemerovo, Russian Federation

[2] Kemerovo State University, Krasnaya Street, 6, 650000, Kemerovo, Russian Federation

Abstract. In view of changing educational paradigm (adopting of two-tier system of higher education concept - undergraduate and graduate programs) a need of using of modern learning and information and communications technologies arises putting into practice learner-centered approaches in training of highly qualified specialists for extraction and processing of solid commercial minerals enterprises. In the unstable market demand situation and changeable institutional environment, from one side, and necessity of work balancing, supplying conditions and product quality when mining-and-geological parameters change, from the other side, mining enterprises have to introduce and develop the integrated management process of product and informative and logistic flows under united management system. One of the main limitations, which keeps down the developing process on Russian mining enterprises, is staff incompetence at all levels of logistic management. Under present-day conditions extraction and processing of solid commercial minerals enterprises need highly qualified specialists who can do self-directed researches, develop new and improve present arranging, planning and managing technologies of technical operation and commercial exploitation of transport and transportation and processing facilities based on logistics. Learner-centered approach and individualization of the learning process necessitate the designing of individual learning route (ILR), which can help the students to realize their professional facilities according to requirements for specialists for extraction and processing of solid commercial minerals enterprises.

1 Introduction

The training of highly qualified specialists in logistics for mining enterprises is one of the actual problems in our country. Nowadays such definitions as “expertise”, “undergraduates’ competence” appear instead of “competence”, “knowledge”, “the common culture”. According to recent requirements the graduates of redbrick universities must have formed

[*] Corresponding author: oela@yandex.ru

DOI: 10.1051/e3sconf/20172104001

organizing abilities, make decisions and take responsibility for their consummation, have all necessary communication skills, be able to define social processes and the role of professional engagement in them, know how to improve it etc.

This problem solution is connected with the content standards modernization, optimization technique of educational process. To join to the World Education Space is necessary to adopt new requirements for the level of education, the quality of training specialists in logistics. In turn it requires significant changes in pedagogical theory and in the practice of education process.

The improvement of the quality of specialists training in logistics for mining enterprises, their academic qualification can be achieved with the help of new training forms, using modern learning and information and communications technologies when carrying out a learner-centered approach.

Moreover it is absolutely necessary to gain experience of individual work, to have skills to define problems, to work with up-to-date resources, to find out-of-the-box solutions etc. Thus centered orientation is recommended for modern student. It is supposed to have individual, comfortable programmers.

One of the technologies which can improve the effectiveness of individual work is individual learning route. It is supposed to found special pedagogical conditions to choose means, forms and methods of training which can keep up different educational and scientific interests.

Individual learning route is defined by educational needs, individual student's skills and capability (the level of readiness to complete program me) and forms competences defined by Federal State Standards of Higher Education.

2 Materials and methods

Individual learning route (ILR) is a purposeful, designed, graded educational program me when a student can take the position of the subject of choosing, making, carrying-gut of an educational program me with support of a lecturer in professional identity and self-fulfillment.

We can mark out some types of individual educational routes inside the system of higher education, let us name the most important of them, in terms of their application for a specialist preparation in logistics sphere for mining enterprises:

-Self-oriented individual educational route (this route is oriented on self-knowledge, personal abilities, personal resources);

-The route, oriented on receiving knowledge (this route is related to the separation of the system subject areas: transport logistics, warehousing logistics, informational logistics, with the system of supply control);

-The route, oriented to the student's self –creation as a specialist to be (this route is related with undergoing of practical training);

-The route, related to the students orientation on the scientific activity, it is characterized by the definite aim dependence. This route is initially planned and it is aimed at the receiving of concrete results (in research sphere, the theme and the problem of the scientific research work). The leading direction of the development of this route is knowledge direction and personal growth direction.

Modern computer technologies allow students to establish order and intensity of studies themselves according to the individual educational route; while realizing the possibility to work at home using PC with University electronic library system, with information materials, with electronic learning and teaching resources, with Internet technologies usage, and the including of students PC into the local University network.

 DOI: 10.1051/e3sconf/20172104001

3 Results and discussion

Scientific research work takes up the main position in the student individual work. But the first year students are not often involved in this activity. There are some reasons:

-First year students don't have necessary skills and abilities;

-There is low level of the teaching staff and curator's activity to coordinate and find students who are interested in the scientific research work.

As a result we can see the progressive reduction of the student's cognitive activity that can have negative result at their further professional activity.

However, it is possible to organize scientific research work while training subjects of general education, based on the individual educational route organization during the first and the second years of education.

Let's study the individual educational route scheme of the first year student scientific research work specialization «Transport logistics», department of automobile transportation, the Institute of information and communications technology, machinery and automotive transport, Kuzbass State Technical University, named after T.F. Gorbachev.

According to the degree course scheme, « Scientific research basics» course is studied on the third year of education, in view of this, it is possible to make the individual educational routes related to the scientific research work for the first year students who study this specialty. Let's name the individual educational route – the map-navigator of scientific research work.

The realization of the individual educational routes is impossible without modern educational resources usage and information and communication technologies. They include:

-Electronic Libraries: scientific and technical library of Kuzbass State Technical University. a T.F. Gorbachev, PC-based library system.

-AUTOSTAT analytic agency has been assisting top managers and key specialists of over 2,000 companies in Russia and abroad in these processes for more than 12 years. The agency carries out surveys of car owners and automotive business experts, prepares marketing reports on different segments of the Russian automotive market, organizes conferences and training seminars. They are the leading analytic agency in the Russian automotive business statistics and analytics and provide complete and actual information on it.

-The official site ofthe « Mining engineering» magazine.

A national specialized scientific/technical/industrial journal about achievements in mining operation technologies, mining equipment and mineral processing, state- of- the- art computer- based systems, R&D, markets of mineral resources, trends and case studies of investments into mining industry in the former Soviet Union and other foreign countries.

The journal presents reviews, research- and- technology, feature, and analytical articles, advertising materials, information on the leading mining and machine- building companies and new products of domestic and world manufacturers.

Mining Industry is a regular participant and media sponsor of major international and regional professional exhibitions specializing in mining, geology, road construction and other related topics.

The journal is circulated among visitors and participants of many conferences and exhibitions on mining in Russia and abroad, such as Mining World Russia (Russia, Moscow), Coal of Russia & Mining (Russia, Novokuznetsk), Expo Coal (Russia, Kemerovo); Mining World Central Asia (Kazakhstan, Almaty), Mining World Uzbekistan (Uzbekistan, Tashkent) and others

-Desktop applications: Microsoft Office, Open Office, Libre Office

 DOI: 10.1051/e3sconf/20172104001

-E-mail technology and provided services in sending and receiving messages (electronic messages) through the distributed (including global) system.

The scheme of the map-navigator of the scientific research work of the first year student of the Kuzbass State Technical University

Table 1. The scheme of the map-navigator of the scientific research work of the first year student.

The educational program specialization 23.03.01.03 « Transport logistics»					
Research advisor					
Stages of the scientific research work	Proceeding	Advisor, consulter	Projected period	Reporting form	Actual period
Organizational aspect	Student's scientific research work review in KuzSTU.	Learning circle advisor	September	Conference	
Course selection	Under the frame of « The introduction in specialty» course	Course lecturer	During the first term	Report	
Problem statement	The urgency of the research statement	Research advisor	October	Conference	
Literature sources work	The view and theoretical survey of scientific literature	Research advisor	November	Report, reference list	
Choice of the object and the subject of the research	The development of the theoretical constructs of scientific research work	Research advisor	December	Conference	
Practice	Empiric study	Research advisor	January-February	Conference, Report	
Presentation of the results	The materials handling, conclusion statement	Research advisor	March	Report, Presentation	
Preparation for the presentation	The scientific research work defense. The Preparation of the scientific research work materials for conferences, roundtables.	Research advisor	April	The proceedings of the report, the presentation at the Russian research-to-practice conference of young researchers« The young of Russia»	

4 Conclusions

In view of changing educational paradigm of higher education, it is necessary to use new educational and information communication technologies applying executing learner-centered approach during highly –qualified specialists preparation for enterprise for solid commercial minerals extraction and recycling.

One of the technologies directed at the improvement of the effectiveness of the student's individual work is the individual educational route.

The scientific research work takes the principal place into the student's individual work.

The individual educational route development of the scientific research work of the first year students will allow them to form scientific research competence, to develop organizing skills, to learn how to make decisions and take responsibility for their achievement, that are corresponding the requirements which are demanded for highly – qualified specialists at enterprise for solid commercial minerals extraction and recycling

References

1. L. Alexitch, The Journal of College Student Personnel, **43:1**, 5 (2002)
2. A. W. Astin, The Journal of College Student Personnel, **25:4**, 297 (1984)
3. V. Baydenko, *Competence-based approach to the design of the state educational standards of higher education (and methodological issues) Toolkit*, 114 (Research Center of quality problems of training, 2005)
4. P. Biner, N. Barone, K. Welsh, & R. Dean, Distance Education, **18:2**, 318 (1997)
5. J. Caulfield, *How to design and teach a hybrid course: Achieving student-centered learning through blended classroom, online, and experiential activities* (Stylus Publishing, 2011)
6. M. Coetzee, & R. M. Oosthuizen, Journal of Psychology in Africa, **22:3**, 315 (2012)
7. D. Gambetta, *Were they pushed or did they jump?: Individual decision mechanisms in education* (Cambridge University Press, 1987)
8. J. Isaksson, R. Lindqvist, E. Bergström, European Journal of Special Needs Education, **22:1**, 75 (2007)
9. C. B. Kandiko, & M. Mawer, *Student expectations and perceptions of Higher Education* (London: King's Learning Institute, 2013)
10. E. Ouweneel, P. M. Le Blanc & W. B. Schaufeli, The journal of positive psychology, **6:2**, 142 (2011)
11. Yu.N. Pak, D.Yu. Pak, Higher education in Russia, **2**, 130 (2012)
12. E. T. Pascarella, Journal of College Student Development, **47:5**, 508 (2006)
13. W. Schaufeli, A. Bakker, Work engagement: A handbook of essential theory and research, **5**, 345 (2010)
14. R. Snir & I. Harpaz, Human Resource Management Review, **22:3**, 232 (2012)
15. T.W. Taris, W. Schaufeli, A. Shimazu, Work engagement: A handbook of essential theory and research, **39**, 567 (2010)
16. M. Tims, A. Bakker, D. Xanthopoulou, The Leadership Quarterly, **22**, 121 (2011)

Sustainable Development vs. Post-Industrial Transformation: Possibilities for Russia

Sergey Zhironkin[1,2*], *Magerram* Gasanov[2], *Galina* Barysheva[2], *Eyvaz* Gasanov[3], *Olga* Zhironkina[4], and *Gennady* Kayachev[5]

[1]T.F. Gorbachev Kuzbass State Technical University, Belovo Branch, 652644 32 Ilyicha st., Inskoy, Kemerovo oblast, Russia
[2]National Research Tomsk Polytechnic University, Department of Economics, 634050, 30 Lenina st., Tomsk, Russia
[3]Khabarovsk State University of Economy and Law, Economics Department of International Economic Relations Faculty, 680042 134 Tikhookeanskaya st. Khabarovsk, Russia
[4]Kemerovo institute (branch) of Plekhanov Russian University of Economics, Department of Humanities, 650992, 39 Kuznetsky av., Kemerovo, Russia
[5] Siberian Federal University, Department of Economy and Business Processes Management, 660041, Krasnoyarsk, 79 Svobodny av.

Abstract. Today the theory of postindustrial society is one of the most widespread concepts which allow adequately comprehending the large-scale changes that have occurred in the ecological consciousness of Western societies for the last thirty years. Offered in the late 1960s and early 1970s by American and European researchers in the field of economics, social philosophy, and ecology, the integrated idea of sustainable development in postindustrial era incorporated the best elements of the scientific tradition dating back to the Age of Enlightenment. The article emphasizes that the key to modern social progress is the rapid technological development based on the transformation of science into a direct productive force. The measure of such progress is a shift from pure economic growth to the sustainable development. The authors describe the ways of changing Russian Government's attitude to economy regulation in postindustrial development to achieve the goals of sustainable development.

1 Introduction

Being a humanistic theory by its very nature, the concept of postindustrial society did not focus on the separation of opinions by economists and ecologists regarding the use of irreplaceable resources to ensure economic growth typical for almost the entire XX century. Assuming the possibility of convergence of two systems (economic and ecological), it was subjected to sharp criticism by orthodox economists of the market direction and anathematized by leading technocratic scientists.

* Corresponding author: zhironkin@inbox.ru

DOI: 10.1051/e3sconf/20172104002

In the 1990s, the most difficult and contradictory stage of recent Russian history, the fate of Russian interpretations of post-industrial theory also developed very dramatically. On the one hand, a significant number of economists and politicians who rejected the communist ideology declared themselves to be supporters of the neoliberal model of development, which became the basis of market transformations that have been carried out for the last decade [1-5]. Having put forward the thesis on the possibility of building a medium-developed capitalist society in our country in a short time, they were objectively not interested in a detailed study of either postindustrial theory or the concept of sustainable development. The reason for this is the unequivocal conclusion that the start of sustainable development in Russia is not feasible, a country oriented to the primary sector of the economy and obviously far from the environmental standards of modern society. On the other hand, the researchers who have not changed their beliefs so radically have adopted the theory of sustainable development as an insignificantly modified doctrine of the postindustrial society and tried to hide behind their environmental rhetoric the desire to restore industry on the old technological basis, without deep technological modernization and changing the paradigm of accelerating economic growth to the ideas of sustainable development. It is understandable that in this case too, there can be no question of a deep penetration of the postindustrial concept into the internal logic of Russia's sustainable development.

2 Materials and Methods

The formula of Russian economy postindustrial development is concretized: neoindustrialization plus vertical integration [6-8]. This is the formula for the rise of the country to the state-corporate stage, to which the most developed industrial countries of the West and the East have now emerged. This is also the formula for entering a path of sustainable development, overcoming systemic and intermediate backwardness.

Solving development problems in accordance with this formula requires the implementation of a national program for the vertical integration of the economy, strengthening environmental responsibility of the state and corporations. The basis for this should be created by high-tech corporations with the government participation, which structure should be complemented by technologically breaking-through and rationally specialized production of the latest equipment capable of implementing technologies that have minimal impact on the environment.

Let's note the fundamental requirements from the part of sustainable development to the state-corporate economic system. It should ensure the neo-industrial transformation of Russia aimed at shaping the national technotronic productive forces, through the effect of vertical integration, the effect of macroeconomic planning, the organization of work on a "just in time" basis, the deep automatization and unmanned technologies embedding into agriculture, transport and mineral resources extracting. Considering vertical integration as a key possibility for improving the foundations of the state-and-corporate relations, the new type of macroeconomic planning should start without delay.

In the field of productive forces development, the new type of national macroeconomic planning should give the priority to the organization of technologically advanced production of microprocessors, robots and engines, which could significantly reduce mineral resources consumption and environmental damage in the form of emissions, air and water pollutions, including radioactive.

The key to success and neo-industrial creation is the curtailment of profit-taking from raw material extracting and semi-finished products for export with simultaneous deployment of effective state-and-corporate system of maximizing value added through the production of high-tech products of final demand.

In order to confidently enter the mainline of sustainable development, it is time for Russia to solve the fundamental issue of our era - the frivolous attitude of corporations to the negative impact on the environment, in accordance with the law of vertical integration, based on the economy of multi-industry corporations.

3 Results and Discussion

Barrier institutions and constraining factors of the structural and technological development of the Russian economy necessary for entering the path of sustainable development are technological multi-wayability, manifested in the functioning of the newest technological structures and relics. Its presence is due to the course of technological development evolution, the peculiarity of Russian Government structural and technological policy and the consequences of market transformations of the economy structure. The high level of fixed assets depreciation, the use of technologies of the third and fourth technological layers, lack of a developed industrial base capable of implementing environmental ideas and technological innovations, the lack of effective institutions conducive to the accelerated development of environmentally friendly technologies, characterized as the deindustrialization, are also limiting technological development.

The basic conditions for the sustainable development of Russia are structural modernization oriented institutional policy and its effective implementation, which has a key impact on the processes of technological and structural shifts and economy development as a whole, since it determines the level of transaction costs. An effective structuring mechanism in the formation of Russia's innovation economy is the clustering (network) structuring principle that ensures more effective implementation of competitive advantages and facilitates the replacement of the resource model, form the dissemination of new knowledge, technologies, innovations and network links between the cluster entities. However, the leading role in managing the processes of structural shifts is assigned to the state as a subject of budget investments and a generator of structural modernization.

The formation and functioning of a structural model meeting the requirements of sustainable development and the real needs of society require highly concentrated state structural policy mechanisms aimed at transforming the basic macroeconomic and sectoral proportions that contribute to overcoming the structural crisis and the sustainable innovation development of economy. In the conditions of financial and economic crisis, the Neo-Keynesian models of social and economic development, based on the methodology of strengthening state regulation, acquired particular relevance, having caused some return to conservative models of economic policy. The use of the mechanism of state-and-private partnership is becoming more effective. Within this framework high-tech projects, joint venture funds, start-up companies, small satellite companies are financed. There is an urgent need for a significant expansion of the range of financial assistance to companies that minimize environmental damage through emitting government's guarantees to innovative companies, subsidizing part of their R&D expenditures, granting them tax benefits and zero-interest loans, carrying out public purchasing of environmentally friendly products.

Practice clearly demonstrates the effect of such conformity: neo-industrial production of fixed assets is carried out by vertically integrated corporations. One naturally suggests another. Outside vertically integrated forms of business, technotronic production is not developed and does not work. And the turning point came in the mid-1970s, when inter-sectoral TNCs took command of the economy in the US, concentrating over 52% of total capital.

Vertical integration is a reproductive synthesis of science, mineral extracting and processing industries, with specialization in the production of specific types of finite science-intensive products: microprocessors, computers, technotronic engines for various applications, aircraft, ships and ships, digital communications and digital television, household appliances, etc.

Let us list the essential features of vertical integration.

The first of them is the inter-branch consolidation of industrial enterprises into a complete and unified cycle of reproduction of specific high-end products of final demand.

The second is guaranteed providing with the necessary material resources of postindustrial quality, starting from raw material redistribution, promotion from extraction minerals to their full processing and utilization of the waste products. In other words, the optimal consumption of the necessary material resources and the optimal proportion between extractive and processing operations are determined and maintained by the most vertically integrated structure.

The third is the target product specialization of a vertically integrated structure. It produces goods of technotronic quality, high-tech, capable of satisfying both intermediate and final demand, without increasing the load on the environment.

The fourth is the unification of technologically complementary extractive and processing industries, more precisely, the unification of manufacturing the means of production and production of end-use items. If a corporate structure needs new machine or assembly lines or modules, new laboratory or test equipment, it either produces it itself, or places orders on a competitive basis, or establishes the contractual relationship through the temporary and flexible connection of a particular production link, center, complex. All this makes it possible to use irreplaceable resources as new requests come from the market and, thus, reduce the overall mineral resource consumption.

The fifth is a flexible organizational structure, mobile in time and space. The corporation quickly gets rid of obsolete links, creates or adds new ones as it is developing, which ensures faster embedding of new technologies.

The sixth is the maintenance of relations between both direct (formal) and indirect (informal) economic dependences of key suppliers. So, it is not necessary to directly include a particular raw material enterprise in the formal structure of the corporation, so that it ensures a common end result.

The seventh is the concentration of industrial research in the structure of corporations and universities (academic science – outside of them). Today in Russia there is an opposite situation. The universities have largely lost their research base and reduced scientific staff. The former Soviet Union industrial science is scattered. Only in the defense industry the modern R&D are still more or less preserved. As we see, there is no inter-branch corporations - no proper industrial research and development. A separate enterprise, outside the corporation, outside of vertical integration, cannot stimulate innovation activity from fundamental to applied researches. All this puts unavoidable barriers between the ideas of sustainable development in the post-industrial era and their implementation in Russian practice.

For a vertically integrated corporation, scientific and research activities become an internal factor. At the time, it was not in vain to argue about the nature of technological progress in relation to fixed assets investment: endogenous or exogenous. When in the 1950s the corporate form of industries only appeared and was still sporadic, such a question was topical. But now it is proved by the life itself, that the formation of a vertically integrated corporation as a universal form of industry organization in postindustrial era.

So, the vertical integration transforms the interdepartmental corporation into the main subject of postindustrial economy. The key criterion of the multi-industrial corporation, alas, is often underestimated: it unites technologically interconnected enterprises of several sectors of economy, i.e. production of final demand items and the means of production. It was in the inter-industry corporation where the practice found the main form of integration of two reproduction divisions: the production of materials and equipment, and the end-use goods.

Unfortunately, as evidenced from the concept of long-term development of the Russian economy until 2030, the government still does not expect the effect of vertical integration.

However, without strengthening the effect of vertical integration, it is impossible to bring Russia to the trajectory of sustainable development, using the technological and logical breakthroughs of the emerging post-industrial world. Moreover, without vertical integration, the transition from raw material growth to an innovative type of development is certainly impossible. The goals outlined by the Russian government are simply unachievable with the preservation of the current economic system, the basis of which is oligarchic property, the dominance of raw material capital over technological and innovation, internal rupture of unified technological chains by redistribution. The processing industries have insufficient international competitiveness, because resources and semi-finished products are exported abroad. Existing corporations extract the maximum profit from the intermediate production, and this is why the manufacturing industry stays uncompetitive.

The transition from commodity growth to sustainable development, innovative diversification of the economy, the widespread lean production implementing in key sectors of the economy by 2020, are the interlinked benchmarks. The system of raw material oligarchy must be transformed into high-tech lobbyism, and as soon as possible. It is vertical integration with its postindustrial organizational and economic mechanisms that puts an end to the whole system of extraction profits from raw-materials and intermediate production. With vertical integration, the state no longer has to pedal the administrative, tax, customs, tariff and other methods of allocating resources between domestic and foreign markets. Multisectoral corporations, in which the intermediate and final productions are integrated, will maximize corporate profits, thereby increasing the output of final demand goods.

The practical solution is unique: the urgent integration of raw material and processing sectors of economy with the primacy of the latest one. Therefore, the process of integration of technologies, labor and capital of both sectors should be initiated and carried out by the state.

4 Conclusion

Thus, entering of Russian economy the way of sustainable development is possible when a set of dominant problems of postindustrial structure changes in productive forces and institutions is solved; when a combination of direct, indicative impact and market self-regulation are taken into account; when the strategic planning and targeted programming of mutual economic and ecological development are fulfilled. Therefore, it is necessary to study further the content, trends, instruments and institutions of the postindustrial transformation of the structure of Russian economy as a key to the sustainable development.

References

1. A. Auzan, *Economics of everything. How the institutes define our life* (Moscow, Mann, 2013)
2. G.A. Barysheva, E.G. Novoselova, App. Mech. Mater. **682**, 550-554 (2014)
3. S. Glaziev, Rus. Econ. J., **2**, 3-6. (2012)
4. I. Kurnysheva, S. Lykov, A. Idrisov, Economist, **10**, 39-47 (2008)
5. V. Volchik, *Institution and evolution economy* (Rostov, UFU, 2011)
6. S. Zhironkin, V. Guzyr, Eur. Proc. Soc. Beh. Sci. **VII**, 160-166 (2016)
7. S. Zhironkin, M. Gasanov, O. Zhironkina, Eur. Proc. Soc. Beh. Sci. **VII**, 124-131 (2016)
8. E.A. Gasanov, N.V. Gasanova, A.V. Zhavoronok, Eur. Proc. Soc. Beh. Sci. Eur. Proc. Soc. Beh. Sci. **VII**, 323-336 (2016)

Neo-Industrial and Sustainable Development of Russia as Mineral Resources Exploiting Country

Marina Prokudina[1,*], *Olga* Zhironkina[2], *Oksana* Kalinina[3], *Magerram* Gasanov[4], *Felix* Agafonov[5]

[1]Kemerovo institute (branch) of Plekhanov Russian University of Economics, Department economic security, accounting and auditing, 650992 39 Kuznetsky pr., Kemerovo, Russia
[2]Kemerovo institute (branch) of Plekhanov Russian University of Economics, Department of Humanities, 650992, 39 Kuznetsky av., Kemerovo, Russia
[3]T.F Gorbachev Kuzbass State Technical University, Department of History, Philosophy and Social Sciences, 65000 Kemerovo, 28 Vesennya st, Russia
[4]National Research Tomsk Polytechnic University, Department of Economics, 634050, 30 Lenina st., Tomsk, Russia
[5]T.F Gorbachev Kuzbass State Technical University, Post-Graduate Study, 650000 Kemerovo, 28 Vesennya st, Russia

Abstract. In the Russian economy, the world leadership in the extraction of different mineral resources is combined with the potential for their processing and a significant scientific sector. Innovative development of raw materials extraction is impossible without the parallel technological modernization of the high-tech sector. In general, the complex of these processes is a neo-industrialization of the economy. Neo-industrially oriented transformation of the economy reflects complex changes in its structure, the transformation of established stable relationships between various elements of the system of social production that determine macroeconomic proportions. Neo-industrial transformations come along with the modification of economic relations associated with investments, innovations, labor and income distribution, with the process of locating productive forces and regulating the economy by the government. Neo-industrialization of economy is not only significant changes in its technological and reproductive structure (the development of high-tech industries, the integration of science and industry), but, above all, the implementation of a system structural policy of innovative development of raw material industry and the recovery of manufacturing industries on a new technological basis.

1 Introduction

At the level of economic reality of Russian market reforms, neo-industrial economic transformations mean a transition from an export-raw model of the reproductive system to a

* Corresponding author: prokudina63@mail.ru

 DOI: 10.1051/e3sconf/20172104003

competitive-processing one. In it, the commodity sector plays a crucial role as a source of investments and material resources for the development of export-oriented manufacturing industries, while the high-tech and R&D sectors form their technological basis [1-3].

The methodology of neo-industrial transformation is based on the analysis of approaches to the study of forms and directions of complex, purposeful structural changes in the economy, in the process of regulating economic relations, institutions, and the development of productive forces [4-5]. The use of this methodology in the process of developing the theoretical fundamentals of the economy neo-industrialization implies the development of a multi-level, multi-object and multi-subject impact on the national economic proportions. This influence should combine normative, indicative, program-targeted regulation and market self-regulation, alternative attractors of structural changes (for Russia - raw-specialized and neo-industrial). The application of the theoretical foundations of neo-industrialization in the practical activities of the state is determined by the vector of its structural policy, ensuring the movement towards the neo-industrial attractor in the process of forming an institutional environment, regulation of economic relations [6-9].

2 Materials and Methods

The methodological fundamentals of neo-industrially oriented economic transformations include an analysis of their principles and conditions.

The principles of neo-industrially oriented economic transformations are the basis on which theoretical positions, institutions, forms and instruments are formed, practical recommendations. The set of these principles includes: the priority of the state as a guarantor of neo-industrialization and regulator of the economy structure; systemic and multilevel character of its structural transformations; a combination of normative, indicative regulation and market self-regulation, the presence of a neo-industrial attractor; goal-setting and strategic programming taking into account the depth of structural problems; development of import substitution, neo-industrial state order, initial state investments, public-private platforms, target social group.

Based on these principles, the social-and-economic conditions of neo-industrialization are defined:

- institutional conditions - the establishment of institutions for interaction between business and the state that can overcome the inertia of the commodity production dominance, social and budgetary dependence on obsolete technologies and raw materials exports;
- technological conditions include the formation and development of convergent technologies - the result of interpenetration and a combination of various technological innovations, united by a common digital form of design, replication and diffusion;
- reproductive conditions consist in accelerating the renewal of the fixed capital of the manufacturing industries, in raising the rate of accumulation, in attracting investments in the branches of the 6th technological mode, in reducing the depreciation of industrial equipment, in mass investment in the digitalization of production and the launch of microprocessor equipment, new materials;
- cluster conditions – the encourage of agglomeration of new technologies as the next stage in the development of innovative clusters, based on the integration of the resource base of industrial enterprises, the innovative potential of scientific organizations and the military-industrial complex;
- market conditions – the promotion of innovations in mass production through public-private partnerships, state guarantees and state loans, subsidizing interest rates and tax incentives, regulating the interaction of innovative firms and natural monopolies;
- social conditions – target development of a stable social group necessary for the diffusion of new technologies into mass production.

The principles and conditions of neo-industrial transformation are common to the economies of the transformational type of countries that are undergoing the process of market transformations. On their basis the theorize of neo-industrially oriented economic transformations, their mechanism, the complex of structural-transforming factors and institutional provision are being formed.

3 Results and Discussion

The mechanism of neo-industrially oriented economic transformations is called to unite multi-level, multi-object and multi-subject interactions of the state and business into a complex of structure-transforming neo-industrially oriented processes in the system of regulation of national reproduction. The impact of the mechanism of neo-industrially oriented reforms on the structure of economy is coordinated with the factors of its change. These factors are the forces of the macroeconomic (external market, aggregate demand, scientific and technological progress and innovation, investment, competition, macro-stabilization, supply of raw materials, labor market, institutions and ideology) and microeconomic (financial condition of enterprises, their links with commercial banks and the vertical integration of science and production) impact on national economic proportions.

Such intensive factors as scientific and technological progress, innovations and investments, market competition, financial recovery of firms greatly contribute to positive structural changes. On the contrary, extensive factors, such as international, aggregate demand, raw material availability of economy, are able both to improve the structure of economy in the process of neo-industrialization, and to strengthen the inertia of structural problems embarrassing innovation development.

Neo-industrialization of economy is impossible without strengthening the structure-transforming factors of market competition, vertical scientific and industrial integration, innovation and investment interactions among the government, banks and enterprises. Conditions for the replacement of short-term and extensive factors with long-term intensive ones (such as scientific and technological progress, vertical integration, investment of innovations, the formation of a stable social group) are defined: the diffusion of innovations from high-tech industries to processing and raw materials; state tax, credit incentives for the links of their subjects; formation of a stable social group.

The prospects for realizing these factors of neo-industrial transformation in the Russian economy are determined by their obstacles, which, in turn, reflect the specific country conditions of market transformation and the depth of structural problems.

The concept of neo-industrially oriented structural transformations unites their methodological and theoretical fundamentals, and is based on the following provisions:

- on the system-and-structural and reproductive methodology, which implies a combination of targeted government regulation and market self-regulation of macroeconomic proportions at several interconnected levels of the economic structure with maximum use of economic potential of the commodity sectors, investment resources of industrial clusters, within the maximum benefits for the manufacturing sector;
- on the concentration of the government's efforts to bring the proportions of production factors and forms of ownership, institutions, sectors and branches of economy, market models, clusters and social groups into the line with the need for innovative development of the basic sectors of the Russian economy and, on their basis, high-tech industries;
- on priority principles of neo-industrial import substitution and state procurement, parity participation of the government in the development of technological platforms and innovation clusters, targeted programming of changes in innovation, investment and social spheres;

 DOI: 10.1051/e3sconf/20172104003

- on the use of resource potential of the basic sectors of the Russian economy in the development of the manufacturing sector by integrating innovative, raw material, manufacturing enterprises, banks and financial companies into high-tech holdings with the government participation;
- on the integrated application of direct framework and indirect investment, credit, tax, institutional instruments of neo-industrial transformations of economy;
- on the institutionalization of investment-and-credit, tax-financial, innovation-technology, social interactions between the state and business.

The obstacles to the neo-industrial transformations of the Russian economy are stable to the regulatory impact, and their inertia is reinforced by the institutional trap of its rental raw material model. It consists in the fact that the existing structure of the Russian economy allows achieving short-term budgetary, macro-stabilization goals, but does not contribute to neo-industrialization and economic development through a consistent change in technological structures. In such an institutional trap, the innovative development of the Russian economy is embarrassed by the stable destructive rules - the institutions of interaction of firms with the state and with each other, formed by the groups interested in the dominance of the export raw material model - the raw material and financial oligarchy and associated officials.

The rules destructive for the neo-industrialization of the Russian economy include the neglect of the public interests of innovative development, the consolidation of the fulfillment of the state's budgetary obligations over the world raw material market, the integration of Russian companies into the production chains of international corporations as a donor of raw materials and the importer of the processed products, etc. As a result neo-industrialization start of the Russian economy is constrained by the dominance of investments with a short-term effect, allowing the interest groups to profit from the situation in the commodity and financial markets, and by the lack of investment, resulting in the long term to a change of technological modes.

Overcoming the de-industrialization of the Russian economy and its move to the neo-industrial way of innovative development require the institutionalization of investment, innovation, credit, tax and financial connections between the state and business, strategic planning and targeted programming of vertical processes of raw materials, processing and high-tech manufacturing, the formation of priority social groups, the development of import substitution.

The integrator of institutional measures to ensure the neo-industrially oriented transformation of the Russian economy is their strategy. It is a program of long-term changes in the structure of the Russian economy, an integrating mechanism, and the factors of neo-industrialization, its conceptual fundamentals and institutions. The main strategic directions of neo-industrial structural policy in Russia are defined as the following:

A) The emphasis on the use of basic (natural, energy, accumulated intellectual capital and educational level of the population) and domestic (funds of the state budget and extra-budgetary funds, capitals of private investors and banks, innovative potential of clusters, special economic zones, techno-parks), rather than external resources (foreign financial capital, international marketing). To do this, it is necessary to create appropriate financial incentives for the state, promote the integration of R & D, high-tech, manufacturing and raw materials industries.

B) The replacement of "participation in authority" (the conditions for the long-term effectiveness of Russian business that appeared in the 2000s) to "participation in innovations" through the active use of neo-dirigisme (the full use of indirect economic regulators by the state), both in financial and innovation, investment spheres, in industrial production, in the labor market. This requires the government to support small innovative entrepreneurship, reduce transaction costs of high technology production - at the expense of state guarantees, subsidies, and insurance.

C) Import substitution of high-tech products, which requires a reduction of the share of imports of late-industrial products - computer manufacturing, ultra-light and ultra-strong materials, microelectronics, and robotics. Neo-industrial import substitution is a necessary condition for innovative recovery of manufacturing industries and the creation of a resource base for the development of new industries. For this, a neo-industrial state order is needed – a set of state contracts for carrying out import-substituting R & D, high-tech products for the needs of state corporations, state guarantees and state export crediting of at least 15% of this output. It is also important to carry out selective state support for projects of neo-industrial import substitution, taking into account the breakthrough nature of innovations, to conduct advanced imports of technologies and highly qualified specialists. To integrate these measures, a state investment program for neo-industrial import substitution is needed.

D) The development of scientific potential, import and adaptation of technologies instead of import of high-tech products and equipment. For this purpose, it is necessary to develop technological platforms and public-private partnerships in innovative clusters, transforming them into vertically integrated companies – high-tech holdings that unite research and development organizations, experimental and innovative firms, manufacturing enterprises, commercial banks, and financial companies. The core of such holdings should become commercial banks and technology platforms.

The most significant form of transforming property relations in the process of vertical integration is the creation of high-tech bank holdings with the participation of the state. Their financial centers - commercial banks - should ensure a long-term capital inflow for lending innovative projects under the state guarantees and on the security of intellectual property (high-tech mortgage).

The growth of social wellbeing of the population and the overcoming of structural desocialization of the Russian economy is the ultimate goal of neo-industrial changes. The main condition and the "conductor" of the influence of neo-industrial transformations on the social wellbeing should become the neo-industrial social group, which include:
- students – future specialists in information technologies, biochemical, radio-electronic and other high-tech industries; employees of aircraft-building industries, instrument-building enterprises; employees of military-industrial enterprises using dual-purpose technologies being diffused in the civil sector; pedagogical workers of universities participating in scientific research; private venture investors. The mechanisms for the formation of sustainable neo-industrial social group include the following:
- the state order for mass training of Russian specialists in the leading foreign universities in the areas of critical and breakthrough technologies with mandatory subsequent employment in companies residing in innovation clusters and techno-parks;
- the development of state guarantees for the employment of highly qualified employees and graduates of higher education institutions in specialties corresponding to critical technologies;
- the exemption of the payroll budget of employees in innovative firms, R & D organizations from all taxes, and their wages - from the personal income tax;
- inclusion of the mandatory requirement for training and retraining of highly qualified personnel in investment projects supported by the Russian venture company and other state funds investing in innovations.

4 Conclusion

Thus, carrying out of neo-industrial transformations of the Russian economy structure is possible when taking into account a set of dominant problems of transforming economic relations and changing the structure of the productive forces and institutions; a combination of measures of direct, indicative impact and market self-regulation; the framework of strategic planning and targeted programming. Therefore, it is necessary to conduct a further study on the content, directions, tools and institutions of neo-industrial transformation of the Russian economy structure.

References

1. O. Krasilnikov, *Structural shifts as a factor of economic growth in modern Russia* (Moscow, Thesis, 2003)
2. S. Glaziev, Rus. Econ. J., **2**, 3-6. (2012)
3. S. Zhironkin, M. Gasanov, O. Zhironkina, Eur. Proc. Soc. Beh. Sci. **VII**, 124-131 (2016)
4. D. Bell, *The coming of post-industrial society. Venture in social forecasting* (New York, Collman Pub., 1973)
5. R. Reich, *The work of nations. Preparing ourselves for 21st Century capitalism* (New York, Univ. of N.Y. Pub., 1992)
6. A.A. Baryshev, G.A. Barysheva, Eur. Proc. Soc. Beh. Sci. **VII**, 352-365 (2016)
7. 7. G.A. Barysheva, E.G. Novoselova, App. Mech. Mater. **682**, 550-554 (2014)
8. I. Kurnysheva, S. Lykov, A. Idrisov, Economist, **10**, 39-47 (2008)
9. V. Volchik, *Institution and evolution economy* (Rostov, UFU, 2011)

 DOI: 10.1051/e3sconf/20172104004

Viral Management as a New Type of Enterprise Management in Coal Industry

Olga Garafonova [1*], *Svetlana* Grigashkina[2], and *Anna* Zhosan[3]

[1] Kyiv National Economic University named after Vadym Hetman, 03057, Prospect Peremogy, 54/1, Kiev, Ukraine

[2] T.F. Gorbachev Kuzbass State Technical University, 650000, 28 Vesennyaya St., Kemerovo, Russia

[3] Kherson National Technical University, 73008, Berislavskoe shosse, 24, Kherson, Ukraine

Abstract. The article considers the evolution of the concept of "management". The Internet analysis of the concept of "viral management" was conducted, the results of which testify to the positive trend, the growing interest in scientific circles, the relevance of the chosen direction for further research and the increasing popularity of the viral management among business leaders. The indices of coal and brown coal extraction in Ukraine and Russia for 2010-2016 are analyzed. Among the problems that exist in the coal industry are the following: mine assets have a high degree of deterioration; the equipment is rather obsolete and does not correspond to the world level; among the existing face equipment, the specific weight of longwall mining equipment, road headers, loading machines, and belt conveyors of the new generation is only several percent. Five steps of introduction of the virus management at the enterprise are identified: opening, development, participation, diffusion, sustainability. The principles of introduction of the viral management in the enterprises of the coal industry are offered. It is established that the main idea of the viral management is to "infect" the organization and, mainly, employees with one or another "virus" in the form of a common idea or goal. It is indicated that the viral management assumes a certain automaticity of changes, internal "obsession" with this or other innovation, involvement of an informal personal factor.

1 Introduction

In modern economic, political, social, and other conditions for the functioning of an enterprises, managers should respond extremely quickly to all changes that occur in the external and internal environment. Often the further activity of the enterprise depends on the speed of reaction. In the 21-st century, most enterprises whether they are big, medium or small, including operations in the coal industry, have already tried everything to improve their activity, including restructuring, reengineering, reorientation, redesigning, and yet again the above techniques do not work as efficiently as efficiently as we would like. One of the alternatives and a new turn would be behavioral changes embodied in the emergence

*Corresponding author: gva.pm@kuzstu.ru

of the viral management that may provide conditions for sustainable large-scale behavioral and cultural changes. Of course, this may require changes in structures and processes, but the final real changes will be behavioral ones. Cultural transformations are formed by the behavior of separate people and small groups, and not from the management downwards. Changing and mobilization of people are the effects of stretching, social infection, social movement, mechanism of leadership at the low level, and support from the performers, and otherwise. For this, we need the viral management, i.e. infection with a common single idea.

2 Materials and methods

The development of management practices leads to a corresponding change in the development of management theory. Humanization of management increasingly shifts attention from technological management to the management of human behavior at the enterprise. At the same time, the development of the theory in its turn affects the practice of management. Thus, rational management requires a rational vision and understanding, both theory and practices of management. In turn, the modern development of management in Ukraine and Russia is influenced by the transition from the traditional management to a modern one.

A feature of the modern management is that its main function is entrepreneurial. Entrepreneurship turns into the core of management. Successful entrepreneurial activity is impossible if the enterprise does not introduce innovations. The state of economic matters in the country can be improved by improving the management and introducing its latest achievements into practice. New realities of the modern management need to be implemented as soon as possible.

Table 1 presents the approaches to the notion 'management' by various scientists [1,2,3,4,5,6,7,8].

Table 1 Evolution of the concept of "management".

N/n	Year	Definition's author	"Management" concept definition
1.	2001	A. G. Ivasenko	Effective use and coordination of resources such as assets, buildings, materials and labor to achieve specified goals with maximum efficiency.
2.	2003	G. Ya. Goldshtein	Influence of one person or group of persons on other persons for inducing actions corresponding to the achievement of the set goals in assuming responsibility for impact efficiency.
3.	2003	V. P. Sladkevitch, A. D. Chernyav'skiy	The integration process through which professionally trained managers form and manage organizations by setting goals and developing ways to achieve them.
4.	2004	I. D. Ladanov	The process of managing an individual worker, a working group, working collectives, is also a process of achieving the goal of organization by the hands of others.
5.	2005	N. V. Komarova	On the one hand, it is activities aimed at planning, organization and effective use of material, labor, financial and information resources; and on the other hand, a set of principles, methods and forms of

			management used to improve the efficiency of the organization performance and increase profit.
6.	2005	V. I. Rudenko	Ability to achieve goals, use labor, intelligence, motives of other people's behavior. Function, the type of activity in managing people in a wide range of organizations. It is also an area of human knowledge helping to carry out this function.
7.	2008	I. P. Derev'yago	An independent type of activity which requires a subject – professional - manager, whose work is aimed at the object - the functioning of the organization as a whole or its specific sphere. The function of management, i.e. the type of activity for the management of people of any branch of the economy and in any field of activity, if it is aimed at making a profit.
8.	2008	V. M. Didenko	A set of certain organizational and economic methods of managing all stages and types of adequate processes and phenomena at the level of various economic entities.

If we consider the concept of "virus", then we should turn to the definition from the biological sciences, where the virus is a non-cellular form of life, which can only be distinguished by an electron microscope. These are intracellular parasites. Outside the cell, they do not show their properties and have a crystalline form.

If we consider a computer virus, it is a type of malicious software that can create copies of itself and be embedded in the code of other programs, system memory areas, boot sectors, and distribute copies of itself across various communication channels [9].

The main purpose of the virus is its distribution, and the violation of the operation of software and hardware packages such as deletion of files, disabling of data placement structures, blocking of users' work, etc. is often its related function. Even if the author of the virus has not programmed malicious effects, the virus can lead to computer crashes due to errors, unaccounted intricacies of interaction with the operating system and other programs. In addition, viruses tend to take up space on storage devices and consume the system resources [9].

Thus, by combining these two concepts, a new "viral management" is formed. The main idea is to "infect" the organization and, mainly, employees with one or another "virus" in the form of a common idea or goal. For example, it is possible to "infect" with the idea of achieving "over-result" by a certain indicator or the idea of creating a corporation as a united, friendly "family", etc. It should be noted that viral management presupposes a certain automaticity of changes, internal "obsession" with this or that innovation, involvement of an informal personal factor. And the main task is to ensure that the "virus" introduced into the organization is ultimately beneficial, not harmful, so that, after "overcoming the virus", the organization become stronger, more powerful, more effective.

The popularization of the concept of "viral management" is proposed to be proved using the Internet-analysis method [14]. This study determines the proportion of links on the Internet to the definition of "viral management" and "social responsibility of the enterprise" in the period from 2000 to 2016. (Table 2). The data are obtained experimentally and then standardized for better visualization and comparison [15].

Table 2 Proportion of links to the 'viral management' definition in the total amount of links (2000-2013), %.

Year	2000	2001	2002	2003	2004	2005	2006	2007	2008

"Viral management" definition	7,36	4,97	4,99	5,49	5,11	5,77	6,48	6,92	7,58
Year	**2009**	**2010**	**2011**	**2012**	**2013**	**2014**	**2015**	**2016**	
"Viral management" definition	7,62	8,58	9,06	9,87	10,20	11,01	11,82	13,35	

The forecast of the popularity of the definition "viral management" in 2017 is made (Fig. 1).

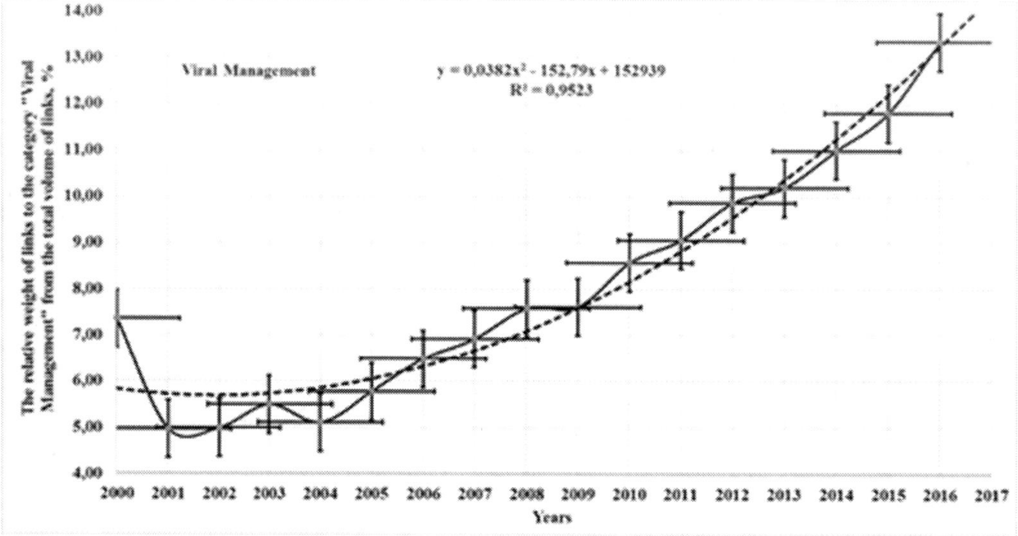

Fig. 1. Forecast of the proportion of links to the "viral management" definition (2-nd order polynomial), %.

The accuracy of the approximation is > 0.95, this confirms a reliable forecast for further increase in the popularity of viral management among business leaders and indicates the feasibility of new research in this area. In 2017, it is planned to increase the proportion of links on the Internet to 15% or more.

Concerning this study, the forecasting of the demand for the study of viral management was carried out based on the method of Internet analysis. So, according to the studies, we can conclude that over the past 16 years the popularity of the concept of "viral management" has grown (Figure 1). This indicates a positive trend, a growing interest in the scientific community, the relevance of the chosen direction for further research and an increasing popularity of viral management among business leaders.

3 Results and discussion

The coal industry is one of the basic for the economy of Ukraine. The work of adjacent basic industries – power engineering and metallurgy as well as the provision of fuel and energy to the domestic sector of the country depends on its stability. The coal industry undergoes restructuring which is driven both by the general processes of the economic transformation of Ukraine and its entry into the world economic space, and by the crisis developments that have been growing in the Ukrainian coal industry in recent years (Table 3).

DOI: 10.1051/e3sconf/20172104004

Table 3. Indices of bituminous and brown coal extraction in Ukraine and Russia for 2010-2016 (in per cent to the corresponding period of the previous year)*.

Year	Production of bituminous and brown coals	
	Ukraine	Russia
2010	99,3	106,9
2011	113,7	104,1
2012	104,5	105,3
2013	97,6	99,3
2014	69,5	101,7
2015	61,9	104,2
2016	104,8	103,2

*compiled by the authors on the basis of statistics [10, 11, 12]

Among the problems that exist the following can be highlighted: mine assets have a high degree of wear; the equipment is rather obsolete and does not correspond to the world level; among the existing face equipment, the specific weight of longwall mining equipment, road headers, loading machines, and belt conveyors of the new generation is only several percent.

To solve these problems, scientists believe it is necessary to accelerate decommissioning, mainly through the conservation of the most unprofitable low capacity mines, that require large subsidies which allow us to improve social protection of released workers (for example, early retirement, retraining, etc.); continue to initiate with appropriate information and advertising support and carry out denationalization (privatization, lease and concession) of mines to promote the practice of joint development of coal deposits by public and private enterprises [13].

The authors consider that it is necessary to apply more innovative approaches to solving the emerging problems at the level of small, medium, large enterprises, incl. in coal industry. One such approach is viral management, consisting of 5 steps, the passage of which will not only involve all the proposed standard solutions to emerging problems in the correct sequence, but also identify new ones that improve the ultimate effectiveness of the actions undertaken (Fig. 2).

Fig. 2. Five steps of introduction of the virus management at the enterprise were identified *the authors' development.

Upon study of the proposed steps, it is clear that the change in the process / system occurs through a change in the behavior of workers. Changes must come from above and be filtered, while serious changes require serious actions. Viral management provides a new basis for changes and shows how the combination of the right language and framework, a small set of necessary actions (distributed by a small number of champions) and the creation of "turning points" creates long-term cultural changes at the enterprise. Unlike conventional methods of change management, viral management can work faster, more effectively increasing organizational effectiveness in everyday organizational life (not only during change initiatives).

4 Conclusions

Thus, the viral management can be called a new field of enterprise management, based on changing employees' behavior, infecting them with a new idea aiming at bringing the enterprise to the desired result. The basic principles of introducing the viral management in an enterprise include: an organization or an enterprise as a single "organism"; multicenter changes and distributed leadership; lateral influence is more important than hierarchical, when it comes to culture; focus on the behavior and behavioral DNA of the enterprise are necessary to support new processes; infection of ideas and behavior; social movement; activism: "I'm doing" .... ", But what about you?"; correctly designed informality; be more silent, act more often invisible (act> speak); "The butterfly effect": small changes are multiplied a thousand times by the community of champions; Internal leadership engine (champions) support the whole team. Thus, the viral management can become an effective alternative to the old traditional "change management" in enterprises of any field of activity, including the coal industry.

References

1. Górny, Advances in Intelligent Systems and Computing, **605,** 92 (2018).
2. F. Voskoboynikov, Advances in Intelligent Systems and Computing, **586,** 380 (2018).
3. K. Klockner, Advances in Intelligent Systems and Computing, **604,** 115 (2018).
4. M. Blašková, R. Blaško, J. Rosak-Szyrocka, R. Ulewicz, Polish Journal of Management Studies **15(1),** 28 (2017).
5. Decision Management: Concepts, Methodologies, Tools, and Applications, **1-4** (2017).
6. A. D'Angelo, A. Majocchi, T. Buck, Journal of World Business, **51(4),** 537 (2016).
7. A. Mukhacheva, A. Sokolov, Economics of the region, **2:13**, 525 (2017).
8. Thomas J Holt, Adam M Bossler, Kathryn C Seigfried-Spellar, Cybercrime and Digital Forensics: An Introduction. – New York: Routledge, 103, (2015).
9. S.V. Kavun, I.V. Mykhalchuk, N.I. Kalashnykova, O.G. Zyma, Intelligent Decision Technologies. Smart Innovation, Systems and Technologies. **15:1**, 38, (2012).
10. L. Korchevska, G. Zhosan, S. Kavun, Journal of Finance and Economics, **4**, 98, (2013).

The Regional-Matrix Approach to the Training of Highly Qualified Personnel for the Sustainable Development of the Mining Region

Evgeny Zhernov[1*], *Evgenia* Nehoda[2]

[1] T.F. Gorbachev Kuzbass State Technical University, 650000, 28 Vesennyaya St., Kemerovo, Russia
[2] National Research Tomsk State University, 634050 36 Lenina st., Tomsk, Russia

Abstract. The state, regional and industry approaches to the problem of personnel training for building an innovative knowledge economy at all levels that ensures sustainable development of the region are analyzed in the article using the cases of the Kemerovo region and the coal industry. A new regional-matrix approach to the training of highly qualified personnel is proposed, which allows to link the training systems with the regional economic matrix "natural resources - cognitive resources" developed by the author. A special feature of the new approach is the consideration of objective conditions and contradictions of regional systems of personnel training, which have formed as part of economic systems of regions differentiated in the matrix. The methodology of the research is based on the statement about the interconnectivity of general and local knowledge, from which the understanding of the need for a combination of regional, industry and state approaches to personnel training is derived. A new form of representing such a combination is the proposed approach, which is based on matrix analysis. The results of the research can be implemented in the practice of modernization of professional education of workers in the coal industry of the natural resources extractive region.

1 Introduction

The purpose of the study is to define a modern theoretical approach to the system of training of highly qualified personnel as an important element of the sustainability of the economic systems of the Russian economy at the level of the resource-type region. The scientific novelty of the study is in the application of the regional-matrix approach to the training of highly qualified personnel in the coal industry of the natural resources extractive region, taking into account the features of the modern innovative stage of economic development. The scientific significance is in identification and systematization of the theoretical (conceptual) approaches to the system of training of highly qualified personnel as illustrated by the Kemerovo region - an important mining region in Western Siberia.

The practical significance of the study is related to the possibilities to use the proposed regional-matrix approach for justification, formulation and adoption of solutions on strate-

* Corresponding author: zhee.eti@kuzstu.ru

gic issues of sustainable development of the system of training of highly qualified personnel in the natural resources extractive regions of the Russian Federation. This goal is achieved by means of scientific substantiation of the necessity, opportunity and feasibility of linking the systems of training of highly qualified personnel with the regional economic matrix previously developed by the author (in co-authorship), into which they fit as parts of the economic systems distributed across the quadrants in the coordinates of natural resources - cognitive resources.

The relationship of this study with the previously published works. The existing approaches to personnel training, which seem relevant to the research topic, can be conditionally divided into three groups. The first group, a nationwide (state) approach in a program form, first of all concerns the development of federal educational and professional training standards. The second group is an industry approach, which today expresses mainly the private interests of owners of privatized primary commodity companies. The third group is a regional approach designed to integrate the interests of local authorities and the local community. Let us briefly characterize these approaches.

The state approach. Training of qualified personnel for high-tech industries and the primary sector of the economy as an indispensable condition for the successful innovative development of the knowledge economy is the most important task of the state. The relationship between the quality of public administration and innovations in the regions (in the European case) was investigated in [1]. It is necessary to recognize the opinion of experts that in the modern world the quality of the functioning of the national education system of a particular state can be regarded as a strategic state resource (in analogy with strategic resources of fresh water, energy resources, etc.). In accordance with the Strategy for Scientific and Technological Development of the Russian Federation until 2035 (hereinafter referred to as Strategy 2035), at the first stage (2017-2019) an integral system of sustainable reproduction and recruitment of personnel is being formed. Strategy 2035 is the basis for the development of state programs of the Russian Federation, state programs of its regions, industry documents of strategic planning in the field of scientific and technological development of the country, as well as planning and program-target documents of state corporations, state companies and joint-stock companies with state participation. In doing so one can take into account, for example, the English experience of integrating sustainable development in planning at the regional level described in [2].

In accordance with the Russian official document among the priorities and prospects for scientific and technological development of the country there is an increase in the efficiency of extraction and deep processing of hydrocarbon raw materials. Hence the need to train specialists in this field, and the Federal State Educational Standard of General Education, developed by the Ministry of Education and Science of Russia, should be coordinated with Strategy 2035 and the Program for the Development of the Coal Industry of Russia for the period up to 2030 (hereinafter referred to as Program 2030). In the latter official document, the interests of the state, regions, professional educational organizations, educational organizations of higher education, specialized scientific organizations and owners of enterprises in the process of improving vocational education are interlinked on the basis of federal state educational standards. Thus, the state program approach has been implemented to the coal industry located on the territory of a number of regions of the country, including the Kemerovo Region.

The industry approach presupposes the formation of an industry-specific training system, which includes state-private partnerships in education and science, vertically and horizontally integrated corporate university complexes, basic departments and laboratories. This approach reorients the centralized financing of education and science "in general" to support specific scientific and educational complexes and projects of the real sector of the economy. For example, the role of small firms in the presence of large laboratories in regions with sustainable innovative development is disclosed in [3]. Industry university complexes are designed to become points of growth in the country's economy based on new innovative principles of development. Recommendations for the training, professional retraining and advanced training of engineering and technical personnel for mining enterprises in accordance with the Concept of Improving the System of Training, Professional Retraining and Professional Development of Personnel for Coal Industry Organizations are given by Russian specialists. Among the most significant problems in the development of human resources are the lacks of industry-wide professional standards recognized at the national and international levels.

The regional approach. Scientists note the regionalization of the training of highly qualified personnel as a trend in the formation of the knowledge economy. A normative legal instrument that is able to ensure the unity of requirements in the professional sphere and to create conditions for professional growth of personnel, taking into account the state requirements for the education system at the federal and regional levels, is a professional standard. The regional component of the professional standard is based on the requirements for personnel taking into account the local socio-cultural, socio-economic and historical features of the development of the region.

The proposed regional-matrix approach to the training of personnel develops such paradigms of the region as the region-quasi-corporation and the region-society. Representation of the region as a quasi-corporation allows researchers to use economic categories for analysis. In this capacity, regions are considered as subjects of economic activity competing with other regions-corporations in the market of educational services. The approach to the region as a society - the community of people living in a certain territory - actualizes the issues of the reproduction of social life. Therefore, the paradigm of the "region-society", on the one hand, is more general than the economic one, on the other hand, requires the consideration of more factors, which complicates its application. The consideration should include not only educational, but also cultural, medical, socio-psychological, political and other aspects of the life of the population of the region. The studies [4, 5] show that, without the growth of human capital, sustainable economic growth in the region is not possible, moreover, the former generates the latter. This fact is taken into account by the proposed matrix: the separation and analysis of the mutual positions of the regions in the matrix are conducted by an integral set of criteria for the knowledge economy. The comparison of existing and proposed approaches in essence, which, we believe, is in the quality of training, is presented in Table 1.

Table 1. The comparison of existing and proposed approaches to personnel training.

An approach	The essence of the approach
State (federal)	The quality of training is determined by the degree of compliance with state standards
Industry	Criteria for the quality of training are specific to each industry, focused on preserving the best industry traditions in the field of training of highly qualified personnel
Regional	The quality of training is determined by the degree of compliance with the set of requirements developed by administrative structures that exercise control in the regional education
Regional-Matrix	The quality of training is determined by the degree of satisfaction of consumers, expressed in market categories of profitability of the educational organization and profitability for the consumer due to the constant improvement of all components of the vocational education process

Thus, the shortcomings of existing approaches have been identified and a new approach, favoring the sustainable development of the region, has been proposed, which is based on the dependence of the analyzed process on the specifics of the regional economy.

Theoretical foundations of the regional-matrix approach. An important factor in the modern training of highly qualified personnel, advanced training and retraining of specialists for mining and processing of solid minerals is their implementation in the conditions of the formation of a new economy. The use of the system of recurrent training and retraining of highly-demanded and highly-qualified personnel within the innovation cycle is the basis for the formation of a sustainably developing "new economy" or "knowledge-based economy". According to Nigerian scientists [6], knowledge economy is a "panacea" for sustainable development. The principles of re-education and retraining of professional personnel for the sustainable development of the economic systems of the regions are: 1) focus on the advanced areas of work; 2) consideration of the available educational potential and maintenance of continuity of training; 3) flexibility of forms of advanced training; 4) the advanced nature of teaching new progressive technologies; 5) optimization of costs for professional training and retraining through multi-channel financing.

The proposed economic matrix of the regions makes it possible to identify the distinctive features of the formation of a new knowledge economy at the mesolevel and to reveal the originality of this process in the mining region. We place Kemerovo Oblast as a typical mining region in the "natural wealth" quadrant of the proposed matrix. It characterizes the raw material economy of the region with abundant natural resources and with cognitive resources below the national average. Such an economic system is characterized by a narrow specialization based on the exploitation of non-renewable natural resources, the predominance of the raw-material orientation of the industrial-type economy. It has a high labour intensity of production with a relatively lower level of equipment and technology, limited investment in the development of the potential of an employee. As a result, human development indicators in the region, including the level of training, are below the national average. To ensure sustainable development of the region and its transition to the "harmonious wealth" quadrant, representing the knowledge economy in the natural resources extractive region, it is extremely necessary to improve the quality of training of innovative personnel in the leading sectors of the region.

For the methodological basis of our study, we adopted the UNESCO experts' statement that in modern conditions an important role plays local knowledge about the methods of using resources and social interactions that are the basis for sustainable development at the local level [7]. At the same time, Russian scientists rightly point out that along with local knowledge and skills there are also general universal knowledge and skills that are necessary for the development of natural resources. It is from the statement about the interconnectedness of general and local (industry, regional) knowledge follows, in our opinion, the understanding of the need for a combination of regional, industry and state approaches to training of personnel. Next, one should choose the form of representation of such a combination. A required new form is the proposed approach, based on matrix analysis. In the case of natural objects, location and geographical space are indeed key factors in explaining cognitive factors (knowledge, innovation). The proposed economic matrix of the regions shows this point clearly. Therefore, in order to analyze and evaluate the role of the processes of creating and disseminating professional knowledge in a particular industry (the coal industry in our case), we need to take into account the spatial (regional) features of these processes. The identification of spatial dependence in the analysis of regional economic development is presented in [8]. The same goes for firms, because they are located in a certain territory, where local knowledge is in demand above all. It is natural to assume that this knowledge is created and spread in the same place. Consequently, the role and place of the carriers of local knowledge transferred in the personnel training vary depending on the type of region (according to the matrix).

Thus, the proposed regional-matrix approach unites three approaches - state, industry and regional, those are implemented in a complex manner in separate territories of a single country.

2 Results and Discussion

The result of this study is the understanding that for sustainable development it is necessary to take into account the state, industry and regional interests in the training of personnel on the ground and to transfer knowledge of all levels, with the domination (d) of interests and knowledge of the corresponding system level (Table 2).

Table 2. Domination of types of interests and knowledge at a certain level of training of highly qualified personnel in the knowledge economy.

Interests and knowledge	System levels		
	state (federal)	**industry, including intrafirm**	**regional**
1. Interests			
1.1. Common	d		d
1.2. Private		d	d
2. Knowledge			
2.1. General	d		d
2.2. Local		d	d

This understanding is possible on the basis of the matrix, which determines the essence of the proposed regional-matrix approach to personnel training. The state-private and state-public partnerships in the field of personnel training based on the economic policy of transferring general and local knowledge and technologies to the industry of the region are the modern forms of the synthesis under consideration. This trend is typical for OECD countries [9].

The following propositions are the most important moments for organizing the regional-matrix training of specialists of the Russian Federation as illustrated by the natural resources extractive region. 1. The specific nature of the extractive region of the country in terms of natural and cognitive resources makes special demands on the training system. Among its most important elements, in addition to professional knowledge, are compulsory practices at local unique enterprises, the ability to act in extreme (emergency) situations, to comply with the rules of using regular means of self-salvage in accordance with international requirements. On the skills for sustainable development and the corresponding transformation of vocational education, see [10]. 2. The development of a scientific and innovative environment in regional universities, which is part of the "know-how carousel" or "know-how hub" [11], reflecting the current level of the world's coal mining machine building and fuel-energy complex, which allows to train personnel in Russia that meet the requirements of the world's leading power. Higher education, forming the intellectual community of the regions of the country, should not only revive the best domestic traditions of training of mining engineers, but also should become one of the attractive sides of Russia regarding the export of education. 3. Development of the educational system for the mining equipment industry and accelerated modernization of the technological base of the coal industry, development of new technologies for extraction, processing and supply to the domestic and world markets of coal products should proceed synchronously. This will generate a synergistic effect of the development of the interconnected industries of the region. 4.

Future development of large multi-level university complexes that unite technical higher and secondary educational institutions, small enterprises at universities, basic departments and laboratories at the leading enterprises of the region; development of multi- and single-industry colleges for the training of highly skilled workers. Such complexes are a source of general and local knowledge for local enterprises [12]. 5. Personnel policy, according to the regional-matrix approach, ensures the selection of personnel on the basis of professionalism and the conformity of education to the profile of the industry (agency), taking into account regional specifics. Without a radical restructuring of the training system, the economic systems of the regions of Russia will not be able to develop sustainably and to solve the socio-economic tasks entrusted to them by society.

3 Conclusion

This study has defined the basic theoretical propositions of implementation of the regional-matrix approach to the training of highly qualified personnel in the coal industry of the mining region of the Russian economy, taking into account the specifics and requirements of the modern innovative stage of sustainable economic development. These propositions, in our opinion, can be taken as a basis for justifying the system of training of innovative personnel of industrial enterprises in the mining region for its sustainable development. The training segment is the foundation for the formation and development of all spheres of the knowledge-based society, including the knowledge economy. Therefore, the sustainable development of the region as an economic entity of the new economy depends on the quality of personnel training. It is proved that the new system of training of highly qualified personnel of the leading coal-mining complex in Western Siberia can and should become an impetus to accelerate the innovative development of its economic system. The proposed development of the modern regional-matrix system of personnel training in the context of the knowledge economy of the leading industries of the region is based on a comprehensive approach to the modern organization of the educational process. It includes the establishment of vertically and horizontally integrated university complexes, state-private and state-public partnerships in science and education. It will ensure the targeted training of innovative personnel, demanded by the leading sectors of the economy of the mining region for its sustainable development.

References

1. A. Rodriguez-Pose, M. Di Cataldo, Journal of Economic Geography, **15**, 673 (2015)
2. D. Counsell, G. Haughton, Geoforum, **37**, 6, 921 (2006)
3. A. Agrawal, I. Cockburn, A. Galasso, A. Oettl, J. of Urban Economics, **81**, 149 (2014)
4. A.A. Batabyal, P. Nijkamp, J. of Evolutionary Economics, **23**, 909 (2013)
5. A. Barnir, Management Decision, **50**, 399 (2012)
6. A. Ogundeinde, O. Ejohwomu, Procedia Engineering, **145**, 790 (2016)
7. *What is Local and Indigenous Knowledge? United Nations Educational, Scientific, and Cultural Organization*, (Washington, UNESCO, 2017)
8. M.B. Pietrzak, J. Wilk, T. Kossowski, R. Bivand, Proceedings of the 8th Professor Aleksander Zelias International Conference on Modelling and Forecasting of Socio-Economic Phenomena, 135 (2014)
9. *OECD Science, Technology and Industry Scoreboard 2015: Science and Research Today*, (2015)

10. S. McGrath, L. Powell, International J. of Educational Development, **50**, 12 (2016)
11. J.G. Wissema, *Towards the Third Generation University: Managing the University in Transition* (2009)
12. H. Pinto, M. Fernandez-Esquinas, E. Uyarra, Regional Studies, **49**, 1873 (2015)

Improvement of the System of Training of Specialists by University for Coal Mining Enterprises

Vadim Mikhalchenko[1,*], and *Irina* Seredkina[1]

[1]T.F. Gorbachev Kuzbass State Technical University, Department of Economics, Vesennaya str., 28, Kemerovo, Russia, 650000

Abstract. In the article the ingenious technique of the Quality Function Deployment with reference to the process of training of specialists with higher education by university is considered. The method is based on the step-by-step conversion of customer requirements into specific organizational, meaningful and functional transformations of the technological process of the university. A fully deployed quality function includes four stages of tracking customer requirements while creating a product: product planning and design, process design, production design. The Quality Function Deployment can be considered as one of the methods for optimizing the technological processes of training of specialists with higher education in the current economic conditions. Implemented at the initial stages of the life cycle of the technological process, it ensures not only the high quality of the "product" of graduate school, but also the fullest possible satisfaction of consumer's requests and expectations.

1. Introduction

The modern development of the world economy demonstrates an increasingly growing competition, the quickening pace of changes and innovations based on the latest achievements of scientific and technological progress, which are inherently continuing in nature. The demand for new, the most sophisticated products is met with increasing speed. At the same time, consumer, showing selectivity, special preferences and expectations, insistence to the standard of goods or services provided, has become a key factor for the modern business success [12]. As noted by W. Scherkenbach, one of the Chief of "Ford" company, "The business process starts with the consumer. If the business was not started with the consumer - it will end with the consumer!"[11].

Strong manufacturers are crowding out the weaker one by offering products of higher quality at the best price and with better service to customers. At the same time, consumers may not be particularly retained even by those suppliers that fully comply with their level of requests. In the face of increasing competition, there is always a chance to get something that can be even better.

* Corresponding author: V.Mikhalchenko@mail.ru

Thus, the strongest positions in the market are received by those producers who have more complete knowledge about the consumer and, ahead of his expectations, continuously improve their production, managerial and social systems, thereby creating competitive advantages for themselves and ensuring long-term stability of their business. At the same time, the level of employees' professional qualification is one of the most important key factors in achieving success and strategic goals of the organization's development.

Under these conditions, in the system of relations with the labor market, the university should position itself not only as a "forge of staff", as a supplier of highly skilled human resources, but also as an active business partner what continuously "improves" its product and ensures full compliance with the level of training of young professionals according to the current and prospective needs of employers.

At the same time, as the results of a sociological survey showed, the employer, for example, a large coal mining company of the Kuznetsk Coal Basin of the Russian Federation, is not satisfied with the quality of training of specialists with higher education by universities. The consumer of the products of higher education institutions does not receive on a permanent basis that "product" that would fully correspond to its ideas about the high level of training of specialists in many aspects of quality which is necessary for the implementation of the organization's strategic goals. At the same time, in the "university-enterprise" system, there is often no direct interaction between the parties to solve the problems of the quality of training of specialists. On the one hand, the employer is not burdened with obligations, not interested and does not consider it necessary to notify universities about what they should pay attention to when forming the required qualities in the process of training of specialists with higher education; on the other hand, universities practically do not make research and do not take into account the needs and requests of real consumers of their products, and in the process of developing their curriculam and work programs rely, to a greater extent, on the compliance of the latest requirements of state educational standards and their capabilities [1].

In this regard, the task of building a system, within which the customer's requirements will be converted into the quality parameters of the product expected by him and, accordingly, into the quality parameters of the training of young specialists by the university, is particularly relevant. At the same time, this system should be able to react flexibly to rapid changes in the business environment, literally "pulling the requirements to the product" from the consumer and reconfiguring the labor, material resources and technological process of training of specialists according to the changing trends of the external environment, which fully corresponds to the most effective and recognized in world practice approaches in building production systems in modern economic conditions and, in particular, to the principles and concept of lean-production [13] and "active" (agile manufacturing) [6] production.

2. Theory

One of the effective tools that has been distributed in recent years and used by many successful manufacturing companies in the world in addressing the above issues is the Quality Function Deployment method [7, 9].

This method was used in 1972 in Mitsubishi Company for the first time, and since 1977 is actively used in Toyota production system (TPS). It is based on the step-by-step conversion of customer requirements ("consumer votes") into specific engineering, technological specifications of the product, organizational, functional and other components of the company itself.

Within the framework of this method, it is necessary to answer the question "How to do it?" in order to implement the list of customer wishes ("What to do?"). In other words, to deploy the quality function means to identify a set of technological parameters, factors and production conditions that radically determine the result (product) of productive activity in such a way as to satisfy the consumer completely.

It is possible to avoid (or at least minimize) adjusting the quality of the product after its appearance on the market by deploying quality at the initial stages of the product life cycle in accordance with the consumer's needs and wishes, and therefore to provide high value and at the same time a relatively low cost product (by minimizing the costs of repairing the spoilage) [2]. The methodology of the Quality Function Deployment has shown its high efficiency for companies operating in a volatile and highly competitive external environment and, of course, can be successfully applied to the conditions of the educational process.

3. Discussion

The product of a university is a specialist with the necessary knowledge to ensure the realization of the company's goals. The process of training of specialist passes through certain stages and can be considered as a technological process with a certain set of technological parameters, conditions, functions. It is possible to get a product that fully meets the employer's needs by managing these components, constructing a meaningful set of elements of the studying process. This task is especially relevant for the companies operating in the field of knowledge. Currently the conversion rate is so high that, according to some statements, total amount of knowledge doubles on average over ten years (at the time of Lomonosov over 150 years), and 30% of knowledge obtained in university becomes obsolete immediately after graduation. In this regard, it is necessary to make changes continuously in the educational process and constantly "be in touch" with the consumer.

The present study proposes to use the following algorithm for improving the system of training of specialists with higher education by university.

At the first stage of the Quality Function Development ("market research"), the team of experts (the quality management team from the university) should identify key consumers and, with the help of selected methods (survey, analysis of statements in the media, complaints, etc.) formulate and systematize their main requirements ("consumer voice") to the quality of training of specialists with higher education. There is a "pulling" of the value of the product that is required by the consumer at the moment. As an example of such work, one can cite the "quality profile" of training of specialists with higher education for the enterprises of the coal mining complex, formed as a result of an earlier sociological study [1]. With the further deployment of the quality function from the whole variety of obtained quality parameters it is necessary to select the most important for the consumer. The rating of the quality parameters can be generated using the results obtained during the application of the Kano model [3, 4], ABC method, construction of the Pareto diagram [10]. The most important quality parameters are assigned the highest score of the rating, the least important quality parameters - the lowest score.

At the same time, a comparative assessment of university is conducted with its main competitors relative to the level of implementation of the identified quality parameters. This work is necessary to clarify the situation according to which consumer requirements the position of the university is strong and it is not required to spend time and resources on their satisfaction at this stage.

In further work on the deployment of the quality function, it is necessary to establish how much the supporting processes affect the level of implementation of the quality parameters. Evaluation of the communication tightness can be based on the experts experience, consumer responses or results of specially conducted experiments. The technical part of this work is the construction of the "product design" matrix (Figure 1) [8].

Figure 1. Product Design Matrix.

		Supporting processes				Evaluation of competitors
Quality parameters – "consumer voice"	Rating of importance for the consumer	Usage of active and interactive forms of learning	Involvement of students to participate in research	Practical training of students	Self-preparation of students	Competitor 1
Professionally significant qualities of personality	5	⊙	⊙	⊙	⊙	2
Personal qualities	5	⊙	⊙	O	⊙	5
Special knowledge	5		⊙	⊙	O	4
Practical training of specialists	4	O	⊙	⊙		3
Absolute weight of the significance of the relationship		102	171	141	105	
Relative importance weight of the relationship, %		19.7	32.9	27.2	20.2	
Key indicators of effectiveness		Specific weight of hours of classes, conducted in active and interactive forms	Proportion of students participating in the research	Proportion of hours spent on internships (as a separate subject)	Specific weight of hours of students independent work	
Unit of measurement		%	%	%	%	
Current value of key indicators of effectiveness of the supporting processes in the university		15	10	1	30	
Values of key indicators of effectiveness of the supporting processes in a university-competitor		25	32	1	30	
Objectives of implementing the supporting processes of the university		25	20	15	30	
Estimation of the implementation complexity of the target installations of the supporting processes		6	6	8	1	

⊙ - strong - 9 points, O – average – 3 points, Δ - weak – 1 point

In the future, the most critical (priority) processes are selected from the whole variety of processes which ensure the realization of quality parameters. It is achieved by calculating the indicator of the *"significance of the relationship"* and the choice of processes with the highest significance of the relationship value:

$$Z = \sum_{i=1}^{n} strength\ of\ relationship \times customer\ rating\ of\ quality\ parameters$$

In the shown example, the involvement of students to participate in research and practical training of students are the university processes that highly influence on the quality formation of specialists with higher education which are valuable for the employer.

From the example (Figure 2), it can be seen that improving the process of practical training of students may require the development of basic educational programs in terms of planning students' training at real workplaces in conditions of real corporate life, methodological and consulting support for such training by faculty, building relations with companies in order to organize on their basis a long-term internship and thereby provide students with the training which meets current industry standards.

At the **next stage** – "production design" - tasks, rules, procedures, instructions, assignment are formulated, job descriptions, regulations of educational, administrative and other processes of the university are developed, equipment, mode of operation, methodological materials, quality control tools are selected.

Figure 2. Process Design Matrix.

		Supporting processes				Evaluation of competitors
Supporting processes	Rating of quality parameters	Creation of the database with companies which are ready to provide students with long-term internships	Conclusion of agreements with companies to provide long-term internships to students	Selection of students for long-term internships in the companies	Inclusion of long-term internship hours into basic educational programs	Competitor 1
Practical training of specialists	0.272	O	⊙	O	⊙	3
Values of relationship		0.816	2.448	0.816	2.448	
Parameters of subprocess		Number of companies providing internship placement	Number of concluded agreements on the provision of internship placement	Amount of students selected for internship	Proportion of hours spent for internship in the total amount of training hours	
Unit of measurement		unit	unit	unit	%	
Current value of sub process parameters		270	10	1	1	
Objectives for implementing sub processes		400	100	30	3	
Complexity level of implementing target installations of sub processes		4	5	7	7	

4. Conclusion

Conversion of the customer requirements and expectations into specific parameters and characteristics of the "product" of higher education through continuous organizational and functional transformation of the university itself is a relevant task of the modern period of world economy development.

An effective tool for solving this problem is "Quality Function Deployment" method, which, in the context of continuous changes in the parameters of the external environment, allows not only "to extract value" from the consumer, but also to determine the directions for improving the technological process of creating the product required by the consumer, and includes four stages: "product planning", "product design", "process design", "production design".

Success in improving the system of training of specialists with higher education for coal mining enterprises and various sectors of the economy depends to a large extent on how in the university:
- all its functions will be interacted;
- communications are organized between participants in the technological process of training of specialists;
- organizational management system of the university is built;
- system of staff motivation has been formed,
- technology of knowledge transfer will be developed and improved.

The efficiency and competitiveness of the university in the current economic conditions will depend on how effectively the university is able to solve the examined problem.

References

1. V. Mikhalchenko, I. Seredkina, E3S Web Conf.**15**, 04005 (2017)
2. AkaoYoji, Ohfuji Tadashi, Naoi, *Tomoyoshi, Survey and Reviews on Quality Function Deployment in Japan*, (Tokio, JHHG, 1987)
3. N. Kano, *Guide to TQM in Service Industries* (Beijing, Asian Productivity Org, 1996)
4. J. Moorman, Proceedings, 882 (2012)
5. J.R. Evans, *Total Quality Management, Organization, and Strategy 4th edition* (South-Western, part of the Thomson Corporation, 2012)
6. P.M. Noaker, Manufacturing Engineering, **113**, 40 (1994)
7. L .P.Sullivan, Quality Function Deployment, **2**, 39 (1986)
8. J. Marsh, *The Continuous Improvement Toolkit* (B.T. Batsford LTD, London, 1998)
9. James L. Brossert, *Quality Function Deployment* (Milwaukee, WI: ASQ Quality Press, 1990)
10. R. Koch, *The 80/20 Principle: The Secret of Achieving More with Less* (London, Nicholas Brealey Publishing, 1998)
11. H. R Neave, *The Deming Dimension* (London, Willey, 2016)
12. Kenichi Ohmae, *The Mind of the Strategist the Art of Japanese Business* (Tokio, JHHG, 2015)
13. J.P. Womack and D.T. Jones *Lean Thinking. Banish Waste and Create Wealth in Your Corporation* (New York , Free Press, 2003)

 DOI: 10.1051/e3sconf/20172104007

The Concept of Resource Use Efficiency as a Theoretical Basis for Promising Coal Mining Technologies

Vadim Mikhalchenko[1*]

[1] T.F. Gorbachev Kuzbass State Technical University, 650000, 28 Vesennyaya St., Kemerovo, Russia

Abstract. The article is devoted to solving one of the most relevant problems of the coal mining industry - its high resource use efficiency, which results in high environmental and economic costs of operating enterprises. It is shown that it is the high resource use efficiency of traditional, historically developed coal production systems that generates a conflict between indicators of economic efficiency and indicators of resistance to uncertainty and variability of market environment parameters. The traditional technological paradigm of exploitation of coal deposits also predetermines high, technology-driven, economic risks. The solution is shown and a real example of the problem solution is considered.

1 Introduction

The current development of the world economy is accompanied by ever-increasing demands for energy. At the same time, coal, as the energy base of the first industrial revolution, more and more is losing its ground to oil and gas, as more technologically sophisticated, cheaper and more environmentally friendly sources of hydrocarbon energy. Since the middle of the XX century, when the share of coal in the fuel and energy balance of Russia reached 65%, it has been steadily declining and currently it is just only 12-13%.

Add to this the modern period is characterized by an intensive quest for solutions to create efficient power plants based on alternative hydrocarbon sources. A number of forecast estimates indicate the near completion of the "era of hydrocarbons". According to the forecast of the famous futurist Ray Kurzweil, technical director of Google, by 2028 the "green" energy (sun, wind, water streams, geothermal sources) will become one of the key technologies for the development of all others. It will be approved as the dominant concept [1].

Expected scientific and technological breakthroughs in solar energy will allow to get a sharp drop in prices for solar panels by 2020 and reduce the turnkey price of solar power from $ 2.5 to $ 0.8-1 per watt, which will make possible to generate a "green" electricity at a lower price than currently it is possible to get the cheapest one from coal-fired power plants [2].

* Corresponding author: V.Mikhalchenko@mail.ru

In this regard, the problem of increasing the efficiency and, consequently, the competitiveness of coal mining production, acquires an extremely high relevance and requires its solution.

2 Materials and methods

An analysis of the underlying causal factors that limit the opportunities for increasing the efficiency of open coal mining enterprises developing steeply falling, inclined and gently sloping coal deposits using technology with the use of a deepen-longitudinal development system (the currently widely used in open coal mining technology) has revealed the following two main prerequisites, which are underlying principles in decision-making, both at the design stage, and in practical activities in mining operations during the exploitation of coal deposits.

The first is the prerequisite of the "transfer of mined rock". According to which it is supposed that the technology includes extracting the rock mass, transporting it for a considerable distance and storing it outside the open-pit mines.

The second is the prerequisite of the "rigid geometry" of the quarry. In this case it is assumed that the quarry is formed as a certain geometric volume with a rigidly predetermined form and relations between the parameters of this volume. The deterministic development model is based on a rigid spatial-temporal relationship between the decisions of the current time period and the possible parameters of the quarry and coal mining technology in the very distant future, up to 50 years or more.

It is these two prerequisites, which implicitly present in most optimization tasks, form the structure of the relationship between the target characteristics (efficiency criteria) and the technology parameters. If to track the connections of these prerequisites, the following closed logical structure is revealed: the principle of "transfer of mined rock" is predetermined by the principle of "rigid geometry" of the quarry; in turn, the "rigid geometry" of the quarry is a consequence of the requirement to maximize the extraction of balance reserves; the estimation of reserves is carried out, proceeding from the precondition of technology application based on principles of "transfer of mined rock" and "rigid geometry" of the quarry.

The realization of these prerequisites generates a high resource use efficiency of coal production (the cost of resources reaches - fifty percent or more in the structure of costs in the coal mining enterprises), its high inertia and high technologically caused economic losses and risks, and as a result, it limits the realization of the possible potential of the open coal mining efficiency.

Among the noted characteristics, it is necessary to highlight such an important in the present circumstances economic parameter as a risk.

In the management literature, the risk is interpreted as the possibility of undesirable deviation from the goal, which the management entity seeks and for the sake of which it invests its resources in the business.

The definition of uncertainty is closely related to the definition of risk. Uncertainty is described as incompleteness of knowledge about the factors that affect the possibility of achieving the goal. In this interpretation, uncertainty acts as a causal factor, as a prerequisite for the emergence of risk. Or, if to use a figurative definition, uncertainty is a habitat for risk. It is obvious that the higher the level of parameters variability of the economic environment, the higher the uncertainty, and, consequently, the risks.

At the design stage of the future enterprise, the initial information uncertainty has the highest possible values, i.e. our knowledge of the parameters of the market environment in the future is zero and, consequently, the probability of a wrong decision is exceptionally high. In this case

 DOI: 10.1051/e3sconf/20172104007

decision making will be similar to random actions, that is, the possibility of making the wrong decision is very high, and thus the risk of error with negative consequences is very significant.

This is very important for the development of the theory and practice of open coal mining. The fact is that economic losses and risks start to form at the stage of project development of the future enterprise, when a set of components of the production system is determined and connections between them are established, i.e., the basic constructive decisions are made and the structure of the future enterprise, its main parameters and characteristics are determined. At the same time, the basis for theoretical approaches to determining the parameters of currently operating enterprises is based on the above prerequisites in the form of a basic methodological principle of an "one-off" optimization of all constructive and production solutions distributed throughout the life cycle of an enterprise, the duration of which, based on the size of industrial reserves and the normative values of production power, is generally 50 or more years. From the point of view of optimization theory, the use of this principle can be interpreted as an attempt to build an optimal trajectory of the enterprise for many years ahead with a rigid fixation of the final state of this trajectory. At the same time, when it comes to the development of gently sloped, inclined and steep deposits, the ultimate depth of open development is the fundamental parameter that determines the entire set of cause-effect technical, technological, organizational, managerial and economic relations of an open coal mining enterprise.

Thus, we are talking about a "rigid" spatial-temporal fixation of the boundaries of the quarry at the time of mine abandoning, determining the parameter of which is the final depth of the quarry, is chosen "optimally" at the design stage for a very remote perspective, based on the condition of the maximum net discounted income for the whole life cycle of a coal mining enterprise.

Decision making on the final boundaries of the quarry, which should be achieved, as already noted in 50 years or more, determines and limits within the whole life cycle such structural relationships and economic parameters of the enterprise as the dependence of the transportation distance of the mined rock and the current stripping ratio, variable costs and production assets of the enterprise from the current and maximum depth of the quarry.

The fact is significant that the "rigid" fixation of the boundaries of the quarry along the surface across the strike and, as a result, the "rigid" spatial fixation of the overburden dumps location is a source of significant losses, the appearance of which is determined by design decisions. These decisions predetermine the occurrence of additional operating costs for moving host rocks beyond the outline of the final position of mining operations at the initial and medium stages of the enterprise's life cycle, i.e. for many decades, in the interests of coal mining in very remote periods of exploitation in the development of steep and, in the case of placement of rock dumps from the hanging side of the deposit, inclined and flat coal seams.

In addition, the specificity of forming the parameters of the quarry working area, predetermined by the need to withstand the angle of the slope of the working side within 15–20°, is such that it is necessary to make additional current investments in capital mining and overburden work carried out in the interests of "future periods". This predetermines an increase in these losses, in the form of an additional expenditure of resources, as well as a high inertia of the production system, which, under conditions of low reliability of the forecast of economic parameters of the external environment, generates significant economic risks.

As a result, the effect of "high resource use efficiency" and "technologically conditioned" capital linking is realized, and thus the conflict of "effectiveness-risk" criteria proves to be predetermined and embedded in the production system long before its appearance, even at the stage of its design, at the stage of making decision on final boundaries of open-pit mining, production capacity, parameters and direction of development of the working area and, as a consequence, parameters of production processes [3].

 DOI: 10.1051/e3sconf/20172104007

The use of the above methodological approach at the design stage of the enterprise is predetermined by the above, as well as by the economic prerequisites typical for the conditions of a centralized planned economy, which are basic: high stability and predictability of business environment parameters, guaranteed sales and payment for all goods produced, low value of economic criteria, availability and low cost of capital use.

However, at the present stage of the development of a market economy, it is necessary to review historically established "rules of the game", i.e. the paradigm that determines the rules and principles of open mining. At the same time, the most promising search directions are based on the revision of historically developed approaches to designing the main parameters of quarries and the rejection of the prerequisites such as "transfer of mined rock" and "rigid geometry" of the quarry.

3 Results and discussion

Consideration of current trends in the development of open-pit mining technology confirms what mentioned above. First of all, it concerns the revision of the principle of "transfer of mined rock". In the literature, technological schemes that are based on limiting the operation of this principle have been described for a long time back. It was most clearly reflected in technological schemes based on the placement of stripping within the open-pit (a block method for the development of coal strip mine and a transverse development system). However, these technologies have not been disseminated in practice, because could not provide a qualitative improvement in the indicators of environmental and economic efficiency in the initial period of coal field operation. Analysis of recent publications reveals a clear revival of the interest of scientists and practitioners to this type of technological solutions.

It is characterized by studies aimed at eliminating shortcomings inherent in these schemes, due to the simultaneous refusal of the prerequisites of the "transfer of mined rock" and "rigid geometry" of the quarry. In particular, the deep-solid transverse development system, protected by the copyright certificate [4], is based on the transition to the principle of "flexible", adaptive tactics of forming the geometry of the quarry, with a complete rupture of the rigid connection between the future quarry parameters and current period decisions, which makes it possible to implement internal dumping since the first years of coal field operation, to significantly reduce the distance of overburden transport and, as a result, to provide a significant reduction the resource intensity of production and the increase in economic and environmental efficiency.

The essence of technology is as follows. The development of a coal seam or series starts from the flank of the open field. In this case, the area of overburden and mining operations are located across the deposit bedding, move it along the line of its strike in the direction of the opposite flank of the quarry field (Figure 1).

The overburden from the first stope is laid on the surface in front of the working side 3 quarry, and after the coal is extracted and the initial capacity is created, the stripping rocks and the temporary dump are placed in the formed goaf 4, filling it.

The formation which is suitable for reclamation is removed from the territory which is spoiled by mine workings and applied to unused surfaces of the internal dump 5.

Deepening of mining operations is carried out step by step through the cutting of new horizons through gaps that provide the formation of the bottom of the 6 quarry with the front-to-back slope $\alpha_д$. The front-to-back slope of the quarry bottom is determined from the condition of ensuring the stability of the internal dump.

Before the beginning of the backfilling of the internal dump, a slope of the shaft 7 is constructed on the specially created path on the bottom of the quarry from the soil side of the formation, later blocked by a dumping massif. Underground preparatory workings 9 are carried out to extract the conserved reserves of minerals 8 from under the dump from the shaft. The coal is extracted by technology with the laying of the worked out space, which prevents deformation of the reclaimed surfaces of the internal dump.

Crushed stripping rocks can be used as a packing material. The minerals mined by the underground method are transported by conveyors through workings 9 and shaft 7 to

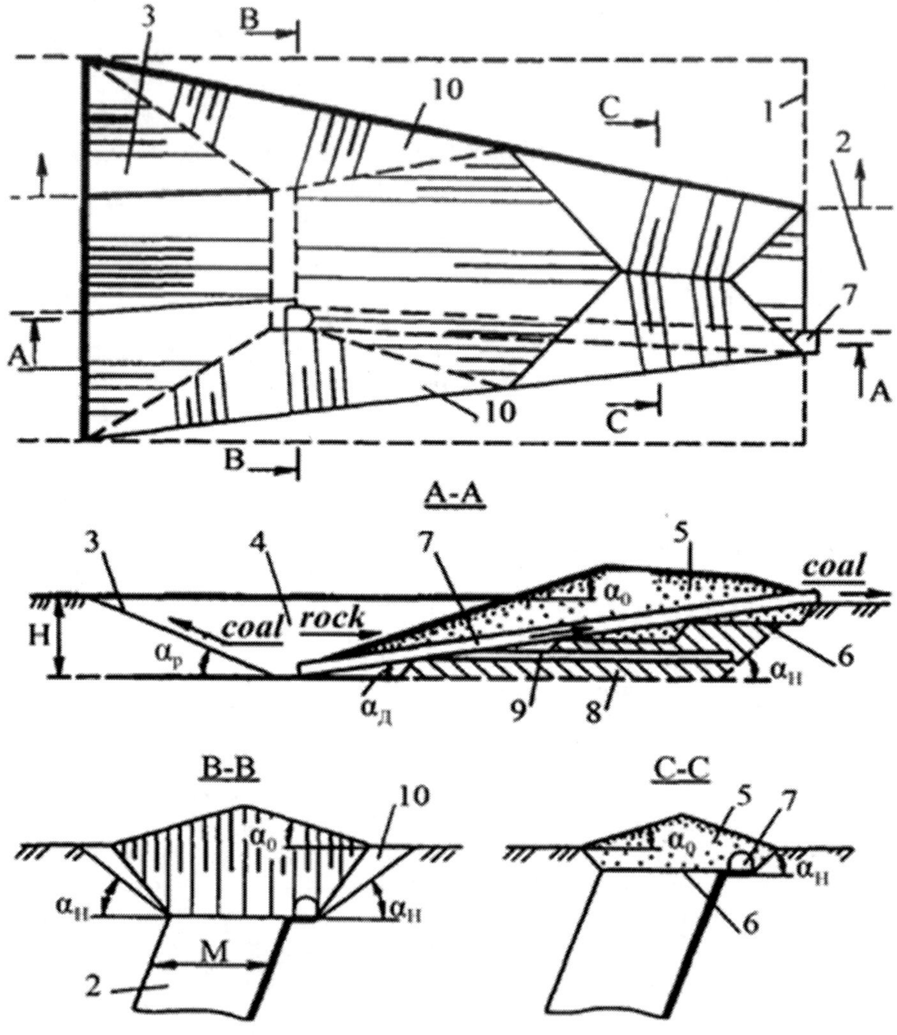

Fig. 1. Technological scheme of mining the deposit with the deepening of mining operations along the line of its strike and the continuous reclamation of the surface.

the surface. With moves forward and deepening of open mining operations, the length of the slope of shaft is increased, and new excavation fields are cut at a greater depth.

The overburden removal from working horizons of the quarry is carried out by dump trucks through haulage bench along the non-working sides 10 to the corresponding stage of the internal dump. The temporary automobile exits are being built for the moving of minerals from the quarry on the working and non-working sides. The processing of lower ledge is worked by a backhoe type excavator with its placement and vehicles on the upper platform, which excludes finding equipment at the bottom of the quarry and the danger of its damage by rock slipping down the slope of the lower stage. Lowering of mining operations is carried out to the final depth H, after which the quarry operates with constant parameters of the working zone. At this stage the technogenic surface acquires a smooth terrain and is lowered practically to the level of the surrounding territories, which makes it suitable for creating highly productive agricultural lands. In case of revision of the lower boundary of effective development of a layer by an open method, a further depth of the quarry is made by cutting the subsequent horizons.

The reduction in the transportation distance of the mined rock due to a change in the direction of cargo flows is typical for "adaptive" development systems, since in this case a technological scheme with practically 100% internal dumping is provided, which gives a significant increase in the dynamic characteristics of the production system. The duration of the production cycle in comparable geotechnological conditions with the use of "adaptive" systems can decrease several times in comparison with the currently used deepen-longitudinal development system. The presence among corporate associations of coal enterprises that use "flexible" and "adaptive" development systems will allow servicing more than 90% of market demand in the "just in time" mode.

In addition, it is necessary to emphasize the fact that increasing the efficiency of open coal mining will create the conditions for revising the decision on the final boundaries of the quarry towards their significant increase in depth and, thus, expanding the resource base of open coal mining.

The conducted studies have shown that the considered option of the deep-open transverse system of open development ("flexible" technology) also allows eliminating the conditions for the emergence of high technologically-caused economic risks, stabilizing the mining regime, significantly reducing the negative impact of mining production on the environment and very significantly, in connection with a multiple reduction in mass transfer, reducing the resource intensity of production.

This is fully in line with the worldwide trends in the development of various industries, as, in particular, one of the reports to the Club of Rome [5], which justifies the concept of "resource productivity" and the relevance of issues of significant reduction in resource consumption. In addition, the proposed approach coincides with the dominant direction of development of modern production systems in various sectors of the economy and, in particular, with the concept of creating "lean" production systems [6, 7].

The noted positive characteristics of the deep-continuous transverse development system make it possible to consider it as a basis for the creation of coal-mining enterprises with a higher level of efficiency, ensuring the transition of the industry to the implementation of the strategy of resource and nature conservation, the strategy of economically efficient adaptation to the changing environmental conditions. At the same time, it should be emphasized that, in accordance with the fundamental principle of "correspondence of the level of complexity of the management system to the degree of variability of the management entity and its environment", effective implementation of the proposed "adaptive technologies" will require a transition to a qualitatively higher level of organization and management of coal mining.

It should also be noted that the deep-continuous transverse development system is only one of a large number of potentially possible options for implementing the principle of adaptive, evolutionary development of production coal mining systems, flexible development of the quarry. In accordance with the methodology of technical knowledge, the above result is achieved through transformations based on a change in the design of the production system, however, with the preservation of the structure of the technological chain, that is, preserving the traditional composition of the production processes of open coal mining.

However, an even greater effect can be achieved through transformations based on changes in the principles of building and operating coal mining systems. The transition to an open-underground coal mining method may be an example of such a solution [8]. Enlarged economic estimates show that in this case the efficiency can increase by more than 4 times.

Obviously, under the influence of high competition in the energy market, the search for solutions and approaches that are enable further enhancement of the efficiency of open coal mining will continue.

4 Conclusion

The viability of socio-economic systems, as well as biological ones, is determined by their ability to adapt to the changed conditions of the external environment. With regard to production systems of open coal mining operating in uncertain and volatile conditions of the market environment, the transition to "flexible" adaptive models for the development of quarry will allow to come close to realizing the idea of resource saving and the concept of "lean production" and, as a result, to ensure a substantially higher profitability of production and the level of its environmental safety, which, undoubtedly, will enhance the competitive advantages of the coal-mining enterprises in the world energy market.

References

1. R. Kurzweil, *The Six Epochs. In The Singularity Is Near: When Humans Transcend Biology.* (New York: Viking, 2005)
2. D. Rifkin, *The Third Industrial Revolution: How Lateral Power Is Transforming Energy, the Economy, and the World,* 304 (St. Martin's Griffin, 2013)
3. V. V. Mikhalchenko, Y.T. Rubanik, IOP Conference Series: Earth and Environmental Science, **45**, 012012 (2016)
4. V.V. Mikhalchenko, S.A. Prokopenko, V.G. Orlov, A.V. Koksin, *Method for the development of stretched along the strike inclined and steeply falling deposits of high power,* **25** (1992)
5. *Factor Four: Doubling Wealth, Halving Resource Use: A Report to the Club of Rome,* 320 (London: Earthscan, 1998)
6. V. Mikhalchenko, Y. Rubanik, N. Osokina, A. Mikhalchenko, The 8th Russian-Chinese symposium coal in the 21st century: mining, processing and safety, 10 (2016)
7. J. P. Womack, D.T. Jones, *Lean thinking. Banish waste and create wealth in your corporation,* 315 (New York, Free Press, 1996)
8. V.V. Mikhalchenko, A.A. Atrushkevich, *The method of open-underground mining of stretched along the strike inclined and steeply falling deposits of high power,* **35** (1999)

DOI: 10.1051/e3sconf/20172104008

Ideological Paradigms and Their Impact on Environmental Problems Solutions in Coal Mining Regions

Vladimir Zolotukhin[1,*], *Natalia* Zolotukhina[2], *Maria* Yazevich[1], *Alexey* Rodionov[1], and *Marina* Kozyreva[1]

[1] T.F. Gorbachev Kuzbass State Technical University, Department of history, philosophy and social sciences, 650000 28 Vesennya st., Kemerovo, Russia
[2] T.F. Gorbachev Kuzbass State Technical University, Department of chemistry and technology of inorganic substances, 650000 28 Vesennya st., Kemerovo, Russia

Abstract. The work presents the analysis of the influence of the ideological paradigms of technology in solving environmental problems in the conditions of modernization of production, where a significant role is given to the protection and preservation of the natural environment and the protection of the rights of a citizen to a favourable environment. The attention is focused on the fact that in a civilized society, the needs of the individual and society are formed within the right social and cultural fields. The main importance of the regulation of everyday reality allotted to the phenomenon of law as a certain ideological paradigm. Stressed that at different stages in the coexistence of ideology and power detected General forms and principles of their influence on social practices, and on consciousness. The article substantiates the position that any interaction of subjects of economic activities leads to the existence of "constructive tension".

1 Introduction

In the conditions of social transformation, a significant role is assigned to the protection and preservation of the natural environment [1-3], the protection of the rights of a citizen to a favourable environment especially for the problems associated with the introduction of new technologies for modernization of industrial enterprises [4-5].

The solution to these problems should be accompanied by a change of legislation at both the Federal and at the regional level, moreover, having the status of "coal producing regions". A well-developed legal framework in the sphere of protection of the environment, within a particular ideological paradigm. On the one hand, the fact that recognition of the possible coexistence of ideological paradigms and the multidimensionality of the social environment allow to speak about existence of the boundaries of the regional society awareness of environmental issues. On the other hand it is accompanied by certain institutional bodies implementation and monitoring of government regulations.

* Corresponding author: zvm64@mail.ru

2 Materials and Methods

To study the influence of the ideological paradigms of technology in solving environmental problems was used to work C. Taylor, M. Weber, Parsons, M. A. Barg, etc. To identify specific relations of ideology and power, ideology and knowledge, as well as the role and importance of it in society contributed to the work of J. Lyotard, M. Foucault, B. Latour, J. Habermas. Analyzed the works and normative and legal acts in the field of environmental protection and the protection of human health and helped to identify the most important social phenomena and methodological characteristics specific to current research in this area. In describing the modern approaches to the consideration of ideological paradigms, their influence on the solution of ecological problems and to identify characteristics of the specific changes of modern technologies, methods were applied: comparative and rational-logical analyses, structural-functional approach, conceptual analysis, historical and logical.

3 Results and Discussion

In a civilized society, the needs of the individual and society are formed within the right and socio-cultural fields, In this regard, corporate ethics and rules of social behavior are the regulators of this interaction. The main importance of the regulation of everyday reality allotted to the phenomenon of law as a certain ideological paradigm.

In the study, under the ideological paradigm (ideology) in the tradition of Karl Marx and Karl Mannheim we understand the illusory form of consciousness, which expresses the interests of certain social groups. The essential features are its irrational and symbolic in nature, dominated by emotional, fantasised and figurative and symbolic components, which makes the ideology is largely similar to the myth and religion (E. Taylor, M. Weber, Parsons, M. Barg, S. Zhizhek, W. Matts, A. Shils, etc.).

In works of J. Lyotard, M. Foucault, B. Latour, A. Kravets, E. Mamchur and others, focuses on the relations between ideology and power, ideology and knowledge, ideology and policy, questions about the role and importance of it in society. It should be emphasized that at different stages in the coexistence of ideology and power can be used to detect a number of General forms and principles of their influence on social practices. The use of certain means and methods of influence on social consciousness [7], the promotion of their own interests, as well as ways to achieve economic, legal and political goals.

Draws attention to the fact, that the classical ideology of liberalism is the result of certain philosophical doctrines, social needs and economic processes. It is based on the principles of pragmatism, individualism, which contributes to the implementation needs of the community of entrepreneurs in the conditions of emerging capitalism in the tradition of K. Marx, M. Weber, K. Mannheim and others, which is the formation of a special installation relations of man to the world, where it manifests its desire to "set the measure of things" [6] and to control reality.

The recognition of ideological pluralism and multidimensionality of the social environment [9] in various aspects of its manifestation (socio-economic [18], environmental [10, 11], technology [8, 17], management [11, 15], etc.) with the existence of differences and contradictions (defining the ontological possibility of conflict) allow to speak about presence of edges in the awareness of environmental issues (environmental space) as a measure of their certainty. This is due to the presence and/or absence of environmental consciousness, the need of defining the boundaries of "Green ecology" in the framework of the mutual interest of the different actors in its formation from the point of view of manifestations of citizenship with the balance of incentives and penalties.

Having distinctive values and semantic units, each ideological paradigm (ideology - liberal, conservative and socialist (communist), etc.) is focused on creating specific to the state and/or territorial entity rational-logical concept descriptions of social life, in particular specific paradigm of ecological space. This paradigm requires the availability and the possibility of using not only theoretical and conceptual apparatus, but also of modern technologies in the field of ecology. The reality is not only knowable, but "used" by man. Of particular importance here is the "historicity" of human existence, the certainty of his place and time – the "situation" in which man finds himself, including at the national and ethno-geographic levels.

Every interaction of subjects of economic activities leads to the existence of "constructive tension" (A. Akhiezer). In this society, including its environment (natural, technological, etc.) should be implemented the possibility of its preservation against the flow of entropy processes. In this case actualizarea the need of seeking consensus as a form of activities on the basis of a recognised (accepted) by all of the rules in the dialectical opposition. Consensus as the absorber of social tensions is implemented in socio-economic activities and contributes to the compromise of all interested parties (economic entities, bodies of state and municipal government and administration, citizens, civil society, etc.). A compromise based on the recognition of certain common interests and strategic objectives is achieved through consensus as a specific form of resolution of environmental conflicts. Consensus involves understanding that the satisfaction of their own interests possible in light of the interests of the "other". In the foreground the anthropological nature of the struggle for the preservation not only within society but also the environment of man himself and his entourage. On this basis, actualizarea the problem of choice of values in one (liberal) or more ideological paradigms.

The preservation and reproduction of values is possible within the socio-cultural space, including the legal culture. Under consideration of the relationship of social injustice in society, it deals with "cultural or symbolic" (Frager N.) injustice. The latter expresses the prevailing cultural domination of one (the forms) of culture on the other, the non-recognition of a particular cultural tradition in terms of the dominant culture and finally contempt, i.e., the systematic negative stereotyping and discrimination of a particular cultural practice in the field available to the mass culture [18]. The elimination of this type of injustice linked with transformation of a common cultural environment, transformation, which is based on the principles of recognition and support of cultural diversity. Within reach mutual consensus, in the words of M. T. Stepanyants "played a significant role, a sober based on the survival advantage of both the contending parties by refraining from the usual at "meetings" of civilizations strategy for the physical destruction of "hostile parties" and the violent eradication of its culture. There was a preference for gradual reconciliation through tactics of inclusion" [12]. That is another proof of the idea that there are no rules and regulations cannot be taken in finished form - they must be checked, missing person through itself, through its constructive experience.

Kemerovo region takes a leading position in coal production in Russia, which, ultimately, indicates the high ecological fragility of the region. It should be noted that a high level of development of the coal and mining industry has no less negative impact and the landscape of Kuzbass. We are talking primarily about the "ridges" dumps, runs through the whole territory of Kemerovo region. According to the 2014 disturbed area in the given subject of the Russian Federation is not less than 100 thousand hectares, which is 10% of the territory of the Kuzbass coal basin. In addition, the resulting tailings mountain parody pollute the air considerably as a result of weathering. One of the methods in the current technological situation can be land reclamation, namely, artificial reforestation of the resulting coal dumps [17].

Today, the reclamation of "carbon ranges" as the method of recovery of disturbed ecosystems, in this applied subject. In the Kuzbass, attempts were made to create a structure dedicated to the management and control of the reclamation of land disturbed by coal mining. So, was organized by the Fund of assistance to the coordination Council for development of the coal industry, labor protection, industrial and environmental safety in the Kemerovo region [15], but in October 2013 the Fund was liquidated.

Land reclamation is carried out at the expense of businesses engaged in coal mining (in 2016 - 126 million rubles., in 2017 it was planned for approximately 400 million rubles). It is also planned to apply new methods of remediation that will definitely improve the quality of the human environment.

Thus, under the decision of environmental problems in coal mining regions is important to consolidate the legal regulation of the ecological situation in coal mining regions, to create funds for the promotion of reclamation of disturbed lands within the existing Association of coal enterprises, to amend relevant regulations in the Russian Federation.

The notion of strategic action Habermas combines the concept of action in accordance with the economic interests and the concept of purposeful rational action. In its framework people focus solely on the success and the results of their actions, they seek to exert economic pressure and other means of coercion on others. Strategic action involves the categories of power, wealth, institutions, purposes. Communicative action is oriented to reconciliation, mutual understanding and recognition in the negotiating process. This idea is very relevant for Russia, where democracy is often understood as "majority rule". Meanwhile, as he wrote Titus Livy: the "Free state, one in which... all citizens are equal participants in the rule of law. This is a self-governing community in which the greater Empire of laws the Empire of man" [14]. Protection of minorities is the protection of their competition (political), and maintain consensus on major issues as a minority - one of the sources of public maneuverability, mobility.

In the annual report of the Federal service for supervision of consumer rights protection and human welfare provides information about the state sanitary-and-epidemiologic wellbeing of the population in the Russian Federation, including focuses on the fact that in recent years the workers of enterprises on extraction of mineral resources noted the highest weight was registered for the first time occupational diseases – 43.03 %. Indicators of occupational morbidity, exceeding the figure for the Russian Federation (of 1.47), as in previous years, marked in the Kemerovo region (of 13.24). It is also noted that, if the whole of the Russian Federation of managerial decisions in the field of environmental protection and the evaluation of health risk was 4.7 % (118), in relation to the Kemerovo region as one of the subjects it is 12 % [12].

As an example, it can be noted that in the framework of the "Charter of the Kemerovo region" with the implementation of the rights of citizens to a favourable environment, the public authorities of the Kemerovo region carried out a number of measures: adoption of laws and other normative legal acts of the Kemerovo region in the field of environmental protection; the establishment of standards of environmental quality; the adoption and implementation of state programs of the Kemerovo region, etc. The division of powers between bodies of state power of the Kemerovo region in the sphere of environmental protection defined in the Law of the Kemerovo region from 18 January 2007 "On the delimitation of powers between public authorities of the Kemerovo region in the sphere of environmental protection". In the framework of the state program of the Kemerovo region "Ecology and natural resources of Kuzbass" for 2014-2018 and "Ecology, subsoil use and water management" in 2017-2019 find the resolutions for the issues of sustainable economy mineral resources and geological information on mineral resources; improve the environmental safety and the preservation of natural systems, etc.

At the level of methodical recommendations establish uniform approaches to development and General requirements to the content and formatting of draft standards for waste and limits on their placement, the Regulations of the Board of administration of the Kemerovo region (03.12.2012 No. 534, 16.09.2014 No. 359) defines the General rules of conducting of works on regulation of emissions of harmful (polluting) substances in atmospheric air during the periods of adverse meteorological conditions; the procedure for the exploration and production of common mineral resources by the subsoil users. Also defined a list of normative legal acts containing mandatory requirements, evaluation of compliance with which is subject to state control (supervision). [12]

It should also pay attention to the fact that within the "paradigm of directed evolution of the biosphere" (A. V. Yablokov) [16] includes not just the awareness of the catastrophic situation that is associated with the environment, but also the development of specific activities and technologies contribute to overcome the key factors, which are the carriers of ecological contradictions. From a number of factors can be identified such as: "the impact of the accumulated anthropogenic high-frequency radiation, chemical substances that are not inherent in nature, and, as a consequence, the growth in the number of people with pathological disorders" and "the speed of concentration and commissioning resources in solving problems of national scale". Timely response to the emergence and diagnosis of dangerous situations, makes it possible to timely prevent, catastrophic development and education of local man-made disasters.

4 Conclusion

Thus, the influence transmitted through the public authorities at both the Federal and territorial levels of ideological paradigms related to the solution of environmental problems, the processes of enterprises modernization and introduction of modern technologies, in particular, contributing to the preservation and restoration of the environment. Emphasized the role and importance of legal culture at all stages of enforcement: discussion, adoption of laws and normative legal acts, their application in various fields, as well as forms and methods of monitoring their performance from the interested in this subject not only economic activities, but also society as a whole.

References

1. M.I. Agienko, E.P. Bondareva, G.V. Chistyakova, O.V. Zhironkina, O.I. Kalinina, IOP Conf. Ser.: Earth Environ. Sci., **50:1**, 012022
2. M. Tyulenev, E. Garina, A. Khoreshok, O. Litvin, Y. Litvin, E. Maliukhina, IOP Conf. Ser.: Earth Environ. Sci., **50:1**, 012035 (2017)
3. M. Tyulenev, S. Zhironkin, K. Kolotov, E. Garina, Poll. Res. **35:2**, 221-227 (2016)
4. E. Dotsenko, N. Ezdina, E3S Web of Conf., **15**, 04012 (2017)
5. V. Trifonov, O. Loyko, D. Nesteruk, S. Zhironkin, E. Strekovtsova, AIP Conf. Proceed., **1800**, 050009 (2017)
6. I.V. Chernikova, *Epistemological schemas and their role in scientific knowledge. Epistemology: the main problems and the evolution of approaches in philosophy of science* (Kemerovo: Kuzbass state technical University, 2007)
7. M.T. Stepanyants, Voprosy Filosofii, **8**, 20-30 (2017)
8. A.B. Efremenkov, A.A. Khoreshok, S.A. Zhironkin, A.V. Myaskov, IOP Conf. Ser: Earth Envir. Sci. **50:1**, 012009 (2017)
9. N. Fraser, *From redistriibution to recognition Dilemmas of justice in a post-socialist age* (London, New Left Review, 1995)
10. T.V.Galanina, M.I. Baumgarten, V.G. Mikhailov , T.G. Koroleva, G.S. Mikhailov, IOP Conf. Ser: Earth Envir. Sci. **50:1**, 012030 (2017)

 DOI: 10.1051/e3sconf/20172104008

11. T.V Kiseleva, V.G. Mikhailov, V.A. Karasev, IOP Conf. Ser: Earth Envir. Sci. **45**, 012013 (2016)
12. S.S. Nipa, The almanac of the Center for social Sciences and the faculty of Economics of MSU named after M. V. Lomonosov, **4:112**, 57-63 (2017)
13. *On the state sanitary and epidemiological wellbeing of the population in the Russian Federation in 2016: State report* (Moscow, Federal service for supervision of consumer rights protection and human welfare, 2017)
14. Q. Skinner, *Liberty Before Liberalism* (Cambridge, CPress, 1998)
15. V. Trifonov, O. Loyko, D. Nesteruk, S. Zhironkin, E. Strekovtsova, AIP Conf. Proceed. **1800**, 050009 (2017)
16. A.V. Yablokov, V.F. Levchenko, A.S. Kerzhentsev, Philosophy & Cosmology, **14**, 34-45 (2015)
17. S.A. Zhironkin, K.A. Kolotov, A.E. Genin, F.V. Agafonov, S.A. Kovalevsky, IOP Conf. Ser: Earth Envir. Sci. **50:1**, 012011 (2017)
18. V.M. Zolotukhin, V.A. Gogolin, M.Yu. Yazevich, M.I. Baumgarten, A.V. Dyagileva, IOP Conf. Ser: Earth Envir. Sci. **50:1**, 012027 (2017)

 DOI: 10.1051/e3sconf/20172104009

The Prospects of Accounting at Mining Enterprises as a Factor of Ensuring their Sustainable Development

Tatiana Tyuleneva[1,*]

[1]T. F. Gorbachev Kuzbass State Technical University, 650000, Kemerovo, 28 Vesennyaya street, Russian Federation

Abstract. One of the problems of sustainable development of mining companies is attracting additional investment. To solve it requires access to international capital markets, in this context, enterprises need to prepare financial statements with international requirements based on the data generated by the accounting system. The article considers the basic problems of accounting in the extractive industries due to the nature of the industry, as well as evaluation of the completeness of their solution in the framework of international financial reporting standards. In addition, lists the characteristics of accounting for mining industry, due to the peculiarities of the production process that need to be considered to solve these problems. This sector is extremely important for individual countries and on a global scale.

1 Introduction

The mining industry is directly connected with the use of natural resources - minerals, therefore, this sector is extremely important for individual countries and on a global scale. The technological revolution that swept the developed world, has led to a increase in the consumption of diverse minerals. Therefore, there is the difficulty of securing mineral and energy resources of the global economy. For most mining companies complex task of cost reduction is not novelty: in recent years, the industry has tried to solve a number of problems of sustainable development, one of which was the need of investment in infrastructure development [1].

Russia's accession to the WTO greatly enhances the investment, but at the same time increases competition from foreign organizations, which in turn determines the necessity of creation of high-quality reporting according to international standards, which will act as a source of information for assessing the efficiency of investment [2]. In these circumstances, it is important to identify and solve the problem of accounting in the Russian mining companies as a source of information for managerial decision-making and enable sustainable development [3].

2 Theory of experimental research

* Corresponding author: kta.bua@kuzstu.ru

 DOI: 10.1051/e3sconf/20172104009

Analysis of existing normative documents of different countries and scientific developments in the field of accounting in the mining sector indicates the absence of an integrated approach that takes into account all the peculiarities of this sphere. Companies use either the uniform standards of accounting and reporting, or separate national standards and recommendations which do not cover the whole range of problems [4, 5]. This situation is typical for the UK, Australia, Nigeria, Canada [6]. Even in the accounting system of the USA – the only national system, which now has quite extensive experience in accounting in mining companies – has not been sufficiently considered, taking into account the specificity of activity of mining enterprises [7].

Traditionally, IFRS` financial statement of companies was distinguished by the presence of such characteristic in the balance sheet as proven reserves resources in the section "Non-current assets" and "Liabilities on the withdrawal of assets from operation".

The main problems of standardization of accounting in the mining sector arise in connection with such key features as the uncertainty of prospects of development of deposits, a significant amount of the necessary costs at the initial stage, the time from mineral exploration to production and implementation, a significant amount of financial obligations for reclamation and for the withdrawal of production assets and infrastructure of the operation, difficulty of assessment and uniformity in the classification of reserves, most of the production facilities at the end of the period of development of the field have no residual value and a number of others [8].

Despite the fact that mining is an important industry for the global economy in recent years, standard-setters have paid little attention. International financial reporting standards before the end of 1990-ies did not contain any specific standards or review to the accounting in the mining industry [9]. The scope of the IFRS 6 "Exploration for and evaluation of mineral resources" is rather limited and covers the costs incurred by the company after obtaining the rights for the implementation of complex of works on search, evaluation and exploration and to confirm the economic viability of field development. Improvements in accordance with the provisions of IFRS 6, are quite limited and involve a lot of conventions [10].

All costs incurred during exploration and evaluation of mineral resources can be reflected in the: statement of profit and loss as an expense; balance sheet in the form of an asset.

The initial cost of the assets on the balance sheet is formed in the capitalization of costs incurred. There are 6 examples of such costs in the IFRS standard (IFRS): acquisition of exploration rights; topographical, geological, geochemical and geophysical studies; exploratory drilling; digging of the pit; selective control; work related to evaluating the technical feasibility and commercial validity of extracting mineral resources.

Of course, the above list is not exhaustive and an entity may identify other exploration for and evaluation that meets the criteria for capitalization as assets. This criterion is considered as the degree of affection of a certain type of costs to specific mineral resources. For example, the costs of obtaining seismic data may not represent a close connection with a certain resource, sufficient for their capitalization as an asset. Emergence of IFRS (IFRS) 6 divided the cost accounting for three phase boundary between given activities and the activities for exploration and evaluation is considered the moment of obtaining legal rights to explore a specific area. Only in rare cases, the cost of exploration and evaluation may include costs incurred in the implementation of the activities preceding the exploration phase and evaluation. As a rule, during this period costs cannot be capitalized, because in general, mineral resources are hypothetical in nature (IFRS 6).

But in some countries the process of obtaining a license is less formal character than in Russia, for example, in some African countries is often a substantial part of the activities of exploration begins already during the completion of the necessary formalities for obtaining a license on the basis of reasonable expectations. In such cases, the entity will consider whether this type of cost recognition criteria as an asset at "the Conceptual framework for IFRS".

The boundary between exploration and development passed the time when it becomes apparent technical feasibility and commercial viability of extracting resources. The term "design" has a traditional association with IAS 38 "Intangible assets", defined as "the application of research results or other knowledge in a plan or design for the production of new or substantially improved materials, components, products, processes, systems or services before the start of commercial production or use" (IAS 38). In the context of industry practice development means a phase when an identified mineral resource prepared for industrial production. For example, the construction of railways (access) to mine the mineral resource.

Develop and evaluate assets should be classified based on the nature, plant and equipment and intangible assets (IFRS 6). Some assets will include fixed assets (vehicles, drilling rigs), while others are clearly intangible assets (license). Also there will be other assets which are difficult to classify, and probably a significant portion of these other assets would consist of costs incurred in the process of construction of exploration wells or mines.

In such cases, the asset classification may be useful to answer the following question: in creation of what object sent to the specified cost object having a physical form, or to obtain information (e.g., exploration well likely to be an intangible asset, because it leads to certain knowledge). Based on the classification of assets, an entity selects accounting policies of the valuation of assets after their recognition and the disclosure of relevant information.

Unclear to date is the recognition of borrowing costs, resulting in a period of exploration and evaluation. From our point of view, borrowing costs are not the costs of exploration and evaluation at its core, therefore, it is necessary to follow IAS 23 "Costs of borrowings". A company can capitalize borrowing costs in the cost of qualifying assets. Qualifying asset is an asset whose preparation for intended use or sale requires significant time. Generally, these are assets that can be created as a result of large-scale development projects and construction.

Borrowing costs that are directly attributable to the acquisition, construction or production of a qualifying asset may be capitalized only if there is a high probability that as a result of these costs will occur future economic benefits. However, exploration is quite problematic to demonstrate the probability of future benefits.

The formation of fixed assets, intangible assets in the exploration and evaluation of resources in its content can be treated as qualifying assets. Borrowing costs that can be capitalized are those costs, the occurrence of which the company might otherwise avoid. Therefore, while there is no General exception to the borrowing costs (in relation to the costs of exploration and evaluation regarding the demonstration of probability of future benefits) it is better to capitalize only the interest on borrowed funds that are raised specifically for the acquisition of a qualifying asset (and costs on other loans that could be repaid, had it not been for the cost of the asset).

After the initial recognition of assets an entity will apply either the model of the actual cost, or the revalued amount model. We believe the first model as a priority, given that in determining the fair value of tangible assets may be missing evidence of the market value of these assets because they usually are highly specialized in nature. According to IAS 16 in the absence of evidence to support fair value, you can use the amortized replacement cost (ASP) or the profitability of the object.

Income, which an asset can generate in the future is highly uncertain, so this approach can not provide a reliable estimate of the fair value of the tangible asset. The same difficulty may arise when the revaluation to fair value of intangible assets, because an active market exists when certain conditions are met.

Assets are assessed for impairment only in cases where the facts and circumstances suggests that the current amount of the asset may exceed its recoverable amount. According to IAS 36: the recoverable amount of an asset represents the greater of the two values: value in use of an asset and the fair value of the asset less costs to sell; the value in use represents the discounted amount of forecasted future cash flows that should arise from the continuing use of the asset or from its disposal at the end of useful life.

The requirements for testing for impairment are applied, if possible, to each asset separately. If this is not possible, the assets typically tested for impairment in groups, called units, generating cash flows.

However, IFRS 6 provides for the possibility of some derogations from the General rules in identifying the presence of indicators of impairment:

in contrast to the procedure applied to other assets, the company is not obliged to analyze the situation to determine the presence of indications of impairment as at each reporting date until, until enough information to come to a conclusion about the technical feasibility of the extraction of mineral resources; does not oblige to test impairment at a low level (a particular geological structures, for example adjacent ore bodies); permits an entity to test for impairment, to combine (merge) one or more cash generating units (provided that the group of merged units will not be higher than the level of the union as part of the reporting segment to which the data belongs it.

Identification of cash generating units or groups of units requires the use of professional judgment and is one of the most complex areas of the audit of assets for impairment. For example, the requirement that each CGU (or group of units) was not larger than the corresponding segment should not be considered in absolute terms. Instead, consider a more objective focus on the fact that the test of impairment will not be conducted on the level at which the result of this test will be a group of assets or cash generating units included in different segments. Also, in practice, may be the preferred option to a level at which the monitored signs of impairment were consistent with the level at which estimated the recoverable amount of assets in the exploration.

The most conclusive indicator of impairment is that the rights of the company to conduct exploration work in a certain area has lost power (due to the expiration of the term), or will lose force in the near future, and thus it is not expected that these rights will be renewed (prolonged). The standard lists other symptoms involving impairment before the specified limit:

the budget is not foreseen and not planned to substantial costs for the continuation of exploration activities and evaluation in this area; the company has not discovered commercially viable deposits of mineral resources as a result up to the present time of exploration and evaluation in the area and decided to stop these operations at this site; even if the probability that it will start development of the field, is high, the company has sufficient data indicating that the carrying value of the asset is not likely to be recovered in full from successful development or sale of the asset.

The signs are determined based on the management information. If the testing for impairment of developed and assessed assets is inappropriate for the comparison of the previous reporting period, the enterprise should disclose that fact.

An entity should disclose information that reflects and explains the amounts recognised in the financial statements and related to the development and evaluation of mineral resources. Disclosure should include the accounting policies and the level at which impairment is assessed developed and assessed assets.

Many enterprises in the extractive industries has obligations to restore the site and the withdrawal of assets from operation as a result of exploration works. These commitments should also be reviewed for impairment. The obligation can either be legal (due to legal requirements), or constructive (reasonable expectations of third parties due to his past activities or public statements that it will perform this action IAS 37). With respect to the duty to withdraw the asset from service or the restoration of the site recognizes the reserve obligation in the full amount immediately after the occurrence of an event constituting the ground for the emergence of the obligation. For example, the relevant provision is an obligation for the anticipated cost of dismantling a test drilling rig is recognized when it is mounted.

Mining companies must apply the requirements of the standard to the costs that they have in connection with the exploration for and evaluation of mineral resources. However, for most Russian companies the most relevant stage of development.

3 Results and discussion

The most difficulties in applying IFRS are as follows [11].

1. Accounting for inventory and resources. In most cases, mining companies take into account only partially confirmed and proven reserves, because they believe that the unaccounted stocks associated with very large uncertainties. This decision may impact the situation when mining without significant impact to the current production plan can defer further work is required to transfer resources to reserves. Mining companies must create the conditions for consistent application of accounting policies and also demonstrate high confidence in the transfer of resources to reserves in the future.

2. The treatment of deferred costs in the open field development. Often as a result of the stripping operations company may have a double benefit: extraction of the mineral, which can be used to produce inventory and improved access to further quantities of minerals that can be mined in future periods. In cases when after the stripping activity mining company gets benefit in the form of reserves, the related costs are recognized in accordance with IAS 2 "Inventories". In cases when after the stripping activity mining enterprise benefits from improved access to deposits in the future, the related costs are recognised as an asset associated with stripping activities [12]. The experts of financial services may require additional information from the production department (project engineers, geologists, surveyors) for a deeper analysis of the costs incurred for overburden works [13]: identification of components in the composition of minerals in accordance with the plan (plans) of development of the field; unit production for the distribution of the total amount of Stripping costs between the cost of inventory produced in the current period, and the capitalized asset; - the definition of the period of beneficial use of each component mineral in order to calculate the depreciation of assets of stripping activity.

3. Long-term storage`s dumps accounting. If mining company has plans for the processing of this raw materials, the costs associated with its production, must be attributed to future periods (non-current reserves). However, this solution depends on whether management evaluates profitability using the price forecasts (at the time of processing of raw materials) or current market prices. In accordance with the international standards for determining the value of other non-current assets is usually used the principle of the time value of money.

4. Accounting of expenses on the excretion of exploitation and land reclamation. One of the specific items in financial statements of mining companies is a reserve for liquidation of fixed assets and restoration of the landscape. However, in IFRS there is no detailed guidance on its calculation, so questions on the calculation of the reserve and its annual revision, become even more relevant. Often, licenses for exploration and development of deposits do not contain clear requirements for action mining to end use of mining land, which complicates the evaluation of reserve, however, IAS 37 "Reserves, contingent liabilities and contingent assets" provides a possibility of making reliable estimates of future costs associated with the dismantling of its assets and the restoration of the deposits [14]. Use of estimates has a significant impact on the reliability of the financial statements. In the case of using the experience of other mining companies it is important to ensure the similarity of

the mining fields, as the costs for dismantling vary depending on the production technology, the type of structure deposits, and other factors that affect the production process.

4 Conclusion

In the current economic circumstances, the accounting in the mining industry has the challenge - with all the features of this sector of the economy securely and reliably reflect the assets, liabilities, the facts of economic life, sources of funding, income and expenses in the financial statements. The efficiency of economic decisions taken by users of these reports in the first place, and potential investors, depends on it. Accordingly, the inflow of investment will ensure dynamic and sustainable development of the mining sector.

Currently in accounting in the mining industry, there are still many unresolved issues. It is necessary that in the framework of the IFRS be created a unified global standards for the training and reporting in the mining industry, taking into account all aspects of the industry. For countries committed to reforming their national systems in accordance with IFRS becomes relevant active part in the development of international standards. This will ensure the most efficient and high-quality preparation and presentation of the information in the financial statements, which will serve as one of the factors of sustainable development of the industry.

References

1. T. Carlin, N. Finch, Macquarie Graduate School of Manag. Work. Paper, **13**, 4 (2008).
2. T. A. Tyuleneva, Europaische Fachhochschule, **6**, 89 (2014)
3. E. V. Kucherova, T. A. Tyuleneva, N. A. Cherepanova, *Coal in the 21st Century: Mining, Processing and Safety*, 130 (2016).
4. I. Tsalavoutas, L. Evans, Managerial Auditing Journal, **25(8)**, 814 (2010)
5. N. H. Laili, Journal of law and financial management, **7(2)**, 8 (2008)
6. J. Kranc, Financial executive, **7-8**, 66 (2011)
7. J.M. Langmead, J. Soroosh, The CPA Journal, **8**, 24 (2012)
8. D.W. Taylor, Asian Review of Accounting, **17(1)**, 40 (2009)
9. E. Kvaal, C. Nobes, Accounting and Business Research, **40(2)**, 173 (2011)
10. G. Guggiola, The International Business & Econ. Res. Journal, **9(12)**, 99 (2010)
11. A. Rahman, H. Perera, Abacus, **38**, 46 (2002)
12. B.J. Epstein, E.K. Jemarkovich, *Interpretation and application of International Financial Reporting Standards* (John Wiley & Sons Inc., Hoboken, New Jersey, 2010)
13. C.M. Graziano, E.M. Heffes, Financial Executive, **25(1)**, 17 (2009)
14. A. Jaruga, J. Fijalkowska, M. Jaruga-Baranowska, M. Frendzel, Accounting in Europe, **4(1)**, 67 (2011)

E3S Web of Conferences **21**, 04010 (2017) DOI: 10.1051/e3sconf/20172104010
The Second International Innovative Mining Symposium

Tools of Realization of Social Responsibility of Industrial Business for Sustainable Socio-economic Development of Mining Region's Rural Territory

Tatyana Jurzina[1], *Natalia* Egorova[1*], *Natalia* Zaruba[1], and *Peter* Kosinskij[1]

[1] T.F. Gorbachev Kuzbass State Technical University, 650000, 28 Vesennyaya St., Kemerovo, Russia

Abstract. Modern conditions of the Russian economy do especially relevant questions of social responsibility of industrial business of the mining region for sustainable social and economic development of rural territories that demands search of the new strategy, tools, ways for positioning and increase in competitiveness of the enterprises, which are carrying out the entrepreneurial activity in this territory. The article opens problems of an influence of the industrial enterprises on the territory of presence, reasons the theoretical base directed to the formation of practical tools (mechanism) providing realization of social responsibility of business for sustainable social and economic development of rural territories of the mining region.

1 Introduction

Regional policy of the development of any territory includes a set of directions and objectives, defining the vector of priorities in various fields and areas of life. In our opinion, areas for economic and social development of the industrial region should become the most actual in modern conditions.

In this regard, we consider it appropriate to draw the attention of research to the SOCIAL RESPONSIBILITY of BUSINESS, which in the framework of its application for the industrial region is of particular importance.

It should be clearly understood that the social responsibility of business primarily of industrial enterprises for different areas plays not just a significant role but, in our view, it should contribute to sustainable socio-economic development of the region.

It is important to note that the industrial enterprises in the area of its activities have various impacts on the territory of presence, particularly on economic, environmental, social and governance ones.

You need to understand that sustainable development of society, individual locations or organizations is not so much a goal, as the process in which the changes must be coordinated with real needs and taking into account future development. Experience shows that to-

[*] Corresponding author: egorovann@list.ru

DOI: 10.1051/e3sconf/20172104010

day, the degree of elaboration of the problems identified is very insufficient, despite the existing research in this area [1-9, 13, 14].

To create conditions and prerequisites for sustainable socio-economic development of the industrial region it is needed to understand that only full interconnection of all spheres and sectors of the regional economic structure, their uniform distribution will be a possible and rational solution to the problems identified.

However, to achieve optimal results, in our view, is possible only by understanding the role and importance of rural development for the industrial (mining) region.

The purpose of this article is the proposal of a mechanism of practical implementation of social responsibility of industrial enterprises for rural areas taking into account the characteristic features of the mining region.

2 Materials and methods

In the concept of sustainable development of rural territories of the RUSSIAN FEDERATION "sustainable development" seems as stable socio-economic development, not violating their natural basis and ensuring the continued progress of society. The transition to sustainable development means the gradual provision of purposeful self-organization of society in economic, social and environmental spheres. This formulation interprets sustainable development as economic efficiency, ecological safety and social justice [10].

Thus it is necessary to consider rural areas as an integral part, which is included in the national strategy for sustainable development and implies the optimum compatibility of the population - local government – economic-industrial complex – the social responsibility of business.

In this case, the priority of economic development improves the life quality of the population of rural territories of the industrial region on the basis of the social responsibility of business.

A separate sustainable development of rural areas requires a search for relevant new strategies, tools, techniques primarily at the level of industries or enterprises, conducting entrepreneurial activity in the territory.

It should be noted that the company is a fundamental element in the pyramid of sustainable development strategy for specific rural areas.

The studies have shown that the implementation of sustainable development principles at the level of the individual enterprise is through the social responsibility system basically only of agricultural enterprises, not businesses, carry out their business activities in the territory. This shows the relevance of the designated problem again.

Today business is ready to move from simple charity to systemic interaction with society to solve problems it faces in carrying out activities in the territory of presence.

An arisen need of business, primarily, of industrial enterprises, in the acquisition of social recognition and, consequently, increase of the competitiveness, at present, is an incentive for modifications of the interaction of business and territories.

Thus, the problem of sustainable socio-economic development of rural areas in the context of social responsibility of business is highly relevant. It is supported by the fact that in recent times in relations between the state, civil institutions and business structures changes have taken place, which are contributed to the emergence of new forms of social interaction, particularly the social responsibility of business.

Today business operating in rural areas, has an impact on its development, and in this case it is important to guide the direction of social responsibility in the direction of the positive dynamics of rural areas development, its socio-economic indicators.

These problems are faced by the companies very strongly, for which changes in the competitive landscape are accompanied by significant transformations at the level of basic business principles. So, for example, such enterprises should include the Russia's largest coal company called SUEK — Siberian coal energy company.

3 Results and discussion

It should be noted that social responsibility of business is a complex, multidimensional concept, which should specify, for example: sustainable socio-economic development of the territory, condition the possibility of expanded reproduction, promote the growth of business potential, increase of competitiveness, improvement of the status of the company.

We believe that social responsibility should be considered from the point of view of its basic functions: ecological, economic, social, managerial. They are part of the strategy, define the essence of the concept and aimed at solving the acute problems of rural territories of the industrial region, such as: pollution of land and water resources, atmospheres, the depletion of natural resources, change of landscape, loss of biodiversity, hazardous waste, etc.

These and other problems are solved by ecological function of social responsibility of business which causes the search for new projects in the environmental field for the development of rural areas. It should be noted that this feature is key for businesses in the area of solving problems in entrepreneurial activity in the territory of presence, in our case it is in an industrial region (the Kemerovo region).

In this regard, we note that in the Kemerovo region three-quarters (63%) of total primary production is the extraction of minerals, this indicates a significant impact on the rural territory of the region for the enterprises of the coal complex, including Siberian coal energy company SUEK. [15]

In addition, it should be noted that as a result of intensive development of the coal industry agriculture annually losses land areas under cultivation of crops. Despite the overall positive dynamics of some key socio-economic indicators of rural areas of the Kemerovo region, for example, the output growth of agricultural production in 2016 compared to 2015 amounted to 115%, it is still the risk of deterioration of indicators such as: investment in fixed capital - 67%, expenses of the regional budget on agriculture -59%, a slight share in the gross regional product in different years varies from 3.8-4%. Also in agriculture of the Kemerovo region there is a high depreciation of fixed assets, increasing the amount of disturbed land over the last 5 years by 19%, from 63.8 thousand hectares to 76.3 thousand hectares, and as a consequence reduced the acreage under the crops in 19, 5% [15]

Therefore, the use of the designated functions of the social responsibility of business, particularly to SUEK, would be a positive change in the problem, in a so-called "weak point" in the field of rural development.

So, the economic function needs to be aimed at achieving the efficient use of resources, including labor, business development, creating new jobs, etc..

Social function has a rapid effect, manifesting itself in a number of actions aimed at the interests of the personnel of the enterprise and the interests of the local community, particularly in professional development of staff, improvement of infrastructure of rural territories, increase in the quality of life, support community organizations, sponsorship of social projects, charity, volunteering, increased employment, development of the presence regions, etc.

It is important to emphasize that social responsibility of business provides legal obligations imposed by the applicable law, ranging from safety regulations to employment law, the law on environmental protection, and respect for human rights.

The features of social responsibility of a business mentioned above are interdependent. They should complement each other and been carried out by enterprises in the process of their activities. Since today the business is the main actor in the market, ensuring the wel-

 DOI: 10.1051/e3sconf/20172104010

fare of society, its social responsibility is a paramount concept in the control system. It must be emphasized that in this case the social responsibility of business is positioning the enterprise in the first place, in terms of the benefits that it provides in the field of corporate image of the company and involve its stakeholders (suppliers, customers, shareholders, employees, governments, local communities, non-governmental, international and other organizations involved in the activities of the enterprise).

However, we believe it is necessary to understand that social responsibility of business in the framework of the mentioned above should be based on the implementation mechanism, presented in the Figure.

It is the tools of the mechanism of implementation of social responsibility of business would allow basic functions that ensure the sustainable socio-economic development of rural areas, which are mentioned above.

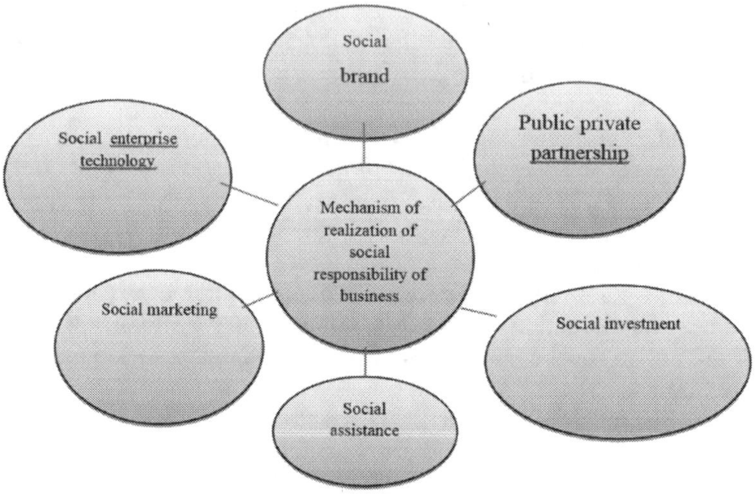

Fig. 1. Tools of mechanism of realization of social responsibility of business.

A distinctive feature of social responsibility of business of enterprises in the industrial region for the development of rural areas is not just a combination of the identified functions, and the allocation of priority to resolve urgent problems, their ranking, but given the priorities facing the area presence of an industrial enterprise.

4 Conclusions

Socio-economic and environmental issues in the mining region is particularly felt today in the modern "sanction state" of the economy, for which it is necessary to use a number of preventive measures, mainly the use of the instruments of implementation of social responsibility of business. Subject to comprehensive, systematic application of the proposed mechanism for the possible increase of efficiency of activity of the enterprise, using social responsibility of business as a fundamental postulate to the execution of strategic decisions. In addition, such an approach in the activities of business structures will allow the rural areas to go on a distinct trajectory of its socio-economic development.

References

1. *Business for Social Responsibility. Overview of Corporate Social Responsibility.* (Geneva, YHHG, 2003)

2. A. B. Carroll, Business and Society, **38:3,** 268 (1999)
3. *The Challenges of Corporate Social Responsibility. Towards Constructive Partnership – Brussels*, (Belgium, BCFFS, 2000)
4. A. Crane, and D. Marten, *Business Ethics A European Perspective*, 32 (Cambridge, Cambridge University Press, 2004)
5. K. Davis, California Management Review, **2:3**, 70 (1960)
6. G. Zwetsloot, M. van Marrewijk, Social Dimensions of Organizational Excellence, (2003)
7. M. Friedman, *The social responsibility of business to increase its profits, in New York Times Magazine*, 32 (New York, Lomann, 1970)
8. B. Hunt, *The New Battleground for capitalism". Financial Times (*Manchester, Management, 2000)
9. M. Weber, *Economy and Society: An Outline of Interpretive Sociology* (New York: Bedminster, 1968)
10. J. Wootliff, and C. Deri, Corporate Reputation Review, **4:2**, 157 (2001)
11. N. Zaruba, N. Egorova, P. Kosinskij, E3S Web Conf. 04003 (2017)
12. N.A. Zaruba, N.N. Egorova, Coal in the 21st Century: Mining, Processing and Safety, 39 (2016)

E3S Web of Conferences **21**, 04011 (2017) DOI: 10.1051/e3sconf/20172104011
The Second International Innovative Mining Symposium

Score Mining Rents in Terms of Investment Attractiveness of Peat Mining

Gennady Alexandrov[1], and *Alexander* Yablonev[1,*]

[1]Tver State Technical University, 170026, Af. Nikitin Street, 22, Tver, Russia

Abstract. In this article, as determinants in the system factors underlying the investment attractiveness of the peat industry is considered a rental factor, which predetermines the significant differences and peculiarities of the investment climate in the mining business and, in particular, in the sphere of peat mining. In contrast to modern studies treated the essence and role of rents in the economic mechanism, is proposed for a new approach to solving the problems of its formation. Our approach differs in that it, firstly, adequate rental relations, objectively in extractive industries, secondly, provides consensus in the interests of the owner of peat deposits and entrepreneurs, businesses in these deposits and, thus, thirdly, contributes to the creation of a favourable investment climate in the peat extraction industry. In practical terms, in accordance with the proposed approach, we have proposed specific allocation algorithm of mining rents from the profits of peat extraction enterprises.

1 Introduction

The relevance of the further development of peat mining in the world, and its revival in Russia, mainly due to the fact that, firstly, there is a clearly expressed the trend deteriorating geological conditions of oil and gas extraction. Secondly, with regard to the emerging ecological situation, ecological standards and regulations on pollution ambient environment are gradually tightened, the degree of which is largely conditioned by the quantity and quality of fuel burn. It should be borne in mind and the fact that, in practical terms, in the year of 2009, the EU adopted energy program, called "20:20:20", providing for the period up to 2020 year reduced electricity consumption by 20%, increasing the share of renewable energy of total consumption to 20% and reduce greenhouse gas emissions 20% also [1]. Thirdly, greater significance of energy conservation issues, the solution of which is associated primarily with the use of alternative energy sources, especially taking into account the strong upward trend of increasing the prices on not renewable hydrocarbon

* Corresponding author: alvovich@mail.ru

raw materials. In this regard, the peat, which is positioned as partially or even entirely renewable energy source, in macroeconomic plans may well be regarded as strategic. In particular, only in Russia around 60% of the total stock of peat on technological factors appreciated as potentially recoverable. Annual extraction in 300 million t, existing stockpiles, even without taking into account their renewability, enough for production for 102 years. Fourthly, it is equally important to use peat as a domestic fuel to meet the challenges of energy and environment (including fire) security and improve their energy efficiency. The Russian experience shows, as well as the experience of some countries, in which the peat industry in the beginning of this century has progressed significantly in its development, the peat industry has all the prerequisites to take its rightful place in the global economy [2]. Especially when you consider the fact, that in the world, about 65% of mined peat is burned to generate heat and electricity. However, it should be noted that, to date, there are discussions about whether or not, in general, to develop production of peat, to what extent and for what purpose to develop, and whether it's on economic conditions. It is therefore not surprising that in publications on this topic are invited to pay greater attention to the serial address fundamental issues of creating economically favourable conditions for the development of the peat industry [3]. As for the Russian business community, its representatives say, that they are in no hurry with their investments "to include in the peat industry is due to the fact that it is not formed the rules of the game". It should be noted that the lack of motivation and incentives for private business to invest in extractive industries – is a common problem [4].

However, formation of adequate "rules of the game" is possible only on the basis of an analysis of the investment climate and, above all, to identify the factors contributing to its attractiveness, taking into account the fact, that these factors have, in one way or another, relate to the rental relationship, folding in the mining sector and, in particular, in the field of peat extraction.

Given the size of this article, we will stop it is not at all the factors influencing the investment climate and the level of its attractiveness, but specifically only on rental factor, so much, so that in some earlier works had already been introduced to a new, so-called "matrix" approach to investment climate assessment [5, 6] and the peculiarities of its use including peat spheres of business.

2 Materials and methods

The important thing is what we would like to focus, is firstly – on the methodology for the analysis of rents and rental relations in terms of their impact on the investment attractiveness of peat mining. Secondly – on the justification copyright value-natural profit sharing method, derived from peat activities on the part of it that is in the form of a mining (natural) rents, to be withdrawal in favour of the owner of the peat deposit, and another part of it, in the form of business income, attributable to an entrepreneur, managing on this deposit. And thirdly – on the principles of formation of organizational-economic mechanism of peat mining, which combines competitive regulation and tax administration. The use of this mechanism will result in permission objective contradictions arising between the peat extraction subjects and, thus, will contribute to the improvement of the investment climate in the peat industry.

Appeal to the methodology for the analysis of economic relations in extractive activities, in general, and peat, in particular, due to significant differences in the proposed withdrawal mechanisms mining rents, proposed by investigators. Firstly, that draws attention to itself, although these proposals have some differences, but are characterized by mostly fiscal direction. This is despite the fact that fiscal approach, by definition, is severely restricted in carrying out motivational and incentive functions. Nevertheless, many economists are inclined to that rental nature of subsoil must be taken into account, but offered to do solely within the framework of improvement of tax administration.

With these same positions do not stand up to criticism and proposals for improving the mining tax for taxation of extraction of solid minerals, by introducing specific tax rates linked to the real financial results of mining and mineral prices on the world market [7]. Besides, it is not clear, how these rates reflect the natural quality of deposits, if real financial results of extractive enterprises depend not only on the quality of deposits. As for the dynamics of world prices, its fluctuations, as conclusively proven, in general have nothing to do with the realities of subsoil use [8]. Proposals of this kind can only strengthen the fiscal, destimulation character of tax administration. The same problems faced by many other countries, which also concluded that existing payment systems in subsoil use require reforming, based not on the change of tax administration, but on the using of rent regulation, resulting from the differentiation of deposits on natural, including mining-technical conditions [9, 10].

Secondly, and most importantly, mechanisms proposed are aimed not at the general that brings together all types of extractive activities, and on the sectoral specifics of each of the extractive industries on their own special interests. This is especially true of leading productions of fuel and energy complex: oil, gas, coal, etc., which not only occupy a monopoly position on the market, but they are budget-making enterprises. It is in orientation on their, in Russia, the tax administration system was established, with a pronounced fiscal, "robbing-distribution" type, practically excepting the possibility of using the rent regulation of extractive enterprises. Inclusion in this system peat mining is a dead-end path, not making it attractive for investment. The more so that this activity is unlikely to be able to hopefully in the foreseeable future to a decent level of state support, the peculiar natural monopolies and could therefore rely mainly on private investment.

It should be noted that in developed Western economies also dominated by tax, largely fiscal, rent withdrawal methods. It is difficult to understand, and therefore, for example, in the UK and Northern Ireland corporate profit tax is complemented by another tax on petroleum extraction and extractive industries in Norway provides for taxation in two forms: special tax and general profit tax.

It seems that the current taxation practice suit all subjects of extractive activities. State is the owner of the subsoil, bearing in mind the interests of corporations, establishes a relatively moderate tax payments, which contain only the part that is created in the course of subsoil use incremental profits. And big Russian mining companies unnecessarily, according to mind many experts, a significant portion of the incremental profit reserves.

However, all members of society are entitled to the balance of interests. Attraction for entrepreneurs will depend on whether he can expect to receive a minimum normal (average) profits on invested in business capital. The State as the owner of the subsoil and the representative of the people, to meet its interests must receive that portion of profits (in the form of rent), which is due solely to the natural quality of his deposits. There is a contradiction, whose permission must be sought in the optimal combination of tax administration and rental management. And the State as the owner of the subsoil, can and must perform a dual role: to regulate process of rent withdrawing from best quality natural deposits, and administer the taxation of business income, attributable to the user of deposit, after rent paying from the gross profit.

It only remains to solve the problem of the division of profits from extractive activities on rents, that is, on the part of it that is subject to rent regulation, and the second part of entrepreneurial income, which is the base for tax administration.

 DOI: 10.1051/e3sconf/20172104011

3 Results and discussion

In addressing the challenge many scholars lean towards, that starting point is normal (average) profits return on capital of the entrepreneur, business in the extractive industry. In other words, mining activity, including the peat industry in accordance, with the principles of inter-branch competition must ensure business entrepreneur (ceteris paribus) profit on capital, according to the same norms, which exists in other fields of activity. And rent, as, for example, for mind J. Schumpeter, is adding to the average profit thanks to cost savings, increased productivity and a higher return capital on the best lands [11]. As a result, it is proposed to calculate rent as the difference between the gross profit from the sale of the product, extracted from the depths, and the normal, average profit. As you can see, one way or another, the definition of rent proposes essentially a purely technical approach, based on the so-called residual principle in relation to mining rent. However, a more logical is a judgement (as, for example, found A. Smith) that rent relations led by the interests of the owner of the land. It is therefore understandable that preparing and signing the agreement, the landowner is seeking to leave land user only such proportion of the product which is sufficient for recovery of capital spent by them for the acquisition of production factors, and to receive regular (normal) profit on invested capital [12]. Agrees with him Karl Marx, who pointed out that no profit is the boundary of the rent and the rent is the boundary of profit [13].

In the end, if the employer conducts business with more or less efficient than the industry average, in first case he receives during the period of validity of the user agreement for more profit, and in the second, his entrepreneurial income after payment of rent less than average profit. Actually, in this action and specific mechanism of intra-industry competition in the mining production, caused by the presence of such an additional factor, as minerals, peat deposits, and, accordingly, the emerging rental relations.

So, we have two pieces created in the production of surplus product. One part of it is due to the natural quality of the deposits and is incremental surplus product and, accordingly – the substance incremental profit (or, as it is called, super-profits), that alienation takes the rent form. Another part of it caused by capital, is the substance mainly average profit, and when implemented, takes the form of entrepreneurial income of subsoil users.

Thus, the initial point in solving the problem of the division of profits from sales of produced products on the rent and entrepreneurial income and, consequently, determine the amount of rent, intended to withdraw, is targeting on the natural deposits of the worst qualities, whose exploitation does not bring incremental product and hence incremental profits (and, hence – rents). On this occasion, J.S. Mill wrote, that the final cost on worst of the land – is a certain measure of allowing to assess the amount of rents will bring all other lands. Any plot land brings so much more regular profit on the capital, how this plot gives more than the nastiest of the cropped plots [14].

The above, allows you to offer theoretically adequate approach to the definition of the mining rents as an extension product (natural form), and, as an extension of income (value form). As you can see, it combines natural and valuation estimations, while in most cases it is proposed that all calculations made on the basis of cost estimates [15]. This approach, let's call it natural-value, is based on the following principles and assumptions:

– produced products are sold at cost, when this extra income or loss occurring when selling products at market price, reflects only on the largest business income;

– the cost of the entire extracting product produced per unit area of the field, as well as additional income, adequate socio-normal, average production conditions, when this deviation from these conditions is reflected to the same extent and subject to the largest business income;

– at worst natural conditions fields where incremental product and, as a consequence, the natural rent is not formed, the public value of the extracted product is seen as production cost, which provides an average profit to which the entrepreneur can expect;

– in the best areas of peat deposits (but with the average production conditions) formed an extension product, acquiring the shape of incremental (over medium) profit, which is the basis of natural rents and which is defined as the difference between the product of two identical amounts of capital and labour. As a result, the incremental profits – is the basis of the withdrawing part of the income in the form of land rents;

– the size of the withdrawing natural rents depends entirely on the purpose and nature of the State policy of its withdrawal. But it should not be higher than the value of incremental product resulting from the best natural conditions, and should not be to renegotiate the Treaty to "penetrate" into the entrepreneurial income, which has entrepreneur as a result of greater efficiency.

Implementation of this approach, in accordance with its logic may be presented in the form of an algorithm involving a number of consecutive actions (steps).

Step 1. Form a block of natural-value indexes of enterprises activity, businesses on the deposits that are included on the basis of the homogeneity of the natural conditions and natural quality in a certain i-based group when the total number of such groups – n. Thus, each of the groups will be represented by a block of indicators such as: area of mines (ha), gross harvest peat (thousands of tons) and, accordingly, the collection of peat per hectare (t/ha).

Step 2. For each group, we determine the amount of incremental product per hectare (t/ha), which is formed relatively group with worse on natural quality of deposits. Naturally, the incremental product for this group is zero.

Step 3. Its necessity is dictated by the fact that, as noted above, production of incremental product is determined not only by the quality of the deposit, but also the technical level of production, which is a factor of entrepreneurial activity (capital). Individual level can vary from one side to another from the socio-normal, average. As a result, the group where individual level above, incremental product contains and that part which is obtained at the expense of the capital factor and it must objectively be attributed to entrepreneurs in the composition of the business income. Conversely, where the technical level of production is lower, then socio-normal, incremental product turns out to be low and in fact wounds rental income and thus overstates the entrepreneurial income. Therefore, the incremental value of product per 1 ha in each of the i-ies groups shall be adjusted on the k_i factor, reflecting a deviation of individual production conditions in one direction or another from socio-normal, average conditions. We assume that the level of technical equipment is expressed by the organic structure of capital value, that can be represented in the natural-value form of capital endowment labour. For its calculation for each of the i-th group, along with data on the largest using mines take data on average annual largest used capital in thous. rub. (except for the passive part) and the number of employees (persons). Then from the relationship of these data (per one hectare of peat deposits) define an individual, for each of the i-th group, capital endowment level. Completing the third step in the calculation of the average, i.e. socio-normal for all groups, the level of labour capital endowment and already regarding on this average size determine for each of the i-th group correction factor k_i.

Step 4. Adjust in each of the i-th group values of incremental product by multiplying it on the corresponding correction factor this group k_i. Thereby, we cite the magnitude of incremental product in each of the groups of deposits, differing from each other by natural quality, to socio-normal, average production conditions. Thus, in those groups where conditions of production, the creation of which is a function of managing the peat deposits of the entrepreneurs, lower, then socio-normal levels, excluded natural rent should be set at a larger size than actually produced a product extension. In other groups where technical level of production is higher than the industry average, by contrast, the excluded portion of the profits in the form of natural rents should be reduced relatively incremental produced product, because it contains a portion of the surplus product, created at the expense of the entrepreneur's activities more effectively.

As you can see from the above, the proposed approach, as we see it, seems quite adequately reflects, firstly – the role of natural conditions in the production of incremental product, which is the substance of the rent. And, secondly, another part of the surplus product, which is the substance of business income, which may include, as the average profit so and excess profits that are attributable to the best individual production conditions compared with the average, socio-normal. But if, on the contrary, individual production conditions will be below the socially required level, the entrepreneurial profit after deducting rent naturally turns out to be even lower than the average profit. In this, actually, is the mechanism of intra-industry competition. And its implementation in the proposed approach – is a pledge that it would become a powerful factor in the attractiveness of peat mining for investors.

4 Conclusion

In conclusion we wants to note, that the practical feasibility of the proposed us approach requires addressing such important problem, as classification and distribution of all intended for the development of peat deposits into groups identical based on homogeneity of natural conditions and natural quality. As well as the problem of forms and methods of rent withdrawal by the State as the owner of mineral resources, taking into account and what each deposit is inherent in the objective life-cycle and its exploitation lead to exhaustion ("wear out") that essentially prejudge the law of diminishing profitability of entrails. In this regard, the rent rate withdrawals should be changed as the deterioration of the conditions using the peat deposit [16].

References

1. Energy 2020. A strategy for competitive, sustainable and secure energy. Communication from the Commission to the European Parliament, the Council, the European Economic and Social Committee and the Committee of the Regions. Brussels, 10.11.2010, URL: http://ec.europa.eu/energy/strategies/2010/2020_en.htm (accesssed 18.07.17).
2. G.A. Aleksandrov, T.B. Yakonovskaya, I.V. Vyakina, G.G. Skvortsova, Eur. Sci. Rev., **11-12**, 96-100 (2014)
3. J. Korpi, Proceed. 13-th Int. Peat Cong., Tullamore, Ireland, **1**, 120–122 (2008)
4. C.P Jones, *Investments: Analysis and Management* (John Wiley & Sons Inc., 2012)
5. G.A. Alexandrov, I.V. Vyakina, G.G. Skvortsova, Act. Prob. Econ., **3:165**, 415-428 (2015)
6. G. Alexandrov, I. Vyakina, G. Skvortsova, Asian Soc. Sci., **2:13**, 10–20 (2017)
7. Ya.I. Yukhimov, Ya.Ya. Ivanova, Min. J., **9**, 59–63 (2012)
8. G.E. Ortega, A. Pugachevsky, G. Walser, *Mineral rights cadastre. Oil, Gas and Mining Policy Division of the World Bank* (Washington, DC, USA, 2009)
9. V.V. Matyukha, Min. J., **7**, 60–62 (2014)

DOI: 10.1051/e3sconf/20172104011

10. M. Nieć, K. Galos, K. Szamalek, Resources Policy, **42**, 93–103 (2014)
11. J. Schumpeter, *A History of Economic Analysis* (Oxford University Press, 1954)
12. A. Smith, *An Inquiry into the Nature and Causes of the Wealth of Nations* (Colin Muir and David Widger, 2009)
13. K. Marx, *Capital. Critique of political economy* (Hamburg, 1885)
14. J.S. Mill, *Principles of Political Economy* (Oxford University Press, 2008)
15. N.V. Galtseva, O.A. Sharipova, I.S. Golubenko, I.N. Grigoryeva, Min. J., **3**, 27–32 (2016)
16. G.A. Alexandrov, A.L. Yablonev, Min. J. (to be published) (2018)

Mastering Foreign Language Competence of Ecology and Environment Managers for Mining Industry of Kuzbass

Oksana Greenwald[1,*], *Roman* Islamov[1], and *Tatyana* Sergeychick[1]

[1]Kemerovo State University, Institute of Philology, Foreign Languages and Mediacommunications, Department of Foreign Languages, 650000, 6 Krasnaya st., Kemerovo, Russia

Abstract. The necessity to solve nature conservation problems of Kuzbass mining industry demands from postgraduate education institutions to train highly qualified specialists in ecology and environment management. As 21st century education is competence-based one, the article clarifies the concept of competence in education, focuses on key competences, namely foreign language competence and its relevance for specialists in ecology and environment management. Foreign language competence is acquired through the course of "Foreign Language" discipline which covers the following aspects: academic reading, academic writing and public speaking. The article also describes the experience of organizing students' individual work taking into account their motivation and specific conditions of the discipline as well. Thus, both the content of the discipline and the approach to organize students' learning contribute to mastering foreign language competence of ecology and environment managers as inherent condition of their professional efficiency for solving ecological problems of mining industry in Kuzbass region.

1 Introduction

Highly qualified specialists in ecology and environment management are vitally urgent for coal-mining enterprises of Kuzbass. They are to solve the following problems of the mining industry:

- assessment of design and production activity of coal-mining enterprises of Kuzbass (in the view of their impact on environment);
- ecological monitoring of regional fauna and flora (biota);
- environment management and expert activity of coal-mining enterprises;
- rational environment management and nature conservation;
- scientific research of regional industrial environmental issues.

Consequently, training qualified specialists in ecology and environment management for a coal-mining region is an important task of the regional higher education system. Modern postgraduate education is competence-based one; it aims at formation and

*Corresponding author: greenwald.ru@rambler.ru

DOI: 10.1051/e3sconf/20172104012

mastering a number of competences of a future specialist, foreign language competence is among them.

2 Analysis

First the term *competence* appeared in scientific and psychological literature in the works of German scientist Jürgen Habermas [1], who used this term in the context of the theory of speech communication. In 1959 Robert White introduced the term *competence* into the theory of management to describe the features of individuality which are most closely connected with effective implementation of work and high motivation [2].

More attention to competence in management and strategy of supervision, the researches of scientists aimed at determination of personal qualities promoting effective functioning in professional area, and requirements of employers became a prerequisite of emergence and development of competence-based approach in education. It is commonly thought that competence-based approach in education appeared in the USA due to the publication of an article by David McClelland *"Testing for competence rather than for intelligence"* in 1973 [3].

Today professional education in the majority of developed countries, in spite of their cultural and national variety and specifics of economic development, unite the description of qualifications in terms of professional competences.

In English, the word *competence* is used along with the word *competency*. According to defining dictionaries they may be considered as synonyms. For example, in Collins Cobuild Advanced Learner's English Dictionary (New Digital Edition, 2008) it is written: *"Competency means the same as competence"*. Besides, the analysis of a number of works on this subject revealed the usage of both words in an identical collocation:

- competence debate – competency debate [4, 5];
- core competence – core competency [6, 7];
- competency-based approach – competence-based approach [8];
- competence movement – competency movement [9].

In summary, we may draw a conclusion that the word *competency/competence* is used to denote what a person, working in a certain professional area, must be capable to do. In the education system, *competency/competence* is understood as a result of training a graduate for the performance of an activity in certain professional areas [4-9].

The analysis of publications devoted to *competence* carried out by Françoise Delamare, revealed that Americans tend to connect *competence* with basic characteristics of a person, which cause one's outstanding achievements in certain kinds of activity. In Great Britain, there is a prevailing opinion that *competence* should be defined as compliance of the results of individual's activity with the authorized standards. Continental Europe countries and Russia adhere to a dominating multidimensional concept of *competence* uniting behavioral and functional concepts [10].

From the classifications of competences, it becomes obvious that researchers in the field of competence-based education allocate key competences also called core competences, basic competences, meta-competences as a separate category of competences. They are considered as competences of a broad scale, not specific to any professional area, and are characterized as universally applied [11-14].

The European Council identified key competences that are fundamental for each individual as a part of one's lifelong learning strategies. The second key competence stipulated by the European Council is a competence in foreign languages that is defined as "communicating in a foreign language: ability to express and interpret concepts, thoughts, feelings, facts and opinions both orally and in writing, also includes mediation skills (i.e.

summarizing, paraphrasing, interpreting or translating) and intercultural understanding" [15].

According to Russian state higher educational standards students are supposed to acquire key and professional competences specific for their specialty. Foreign language competence is included in the list of key competences for all specialties taught at higher educational institutions.

Relevance of foreign language competence for any specialist aspiring to gain professional success in a career (among them ecology and environment managers) is evident. Lack of foreign language competence of a specialist hinders one's international professional communication and studying cutting-edge achievements in the field of solving professional problems, makes impossible academic mobility and conducting cooperative scientific research with international colleagues.

Consequently, the list of competences specified by state education authorities for postgraduates, who are getting Master's degree in ecology and environment management, includes foreign language competence as an essential means of scientific and professional communication. Knowledge, skills and experience related to foreign language competence for postgraduates, who are getting Master's degree in ecology and environment management, requires: knowledge of vocabulary and functional grammar of scientific texts and documents; skills to understand and produce business, academic or scientific written and spoken messages, to initiate, sustain and conclude conversations concerning the field of scientific research; experience in public speaking with the results of their scientific research.

3 Results and Discussion

The aforementioned foreign language competence is acquired through "Foreign Language" discipline. The program of the discipline includes all the aspects of foreign language competence and consists of the following sections: *My Master's Course, Characteristic Features of Scientific Discourse, Academic Reading, Academic Writing and Public Speaking*. It should also be mentioned that, while creating the program of the discipline, we had to take into account the following facts.

Firstly, the majority of time designated to "Foreign Language" discipline is assumed to be individual students' work (viz. 76 hours out of total 108). In addition, nearly all postgraduates are employed and often have no opportunity to attend classes and tutorials in person. As a result, the main task was to organize individual students' work and to provide them with methodical help and possibility to get private tutorials on-line.

Secondly, the success in mastering foreign language competence by the postgraduates, who are getting Master's degree in Ecology and Environment Management, could be partly hindered by the level of the competence acquired at the previous stage of their education and lack of motivation to get the competence. The reasons of those are the following: (i) according to the state educational standards, at the previous stage of education, leading to Bachelor's degree, foreign language competence is formed during the first two years of studies. So, by the time of postgraduate course students have at least 3-4 years gap in foreign language training; (ii) some postgraduates appear to be qualified specialists in the field of environment management and they are highly estimated by their employers. Sometimes their low level of foreign language competence makes them feel psychological discomfort and causes lack of motivation in acquiring the competence. As a result, we have to persuade students in their ability to master foreign language competence by showing and explaining strategies of individual work, by facilitating their getting through the tasks, by creating assignments they can cope with.

The authors' long-term practice in foreign language teaching proves that positive motivation helps to overcome lack of foreign language skills. In our opinion, creating positive motivation in students is one of the main duties of any foreign language teacher.

The first section of the discipline, called *My Master's Course*, contributes to enhancing students' motivation to master foreign language competence. It opens with the text *Why Get a Master's Degree?* and proceeds with the description of Master's courses in the similar field from the sites of the leading universities in the world. While working on the text, students get the vocabulary necessary to make a presentation *My Master's Course*. The presentation should include personal educational background, the description of one's Master's course (specialty title, time schedule, disciplines, scientific research area, etc.) along with the information of one's career plans.

The section *Characteristic Features of Scientific Discourse* is devoted to grammar, vocabulary, stylistic and syntactic peculiarities of academic and scientific discourse in English. While working on this part, postgraduates are referred to scientific journals on Web portals (e.g. https://danieljhocking.wordpress.com/links/oa-journals/) to find an article in English that discusses a topic similar or close to their scientific research.

To acquire and develop experience in understanding scientific discourse, students are to analyze the language of the article (scientific and professional terminology, grammar constructions, etc.), practice their skills in translating and summarizing with the prospect to refer to the article in their Master's paper.

Students are supposed to work individually on the sections of the program *Academic Writing* and *Academic Reading*. The purpose of the section *Academic Reading* is to master postgraduates' skills in different strategies of academic reading: intensive or extensive reading, skimming and scanning. Intensive reading is decoding the message of a text. It deals with the study of grammatical and lexical features. The major objective of extensive reading is developing the ability to extract the necessary information from a text. Skimming is a strategy of previewing an article or a book when it is read quickly with noting only chief points. Scanning is reading for specific information.

Working through the tasks of the section students learn to look for such signals as headlines, introductory and concluding paragraphs, repeated key words and key terms, pictures, names, places and numbers, division into paragraphs, transitional words. Students are taught to see the idea of each paragraph, to find unknown words, terms which can be interpreted by the author of the article. Sometimes students may refer to explanatory or defining dictionaries, encyclopedias. At the same time, there is no need to look for each word in dictionaries as a meaning of some words is deduced from a context. Therefore, it is very important for a teacher to explain the strategy to work with vocabulary of a scientific text.

The requirement to develop skills in academic reading is caused by necessity to work with scientific texts which are notable for their general difficulty of perception, long paragraphs, special vocabulary and complex sentences. The ability to see similar features of a scientific text through its organization becomes a paramount task for a student.

The task of a teacher is to show typical structural features of these texts and to deduce their template character. For example, a scientific article has a uniform structure or a form which is appropriate for texts of any scientific field (introduction, analysis, results, conclusion); books, textbooks or guides are always divided into chapters, sections, etc.

The purpose of the section *Academic Writing* is to prepare postgraduates to write different kinds of scientific texts in English. For this purpose, postgraduates are to read the text *Writing an article* and to make a list of recommendations for writing a scientific article; to read the text *Writing a Conference Report* and to make a list of recommendations for writing a report for a conference; to define a difference between composing a scientific article and composing a report for a conference; to read the text *Writing a Thesis* and to make a list of recommendations for writing a thesis. It should also be noted that the language of the instructions itself serves the aim to master students' foreign language competence.

Special attention is paid to mastering the ability to write an abstract to an article in English since any person, who wants to publish the results of a scientific research in scientific publications, faces this problem. Hence, postgraduates are provided with the examples of articles and corresponding abstracts to carry out their detailed analysis. As a result, students are to prepare an abstract to their own article or some offered articles in English with the use of corresponding speech clichés.

The section *Public Speaking* contains clues and recommendations how to prepare a presentation, what information is to be included and excluded, how to use demonstration equipment, the speaker's appearance and formulas of business etiquette as well. Students are assumed to study the recommendations and to answer the questions grouped according to the sections of a thesis: *Field of Science and Research, Research Problem, Historical Background of Research Problem, Purpose and Methods, Results and Conclusion.* On the basis of their answers they make up a presentation *My Scientific Research* done in compliance with the recommendations offered, and also answer questions of the audience on their research subject.

The presentation *My Scientific Research* gives the students initial experience of future professional dialogue and interaction with colleagues and partners. It teaches how to select material for discussion proficiently and to state one's own thoughts, how to formulate one's own point of view correctly and to advocate one's personal interests as well as the interests of one's company. Besides, such activity, being correctly organized by a teacher, will remove psychological barriers and typical fears of contact with audience.

Considering that students, who are getting Master's degree in Ecology and Environment Management, have to take their examination on the "Foreign language" discipline at the end of the 3rd semester, they have already conducted certain research and at least intermediate results can be presented by that time. It should be noted that the general scope of all postgraduates' research is connected with the environmental issues of the native region incorporated in the global problems of environment. The research is based both on practical experience, and on extensive theoretical material received at the lectures and seminars. It allows creating harmonious conditions for incorporating the process of mastering foreign language competence in a general course of the Master's educational program. Besides, bilingual vision of the scientific project allows students to understand their research subject more deeply.

4 Conclusions

Qualified specialists in ecology and environment management are highly needed for mining industry to solve ecological problems and to prevent ecological disaster that may result from its work. Modern society demands from a specialist to possess competences, and foreign language competence is among the key ones. The content of the "Foreign Language" discipline and the approach to organize students' individual work for Master's course in ecology and environment management contribute to mastering foreign language competence of postgraduates as inherent condition of their professional efficiency and career development. It serves as "a bridge" for establishing foreign contacts, creates conditions for professional experience exchange. The result of the process is the integration of regional coal-mining enterprises of Kuzbass, possessing the staff of such employees, into world industrial community.

 DOI: 10.1051/e3sconf/20172104012

References

1. J. Habermas, *Moralbewusstsein und kommunikatives Handeln* (Suhrkamp, Frankfurt am Main, 1983)
2. R.W. White, Psy. Rev., **66** (1959)
3. D.C. McClelland, Amer. Psy., **28**, 1-14 (1973)
4. R.B. Brown, Pers. Rev., **22**, 25-36 (1993)
5. R.B. Brown, Manag. Learn., **25**, 289-299 (1994)
6. A. Snyder, H.W. Ebeling, J. Bus. Strat., **13**, 26-32 (1992)
7. H. Scarborough, Brit. J. Manag., **9**, 219-232 (1998)
8. J. Burgoyne, *Competency Based Approaches to Management Development* (Lancaster: Centre for the Study of Management Learning, 1988)
9. I. Bates, Stud. Sci. Educ., **25**, 39-68 (1995)
10. F. Delamare Le Deist, Hum. Res. Dev. Int., **8**, 27-46 (2005)
11. D. Gibbons-Wood, Education + Training, **42**, 24-32 (2000)
12. P. Humphreys, Journal of European Industrial Training, **21**, 63-69 (1997)
13. S.C. Schoonover, *Implementing Competencies: A Best Practices Approach* (London: Schoonover Co, 1998)
14. S. Shaw, *Development of Core Skills training in the Partner Countries* (ETF, 1998)
15. 15. Official Journal of the European Union. Recommendation of the European Parliament and of the Council of 18 December 2006 on key competences for lifelong learning. URL: http://eur-lex.europa.eu/legal-content/EN/TXT/?uri=celex:32006H0962

NBIC-Convergence as a Paradigm Platform of Sustainable Development

Elena Dotsenko[1,*]

[1]Plekhanov Russian University of Economics, Department of Political Economy and History of Economic Science, 117997, Moscow, 36 Stremyanny lane, Russia

Abstract. Today, the fastest rates of scientific and technological development are typical for the spheres of nano-systems and materials industry, information and communication systems, as well as spheres of direct human impact on environment – power industry, urbanization, and industrial infrastructure. Accelerate replacement of a human by machines and robots, the construction of megacities; the transportation of huge volumes of environmentally hazardous goods takes place against the background of intensive generation of knowledge, the transition of the results of fundamental research into specific production technologies. In this process, on the one hand, a fundamentally new format for technological restructuring of the world economy is being developed. On the other hand, a new platform for human-environment interaction is being formed, where both positive and negative environmental impacts will be determined by unstudied factors in the near future. The reason for this is in the forthcoming replacement of the technologies that are familiar to us, although dynamically developing, by fundamentally new - convergent. Entering the front line of technological development - NBIC-convergence - requires a new paradigm of sustainable development.

1 Introduction

The rapid development of bio- and nano-technologies is going faster than computer technology. The development of science, technology and the actual process of cognition, as a higher level of human intellectual development, were in a state of relative balance and harmony. However, with the rapid development of high technologies, the process of cognition cannot keep the evolutionary principles of development. The giant speed of the emergence of fundamentally new information, its scale and modification justifies the creation of a new format and principles for responding to the results of scientific and experimental activities. The procedure of analytical generalization and operational decision-making on the reduction of dangerous consequences of risks, the achievement of the desired effectiveness of symbiosis of science and technology, should be presented by electronic data processing complexes in the format of information technology and analytical standards.

[*] Corresponding author: ktyf110372@rambler.ru

This trend of development of scientific activity is realized in a relatively young sphere of social development - cognitive sciences and technologies. It is these technologies that will determine the form, content and intensity of human impact on the environment in the near future. And being the main engine of not only scientific and technical, but also economic, social progress, NBIC-convergent technologies will become a key for the sustainable development of the mankind in the 21st century.

2 Materials and methods

In the dialectic of cognition, the development of cognitive sciences represents a new stage in the development of meta-sciences, their consolidation into the system of a scientific worldview, taking into account a fundamentally new scientific and technological basis and the level of innovation development. It has a special importance for mining regions where sustainable development is constrained by technological [1-2], economic [3-6] and environmental [7-9] challenges. The development of cognitive sciences and technologies will be carried out at an accelerated pace and cover practically the entire spectrum of social and natural science disciplines, fundamental and applied research and development [11-13].

Biotechnologies and nanotechnologies, information and communication technologies and cognitive sciences and technologies closely interact with each other and in their totality form a system of converged NBIC technologies that predetermine the main directions of the progress of science and technology at the present stage of social development.

The effectiveness of such convergence is justified by the realities of the synergetic effect of the system development of basic technologies and science. Converged NBIC-technologies are characterized by many distinctive features. Here, first of all, there are basically new functional qualitative features:

- science is the dominant factor in the development of basic and converged technologies;
- the methodology of science, remaining the basis of knowledge, in the conditions of the development of cognitive sciences and the development of NBIC-technologies passes to a higher level of its development - the technology of cognition. As a result, the interpenetration of sciences and the integration of scientific knowledge into the overall system of the scientific worldview take place at the level of cognitive technologies;
- the technological nature of cognition procedures makes it possible to carry out strategic planning of scientific activities, to evaluate the realism of fantastic views, to foresee risks and to achieve the desired effect of scientific research and development;
- science as a system-forming link of the whole complex of NBIC-technologies forms the technological image of the whole sphere of production of goods and services.

The latter circumstance is extremely important for understanding the quality of sustainable development - progress or regress, the speed and nature of innovations, the scale of structural and dynamic changes.

3 Results and Discussion

So, if the current system of production, economic and social and ecological relations perceives the challenges of NBIC-technologies, receives real support in the form of demand and proposals from the production and economic environment and civil society, a harmonious sustainable development will start with the realization of environmental potential in the form of converged technologies. If this balance is broken, then the prospects for the development of NBIC-technologies will be depressing, regardless the scenario for their further development. They can go in isolation from the imperative of sustainable development. NBIC-technologies have enough internal motivation for this subject. However, having become a goal in themselves, they will lose environmental constraints. The results of scientific and experimental activities, irrespective of the needs for sustainable development, will lead to their universal human futility. The feasibility of the development of NBIC-technologies will be in doubt and in the case of weak public susceptibility of the synergy of social and economic and environmental effects. In this version, the converged technologies will not be able to develop due to contradictions with the motivational incentives contained in them.

It can be argued that NBIC-technologies are an objective reality, and not a product of virtual intellectual constructions. In the format of these technologies society receives a subtle instrument and at the same time a powerful impetus to its sustainable development.

Direction to the results of research and development achieved in the system of basic technologies becomes the vector of development of the economies for all countries. Thus, the list of Critical Technologies of the Russian Federation is approved by the President of the Russian Federation. It includes 34 critical technologies. In addition, eight directions for the development of science and technology in the Russian Federation have been approved, which also cover critical technologies. The Government of the Russian Federation also established a rule for the establishing, adjustment and implementation of priority areas for the development of science, technology and technics in the Russian Federation and a list of critical technologies of the Russian Federation. In other countries, critical technologies have become less used to manage the strategic development of the economy. For example, the United States, France, Japan and other countries, by the beginning of this century, gradually moved away from the idea of creating the list of critical technologies. The USA were the pioneers of building such lists. In the framework of the famous RAND Corp. even in the second half of the 1990s, the Institute of Critical Technologies preparing such lists was closed. There was an opinion that a large list of critical technologies needed for the economy sprayed budgetary funds, did not allow to concentrate resources on the latest breakthrough technologies. In addition, it restricted the initiative of scientists and engineers to choose future, challenging disputes, sometimes "futuristic" trends of technological development at this stage. In addition, in the XXI century science is changing so rapidly that the choice of specific priority technologies can serve as a certain deterrent.

At the same time, no matter what the attitude to critical technologies was, there is no doubt that they are significant for concretizing the choice of directions for development. The availability of the list of critical technologies is especially important for the Russian economy, where, against the background of terminological confusion in terms of priority areas and national priorities, the vector of scientific and technological development of the country for the foreseeable future has been worked out only in the draft version. It is necessary to start from something and, gradually enlarging (by means of generalization) the number of directions, to choose one of the following qualitative trends of scientific and technological development:

First, the progress in the field of NBIC-technologies with the subsequent access to a fundamentally new trajectory of sustainable development of the nation and the whole of mankind, with the provision of potential leadership positions of the most innovative countries;

Secondly, the elimination of the technological backwardness of the Russian economy and the consequent transition to a high-tech basis for the economic potential development. Such a quality of economic development is associated with a small gap from the world scientific and technological level and the possibility of accelerated development of certain NBIC technologies.

Whatever the objective function of sustainable development has been chosen as a priority, the list of critical technologies will be refined and brought into line with the tasks of scientific and technological development that are at the specific time. It is important that the list of critical technologies is not declarative, but is a real tool for managing science and technology.

It is obvious that the real assessment of the resource potential, the degree of progress of the scientific and technological directions, and the image of the country's economy in the medium and long term run should become the criteria for selecting qualitative parameters for economic development.

The revolution of NBIC-technologies, global challenges for sustainable development make changes in the traditional innovation policies and strategies in the developed countries all over the world. This was demonstrated with particular conviction by the global financial and economic crisis at the end of the first decade in the 21st century. Mankind is not protected from subsequent global challenges and crises, such as energy, environmental, demographic and others, which are difficult to predict today. At the same time, the convergence and synergy of NBIC-technologies can largely support governments in overcoming and mitigating the global challenges and crises of the 21st century, in particular changing the structure of the world trade, which will increasingly be saturated with innovative goods and services.

Another major trend of global development in the XXI century is the birth, at this stage, of various social structures or societies emerging as a result of the large-scale development of certain global technologies, which include NBIC-technologies. As a result of penetration of these technologies into the various political, social and economic and cultural subsystems of modern post-industrial civilization, the "information society", the "bio community", the "nano- society" and, in the long term, the "cognitive society", which in the world scientific literature is conceptually united under the general term "knowledge society», are being formed. However, convergence and synergy of NBIC-technologies will ultimately lead to social and economic, political and other types of convergence and synergy, forming a new "sustainable innovative society", and ultimately a stable and harmonious civilization of the 21st century.

Started in the late XX - early XXI century the active process of innovation and technological convergence means not only mutual influence, but also interpenetration of technologies, when the boundaries between individual technologies are blurred, and the final results appear in the framework of interdisciplinary research at the intersection of various fields of science and technology. Technological convergence is clearly manifested at the intersection of technologies that are part of the "bundle" of NBIC- technologies, which represent the merging of various sciences and technologies into a single scientific and technological area of knowledge. Such a scientific field will include almost all levels of organization of substance: from the atomic-molecular nature (nano), to the nature of life (bio), the nature of mind (cogno), and the processes of information exchange in material structures (info). Apparently, as a result, a meta-area of knowledge may arise.

At the international political level, the concept of technology convergence was developed by experts from a number of industrialized countries that are members of the EU and G-8, as early as the beginning of the 21st century. This concept is based on the materials of four conferences held in the USA (2001-2005) and reports of the European High Level Expert Group (HLEG).

In the US, these conferences were held under the auspices of the "National Nano-Initiative" of the US President and the NSF, and in Europe - under the auspices of the EU. Based on the results of the discussions, the principle of the "synergetic combination of the four scientific and technological areas", developing at an exceptional rate, was taken as a basis for the concept of technological convergence:

1) Nano-science and nanotechnology.
2) Biotechnology and biomedicine, including genetic engineering.

3) Information technologies, including the latest computer and communication technologies.
4) Cognitive sciences, covering neurosciences and cognitive technologies.

These areas have received the generally accepted and currently used in the world practice term "NBIC-science and NBIC-technologies". The technological convergence scheme was proposed in 2003 by American scientists M. Roco and U. Bainbridge from the National Science Foundation of the United States.

The development of NBIC-convergence forms a new living space for mankind, its new ecology. Atoms, molecules, information bits and neurons, which are fundamental elements of NBIC technologies, interact with each other in various ways in the process of convergence and synergy of these technologies. An important element in this conceptual system is the national and global policy of the industrialized countries, which accelerates or stagnates the development of NBIC technologies, and consequently, the entire process of the formation of the innovative economy of the 21st century. In accordance with the vector of these processes, a new policy of environment protection and improving living conditions is being formed.

Thus, in the line of the European High Level Expert Group (HLEG), the focus in the development of the EU innovation policy is not on convergence areas but on specific R & D areas that need to be implemented in order to move to the creation of converged technologies and, on their basis, to develop the corresponding marketable products by 2020-2030 years. This group has identified five main areas of research, in which the development of specific converged technologies and related devices and other promising innovative products are provided.

So, for example, the direction number 1 "Healthcare" includes the creation of the following objects that can massively change the habitat and lifestyle of people:
- technologies and so-called laboratories, mounted in chips for the purpose of conducting a quick scan of the health state and determining the diagnosis of the patient, including a patient remote from the medical facility;
- "smart prostheses", which are capable of responding to human brain signals and transmitting relevant sensory information back, and others.

Direction number 4 "Ecology" and direction number 5 "Power energy" include the development and creation of:
- new energy carriers and technologies for their storage, imitating natural systems;
- renewable energy sources, including photoelectric systems, hydrogen, geothermal and solar energy technologies;
- autonomous "smart homes" and "smart energy systems";
- biomimetic technologies that simulate natural processes and integrate photoelectric systems into "smart materials";
- environmental sensors, incorporated into the information and management systems.

It seems that the application of technologies in directions numbers 1-5 will have a generally positive impact on the environment. However, the high cost and inaccessibility of some equipment and technological innovative products for all members of society can, from the point of view of the principles of technological synergetic, cause certain social fluctuations and chaos in the society. The main emphasis in the innovation policy of industrialized countries is made on R & D aimed at strengthening the neuro-systems and potential capabilities of the human brain, as well as on the development of cognitive and computer neuroscience, nano-medicine, artificial intelligence, nano-electronics, biomedicine, enhancing the physical capabilities of a human (for example, creating exoskeletons) and etc.

4 Conclusion

 DOI: 10.1051/e3sconf/20172104013

Analysis of global trends and forecasts of the EU and the US shows that at the moment some converged technologies from the group of NBIC-technologies are still at the embryonic stage of development. However, in the future until 2030, these technologies should take a leading position in the global innovation economy and become a priority for sustainable development of G-8 countries, as well as some countries in Asia. Convergence and synergy of NBIC-technologies will lead to the formation of new trends of the "green economy" - nano-economy, bio-economics, information economy, cognomics (economy using cognitive technologies). Under their influence, a transition to new forms of social development aimed at harmony of the interests of mankind and environmental protection is expected. Such a transition will be accompanied by a change in ecological culture, values in society, a new psychology of social development, the emergence of new moral and ethical norms, which is already taking place now. Synergy of NBIC-technologies has a direct or indirect impact on people's ecological thinking and their readiness to develop environmentally safe products, create demand for it in local and global markets. NBIC-technologies will affect the transformation of people's thinking, which may entail a change in the content and structure of modern civilization or lead to a transition from a post-industrial to a harmonious with nature society.

References

1. A.A. Khoreshok, L.E. Mametyev, A.Y. Borisov, A.V. Vorobyev, IOP Conf. Ser.: Mater. Sci. Eng., **127:1**, 012039 (2016)
2. A.A. Khoreshok, L.E. Mametyev, A.Y. Borisov, A.V. Vorobyev, IOP Conf. Ser.: Mater. Sci. Eng., **127:1**, 012014 (2016)
3. S.A. Zhironkin, Ugol', **6,** 62-63 (2001)
4. S.A. Zhironkin, Ugol', **4**, 47-49 (2002)
5. V. Trifonov, O. Loyko, D. Nesteruk, S. Zhironkin, E. Strekovtsova, AIP Conf. Proceed., **1800**, 050009 (2017)
6. E.V. Kucherova, T.A. Ponkratova, T.A. Tyuleneva, N.A. Cherepanova, Economics and Innovation Management, **1**, 76-89 (2017) DOI: 10.26730/2587-5574-2017-1-76-89
7. N.N. Golofastova, V.G. Mikhailov, I.V., Seredyuk I.V. Economics and Innovation Management, **1**, 66-75 (2017) DOI: 10.26730/2587-5574-2017-1-66-75
8. A.V. Myaskov, A.S. Ilyin, S.M. Popov, Gornyi Zhurnal, **2**, 51-56 (2017)
9. O. Erdem, T. Güyagüler, N. Demirel, J. Sout. Afr. Inst. Min. Met., **112:5**, 405-412 (2012)
10. S. Zhironkin, M. Gasanov, G. Barysheva, K. Kolotov, O. Zhironkina, E3S Web of Conf., **15**, 03012 (2017)
11. A. Balabanova, V. Balabanov, E. Dotsenko, N. Ezdina, E3S Web of Conf., **15**, 04013 (2017)
12. S.A. Zhironkin, K.A. Kolotov, O.V. Zhironkina, Economics and Innovation Management, **1**, 4-16 (2017) DOI: 10.26730/2587-5574-2017-1-4-16
13. Z. Jurkasová, M. Cehlár, S. Khouri, Int. Conf. on Engin. Sci. and Prod. Man., ESPM, 409-412 (2015)

DOI: 10.1051/e3sconf/20172104014

Sustainable Development Strategy for Russian Mineral Resources Extracting Economy

Elena Dotsenko[1,*], *Natalya* Ezdina[1], *Angelina* Prilepskaya[2], and *Kirill* Pivnyk[3]

[1] Russian University of Economics, Department of Political Economy and History of Economic Science, 117997, Moscow, 36 Stremyanny lane, Russia
[2] Russian University of Economics, Foreign Languages Chair No.2, 117997, Moscow, 36 Stremyanny lane, Russia
[3] ZAO Comenergo, 127591, Moscow, 77 Dubninskaya Street, Russia

Abstract. The immaturity of strategic and conceptual documents in the sphere of sustainable development of the Russian economy had a negative impact on long-term strategic forecasting of its neo-industrialization. At the present stage, the problems of overcoming the mineral and raw material dependence, the negative structural shift of the Russian economy, the acceleration of the rates of economic growth, the reduction of technological gap from the developed countries become strategically in demand. The modern structure of the Russian economy, developed within the framework of the proposed market model, does not generate a sustainable type of development. It became obvious that in conditions of the market processes' entropy, without neo-industrial changes, the reconstruction of industry on a new convergence-technological basis and without increasing the share of high technology production the instability of macroeconomic system, the risks of environmental and economic security of Russia are growing. Therefore, today we need a transition from forming one industry development strategy to the national one that will take into account both the social and economic and environmental challenges facing Russia as a mineral resources extracting country.

1 Introduction

The strategy of sustainable development is considered to be the program of long-term changes in the structure of the Russian economy, integrating the mechanism of interaction between the state, business, ecologically significant subjects of the economy and the population. The implementation of the sustainable development strategy should include a number of steps that must be followed by certain strategic directions - the components of Russia's social and economic development associated with the neoindustrialization of the country's economy.

2 Materials and Methods

*Corresponding author: ktyf110372@rambler.ru

DOI: 10.1051/e3sconf/20172104014

Delaying of sustainable development problem's solving in Russia forced the discussions among domestic and foreign researches. These discussions concern mining technologies development [1-5], creating basis of economic growth and improving socio-economic situation [6-9], finding new possibilities of lowering the pressure of mineral resource extraction on environment [10-16]. As a result of these discussions the new strategy of sustainable development must receive its fundamental bases.

The stages of programming and implementation of sustainable development strategy include the following:

At the first stage, the implementation of the sustainable development strategy should completely change the raw material production of the 4th technological mode. This implies the automation and robotization of coal, oil and gas, metal ores extraction and an increase in labor productivity in these industries by 1.5-2 times (up to the level of the US and EU companies), the transfer of these industries to the full cycle of production of modern products of organic chemistry, modern alloys and carbon plastics, full supply of the latest materials of domestic engineering, construction industry. Technological base of the first stage should become available technological platforms, and investors - high-tech holdings. With regard to strategic changes in the structure of property relations and institutions, at the first stage, the state needs to initiate vertical integration of the subjects of the raw, processing and sectoral sectors, legislatively establish the most attractive conditions for investing innovations that increase the technological level of raw material extraction and processing, creating conditions for the start of environmentally friendly production.

At the second stage, it is necessary to ensure the modernization of the processing enterprises of the 5th technological mode: auto, aircraft, shipbuilding, and energy. It is these sectors that should ensure the improvement of the ecological level of deep processing of raw materials (the 4th technological mode), which will expand the scope of application of environmentally friendly technologies and enhance their international competitiveness. We believe that it is impossible to do without imports of modern environmentally friendly technologies and their adaptation in existing technological platforms and innovative clusters, with investment support for existing in Russia commodity holdings, state guarantees and insurance. That is, the role of the state at the second stage is to stimulate the transition from the export of raw materials to its processing, large-scale investment of environmentally friendly technologies in innovative clusters, and the formation of a powerful environmental lobby.

At the third stage, the development of the 6th technological mode processes is connected with convergent technologies and the emergence of new industries, in which generated environmental innovations can find demand in the raw and processing industries. These are such convergent branches of the 6th technological mode as bioenergy and the production of environmentally friendly fuel, industrial distributed computing programming, and nanomaterial engineering. The technological basis of the third stage should be formed on a cross-platform principle, in network clusters, with large-scale government support. The state at the third stage of implementation of the neo-industrial strategy should concentrate on the development of inter-cluster and cross-platform interactions, maximum benefits and investment in the development and implementation of environmentally friendly technologies, in the creation of network clusters for the production of environmentally friendly materials.

It should be noted that, without government support, high risks of investing in R & D and industries in the sphere of environmentally friendly technologies discourage large Russian and international investors. At the same time, in Russia there is no real support for the neo-industrial maneuver necessary for sustainable development either in the industrial sector or in the strategic documents of the Government. The share of high-tech products of the 6th technological mode in gross Russian exports does not exceed 1.5%, while the share of innovative enterprises is below 10%. The task of creating 25 million high-tech jobs by 2020 and a general increase in labor productivity in one and a half times that of 2010 (due to the growth of investments and technological renewal of industry, state support for R & D and the upgrading of employees' skills) is a demonstration of ambitions that are not backed by any strategic documents. Therefore, if in 2012 the share of products of the 6th technological mode was about 5% in the US GDP, in Russia it was 0.3%, which corresponds to the contribution of fundamental science to the production of the final product of civil industries.

3 Results and Discussion

In order for the structural transformation of the Russian economy to take place in the form of neo-industrialization, rather than the consolidation of raw material specialization, it is necessary to develop a strategic union of the state and business. Today, they cannot separately initiate the sustainable development, because the state does not have corresponding institutions that determine the conditions for the inter-sectoral redistribution of financial, labor, intellectual, production resources. In turn, Russian business does not show interest in mass investment in environmentally friendly technologies.

Therefore, the neo-industrial strategic union is understood, first of all, as an institutional solution to the problem of interaction between business and the state, that is, as a definite long-term plan. On the other hand, such an alliance is a combination of forms of interaction between the state and business - integration, investment, credit, emission, and social. The following strategic directions of the structural policy of sustainable development can be identified:

1. *The emphasis on the use of basic* (natural, energy, accumulated intellectual capital and educational level of the population) and *domestic* (funds of the state budget and extra-budgetary funds, capital of investors and banks, market potential of special economic zones, technological parks, interregional sales networks), but not *external resources* (funds of international funds, foreign capital, international marketing). This means involving the financial capital of the banking system, the stock market, the largest national corporations in the process of investing innovations, manufacturing the products of the 6th technological mode industries for the domestic market and its promotion abroad, by creating appropriate tax, credit incentives for the state, promoting the integration of R & D and high-tech processing, and also raw material productions, by using the market potential of IT, biochemical, nuclear-energy clusters.

2. *Replacing* the established after the "return of the state" in the 2000s term of long-term effectiveness of Russian business "*participation in authority*" on "*participation in innovation*". To do this, it is necessary to actively apply neo-dirigisme that is the full-scale use of indirect economic regulators by the state, not only in the financial sphere (in the sense of monetarists), but also to stimulate the inter-sectoral flow of investment, labor, intellectual capital, technology transfer, the development of regional clusters of productions 6th technological mode and small innovative entrepreneurship. To do this, it is important, first of all, to create the required institutional environment, which includes both state institutions that fully encourage innovation and the priority of breakthrough technologies in state support of industry, as well as business institutions that can reduce its transaction costs when investing innovations.

3. *The creation of long-term conditions for sustainable growth of wellbeing* in the process of formation of environmentally friendly production in processing industries on a new technological basis, and active support of the state for environmental entrepreneurship. As a key, long-term goal of structural reforms in the Russian economy, the growth of wellbeing is impossible without reducing the number of loss-making industrial enterprises (by 2014 - up to 40%), expanding employment in manufacturing and high-tech industries. At the same

time, raw materials extraction remains the most profitable type of business in the Russian industry. The narrow layer of innovative firms is unstable, and most of these business entities - enterprises, banks, and venture funds - are in fact not engaged in innovations, but import of finished products and shallow processing of raw materials, violating the ecological balance.

Therefore, the union of the state and business, aimed at changing the most stable elements of the structure of the national economy - formal and informal institutions, means of production, mechanisms for inter-sectoral and inter-regional distribution of factors of production and resources - should become the first step towards sustainable development in the process of neo-industrial structural transformations. That is, by influencing on the most stable elements of the structure of the national economy, the state can initiate a reduction of obstacles to sustainable development by stimulating innovative activity in manufacturing industries, involving large businesses in investing innovations. To do this, it is important to develop legislation concerning public-private partnerships, high-tech holdings, tax incentives, government loans and guarantees for innovative firms, to initiate a neo-industrial lobby. This will give impetus to the change in technological, cluster structure, due to redistribution of resources between environmentally hazardous and safe production, between raw materials and innovative clusters.

4. *The development of own scientific potential, import and adaptation of* nature-saving *technologies* instead of import of production equipment. It is necessary to take into account the experience of Argentina in the implementation of the strategy of "neo-nationalism" (comprehensive support of the national industry and domestic investors). Due to the fact that after the economic crisis in the late 1990s, the manufacturing industry was developing at a faster pace, the Argentine economy experienced a positive structural shift (the share of the modern 5th technological mode in 1997-2007 increased from 15 to 35% of GDP). This was preceded by a five-year economic policy of the state, including the return of the natural rent to the state, the creation of state-owned scientific and industrial enterprises (mainly in agriculture), large tax breaks to R & D organizations, partial control over raw materials prices, the use of state investments to increase the capitalization of the largest manufacturing companies, transfer of the latest technologies as a prerequisite for multinational corporations' foreign direct investment. As a result, Argentina overcame the dominance of agriculture and became one of the industrial countries that came close to the beginning of sustainable development.

Realization of the potential for sustainable development of the Russian economy is impossible without the development of technological platforms and public-private partnerships in innovative clusters. Public-private partnerships in Russia play the role of attracting business entities to maintenance, operation, reconstruction, modernization or new construction of public infrastructure facilities on terms of sharing risks and responsibilities. But the state does not take responsibility for fulfilling the main condition of sustainable development - changing the institutional and reproductive structure of the economy in such a way as to ensure a massive inflow of investments in environmentally friendly technologies and speed up the replacement of existing production facilities with new ones that take into account the growing requirements of environmental safety.

Neo-industrial structural transformations require an early overcoming of the existing disintegration of science, production and finance in the conditions of reforms. Today, scientific institutions, small innovative enterprises, development companies practically have neither a financial base nor an opportunity to produce an innovative product, since their connections with large industrial enterprises were not established during the reform period.

Considering this, for the development of resource-technical and technological factors of sustainable development of the Russian economy, the following tools can be offered:

 DOI: 10.1051/e3sconf/20172104014

1. *Neo-industrial state order*, which should become a means of changing the trajectory of the development of relations between innovation and production business and the state towards sustainable development. We understand it as the development of market demand for environment friendly technologies, means of production and ready-to-introduce technologies for environmental protection by the state. The economic mechanism of such state order should consist of three parts:

a) an order for R & D projects on environmentally friendly technologies in the framework of the appropriate sustainable development strategy, which the Russian government should adopt;

b) an order for the production of products capable of reducing environmental damage from the activities of extractive enterprises dominating in the Russian industry;

c) state guarantees and state export crediting not less than 15% of environmentally friendly technologies produced within the framework of the neo-industrial state order, which should serve as a guarantee of their high international competitiveness.

As a promising organizational form of neo-industrial state order, a public-private partnership is considered. Within it an investment fund and venture capital funds, specialized tax and customs privileges should be formed.

2. *Selective investment support for innovative projects* of environmental-friendly technologies. When examining them, it is necessary to take into account the technological level of the replaced products, in order to concentrate government investments on the most promising for sustainable development of the Russian economy projects. To do this, it is proposed to classify enterprises - potential beneficiaries of state investments, in the following way: users of foreign technologies (not implementing the full cycle of R & D), converters of environmental technologies (adapting them to Russian production conditions using their own scientific capabilities), effective innovators (using R & D sporadically for individual projects) and breakthrough innovators, for which the creation of environmentally friendly technologies is the main factor of competitiveness.

For the least developed in Russia manufacturing and high-tech industries, such as radio electronics, instrument making, computer building, production of ultra-light and super-strong materials, microelectronics, robotics (in which the share of imports reaches 95-100%), it is advisable to provide the state orders and state investments both to breakthrough and effective innovators, as well as to the firms-converters of environmentally friendly technologies.

In turn, in industries with 60-80% import share (chemistry of semiconductors, aircraft construction, electro-instrument engineering, machine-tool building, satellite engineering, etc.), the investment state support is appropriate for efficient and breakthrough innovators. Finally, for industries with an import share of 40-60% (organic chemistry, vehicles), the state order and state investments are needed only for breakthrough innovators.

3. *The import of environmentally friendly technologies and highly qualified specialists*, which should eventually replace the import of the finished product. Along with this, it should be emphasized that there is a need for a comprehensive use of investment, organizational, institutional, financial and credit instruments of structural transformation of the Russian economy for the development of non-industrial import substitution in it.

4 Conclusion

The mechanism of development of neo-industrial structural transformations in the Russian economy aiming to enter the way of sustainable development reflects a multiplicative effect. It consists in the consistent modernization of the extractive and processing industries based on environmentally friendly technologies, in systematic interaction between the state and business in the development of the necessary institutions, in the formation of a neo-

 DOI: 10.1051/e3sconf/20172104014

industrial state order, involving technological platforms and innovation clusters, implementing interest rate policies, state guarantees and investments.

References

1. A.A. Khoreshok, L.E. Mametyev, A.Y. Borisov, A.V. Vorobyev, IOP Conf. Ser.: Mater. Sci. Eng., **127:1**, 012014 (2016)
2. A.B. Efremenkov, A.A. Khoreshok, S.A. Zhironkin, A.V. Myaskov, IOP Conf. Ser.: Earth Environ. Sci., **50:1**, 012009 (2017)
3. A.A. Khoreshok, L.E. Mametyev, A.Y. Borisov, A.V. Vorobyev, IOP Conf. Ser.: Mater. Sci. Eng., **127:1**, 012039 (2016)
4. M.A. Tyulenev, S.A. Zhironkin, O.I. Litvin, E.A. Tyuleneva, O.V. Zhironkina, S.O. Markov, Geotech. Geol. Eng. **35:5**, 2065–2077 (2017)
5. S. Zhironkin, M. Gasanov, G. Barysheva, K. Kolotov, O. Zhironkina, E3S Web of Conf., **15**, 03012 (2017)
6. S.A. Zhironkin, Ugol', **4**, 47-49 (2002)
7. S.A. Zhironkin, Ugol', **6,** 62-63 (2001)
8. V. Trifonov, O. Loyko, D. Nesteruk, S. Zhironkin, E. Strekovtsova, AIP Conf. Proceed., **1800**, 050009 (2017)
9. E.A. Gasanov, M.A. Gasanov, Economics and Innovation Management, **1**, 30-38 (2017) DOI: 10.26730/2587-5574-2017-1-30-38
10. N.N. Golofastova, V.G. Mikhailov, I.V., Seredyuk I.V. Economics and Innovation Management, **1**, 66-75 (2017) DOI: 10.26730/2587-5574-2017-1-66-75
11. A.V. Myaskov, A.S. Ilyin, S.M. Popov, Gornyi Zhurnal, **2**, 51-56 (2017)
12. O. Erdem, T. Güyagüler, N. Demirel, J. Sout. Afr. Inst. Min. Met., **112:5**, 405-412 (2012)
13. A. Balabanova, V. Balabanov, E. Dotsenko, N. Ezdina, E3S Web of Conf., **15**, 04013 (2017)
14. S.A. Zhironkin, K.A. Kolotov, O.V. Zhironkina, Economics and Innovation Management, **1**, 4-16 (2017) DOI: 10.26730/2587-5574-2017-1-4-16
15. Z. Jurkasová, M. Cehlár, S. Khouri, Int. Conf. on Engin. Sci. and Prod. Man., ESPM, 409-412 (2015)
16. T.V. Kiseleva, V.G. Mikhailov, V.A. Karasev, IOP Conf. Ser.: Earth Environ. Sci., **45:1**, 012013 (2016)

 DOI: 10.1051/e3sconf/20172104015

Humanity and Environment Co-influence in the Shadow of Technological Convergence

Natalya Ezdina[1,*]

[1]Plekhanov Russian University of Economics, Department of Political Economy and History of Economic Science, 117997, Moscow, 36 Stremyanny lane, Russia

Abstract. Today one of the most important issues is the search for controlled ways of sustainable development of civilization, society, and a human. Previously, the development of society was determined by natural and evolutionary factors based on the principle of natural selection of living organisms and populations. However, as social institutions are developing, the technological factors of development, determined by the ideas about this phenomenon – the special arrangement of productive forces and the need to protect or change the environment, began to appear. The end of the 20th century was marked by the realization of a new, universal form of economic and technological synthesis - convergence, which, in the process of developing science and improving technology, became so comprehensive and widespread that today it can be called the force that transforms the environment. In this regard, it is convergent technologies that become the focus and the basis for research of the processes uniting the main spheres of human activity - science, society and technology, which will lead to a new form of interaction with the environment, conditioned not only by scientific and technological changes and social factors, but also new organizational forms of their interaction.

1 Introduction

Convergent technologies arise at the intersection of different technologies, for example bio- and nanotechnologies, information and cognitive technologies, etc., which requires inter-disciplinary and trans-disciplinary research of the science and technology. For example, for Russian economy if means technological alley between mining, machinery, chemistry, bio-technologies [1-4]. Huge speed of development of the technogenic world, the emergence of global information and communication technologies, a giant network covering the whole world, modify modern spirituality in its various ways. Hypermedia transforms not only the reality itself; it changes the public consciousness, the consciousness of a person, his identity. The expansion of modern technologies forms their risks and dangers, a threat to human and his existence on the earth. But there are also optimistic forecasts of technospheric genesis, promising to improve the environment, the very essence of a human, to create his har-

* Corresponding author: ktyf110372@rambler.ru

 DOI: 10.1051/e3sconf/20172104015

monious ties with nature. This is the essence of the new directions in science – transhumanism and immortology. In this connection, it becomes necessary to study all the features of this process, to promote its positive results and to counteract the negative. It is necessary to understand the causality of the mutual influence of technologies, their desire to converge not only among themselves, but also with the environment.

In the 21st century, the basis of all social changes is a science. The instrument of all these changes is technology - scientific, industrial, social, social and cultural. The latter include the actual technologies of influence on people - individually and massively: advocacy of innovations, organizational and self-organizational projects, change of social and individual consciousness and culture. In fact, this is a new kind of social practice aimed at sustainable development.

2 Materials and Methods

The meaning and goal of positively oriented social and cultural phenomena is to set such a way of changing social structures in which the processes of the formation of the new - integrative (convergent) and disintegrative (divergent) - go purposefully and in an organized way - from the lowest to the highest levels upward the whole hierarchy.

Therefore the term "convergence" sets a new conception of the new reality of co-organization of society and science, which requires philosophical reflection and analysis. Although the phenomenon of convergence has become widely known due to the so-called NBIC-convergence (nano, bio, info and cognitive technologies) it should be noted that convergence appeared unexpectedly in many areas - in science, technology, and society [5-6]. The idea of universal convergence aroused a wide resonance in society and scientific circles, took upon itself the dominant role of the development of science, technology and the society itself. Convergence has gained a dominant position in the sphere of interaction of technology, science and environment, as well as the whole society as a whole. It also requires new standards of corporative relations [7-8[and cross-cultural communications [9].

The rapid development of science-intensive technologies that occurs in the 20th century, or rather in its second half, has created a number of problems for the solution of which it is necessary to actualize the origin of synergies and convergence of not only technologies and knowledge but also the environment of the person's life with its values and ethical imperatives, in relation to the values of sustainable development, which also requires a philosophical understanding and justification. Convergence acquired leadership status in the development of science, technology and the society itself, began to dominate in the processes of their interaction, and this required its inclusion in the field of sustainable development.

The first school of thoughts is represented by the theories of the industrial society (R. Aron [10]), the postindustrial society (D. Bell [11]), the "growth stages" (W. Rostow [12]), who proposed the "optimism" of technological determinism as the most adequate to the spirit of modernity. In the concepts of this direction, modern technologies act as locomotives for techno-social development. It is scientific and technological knowledge in the form of technology that moves the transformation of industrial society into a postindustrial one, and also helps to smooth out ideological, social and economic differences.

The second school of thoughts can be called "pessimistic". It represents the negative impact of a technogenic society on the transformation of modern society and its criticism. This direction was represented, first of all, in the 1960-70s of the XX century in the reports of the Club of Rome (D. Meadows etc. [13]). In their writings it was noted that the domination of technology, not least of information, "leads to the loss of freedom, alienation of the individual, the dehumanization of social relations and, ultimately, the death of civilization."

The third school of thoughts is connected with the works made by M. Castells [14], A. Toffler [15] and others which appeared in the early 1970s. These concepts are suitable for the social consequences of revolutions: scientific, technological and informational - soundly and balanced. Such a broader approach allows us to consider scientific and technological development in co-evolution with the development of changes in the environment, generating both positive and negative consequences and problems. Emerging technologies free a person from heavy physical, routine mental work, give hope for improving the functionality of a human being, including his corporeality, ability, way of life. At the same time, these concepts mark the emergence of ecological consequences of the use of new technologies.

The methodology of the convergence concept was further developed in the works of Helga Novotny, a modern sociologist and philosopher. Helga Novotny [16] introduces the concept of an emergent interface. The interface is the basis of interaction, the means of communication of systems among themselves.

The concept of interface has two meanings. The first meaning is the division, the partition, and the second meaning is the interaction of any two systems for information exchange. Its peculiarity consists in the fact that the interface generates completely new effects or properties that are different from the effects or properties of interacting systems. The system suddenly has qualitatively new, emergent properties that are not equal to the sum of the properties of its parts. H. Novotny extended the idea of the interface to the field of knowledge, where different forms and areas of knowledge intersect. Such boundary areas of knowledge can become messengers of future breakthroughs that reduce the level of complexity that can optimize them. This situation of "unexpected meeting" of various fields of knowledge, the emergence of emergent interface, gives impetus to the convergent technological development of society. The concept of "emergent interface" makes a new contribution to the development of the idea of unification, "fabric", intertwining all sorts of activities: scientific and technological. As a result, the concept of the interface and the concept of an innovative, environmentally friendly idea arise within the scientific community.

3 Results and Discussion

Convergent technologies are not only a purely technical tool of activity. They should have a broad interpretation such as the attitude of a human being to the environment, which includes the activity and reflexive components, which allows taking into account the ecological component in convergence. Consequently, the NBIC-convergence model is not sufficient to answer the challenge of modern environmental requirements, since it carries a reflexive component that can only be tested with the help of SNBIC-convergence, which includes an environmentally-oriented consciousness of people.

In the era of global crises and changes, the specification of the scientific and technological development of mankind changes which requires re-evaluation and reassessment. There are acute social problems such as the threat of man-made disasters, pollution of the environment, Russia's position in the world, the situation in the sphere of education, energy security, etc., which ecology should study. As its purpose is to embrace the reality from the top, see the depth of dangers and risks, to build a picture of a new reality, as a plan to overcome routine, break stereotypes and anticipate, i.e. preparing and organizing the coming changes. The economics of nature use as modern science possesses cognitive methodological tools that allow not only to comprehend the problems of modern development of society, but also to construct the future, since it immanently exists in the present.

Likely crisis of the global social and economic system, the danger of environmental disasters, the transition to a new technological order requires new tools, algorithms, sustainable development technologies, and a new paradigm not only in science and technology but also in the environment. Rapid progress in the processes of convergence and the development of convergent technologies has become a trend of the times, a prerequisite and dominant of the development of modern society, allowing us to talk about the future unification of science, modern technology, society and a man into a single integrated complex. The starting point of our study is the hypothesis of M. Castells "that all the main trends of changes that make up our new, confusing world are correlated, and we can make sense out of their relationship."

The ground of our research was also the placement of a process of revolutionary technological change, such as the emergence of technology convergence, into a context of sustainable development without which it cannot take place and be formed.

Our hypothesis is the following: earlier science as such, including ecology, discovered new phenomena of nature, analyzed them, built on this basis theories, on which then engineers made inventions, created technologies and new models of technology, which, in turn, became the basis of new types of production. And all this happened spontaneously, producing a negative impact on the environment. Today, the question about the integration of all these stages into a single conceptual and organizational system is raised. It will be the basic project of sustainable development, which initially assumes mutual interest and mutual coherence of all these stages, i.e. immediately eliminating occasional useless deviations and related misunderstandings and conflicts, as well as unnecessary costs for unpromising attempts to improve the relationship between a man and nature.

Therefore, on the one hand, convergence appears as a general regulatory mechanism (configurator) of modern development of society. On the other hand, we define a convergent paradigm as an aggregate of technologically and socially acceptable ways and methods of sustainable development of mankind. Representation of convergence as a paradigm has arisen not casually. The unifying tendencies of the scientific and technological development of the society were previously characterized by various paradigms, for example, interdisciplinarily, trans-disciplinarily, integrativity, synthesis. In modern technogenic world, the technological paradigm, built on the convergence of science, technology and society, has become a semantic factor

At the present stage of society development the paradigm of convergence arises as a combination of technological, organizational, ecological innovations. The advantages of this are not only in the production of new products and systems, but most of all - in the dynamics of the regulation of the structure of various systems, as a result of their spontaneous interaction with the obtaining of a new synergistic effect - the accidental emergence of new beneficial results that have arisen on the basis of systemic quality. "The key factor" of this paradigm is not a specific investment of resources - cheap energy or cheap information, derived from the successes of power industry or microelectronics and telecommunications technology, as it was in previous technological paradigms, but collective, intra-system interaction. As a result, a new scientific and technological result is obtained, which inevitably leads to a drop in relative costs and universal availability of production, as well as generating the corresponding social transformations. The difference of the convergence paradigm from the modern technological paradigm lies in the transition from technology based on cheap sources of energy, information, to technology based on a new tool, a new regulator, a new form-maker leading to the unification of systems, with a new result. As such systems, firstly, the technologies that form clusters can be used, by which we mean the unification of several homogeneous elements, which can be regarded as an independent unit with certain properties. And, second, science and technology, forming a new complex – techno-science.

The concept of convergence has a complex structure: it is based on a holistic paradigm (the idea of absolute all-embrace and integrity); the nature of its generality (metaphoricity) is based not on the similarity of objective features, but on the similarity of the structural qualities of systems of any kind - from scientific and theoretical to cultural ones; it has a mirror side - a divergence, which indicates its spontaneously dialectical nature; synergetic based on the mechanism of extrapolation of conclusions and understanding of the nature of development as self-organization on the basis of nonlinear and bifurcation processes of the emergence of a new in science, technology and society.

The development of innovative technologies, a new science and their widespread introduction into all fields of human activity, sets before the researchers exclusively new classes of cognitive tasks, for the solution of which it is necessary to comprehend convergent technologies that bring about the convergence and interpenetration of science, technology and a man. The sixth wave of technological development, the beginning of which was expected from 2010, many scientists associate with the development of convergent technologies that interlink, unite and generate new fields of knowledge, at the intersection of such areas as biotechnology, nanotechnology, life design, human change, new nature use, robotics, new medicine, high humanitarian technologies, projecting the future and managing it, etc.

The combination of cognitive, intellectual technologies with information and network ones creates a sphere of convergence - the formation of new social systems that can harmonize the human impact on nature - institutions, laboratories, industries, from politicized groups to parties, etc. - on the basis of free information exchange, the conclusion of contracts and transactions. The general scheme of such a connection is called "SNBIC-convergence", and its global use allows us to speak about management of development of all social spheres, social institutions, forms and activities and organization, i.e. society in general and, potentially, the environment.

Prospects for environmentally oriented deployment of SNBIC-convergence are based on such socially significant innovations as nanotechnology (revolution in material production), biotechnology (from genetically modified products to human genotype change), information technologies (from the global Internet to artificial intelligence), and cognitive technologies. These technologies, on the one hand, allow eliminating any obstacles to the spread of convergent ideas, but on the other hand, they give opportunities for abuse of irresponsible individuals and organizations seeking profit or power at any cost (from advertising unproven products to projects of new global wars). This creates a problem of control over processes of technological convergence.

Prospects of social and cultural unfolding of "NBIC-convergences" in the concept of sustainable development concern not only the society, but also the person himself: his body (cyborgization), thoughts (chipping), character (psycho-programming), evolution (unlimited development), his values (unlimited enjoyment – paradise-engineering – immortality); his capabilities (the extropian ideology of unlimited self-perfection of a person's abilities), etc. That is, convergent technologies can become, on the one hand, the engine of sustainable development, on the other hand, a means of transforming a person into a post-human, in the face of technological and cognitive singularity. However, the scale of these ideas is certainly superior to the person's available opportunities - both in the implementation of these ideas, and in anticipation and prevention of possible negative consequences. This creates the problem of a new definition of human nature and the criterion of humanity within the framework of the environmental responsibility ethics.

4 Conclusion

Thus, the basis of convergent technologies is post-non-classical cognitive methodology of the formation of new ways of working and thinking tools, the value of which has to be justified specifically and separately from science – in the field of economic growth and sustainable development. The socialization of this practice generates the eco-techno-economy – a combination of technological, organizational, economic and organizational innovations. Its expansion through information and network structures sets a new foundation for sustainable development and a new hierarchy of social structures, the focus of which is nanotechnology – the totality of the newest means of material production.

DOI: 10.1051/e3sconf/20172104015

References

1. A.B. Efremenkov, A.A. Khoreshok, S.A. Zhironkin, A.V. Myaskov, IOP Conf. Ser.: Earth Environ. Sci., **50:1**, 012009 (2017)
2. S. Zhironkin, M. Gasanov, G. Barysheva, K. Kolotov, O. Zhironkina, E3S Web of Conf., **15**, 03012 (2017)
3. A.A. Khoreshok, L.E. Mametyev, A.Y. Borisov, A.V. Vorobyev, IOP Conf. Ser.: Mater. Sci. Eng., **127:1**, 012039 (2016)
4. A.A. Khoreshok, P.V. Buyankin, A.V. Vorobiev, A.A. Dronov, IOP Conf. Ser.: Mater. Sci. Eng., **127:1**, 012014 (2016)
5. S.A. Zhironkin, K.A. Kolotov, A.E. Genin, F.V. Agafonov, S.A. Kovalevsky, IOP Conf. Ser.: Earth Environ. Sci., **50:1**, 012011 (2017)
6. F. Agafonov, A. Genin, O. Kalinina, O. Brel, O. Zhironkina, E3S Web of Conf., **15**, 04011 (2017)
7. S.A. Zhironkin,.Ugol', **4**, 47-49 (2002)
8. E.V. Kucherova, T.A. Tyuleneva, N.A. Cherepanova, Economics and Innovation Management, **1**, 45-51 (2017) DOI: 10.26730/2587-5574-2017-1-45-51
9. I. Pevneva, O. Gavrishina, A. Rolgayzer, M. Agienko, A. Myaskov, E3S Web of Conf., **15**, 04016 (2017)
10. R. Aron, *Démocratie et totalitarisme* (Calmann-Lévy, Paris, 1965)
11. D. Bell, The coming of post-industrial society. Venture in social forecasting. – N.Y.: Collman Pub., 1973. – 884 p.
12. W. Rostow, *The Process of Economic Growth* (Nation, New York, 1952)
13. D. Meadows, J. Randers, *Limits to Growth: The 30-Year Update* (Ventura Pub., Los Angeles, 2004)
14. M. Castells et al., *Mobile Communication and Society: A Global Perspective* (MIT Press, Cambridge MA, 2006)
15. Toffler A. Powershift. Knowledge, Wealth and Violence at the Edge of the 21st Century. - N.Y., 1990. – P.12.
16. H. Novotny, *The Cunning of Uncertainty* (Polity Press, Cambridge, 2015)

Modern Trends of Additional Professional Education Development for Mineral Resource Extracting

Olga Borisova[1], *Victoria* Frolova[1], and *Elena* Merzlikina[2]

[1]Financial University under the Government of the Russian Federation, 125993 49 Leningradsky Prospekt, Moscow, Russia
[2]Moscow Polytechnic University, 107023 38 Bolshaya Semenovskaya str., Moscow, Russia

Abstract. The article contains the results of development of additional professional education research, including the field of mineral resource extracting in Russia. The paper describes the levels of education received in Russian Federation and determines the place and role of additional professional education among them. Key factors influencing the development of additional professional education are identified. As a result of the research, the authors proved the necessity of introducing additional professional education programs on educational Internet platforms for mineral resource extracting.

1 Introduction

In Russia, the market of educational services in the sphere of additional education is being transformed. Until recently consumers of educational services market came to universities for a higher national diploma. Since September 1, 2013, the Ministry of Education of the Russian Federation has abolished these diplomas for programs related to additional education. In addition, in recent years, the system of professional standards of Russia, included in the register of the Ministry of Labor of the Russian Federation, has been created, which contributes to the definition of skill levels, training and practical implementation, for example, Standard of Mining Engineer. In international practice, international of professional standards are widely applied, for example, international management accounting certification (CIMA), international certification of financial management (IFA). Since 2011, Russian market for educational services has been significantly influenced by the merger of a number of large universities, which provide educated stuff for many sectors of the Russian economy. Today about 50 universities in Russia are training specialists in the field of mineral resource extracting. It is being conducted according to the bachelor's, master's, postgraduate programs, as well as a number of retraining and advanced training programs.

Currently, additional education programs in many universities, including national mineral resource institutions, need to be modernized and attractive for potential clients. The purpose of the study was to identify effective areas for additional professional education development. Object - additional professional education. Subject - problems arising in implementation of programs for additional mineral resource extracting education. The practical significance of research is proved by applying its results in the process of additional

 DOI: 10.1051/e3sconf/20172104016

professional mineral resource extracting education, in accordance with current trends.

2 Materials and Methods

The ground of the research is based upon the papers of the scholars which observe the issues of additional professional education. Basic moments of mineral resource extracting higher school development in Russia are shown in the works of some domestic scholars [1-2]. The issues of modern knowledge estimation system formation researched by V. Frolova, O. Borisova [3-5]. The matters of pedagogical personnel training described by many Russian and foreign scientists [6-8]. Regional and sectorial peculiarities of Russian education system development are the objects of peculiar interest of Kuzbass researchers [9-10]. The main directions of specialists' preparing for mineral resource extracting are shown in the some papers [11-12].

Today, in Russian Federation, the model of national educational system is built in accordance with the Federal Law "On Education in Russian Federation" [13]. In this normative document, the levels of education that exist on the territory of Russia are indicated: general (preschool, primary, basic secondary general) and professional (secondary, higher – bachelor, special, magistracy, and research personnel training). Consequently, additional professional education is not allocated to a separate level of education. Educational services in this area can be provided by any educational institution.

3 Results and Discussion

Additional professional education should be understood as the kind of vocational education, which is obtained in addition to secondary or higher education. The basic rules regulating the process of additional vocational education are specified in Art. 76 of Russian Federation Federal Law "On Education in the Russian Federation".

Vocational professional education in Russia is divided into types, which include programs of professional development and professional re-training. The main criterion for training under these programs is the availability of professional education. These programs allow to raise the professional level to citizens who have a qualification or to master a new kind of professional activity.

Each industry has specific professional requirements, meanwhile, the general tendencies in education development testify to the limitation and closure of economic, managerial and legal specialties in Russian higher education institutions. Nevertheless, higher and additional professional education in mineral resource extracting area is expands in Russia. According to the hh.ru site, more than 50% of vacancies in the mineral resource extracting segment of labor market are lawyers which indicates the need for special training or retraining of personnel for this direction. Therefore, there is a need in "re-graduation" of a graduate from economic or legal university, which, in fact, received only general basic knowledge which is not always matching the specificity of mining industry. As a result, the employer faces the priority task of creating in the company a system of training or retraining employees at the sectoral level, in the parent organization, industry associations, especially created professional training and retraining centers. It should be noted that there is no training of individual specialists in sectoral universities in Russia existing. For example, none of the industry-specialized higher education institutions prepares financiers, however, in the current conditions of management, limited resources and instability of the business environment, the role of financial management of mineral resource extracting companies is growing permanently. And in order to competently manage finances, you need a clear understanding of the production process carried out by mineral resource extracting companies.

Currently, the universities are developing two types of services for additional professional education. Let us analyze them in more detail. The first type of service is professional retraining. It involves a 900-hour training course. Depending on the educational institution, the amount of classroom hours varies on average from 250 to 500 hours. The rest of the time is allocated for self-training. Based on the results of the program, students receive a diploma of retraining. At the same time, among the studied disciplines, only one discipline in management and economics is encountered among experts in the field of mining or surveying. In some universities, a certificate of qualification is issued. In this case, the students get the right to perform a new type of professional activity.

The analysis of the market of additional education for mining specialists, carried out by us, indicates a wide range of programs being implemented. For example, professional retraining in the field of mining education is carried out in the areas of intra-industry bases, for example: "Mine Surveying"; "Engineering and geodesic surveys for construction"; "Rational use and protection of mineral resources"; "Geology"; "Land management and land cadaster"; "Industrial safety of hazardous production facilities"; "Organization of Personnel Service and Personnel Management in Mineral Resource Extracting"; "Environmental protection in mineral resource extracting".

The second block of programs is the programs of professional development. Their main task is to improve the educational level of employees. The target audience of such programs is the specialists who are required to undergo professional development every three years. The main goal of these programs is to improve skills and experience of industry workers. These programs involve training with a time budget of 16 hours or more. They allow improving professional knowledge in the field of mining.

It should be noted that for qualified management, including the sphere of mineral resource extracting these programs will not be enough. Specialists of the middle and senior level can improve their managerial effectiveness using programs of business education. Their training duration and list of activities studied differ from the two previous types of programs. As a rule, they are in demand among specialists who have a diploma of higher education. Some of the programs are selected for admission (for example, to the "Presidential program"), which is absent in the two previous types of programs.

As a result, the conducted research allows asserting that it is expedient to subdivide programs of additional professional education for specialists in the field of mineral resource extracting into three types. These programs are in demand by different specialists working in the industry. Some of them are in demand by land-surveyors, geologists, etc., the other - economists, financiers, lawyers:

1. Professional retraining programs.
2. Business education programs (seldom take into account the specifics of the industry): DBA, EMBA, MBA, MPA; programs on international certification; corporate educational programs; state certification; presidential program for the training of management personnel.
3. Professional development programs - seminars, trainings, etc.

Today these programs are implemented both on the basis of industry sectorial universities and consulting agencies. When implementing a program at the university, it is advisable to take into account the differences in methods, forms, approaches to the organization of the educational process and the assessment of knowledge of the students. First of all, along with theoretical issues, in-depth practical aspects of the chosen topic should be considered. Secondly, the methodology for testing knowledge should be built taking into account the psychological characteristics of adult learning. Thirdly, it is expedient to conduct classes in an interactive form, which helps to determine the degree of mastering the material of each participant in the learning process.

DOI: 10.1051/e3sconf/20172104016

A study of the forms of additional professional education programs implementation showed that in most universities and training centers a remote form is used. Programs in the field of mineral resource extracting are no exception.

Many universities develop their own portals and give their students the remote access to them. On the portal you can get acquainted with the materials on a certain problematic, which can represent both lectures and scientific developments on the theory and history of the issue under study. However, almost all educational institutions provide these services for a fee. Portals of distance education are actively developing, on the basis of the active use of computer technologies. They have already laid out the courses of leading universities on "Engineering Geology", "Hydrogeology", "Mining Engineering". Some of these courses are free. However, as can be seen from the list, this is not enough for full qualification improvement. At the same time, there are no courses on the economics of the industry, industry finance or mining legistation.

Our study of online educational projects has shown that almost all of them are made not in Russia. The name of the projects and the specificity of the material provided for them are given in Table 1.

Table 1. Remote portals used for obtaining professional skills, including the field of mineral resource extracting engineering and their content.

Project name	Materials tutoring on the educational platform
Coursera, Study the new	Courses on a wide range of subjects
Zillion	Articles and webinars on doing business, realizing business ideas and career building
Universarium, Open Education	Courses of lectures on the disciplines
Eduson	Educational courses on doing business in Russian and English
Uniweb	Courses and educational programs of additional professional education in the field of economy, finance, development of e-courses, etc.
DigitalOctober	Courses, translated into Russian, "Open Laboratories"
Univer.tv	Lectures, educational movies, speeches at conferences.
Intuit.ru	Lecture courses for bachelor's degree and advanced training courses mainly in IT
Hexlet (hexlet.org)	IT courses
Lektorium.tv	A library of video lectures on various topics
Businesslearning.ru	Business education courses
Netology (netology.ru)	Courses and webinars on IT topics
University without borders	Courses on the methodology

Distance courses on most platforms are not fee, that allows them developing actively and depriving a part of the educational institutions of listeners. Consequently, the future of Russian additional education is the portals of distanced education with a variety of courses on sectoral subjects.

4 Conclusion

The conducted research showed that the programs of additional education in mineral resource extracting should be divided into programs of professional retraining, advanced training and business education, since the requirements for each of the programs listed are different. These programs are in demand in the market of educational services in Russia. At the same time, there is an increasing demand for distance learning programs. However, the emergence of educational platforms and courses in the result may adversely affect the traditional forms of education in universities and training centers. Therefore, even today, for the purposes of advertising, it is necessary to introduce new courses in mineral resource extracting area to remote platforms, as well as for economics, finance and jurisprudence, taking into account the industry specific features of specialists training.

References

1. M.A. Gasanov, K.A. Kolotov, K.A. Demidenko, E.A. Podgornaya, O.V. Kadnikova, IOP Conf. Ser.: Earth Environ. Sci., 50:1, 012025 (2017)
2. I. Pevneva, O. Gavrishina, A. Rolgayzer, M. Agienko, A. Myaskov, E3S Web of Conf., 15, 04016 (2017)
3. V.B. Frolova, O.A. Kalugina, L.S. Artamonova, A.I. Boykov, IEJME: Math. Educ. **11:5**, 1185-1193 (2016)
4. O.V. Borisova, D.G. Vasbieva, V.B. Frolova, E.M. Merzlikina, IEJME: Math. Educ. **11:7**, 2483-2491 (2016)
5. O.V. Borisova, D.G. Vasbieva, N.I. Malykh, S.A.Vasnev, IEJME: Math. Educ. **11:5**, 1175-1184 (2016)
6. F. Agafonov, A. Genin, O. Kalinina, O. Brel, O. Zhironkina, E3S Web of Conf., **15**, 04011 (2017)
7. S.A. Zhironkin, K.A. Kolotov, O.V. Zhironkina, Economics and Innovation Management, **1**, 4-16 (2017) DOI: 10.26730/2587-5574-2017-1-4-16
8. S. Anyona, B.K. Rop, Economics and Innovation Management, **1**, 17-29 (2017) DOI: 10.26730/2587-5574-2017-1-17-29
9. V. Trifonov, O. Loyko, D. Nesteruk, S. Zhironkin, E. Strekovtsova, AIP Conf. Proceed., **1800**, 050009 (2017)
10. S.A. Zhironkin, K.A. Kolotov, A.E. Genin, F.V. Agafonov, S.A. Kovalevsky, IOP Conf. Ser.: Earth Environ. Sci., **50:1**, 012011 (2017)
11. M. Cehlár, J. Janočko, Z. Šimková, T. Pavlik, E3S Web of Conf., **15**, 01019 (2017)
12. S.A. Zhironkin, Ugol', **4**, 47-49 (2002)
13. *Federal Law of Russian Federation "On Education in Russian Federation"*, December 29, 2012, No 273-FZ

E3S Web of Conferences **21**, 04017 (2017) DOI: 10.1051/e3sconf/20172104017
The Second International Innovative Mining Symposium

Corporate Social and Ecological Responsibility of Russian Coal Mining Companies

Nikita Ravochkin[1,*], *Vladimir* Shchennikov[2], and *Vasiliy* Syrov[3]

[1]Kemerovo State Agricultural Institute, 650056 5 Markovtseva Street, Kemerovo, Russia
[2]Kemerovo State University 650000 6 Krasnaya Street, Kemerovo, Russia
[3]National Research Tomsk State University, 634050, 36 Lenina prospect, Tomsk, Russia

Abstract. Based on the provisions of corporate social responsibility and taking into account the specifics of Russian mining enterprises, the authors attempt to understand theoretically the corporate social and environmental responsibility in this paper. The study shows that the essence of the principles of socially responsible behavior has ancient roots, while the consumer's attitude towards nature begins only in the era of modern times. The genesis, evolution and transformation of social responsibility in Western countries in the twentieth century are traced. The necessity of taking into account the national social and cultural specifics of the domestic economy is substantiated instead of blind copying of foreign management practices. The difference in the formation of corporate social responsibility (CSR) abroad and in Russia is shown. The list of facts and factors contributing to the formation of CSR in Russian realities is given. With regard to the coal industry enterprises inconsistencies have been identified. Their overcoming will allow the enterprises formulating strategies for corporate social and environmental responsibility. The advantages of social and environmental responsibility in comparison with the legal one are presented. In conclusion, the authors summed up the theoretical interpretation of the object claimed in the introduction.

1 Introduction

Uncontrolled expansion of a human being into the space of nature, which led to the global environmental problems, we associate with the proclaimed by F. Bacon imperative "Knowledge is power" [1]. Any statement can be understood differently, however, it was this phrase that was aimed at increasing man's power over nature. First of all, the use of nature was taking place in connection with the implementation of economic activities by economic agents. The main goal of the capitalists was to obtain the maximum profit without taking into account any ethical aspects of conducting entrepreneurial activity. Such consumerism, on the one hand, has opened to the mankind unlimited possibilities for the transformation and use of nature in its own interests, on the other, increased its negative impact on it. The industrial revolution led to a rapid economic growth and simultaneous degradation of the environment: the maximum speed and efficiency of the use of natural resources

* Corresponding author: nickravochkin@mail.ru

caused a rapid destruction of the environment. Natural resources, which seemed unlimited, could not withstand their outstripping use by entrepreneurial structures.

At the turn of the XIX-XX centuries the tendencies that testify to the system and interdependence of the processes taking place in all spheres of public life intensified. For the first time, an emphasis is placed on the environmental consequences of capitalism. Beginning with the problems of individual industrial cities, the current economic, environmental and social contradictions have clearly regional and, more recently, global character, presenting one of the problems of our time. For example, meeting fast changing needs of the population in developed countries can lead to the simultaneous death of people in the Third World countries due to environmental pollution, poor medical care, food problems, etc. In other words, the problem of favorable living conditions for future generations is treated just as an ecological crisis.

In modern economy, the outlines of globalization are ever more clearly emerging, setting the single vector of development and unifying the standards of doing business. However, it is true that not all subjects of the market economy take effective measures to respond to the fact of ecological crisis. The authors attribute this mainly to the mental processes and the remaining intentions of profit maximization for the management of most companies and the unwillingness to take the responsibility for the consequences of their own activities.

Russian enterprises are the most vivid examples of so-called "irresponsible corporate behavior". The lagging behind of the domestic economy from the West is due, in our opinion, ideological isolationism, which contributed to the later realization and implementation of the world business standards. The Russian coal enterprises have been least adapted to socially responsible operations, as only the first of three interrelated components (economic, social, ecological) is taken into account. These companies are potentially dangerous for the environment. Domestic practice has shown that in the operations of these enterprises the most difficult is to make effective management decisions that determine their integration into the global economic environment. Impacting the environment, coal industry enterprises have a complex negative influence on it. The problem of this study is related to the fact that more than half of Russia's population lives in unfavorable environmental conditions. At the same time, their ignoring the consideration of the interests of society and sustainable development sets the urgency of the research and allows formulating its object. It is a theoretical comprehension and identification of the state of corporate social and environmental responsibility of Russian coal industry enterprises.

2 Materials and Methods

Having analyzed the possible irreversible consequences of the consumer attitude toward environment, the Western business came to the need to identify its own functions and role in the social development and interaction in the realities of the conflicting interests of various social actors. Initially, the boundaries of corporate social responsibility (CSR) were blurred and narrowed down a trivial abstract view: "Is it possible to maximize profits and make the world better?" G. Bowen [2] was the first who formulated the main provisions and directions of socially responsible activity of entrepreneurs. He pointed to the need for the business to participate in values and goals of the interested parties (stakeholders), the main of which were society and the state. It is curious that similar ideas of following the moral duty and free will that determined the responsibility of human behavior in this world could be found in the first philosophical teachings of Indian, Chinese and Greek thinkers. However, we observe that their introduction into the world business practice occurred much later - in the middle of the twentieth century. Within the framework of our research, we note that the first western socially responsible models of behavior of market economy subjects took into account the aggravation of social and economic and environmental problems.

Later, in the works of H.I. Ansoff [3], T. Donaldson [4], P. Drucker [5], Ph. Kotler [6], A. Carroll [7, 8], M. Porter [9], R. Freeman [10] and other researchers, there was the evolution and transformations of the concepts of CSR. Their essence lies in the expansion of the circle of stakeholders, the substantive concretization of their relations with them in the conditions of growing competition, the first systemic formulations of value expectations that contribute to social progress, social forecasting and response to the constant changes and requests for the world development. It should not be forgotten that the image of a socially responsible company is developed not only in the eyes of the public, but also in the minds of its own personnel. The future of the enterprise depends on the integrated accounting of the views of stakeholder groups. To do this, it is necessary to maximize benefits for all stakeholders (utilitarianism for the greatest number). Note that the Western CSR practice was also developed under the influence of the ideas of civil society.

The theoretical analysis of the works of the twentieth century revealed that, despite the recognition of universal responsibility, the main provisions of CSR only episodically describe the synthesis of environmental and social responsibility [11]. This indicates the need to take into account environmental values and factors that are now actively presented in the social reports of large industrial enterprises for which the greening of operations is mandatory. It is obvious that today the activities of industrial organizations should be programmed in accordance with economic, social and environmental aspects, giving a synergistic effect.

According to the author's opinion, corporate social and environmental responsibility is the sum of voluntary investments by enterprises to solve problems of social injustice and reduce tensions in society in order to avoid economic losses and to enhance the reputation effect.

3 Results and discussion

Russia, unlike the West, had no institutional foundations for CSR formation until the early 1990s. The main ideas of responsible business were borrowed from the most successful foreign companies, without taking into account the specifics of the actual Russian realities. Economic responsibility, i.e. the company's performance of its basic functions, traditionally was the only one for Soviet companies. Activities aimed at addressing social problems and the achievement of the public benefit, was practically absent [12, 13]. CSR in Russia was formed not under the influence of civil society structures and public policy, but in view of the following facts and factors:

- ambiguous attitude towards the results of privatization reform;
- lack of rational arguments regarding the origin of large capitals and the relationship between the increase in the value of companies and socially responsible strategy;
- uneven concentration of the capital of enterprises in the regions of the country (climatically unfavorable and undeveloped);
- the growing gap between the poor and the rich in the regions;
- mental features of the population of the country with simultaneous high level of expectations, pessimism and low level of social activity;
- traditionalism in the methods of management and practice of labor relations (double standards);
- the variety of social problems which multiplication is observed in the Russian society up to the present time.

It is therefore natural that today most large industrial enterprises for a number of reasons are in a transit from the socialist paternalistic views to the universal practice of CSR. It should be noted that a number of industries integrated into the global economy (usually

 DOI: 10.1051/e3sconf/20172104017

dealing with the Western partners) are already successfully implementing a socially oriented business strategy, trying to take into account the interests and expectations of stakeholders.

Russian coal industry has long been considered a subsidized industry. Despite this, coal mining enterprises face a number of factors constraining the development of the industry. This can include technical, technological, economic factors, but the main ones, we believe, are managerial and environmental. The impact of this industry goes in a negative way, affecting the living standards of the population, its expectations and conditions of existence in general. There are ecological-and-economic and social-and-ecological contradictions, formed by the trajectory of the development of operations of the coal industry enterprise, which do not coincide with the values and even natural rights of the citizens.

The latter is very often observed in single-industry cities, as the relation of the coal mining enterprises to nature, like their organization, has not fundamentally changed for a long time. The coal mining enterprises in the settlements objectively worsen the living conditions. As a rule, the coal industry enterprises are organized in such a way that their personnel live in mining towns and villages in close to coal deposits or within the boundaries of mining fields. All this creates a number of environmental problems inherent only in the coal regions. These include increased standards for the permissible load of harmful substances on all layers of the biosphere, the need to extinguish burning rock dumps, the widespread communication of occupational chronic and acute diseases, problems with waste disposal, etc.

At the same time, the current environmental crisis in the Russian coal regions contributed to the first alarmist estimates by the local communities. Restructuring of the coal sector is one of the characteristic features of the promising development, since it directly points to the renewal and modernization of its organizational and production structures. The consideration of the environmental aspect of social responsibility in the coal industry presupposes a gradual reduction of external negative effects of the industry enterprises' operations in the synthesis with environmentalism and preventive measures that would reduce the burden on the environment.

The Russian practice of entrepreneurial activity of coal mining enterprises reveals a number of inconsistencies, which indicate that the strategies of social and environmental responsibility have not yet received wide circulation and there is a need for overcoming the "socialist" vision of entrepreneurial activity:

- the wide use of command methods that lead to unavailability for conscious and active initiatives;
- the absence of special structural units and a clear understanding of objectives and essence of social and environmental responsibility and the reluctance to implement social investments leads to the lack of motivation for the "responsible" activities and a dual attitude towards CSR concepts;
- relatively low level of "civic self-awareness" among management;
- single and forced nature of ongoing initiatives and lack of social reporting;
- inability to anticipate the expectations of stakeholders, guided by the Soviet standards of planning the enterprises' operations.

Nevertheless, it is obvious that social responsibility is not absent in the domestic coal industry. Modern age does not correspond to the realities of the USSR, and the established management paradigm is approaching its regular change and subsequent differentiation. The efficiency of the coal industry today is measured not only by making a profit, but also by its correlation with the environment. The management of enterprises is aware of the need to implement a strategy of social and environmental responsibility, focusing on those enterprises of the sector that belong to multinational and national holdings, and consequently - implement the principles of socially responsible behavior. In positive modality, it can be estimated that the development of corporate social and environmental strategies of coal mining enterprises does not indicate a simple transfer of Western practice to Russian conditions, but rather demonstrates the creation of their own methodology for taking into account the expectations of specific stakeholders.

The social and environmental responsibility of coal enterprises, unlike legal responsibility, is estimated by us as the most effective. First, it is connected with the voluntary response of enterprises to emerging contradictions. Secondly, social and environmental initiatives go beyond the current legislation on environmental protection. Thirdly, forward-looking and conducting business according to the global environmental standards is assessed as a conscious activity, which meets the support in the eyes of stakeholders. Fourthly, the above-mentioned Russian environmental legislation is characterized by the lack of compliance, the absence of real measures to bring to responsibility and the weakness of social control.

The Russian enterprises of coal industry, unlike foreign ones, have not yet reached a balance at the level of "economy-society-ecology", since there are no clear ideas about what economic efficiency should be (especially guided by various technological and environmental standards of operation). In addition, there is a question of what is the fairness of the distribution of limited resources and the regularities of building up the reputation of the enterprise. Finding out points of positive interconnections with ambiguous economic, social and environmental aspects of the management strategy contributes to the achievement of synergies and sustainable business competitiveness in the short, medium and long term, opening up new horizons for social and environmental initiatives. In addition, for the public to understand these perspectives, it is necessary to observe the principle of openness of information about the company's activities. Foreign standards of non-financial reporting are universal, therefore, we will only point out those sections of the reporting by which the interested parties, taking into account the specifics of the coal industry, can give their assessment:

1) "Economy": general coal mining, business contacts.
2) "Social activity" in relation to the population, public, municipal and non-profit organizations and associations.
3) "Ecology": environmental information on the operations of the coal company.

As a theoretical core of the concept of corporate social and environmental responsibility, we proposed the synthesis of the social ecology laws of B. Commoner [14] with the generalization and systematization of data on the social and ecological problems of coal enterprises, the creation and implementation of social and environmental standards to modernized enterprises and technologies and the subsequent congruent development of technologies for the development and processing of deposits. Increasing utilization, transforming waste into useful forms, reducing energy costs, saving transportation costs, processing and disposal of garbage, improving product quality, environmental innovation - these are just some areas of the social and environmental responsibility of coal enterprises.

4 Conclusion

Every year Russian coal industry enterprises increase the degree of integration into the global economic processes, which makes them move along the path of increasing corporate transparency, social responsibility and efficiency. It is obvious that the introduction of the provisions of social and environmental responsibility in the Russian coal companies will become the norm of doing business in the near future. The permanent open dialogue with stakeholders and response to the social demands become the philosophy of the coal industry enterprises, helping them to determine and be informed about the true meanings and values of human existence, and the industry itself - to increase its competitiveness.

 DOI: 10.1051/e3sconf/20172104017

References

1. F. Bacon, *The New Organon* (Cambridge University Press, Cambrige, 2000)
2. H.R. Bowen, *Social Responsibilities of the Businessman* (Harper and Row, London, 1953)
3. H.I. Ansoff, Strat. Man. J. **1:2**, 131-148 (1980)
4. T. Donaldson, T.Dunfee, *Ties that Bind: A Social Contracts Approach to Business Ethics* (Harvard Business School, Boston, 1999)
5. P. Drucker, *Innovation and Entrepreneurship* (Harper, New York, 1985)
6. Ph. Kotler, N. Lee *Corporate Social Responsibility – Doing the Most Good for Your Company and Your Cause* (John Wiley and Sons, New Jersey, 2005)
7. A.B. Carroll, Bus. Soc. **28:3**, 268- 295 (1999).
8. A.B. Carroll, Acad. Man. Rev. **4:4**, 497-505 (1979)
9. M. Porter, M. Kramer, Harv. Bus. Rev. **80:12**, 56–69 (2002)
10. R.E. Freeman, Calif. Man. Rev. **25:3**, 88-106 (1983)
11. R. Barth, F. Wolff, *Corporate Social Responsibility in Europe: Rhetoric and Realities* (Edward Elgar Publishing, New York, 2009)
12. H. Söderbaum, *Corporate Social Responsibility in Russia* (Uppsala University, Helsinki, 2006)
13. M.S. Fifkaa, M. Pobizhan, J. Clean. Prod. **82**, 192-201 (2014) https://doi.org/10.1016/j.jclepro.2014.06.091
14. B. Commoner ,*Making Peace With the Planet* (Pantheon Books, New York, 1990)

Diversification of the Higher Mining Education Financing in Globalization Era

Victoria Frolova[1], *Olga* Dolina[1], and *Tatyana* Shpil'kina[2]

[1]Financial University under the Government of the Russian Federation, 125993 49 Leningradsky Prospekt, Moscow, Russia
[2] Capital Financial and Humanitarian Academy, 109383 90-17 Shosseynaya st., Moscow, Russia

Abstract. In the current conditions of global competition, the development of new mining technologies, the requirements to labor resources, their skills and creative potential are increasing. The tasks facing the mining industry cannot be solved without highly qualified personnel, especially managers, engineers and technicians, specialists who possess the knowledge and competences necessary for the development of science and technology of mining, and ensuring mining industrial safety. The authors analyze personnel problems and financing of mining higher education, conclude that there is a need to develop social partnership and diversify the sources of funding for training, advanced training and retraining of personnel for mining and processing of solid mineral deposits.

1 Introduction

The lack of specialists of the appropriate qualification restrains the development of mining industry in Russia. The solution of tactical and strategic tasks of the industry development cannot be carried out without an effective innovative system of training and development of highly skilled personnel.

Modern personnel problems at the enterprises involved in the extraction and processing of solid minerals are due to the following factors:
- the lack of a close relationship between production, education and science;
- insufficient financing of research and development (R&D);
- shortage of highly skilled personnel to ensure innovative development of industry enterprises.

The specific features of the mining industry are also stipulated by the additional specific training problems, such as:
- the mining profession is not currently prestigious for the younger generation;
- the graduates of universities do not want to work in dangerous and harmful working conditions and move to the remote sparsely populated villages.

It is not possible to solve the personnel problems in the industry only by the increase in wages: large businesses traditionally spend most of the profits for dividends to the owners of the company, and small businesses simply cannot afford to increase their production cost by the wage increase for employees in conditions of severe internal and external competition.

2 Materials and Methods

Extracting and processing fuel, energy and mineral raw materials enterprises are the budget-forming in the Russian Federation. They provide more than 50% of Russian exports, in the labor market it is several million jobs [1-2]. At the same time, science costs are less than 1-2% of profits (for comparison, in developed countries, about 30-40% of profit is spent on research and development in similar spheres of activity) [3-4]. Recently in the large mineral resource companies there has been an active reduction in spending on scientific research: approximately 80% of scientific personnel have been cut by "Rusal", "Norilsk Nickel" and other mineral resource companies are reducing the funding of their scientific departments. To change this situation in the context of globalization, to ensure future competitiveness of Russian mining enterprises, it is necessary to increase investments in the human factor, the costs of science and education, retraining of personnel [5].

Financing is one of the most serious problems of improving mining industry in Russia.

Currently, the training of personnel for mining industry within the framework of per capita financing does not allow training a qualified specialist taking into account the specifics of the industry and the need for additional transport and field costs during on-the-job training in remote regions of Russia. Education in educational institutions of various organizational and legal forms and in-house training also require significant investments in the training, retraining of teachers, advanced training, new information technologies, the acquisition of special literature, etc. In future insufficient financing can negatively affect the professionalism of engineering personnel. Additional costs in mining education are also associated with the acquisition of new imported equipment, instruments and mechanisms for the extractive industry that do not have Russian counterparts.

Budget financing of mining education in conditions of limited budget is not able to ensure an effective modernization of the industry. Due to this fact, it is necessary to search for and use additional sources of financing for the mining education. The professional mining community is interested in finding a means for non-state financing of educational institutions. To prepare high skilled specialists the part of the costs for education should be invested by enterprises, private investors, and entrepreneurs.

In the conditions of external competition mining enterprises need investments for their development - they are forced to invest in the introduction of innovative technological programs and installation of equipment in order to increase production efficiency, while the costs of labor remuneration for employees and their retraining are reduced. Now it is not possible to attract investments into the extraction and processing of solid minerals from foreign sources. As the analysis showed, according to the results of 2016, the investment flow to Russia totaled $ 1.14 billion (in 2015 this figure was $ 208 million). Taking into account the problems mentioned above, the expenses for the government program "The development of industry and increase of its competitiveness" in the amount of 154.4 billion rubles were included into the budget of 2016. The financing of industrial enterprises due to this program caused a slight improvement in their activities, as these funds are insufficient for the development.

3 Results and Discussion

We consider the fuel and energy sector as the key industries which are promising today for attracting foreign investments. At the same time the outflow of investments is taking place from other extractive industries, including those engaged in the extraction and processing of solid minerals. In Russia, unfortunately, the financing of business ideas from "business angels" is not developed, as it is the case in foreign practice.

In such difficult financial conditions, we consider it expedient to develop and strengthen the system of social partnership taking into account the interests of all interested parties. Social partnership can be implemented in the form of interaction between universities, employers, academic and industrial institutes, through the establishing of educational and scientific-educational centers, departments of training and retraining of enterprise personnel, training and workshop complexes, as well as directly at the places of work with the involvement of specialists in the form of group and individual training. At the same time, it should be kept in mind that, in global competition, the training of personnel must be business and practice-oriented, the quality of personnel must meet the requirements of the most progressive technologies and innovations. Educational programs must be developed with the participation of representatives of production. Special grants of interested ministries and departments could contribute to improving the quality of education through the organization and conduct of regular training workshops, contests for employees, students and teachers. To ensure the staff meeting the global requirements, special attention should be paid to training technical specialists in English.

Mining companies should carry out training and retraining of indigenous workers and provide opportunities for the development of labor force in the region and in the country as a whole in order to meet the production tasks and reduce the shortage of high skilled personnel. This approach corresponds with the needs of local society and helps to increase the indicator of labor mobility.

Many organizations solve the problem of staff shortage by improving the skills of their own personnel, which makes it possible to prepare the employees for the solution of new tasks and perform new labor functions.

Advanced training and retraining of enterprise specialists on the basis of professional training and retraining of personnel is a system of measures aimed at improving the quality of labor resources and, accordingly, the labor potential of the whole enterprise.

The retraining personnel system allows obtaining not only an overall knowledge, but a complete training of skills in a given professional sphere. At the advanced training courses, the specialist will have an opportunity to learn about new technologies and techniques that are being introduced at the leading enterprises in Russia and all around the world.

The special programs for skills improvement and retraining of specialists in mining include studding new technologies and developments in the extraction and processing of solid minerals, underground and open pit mining, labor safety, and also consider the economic aspect of mining.

Mining specialist requires a special attention. This is the sphere of labor in which the theoretical knowledge obtained in training at higher education institutions, as a rule, is not enough. We need personal experience and continuous improvement of skills and knowledge.

In the period of scientific discoveries and the digital revolution, it is necessary not only to be well-educated in the latest technologies of coal mining and processing, but also to be able to apply them in practice. Only through the introduction of innovations and new technologies the efficiency of the enterprise can be improved. Competent specialists are the key factor that influences the effectiveness of the entire industry. The qualifications of both workers and management personnel must be constantly updated [5]. This can be seen from the production index based on the type of economic activity "Extraction of minerals", which for the last 10 years has not changed much.

The data shows a decline in the production index in 2009 compared with 2008 to 97.2%, which was due to the economic crisis. The highest value of the production index was observed in 2007 and in 2010. In 2016, the production index was 102.5%, which indicates a slight increase in production and processing of solid minerals in comparison with 2015. In the long term, to improve the efficiency of the industry, it is necessary to introduce high-precision and high-tech equipment, which must be produced, and highly skilled personnel are also needed for this.

As the experience of Moscow Higher Engineering School, which prepares managers and engineers for solving urgent problems of developing industrial enterprises and increasing labor productivity, show the best form of staff retraining has been the creation of a comprehensive program composed of separate modules. In addition, short courses and expert seminars are offered. If desired, one can choose either the entire complex program, or one of its parts or individual modules or courses.

The peculiarity of modular programs is that they provide interdisciplinary competences, first of all, managerial, economic and engineering.

Work experience of some large enterprises showed that at first a small group of senior staff headed by the Deputy Director for Development had been trained by individual modules. After that, the management of the enterprise made a decision to train the group of heads of shops and project managers by the special program directly at the plants of the corporation. Being trained directly at the enterprise allowed them, and, if necessary, to make managerial decisions.

The practice of staff development and retraining when teams of 2-4 people (for example, a technologist, production director, chief engineer, and economist) from one enterprise have been trained is of much interest. Within the framework of the program, under the guidance of the curator, they work out various concepts and make a feasibility study, which is of practical value and will increase the efficiency of the enterprise's activity.

We believe that in order to develop advanced methods for improving the process of coal mining, it is necessary to improve the skills of management personnel and workers. For this purpose, it is advisable to use comprehensive programs for professional development and retraining, since they have shown their effectiveness.

The algorithm of the comprehensive program for advanced training and retraining is following:

1. Analysis stage
2. Conceptual stage
3. Design stage
4. Implementation phase
5. The stage of analysis and formation of technical task
6. Conceptual planning stage
7. The stage of detailed study and implementation
8. Launching and operation of new technology, models, equipment, etc. stage.

In modern conditions, in order to increase the efficiency of a mining enterprise, it is also necessary to manage properly the capital structure, financial stability, and financial solvency of the enterprise. All these matters are in the competence of the financial manager. Along with the training of management personnel and workers, one should not forget about the improvement of skills of the company's specialists. With their help the company can not only rationally use the available funds, but also optimize the company's expenses, which is important in the period of recovery from the crisis.

The authors conducted a study of motivation and assessment of staff development effectiveness (See Table 1).

Table 1. Motives for professional development.

Motivation for professional development	**Interview results, %**	**Effectiveness of professional development**
Employer's requirement	28	There is a formal approach to the advanced training, and sometimes denial of offered knowledge on the basics of personal experience

Professional growth	21	As a rule, there is an interest in learning process
Mastering modern professional methods in work	18	There is active work, interest, complete fulfillment of all tasks
Waiting for salary growth	18	As a rule, an employee's professional development is not grounds for the employer to increase the salary of an employee, if this does not mean a change in activity
Desire to improve the position of the company	15	The top manager wants to raise his level of knowledge and professional skills in order to bring the company to a new level, to develop new areas of financial and investment activity, and to increase their effectiveness

As shown by the survey, the leaders most often motivate the employee's professional development, and senior managers tend to increase knowledge and professional skills with greater desire. At the same time, it should be noted that the greatest result of staff development and retraining can be achieved by the synergistic effect of several motives. For example, if the financial manager is the owner of a mining enterprise and he has much interest in the growth of its activities' results, and consequently the profit growth.

From the point of view of the employer, some problems concerning the possibility for an employee to attend training classes depending on their forms arise (see Table 2).

Table 2. Forms of training.

Form	Advantages	Disadvantages
On-site training on the job	The possibility of combining professional and educational activities	Not enough time to study and consolidate the knowledge
On-site training off the job	Complete involvement in the educational process	The possibility of losing a part of wage/salary
Remote	The choice of the time of classes	The impossibility of conducting interactive classes

In case of the on-the-job training, the level of material development is lower than in off-the-job training, but it is often necessary for the financial manager to perform his duties even when studying off-the-job, which also has a negative impact on the level of mastering the training program. An important problem is the issue of payment for training. The employer is not always ready to bear the costs, and all expenses are assigned to the employee improving the qualification. It should be noted that the cost of remote training is less, as a rule.

From the point of view of organizing the learning process, an important issue is the organization of current, intermediate and final forms of control (see Table 3).

Table 3. Forms of control.

Form	Advantages	Disadvantages
Testing	The possibility of using computer processing, the absence of a subjective attitude of the examiner	A variant of a random correct answer is possible. Subjectivity of opinion of the author - the compiler of test tasks
Oral exam	The possibility to identify the true knowledge more accurately	Human factor
Creative work	The development of an individual approach to solving the task	Does not reflect the whole set of knowledge, skills and abilities

Thesis defense	Study and consolidation of theoretical knowledge and their practical implementation in the paper	Preparing a thesis on a specific issue does not allow evaluating the result on all subjects

Obviously, in conducting training classes for all categories of employees, including management, it is necessary to apply various forms of motivation, training and control to achieve the best result.

At the same time, the spread of digital production, the emergence of completely new technologies make both employers and employees constantly improve not only their skills, but also be able to learn and retrain quickly, including directly during the production process.

However, it should be borne in mind that practical training directly in production process is very dangerous. Young professionals or newly recruited staff which do not have the skills and work experience cannot be admitted to the work on hazardous equipment. But they need to learn how to work on complex equipment in the process of specific production. To solve this problem, we can recommend the creation of training grounds equipped with special equipment that is safe for trainees.

We consider it expedient in the process of strategic and tactical planning of the development of the extraction and processing of solid minerals enterprise to provide the costs for training and upgrading the skills of personnel, as well as to calculate the effectiveness of personnel development activities (EF) according to the following formula:

$$EF = PR - CE \times EXP \quad (1)$$

where: PR – balance profit of the organization for the reporting period;

CE – the efficiency coefficient;

EXP – actual expenses for personnel development in the reporting period.

The magnitude of the efficiency factor determines the management of the enterprise, based on the tasks assigned to the personnel development programs.

4 Conclusion

Thus, in the new model of economy, there will be a need for people who are able to modernize production, solve problems and make decisions in non-standard situations, create new models and business processes, process and analyze data. All this is possible on condition of retraining and constant improvement of the professional skills. Consolidation of efforts of educational institutions, scientific organizations and industrial enterprises will ensure a high level and quality of education in conditions of limited funding.

References

1. Zhironkin S.A., Kolotov K.A., Zhironkina O.V., Institutional traps and externalities of innovative neo-industrial import substitution, Economics and Innovation Management, **1**, 4-16 (2017) DOI: 10.26730/2587-5574-2017-1-4-16
2. F. Agafonov, A. Genin, O. Kalinina, O. Brel, O. Zhironkina, E3S Web of Conf., **15**, 04011 (2017)
3. M.A. Gasanov, K.A. Kolotov, K.A. Demidenko, E.A. Podgornaya, O.V. Kadnikova, IOP Conf. Ser.: Earth Environ. Sci., **50:1**, 012025 (2017)
4. M. Gasanov, S. Zhironkin, M. Hellmer, Proc. Soc. Behav. Sci., **19**, 745-752 (2017)
5. A. Zhavoronok, M. Gasanov, S. Zhironkin, SHS Web of Conf., **28**, 01144 (2016)

Andragogical Model in Language Training of Mining Specialists

Evgeniya Bondareva [1,*], *Galina* Chistyakova[2], *Yury* Kleshevskyi[3], *Sergey* Sergeev[4], and *Aleksey* Stepanov[5]

[1]Plekhanov Russian University of Economics, Kemerovo Institute (branch), College of information technologies, economics and law, 650992 Kuznetsky Av. 39, Kemerovo, Russia; Ağrı İbrahim Çeçen University, Department of Russian language and Literature, 04100 Erzurum Road 4 km, Ağrı, Turkey
[2]Plekhanov Russian University of Economics, Kemerovo Institute (branch), Department of Humanities, 650992 Kuznetsky Av. 39, Kemerovo, Russia
[3]Plekhanov Russian University of Economics, Kemerovo Institute (branch), Department of Management and Business technologies, 650992 Kuznetsky Av. 39, Kemerovo, Russia
[4]Ağrı İbrahim Çeçen University, Department of Russian language and Literature, 04100 Erzurum Road 4 km, Ağrı, Turkey
[5]Plekhanov Russian University of Economics, Branches Center, 117997 Stremyanny Lane 36, Moscow, Russia

Abstract. Nowadays foreign language competence is one of the main professional skills of mining engineers. Modern competitive conditions require the ability for meeting production challenges in a foreign language from specialists and managers of mining enterprises. This is the reason of high demand on foreign language training/retraining courses. Language training of adult learners fundamentally differs from children and adolescent education. The article describes the features of andragogical learning model. The authors conclude that distance learning is the most productive education form having a number of obvious advantages over traditional (in-class) one. Interactive learning method that involves active engagement of adult trainees appears to be of the greatest interest due to introduction of modern information and communication technologies for distance learning.

1 Introduction

Modern competitive production conditions require a high level of training, including knowledge of a foreign language, from specialists and managers of mining enterprises for solving many issues at the international level. Adult trainees being professionals in their working area have a critical need for learning a foreign language that is connected with international processes of society liberalization, expansion of international trade and production cooperation, rapid development of new production decisions, effective cooperation with foreign partners in mining sphere.

* Corresponding author: bondarevaep@list.ru

Many centres offering language training and (or) retraining programmes for engineers who actively learn a foreign language as a means of professional communication do not take into consideration some factors that impede quality linguistic education. The fact that an adult needs to be taught in a different way than a child and / or an adolescent has become clear for a long time. The 20th century gave a powerful impulse to the developing of andragogy, a new branch of pedagogy. In the 1970s, American scientists M. Knowles, P. Jarvis, et al suggested the first theoretical theses describing the specific behaviour of adults in learning process. They dwelt upon the procedure model of education as opposed to the content one used in children teaching [1]. Andragogical model involves *lifelong learning*. In the world practice, the terms *adult education, adult learning, adult training, continuing education* are used as synonymic expressions.

Research works of many scientists and practitioners are devoted to different aspects of andragogy: general andragogical issues [2, 3, etc.]; organization of courses for adults [4]; methods, technologies and tools used in continuing education [5, 6, etc.); motivation appeal of adult learners [7, 8, etc.]; training of teaching staff for vocational education [9, etc.]. The issues of language training/retraining of adult learners are discussed in [10, 11, 12, 13, etc.].

Nowadays adult educators should not strive for using as many diverse techniques and technologies as possible. To achieve maximum efficiency, organizers of language courses for adults are to design the learning process by combining the most effective learning methods with capabilities of modern information and communication technologies. Besides, they are to pay special attention to the requirements imposed on their employees by professional communities. In our opinion, it is necessary to offer a clear framework of andragogical learning model, to choose the most effective training methods, and to make practical orientation the cornerstone of adult learning process.

2 Materials and methods

There are several reasons for mining professionals to attend language courses. Firstly, to communicate on professional topics effectively one should keep mastering his or her foreign language all the time. If the work process does not involve any tasks that require a foreign language to be completed, linguistic skills become unproductive. Consequently, graduates with Upper-Intermediate or Advanced Level of English cannot communicate fluently as they used to while studying at higher education institutions. Secondly, some mining specialists need to change their specialization area due to some working conditions. So, their aim will be to enlarge vocabulary in some specific professional area. Thirdly, foreign language level of many graduates in Russia leaves much to be desired. Students often have low motivation for studying foreign languages caused by counterproductive combining of learners into training groups (the problem of mixed-abilities groups and the problem of overcrowding groups), the choice of ineffective teaching methods, improper organisation of out-of-class independent work, and failure to use information and communication technologies in the educational process. When former students start working at some enterprises they are often demanded to be able to solve production tasks both in their native and a foreign language. This imperative as well as the desire to get higher salary and to have some career development form steady motivation to improve the foreign language studied before (or even to learn one more). Thus, irrespective of the reason for which learners come to language courses all of them need mastering a foreign language to become professionally and financially stable.

The problem of mixed-abilities groups and the complex of challenges resulted in low motivation are equally essential for language training/retraining courses. The aim of language training centers is quite transparent. They are to make all adult learners successful in solving professional tasks in a foreign language regardless their language level at the beginning of the course. This article describes one of the ways of designing language training/retraining courses for adult mining specialists.

While choosing a working pattern suitable for trainees / retrainees of language courses one should take into account that they are usually people facing with time and location constraints. To organise traditional in-class lessons with a strict timetable at a definite place is not always possible. Therefore, it is necessary to elect another education form. The best option is believed to be distance learning. The development of distance learning technologies ensures the possibility of providing educational services without being tied to a specific geographical location (Open and Distance Learning). The mixed (distance and in-class) form of education can become a good alternative if the trainees of one course are not too distant-located and are able to meet in class from time to time. This position is supported by the fact that educational results in groups with mixed education forms are the highest as compared to the traditional one [14].

One cannot ignore the fact that modern educational process, organised in distance form, is inconceivable without the use of information and communication technologies (ICT). The supply of educational material through electronic means makes education process fascinating and keeping pace with the times. The reference to the materials of the Internet, the organisation of webinars, the use of interactive multimedia and others are already an indispensable element of learning process. Learners should be provided opportunity to use multiple resources, thus favoring flexibility in the acquisition of knowledge [15].

Modern ICT instruments include MT (machine translation), CAT (computer-assisted translation) and HAMT (human-assisted machine translation) systems aimed to facilitate and speed up translation process. Some MT systems (Yandex translator, Google translator, Translate.ru, etc.) are very popular among learners of all ages. However, in practice, many adult trainees are not aware of the whole sets of capacities provided by the systems used. Similarly, they are not quite skilful in using online and offline dictionaries offering a wide range of opportunities for thematic translation (mining, engineering, geology, etc.). Thus, it is advisable to develop MT, CAT and HAMT skills in the course of training programmes for adult professionals.

Mastering listening and speaking skills is proved to be a highly complicated task for adult learners. The challenge of becoming proficient in these activities is emphasized by many scientific laboratories, including the American Educational Research Association [16]. Consequently, for adult trainees effective foreign language instruction should include less direct teaching and systematic practice involving rules and grammar, and plenty of opportunities for conversation. The educator needs to provide more occasions for communication between trainees themselves. While communicating with each other, adult learners can practice foreign language skills and share their professional experience, discuss some business matters equally important for all of them. This opportunity of professional communication with colleagues throughout the world can become an additional motive for attending language classes. Thus, it is practical professional orientation of language training that must become the cornerstone of continuing language education.

In the context of implementing distance learning technologies and ICT into language training/retraining of mining engineers, the choice of the most effective learning forms and methods is crucial for the process of educational design (specifically targeted planning of training courses). These technologies should encourage adult learners to self-learning alongside with the training programme. Among the innovative methods of foreign language training, the interactive approach appears to be the most promising in terms of distance technologies. It is based on active interaction of learners. Educators act as consultants and coordinators of trainees' educational activities. They step aside while trainees' work intensifies and dominates the process. In combination with ICT, this approach contributes to easier perception of learning and professional information, thus, developing intrinsic motivation for independent educational activity and demonstrating maximum efficiency in developing professional linguistic competence of adult learners [17]. So, considering professional orientation of language training/retraining courses for mining specialists, the most effective training technology is believed to be interactive learning method.

3 Results and discussion

When summarizing authors' long-term teaching practice as well as data from modern scientific and methodological literature, it can be stated that there are some fundamental differences between adult and children education shown in Table 1.

Table 1. Criteria of adult and children education.

Criterion	**Young learner** (pedagogical model)	**Adult learner** (andragogical model)
Aim of learning	Vague, defined by the third party	Clear, defined by the learner
Motivation	Often low	High
Leading role	Educator	Learner (wants to study and gain a skill)
Independence of learning	Often low	Often high (wants to manage the learning process himself)
Attitude to learning process (selected methods, educational literature, etc.)	Often worriless	Often critical (evaluates the learning process in terms of his previous educational experience)
Practical orientation of the skills developed	No	Yes
Self-discipline	Often low	High
Attitude to timescale of learning process	Often worriless (ready to learn for a long time)	Conscious (ready to learn for a limited time period)
Content of learning material	Often general	Specific and individual

Considering the mentioned characteristics of andragogical educational model, it is expedient to adhere to the principles below, when organising academic work with adults:

- using personal and professional experience of language courses trainees;
- exercising individual approach, flexibility in curriculum development;
- prioritising self-learning;
- practical applying the knowledge acquired and skills developed.

In addition, children learning process and adult learning process differ by the relations "teacher-student". In andragogical model, a teacher is a tutor, a designer of the learning process. His aim is to provide the educational process methodically and only then to act as a source of knowledge (*student-centered approach / learner-centered approach*). As V. Singh and P. Kharb mention, educators no longer serve chiefly as the dispensers of the subject content, but act as facilitators of learning and evaluators of competency [15]. This essentially different function is often not taken into account by educators of adult training/retraining courses, still exercising *teacher-centered approach*.

Thus, when training mining specialists for meeting professional challenges in a foreign language, it is necessary:

- to provide adult learners, non-humanitarian specialists, with the opportunity of quick developing skills for solving production issues in a foreign language with minimum time and finance costs;
- to use modern ICT when designing the learning process,
- to organise a continuous communicative interaction in a foreign language between the trainees.

Further in the article one of possible scenarios of designing language training courses for adult professionals is described. To start with, it is rational to conduct pre-testing of trainees in all language activities (listening, speaking, reading, and writing) to assess their language level. Along with stylistically neutral vocabulary, the tasks should include widely used professional words (e.g. *employee, employer, to dismiss, to hire, personnel department, salary*, etc.) as well as mining terms (e.g. *openings, workings, mine foreman, coal-breaker*, etc.). The pre-testing aims to divide trainees into language groups.

The first group is to join trainees with the levels up to Pre-Intermediate. Before the beginning of the main course, such trainees should be offered an intensive language course (not less than 20 hours long) aimed at enlarging active professional vocabulary (stylistically neutral words as well as a minimal set of professional words) and mastering simple grammar constructions necessary for making phrases. The main purpose of the second group of trainees (Intermediate level and upper) at the preparation stage is refreshing their listening and speaking skills (not less than 10 hours long). These trainees will take leading roles in the main course. The work of both groups can be organised in distant and/or mixed forms.

To organise a continuous communicative interaction in a foreign language between the trainees (in combination of the other tasks mentioned above), it is advisable to use interactive learning method. In terms of distance/mixed learning, among interactive teaching techniques the main interest is aroused in: brainstorming, round-table discussion (debate), case study (situational analysis), business game, role play, master classes, etc. This article describes the "Role play" technique as a technology successfully used at language courses for adults, including distant training/retraining language courses for mining specialists.

The "Role play" technique is successfully used in teaching children and adolescents at school, students at colleges and higher educational institutions as well as adults at language training/retraining programmes. There are many kinds of role plays: strategic, educational, organisational, etc. The applicability of the techniques for adult learners is easily explained by the advantages it has. Role plays let to simulate various situations of business interaction; to act in various professional roles; to master foreign language skills in real-like professional situations. The "Role play" technique can also include a number of other technologies involving an active interaction of participants: discussion, brainstorming, manipulative influence, and the search for a manipulation response, etc. In addition, adults are pleased to be included in game-based methods of learning material.

Disadvantages of role plays as a learning technology are reported to include the workload at the preparation stage and the lack of skills in playing a specific role in a real professional interaction. These disadvantages, in our opinion, are fully neutralised in andragogical educational model for two reasons. Firstly, the preparation for the role play is associated with developing foreign language skills that are necessary for real professional and gaming interaction, which corresponds to the main aim of language training/retraining courses. Secondly, adult learners are quite experienced in the professional sphere intended for practicing during the play. So, they can easily perform different professional roles of the area studied.

The basis of the "Role play" technology is an approximate scenario. It is reasonable to discuss some possible variants of scenario as this technology presupposes elements of "free speaking" (in contravention of the scenario discussed). The topic for the role play can be offered by the trainees themselves. While practicing foreign language skills they will are able to discuss some real production challenges they encounter with at their enterprises. Thus, the task "practical orientation of adult learning process" can successfully be accomplished.

At the preparation stage the educator is advised to act in different roles: a tutor while giving instructions, a group member while discussing a scenario, an advanced foreign language speaker while working on the wording. The educator can actively help while working with essential vocabulary on the topic (prepare some lexical minimum and some lexical exercises to practice it) and giving parts in the play (as it is he who knows foreign language abilities of all trainees). His main role at this stage is to assist (not to manage).

At the presentation stage the educator's role changes. He cannot assist but only observe the course of the play and caution the participants in case they turn to using their native language. It is necessary to achieve maximum self-regulation of participants. While observing, the educator should record language (speech, grammar, stylistic) and logical mistakes for the following individual and/or group analysis. It is also advisable to record the play to use the video at the evaluation stage.

At the evaluation stage trainees should also play the leading role: they are to be the first who give the language and professional feedback of the play. They are to assess their own language success / failure and that of their group mates. Besides, they can have a kind of round-table discussion on the professional matter discussed during the role play. The role of the educator here is an advanced foreign language speaker who is able to comment on the language and logical mistakes made. He can also offer some after-activity language practice and/or testing.

The core demand of the technique described is speaking the foreign language at all stages of the role play. The only exception can be made for the preparation stage, when vocabulary is being developed. And this is the main duty of the educator to monitor meeting this requirement.

At all stages of the role-playing activity, an extensive use of ICT is expected: online dictionaries and translators when working with vocabulary, video conferences for preparation, presentation and evaluation stages, messengers for the exchange of written documents and visuals, etc.

It is quite obvious that this learning pattern requires from the adult educator to meet the challenge: be competent as a language instructor, and advanced ICT user, an expert in business negotiations, and a specialist in the professional field discussed. So, this is the reason why adult educators should specialise in those professional spheres his or her trainees work at.

4 Conclusions

While designing language training/retraining courses for mining specialists, adult educators should follow the principles of andragogical educational model presented below:

- It is necessary to take into consideration the main criteria of adult learning: using trainees' professional experience; exercising individual approach; self-learning; practical applying the knowledge acquired and skills developed; treating the educator as a facilitator of learning and evaluator of language competency.
- Distance (mixed) form of education is the most suitable work pattern in andragogical model. It has a number of advantages of both existential and professional content.
- The extensive use of ICT should become an indispensable element of foreign language learning process.

- Interactive learning technology is the most productive method of language education for adult trainees.
- Using the "Role play" technique at language courses for mining specialists can meet two challenges: develop foreign language skills and exercise practical orientation of adult training process.
- Other techniques of the interactive learning method can also be effectively used within a framework of andragogical learning model [17].

References

1. M.S. Knowles, E.F. Holton, R.A. Swanson, *The Adult Learner: The Definitive Classic in Adult Education and Human Resource Development* (Elsevier Butterworth Heineman, London, New York, 2005)
2. S.B. Merriam, New Directions for Adult & Continuing Education, **89**, 3-14 (2001)
3. E.F. Holton, R.A. Swanson, S.S. Naquin, Perf. Improv. Quart. **14:1**, 118-143 (2001)
4. J.K. Dhillon. Stud. Educ. Ad. **48:1**, 118-119 (2016)
5. A. Gitterman, J. Teach. Soc. Work, **24:3/4**, 95-112 (2004)
6. K. Yang, R. Lawrence, *Participatory Visual Approaches to Adult and Continuing Education: Practical Insights* (CA: Jossey-Bass, San Francisco, 2017)
7. R.K. Sealana, *Examining the Efficacy of Adult Learning of Government-Mandated Content Using Andragogical Delivery Methods versus Traditional Pedagogical Delivery Methods. Doctoral Dissertation* (The University of San Francisco, 2014).
8. A. Jenkins, J. Educ. Work, **30:4**, 445-455 (2017)
9. K.P. King, P.A. Lawler, New Directions for Adult & Continuing Education, **98**, 5-13 (2003)
10. O. Tarnopolsky, Cog. Educ,, **3**, 1135771 (2016)
11. L. Arthur, F. Beaton, Lang. Lear. J., **21:1**, 31-36 (2000)
12. M. Lukes, Multil. Res. J. **5:1**, 19-38 (2011)
13. T. Deveci, New Horiz. Adult Educ. Hum. Res. Dev.t, **21: 3/4**, 16-28 (2017)
14. L.-T. Lee, J. Hung, Hum.-cen. Comp. Inf. Sci. **5**, 13 (2015)
15. V. Singh, P. Kharb, J. Anatom. Soc. India, **62:1**, 84-89 (2013)
16. R.M. DeKeyser, Am. Educ. Res. Assoc., Res. Poin., **4:1** (2006)
17. G. Chistyakova, E. Bondareva, K. Demidenko, E. Podgornaya, O. Kadnikova, E3S Web Conf. **15**, 04010 (2017)

E3S Web of Conferences **21**, 04020 (2017)
The Second International Innovative Mining Symposium
DOI: 10.1051/e3sconf/20172104020

Improving Occupational and Industrial Safety Management System at Coal Mining Enterprises

Svetlana Smagina[1,*], *Oksana* Kadnikova[2], *Ksenia* Demidenko[2], *Galina* Chistyakova[2], and *Anastasia* Rolgayzer[2]

[1] Kemerovo State University, Department of Psychology, 650000 Krasnaya St. 6, Kemerovo, Russia
[2] Plekhanov Russian University of Economics, Kemerovo Institute (branch), Department of Humanities, 650992 Kemerovo, Kuznetskiy Av. 39, Russia

Abstract. The article highlights the problem of mine workers occupational safety. The foreign and domestic experience of the field is analyzed. Some areas, activities and methods that meet modern requirements for ensuring occupational safety and production development, which are aimed at improving the quality of occupational safety management system, are described. The possibilities of their use in industrial safety of Kuzbass coal mining enterprises are considered. The proposed methods were implemented by JSC “Shakhta Yuzhnaya”, “Kiselevsky Razrez” Ltd., the coal mining enterprises of the Kemerovo region.

1 Introduction

Ensuring industrial safety and labour protection of coal mining enterprises employees is one of the most difficult social-economic and moral challenges of Kuzbass coal industry.

The number of accidents and the rates of fatal injuries in Kuzbass are significantly higher than at European coal mining enterprises with comparable volumes of coal mining [1, 2]. At the same time there are practically no technical and technological reasons that could lead to major accidents and emergency events at Kuzbass mines.

The basic factors of emergency events and accidents with severe consequences are the following: violation of technology; non-compliance with mandatory safety requirements when performing technological operations; commission of faulty, dangerous, forbidden actions by employees of all levels – workers, foremen, mine overseers, first-line managers, middle and top managers. Industrial safety requirements are violated by both contractors as well as technical and engineering employees [3].

Thus, the current level of fatal, severe, minor injuries is a consequence of the so-called “human factor”. Personal reasons such as faulty, careless or dangerous methods of work, deviation from job descriptions and technological instructions, non-compliance with special safety measures, violation of labour discipline, aspiration to overfulfillment of the production plan at the expense of evasion from safe production rules are the causes of 93-97% of industrial injuries of different types (including fatal cases) in comparison to 3-7%

* Corresponding author: kennedy22.1963@yandex.ru

caused by dangerous working conditions (mining-and-geological, natural and technical-and-technological factors).

2 Materials and methods

The methodological basis for the effective functioning of a complex occupational, industrial and environmental safety management system is domestic and international standards: GOST R 12.0.006-2002 "Occupational safety standards system. General requirements on occupational health and safety management in organisation system", the international standard OHSAS 18000 "Occupational Health and Safety Management System", the project National standard "General requirements on occupational safety and industrial safety management system", where occupational, industrial and environmental safety management system is presented as a course of actions, regulated in the form of local documents: job descriptions, occupational safety instructions, etc. The key conditions for the effective functioning of the system are competence and professional literacy of staff, interaction and partnership of enterprise management with public organisations, state bodies of supervision and control, municipal labour authorities.

Nowadays, the modern occupational and industrial safety management system at coal mining enterprises of Kuzbass represents a relatively wide range of measures aimed at preserving the life and health of mine workers, preventing accidents and their consequences. Traditional methods of organising work on labour protection correspond to the current level of management and development of production, though, they are not effective enough. The disadvantages of the existing safety management systems at mines include: the system is redundantly formalized, the staff is not sufficiently interested in meeting the safety requirements, in some cases employees want to overfulfill the targets by violating safe production regulations, there is a weak feedback from employees on safety issues, as well as the invalidity of motivational appeal for the staff to meet safety requirements.

3 Results and discussion

The problem of production safety increase at coal mines can be solved not only by expanding technical, technological and organisational measures, but also by exploiting the "human factor". If the so-called "human factor" is universally recognized as the main cause of production injuries, why not to use it to deal with the problem mentioned. There is a critical need to find some new measures meeting modern conditions and requirements that would allow further reduction of injuries and accidents at Kuzbass coal mining enterprises.

Along with the traditional methods of organising occupational safe and health activities, it is legitimate to introduce new approaches and directions that would correspond to the current level of industrial safety management and production development. Active involvement of personnel in safety management process by forming internal motivation for unconditional compliance with labor protection requirements, as well as appropriate behavior in hazardous industrial situations is one of the most progressive approaches that can be implemented in two directions:

- the increase in the level of communicative and psychological competence of managers and specialists in the direction of teaching methods of information and psychological impact on group and individual professional consciousness of the enterprise staff;
- the combination of material incentives and disincentives for compliance with safety requirements.

The main psychological regulators of safe professional behavior of employees are not only individualized personal characteristics that were formed during ontogenetic development (caution, risk tolerance, impulsiveness, discipline, courage, self-confidence, inattention, etc.), but also coordinated motives and attitudes (willingness to violate or readiness to comply with), group and individual perception of norms, rules, stereotypes of professional activity ("as it should be", "as it is correct", "like everyone does", "as it has always been done") including those related to health and safety rules applied while conducting mining operations and other technological processes; the need and accuracy of their observance, possibilities of evasion, etc., which were developed in a definite labour collective.

The degree of maturity of motives, attitudes, concepts provides and allows predicting the direction of professional behavior with a high degree of probability, including those aspects that are related to risk tolerance and meeting safety requirements [4].

The considered course of action suggests educating managers and labour protection specialists on alternative methods of providing subordinates with information on safety, which include specific speech strategies designed to increase the effectiveness of psychological impact on subordinates. It will allow building intrinsic motivation and determination of staff to comply with labour protection requirements in the course of the communication process regardless of the management hierarchy level. Exerting pressure in the right area and employing labour behavior management that reduces labour riskogenics is only possible, if employees themselves are aware of the need to comply with safety regulations, use safe techniques for work performance, and are willing to act accordingly [5].

Methods of psychological impact on personnel, including manipulative ones, are the most important methods of personnel management, which dynamically combine all the necessary and legal techniques and ways of influence (suggestion, infection, pressure, persuasion, placebo, purposeful dissemination of rumors; prohibition, hints, compliments, requests, etc.), induce or restrain employees, form certain moods and opinions at the workplace.

By using arguments, manipulating information, and applying persuasion all leaders and specialists involved in labour protection management as well as industrial and environmental safety management at mines in the course of promoting labour protection should strive not only to broadcast information through interpersonal communication, but also to inspire intolerance towards violators of safety requirements, motivate unconditional compliance with such requirements as well as reinforce positive stereotypes of safe and professional behavior at a particular workplace.

Employees tend to simplify their operations while performing work, as a consequence they fail to take certain factors into account, lose attention, show carelessness, excessive self-confidence, and risk taking. Change in habitual behavior, not only of an individual employee, but also of most workers, is merely possible if external information becomes "internal property", an element of personal experience, a persistent inner conviction.

Supplying employees with information on safety issues does not have to be limited by "passive forms" such as instructing, lecturing, providing reference books, inviting to thematic exhibitions.

Using manipulative tactics and techniques presents wide possibilities and can be applied at various levels, but the effectiveness of its impact largely depends on communicative competence (communicative qualities, communicative skills, applying different approaches to employees). In particular, the works of A. Cudworth and Mohammad Abu-qulah justify the use of the following manipulative techniques that can enhance the leader's persuasion of employees:

- the "Cards Juggling" technique is selecting and demonstrating factual nuances of an event or erroneous or dangerous actions that led to an injury or accident, at the same time ignoring or concealing some details of the event in order to impose a certain attitude towards the person or event;
- the "Simplification" technique is based on the fact that along with a number of general truths, an observation/assertion, that does not correspond to reality and is perceived by subordinates with no criticism, is proclaimed;
- the "Label Sticking" technique is identifying an event (which resulted in an industrial accident, emergency or prevented injury, accident) or an employee (who violated safety rules once/has constantly violated safety rules) by a word or expression that originally contains an acute negative tone in subordinates' minds;
- the "Pretending to be One of the Team" technique is a method of winning employees, shortening the distance between a manager and his subordinates by identifying with them through declaring "I am the same as YOU are", "I am one of YOU", "I feel the same as YOU do";
- the "Witness" technique is a reference to a positive or negative evaluation of an employee's idea, action or professional behavior, presented by a reputable or a highly competent person, for example, a reference to the opinion of the well-respected CEO of a coal mining enterprise in the course of introducing a new polymer resin based roof strengthening technology;
- the "Insinuation" technique is based on objectively existing differences that subordinates are aware of;
- the "Follow Everyone" technique is expressed in the manager's or the labour protection specialist's persuading the team that the overwhelming majority of the workers on the site (the mine as a whole) share a certain opinion, consequently, others have no alternative, but to join this opinion and behave correspondently;
- the "Axiom" technique means supplying safety requirements information in the form of indisputable truth, for example: "Employees of our coal mining enterprise must comply with all safety requirements unquestionably and unconditionally";
- the "Aggressive Sale" technique is attributing or subjectively accentuating causes of emergency/traumatic situations, accidents as a preventive measure;
- the "Transferring" technique is a projection of an approving attitude from something well-known to something new and incomprehensible; the technique is especially effective when innovations are to be introduced in safety regulations, since it is intended for workers' noncritical perception. The main difficulty is to find a suitable image or an example [6-9].

According to Pink D.H., in order to improve the efficiency of labour protection and industrial safety in the course of instructing employees it is legitimate to use information that causes fear and is supported by statistical (or "pseudo-statistical") data. Frightening information is distributed for the purpose of changing workers' attitude towards risks associated with hazardous industrial situations (for example, for encouraging people's active use of personal protective facilities as an effective way of preventing dangers) [10].

If the communicator's implicit purpose is to "intimidate" employees, it is advisable to adhere to the following recommendations when transmitting information:

- the message should express concern about possible consequences of erroneous, risky or careless actions of employees;
- at the same time the message should contain an obvious positive evaluation of the recommended "correct" actions and measures;
- the message should contain both logical argumentation and value judgments based on common sense and the logic of the common mind (undeniable truths);
- the actions recommended should be specific, detailed and elabourated, as well as be perceived as effective means of preventing potential danger;

- the message should correspond with the level of awareness and competence of the employees who perceive information;
- the information should contain specific examples in which the employee who systematically violates labour protection requirements inevitably experiences more losses than acquisitions including financial ones, and vice versa, strict compliance with labour protection requirements makes the employee more successful in the production process;
- the communicator's level of authority is essential (the higher the level, the less critical the perception of the recipients).

According to Drupsteen L., Groeneweg J., the formation of a company's safety culture begins not only with imposing high responsibility for accidents on the company management, but also with the ability of the management to communicate with the personnel, make it a personal concern for everyone and convince every employee that compliance with all safety requirements and regulations is necessary and mandatory. This is "human factor" management, which makes it possible to achieve impressive results [11]. Communicative and psychological competence of the personnel: master, mine overseer, engineer, operations manager, headmaster and other specialists (not only in the field of expertise and Process Safety Management) can influence the level of traumatism at coal mining enterprises. Highly developed intrinsic motivation, determination to comply with safety regulations and self-control act as significantly more powerful regulators of workplace behavior, which can exclude violations of occupational safety requirements more effectively then administrative pressure and operating control [12].

The analysis of foreign practice shows that one of the directions of improving the quality of HSE Management System of coal producers is a combination of financial incentives and disincentives for the personnel to meet the requirements of individual and collective safe conduct [13].

In particular, M. Evans states that the amount of incentive payments should correspond to or outstrip the period of accident-free work, and economic sanctions should be perceived by employees as justified and fair measures increasing pressure in the area demanded [14].

Being one of the most effective methods of influencing employees labour behavior and motivation, financial encouragement of the personnel is aimed at eliminating the factors which contribute to violating labour protection rules and neglecting personal safety, as well as increasing the interest of the staff in ensuring safe labour conditions at the workplace in particular and at the enterprise as a whole). The standard of financial incentives for mine employees' complying with the requirements of individual and collective safe conduct includes a system of economic measures (sanctions and incentives) [15].

The leading form of encouragement is bonus, which is extremely effective, being gradual, meaningful and corresponding to the principles of "graduality" and "tangibility". Thus, payments for compliance with / not violating security requirements should not be regular, otherwise it may cause habituation and relaxation of vigilance (the employee is clearly aware of the linear relationship between specific professional actions and the financial encouragement); a large one-time bonus contributes to the formation of increased remuneration expectations of the staff. Bonuses should comprise at least 15-20% of the average month, quarter or other reporting period's earnings. Publicity is also a way of intensifying the effect of financial rewards [16].

Conclusion

The analysis of the efficiency of the occupational safety management system applied in the coal mining industry of Kusbass shows that the solution to this problem is impossible without active involvement of the coal mining industry personnel in the process of ensuring safe mining operations. It is necessary to emphasize that the direction considered in the

DOI: 10.1051/e3sconf/20172104020

article reflects some principles of the "Zero Accident Vision" concept, applied to corporate programs of the metallurgic and mining companies in the European Union countries, Finland and the Netherlands in particular [17]. It is recommended in order to form psychological attitudes to meeting safety requirements. In combination with financial incentives, these actions will provide effective management of the personnel's motivation. The proposed methods were tested and received approval at the coal mining enterprises of the Kemerovo region, OAO "Shakhta Yuzhnaya", OOO "Kiselevsky Razrez".

References

1. J.L. Hawksley, Haz. Mat., **65 (2)**, 109-121 (1999)
2. M. Cehlár, J. Janočko, Z. Šimková, T. Pavlik, E3S Web of Conf., **15**, 01019 (2017)
3. M. Prekopová, J. Janočko, Geol. Carp., **60:6**, 485-494 (2009)
4. G. Wittenberger, M. Cehlár, Z. Jurkasová, Acta Montanistica Slovaca, 17:4, 241-246 (2012)
5. J. Song, Y. Zhu, F. Dong, Cross Strait Quad-Regional Radio Science and Wireless Technology Conference **24**, 01124 (2011)
6. A. Cudworth, Strat. Com. Man., **14 (1)** 16-19 (2009)
7. Mohammad Abu-qulah, Int. J. Man. Fin. Educ, **24 (6)** 42-55 (2014)
8. W. Wen-Qing W, Z. Lei, Z. Wei-hua, Proc. Engin.,**4,** 24-52 (2013)
9. F. Mingchao, S. Bin, Sec. Int. Conf. Dig. Manuf. Autom, **25**, 03124 (2011)
10. D.H. Pink, *Drive: The Surprising Truth about What Motivates Us* (Canongate Books Ltd., Great Britain, 2010)
11. L. Drupsteen, J. Groeneweg, G. Zwetsloot G., J. Occup. Saf. Erg., **19 (1)** 63-77 (2013)
12. G. Chistyakova, E. Bondareva, K. Demidenko, E. Podgornaya, O. Kadnikova, E3S Web of Conferences, **15**, 04010 (2017)
13. G. Pinigina, I. Kondrina, S. Smagina, E3S Web of Conferences, **15**, 04017 (2017)
14. M. Evans, J. Manag., **12 (2)**, 203-222 (2006)
15. Zwetsloot, M . Aaltonen, J. Wybo, J. Saari, P. Kines, R. Op De Beeck, Saf. Sci., **58(3),** 41-48 (2013)
16. K. J. Foster, Risk Anal. **17 (2)**, 215-225 (1997)
17. K.M. Dange, R.T. Patil, Int. J. Engin. Sci. Inven., **2(7),** 14-19 (2013)

DOI: 10.1051/e3sconf/20172104021

Education within Sustainable Development: Critical Thinking Formation on ESL Class

Inna Pevneva[1,*] , *Olga* Gavrishina[2], *Anna* Smirnova[3], *Elena* Rozhneva[4], and *Nataliya* Yakimova[5]

[1] Kemerovo institute (branch) of Plekhanov Russian University of Economics, Department of Humanities, 650992 39 Kuznetsky pr., Kemerovo, Russia
[2] Kemerovo State University, Institute of Sciences, UNESCO IT department, 650000 6 Krasnaya ul., Kemerovo, Russia
[3] Kemerovo State University, Department of Germanic and Romance studies, 650000 6 Krasnaya ul., Kemerovo, Russia
[4] T.F. Gorbachev Kuzbass State Technical University, Department of foreign languages, 650000 28 Vessennyaya st., Kemerovo, Russia
[5] Kemerovo State University, Department of Foreign Languages in Professional Communication, 650000 6 Krasnaya ul., 6, Kemerovo, Russia

Abstract. The article is devoted to consideration of the critical thinking formation in course of foreign language teaching within the education for sustainable development as a crucial skill of perspective employee and a future leader of Russian employment market. The necessity to include the component of problem education and critical thinking methodology in course of the foreign language class is justified along with analysis of the basic principles of critical thinking and certain strategies that can be implied in class. This model targets communicative language competences of students as well as critical thinking due to interconnection of various types of cognitive activities in class. The role in personality development of the students is considered along with the formation and enhancing of critical thinking skills within the modern personality-oriented approach.

1 Introduction

Education for sustainable development allows each person to acquire the knowledge and skills and to form the attitudes and values necessary to build a sustainable future. Education for sustainable development involves teaching and learning the main issues related to sustainable development such as climate change, biological diversity, poverty reduction and sustainable consumption. It also implies the use of methods based on participation, motivating students and allowing them to adapt and become active participants of sustainable development. In this regard, education within sustainable development would seek to develop critical thinking ability and qualities that allow graduates to predict certain crucial events in the future and take correct decisions [1-2]. Nowadays we obviously

* Corresponding author: Inna_nag@mail.ru

witness that sustainable development can not be achieved solely through technology, political regulation or financial mechanisms. Mankind needs to change the way of thinking and behavior, thus requiring to provide quality education and training for sustainable development at all levels and independently of social conditions.

Situation on Russian graduate market has its specifics, but in general is close to European and American. Employers increasingly recognize what they want to see in graduates is not so much technical knowledge, but applied skills, including critical thinking and abilities to take justified decisions. According to a report of employers' perspectives, a consortium of more than 400 US employers ranked "critical thinking" as the most desirable skill in new employees. It was ranked higher than skills in "innovation" and "application of information technology" [3].

2 Methods

One of the technologies that present enormous opportunities for solving this problem is the technology of developing critical thinking through reading and writing, developed in the 80s of the 20th century in the USA. Even at the beginning of the twentieth century, D. Dewey spoke of critical thinking identifying it with reflective thinking. He believed that through exercises you can teach a person to think constructively and saw in this the main task of education. The task of education, in his opinion, is precisely to develop intelligence and form a "disciplined" mind [4-6].

Academics and researchers for the most part understand critical thinking as a cognitive strategy, consisting largely of continuous challenging and testing of possible solutions. Other definitions indicate that critical thinking is characterized by constructing logical conclusion as well as creating consistent logical models and taking justified decisions. They refer to the mental activity which is aimed at solving a particular cognitive task [7-8]. The mechanism of critical thinking includes mental operations that determine the process of reasoning and argumentation: setting the goal, identifying the problem, putting forward hypotheses, bringing arguments, justifying them, predicting the consequences, accepting or rejecting alternative options. It includes the ability to apply basic intellectual skills (knowledge and understanding) for the synthesis, analysis and evaluation of complex and ambiguous situations and problems.

There are basic principles that characterize the development of critical thinking. Each of these principles to a certain degree can be considered a mental skill so they should be given great attention in the process of students' education. First, a student can identify and challenge assumptions. It can be verified by posing special questions that clarify the details. Students should study the grounds on which the conclusion was drawn and analyze whether the various sides of this conclusion are facts or opinions. Secondly, it should come to verification of actual accuracy and logical sequence checking the logical coherence and consistency in the argument. The third principle is based on context consideration. Upon considering new information, a person having critical thinking skills will pay attention in what context this information makes sense. He analyzes whether this information can be applied in any situation or only under specific conditions. The last principle infers the studying of options or alternatives, just like all other elements of critical thinking this skill can be formed by certain training [9].

The strategies which are often referred to in course of students' education are brainstorming, problem formulation, narrowing the information and others. All of them to a certain degree are used by teachers of foreign language. The above mentioned principles, when formed, will allow students and graduates to critically comprehend any information received from the mass media, the Internet, routine communication with colleagues and make the right choice. It is important to emphasize that those principles and characteristics of critical thinking are directly interrelated. In the integrated form they are the original "stages" relying on which it becomes possible to use this or that principle of critical thinking when teaching students a foreign language. This correlates to the phases mentioned in other researches: evocation, realization, understanding and reflection [11-12].

Nevertheless, critical thinking doesn't exclude creativity. Visa versa, by critical thinking we mean the ability to think, finding strengths and weaknesses, to evaluate the result and the process of thinking but also includes the ability to understand, analyze, evaluate, synthesize, which, in essence, is a creative process. The creative essence of thinking is revealed at all stages from knowledge to evaluation. The background knowledge is not the same for all students so the phase of referring to this knowledge can take completely various forms and verbal form will largely depend on individual style and personality. The application of methods of developing critical thinking in the process of foreign language teaching has a number of advantages but also may present a challenge to the teacher in terms of control and assessment. Undoubtedly, along with critical thinking students develop their independence, creativity, increase motivation, improve ability to understand context. But difficulties may lie in the problematic evaluation of skills and the complication of control over discipline. Along with the development of skills of reading, it is necessary to develop students' ability to express their thoughts in writing. It is the written speech that is accompanied by the greatest difficulties especially in a foreign language.

Some scientists suppose that the current educational trend to standardize curricula and put emphasis on test results makes it hard for teachers to address critical thinking in the classroom. The focus on "preparation for the test" may mislead the educational process from student-centered instruction and put an emphasis on the content instead [8]. It should be emphasized that the formation of critical thinking skills is linked with the formation of speech skills in the learning process of all types of activity, since the subject of any speech activity is thought. Several researchers thus distinguish between "critical reading", "critical writing", "critical listening" and "critical speaking" [9-10].

3 Discussion

Thus in the process of teaching a foreign language the development of critical thinking can be considered as one of the factors in the formation of communicative competence.

Since the primary goal of teaching a foreign language is the communicative activity of students the teaching material must be problematic and the educational process should be organized in such a way that students would have a permanent desire to express themselves and defend their opinion. The ability to formulate a problem, find answers to the questions posed, draw additional information, form a position on a particular issue, back it up with arguments and facts, draw conclusions – all these is the goal of problem-based learning based on critical thinking skills.

Thus problem methods are both a goal and a means of learning as the work on a problem also involves interaction in a team as participants need to exchange knowledge, share experience, ideas, information received and, as a result, come to a mutual decision. Such methodology is based on the principles of the personality-oriented approach (education in cooperation, discussions, round tables, projects etc.), which corresponds to the paradigm of modern education.

When applied, this methodology can track the dynamics of students personal development which means changes in willingness to pose questions about what is happening and their actions, enhance readiness to refer and apply their own experience and not the external sources (teacher, reference book, etc.). It sufficiently increases self-control and as a result self-organization, students begins to understand what, how, why he does certain things and consequently gets an opportunity to organize and control his own

DOI: 10.1051/e3sconf/20172104021

activities. Students obviously develop abilities to understand phenomena and events and not act according to a prescribed standard.

Haynes and Bailey in the research stress the necessity of posing the right questions to stimulate students' critical thinking skills. In other researches the focus on integrating questioning techniques into class discussions to encourage an environment for learners to reveal and practice critical thinking skills [6-7]. Following the "IDEALS" method in class a teacher can help to integrate methods of critical thinking development and it can be adapted to various level and theme of class:

I – Identify the Problem
D – Define the Context
E – Enumerate the Choices
A – Analyze the Options
L – List Reasons Explicitly
S – Self-Correct

All in all, teaching a foreign language on a university level shouldn't be limited to forming communicative competence, it should also contribute to critical thinking development though the foreign language. Though formation and development of critical thinking skills should be carried out at all stages of teaching a foreign language, at a university level it becomes crucial. This skill becomes a new important component of modern higher professional education, an substantial factor in the professional competence of the graduate.

4 Conclusion

The use of methods of critical thinking in the process of teaching a foreign language contributes not only to the intensification of this process, but also to the development of the motivation for attaining and value attitudes of students. Critical thinking methods allow students to develop the habit of creating their own texts, making adjustments and presenting them to the audience. A student with critical thinking is distinguished by the readiness for planning, flexibility, perseverance, readiness to correct his mistakes, and search for compromise solutions.

References

1. M. Bonnett, Camb. J. Ed., **29:3**, 313-324 (1999)
2. N.S. Kasimov, *Education for sustainable development: Experience of West Europe, Russia and Central Asia* (Moscow, Lomonosov Moscow state university, 2008)
3. Are They Really Ready To Work? Employers' Perspectives on the Basic Knowledge and Applied Skills of New Entrants to the 21st Century U.S. URL: http://www.p21.org/storage/documents/FINAL_REPORT_PDF09-29-06.pdf
4. R.H. Ennis, *The Nature of Critical Thinking: An Outline of Critical Thinking Dispositions and Abilities,* URL: http://faculty. education.illinois.edu/rhennis/documents/TheNatureofCriticalThinking_ 51711_000.pdf
5. J. T. Broadbear, J. Schol. Teach. Learn., **3:3**, 1-8 (2003).
6. T. Haynes, G. Bailey, Business Education Forum, **57,** 33-37 (2003).
7. H. E. Hemming, McGill J. Educ., **35**, 173 (2000).
8. L. G. Snyder, M. J. Snyder, J. Res. Bus. Educ., **50**, 2 (2008).

9. R. Nault, *Teaching Critical Thinking: Selected Readings Compiled, BBA - WLU, Canada, 1987, CELTA - Canada, 2004, MA Development Studies - RUPP, 2010* (TCTV1 July 31, 2014).
10. P. Brodie, K. Irving, Assessment & Evaluation in Higher Education, **32:1**, 11-19 (2007).
11. S. Hannon, H. McBride, B. Burns, Industry and Higher Education, **18:2**, 95-100 (2004).
12. R. Feldman, S. Schwartzberg, *Thinking critically: A psychology student's guide* (New York., McGraw-Hill, 1990).

Some Diversification Factors of Old Industrial Regions' Economy and Transition to the Innovative Development

Olga Tabashnikova[1]

[1] Plekhanov Russian University of Economics, Kemerovo Institute (branch), 650992 Kemerovo, Kuznetskiy Av. 39, Russia

Abstract. The article presents the grounds for the necessity to diversify the mono-economy of old industrial regions and its transition to the innovative development based on the interaction of small and large businesses with the support of municipal, regional and governmental authorities. The examples of the world practice in state regulation of depressed territories of old industrial type and the participation of multinational corporations in their modernization are given. The role of business groups in the diversification of the Kemerovo region economy is described, as well as the importance of supporting this process by the governmental authorities.

1 Introduction

The need to diversify the mono-profile (mono-product, mono-industrial) economy of old industrial regions, including the Kemerovo Region, is due to the objective requirement of their sustainable development and increase of social and economic parameters of the territorial management system. One of the important factors of this process is the effective interaction of small and large business. However, in modern economy, when the existing economic, production, and other links and mechanisms lose their importance and new ones are just being established, to create such interaction and ensure its development the purposeful efforts of the authorities on federal, regional and municipal levels are necessary.

2 Materials and Methods

It should be noted that there is no "exact" (single, universal) definition of the term "old industrial region", despite the fact that many scientists have paid attention to this phenomenon during the past twenty five years in Russia [1-5]. The fact is that the definitions proposed by the researchers are determined by the aims of this or that study. At the same time, a sufficiently extensive scientific material accumulated in this sphere of research allows us making some generalizations.

The English scientist R. Hudson believes that the old industrial regions, which he calls "the world's workshops," lose their importance over time [1]. In the development of the

DOI: 10.1051/e3sconf/20172104022

strategy for the survival of "peripheral areas," to which mining areas exactly belong, according to Hudson, four possible reorganization options should be considered:

1. The development of small and medium-sized enterprises, the search for ways to create new jobs in small business, the development of the so-called entrepreneurial "culture of self-employment" instead of large-scale production.

2. The creation of an updated economic system based on the branches of large, and, if possible, multinational companies attracted to the region, upon condition of propaganda of an attractive image of the territories.

3. Transition to the economy of "service" - tourism, recreation, entertainment, etc.

4. The refusal from the production sphere and the transformation of territory into a zone of "housing and consumption", where the main source of living is state benefits and subsidies.

However, R. Hudson emphasizes, that only local initiatives in implementing any variant or combination of options are not enough to solve the problems of territory restructuring [1].

At the same time, D. Sadler, being a member of the Cabinet of Ministers of Great Britain, drew attention to the fact that the depressive state of some territories leads to the rejection of entire regional population groups from the processes of social and economic development of society. By the example of some old industrial territories of Western Europe one can observe that low labor status, the regressive distribution of incomes, the underdevelopment of social infrastructure; the contradictions of social strata escalate the social and political situation in such areas [2].

3 Results and Discussion

At present in Western Europe there is a certain system of state regulation of depressed territories of the old industrial type. Common principles for all countries of this system are the following [6-8]:

- recognition of the government support as a necessary condition for the recovery of crisis and depressed regions, which should be provided in advance, before the crisis, at the stage of stable development;
- implementing regional policy to have an impact not on the region as a whole, but on the most problematic part of it;
- solving priority social problems by creating new highly skilled jobs for local personnel, developing social infrastructure, preventing the decline in income of the population;
- stabilization of peripheral old industrial areas through the development in them new high-tech industries, business services, recreation;
- orientation to the internal resources of the regions to achieve the stability of their development.

As a rule, the old industrial areas play a special role in the development of economy of the territories, but they have a whole range of problems, from ecological to demographic ones. The main problem of such regions, and this fully applies to the Kemerovo region, is the excess of production capacity, represented by large and medium-sized enterprises with obsolete equipment and technologies.

A basic feature of the Kemerovo region as an old industrial zone is its predominantly resource nature. Herewith the main economy resource in the absolute majority of Kuzbass towns, and in fact the source of their existence, is coal deposits the volume of which is limited, and, according to experts, will provide the operation of coal mines and open pits for the next 50-60 years.

By the level of economic development the Kemerovo region occupies an intermediate place between the relatively developed and moderately developed regions of Siberia.

However, the following trend attracts attention: during recent years the sectoral structure of the gross regional product (GRP) of the Kemerovo region has become more primary. According to the Independent Institute of Social Policy, the share of extractive industry increased from 20.6% to 28.4% from 2004 to 2008, while the share of manufacturing industry fell from 24.5% to 19.8%. The reduced life expectancy and educational level of the population do not allow the Kemerovo region rising above the 31st place by the HDI in the ranking of regions (2008). Earlier the position of the region was even worse (35-42 places in 2004-2007). The region's quality of life index grew in 2002-2005; it increased from the 26th to the 11th place. As for the innovation index Kemerovo region occupies the 30th place. The innovative development of the region is hampered by a small proportion of those employed in science and scientific development. In 2000-2004, by the index of democracy, the region occupied only the 66th place. Thus, according to the integral social indices, in 2008 the Kemerovo Region lagged behind most of the developed industrial regions of the country.

At the same time, the recovery of enterprises of the Kemerovo region also began earlier than in other Russian regions. This was due to the improvement of the metallurgy global market. However, this impetus for growth has dried up in 2010. The decline in 2009 as a whole was slightly less than the total for Russia (9 and 11%, respectively). The production of the most important type of products - steel - decreased by 7% and amounted to 7.6 million tons (13% of all-Russian production).

Unfortunately, in 2016 the value of integral indicator of the development of the Kemerovo region fell even further and amounted to 0.152, as a result of which the region took the 78th place out of the 80th. The economic, social and environmental indices were 0.53; 0.223 and 0.03, respectively.

The volume of production of coal mining industry in 2008 was 178 million tons. Crisis helped to increase the concentration of Russian coal production in Kuzbass: if in 2008 it accounted for 55% of Russian coal production, in 2009 it was 59%. At the same time, the volume of coal mining in Kuzbass is constantly increasing. Thus, the volume of coal production in 2016 increased by 5.4% compared to the level of 2015 and amounted to 227.4 million tons.

The coal mining industry is the main branch of Kuzbass specialization. The most important centers of coal industry are Prokopyevsk, Mezhdurechensk, Belovo, Kemerovo, Novokuznetsk, Osinniki, among the districts are Leninsk-Kuznetsk, Belovsky, Kemerovo, Novokuznetsk and Prokopyevsky.

Leading coal mining companies are owned by large Russian business, mainly metallurgical. Russian metallurgical companies bought mines and pits that extracted coking coal in order to control the entire technological chain of steel and rolled steel production. OJSC "United Coal Company “Yuzhkuzbassugol” and the mine "Raspadskaya"are controlled by the holding company Evraz Group. Kuzbassugol (currently OJSC Coal Company “Severny Kuzbass”) is owned by OJSC “Severstal”, OJSC Coal Company “Yuzhniy Kuzbass” – by metallurgical company OJSC “Mechel”.

The largest company in the Kemerovo Region, OJSC Coal Company “Kuzbassrazrezugol”, which accounts for a quarter of Kuzbass coal produced in 2009, is a part of the OJSC “Urals Mining and Metallurgical Company”. Moreover, this property is not linked by a single technological chain, since the “Ural Mining and Metallurgical Company” is the second largest copper producer in Russia.

In the 90s of the 20th century, coking coals were supplied primarily to the domestic market. Their consumers were mainly metallurgical plants. At that time, the producers of power generating coals were export oriented. However, in the second half of the 2000s the export vector of Kuzbass grew stronger due to the increase in world prices for coal. In 2006, the volume of export of Kuzbass coal exceeded the volume of supplies to the domes-

 DOI: 10.1051/e3sconf/20172104022

tic market for the first time, and in 2016 the shipment of products exceeded 188.9 million tons for the year - an increase of 7.8 million tons, 66% was exported. In 2016, Kuzbass significantly expanded the sales geography by supplying coal to Bosnia and Herzegovina, Chile. At present Kuzbass supplies coal to 55 countries of the world [6].

The enterprises of metallurgical industry provide more than half (almost 60%) of the production of the processing industry in the Kemerovo region. Two full-cycle metallurgical plants are located in Novokuznetsk and since 1999 belong to the holding company EVRAZ Group. One of them, OJSC "Zapadno-Sibirskiy metallurgicheskiy kombinat", is one of the five largest in Russia and in the thirty of the world's largest steel works. Its capacity is 8 million tons of steel and 5 million tons of rolled steel a year. Novokuznetsk Metallurgical Complex (NMC), built during the industrialization period, has less capacity and specializes in the production of rails (about 70% of Russian production). From July 1, 2011, OJSC "EVRAZ Zapadno-Sibirskiy metallurgicheskiy kombinat" merged ZSMK and NMC. In 2016 EVRAZ ZSMK produced 5.4 million tons of cast iron, 6.9 million tons of steel, 6.3 million tons of steel products. Except the enterprises of ferrous metallurgy Novokuznetsk aluminum plant is located in the city. It belongs to the company "Russian Aluminum" and is the fifth largest aluminum producer in Russia.

Among the enterprises of manufacturing industry of the Kemerovo region there are such enterprises as the OJSC "Yurga Machine-Building Plant" (Yurga), the largest machine-building enterprise in the region producing mining equipment, the Kuzbass Railcar Company (Kemerovo), a branch of "Altayvagonzavod" that manufactures cargo platforms and components for wagons, OJSC Holding Company "Sibirskiy Cement" (the headquartered is in Kemerovo, the main enterprise "Topkinsky Cement" is in Topki), which produces cement for housing and road construction, OJSC "Azot" - a powerful industrial complex, which produces more than 40 kinds of chemical products on the basis of modern chemical equipment, etc. All of these companies have financial difficulties in recent years.

At the same time, the experts note that for the Kemerovo region the general tendency of increasing the service sector in economy is not typical: it makes just over 40% of GRP, while in Russia as a whole it is 56%. Among the main threats to the development of the region it is necessary to single out the following:

- mono-branch structure of the economy of the Kemerovo region;
- the dependence of regional budget and, accordingly, municipal budgets on large taxpayers;
- manufacturing of products with low added value;
- the dependence of economy on the market prices for coal and ferrous metals;
- the dependence of economy on transport and energy tariffs;
- high degree of fixed assets depreciation;
- insufficient level of use of modern management and labor management systems;
- infrastructure constraints;
- the lack of skilled labor force;
- the shortage of investment resources.

At the same time, for municipal entities of the Kemerovo region, which have a single-sector economic structure dominated, as a rule, by one type of economic activity, there are a number of factors that determine the opportunities for diversification of economy and transition to the innovative development. Among them are:

1) dominance of the fuel and energy complex as a source of financial resources for the future transformations;

2) the availability of relatively high-tech industrial and human resources, historically formed since the times of the USSR;

 DOI: 10.1051/e3sconf/20172104022

In our opinion the most interesting example of industry diversification on such old industrial area as the Kemerovo Region is the experience of regulating the problems in the Ruhr Area in Germany.

The old industrial Ruhr Area, which played a huge role in Germany's industrialization, unites today in one communal union (the Ruhr region) 11 cities and adjacent areas occupying an area of 4,4 thousand square km with a population of 5.4 million people.

After the Second World War, the region restored its economic power within a decade, based on the coal industry. 50 new mines were built, and coal production reached 112.8 million tons in 1954. However, already in the second half of the 1950s, a crisis started in the Ruhr coal industry due to the imported oil competition. Coal production fell sharply, many mines were closed, the crisis spread to the coal-related industries, and unemployment increased.

The government of the Federal Republic of Germany chose the way of special economic support for the Ruhr Area. Subsidies were granted to the coal industry, the coal consumption, especially to the Central Heating and Power plants, in order to maintain the consumer demand for coal. In 1968, a federal law was adopted to restructure and recover the coal industry and the coal-mining region, according to which the government determined the minimum required volume of coal production in the country, the amount of subsidies, subsidies to the freed-up workers and their retraining costs.

In 1969, the whole Ruhr coal industry was merged into one concern "Rourkole" ("Ruhr Coal") on the basis of the agreement of the German Minister of Economy, the government of North Rhine-Westphalia, the owners of mines and trade unions.

Coal mining enterprises were bound by the long-term contracts with metallurgical plants and power plants. A specific tax - "coal pfennig" - had a significant impact on the stabilization of coal sales. Each extra pfennig (0,01 marks) paid by the consumers extra price of kilowatt-hour of electricity came to a special fund for coal consumption, all of which went to buy German coal instead of cheaper foreign one. The tax was abolished in 1995.

These measures of the government of the Federal Republic of Germany allowed avoiding the collapse of coal production and gaining time for the gradual and controlled restructuring of the entire Ruhr economy.

The main directions of restructuring the region's economy were:
- the modernization of "old" industries;
- the development of higher education;
- the creation of technology centers and, on their basis, high-tech industries;
- improving the environment.

Industrial modernization covered most of the metallurgical and chemical plants, heavy engineering enterprises. Ecologically polluted industries were eliminated, the range of products was radically updated, and industrial territories were reduced. In 1960-1980 in the Ruhr region 5 new universities were founded with the aim of creating training centers, reducing youth unemployment, limiting emigration from the region and attracting new personnel. A significant part of local students studies free of charge, low-income students receive scholarships.

In 1993, the total number of students reached 150 thousand, which was twice the number of mines, the number of scientists and teachers increased multifold. Twelve technological centers were created that contributed to strengthening the scientific and innovative potential of the region, providing premises for entrepreneurs, equipment, logistics, organizational and financial assistance.

Elimination and recycling industrial waste, protection and expansion of the green landscape were coordinated by the "Municipal Union" using various financial sources. The result of this program was the ecological improvement of the entire region, and the most polluted area along the Emscher River was turned into a huge landscape park with a length of 320 km, containing recreation areas and eco-zones, technology museums and innovative exhibitions [6].

The effect of specialization may arise when using highly specific material and / or intangible assets, for example, the interaction of a large construction company "Kemerovopromstroy" with a small enterprise "Finestra", which supplies it with 90% of the production (wood constructions for building) on the basis of long-term contracts. There are a lot of examples of this kind in economic activity, in particular, the transfer to the audit firm a part of the financial planning functions by a large enterprise "Kemerovohimmash". Aaccounting, providing legal support, consulting, recruiting, advertising and other services provided by firms specializing in their implementation to a wide range of small and large enterprises. The need to distinguish specialization as the basis for interaction between small and large businesses is determined by the processes of division of labor, and also by the fact that it is small business that largely determines the viability and effectiveness of new types of business taking innovative risks.

Complementary (additional) effect occurs when using tangible assets (tools, raw materials, materials) and is focused on their full use. An example of a more complete use of production capacity is the provision by the large enterprise "Kemerovo plant of reinforced concrete structures" of existing equipment, which is engaged in the basic production of an average 1.6 shifts, to a small enterprise that produces paving slabs. An example of more complete use of raw materials is the production of household products by a small enterprise "Kuzbasski Scarabei" from the residues left as a result of production at a large chemical enterprise "TOKEM".

4 Conclusion

We emphasize that the desire to achieve such effects in an explicit or implicit form acts as one of the incentives for the interaction of different forms of business. The basis for solving practical problems of territorial-industrial integration development is the theory of clusters. According to it, the production structure of the region should be developed in such direction that allows using the product of one industry for the needs of the others.

Thus, between all the industries represented in the region stable links are being created. It allows supporting these industries and promoting the innovative development of the economy of old industrial region.

References

1. R. Hadson, *Economia Restraction Tendencies and their Implications for Development Strategies in the Peripheries of Europe* (Birmingham, Lommond, 1994)
2. D. Sadler, *Marginality in the New Europe* (Birmingham, Lommond, 1994)
3. H. Burke, E. Hough, D.J.R. Morgan, L. Hughes, D.J Lawrence, UK Land Use Policy, , **47**: 321-331 (2015)
4. M. Cała, A. Ostręga, Arch. Min. Sci., **58:2**, 361-374 (2013)
5. J. Kretschmann, Euras. Min. **2**, 52-56, (2015)
6. J. Kretschmann, Int. J. Min. Sci. Tech. **23**, 659-663 (2013)
7. A. Prossek, H. Schneider, H.A. Wessel, B. Wetterau, D. Wiktorin, *Atlas der Metropole Ruhr –Vielfalt und Wandel des Ruhrgebietes im Kartenbild, Emons Verlag* (Hamburg, Hollmann, 2011)

DOI: 10.1051/e3sconf/20172104022

8. *Gespräch mit Experten des Regionalverbandes Ruhr in Essen am 04.12.2012: Michael Schwarze-Rodrian* (Netzwerke Ruhr, RVR, 2013)

Key Trends in Institutional Changes Under Sustainable Development

Olga Karpova[1,*], *Inna* Pevneva[2], *Irina* Dymova[3], *Tatiana* Kostina[4], and *Sergey* Li[5]

[1]Kemerovo institute (branch) of Plekhanov Russian University of Economics, Department of Economic Security, Accounting and Audit, 650992 Kuznetsky pr.39, Kemerovo, Russia
[2]Kemerovo institute (branch) of Plekhanov Russian University of Economics, Department of Humanities, 650992 Kuznetsky pr.39, Kemerovo, Russia
[3]T.F. Gorbachev Kuzbass State Technical University, Department of Management Accounting and Analysis, 650000 Vessennyaya st., 28, Kemerovo, Russia
[4]Kemerovo institute (branch) of Plekhanov Russian University of Economics, Department of Management and Business Technology, 650992 Kuznetsky pr.39, Kemerovo, Russia
[5]Kemerovo institute (branch) of Plekhanov Russian University of Economics, Department of Information technology and Applied Mathematics, 650992 Kuznetsky pr.39, Kemerovo, Russia

Abstract. The article is devoted to the consideration of the essential problems of accounting institution formation under the sustainable development of the country and the region. The research is based on the key research the field of the intuition economics and considers the trends of institutional changes including incremental, evolutionary and revolutionary. Approaches to the analysis of institutions are presented as well. The first approach states that economic efficiency is guaranteed by newly emerging institutions. The second approach involves certain internal and external incentives for changing institutions. Whereas the third approach insists on considering institutional changes to be the relation of individual economic entities to institutional innovations in terms of the net benefit from their implementation. The conclusion draws the leading role of the state in the process of the emergence and further development of newly created institutions focusing on the fact that not every change leads to greater efficiency. Thus it is crucial to consider the previous background of institutions' development at implementing changes in accounting and control.

1 Introduction

To develop a real sustainable development policy it is vital to take into account the interconnection and interdependence of all components, that is economic, social and environmental. Traditional instruments of economic analysis do not allow organizations to reliably determine the efficiency. Currently we see rather favorable conditions for the further improvement of accounting in the field of sustainable economic development as normative legal acts covering most accounting objects have come into effect.

* Corresponding author: oskarp@yandex.ru

Nevertheless, Russian institutional environment and, accordingly, the region's environment continues its formation. Under the current economic conditions it is very important to develop institutions which will contribute to the growth of the economy. Undoubtedly such are the institute of accounting and the institute of control. The institutional economics theory defines accounting and control as economic institutions that fully correspond to this theory.

The sustainable development of an enterprise depends on the sustainable condition of the enterprise itself as an integral element in the external environment and the stability of its internal processes. Sustainable development of an enterprise is a process of permanent internal and external changes in which the use of all types of resources including investment, scientific and technological development, innovations and new technologies, HR and improving the quality of knowledge capital as well as the development of social institutions are coordinated with each other. They are all aimed at increasing the present and future potential of the enterprise, meeting its needs and achieving strategic goals. Thus, the principles of sustainable development are thoughtful management, harmony of the enterprise and the environment, the development of high technologies oriented to the environmental values.

As the research shows there is a large number of scientific works devoted to the analysis of institutional changes. This is due to the fact that institutional changes are an important condition for highly effective economic development. Traditionally, institutional changes mean the emergence of new rules with corresponding mechanisms for their implementation which substitute the old ones. Changes in institutions are related to the fact that the values that used to ensure their existence change as well come into conflict with other values and institutions [1-5, 11-14].

The analysis of institutional changes is good to start with considering formal and informal rules that determine the system of institutions, namely the institution of accounting and the institution of control. Due to the action of formal and informal institutions a stable institutional structure is achieved. Stability of the institutional structure of society gives us a feeling of comfort and confidence in our actions. However, despite the stability of institutions there is no guarantee that the institutions we rely on are effective.

2 Materials and Methods

Today the problem of the efficiency of the institute is one of the most essential in the field of research of institutional dynamics. Typically, changes in institutions happen due to the fact that over time the values that condition their existence change or they themselves may come into conflict with other institutions, as well as values. Stable institutions are also subject to change, but for the most part development takes place through insignificant increments that are permanent, although at times revolutionary changes in the institutional structure of society occur as well.

There are the following types of institutional changes that are described in the economic literature. Incremental institutional changes that happen due to the consolidation of informal norms and rules in small groups. Evolutionary institutional changes presuppose a gradual consolidation of informal norms as generally accepted in the formal rules of the relevant institution. Revolutionary institutional changes manifest themselves in the exogenous borrowing of the institute or their "import".

Some researchers miss the incremental institutional changes focusing on endogenous and exogenous changes. As mentioned, endogenous changes are carried out through evolutionary changes in existing norms and rules and exogenous institutional changes are radical and manifest themselves when importing institutions. It should be noted that the import of institutions does not always lead to an increase in production efficiency or to an accelera-

DOI: 10.1051/e3sconf/20172104023

tion in the rates of economic growth. Evolutionary changes are studied within the framework of the mainstream theory [6-7] .

3 Discussion

The evolutionary path of change is less painful and more effective as it reduces the costs of updating changes to the relevant institutional environment. Nevertheless, evolution does not guarantee the dependence on the previous direction of development, the constant reproduction of knowledge aimed at maximizing the possible gains precisely in the existing institutional matrix that is a radical change in the institutions necessary for the withdrawal of society from stagnation and economic decline. It can't reduce transaction costs in the economy for such a short time period. Evolution does not reject changes of a qualitative nature, but refers to their mandatory historical dependence.

Changes in institutions are usually considered in the context of the three-levelled institutional organization of the economy. The levels of institutional organization are institutional order, institutional environment and institutional arrangement.

Targeted changes predominate at the level of the institutional structure which is primarily a formal institutional constraint. Changes at the level of the institutional structure of accounting are observed quite often in modern Russia. Infinite amendments to the existing normative legal acts on tax and accounting make it very difficult to focus in formal rules and create a favorable ground for the use of illegal informal rules.

The main reason for the change is the pressure of the institutional environment. The conflict between the institution and the environment in which they are function gives an incentive to activate one or several types of institutional isomorphism to restore conformity.

At present there are different opinions on the issue of institutional changes. We would like to focus on three basic approaches to the analysis of institutional changes.

The first approach ensures that high economic efficiency is guaranteed only by newly emerging institutions, and the prerequisites for their emergence presuppose previous changes in various areas from population dynamics to the technology development. The state as well as political processes in this case are inactive.

The second approach involves certain incentives for changing institutions, both internal and external. The logic of this approach was set out by D. North, who described the following stages. The emergence of new technologies is driven by changes in the level of knowledge; new technologies change relative prices of resources; new price levels create incentives for owners to transform ownership; the emergence of rules allowing to maximize the value of using these rules, lead to new price levels; non-zero transaction costs in the political market prevent possible institutional changes from being realized [8].

The authors of the third approach state that institutional changes are considered to be the relation of individual economic entities to institutional innovations in terms of the net benefit from their implementation [9].

Besides, the institutional economy speaks about the Harold Demsets scheme changes. Demsetz believes that the source of change is external shock which leads to a change in relative prices. The change in property rights is aimed at the effective use of resources and technologies. And most importantly it is assumed that the process of change is cost-free and is not related to the active actions of the state and stakeholding groups and organizations. This scheme of changes, in our view, can refer to the revolutionary type of institutional change. Armen Alchian put forward a "naive" theory of institutional changes holding the view that only effective institutions survive in the course of competition. This theory can refer to the evolutionary type of institutional change [10].

DOI: 10.1051/e3sconf/20172104023

The above mentioned approaches to the essence of institutional changes have one common component - the state, which plays a decisive role in the process of the emergence and further development of newly created institutions.

The institutional changes that restore economic efficiency in the economy or contribute to its achievement include changes in relative prices. Such changes undoubtedly relate to the processes of globalization, the tendency of international integration, the growth of international cooperation and trade and, as a result, the transition to international financial reporting standards through the convergence and harmonization of national accounting systems. New relative prices create new opportunities for income generation by Russian companies, which in turn require appropriate institutional changes.

In addition, new technological opportunities entail the need to develop methods for accounting for new objects and new transactions, which in turn lead to institutional changes. Such objects can the accounting means of automation, a brand, web sites, scientific and technical development, advertising etc. These accounting objects are not a luxury but a necessity of modern companies for successful work and further promotion on the market. Thus, accounting is changing in order to reflect the company's attitude to innovation, the impact of risks and costs, obligations and the degree of liability associated with innovation. In connection with the advent of new technologies in the accounting organization we see changes in tastes and preferences. Lenders and investors today require for decision-making an extensive data and information of a different nature that can be easily and quickly generated through automated accounting systems.

It should be noted that in some cases, institutional changes lead to increased efficiency, in other cases changes play a redistributive role. There is also a case where there is a decrease in economic efficiency. The most common is the redistributive version of institutional changes. Hence, all the institutions can be hypothetically divided into three groups: increasing the efficiency of the economic system, neutral and reducing the efficiency of the economic system. In the framework of neoinstitutional analysis institutions that maximize social wealth are recognized as effective. However in Russia we see a paradoxical situation when the introduction of formal effective market institutions is accompanied by a decline in the efficiency of the economic system functioning. Explanation of this phenomenon should be sought in the mismatch of formal and informal institutions that together represent an institutional environment.

4 Conclusion

Institute of accounting and control can be attributed to institutions that increase the efficiency of the country's economic system. Despite the various institutional changes in Russia, the institutions of the command economy have firmly established themselves in the Russian economic system. Besides, the basic institutions inherent to a planned economy continue to exist today. Undoubtedly, such institutions include accounting and control.

All in all, the state plays a decisive role in the process of the emergence and further development of newly created institutions. Changes in the accounting and control institution are of an evolutionary nature and increase the efficiency of the economic system of the country. Implementing institutional changes in accounting with the aim of building an effective model, it is necessary to take into account the historical background and development of institutions.

References

1. D. Dequech, J. Econ. Iss. **36:2**, 565-572 (2002)

2. G.M. Hodgson, J. Econ. Iss. **34:2**, 317-329 (2000)
3. M. Rutherford, J. Econ. Persp.**15:3**, 173-194 (2001)
4. M. Rutherford, *Institutions in economics: the old and the new institutionalism* (Cambridge University Press, Cambrige, 1996)
5. T. Lawson, Evol. Inst. Econ. Rev. **2:1**, 7-20 (2005)
6. R.R. Nelson, S.G. Winter, *An evolutionary theory of economic change* (Harvard university press, Harvard, 2009)
7. J. Schumpeter, U. Backhaus, *The theory of economic development* (McGraw-Hill, New York, 2003)
8. D.C North, M.C. Brinton, V. Nee, *The new institutionalism in sociology* (McGraw-Hill, New York, 1998)
9. F. Toboso, Institutional arrangements matter for both efficiency and distribution: Contributions and challenges of the New institutional Economics (Springer, Berlin, 2011)
10. A.A. Alchian, H. Demsetz, Amer. Econ. Rev. **62:5**, 777-795 (1972)
11. S. Tang, *A general theory of institutional change* (Routledge, London, 2017)
12. E. Brousseau, P. Garrouste, E. Raynaud, J. Econ. Behav.Org. **79:1**, 3-19 (2011)
13. L.J. Alston, Emp. Stud. Inst. Chan. **10**, 25-30 (1996)
14. G.M. Hodgson, New Polit. Econ. **6:2**, 245-253 (2001)

E3S Web of Conferences **21**, 04024 (2017) DOI: 10.1051/e3sconf/20172104024
The Second International Innovative Mining Symposium

Integration of MOOCs in Advanced Mining Training Programmes

Irina Saveleva[1,*], *Oksana* Greenwald[1], *Svetlana* Kolomiets[1], and *Elena* Medvedeva[1]

[1]Kemerovo State University, Institute of Philology, Foreign Languages and Media Communication, Department of Foreign Languages, 650000 Kemerovo, 6 Krasnaya st., Russian Federation

Abstract. The paper covers the concept of innovative approaches in education based on incorporating MOOCs options into traditional classroom. It takes a look at the ways higher education instructors working with the mining engineers enrolled in advanced training programmes can brighten, upgrade and facilitate the learning process. The shift of higher education from in-class to online format has changed the learning environment and the methods of teaching including professional retraining courses. In addition, the need of mining companies for managers of a new kind obligates high school retraining centres rapidly move towards the 21st century skill framework. One of widely recognized innovations in the sphere of e-learning is MOOCs (Massive Open Online Courses) that can be used as an effective teaching tool for organizing professional training of managing staff of mining companies within the walls of a university. The authors share their instructional experience and show the benefits of introducing MOOCs options at the courses designed for retraining mining engineers and senior managers of coal enterprises. Though in recent researches the pedagogical value of MOOCs is highly questioned and even negated this invention of the 21st century can become an essential and truly helpful instrument in the hands of educators.

1 Introduction

The demands of the modern information age compel higher education graduates and those people who already have working place to be equipped with the 21st century skills. This set of skills will allow professionals to achieve great results in their career and to make a significant contribution to society or industry. According to the Partnership for 21st Century Skills' list the most valuable qualities to a new high school graduate and to employers are flexibility and adaptability, collaboration and teamwork, creativity, imagination, critical thinking, problem solving. They all interact with each other, e.g. creativity and imagination go hand-in-hand with critical thinking and problem solving.

2 Materials and Methods

*Corresponding author: saviren1973@mail.ru

Giving an extensive survey on 21st century skills Thrilling and Fadel note that flexibility and adaptability can be learned by working on progressively more complex projects that make students change decisions and ways when things aren't working well, adapt to new developments in the project, and incorporate new team members on both current and new projects". To develop critical thinking and problem solving skills it is necessary to make judgments and decisions, analyze and evaluate, synthesize and make connections between information and arguments, etc. [1].

"The increase in the number of service-sector jobs, which include high-growth, high-wage and high skilled occupations in new and emerging occupations, requires students to have different skills and a greater familiarity with technology than ever before [2]". Seniors of coal industry also want their managers of different level working for mining companies to gain the skills necessary to participate in a global economy. As CEOs of coal industry all over the world claim, the prior strategies of mining enterprises are as follows: moving collaborative research beyond regional boundaries to global market, to increase competency creating a mining research and innovation projects effectively linking research capability to industry needs [3, 4].

Thus, retraining courses organized for this group of adult learners should both meet the requirements of their current needs and supply the effective teaching schedule developing the set of 21st century skills. Recent researches in the sphere of teaching methods show that application of one or two approaches does not result in high learning outcomes and, therefore, does not shape the set of skills required. Combination of face-to-face and online teaching formats especially used for retraining programmes is the call of the time [4-7]. Incorporating distance learning elements into traditional classes lead to such innovative techniques as "hybrid teaching" or "blended teaching" [8]. The idea of integrating MOOCs into university training programmes is one of the possible solutions. In order for teachers to deliver quality in both face-to-face and online learning environments, it is pivotal that they experience and understand MOOC options [5]. Indeed, education is now at a turning point and this is going to occur rapidly at a global scale and massive courses are just the beginning [9].

3 Results and Discussion

Massive Open Online Course (MOOC) is a technology that provides educational courses via online to a considerable number of people for free of charge or at an affordable cost. It is open to anyone who is interested in receiving knowledge and education. MOOCs are commonly offered via a web-based platform [10]. There are two types of MOOCs – cMOOCs and xMOOCs. Commonly, xMOOC consists of short video lectures, quizzes, self-graded / peer-graded assignments and forum to facilitate the communications and collaborations [11]. Typically, this kind of a course expands 4-8 weeks and could be taken on scheduled dates or self-phased. The other type, cMOOC, uses social media to connect people with the information of Open Educational Resources and stimulates the participants' discussion and collaboration.

The term MOOC originated in Canada in 2008 being coined to describe the online course *Connectivism and Connected Knowledge* designed by George Siemens and Stephen Downes. Then, in 2012, MOOC evolution took place when the University of Stanford organized the Artificial Intelligence course, attracting over 100.000 interested learners. MOOCs were new and they shook the educational establishment [2].

Larger corporations such as Pearson and Google are also planning to move into the sector of higher education as global players and are likely to adopt a MOOC-based approach as a part of their plans. Recently a new company, Futurelearn, has been launched by the Open

University in the UK, to bring together a range of free, open, online courses from leading UK universities for learners around the world [12].

A report for the Becta agency [5] on technology-assisted learning identified 17 essential learning practices that might be delivered or supported by the technology. These include: expository, reflective, networked, tutored, case-based, problem-based, inquiry-driven, game-based, and constructive. MOOCs is to provide teaching by video, with opportunities for learners to pause the presentation and reflect or respond vocally, and for educators to embed formative quizzes into the video. The problem is that, while direct instruction by lecturing works at massive scale to engage learners and introduce big topics, this is a less effective approach to teaching than more active and collaborative pedagogies [13].

Since then a lot of MOOC research and experimenting has been done resulting in a better understanding of how MOOC can help teachers and learners in this new eLearning era. MOOCs are seen to become a framework of an innovative, comprehensive strategy for evolving education to meet the needs of 21st century society [1].

Though in the beginning of MOOCs dissemination at the learning market this platform was exploited apart from traditional education patterns, rather independently, some educators considered massive courses from a different perspective and introduced a new approach to teaching university students borrowing MOOCs options. This was due to an attempt to build a new model of a student centered classroom instead of a teacher centered classroom as well as to change the learning resources by incorporating e-resources into education of any level. Thus, there have appeared some researches where combined teaching principles were claimed, described and justified [5, 7, 8, 11].

Yet pedagogical value and effectiveness of MOOCs was called into question [7, 10, 12, 14]. For instance, Armellini and Rodriguez think that MOOCs as open educational resources are often re-structured as courses with teachers' participation. This traditionally results in mere transmission of content. The researchers conclude that this approach is hardly an innovative pedagogical practice [14]. As for the concept of MOOCs it is often doubted that they really provide education to anyone who has a desire to develop intellectual abilities as they cannot access any course without payment. Some MOOCs are open and massive, but they are not many [12].

According to Inge de Waard, online education does not vary much from sound, face-to-face learning in its pedagogical essence. MOOCs based on the importance of dialogue being present at the core of a MOOC. Never the less, dialogue between participants immediately moves the focus from content-to-learner toward learner-to-learner dynamics. MOOCs, and particularly the connectivist MOOCs lend themselves for problem-based learning, as they rely in part on enquiry based peer interactions and these learning models benefit from more interaction and access to information [5]. As Maria Israel states in her study "incorporating MOOCs in traditional classroom settings was almost equal or slightly better than face-to-face teaching environments [8]".

Certain scholars suggested to infuse the EFL class with some of these ideas in order to optimize learning outcomes by introducing the innovative method of flipped classroom (a type of a lesson where students discuss and analyze content they familiarized with before coming to the lesson, mostly practice and master their skills) [15-17].

Nowadays the two types of MOOCs are mostly presented by Coursera and FutureLearn as the most widespread massive education platforms. The main difference between them being the underlying pedagogy, they both are useful tools to apply for training qualified specialists.

The courses offered by xMOOCs (Coursera) are using the instructive (or transformative) approach. As it was mentioned above they usually contain fragmented video lectures, additional material for reading, multiple-choice quizzes, self-graded/ peer graded assignments and forums for discussion. However, the really working parts of it are the following:

firstly, a video (as it usually provides interesting and significant information); secondly, multiple choice quizzes (as they are not easily answered and require full understanding, analyzing and rethinking of the given material), thirdly, some assignment (as it gives some practical application of the knowledge and skills obtained).

As for the forum is concerned, it does not work to the full and students prefer to communicate and discuss problems in personal contact, face-to-face. Lacking in organizing real massive communication, this kind of MOOCs is still a very useful tool of training a professional in the 21st century as it provides good opportunities for developing necessary skills.

First of all, blending traditional face-to-face teaching with a suitable course (or a part of it) in MOOCs allows the teacher to show the different approaches to the same problem especially in case of the course developed in another country or university because each one has its own peculiarities, national or institutional. The students are asked to enroll for the course and pass it (or at least part of it) before the usual class, then the teacher can organize discussion or some problem-solving group task demanding knowledge and skills obtained at the course to cope with.

It is also important that the students can schedule their doing the tasks of the course according to their own pace as the participants of retraining courses are busy working people and are not able to spend much of their time at the university or any other educational centre. The educational institutions in their turn are also interested in offering the most effective training with as small number of classroom hours as possible offering different ways of organizing self-study.

The courses of the second type (cMOOCs) which can be found at FutureLearn are based on connectivist or constructivist approaches. The first make use of social media connecting people and information thus enabling people to learn more. The second (social constructivist) apply to the theory of learning effectively in conversations with yourself and others. In this case, the content of the course, its questions and given problems serve as a starting point for the participants to build their learning process. The latter can take part in the discussions, add their own resources and produce their contextualized content. Hence, this type of MOOCs is more suitable for developing such skills as collaboration, creativity, teamwork and problem solving. It requires a high level of participant's motivation, though. Taking into consideration that some of our target audience are highly motivated professionals this kind of courses appears to be extremely useful and effective for them as it gives the opportunity to put into practice their knowledge and to solve real problems while studying.

All these views support our idea of integrating MOOCs options into advanced mining training programmes. This was done during President courses aimed to master language skills of Kuzbass mining companies managerial staff. First of all, some MOOCs were selected and partially included into the syllabus. It should be noted that online lessons created by leading British, American and Australian universities and disposed on Coursera, FutureLearn, Edx and Open2Study Platforms provide short specialized open courses that can boast their authenticity in regards to the language presented and the sound ideas outspoken. They offer the courses that provide information and practice in the sphere of management as well as mining and include diverse theme modules and multi-purpose content.

For example, the course "Developing the Opportunity for Corporate Entrepreneurs" disposed on Coursera is designed for managers of different level from junior to senior working for companies interested in learning how to innovate and apply entrepreneurship principles in the corporate setting. Learners will develop knowledge on how to navigate the barriers to creating, developing, and sustain innovative new businesses within large companies. These new businesses rarely fit neatly within well-established systems, processes, and cultures. Learners will develop the skills, and learn the tools and best practices, for identifying and developing the entrepreneurial opportunities, building business models, creating strategies for leading innovation, and financing innovation [18]. The following modules and discussion-provoking topics have been practiced:

- Setting up a formal structure for corporate entrepreneurship.
- Key enablers of corporate entrepreneurship.
- What is strategic decision-making?
- Why study strategic decision-making?
- What is your decision-making style?
- Designing a career path for corporate entrepreneurs
- Do you have a high need for achievement?
- How does individualism influence your decision-making?
- Do you believe that you have control of your own destiny?
- Are you able to effectively focus your time and resources?
- Is the industry advertising intensive?
- Are you trying to enter a concentrated market?
- What is the average size of a company in the industry?

More specific programmes provided by Edx and Open2Study were optionally incorporated into the syllabus of the programme designed for professional training of mining company managers. Edx MOOC runs the course called "The Business of Mining". The content presented includes four modules:

1. Exploration. The students were able to analyse geophysical surveys and geochemical data.
2. Mine Planning. The Round Table discussions were devoted to the following problems:
   - What infrastructure will you build to run your mine?
   - Will you use a local workforce, or a fly-in/fly-out model?
   - What are you planning to do with your mine waste?
   - Can you pay for all of this and still make a profit?
3. Operation. Trainees mostly touched practical points, e.g. "Process your ore and recover as much gold as you can. Sounds pretty simple, until the gold price takes a tumble and suddenly your mine plan needs adjusting".
4. Closure and Post-closure. This module included lectures, pedagogical tasks and tests concerned with scarce supplies: "Nothing lasts forever, particularly in mining. Eventually you're going to run out of gold, or the market will change and your profit margin will disappear. How will you ensure you leave a safe, rehabilitated site?" [19].

If the programme syllabus counted fewer hours of face-to-face in-class study the method of flipped classroom was effectively applied. Typically, flipped teaching format presupposes substantial hours for self-study and gives students the opportunity to be engaged into larger number of practical tasks and project making assignments when working in the classroom.

"Mining Engineering" course created by the University of Newcastle, Australia and logged on Open2Study educational platform allows those who are mastering their professional skills to learn: the steps in the mining process from mineral exploration to closure; fundamentals of Rock Mechanics (stress-strain, rock strength, effect of discontinuities, rock testing, stress in the ground and instabilities); fundamentals of surface and underground mining including terminology, types of operations, economics and safety. Preparing for the class students can watch video content on the mining process, mining geomechanics, surface mining, underground mining. Yet, in the class they are to interpret information they familiarized with, solve different kinds of problems they face at work in both conventional and innovative ways, draw conclusions in groups and individually.

 DOI: 10.1051/e3sconf/20172104024

Combined teaching approaches applied for retraining coal industry managers that exploit MOOCs platforms, content and techniques, format and assessment patterns can be effectively used for developing 21st century skills listed and described above. Using fragments of MOOCs, which are designed exactly for mining company workers, the instructors can enable trainees to acquire professional knowledge as well as to apply creative and critical thinking. While the adult learners can be involved into the process of watching and listening the teacher is able to mediate the learning situations and participate launching the discussions that require the ability of trainees to work in a team and to interact with each other in order to work out a solution (problem solving activities and tasks). Project work based on MOOC content can also help professionals practice and master their team-building and collaboration skills, they get more complex vision and understanding of global technologies and economies.

4 Conclusion

MOOCs as an innovative education framework have been and remain a controversial issue. One of the reasons is that they are still being exposed to the process of evolution.

However, the era of information technology, expansion of learning and extensive use of information sets its requirements for both teachers and learners.

Education of the 21st century tends to apply teaching formats and approaches that integrate learning strategies, digital competences and career abilities and MOOCs implementation into the learning process can serve as such. Using complete MOOCs or integrating parts of them according to the personal needs and abilities of the learners into the process of study allows to develop skills necessary for a 21st century professional.

To sum it all up, MOOCs act as a great stimulus for intellectual activities of professional mining engineers and encourage their desire for further self-education and implementation of obtained skills at the working place. It is crucial that their digital and technological skills are enhanced.

References

1. B. Trilling, Ch. Fadel. *21st Century Skills. Learning for life in our times* (Jossey-Bass, USA, 2009)
2. Hanover Research. *A Crosswalk of 21st Century Skills.* (Washington DC, USA, 2011)
3. J. L. Kohler, Int. J. Min. Sci. Tech. **25:3**, 325- 332 (2015)
4. G. Johnson. *Strategic Change and the Management Process.* (Oxford: Blackwell, 1987)
5. I. de Waard, Revista Mexicana de Bachillerato a Distancia, **7:13**, (2015)
6. R .Williams, R. Karousou, J. Mackness, Int. Rev. Res. Open Dist. Learn. **12:3**, 39-59 (2011)
7. R. Ferguson, D. Clow, R. Beale, A. Cooper, N. Morris, S. Bayne, A. Woodgate, *Design for Teaching and Learning in a Networked World, Lecture Notes in Computer Science* (Springer International Publishing Switzerland, 2015)
8. M. Israel, Int. Rev. Res. Open Dist. Learn. **16:5** (2015)
9. O. Shapira-Lishchinsky, Teaching and Teacher Education, **27** (2011)
10. S. Bayne, J. Ross, *The pedagogy of the Massive Open Online Course: the UK view* (Higher Education Academy, the University of Edinburgh, 2014)
11. J. Daniel, J. Interact. Med. Ed. **3** (2012)

12. I.de Waard, S. Abajian, M. Gallagher, R. Hogue, N. Keskin, A. Koutropoulos, & O. Rodriguez, Int. Rev. Res. Open Dist. Learn. **12:7**, 94-115 (2011)
13. M. Prince, J.l Engin. Educ., **93:3** 223-231 (2004)
14. A. Armellini, B. Rodriguez, J. Interact. On. Learn. **14:1**, 17-28 (2016)
15. Y. Parra Sr., Gist Educ. Learn. Res. J., **7**, 190-208 (2013)
16. M. Bali, Int. Rev. Res. Open Dist. Learn. **10:1**, 44-57 (2014)
17. Ya. Parra, GIST Educ. Learn. Res. J. **7**, 190- 208 (2013)
18. *MOOC Resource, Developing the Opportunity for Corporate Entrepreneurs* URL: https://www.coursera.org/learn/corporate-entrepreneurs-opportunity (Accessed: 08.09.2017)
19. *MOOC Resource, The Business of Mining* URL: https://www.edx.org/course/business-mining-curtinx-tbomx-0# (Accessed: 08.09.2017)

Innovative Technological Development of Russian Mining Regions (on Example of Kemerovo Region)

Evgeniya Shavina[1], *Oleg* Kalenov[2]

[1]Plekhanov Russian University of Economics, Academic Department of Political Economy and History of Economic Science, 117997, Moscow, 36 Stremyanny lane, Russia
[2] Plekhanov Russian University of Economics, Academic Department of Industrial Economics, 117997, Moscow, 36 Stremyanny lane, Russia

Abstract. A characteristic trend of many countries modern development is the transition to an innovative economy. At present, this is the only opportunity to secure and maintain a high standard of living for the population. Moreover, innovative development of Russian can be achieved during technological progress in its regions. In this regard, it is necessary to assess the innovative potential of the region and identify the most actual problems that impede the transition to the trajectory of innovative development. The authors outline several main indicators that help to determine the level of innovation and technological development of one of the largest industrial areas of Russia – Kemerovo region. The special economic role of Kemerovo region as a large territorial old-industrial complex of Western Siberia requires a large-scale work to solve the most acute problems of regional development. It is necessary to find the answer for existing problems through the implementation of a system of state regulation aimed at making the innovation component a leading factor of the regional economy competitiveness.

1 Introduction

To achieve a high level of innovation and technological development of Russia, it is necessary to begin modernization from the regions as the parts of the national economic system. Being the largest industrial, and primarily the mining region of Russian Federation, Kemerovo region requires special attention to activating the process of implementing innovations and modernizing the traditional sectors of regional economy. Its independent scientific, technical and technological support of industrial potential rise is a matter of national security. An analysis of the level of innovation and technological development of Kemerovo region will enable understanding the current situation in the region and identifying the most urgent development problems.

2 Material and method

The methodology of the analysis of regional innovative and technological development is based on the concepts of innovative economy formation, introduced by G. Mensch [1], J. Schumpeter [2], B. Lundvall [3], N. Kondratiev [4], S. Glazev [5] . The problems of innovation and technological development of the regions are examined in the works of M. Golova and A. Sukhovey [6], V. Osipov, A. Zeldner, T. Skryl [7], including the problems of innovative development of the Kemerovo region - in E. Dotsenko and N. Ezdina [8].

To assess the innovation and technological potential of the mining region and the prospects for its development, it is necessary to analyze a number of indicators, such as:

- number of organizations performing scientific research and development,
- number of advanced industrial technologies;
- the share of costs for technological innovations and the share of innovative goods and services in the gross regional product;
- fixed assets investments.

It is the analysis of the above-mentioned indicators that will make it possible to understand how the region is ready for the transition to an innovative development and to assess its technological potential. In addition, it is necessary to analyze existing regulatory documents adopted at the regional level in the innovation field and identify problems that hamper the innovative and technological development of the mining region. For this study, Russia's key industrial region, Kemerovo Region (Kuzbass), was selected. It mines more than 50% of coal in the country, founds a significant part of iron and steel, as well as finished ferrous metals. Proceeding from this, the problem of innovation and technological development here is particularly acute.

3 Results and discussion

To ensure a high level of social and economic development of the region and achieve compliance to the world's advanced mining regions, it is necessary to keep pace with scientific and technological progress. Creation of measures to stimulate innovation activity can not achieve positive efficiency, if the level of regional innovative and technological development is not properly assessed in advance. That is, it is necessary to trace how the region is developed in this area for a long time.

In the structure of the gross regional product of Kemerovo region in 2015, the main type of economic activity was the extraction of minerals - 25.6% (in 2005 - 27.1%, in 2010 - 31.4%). As can be seen from Table. 1, production index for Kuzbass like in Russia as a whole varies abruptly. It reached a peak value of 109.6, in 2010, in 2011 - a sharp decline to 98.1%, then there were fluctuations in the range of 101-104%. In the Siberian Federal District (SFO) has was also an increase of production index in 2010 - 142.2%, and then a gradual decrease to 101% in 2015. At the same time, the situation is more stable in Kemerovo region along with Irkutsk region, than in other mining regions of Siberia. This indicates more favorable conditions for mining industry innovative development.

Table 1. Production index for "Mineral Resources Mining" as a kind of economic activity, % to the previous year [9].

Territory	2005	2010	2011	2012	2013	2014	2015
Russian Federation	101.4	103.8	101.8	101	101.1	101.4	100.3
Siberian Federal District	101.9	142.2	111.5	113.5	109.2	103.4	101
Krasnoyarsk region	102.1	180.4	117.6	120.8	115.9	103	99.6
Irkutsk region	103.5	133.6	152.6	133.2	109.7	110.9	113.6
Kemerovo region	*103.6*	*109.6*	*98.1*	*104.5*	*101.4*	*104.2*	*102.6*
Novosibirsk region	147.6	100.3	104.5	106.2	116.8	94.5	95.3

Let's turn to the index of innovative activity of organizations (Table 2). Unlike other regions, rather low value of the indicator is observed in Kemerovo region. In 2005, it was 7.3% (in the Siberian Federal District - 8.1%, in Russia - 9.9%), in 2010 - 5.9% (in Siberian Federal District - 8.2%, in Russian Federation - 9% 5%), in 2015 - 3.9% (in Siberian Federal District - 8%, in the Russian Federation - 3.9%). The situation is aggravated by negative dynamics. Comparison with world leaders in this direction is also not in Kuzbass favor. In 2015 the value of the innovation activity indicator of organizations was: Australia - 55.8%, Brazil - 38.2%, India and China - 35.4%, Canada - 32.3%, the United States - 14.3%.

Not being an innovative leader, Kemerovo region faces a number of problems in creating, implementing and promoting innovations. The reasons for this lag, both at the level of the region and the country as a whole, according to the authors, lie consist of the following. First, the number of personnel engaged in research and development is being reduced. The low level of remuneration of scientific workers and the prestige of the profession does not ensure the inflow of young researchers. Secondly, the problem of intellectual and business migration, which is directly related to the outflow of specialists in the scientific and research field. Thirdly, low financial activity, including the commercial sector aimed at obtaining new knowledge and its practical implementation. Fourth, lack of support from the state, not only material, but also moral.

Table 2. The share of innovative activity of organizations, % to all activities [9].

Territory	2005	2010	2011	2012	2013	2014	2015
Russian Federation	9.9	9.5	10.4	10.3	10.1	9.9	9.3
Siberian Federal District	8.1	8.2	8.8	8.5	9.1	8.8	8
Krasnoyarsk region	8	10	10.2	9.5	11.2	9.3	8.8
Irkutsk region	9.2	8.7	6.5	6.9	8.7	6.4	7.9
Kemerovo region	*7.3*	*5.9*	*6.4*	*6.1*	*4.6*	*7*	*3.9*
Novosibirsk region	5.3	5.5	8.2	8.6	9.9	9.7	9.4

All this is confirmed by the data presented in Table. 3. In Kemerovo region like in Russian Federation, the number of organizations performing scientific research and development for the period 2005-2015 has changed insignificantly: only in 2015 the indicator grew by 18% (in Russia - by 16%). This can be explained by the intensification of research in response to the sanctions policy of western states, which began in 2014.

Table 3. The share of organizations performed R&D [9].

Territory	2005	2010	2011	2012	2013	2014	2015
Russian Federation	3566	3492	3682	3566	3605	3604	4175
Siberian Federal District	419	404	424	424	428	424	491
Krasnoyarsk region	60	54	53	52	52	52	72
Irkutsk region	35	44	46	49	51	45	52
Kemerovo region	*29*	*27*	*27*	*26*	*27*	*27*	*32*
Novosibirsk region	119	104	111	113	118	120	122

The raw orientation of Kemerovo region economy had an impact on innovative developments (Table 4). Other mining regions significantly overtake it in this direction. In

2015, Kemerovo region has introduced 8 innovative technologies, while the Irkutsk and Novosibirsk regions are more than 20. The main element of Kemerovo region innovation infrastructure is the Kuzbass Technopark, created to fulfill state program for supporting small and medium-sized innovation projects in 2007. The major project of Kuzbass technopark is a global project for the extraction of methane gas from coal seams, which is currently an indicator of the level of innovative development of Siberia. Another important project of residents of the Kuzbass technopark is the production of heart valves, for all regional and federal clinics located to the East from the Urals, as well as many medical facilities in Moscow, St. Petersburg, and the Volga region.

Table 4. Developed modern industrial technologies [9].

Territory	2005	2010	2011	2012	2013	2014	2015
Russian Federation	637	864	1138	1323	1429	1409	1398
Siberian Federal District	68	64	126	151	123	116	92
Krasnoyarsk region	15	6	33	38	24	26	19
Irkutsk region	4	10	7	53	41	31	22
Kemerovo region	*6*	*7*	*14*	*11*	*10*	*3*	*8*
Novosibirsk region	21	23	53	31	26	30	23

Dynamics, presented in Table 4, can be explained by the amount of costs for technological innovation. In 2005, they amounted to 1,845.1 million rubles, in 2012 - 12978.9 million rubles, reaching its maximum. In 2013, the decline was almost in 2 times - 6662.1 million rubles. In 2014 the decline was almost to 1 billion rubles, and then a sharp jump up to 3,899.6 million rubles. A sharp increase in the cost of technological innovation in 2012-2013 was the evidence of the fact that the economy of Kemerovo region was developing an innovative potential for the next innovation cycle, and the "surge" of activity was manifested in the increase in the volume of innovative goods and services. In 2014 there was a "jump" of this indicator almost in 7 times from 3242.9 million rubles to 21346.2 million rubles. In 2015 the volume of innovative products grew by another 30% and amounted to 32,435 million rubles. [9].

The most important factor of innovation-technological development effectiveness is investment in fixed assets. Especially in Kuzbass the problem of deterioration of the basic industrial and infrastructural funds is actual. As can be seen from Table. 5, the dynamics of investment over the past decade is characterized by sharp ups and downs. So in the period 2005-2010 investments in fixed assets increased almost in two times. Positive trends continued until 2013, until there was a sharp decline: compared to 2012, investment fell down by almost 20%. In 2015 the falling was also more than 30%.

Table 5. Fixed assets investment [9].

Indicators	2005	2010	2011	2012	2013	2014	2015
Investments, RUB million	80315	156519	214780	267812	217711	230951	162059
Index, in % to the previous year	100.0	194.9	137.2	124.7	81.3	106.1	70.2

Among the problems faced by the mining region in the transition to the innovative development, it is possible to distinguish primarily institutional ones, such as the deforming of rules in the sphere of interaction between government and business, the significant role of the administrative resource, the incompleteness of institutional transformations in the leading sectors of Kuzbass economy and the lack of state and business institutions, necessary for structural reforms initiation (large venture funds and innovative banks, technology promoting zones, a system of guarantees for investment in innovation, etc.) [8].

According to experts, the lack of infrastructure necessary for technological development, including ensuring interaction between science and business, is a central problem of innovative transformations. To solve it, by order of Russian Federation Government No. 904 in 2007, an industrial park was established in Kuzbass, and in 2008 the law of Kemerovo Region "On Technoparks in the Kemerovo Region" was adopted. There are no other technoparks specialized for coal industry in Russia.

The need for the regulation of private-and-state partnership in innovative development predetermined the adoption of the law "On Innovative Policy of the Kemerovo Region" in 2008 and the development of the state program "Economic Development and Innovative Economy of Kuzbass" for 2014 - 2019 [10]. The implementation of this program will allow concentrating resources on solving key scientific and technological problems, and will also contribute to enhancing the role of science and technology in the development of Kemerovo region. And the implementation of the formed innovation policy in the long term should turn Kuzbass into the leading Russian center of technological support of the mining industry of global importance and dramatically improve the quality of life of the population in Kemerovo region.

4 Conclusions

The solution of existing problems requires the implementation of a system of measures, with the help of programme methods, aimed at making the innovation component a leading factor in the competitiveness of Kemerovo region.

The main areas for strengthening the innovative development of the mining region should be the following:

1. Detailed development of the long-term innovation strategy of the region with the definition of priorities and the development of innovative programs to support young entrepreneurs and innovators.
2. Improving the innovation infrastructure, which helps to create a favorable innovative climate in the region.
3. Establishment of a financing system for innovation.
4. Assistance in training and finding qualified stuff in the innovation area.
5. Development of interregional and international innovative and technological cooperation, in particular with the BRICS countries and the EEMP.

Revival the modern industry of mining regions on the new technological bases is possible only during systematic investing in technical innovations and the formation of structural clusters to maintain industrial growth with information development and research. A positive effect can be achieved only with the initiation and support of the innovative development strategy of mining region by state authorities and coordination of efforts of all participants of the innovation process.

References

1. G. Mensch, *Stalemate in technology: innovations overcome the depression* (Cambridge, Masachusetts, 1979).
2. J.Schumpeter, *The theory of economic development* (Progress, Moscow, 1982).

DOI: 10.1051/e3sconf/20172104025

3. B. Lundvall, *Nation system of innovation.Towards atheory of innovation and interactive learning* (Pinter Publishers, London, 1992)
4. N. Kondratiev, *Large cycles of conjuncture and theory of foresight* (Economics, Moscow, 2002)
5. S. Glazyev, *The economic theory of technical development* (Science, Moscow ,1990)
6. M. Golova, A. Sukhovey, Econ. Reg. **1** (2014)
7. V. Osipov, A. Zeldner, T.Skryl, Bus. Econ. **51:4-2** (2016).
8. E. Dotsenko, N.,Ezdina, E3S Web of Conf. **15**, 04012 (2017)
9. *Official site of the Federal State Statistics Service of the Russian Federation*. URL: http://www.gks.ru/
10. *State Program of the Kemerovo Region «Economic Development and Innovative Economy of Kuzbas» for 2014-2019.* URL: http://docs.cntd.ru/document/412805057

Age-Sex Structure of the Population and Demographic Processes in Environmentally Challenged Mining Region (on the example of Kemerovo region)

Timofey Leshukov[1], *Olga* Brel[1*], *Anna* Zaytseva[1], *Philipp* Kaizer[1], and *Kirill* Makarov[1]

[1] Kemerovo State University, Department of Geology and Geography, 650000, Kemerovo, Krasnaya Street, 6, Russia

Abstract. The main goal of the article is to show the influence of the age-sex structure of the population on the basic demographic processes in the Kemerovo region. During research the authors have established correlation links between the sex-age structure of the population and the main demographic indicators (birth and mortality rate, morbidity rate, migration and others) based on the analysis of official statistical data. The direct influence of internal and external factors on the age-sex structure of the population is revealed. Conclusions about the impact of demographic processes on the sex-age structure of the population of the Kemerovo region are drawn.

1 Introduction

The Kemerovo region is one of the most important industrial centers of Russia which is fairly attributed to the old industrial regions. The effectiveness and dynamics of the development of the region largely depend on its population: the number, growth rate, life expectancy, migration, etc. The demographic situation in the Kemerovo region, as well as the country in general, is characterized by a steady population decline. The Kemerovo region is not considered a depressive territory, but a number of features of the old industrial region are inherent in it (the dominance of the early industrial branches – coal-mining and metallurgy, the underdeveloped social infrastructure in a number of coal-mining villages of the region, a high proportion of people engaged in heavy physical labor, etc.). The social and economic development of the Kemerovo region is substantially determined by factors which are external to the region. They are defined by the overall dynamics of the socio-economic situation in the country and the situation in the commodity markets of the basic sector of the region's economy - extraction and primary processing of raw materials [1].

Demographic processes which are formed from a large number of various components that affect them are typical for each region. These determinants can be usually divided into external and internal factors in their relation to the community of people. External or exogenous factors are caused by the impact of the environment on the people community, form-

* Corresponding author: brel_o_a@mail.ru

ing certain demographic processes. Internal or endogenous factors are associated with internal processes occurring in the community itself.

As a rule, the endogenous factors influencing demographic system are laid in the human population itself and are described by probability of approach of a demographic event. For example, among senior-age people an increased mortality and morbidity from blood circularity system diseases and neoplasms are observed [2]. Proceeding from it in areas where the age-sex structure of the population will have a regressive form, one should expect a higher incidence of morbidity and mortality from these pathologies. Aging of the population should be expected in the old industrial regions. It'll naturally lead to a decrease in the demographic potential of the territory, which in turn will affect the decline in fertility, the increase in mortality and morbidity of cardiovascular system diseases and malignant neoplasms, and will also lead to an intensification of immigration processes in the region.

Internal factors defining the demographic development of the municipalities of the Kemerovo region are the differences in the age and sex structure of their population. Territorial differences in the number of people assigned to different age groups lead to spatial differentiation of demographic events that are pre-characterized for a particular age. The probability of this event increases with the approach to the average age of its onset or decreases with the distance from it (for example, death from a heart attack or stroke is characteristic of senior-age people). These events can be, f. ex.: the average age of death from blood circulatory system diseases and neoplasms, the average age of the mother at childbirth, the age at which the greatest number of certain types of newly diagnosed diseases are observed, the age groups of migrants [3].

Age-sex structure has a serious impact on the level and causes of mortality. The older population will have a higher overall mortality rate. It is also known that the male population, aged 15 and over, has a higher rate of age-related mortality, in relation to the similar rate of female mortality. The main causes that lead to the death of the young population are the external causes of death associated with accidents, murders, car accidents, etc. [4].

2 Materials and Methods

The article deals with the analysis of the main demographic processes (fertility, mortality, migration). Based on the results of the analysis, their relationship to the age and sex structure of the population of the municipal districts and urban districts of the Kemerovo region. According to the results of the analysis, the authors established their relationship with the age-sex structure of the population of the municipal districts and urban territories of the Kemerovo region. All the data for analysis are taken from the official statistical sources. The processing of statistical data and correlation links were made by using the Spearman correlation coefficient. A check of statistical significance was carried out by the Student's criterion at a level not less than 0.05 with the use of the program "Statistic 10.1".

3 Results and Discussion

Distribution of age mortality rates in all countries confirms the increase in probability of approach of a demographic event – death with age [5]. At a high level of life the intensity of increase of this probability decreases that is provided with timely diagnosis of diseases and effective protocols of treatment. As a rule the quality of environment and specialization on "dirty" braches in old industrial regions create conditions for the earlier development of certain types of pathologies, including cardiovascular and pulmonary disease. The study found correlation between the age-sex structure of the population and the overall mortality rate (Tab.1). The strongest link is noted between the total mortality and the number of people over 50-years old. This correlation confirms the fact of an increase in the mortality rate in those regions where more adult population lives. For all the proposed parameters, a fairly high correlation is observed at a high statistical significance.

In the areas where there are more people over the age of 50, the total mortality rate will exceed the average regional rate. In this case the improvement of the demographic situation can be achieved by the prevention and timely diagnosis and also modern treatment of quasi-endogenous and endogenous diseases. These measures will increase the life expectancy of the population and avoid depopulation.

Table 1. Correlation between age and sex structure and total mortality rate.

Index (2015)	Spearman's rank correlation coefficient	*t*-test	p-value
Population over 70 years old, % Total mortality rate, ‰	0.518462	3.429847	0.001683
Population over 60 years old, % Total mortality rate, ‰	0.634967	4.649492	0.000055
Population over 50 years old, % Total mortality rate, ‰	0.716000	5.801907	0.000002
Men population over 50 years old, % Total mortality rate, ‰	0.512040	3.372141	0.001964
Men population over 60 years old, % Total mortality rate, ‰	0.571822	3.942955	0.000411
Women population over 50 years old, % Total mortality rate, ‰	0.666922	5.063129	0.000017
Women population over 60 years old, % Total mortality rate, ‰	0.548429	3.710111	0.000785
Women population over 70 years old, % Total mortality rate, ‰	0.570905	3.933567	0.000422

The most unfavorable situation is observed in the Tyazhinsky and Izhmorsky municipal districts, where from 38% to 40% of the population are over 50 years old. A decline in fertility caused by the output of a large part of the population from reproductive age will also be noted in these areas. It is also worth noting that in these areas there should be an increased incidence of morbidity and mortality from the circulatory system diseases and neoplasms, which requires appropriate measures from health facilities. The most favorable situation is observed in Kemerovo and Taiga urban districts, Mariinsky and Promyshlenniy municipal districts.

The age-sex structure also affects mortality from certain causes, as some pathologies develop significantly more often in older ages, which was found earlier and confirmed in this study (Tab. 2).

Correlation link is found between the number of deaths from tumors and the amount of people over 70 years old. The strongest correlation links were found between the proportion of people with mortality from the digestive system diseases and the proportion of women over 50-60 years of age.

Also a medium strength correlation with mortality from tumors was found among women. It should be noted that the absence of correlations with pathologies most often developing in older ages among men (oncopathology, circulatory system diseases) does not indicate the absence of these diseases, but it indicates higher risks of dying from external causes (injuries, alcohol poisoning, murder, suicide, car accidents) at earlier ages (correlation coefficient 0.42 ($p < 0.05$)). It's commonly known that age-sex structure has a significant effect

 DOI: 10.1051/e3sconf/20172104026

on fertility and is associated with the biological abilities of the body to perform its reproductive function at early ages, and the extinction of this function with age. It is worth noting that the birth of a child in modern society has become not only a biologically determined fact, but also socially and economically determined event. It is confirmed by a decrease in the birth rate in the age groups of 18-24 and an increase in fertility in the ages of 25 to 29 years.

Table 2. Correlation links between mortality rate for a specified diseases and age-sex structure of municipalities' population.

The mortality rate for	Spearman's rank correlation coefficient	*t*-test	p-value
circulatory system diseases per 100 000 people Population over 50 years old, %	0.374179	2.282482	0.029250
neoplasms per 100 000 people Population over 70 years old, %	0.590114	4.134900	0.000239
neoplasms per 100 000 people Population over 60 years old, %	0.505615	3.315170	0.002285
neoplasms per 100 000 people Population over 50 years old, %	0.439147	2.765084	0.009367
neoplasms per 100 000 people Women population over 50 years old, %	0.589808	4.131616	0.000242
neoplasms per 100 000 people Women population over 60 years old, %	0.536634	3.597540	0.001069
neoplasms per 100 000 people Women population over 70 years old, %	0.557720	3.801002	0.000610
neoplasms per 100 000 people Men population over 40 years old, %	0.353400	2.137027	0.040338
neoplasms per 100 000 people Men population over 50 years old, %	0.416043	2.588119	0.014394
respiratory diseases per 100 000 people Population over 60 years old, %	0.402903	2.490228	0.018149
respiratory diseases per 100 000 people Population over 50 years old, %	0.464629	2.968183	0.005633
respiratory diseases per 100 000 people Men population over 50 years old, %	0.412987	2.565184	0.015203
respiratory diseases per 100 000 people Women population over 50 years old, %	0.415737	2.585821	0.014473
digestive system diseases per 100 000 people Population over 70 years old, %	0.536481	3.596102	0.001073
digestive system diseases per 100 000 people Population over 60 years old, %	0.570097	3.925322	0.000432
digestive system diseases per 100 000 people Population over 50 years old, %	0.449538	2.846837	0.007648
digestive system diseases per 100 000 people Women population over 50 years old, %	0.600199	4.244836	0.000175
digestive system diseases per 100 000 people Women population over 60 years old, %	0.617923	4.445857	0.000099
digestive system diseases per 100 000 people Women population over 70 years old, %	0.520131	3.444977	0.001616

Migration processes can be used as indicators of the aggregate competitive advantages of the regions. Migration can signal the adverse social, economic and environmental conditions of residence. Immigration processes are indicators of the competitive advantage of a

DOI: 10.1051/e3sconf/20172104026

single administrative education. Migration data is one of the problem areas of population statistics that is not correctly and systematically maintained. In 2015, a positive migration balance was observed only in the Kemerovo, Novokuznetsk and Berezovsky urban districts. From 2009 to 2015 almost all municipalities, except for Kemerovo, Novokuznetsk, Yurga urban districts and Kemerovo municipal district, showed a negative balance. The Yurga urban district showed a high intensity of migration flows due to the location of the military unit there. This territory can't be regarded as having competitive advantages. The remaining municipalities that have a positive migration balance are permanent centers of attraction of the population due to a higher quality of life. Those municipalities in which industrial monocities such as Belovo, Kiselevsk, Prokopyevsk, are located, are losing their population the most intensively. The age-sex structure of the population influences the migration activity of the population of the municipalities, that is confirmed by the statistically significant correlation ($p < 0.05$) (Tab. 3).

Table 3. Correlation between age-sex structure of the population and amount of people emigrated from the municipalities

Index	Spearman's rank correlation coefficient	*t*-test	p-value
Emigration rate, ‰ Population aged 18-59 years, %	-0.389458	-2.39196	0.022804
Emigration rate, ‰ Population aged 25-59 years, %	-0.470130	-3.01322	0.005022
Emigration rate, ‰ Population aged 18-44 years, %	-0.551413	-3.73908	0.000725
Emigration rate, ‰ Population aged 25-44 years, %	-0.639114	-4.70071	0.000047
Emigration rate, ‰ Population of municipalities aged 18-44 years, %	-0.616099	-3.12872	0.006478
Emigration rate, ‰ Population of municipalities aged 25-44 years, %	-0.717234	-4.11711	0.000807

The greatest correlation link was found in the 25-44 age group as for the whole population of the Kemerovo region, as separately for the municipalities. The statistical links confirm the connection between the migration activity and working age groups. This circumstance should be taken into account when creating a program for the demographic development of the territory. It is also worth noting that the age and sex structure has a great influence on the prevalence of certain diseases, the frequency of their occurrence, and overall the incidence of the population. During the study, correlations were found between the age-sex structure of the population and the frequency of oncological diseases (Tab. 4).

The revealed patterns of demographic processes occurring on the territory of the Kemerovo region, on the whole, correspond to all-Russian trends. However, regional specifics are also noted, due to both internal causes inherent in this region, and external factors (the level of socio-economic development, the state of the environment, etc.). The influence of external factors affects the lower life expectancy of the population of the Kemerovo region compared with the average Russian and world rates. For reducing the significance of external and internal factors that directly affect the age-sex structure of the population, a multilateral system of measures for stabilization the demographic situation in the Kemerovo region is needed. So, modernization and ecologization of the mining industry, ferrous and non-ferrous metallurgy will improve the atmosphere condition, soil cover and surface water, which in turn will contribute to reducing mortality, increasing fertility, decline the level of newly diagnosed diseases among the population of the old industrial region [6].

 DOI: 10.1051/e3sconf/20172104026

Table 4. Correlation between age-sex structure of population and the level of newly diagnosed morbidity with various nosologies

Rate of newly diagnosed morbidity of	Spearman's rank correlation coefficient	*t*-test	p-value
All types of neoplasms per 100 000 people Population over 70 years old, %	0.505271	3.312139	0.002304
All types of neoplasms per 100 000 people Population over 60 years old, %	0.394041	2.425255	0.021118
All types of neoplasms per 100 000 people Women population over 50 years old, %	0.466463	2.983145	0.005423
All types of neoplasms per 100 000 people Women population over 60 years old, %	0.477158	3.071416	0.004325
All types of neoplasms per 100 000 people Women population over 70 years old, %	0.508021	3.336415	0.002160
All types of pathologies per 1000 people Men population over 40 years old, %	-0.353400	-2.13703	0.040338
All types of pathologies per 1000 people Men population over 50 years old, %	-0.443545	-2.79951	0.008603
Neoplasms of respiratory system per 100 000 people Population over 50 years old, %	0.383346	2.347901	0.025225
Neoplasms of respiratory system per 100 000 people Men population over 50 years old, %	0.446600	2.823582	0.008104
Neoplasms of respiratory system per 100 000 people Men population over 60 years old, %	0.361956	2.196461	0.035420

According to Tab. 4, high rates of newly diagnosed morbidity of all localizations neoplasms are observed in groups in which there is a higher proportion of people older than 60-70 years. There is also an increase in the development of oncopathology in groups that have a high proportion of the female population older than 50-70 years. Such a pattern among men was not found. However, this may be due to the high men's mortality in earlier ages.

4 Conclusions

The results of the research indicate the necessity of creating a program for the region's development, taking into account the age-sex structure of the population. The necessity of prevention organization and early cardiovascular system diseases diagnosis and neoplasms is caused by a high proportion of people older than 50-70 years [7]. The development of social infrastructure objects, including the system of preschool and school institutions, is necessary because of a large proportion of 20-29 years people. The probability of childbearing is increase in this age group. A possible solution to the problem will be the development of a program for reducing male mortality, since the average life expectancy of men is 62-64 years (depending on the type of settlement). This indicator is 5.5% lower than on average in Russia. This program should contain activities aimed at reducing mortality from external causes, first of all by improving working conditions in mining enterprises. The problems related to the state of the environment remain unaccounted for. These problems are also relevant for mining enterprises and old industrial regions as a whole. For solutions these problems it's necessary to improve the ecological policy of the Kemerovo region by adopting appropriate laws. The development and implementation of a program of integrated

DOI: 10.1051/e3sconf/20172104026

measures within the framework of environmental and demographic policy will be the key to development of an innovative and post-industrial economy in the Kemerovo region.

References

1. F. Agafonov, A. Genin, O. Kalinina, O. Brel, O. Zhironkina, E3S Web of Conf., **15,** 04011 (2017)
2. V. Aboyans, J. Abrahan, T. Adair, R. Aggarwal, S.Y. Ahn, M. AlMazroa, M. Alvarado, H. Anderson, L. Anderson, *Global and regional mortality from 235 causes of death for 20 age groups in 1990 and 2010: A systematic analysis for the Global Burden of Disease Study 2010*, **380,** 2095-2128 (2012)
3. J. Hobcraft, J. Menken, S. Preston, *Cohort analysis in social research: Beyond the identification problem* (Springer-Verlag, New York, 1985)
4. H. Wang, L. Dwyer-Lindgren, K.T. Lofgren, J.K. Rajaratnam, J.R. Marcus, A. Levin-Rector, C.E. Levitz, A.D. Lopez, Christopher J. L. Murray, *Age-specific and sex-specific mortality in 187 countries, 1970-2010: a systematic analysis for the Global Burden of Disease Study 2010*, **380**, 2071-2094 (2012)
5. C.J. Murray, A.D. Lopez, *Global mortality, disability, and the contribution of risk factors: Global Burden of Disease Study*, **349**, 1436-1442 (1997)
6. J.A. Salomon, H. Wang, M.K. Freeman, T. Vos, A. Flaxman, A.D. Lopez, C.J. Murray, *Healthy life expectancy for 187 countries, 1990-2010: a systematic analysis for the Global Burden Disease Study 2010*, **380**, 2144-2162 (2012)
7. Yu. Lesin, T. Leshukov, IOP Conf. Ser.: Mater. Sci. Eng., **142: 1**, 012124 (2016)

The Distribution of the Informative Intensity of the Text in Terms of its Structure (On Materials of the English Texts in the Mining Sphere)

Ludmila Znikina[1*], and *Elena* Rozhneva[1]

[1] T.F. Gorbachev Kuzbass State Technical University, 650000, 28 Vesennyaya St., Kemerovo, Russia

Abstract. The article deals with the distribution of informative intensity of the English-language scientific text based on its structural features contributing to the process of formalization of the scientific text and the preservation of the adequacy of the text with derived semantic information in relation to the primary. Discourse analysis is built on specific compositional and meaningful examples of scientific texts taken from the mining field. It also analyzes the adequacy of the translation of foreign texts into another language, the relationships between elements of linguistic systems, the degree of a formal conformance, translation with the specific objectives and information needs of the recipient. Some key words and ideas are emphasized in the paragraphs of the English-language mining scientific texts. The article gives the characteristic features of the structure of paragraphs of technical text and examples of constructions in English scientific texts based on a mining theme with the aim to explain the possible ways of their adequate translation.

1 Introduction

In the research field of constructive effects of language discourse-analytical approaches are of interest as the ability of a structured and qualitative analysis of texts. The term "discourse" as it is known, originally began to be used in the sense of "functional style" as a particular type of texts: non-fiction, newspaper, scientific, technical etc., having their structural, lexical, and grammatical features [4]. Now we perceive this concept more widely, considering the discourse as a complex communicative phenomenon that includes the conditions of the actualization of texts. Naturally, these conditions include and extra-linguistic factors, such as the target installation of the sender. This makes them "pragmatic discourse" expressed appropriate communicative strategies in vocabulary, syntax, semantics, etc.

2 Results and discussion

We used discourse analysis to the study of English scientific texts of the coal and mining theme with the aim of identifying characteristics of the subsequent formalization of these

* Corresponding author: znikina@mail.ru

texts. In other words, we consider discourse analysis as a practical receiving adequate transfer of the primary text into a derivative one.

The problem of unoriginality in the text formation and the actual texts, which are called derivatives, are the object of study of many researchers. This interest is quite natural, since there is too high proportion of derived texts in a text continuum in which we live.

As such we consider that a formal text in a summary essay, which is the result of clotting (and also the translation into the Russian language) of the English scientific and technical text. Issues of translation and the further formalization of foreign language scientific and technical texts categorized as theoretical and methodological, have always been a subject of interest, as in its complex, they relate to semantic aspects of language and lexical patterns of structural and systematic organization of the text. The study of these aspects provide, in our opinion, the availability and validity of necessary and sufficient perception of the adequacy of the information of a formal text.

In the theory of information systems semantic information is classified into primary and derivative semantic information [8], that is, the source of information, which is translated, may be called primary information. The relationships between elements of linguistic systems are structured and form a model of information transfer with a certain level of adequacy. To obtain a secondary semantic information – in this case the translation of scientific and technical texts from the English language and its formalization is a necessary lexical, grammatical, semantic and logical analysis of the source. From this point of view, getting derived semantic information is considered as a continuation of the process of cognition of the object and is a model of primary information. The degree of a formal conformance of this model with the source directly depends on the objectives of the translation, as well as the intentions and knowledge of the author (subjective factors). In other words, the relationship between the elements of the language system must be considered at two levels: the preservation of adequacy in the semantic coagulation, firstly, and, secondly, the adequacy of the translation.

Analysis of primary and derived texts is based on the assumptions of the constructive effects of language. So a translation of the abstract is the primary formalization of semantic information. Following to the theory of systems, the transformation scheme of the primary information into the derivative one (the transition from one language system ($Language^1$) to another language system ($Language^2$), it is possible to present a model for elements of these systems – the so-called constructive and neglecting attitude [4]. The first ratio shows the transition from the system with a simple language structure into a system with a more complex structure. Neglecting defines the reverse transition, which in our case corresponds to the formalized translation. Pragmatic factor, that is, a specific statement of the problem, determines the degree of formalization and, in this case, the sufficiency of the language system 2 ($Language^2$).

The translation of a foreign text into another language as a speech unit is complex from both a cognitive and linguistic point of view. And it's not just the lack of the necessary of the vocabulary or knowledge of grammar: when reading the text it is necessary to proceed not only from the lexical set but from the structure of the text as a whole, and from its structure's peculiarities[3]. A sign of linguistic organization is the source for the translation and formalization of scientific and technical texts: they are considered as a rule, from the point of view of availability of semantic and formal connections between text units (lexical units, sentences, with super-phrasal unities, paragraphs) as voice work in a multilevel and diverse aspect [2]. The absence of these links makes the text non-functional.

Naturally, when translating the foreign text becomes relevant to the issue of the adequacy of the semantic information of the language system. In our view, the translation with the specific objectives and information needs of the recipient must be understood under the appropriate translation of scientific and technical text but not its exhaustive accuracy. If the translation is considered to be adequate and it provides the coincidence of the denotative

DOI: 10.1051/e3sconf/20172104027

structure and forms of representationof the information, then the attitude to the language coagulation of this information (a summary, an abstract) suggests, obviously, a change of not denotative unity of the text, but the frame structure of the utterance. So, the translation is carried out by means of grammatical and structural changes. Such goal setting is also an adequate translation. However, it should not be considered pragmatically appropriate, because in this case it is based on a change of the communication setup, inherent in the source.

In any case, the correct solution of the problem of derived semantic information (Language²), is adequate to the primary one (Language¹). Otherwise, this would contradict the structural, analytical and logical transformation of information. The essence of such conversions is not in receipt of the formal copy, is is in preserving the communicative significance.

In this case we are dealing with a super-phrasal unities. Such units include the paragraph, graphically, placing the text and at the same time separating it into information segments. The structure of a paragraph, the scientific and technical text is not very diverse, due to which they can model. The technical text is characterized by the presence of key sentence in a paragraph. In a concentrated form, the key sentence conveys the information contained in the paragraph, and can be in any part of it [6]. However, the vast majority of the paragraphs it is in the initial position and serves as a subtitle of the paragraph (a method of introductory or screening reading of a text is based on it: paragraph by paragraph information search requires the ability to extract the basic elements of the meaning from the original quickly enough) [7].

Informative important segments occupy a certain logical position in the structure of the text.

We consider (here and later) some examples of texts taken from the mining sphere [1]. In the scientific and technical article these segments occupy the vast majority of the starting position: the first paragraph and the first sentences of the middle paragraphs of the text. From the boundary position of the whole text, the priority in terms of informativeness is given to the first paragraph.

For example:

1. The methane content in coal bed can be estimated: a) directly in situ by desorption of the coal steam; b) under laboratory conditions by determination of the gas amount incorporated into the extracted coal.

2. The collected data made it possible to prepare models for the sedimentary environment of the coal beds on two different levels:

- General level, prepared from the study of inorganic/organic macrofacies relations, using mostly results of field studies and borehole lithological analysis;
- Detailed level, prepared from the study of microbotanical and micropetrographic features of coal.

3. As generally known, the problem of prevention for coal self-heating and self-ignition is of great importance. Such scientists as E. Pilarczyk, P. Leonhardt, W. Wanzl, D. Rusezev, K. Markova, N. Oils, X. Dong, D. Drysdale, A. C. Smith, P. Panek, B. Taraba, E. Sebestova and others study the process of coal oxidation and different inhibitors for suppressing of coal self-ignition.

4. In contrast, the outlook for Indonesian coal miners is more positive. Two key trends will develop during 2017 and beyond. Firstly, an increase in production that will outperform expectations as Indonesia seeks to capitalise on growing demand for coal from not only China but increasingly other Asian emerging markets, such as Vietnam, Malaysia and the Pilippines. Secondly, there will be a sustained rise in domestic demand that will underpin higher production levels as the government invests heavily in power infrastructure and ramps upcoal as a share of Indonesia's energy mix.

5. Coal demand, and in particular thermal coal deman,hasincreased significantly from 2012 after Japan'snuclear reactors were switched off following the Tohoku Earthquake in 2011.

This are the first paragraphs of scientific-technical articles. The accuracy requirement led to the emergence of the syntactic feature such as information intensity of the sentences. The need to express a complex system of scientific concepts, to establish relationships between them creates a complex and complicated sentences, which can hold much information. These segments are informative significant, as they clearly state the key ideas of the article provided, for example, using modal structure can be estimated. This clarity is achieved by dividing key ideas into paragraphs in the first and second examples. The emphasis on key words is done graphically through the colon. In addition, the third, forth and fifth examples introduce the phrases As generally known (as you know), in contrast, in particular which make the reader pay special attention to it, as different causal relationships and the relationships as moments of universal interconnection, "communications", "chain of causes" are chiefly important in scientific discourse. It is impossible to execute and perceive complex logical and conceptual relationships of the content of scientific texts without this, as well as strict, accented logical coherence of scientific discourse is aslo not possible [5].

The structural and systematic organization of the scientific and technical text is quite clear: in most cases it consists of thesis, arguments and conclusion with the presence of the constituents to argue, to convince, to raise a scientific interest.

In the distribution of information intensity by means of pragmatic impact to enhance persuasion and argumentation such units as: in addition to, for example, for instance, etc, and so on are frequently used. Modal and cognitive verbs, including a component of "knowledge" (to convince, to persuade, etc.) strengthen the structure.

Let's consider some examples on this piece of information:

1. In addition, while there is a current oversupply domestically, India's coal usage is set to continue to rise as the government seeks to increase power capacity.

2. For example, coal from South Kalimantan mines can arrive in Guangzhou after 7-8 days, while Chinese mined coal from large mines at Shanxi can take a total of 8-9 days to arrive in Guangzhou.

Since the main function of this type of the text is a communicative and evaluative function, the basic modality of its structure always remains the subject (objective) modality.

An example of the structure from an English scientific and technical text representing not only the modalities but also the position of the modal structure is the following sentence:

Therefore we must confine ourselves to a brief survey.

The use of the modal verb "must", in addition to the introductory word "therefore", is taken for greater emphasis of the reader's attention on the key ideas of the article.

Let's consider another example:

The effect of temperature under 1MPa partial pressure of oxygen for 30 min on the solid product yield of the lignite from Gediz area is shown in Figure 2.

This is the first sentence of the second paragraph, which is the focus of an expandable informative article: there are accurate data and a link to the picture.

Thus, it is in the first position of the text, especially in the first paragraph, we observe a saturation of informative sentences.

An approach of search and retrieval of information in the process of discourse analysis of text paragraph by paragraph is key when you minimize or formalize it, including when the information is any text a summary, an abstract, a reference, a review, etc.). If we continue our discussion specifically about the translation of a foreign-language text, the logical partitioning of a paragraph (definition of rheme and theme) suggests the formalization of the text with a sufficient level of adequacy of translation. The term "content" is a set of objects and their characteristics. Under the certain restriction on the lexical and syntactic levels in the process of formalizing the necessary condition for the correct transmission of the content of the text is to identify the key positions in the paragraph, the choice of the adequate level of nuclear structures to the original and correct (relevant from the point of view of pragmatics and language norms) of the translated equivalent.

3 Conclusion

Using elements of a discourse analysis of a scientific text, we investigate compositional and content aspects of the process of formalization of the text from the following viewpoint: studying those segments which are rich in information in correlation with the structural and graphic design of the text (*paragraph by paragraph*); identifying of logical and conceptual relations expressed by lexical features of the discourse, which were essential to the perception and design of the logical coherence of scientific statements and, accordingly, the subsequent adequacy of the text with derived semantic information.

References

1. D. Geeraerts, *Theories of lexical semantics*, 336 (Oxford University Press Inc., New York, 2010)
2. V. A. Kolosova, Modern problems of science and education, **6**, 567 (2013)
3. J. Peregrin, V. Svoboda, Synthese, **14:190**, 2897 (2013)
4. S. Pohorec, M. Verlič, M. Zorman, Proc. of the 14th east European conference on Advances in databases and information systems (ADBIS'10), 445-453 (2010)
5. P.H. Portner, *What is Meaning? Fundamentals of Formal Semantics* (Oxford, Blackwell, 2005)
6. N.M. Solomatin, *Information Semantic Systems* (High School, 1989)
7. L.S. Znikina, N. A. Zaostrovskaya, World of science, culture, education, **4:2**,142 (2011)

Training of Engineering Personnel for the Innovative Coal Industry: Problems and Ways of Solution

Natalya Zaruba[1], *Tamara* Fraltsova[2], and *Tatyana* Snegireva[2]

[1] T.F. Gorbachev Kuzbass State Technical University, 650000, 28 Vesennyaya St., Kemerovo, Russia

[2] Institute of improvement of professional skill of executives and specialists of fuel and energy complex

Abstract. The article is written based on some results of the long-term scientific research of the problem related to the urgent need to find the ways of training personnel for the innovative coal industry in the higher education system. This is due to the fundamental changes in the Russian social and economic conditions: the change in the social system and the owner of the coal industry, the emergence of new technologies in the field of coal mining and processing, and in the management of these processes. At the same time, the system of training specialists for the coal industry in the higher education institutions has largely remained unchanged: technologies and principles of training, scientific approaches and concepts take little account of the changed situation, traditional views of specialists working in the university continue to dominate innovative ideas. Many innovations, especially related to technology and the principles of education, struggle to make their way into the higher education system. The article substantiates the urgency of the problem of training personnel for the innovative coal industry in the higher education system, as well as the importance of scientific analysis of the problem in order to find the ways to solve it.

1 Introduction

The turn of the third millennium was marked for the mankind by the entry into an era of great changes that radically changed all spheres of human life. The new conditions for socio-economic development and the nature of changes in society caused by the two revolutions (scientific and information) contributed to the formation in the leading countries of the post-industrial "new economy", for which knowledge and human capital became the main resource. According to a number of sources, the new knowledge that is used in modern technologies, personnel, in the organization of production of developed countries accounts for 80 to 95% of the growth of the gross domestic product. The researchers found that the knowledge and the level of development of human capital present the main source of economic development and the wealth of the nation in the new era rather than the wealth of natural resources.

 DOI: 10.1051/e3sconf/20172104028

The competitiveness of states, their leading positions in the development of national industry and their place in international markets present an important factor in the economic development of the world's leading countries today. The modern global nature in the supply of a large number of goods and services gives way to tools that help to conquer the markets through active technological competition in the field of innovation.

In the Strategy for Innovative Development of the Russian Federation for the period up to 2020, it is noted that the most important problem in ensuring a high level of competitiveness of the country's economy is the problem of being competitive in everything: a person must be competitive, as well as a city and an industry, in other words, the whole country and all its components need to be competitive [1, 3-5].

In this document, it is noted that in addition to quality education, the person's attitudes to life and models of his/her innovative behavior play an important role in the innovative development of society. It should be recognized that the desire for lifelong learning, innovative thinking and creativity, as well as other personality properties important for innovative development, in general, are not sufficiently developed among graduates of the Russian higher education system. The development of a higher level of innovative activity of the university graduate is an actual problem that requires attention of both scientists and practitioners.

2 Materials and methods

A major role in this is given to the system of education, its integration into the world educational space while preserving its own value. This causes profound transformations in all its spheres and generates a change in the educational paradigm: the "knowledge" model of education is replaced by the "competence" one filled with activity basis. The system of mutual relations of participants in the educational process in the university is being restructured. Formation of the key competencies of engineers, including in the field of innovations that contribute to the mobility of the individual in a changing world and professional success in the innovative economy becomes the basis for the training of professional specialists, including in the coal industry.

In this regard, it is extremely important to actively form an understanding of the special mission of the university in the public consciousness and the state authorities and develop an education system in it that would be oriented towards training specialists possessing innovative thinking necessary for the development of the innovative Russian economy. Training of an innovative type engineer requires the implementation of innovative educational technologies in the higher education system: a university teacher using new technologies is able to train an engineer with innovative thinking.

The urgency of the search for the grounds for the training of engineering personnel in the university for an innovative economy is determined by the following circumstances:

-significant changes in domestic and foreign education systems associated with the Bologna Process;

-a change in the social system in Russia, its transition to a market path of development;

-the needs of the post-industrial economy in the preparation of innovative engineers with innovative thinking.

Such concepts as "innovative economy", "innovative educational technology", "innovative thinking", "innovative type of specialist" are important for our research, so it is important to define the concept of "innovation". In the Russian and foreign literature, this concept is interpreted in various ways, the preferences are based on methodological approaches: if it's the activity approach, then innovation is the result of creative activity; if it's the process approach - that is the process of introducing innovations; if it is functional, it is a system of functioning of innovations.

There is a reference point in the understanding of "innovation". This is a kind of international standard, it is taken as the basis for the development of concepts, programmes and other strategic documents of innovation. In accordance with it, innovation is considered as a novelty introduction, i.e. the final result of creative activity, which is realized in the form of new or improved products sold on the market, or a new or improved technological process used in practical activities [1]

The attempt to find ways of training engineering personnel for the innovative economy, in particular for the coal industry, allowed us to determine, as a direction of the movement, the need to develop the theoretical and methodological grounds for this. The analysis carried out confirmed the importance and necessity of a scientific search in this direction.

For example, the need to train engineering personnel for the innovative industry was identified as the basis for the comprehensive development of the coal industry in Kuzbass. This is reflected in the developed and adopted Strategy of the region until 2025, "... the creation of a multidisciplinary economy, the transition to a more qualitative, innovative stage of its development" [2]. It proposes to master the newest technologies for coal processing, to develop mining machinery building, strengthen the innovative component in each branch of the economy in order to bring products of high limits to the world and domestic markets.

With this purpose in view, the Concept and the integrative - competence model were developed in the course of the study for training the engineering personnel for an innovative economy using the example of the coal industry in the region. Technologies for their implementation were proposed, which were presented to the scientific and practicing community for wide awareness [3-5].

During the study of the problem, some inconsistencies in the training of engineering personnel for the innovative coal industry were identified, including:

-a high level of theoretical studies of the problems of professional education and an insufficient degree of study of the development of modern approaches and theories of training engineering personnel for the innovative economy;

-the need for training of engineering personnel for the innovative economy and an insufficient level of development of its theoretical and methodological justification;

-a high innovative potential of modernization of the education system and a lack of theoretical developments that allow on a scientific basis to carry out training of engineering personnel for the country's innovative economy;

-a significant potential of ideas for innovative development of professional education for engineers and inadequate elaboration of the conceptual and technological foundations for the training of engineering personnel for the innovative economy.

The inconsistencies also testify to the urgency of the attempt to develop the concept of training specialists in the system of higher education for the innovative economy, using the example of the coal industry. The concept is based on the integrative interaction of the components of the scientific, educational, production and innovation spaces, as well as regional, federal and international levels of economic development with the goal of training a specialist, a mining engineer with the competence in innovation.

The development of the concept of integrative-competence training of specialists for the innovative economy, including the mission, goals, objectives, principles, resources, indicators and indicators for assessing the effectiveness of the professional education system development, will contribute, in our view, to the task of forming and developing mining engineers required for development of the innovative sector of the coal industry.

 DOI: 10.1051/e3sconf/20172104028

The main idea of the Concept is the need to create a system of training specialists for the innovative economy, as an integral characteristic of the process, result, the system of professional education, which envisages regulation of the process based on the integration of its main components, as well as factors influencing the final result considered to be the innovative competence of a specialist.

The integrative-competence concept is focused on training specialists for the innovative economy, in particular the coal industry, and is based on the idea of integrating a number of grounds (attributes):

-organizational - integration of the systems of training specialists for the innovative economy with the aim of forming their innovative competence at the following levels of professional education: general, primary, special, higher, postgraduate and additional;

-functional - integration of basic and additional processes of formation of innovative competence of specialists in the system of professional education implemented in the educational system.

The essence of the integrative-competence concept is the integration of the management functions of the process of forming the innovative competence of specialists in various educational processes (education, upbringing and development), etc.

The implementation of the functions of managing the process of forming innovative competence of specialists in certain educational processes based on integration is reflected in the following procedures: introduction of state educational standards, accreditation and licensing of educational institutions, final certification of graduates, certification of pedagogical staff, examination of projects and programs, etc.

The process of formation of the innovative competency of specialists in certain educational processes based on the principle of integration includes, for example, such components as:

-content - is the integration of the content of education at various levels of professional education included in the integrated system of training specialists for the innovative economy; the requirements of the education standard, employers, the labor market and the economy, the achievements of modern science and practice (domestic and world);

-parameter - is an evaluation - criteria model of training a specialist for the innovative economy, which includes such elements as:

- Criterial system for assessing the quality of training specialists for the innovative economy, in the context of the integrative system of professional education, focused on the formation of an innovative and competent specialist;
- Evaluation of the effectiveness of the implementation of educational programs and the quality of training of future specialists in the coal industry, in the context of the integrated innovation-oriented system of professional education.

The concept includes the following components: scientifically substantiated goals and objectives of the integrative - competence approach to training specialists for the emerging innovative coal industry; principles, features, conditions and directions of its implementation; technology of designing and implementation of the integrative-competence approach to the training of mining engineers in the context of the integrative system of professional education.

It also includes methodological requirements for designing a multi-level regional system of training specialists for the innovative economy.

The features of the competence approach to training of specialists for the innovation economy, its components, criterial apparatus, organizational and pedagogical conditions (personnel, regulatory, legal, information) and resources that contribute to ensuring the effectiveness of training specialists for the innovation economy were based on the integration of the system, integrated, interdisciplinary, technological, competence, and situational approaches and models of training specialists for the innovation activity.

The essence of integrative - competence approach to the training of specialists for the innovative economy consists in the integration of organizational, methodical, scientific, personnel, managerial and other efforts in order to achieve a new quality of training specialists for the innovative economy with competence in innovation. Based on the principle of a multi-level and integrated construction of a specialist training system in the field of higher education for the innovative economy introduced into the concept, we propose an approach to designing a regional specialist training system based on integrating interrelated processes and highlighting the stages in the life cycle of the training process. The theoretical and methodological grounds of designing a regional educational system for training specialists for the innovative economy were determined, taking into account the experience and achievements of the modern domestic and world science of professional education. They are presented as a set of scientific approaches (systemic, integrated, interdisciplinary, technological, competence and situational) with the dominance of the competence one. The peculiarities of the process of designing the integrative - competence approach of training a specialist for the innovative economy were identified. The idea of an integrative - competence approach to training specialists with an emphasis on innovative parameters of the development of the coal industry and, the whole economy was implemented. The idea of the integrative - competence concept of training a specialist for the innovative economy presents a systemic integration interaction of its components. A sign of such interaction is the unification of the organizational, functional, content and parametric grounds – this is firstly. And, secondly, the impact of this unification on the process of training specialists, on the group of other sub-processes.

3 Conclusions

The methodological basis of the integrative competence approach to training a specialist for the innovative economy is the integration of a systemic, integrated, interdisciplinary, technological, competence and situational approaches, the implementation of which ensures a real increase in the educational, personal and professional achievements of graduates who meet the requirements of the innovative economy.

As we determined, the basic principles of the integrative and competence approach to the training of specialists for the innovation economy are as follows: compliance with the requirements of the current external environment, potential flexibility, integration of activities and resources, the innovative orientation of scientific and educational activities, the conjugation of educational programs, the optimality and adaptability of management, etc.

References

1. V.I. Vinokurov, Innovations, **4**, 6 (2005)
2. V.P. Mazikin, *Innovations as the basis of the integrated development of the coal industry in the regions of Russia and CIS countries*, (Prokopievsk, 2009)
3. N.A. Zaruba, S.G. Kostyuk, *Training specialists for the innovative economy: the integrative competency approach*, 146 (T.F. Kuzbass State Tech. University, Kemerovo, 2013)

4. N.A. Zaruba, S.G. Kostyuk, *Training specialists for the innovative economy: theoretical and methodological aspect*, 164 (T.F. Kuzbass State Tech. University-Kemerovo, 2012)
5. N.A. Zaruba, S.G. Kostyuk, *Integrative - competence approach to training specialists for the innovative sector of the coal industry*, 309 (T.F. Kuzbass State Tech. University-Kemerovo, 2013)
6. *The Strategy for Innovative Development of the Russian Federation for the period up to 2020* (2011).
7. N. Zaruba, N. Egorova, P. Kosinskij, e3s web of conferences, **15**, 04003 (2017)
8. E.A. Fedulova, A.V. Medvedev, P.D. Kosinskiy, S.A. Kononova, P.N. Pobedash, Foods and raw materials, **4:1**, 154 (2016)
9. E.A. Fedulova, A.V. Medvedev, P.D. Kosinskiy, S.A. Kononova, P.N. Pobedash, Foods and raw materials, **4:2**, 157 (2016)

Innovative Model of Practice-Oriented Training of Employees of the Town-Forming Enterprise in the Mining Region (by the Example of JSC "SUEK-Kuzbass")

Svetlana Kulay[1*], and *Gennady* Kayachev[2]

[1]Prokopievsk branch of T. F. Gorbachev Kuzbass State Technical University, 653039, Nogradskaya st., 19a, Prokopyevsk, Russia
[2]Siberian Federal University, Department of Economy and Business Processes Management, 660041, Krasnoyarsk, 79 Svobodny av.

Abstract. The article proposes to improve the system of training of employees of joint-stock company SUEK-Kuzbass in the educational institutions of the Kemerovo region according to the requirements of the company using practice-oriented training technology. The aim of the work is to substantiate the effectiveness of implementing practice-oriented training, identify priority directions and ways of its development. The main objectives of the study are: to identify the main advantages for the company and students; determine the criteria for the success and practical value of applying practice-oriented training for the company; conduct a comparative analysis of the target and practice-oriented model of training for the company. The real needs of the employer through the dual form of training were taking into account. The expansion of positive experience in training engineering personnel in higher education in technology-based training with the involvement of specialists from other regions of the company and expanding training in mining was also included.

1 Introduction

In currently economic conditions the problems of sustainable regional development are of particular importance in the Russian Federation [1-3]. In Siberia, a significant number of coal industry enterprises in the region are classified as city-forming, however, in the crisis periods, the region particularly feels the shortcomings of mono-profile economic development.

In developed countries, the principles of state regulation of problem regions have been formed, but the role of transnational corporations with significant investment opportunities is also significant [4, 5]. It should be noted the successful experience of modernization of the Japanese city of Kamaishi with a high share of social responsibility of the city-forming enterprises and their close cooperation with local self-government bodies [6, 7].

* Corresponding author: osv-07@mail.ru

DOI: 10.1051/e3sconf/20172104029

The city-forming enterprise in the Kemerovo Region is the SUEK SUEK-Kuzbass, one of the main employers and taxpayers in the region. The company implements a complex of socio-economic and charitable programs with the aim of creating a favorable social environment and raising the standard of living of the population.

The analysis of the company's age structure for the last 5 years shows the relative stability of the ratio of young (under 35, 42-44%) and experienced employees (over 35, 56-58%). Due to the targeted training program implemented in the company, the problem with the deficit of IT personnel has been successfully solved. Inspite of this the number of age groups under 20 and 20-25 years is decreasing (due to a decrease in the number of young workers). There is an acute problem of the shortage of workers in the company.

There are several reasons for dismissal and decrease in the intake of this category of personnel: weak prestige of trade and mining professions; the compatibility between the expectations of young workers for the chosen profession, the unconscious choice of a profession; ineffective adaptation measures for workers, ineffective measures to retain employees in the company, inconsistency of the skills level of preparation for the company's requirements.

The problem of the deficit of highly skilled personnel, especially workers, can be decided only when the employer begins to take a direct part in training, starting from career counseling to the procedure of evaluating of the professional qualifications.

2 Materials and methods

During the research the following methods were applied: interview and questioning, experiment, testing, comparison, risk analysis, expert assessments, induction, study and generalization.

In the course of the study, the first results of the training of employees of the SUEK-Kuzbass joint-stock company on practice-oriented training technology in the branch of KuzSTU in Prokopyevsk, the target training program in JSC SUEK-Kuzbass were taken as a basis.

3 Results and discussion

In foreign practice, the universally recognized leader in the formation of a practice-oriented (dual) system of professional training of personnel is Germany, where this system takes its origin [8, 9]. The foundation high-quality specialists training is the integration of the learning process and practice, which is proved by the experience of educational systems in Germany, Sweden, Japan, Britain, and China [10-12].

There is a positive experience in training the employees of SUEK-Kuzbass, based on the technology of practice-oriented training in the branch of the "Kuzbass State Technical University named after T.F. Gorbachev" in the city of Prokopyevsk.

With the company JSC "SUEK-Kuzbass", the branch signed an agreement, according to which for the group of practical-oriented training a curriculum an individual study schedule were drawn up. The schedule of the educational process provides for each half-year of 2 months of classroom hours within the university with a retention of labor in the amount of 20 thousand rubles. Thus, for an enterprise, the annual costs are about 80 thousand rubles, as well as 4 months of absence of an employee in the workplace. For comparison: a student who is on targeted training costs the enterprise 100 thousand rubles. per year, but the employee is not during the entire training period.

Scientists of the KuzSTU branch conducted a major psychological study. Respondents were trained in this technology. The results are positive, the training technology is recommended for further implementation.

As a result of the technology introduction , the current development trends for the entire coal-mining region have been identified:

1. Using this technology to train young workers and, as a consequence, to choose professional educational organizations as a provider of services, which is confirmed by world experience in applying practice-oriented technology of education;
2. Scale the experience of introducing technology to train mining engineers in other regions of the company's presence and other specialties.

The criteria for the success of the introduction of an innovative model of practice-oriented tning will be: optimization of the terms of staff training, adaptation of graduates in the enterprise, staffing of the company with working personnel, lowering the turnover of workers, increasing staff motivation, increasing labor productivity and product quality.

Economic benefit is of high importance when choosing a training model. Table provides a comparative analysis of the company's costs per person for targeted and practice-oriented forms of education.

Table 1. Costs of the company for training students on targeted and practice-oriented forms of education.

The indicator	The target form of training	The practice-oriented technology of training
Education services, rubles / year	90000	0 (budget places)
Scholarship, rubles / year	48000	0 (paid by the state)
Certificates (for working positions), rubles.	18000	0(usually already available)
Total costs, rubles / year	156000	0

4 Conclusion

To sum up it is advisable to develop this model of education on the basis of positive experience and the existing material and technical basis for training workers in specialized educational institutions. Within the framework of the project it is proposed to study the experience of using this technology on the territory of Russia and Kazakhstan.

It is also proposed to expand the training of mining engineers in a practice-oriented training system by connecting the specialization "Mineral processing" and forming a group of employees from all regions of JSC "SUEK" (the training process and production practice will be carried out in the Kemerovo region, while students return to work their workplaces).

The proposed model of training is a solution to the current problems in the training of personnel for the coal industry by combining training with practice, which allows the student already in the process of learning to learn in more detail the chosen professional activities, team, working conditions, corporate culture of the enterprise, skill level requirements. A student studying on a practice-oriented model of training is more valuable for the company, since during the training is already participating in the creation of products. Thus, the practice-oriented form has a significant advantage in saving money

 DOI: 10.1051/e3sconf/20172104029

spent on the student, productivity during training, and also in the level of quality of specialists.

Mono-dependence is an acute problem, but in the case of practice-oriented learning, it is more a possibility, because only single-industry cities have city-forming enterprises that are related to big business, whose strengths and interests are to support the development of dual education.

References

1. P. Annoni, K. Kozovska, *Econometrics and Applied Statistics Unit* (London, BHHT, 2010)
2. M. Bussiиre, J. Imbs, R. Kollmann, R. Rancire, Amer. Econ. J.: Macroec. **5**, 75 (2013)
3. L.A. Kormishkina, E.D. Kormishkin, N.N. Semenova, D.A. Koloskov Mediter. J. Soci. Sci. **6:4**, 163 (2015)
4. J. Hollander, J. Németh, Housing and Policy Debate, **21:3**, 349 (2011)
5. L. Ping, North-East Asia Academic Forum, **63**, 550 (2013)
6. D. Carson, D. Carson, Scandinavian Journal of Hospitality & Tourism, **14:4**, 460 (2014)
7. D. Littlewood, Journal of Business Ethics, **120:1**, 39 (2014)
8. *How the dual system – practical vocational and academic - works in Germany. – Bonn* (Germany: BIBB, 2012)
9. K.A. Geissler, , *Das duale System der industriellen Berufslildung hat keine Zukunf, Leviathan*, (Hampshire, JOOP, 1991)
10. W. D. Greinert, *Das deietsche System der Berufsbildung* (Baden-Baden, Inner Pub., 1998)
11. A.T. Kapanova, The Third International conference on development of pedagogical science in Eurasia, **29**, 1556 (2015)
12. I. Abdulloev, I. N. Gang, M. S. Yun, The European Journal of Development Research, **26:4**, 509 (2014)

Improvement And Development Of The Motivation System In The Occupational And Industrial Safety Field

Arkhip Pavlov[1*], and *Dmitrij* Gavrilov[1]

[1]JSC "VostNII" 650002, 3 Institutskaya Str., Kemerovo, Russia

Abstract. This paper discusses one of the main problems in labour and industrial management in the occupational and industrial safety field - motivation to work safely. The problem is complex and should be solved by a set of measures, where the assignment of responsibility to employees for the results of their work is absent, including in the field of labour protection and industrial safety. In accordance with the obligatory management principles, employees' work resolves to the strict implementation of the actions prescribed by the regulations. The responsibility for the negative result rests with the person who enacted or instructs employees. Thus, the employee is practically exempt from responsibility for the final result. One of the possible solutions to this problem is to put an assignment of responsibility on the employees for the results of their activities also in the occupational and industrial safety field. This is illustrated by the experience of other states, particularly of Australia. In conclusion suggestions for improvement and development of the motivation system in the field of occupational and industrial safety.

1 Introduction

Article 212 of the Labour Code of the Russian Federation stipulates that "the duties in respect of ensuring labour safety in the organization are imposed on the employer"; next follows a list of many conditions that the employer must ensure.

Article 9 of the Federal Law "On Industrial Safety of Hazardous Production Facilities" begins with the words: "An organization operating a hazardous production facility shall", and then there is a multi-page recitation of duties. That said, in accordance with Article 216 of the Criminal Code, "violation of safety rules during mining, building or any other activities, if this has involved by negligence the infliction of grave injury to human health or major material damage, shall be punishable" either by a fine or by deprivation of liberty.

Here, it is appropriate to express a reservation that both the Labour Code of the Russian Federation and the Federal Law "On Industrial Safety of Hazardous Production Facilities" are not safety rules as such. Such a discrepancy of the legislation on labour protection and industrial safety with the Criminal Code is due to the fact that Article 216, like many other

* Corresponding author: a.zakirova@nc-vostnii.ru

articles of the Criminal Code, was formulated at a time when these legislative acts on labour protection and industrial safety did not yet exist. There were only industry-specific safety rules.

At the same time, Article 216 of the Criminal Code is in force. When an event falling under this article occurs, there arises a problem of justifying the causal relation between the action or inaction of an employee and the accident that caused such "grave injury to human health or major damage." However, when identifying such dependence, few turn to the obligations of the 'employer' stipulated by the Labour Code of the Russian Federation and the duties of the 'organization' contained in the Federal Law "On Industrial Safety of Hazardous Production Facilities", while the Criminal Code is aimed at a particular responsible person.

This is the present-day problem of placing responsibility for "grave injury to human health or major damage" through negligence.

2 Materials and methods

When investigating the circumstances that led to "grave injury to human health or major damage", the picture of the emergence and development of the dangerous event is reproduced in the first place. For this purpose, the network model of the processes of formation and manifestation of the danger in all casual relations is appears most convenient. In this case, all the reasons can be taken into account: technical, organizational, and personal. All persons involved in the analyzed event are identified, and the analysis often covers all periods of the operational life of the production facility, starting from the stage of design.

With such approach, the investigation ends with the identification of the inadequate actions or inaction of particular employees, which resulted in "grave injury to human health or major damage". But in most cases, the investigation covers only immediate performers, middle-ranking managers and specialists or blue-collar professionals. Such an investigation of the accidents often does not bring light the causes of inadequate actions or inactions of the immediate performers and does not ask why the employees did not perform their duties or why they performed their duties improperly.

To find answers to such questions, it is necessary to investigate the production management system and personnel management system, including labour protection and industrial safety, since such answers as 'did not want to' or 'could not' reflect, first of all, the presence of defects in the systems of production management, personnel management and labour safety management, and this is the sphere of motivation and organization of labour that belongs to the senior executives and specialists and the top management of the organization.

However, this is yet to come. As a result, the statistics of criminal responsibility based on the results of the accident investigations still includes only managers and specialists of the middle level of mining management.

The existing domestic judicial practice appears to correspond to the previous state system. Today, the state system is different, many issues are seen in the other way; the legislation on labour protection and industrial safety was adopted. Therefore, different mechanisms for employee motivation in the field of labour protection and industrial safety should also be applied.

Addressing this issue, it must be recognized that judicial practice largely depends on the perfection of legislation, the efficiency of accident investigation and investigative actions, guided by such precedents. Let us consider a precedent from the foreign experience of legal proceedings following the investigation of an accident in a mining company, since the current Russian legislation on labour safety was largely based on the international standards of health protection of personnel at industrial enterprises.

The Gretley Coal Mine Disaster

Case. Gretley management obtained from the Department of Mineral Resources some old plans that purported to indicate the location of the earlier workings of an adjacent closed mine. The Department of Mineral Resources provided copies of the old workings plan, but these plans were wrong.

Based on the department's data, Gretley management organized excavations of a new section of the mine. Miners inadvertently broke through into the flooded workings of an adjacent abandoned mine, and four miners died in the inrush of water.

Such accidents occur in mining. What is of interest here is the judicial decisions made on this case, which give an answer to the question: when do the managers and specialists carry the responsibility? Consider this in more detail [1-5].

Defence. The coal company claimed that the mining operations were carried out according to plans presented by the Department of Mineral Resources, which, as it turned out, indicated the location of the old workings incorrectly.

Judgment. The court rejected the company's claim, arguing that the company may have relied on the information obtained from the Department of Mineral Resources and considered it accurate, but this did not exempt the company from the obligation to safely conduct mining operations; the company had to ensure the accuracy of plans submitted by the Department of Mineral Resources.

Based on the investigation, the court rules that in accordance with the Mining and Quarrying Safety and Health Act of 1999 and the Coal Mining Safety and Health Act of 1999:

1. The company has the right to rely on information received from outside organizations, institutions and individuals, considering it to be accurate, but it does not exempt the company from the obligation to provide and guarantee the safety of works. The company must verify the accuracy of the information received.

2. There are two grounds for assigning responsibility to employees and recognizing their culpability:

- employee's having opportunity to influence the company's policies,
- employee's failure to perform or improper performance of all possible actions to prevent the accident.

3. The grounds for dismissing the accusations towards the employee are the absence of conditions and opportunities for their influence on the management of the company in order to prevent the accident.

4. If the company is found culpable, then the accusation is presented to the chief executive officer.

The chief executive officer of the company is a person who:

- is one of the company's directors,
- takes part in the management of the company, regardless of the name of their position, and is the immediate head of the company.

5. Responsibility must be imposed on those people whose actions or inactions have led to violations of the company's activities, and such actions or inactions should be considered complicity in the activities that led to such violations [6-10].

6. 'To take part in the management of the company' means to bear responsibility for actions or inactions that led to violations of the company's activities.

The actions of a particular person that played a role in the activities of the company may include:

- advice given to the company management;
- participation in decision-making;
- execution of decisions received from the company management.

7. The actions entailing responsibility for the management of the company are:

- participation in the budget planning;

DOI: 10.1051/e3sconf/20172104030

- participation in drawing up the production plan;
- participation in the management meetings;
- control over the implementation of decisions;
- responsibility for certain functions of the company;
- organization and conduct of audits;
- development and approval of the company plans;
- signing of reports.

8. 'Advisor' is deemed involved into the management of the company:

- if the executive officials of the company use their advice to make decisions and when such pieces of advice play a leading role in making corporate decisions;
- if such advisor at least partially implement the decision, affecting the company as a whole;
- if such advisor makes decisions that purport to decisions related to the violation in question, although they may not affect the company as a whole.

9. The statement of any employee about their innocence must be supported by evidence that they had no influence on the prevention of the accident.

10. The specialist is not responsible for the accuracy of the information presented to them (the initial data), but is responsible for the accuracy of any production and technical design document developed by them based on such information.

11. Anyone who was able to prevent the accident but did not make proper effort to do so, is recognized as culpable.

12. The director of the mine is responsible for a wide range of issues related to occupational safety, and it is not difficult to prove that the managers of the mine took immediate part in the management of the enterprise.

The court dismissed the plea of non-guilty presented by two company executives and the mine surveyor, having decided that they took part in the management of the company, since it was not established that they had no opportunity to influence the company's policy, and also that they did not take proper actions to prevent the accident.

The mine surveyor of the Gretley mine claimed that he could not be held responsible for the accuracy of the documents made by the previous surveyor or signed by another surveyor. But, according to the court's opinion, he certified the accuracy of plans drawn up by him or under his direction with his signature, and it was exactly his signature on the plans that made him responsible. The surveyor was able to verify in the mining plans, but he did not take proper efforts to study the location of the old mine workings [11-15].

3 Results and discussion

Firstly, in accordance with the Criminal Code of the Russian Federation, the mechanism and the logic of assignment of guilt in labour protection and industrial safety issues is based primarily on the provisions of the rules for the safe conduct of mining, construction and other works. These rules regulate the activities of performers of mining, construction and other works and are violated mainly by the middle-ranking managers and specialists and blue-collar professionals.

Secondly, both the Criminal Code of the Russian Federation and the Industrial Safety Rules in the sectors of Russian economy were adopted much earlier than the current legislation of the Russian Federation on labour protection and industrial safety, which largely relies on foreign labour protection experience. As a result, considerable obligations were imposed on both the employer and the organization, but a mechanism that would impose responsibility for the implementation of such obligations has not been formed yet. The exist-

DOI: 10.1051/e3sconf/20172104030

ing mechanism of imposing responsibility for violating the Safety Rules appears to be impracticable.

The problem arises from the fact that there is no efficient mechanism to motivate employees to work safely and efficiently in the interest of the whole business. It should be noted that the problem of motivating employees to work safely, as any other problem, cannot be solved by a single action. The solution can be achieved by implementing a package of measures. The problem of motivating employees to work safely demonstrates the absence of one of the motivation methods — the imposition on the employee of all forms of responsibility, including criminal liability, for any negative outcome of their actions or inaction. Virtually having the right of operational management of the production activities on the facility under their control, the employee remains free from liability for any negative outcome of his actions or inactions. This, first, violates the balance of rights and responsibilities, and, second, a rather effective mechanism for motivation as criminal liability is ignored.

Thirdly, when assigning responsibility for non-fulfilment or improper fulfilment of labour protection and safety legislation in the judicial practice of many foreign countries, the courts are guided by the following provisions:

- culpability for the accident lies with the company,
- if the company is found culpable, then the accusations are presented to the chief executive officer,
- a chief executive officer of the company is the person who takes part in the management of the company, regardless of the name of their position, and is the immediate head of the company,
- 'to take part in the management of the company' means to bear responsibility for actions or inactions that led to violations of the company's activities.

4 Conclusions

1. The logic of assigning responsibility is rather simple: so many duties, so much responsibility. If, in accordance with Article 212 of the Labour Code of the Russian Federation, "the duties in respect of ensuring labour safety in the organization are imposed on the employer", the responsibility for ensuring safe working conditions and labour protection in the organization shall also borne by the employer. In accordance with Article 20 of the Labour Code the Russian Federation, the *employer* is an individual or a legal entity (organization) entering labour relations with the employee and having the right to enter into employment agreements. An individual, in this case, usually acts an individual entrepreneur, white the head of the organization acts on behalf of the legal entity.

2. Following the terminology used in the Labour Code of the Russian Federation, the industrial safety requirements stipulated in Item 1 Article 9 of the Federal Law "On Industrial Safety of Hazardous Production Facilities" and addressed to the *organizations*, should be seen as addressed to a *legal person*, employer, head of the organization.

3. The mechanism of imposing criminal liability for violations of labour protection and industrial safety legislation should be streamlined with the current legislation of the Russian Federation on occupational and industrial safety.

4. Harmonization of Russian legislation on labour protection and industrial safety with the international legislation demonstrates the need in improving and updating the mechanisms imposing responsibility for occupational and industrial safety based on international experience.

5. Responsibility should be imposed on those people whose actions or inactions have led to violations of the company's activities, and such actions or inactions should be considered complicity in the activities that led to such violations.

6. The principle that 'implementation of the rules and norms cannot be sufficient grounds of non-culpability for any negative outcome' should be applied. The concept that everyone is obliged to take every possible action of those actions that are necessary, regardless of whether it is stipulated by the rules or not, should be made a legal norm.

7. Terms and definitions:

- Competence is the professional ability to do the assigned work;
- Competency is the right to express professional ability under one's personal responsibility, according to the principle: *if you do, take the consequences*.

References

1. F. Hopkins, *Managing Major Hazards: The lessons of the Moura Mine Disaster*, (Sydney, Allen and Unwin, 1999)
2. J. L. Massie, J. Douglas, *Managing: A Contemporary Introduction*, (Englewood Cliffs: Simon & Schuster Company, 1992)
3. R.J. Schonberger, *Japanese manufacturing techniques: Nine Hidden Lessons in Simplicity*, 55 (New York, The Free Press: a division of Macmillan Publishing Co., 1982)
4. F. Luthans, *Organizational Behavior*, (New York: McGraw-Hill, 1989)
5. W. G. Ouchi, *Theory Z: How American Business Can Meet the Japanese Challenge*, 244 (New York: Avon Books, 1981)
6. *Queensland Mining Industry Health and Safety Conference*, 206 (collection of papers. Townsvill, Queensland, Australia, 2005)
7. K. Yin. Robert, *Case Study Research*: Design and *Methods* (SAGE Publications. California, 1989)
8. I Masaaki, G. A. Kaizen, *Commonsense, Low-Cost Approach to Management*, (New York, McGraw-Hill, 1997)
9. J. Womack, D. Jones, *Lean Thinking: Banish Waste and Create Wealth in Your Corporation* (New York, Simon & Schuster, 2003)
10. M. Rother, J. Shook, *Learning to See: Value Stream Mapping to Add Value and Eliminate Muda*, (Brookline, MA, Lean Enterprise Institute, 1999)
11. J.P. Kotter, *Leading Change*, (Boston, MA, Harvard Business School Press, 1996)
12. B. Emiliani, *Better Thinking, Better Results: Using the Power of Lean as a Total Business Solution*, (Kensington, CT, The Center for Lean Business Management, 2002)
13. J.H. Dyer, *Collaborative Advantage: Winning Through Extended Enterprise Supplier Networks*, (New York, Oxford University Press, 2000)
14. H. Emerson, *The twelve principles of efficiency*, (New York, Engineering magazine, 1912)
15. T. Fujimoto, *The Evolution of a Manufacturing System at Toyota*, (NY, Oxford University Press, 1999)

AUTHOR INDEX

AUTHOR INDEX

AUTHOR INDEX